Human Physiology

An Integrated Approach

About the Author

Dee Unglaub Silverthorn

Dee Unglaub Silverthorn studied biology as an undergraduate at Tulane University and went on to earn a Ph.D. in marine science at the University of South Carolina. Her research interests center on comparative invertebrate physiology; however, she began her teaching career in the Physiology Department at the Medical University of South Carolina. Drawing on her training in both comparative and traditional human physiology, Dee has taught a wide range of students, from those in medical school and college to those still preparing for a higher education. At the University of Texas she lectures in physiology, coordinates undergraduate laboratories in physiology, and instructs graduate students in a course on developing teaching skills in the life sciences. She has substantial experience with active learn-

ing in the classroom, and has given workshops on this subject at regional, national, and international conferences.

Dee is a member of the American Physiological Society and an associate editor of Advances in Physiology Education. She works with members of the International Union of Physiological Sciences to improve physiology education in developing countries. She is also a member of the Human Anatomy and Physiology Society, the Society for Comparative and Integrative Biology, the Association for Biology Laboratory Education, the Society for College Science Teaching, the American Association for the Advancement of Science, and Sigma Xi. Her free time (what little remains after committing to write a book!) is spent creating multimedia fiber art and enjoying the Texas hill country with her husband, Andy, three dogs, and cat.

About the Illustrators

Dr. William C. Ober (art coordinator and illustrator)

Dr. William C. Ober studied as an undergraduate at the Washington and Lee University and obtained his M.D. from the University of Virginia in Charlottesville. In addition to his medical school education, he studied in the Department of Art as Applied to Medicine at Johns Hopkins University. After graduation, Dr. Ober completed training as a resident in family practice. He is now a Clinical Assistant Professor in the Department of Family Medicine, University of Virginia. During the summer, he teaches biological illustration at Shoals Marine Laboratory, where he is part of the Core Faculty. Dr. Ober's professional life now focuses on medical and scientific illustration.

Claire W. Garrison, R.N. (illustrator)

Claire Garrison chose to make medical illustration her career after almost two decades as a pediatric and obstetric nurse. After a five-year apprenticeship, she became Dr. Ober's associate in 1986. Ms. Garrison is also a core faculty member at Shoals. The texts illustrated by Dr. Ober and Ms. Garrison have won numerous awards from groups including the following: Association of Medical Illustrators (Award of Excellence); Chicago Book Clinic (Award for Art and Design); Printing Industries of America (Award of Excellence); and Bookbinders West. They have also received the Art Directors Award.

About the Clinical Consultant

Andrew C. Silverthorn

Andrew C. Silverthorn, M.D., is a physician in private practice in Austin, Texas. He is a graduate of the United States Military Academy (West Point), served in the infantry in Vietnam, and upon his return entered medical school at the Medical University of South Carolina in Charleston. He completed his family practice residency at the University of Texas Medical Branch, Galveston, where he was chief resident, and he is board-certified by the American Board of Family Practice. When Andy is not busy seeing patients he may be found on the golf course, playing golf or running with his black lab, Maggie.

About the Cover

The confocal microscope image on the cover shows neurons (red or green and yellow) and support cells (blue) in the brain after immuno-fluorescence staining. The brain is the primary integrating center of the body.

Human Physiology
An Integrated Approach

Dee Unglaub Silverthorn, Ph.D.
University of Texas

with

William C. Ober, M.D.
Illustration Coordinator

Claire W. Garrison, R.N.
Illustrator

Andrew C. Silverthorn, M.D.
Clinical Consultant

Prentice Hall
Upper Saddle River, New Jersey 07458

Library of Congress Cataloging-in-Publication Data
Silverthorn, Dee Unglaub
    Human physiology : an integrated approach / Dee Unglaub
Silverthorn with William C. Ober, Claire W. Garrison, Andrew C. Silverthorn, clinical consultant.

    Includes index.
    ISBN 0-13-262528-8
    1. Human physiology. I. Title.
    [DNLM: 1. Physiology. QT 104 S587h 1998]
  QP34.5.S55 1998
  612—dc21   97-16184
                                                                       CIP

Executive Editor: David Kendric Brake
Development Editor: Shana Ederer
Associate Editor in Chief, Development: Carol Trueheart
Editor in Chief, Development: Ray Mullaney
Production Editor: Debra A. Wechsler
Editorial Director: Tim Bozik
Editor in Chief: Paul F. Corey
Director, Production & Manufacturing: David W. Riccardi
Executive Managing Editor: Kathleen Schiaparelli
Assistant Managing Editor: Shari Toron
Marketing Manager: Jennifer Welchans
Creative Director: Paula Maylahn
Art Director: Heather Scott
Art Manager: Gus Vibal
Interior Designer: Robin Hoffman, Judith A. Matz-Coniglio
Cover Designer: Tom Nery
Copy Editor: Margo Quinto
Manufacturing Manager: Trudy Pisciotti
Photo Research: Stuart Kenter Associates
Photo Editor: Lorinda Morris-Nantz
Illustrators: William C. Ober, MD; Claire W. Garrison, RN; Dartmouth Publishing, Inc.
Editorial Assistant: Byron Smith
Cover Photograph: U. Blomer and F. H. Gage. Reprinted with permission from Science Magazine.
Copyright © 1997 American Association for the Advancement of Science.

© 1998 by Prentice-Hall, Inc.
Simon & Schuster/A Viacom Company
Upper Saddle River, New Jersey 07458

Printed in the United States of America
10 9 8 7 6 5 4 3 2 1

ISBN 0-13-262528-8

Prentice-Hall International (UK) Limited, *London*
Prentice-Hall of Australia Pty. Limited, *Sydney*
Prentice-Hall Canada Inc., *Toronto*
Prentice-Hall Hispanoamericana, S.A., *Mexico*
Prentice-Hall of India Private Limited, *New Delhi*
Prentice-Hall of Japan, Inc., *Tokyo*
Simon & Schuster Asia Pte, Ltd., *Singapore*
Editora Prentice-Hall do Brasil, Ltda., *Rio de Janeiro*

OWNER'S MANUAL
How to Use this Book

Welcome to *Human Physiology!*

As you begin your study of the human body, you should be prepared to make maximum use of the resources available to you, including your instructor, the library, the World Wide Web, and your textbook. One of my goals in this book is to provide you not only with information about how the human body functions, but also with tips for studying and learning to solve problems. Many of these study aids have been developed with the input of my students, so I think you may find them particularly helpful.

On the following pages, I have put together a brief tour of the special features of the book, especially those that you may not have encountered in textbooks previously. Please take a few minutes to read this section so that you can make optimum use of the text as you study. If you take advantage of features such as the *Graph Questions*, *Concept Checks*, and the *Running Problems*, you will find that you begin to think about physiology in a different way.

One of your tasks as you study will be to construct for yourself a global view of the body, its systems, and the many processes that keep the systems working. This "big picture" is what physiologists call the integration of systems, and it is a key theme in the book. In order to integrate information, however, you must do more than simply memorize it. You must truly understand it and be able to use it to solve problems that you have never encountered before. If you are headed for a career in the health professions, you will do this in the clinics. If you are headed for a career in biology, you must solve problems in the laboratory, field, or classroom. Learning to analyze, synthesize, and evaluate information are skills that you need to develop while you are in school, and I hope that the features of this book will help you with this task.

So find a comfortable chair and take a peek at what lies ahead!

Warmest regards,

Dr. Dee
(as my students call me)

P.S. Join me on the book's World Wide Web site for additional information and activities that could help you be more successful in your physiology course. Who knows, it might even help you on the next test!

http://www.prenhall.com/silverthorn

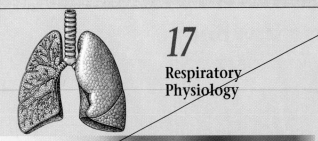

17
Respiratory Physiology

CHAPTER OUTLINE
Each chapter begins with an outline listing the major headings, or topics, for that chapter. You will find a page reference next to each heading. Use this outline to quickly preview the chapter. Do you see any connections between these topics and things you have learned in previous chapters or other courses?

BACKGROUND BASICS
This list is a handy tool at the beginning of each chapter that shows you which important concepts you should have mastered prior to beginning the chapter. If any of these topics seem unfamiliar, use the page references to revisit the subjects you need to review.

Imagine covering the playing surface of a racquetball court (about 75 m²) with thin plastic wrap, then crumpling up the wrap and stuffing it into [a 2-liter (?) drink] bottle. Impossible? Maybe so, if you use [a] drink bottle. But the lungs of a 70-kg [human have an] exchange surface the size of that plastic [wrap crumpled] into a volume that is less than the size [of ...] tremendous area for gas exchange is nee[ded to supply] trillions of cells in the body with ade[quate amounts of] oxygen.

Aerobic metabolism in animal cells [depends on a steady] supply of oxygen and nutrients from [the environment,] coupled with the removal of carbon d[ioxide and other] wastes. In small, simple aquatic animal[s these needs can] be met by simple diffusion across the [skin. However, the] rate of diffusion is limited by distance, [and as a] result, most multicelled animals use [some type of respi-] ratory organs associated with a cir[culatory system.] Respiratory organs take a variety of form[s, but all have a] large surface area compressed into a sm[all space. ...] Besides needing a large exchange sur[face area, ...] other terrestrial animals face an additi[onal ...] challenge, that of dehydration. Their exch[ange ...] be thin and moist to allow gases to pass [...], yet at the same time they must be pr[otected ...]

CONCEPT LINKS
These blue "chain-links" with page references are an extension of the Background Basics, linking you to concepts discussed earlier in the text. They help you find material that you might have forgotten or that might be helpful for understanding new concepts.

CONCEPT CHECKS
The red check marks placed at intervals throughout the chapters act as stopping points where you can check your understanding of what you just read. Some of these questions are factual, whereas others ask you to think about or apply what you have learned. Answers to the Concept Check questions are in Appendix D for instant feedback.

---

abdominal muscles. These muscles are collectively called the *expiratory muscles.*

The internal intercostal muscles line the inside of the rib cage. When they contract, they pull the ribs inward, reducing the volume of the thoracic cavity. To feel this action, place your hands on your rib cage. Forcefully blow as much air out of your lungs as you can, noting the movement of your hands as you do so.

The internal and external intercostals act as antagonistic muscle groups (∞ p. 326) to alter the position and volume of the rib cage during ventilation, but the diaphragm has no antagonistic muscles. Therefore, the abdominal muscles become active during active expiration to supplement the activity of the intercostals. Contraction of abdominal muscles during active expiration pulls the lower rib cage inward and decreases abdominal volume. The displaced intestines and liver push the diaphragm up into the thoracic cavity, passively decreasing the chest volume even more. This is why when you are doing abdominal exercises in an aerobics class, the instructor tells you to blow air out as you lift your head and shoulders. The active process of blowing air out helps you contract your abdominals, the very muscles you are trying to strengthen. The diaphragm stays relaxed during active expiration because contraction would move it downward and would work against the abdominal muscles that are trying to push the diaphragm upward.

Any neuromuscular disease that weakens skeletal muscles or damages their motor neurons can adversely affect ventilation. With decreased ventilation, less fresh air enters the lungs. In addition, loss of the ability to cough increases the risk of pneumonia and other infections. Examples of diseases that affect the motor control of ventilation include **myasthenia gravis,** an illness in which the acetylcholine receptors of the motor end plate of skeletal muscles are destroyed, and **polio** (poliomyelitis), a viral illness that sometimes paralyzes the respiratory muscles.

✓ Scarlett O'Hara was trying to squeeze herself into a corset with an 18-inch waist. Was she more successful when she took a deep breath and held it or when she blew all the air out of her lungs? Why?

✓ Why would loss of the ability to cough increase the risk of respiratory infections? (Hint: What does coughing do to mucus in the airways?)

Intrapleural Pressure Changes during Ventilation

Ventilation requires that the lungs, which are unable to expand and contract on their own, move in association with the contraction and relaxation of the thorax. As discussed earlier in this chapter, the lungs are "stuck" to the thoracic cage by the pleural fluid, the fluid between the two pleural membranes.

The intrapleural pressure, pressure within the fluid between the pleural membranes, is normally subatmo-

spheric. This subatmospheric pressure arises during development, when the thoracic cage with its associated pleural membrane grows more rapidly than the lung with its associated pleural membrane. The two pleural membranes are held together by the pleural fluid bond, so the elastic lungs are forced to stretch to conform to the larger volume of the thoracic cavity. At the same time, however, elastic recoil of the lungs creates an inwardly directed force that attempts to pull the lungs away from the chest wall (Fig. 17-11a ■). The combination of the outward pull of the thoracic cage and inward recoil of the elastic lungs creates an intrapleural pressure of about −3 mm Hg.

You can create a similar situation with a syringe half-filled with fluid and capped with a plugged-up needle. Begin with the fluid inside the syringe at atmospheric pressure. Now pick up the syringe and hold the barrel (the chest wall) in one hand while you try to withdraw the plunger (the elastic lung pulling away from the chest wall). As you pull on the plunger, the volume inside the barrel increases slightly, but the cohesive forces between the water molecules cause the fluid inside the syringe to resist expansion. The pressure within the barrel, which was initially equal to atmospheric pressure, decreases slightly as you pull on the plunger. If you release the plunger, it snaps back to its resting position, restoring atmospheric pressure inside the syringe.

But what happens to subatmospheric intrapleural pressure if an opening is made between the sealed pleural cavity and the atmosphere? A knife thrust between the ribs, a broken rib that punctures the pleural membrane, or any event that opens the pleural cavity to the atmosphere will allow air to flow in down its pressure gradient, just as air enters when you break the seal on a vacuum-packed can. Air in the pleural cavity breaks the fluid bond holding the lung to the chest wall. The elastic lung collapses to an unstretched state, like a deflated balloon, while the chest wall expands outward (Fig. 17-11b ■). This condition, called **pneumothorax** [*pneuma,* air + *thorax,* chest], results in a collapsed lung that is unable to function normally. Pneumothorax can occur spontaneously if a congenital *bleb,* or weakened section of lung tissue, ruptures, allowing air from inside the lung to enter the pleural cavity. Correction of a pneumothorax has two components: removing as much air from the pleural cavity as possible with a suction pump, and sealing the hole to prevent more air from entering. Any air remaining in the cavity will gradually be absorbed into the blood, restoring the pleural fluid bond and reinflating the lung.

Pressures in the pleural fluid vary during a respiratory cycle. At the beginning of inspiration, intrapleural pressure is about −3 mm Hg (Fig. 17-10 ■, point B_1). As inspiration proceeds, the pleural membranes and lungs follow the thoracic cage because of the pleural fluid bond. But the elastic lung tissue resists being stretched. The lungs attempt to pull farther away from the chest wall, causing the intrapleural pressure to become even more negative (Fig. 17-10 ■, point B_2). Because this

FIGURE REFERENCE LOCATORS
The red squares, found next to every figure callout in the text, function as place markers, making it easier for you to toggle between an illustration and the narrative associated with that illustration.

Chapter Summary

1. Aerobic metabolism in living cells consumes oxygen and produces carbon dioxide. (p. 474)

2. Gas exchange requires a thin, moist, internalized exchange surface with a large surface area, a pump to move air between the atmosphere and the exchange surface, and a circulatory system to transport gases between the exchange surface and the cells. (p. 474)

3. The functions of the respiratory system in addition to gas exchange include pH regulation, vocalization, and protection from foreign substances. (p. 475)

The Respiratory System

4. **Cellular respiration** refers to the metabolic processes of the cell that consume oxygen and nutrients and produce energy. **External respiration** is the exchange of gases between the atmosphere and the cells. It includes ventilation, gas exchange at the lung surface and at the cells, and transport of gases in the blood. **Ventilation** is the movement of air into and out of the lungs. (p. 475)

5. The **respiratory system** is composed of the anatomical structures involved in ventilation and gas exchange in the lungs. (p. 475)

6. The **upper respiratory tract** includes the mouth, nasal cavity, **pharynx, larynx,** and **trachea.** The lower respiratory tract includes the bronchi, bronchioles, and exchange surfaces of the alveoli. (p. 475)

7. The thoracic cage is bounded by the ribs, spine, and **diaphragm.** Two sets of **intercostal muscles** connect the ribs. The respiratory muscles are skeletal muscles innervated by somatic motor neurons. (p. 475)

8. The lungs are paired air-filled organs, each contained within a double-walled pleural sac. A small quantity of **pleural fluid** lies between the pleural membranes. (p. 476)

9. Air passes through the upper respiratory system to the **trachea.** In the thorax, the trachea divides to form the two **primary bronchi** that enter the lungs. Each primary

CHAPTER SUMMARY

The end-of-chapter summary has enumerated points and page references, providing you with a quick summary of every major topic in the chapter.

Questions

LEVEL ONE Reviewing Facts and Terms

1. List four functions of the respiratory system.

2. Give two different definitions for the word *respiration*.

3. Which sets of muscles are used for normal quiet inspiration? For normal quiet expiration? For active expiration?

4. What is the function of pleural fluid?

5. Name the anatomical structures that an oxygen molecule passes on its way from the atmosphere to the blood.

6. Diagram the structure of an alveolus and give the function of each part. How are capillaries associated with an alveolus?

7. Trace the path of the pulmonary circulation. About how much blood is found here at any given moment? What is a typical arterial blood pressure for the pulmonary circuit and how does this compare to that of the systemic circulation?

8. List three factors that influence the movement of gas molecules from air into solution. Which of these factors is usually not significant in humans?

9. What happens to inspired air as it is conditioned during its passage through the airways?

10. During inspiration, most of the volume change of the chest is due to movement of the _____.

11. Describe the changes in intra-alveolar and intrapleural pressure during one respiratory cycle.

12. What is the function of surfactant?

13. Of the three factors that contribute to the resistance of air flow through a tube, which plays the largest role in changing resistance in the human respiratory system?

14. Match the following items with their correct effect on the bronchioles:
   (a) histamine
   (b) epinephrine
   (c) acetylcholine
   (d) increased P_{CO_2}
   1. bronchoconstriction
   2. bronchodilation
   3. no effect

15. Define these four terms and explain how they relate to one another: tidal volume, inspiratory reserve volume,

binds to hemoglobin. Which of these four factors is the most important?

18. Describe the structure of a hemoglobin molecule. What element is essential for hemoglobin synthesis?

_____ The centers for control of ventilation are found in the _____ and _____ of the brain. The dorsal and ventral respiratory groups of neurons control what processes, respectively? What is a central pattern generator?

_____ Name the chemoreceptors that influence ventilation and explain how they do so. What chemical is the most important controller of ventilation?

_____ Describe the protective reflexes of the respiratory system. What does the Hering-Breuer reflex prevent? How is it initiated?

END-OF-CHAPTER QUESTIONS

The end-of-chapter questions have been organized into three levels of difficulty to allow you to work your way up to problem solving.

- **Level One** questions review basic facts and terms. Successfully answering these questions gives you a foundation for more conceptual questions.
- **Level Two** questions test your ability to understand and integrate the chapter's key concepts into the big picture.
- **Level Three** exercises are real-world scenarios designed to develop your problem-solving skills. You will be more successful dealing with Level Three questions if you have successfully worked through the first two levels. To check your answers or to find additional help, you can go to the book's Web site.

LEVEL TWO Reviewing Concepts

22. **Concept map:** Draw an alveolus and adjacent capillary similar to that shown in Figure 17-20. Write in the partial pressures of oxygen and carbon dioxide for the arterial and venous blood and the air. Draw arrows showing diffusion of the gases. Relate the four rules governing diffusion to the figure. Indicate how at least three different pulmonary pathologies can interfere with gas exchange.

23. A container of gas with a moveable piston has a volume of 500 mL and a pressure of 60 mm Hg. The piston is moved and the new pressure is measured as 150 mm Hg. What is the new volume of the container?

24. You have a mixture of gases in dry air, with an atmospheric pressure of 760 mm Hg. Calculate the partial pressure of each gas for each of the examples below:
   a. 21% oxygen, 78% nitrogen, 0.3% carbon dioxide
   b. 40% oxygen, 13% nitrogen, 45% carbon dioxide, 2% hydrogen
   c. 10% oxygen, 15% nitrogen, 1% argon, 25% carbon dioxide

25. Compare and contrast the following sets of concepts:
   a. compliance and elastance
   b. inspiration and expiration
   c. intrapleural and intra-alveolar pressures

d. total pulmonary ventilation and alveolar ventilation
e. transport of oxygen and carbon dioxide in arterial blood

26. Neelesh is helping his mother clean a dusty attic. As he watches the dust particles dance in a sunbeam, it occurs to him that he is inhaling many particles like these. Where do these particles end up in his body? Explain.

27. Define the following terms: pneumothorax, spirometer, hypoxia, COPD, auscultation, hypoventilation, hypercapnia, bronchoconstriction, minute volume.

28. The cartoon coyote is blowing up a balloon as part of an attempt to once more catch a roadrunner. He first breathes in as much air as he can, then blows out all that he can into the balloon. The volume of air in the balloon is equal to the _____ _____ of the coyote's lungs. This volume can be measured directly, as above, or by adding what respiratory volumes together? In 10 years, when the coyote is still chasing the roadrunner, will he still be able to put as much air into the balloon in one breath? Explain.

29. Define these terms and explain how they differ: oxyhemoglobin, carbaminohemoglobin, hemoglobin saturation, carbonic anhydrase, polycythemia, erythropoietin.

LEVEL THREE Problem Solving

30. Li is a tiny woman, with a tidal volume of 400 mL and a respiratory rate of 12 breaths per minute at rest. What is her total pulmonary ventilation? Just before a physiology exam, her ventilation increases to 18 breaths per minute from nervousness. Now what is her total pulmonary ventilation? Assuming her dead space is 120 mL, what is her alveolar ventilation in each case?

31. Marco tries to hide at the bottom of a swimming hole by breathing through a garden hose that greatly increases his dead space. What happens to the following parameters in his arterial blood, and why?
   a. P_{CO_2} b. P_{O_2} c. bicarbonate ion   d. pH

32. A hospitalized patient with severe chronic obstructive

lung disease has a P_{CO_2} of 55 mm Hg and a P_{O_2} of 50 mm Hg. To elevate his blood oxygen, he is given pure oxygen through a nasal tube. The patient immediately stops breathing. Explain why this might occur.

33. You are a physiologist on the manned space flight to a distant planet. You find intelligent humanoid creatures inhabiting the planet, and they willingly submit to your tests. Some of the data you have collected are described below.
   a. The graph below shows the oxygen dissociation curve for the oxygen-carrying pigment in the blood of the humanoid named Bzork. Bzork's normal alveolar P_{O_2} is 85 mm Hg. His normal cell P_{O_2} is 20 mm Hg but it drops to 10 mm Hg with exercise.

Anatomy Summary   **Respiratory System**

■ Figure 17-2

(a) Muscles used for ventilation
The muscles of inspiration include the diaphragm, external intercostals, sternocleidomastoids, and scalenes. The muscles of expiration include the internal intercostals and the abdominals.

(b) The respiratory system
The respiratory system consists of the upper respiratory system (mouth, nasal cavity, pharynx, larynx) and the lower respiratory system (trachea, bronchi, lungs). The lower respiratory system is enclosed in the thorax, bounded by the ribs, spine, and diaphragm.

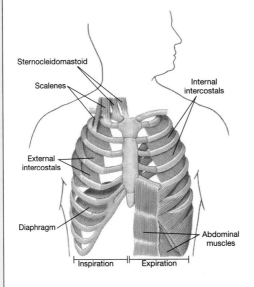

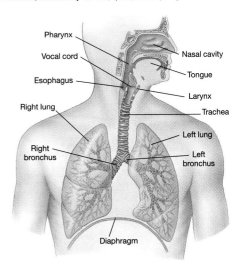

(c) External anatomy of lungs
Externally, the right lung is divided into three lobes and the left lung into two.

(d) Sectional view of chest
Each lung is enclosed in two pleural membranes. The pleural fluid and space is much smaller than illustrated. The esophagus and aorta pass through the thorax between the pleural sacs.

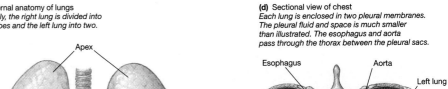

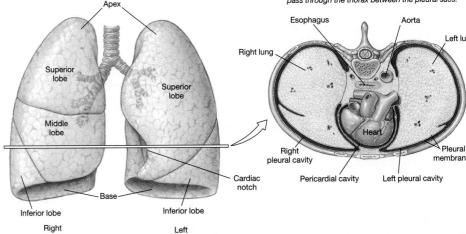

ANATOMY SUMMARIES

In order to understand physiology, you must have a good understanding of anatomy. To help you out, this book includes several Anatomy Summary figures that show the anatomy of a physiological system, from a macro- to micro-perspective, in one large spread. "Zoom" arrows are used to "explode" structures into their component parts, all the way down to the tissue level. These summaries allow you to see all of the essential features of each system in a

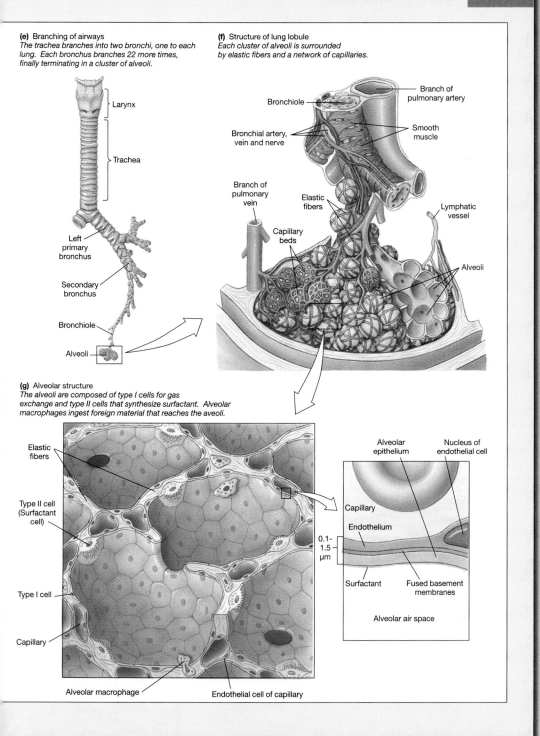

(e) Branching of airways
The trachea branches into two bronchi, one to each lung. Each bronchus branches 22 more times, finally terminating in a cluster of alveoli.

(f) Structure of lung lobule
Each cluster of alveoli is surrounded by elastic fibers and a network of capillaries.

(g) Alveolar structure
The alveoli are composed of type I cells for gas exchange and type II cells that synthesize surfactant. Alveolar macrophages ingest foreign material that reaches the aveoli.

single figure, either to review them or learn them for the first time.

The Anatomy Summaries and many other figures in the book were created by Bill Ober, M.D. and his associate Claire Garrison, R.N. Both Bill and Claire have medical backgrounds and understand what students need to know.

REFLEX PATHWAY AND CONCEPT MAPS

The color-keyed reflex pathway maps and structure/function maps organize material in a logical format to carefully walk you through the processes and organization of physiological systems. You will find it easier to remember complex pathways when you see them organized into maps, and you will be able to create your own maps using suggestions from the end-of-chapter exercises. You will find information on how to map inside the back cover and in Chapter 1.

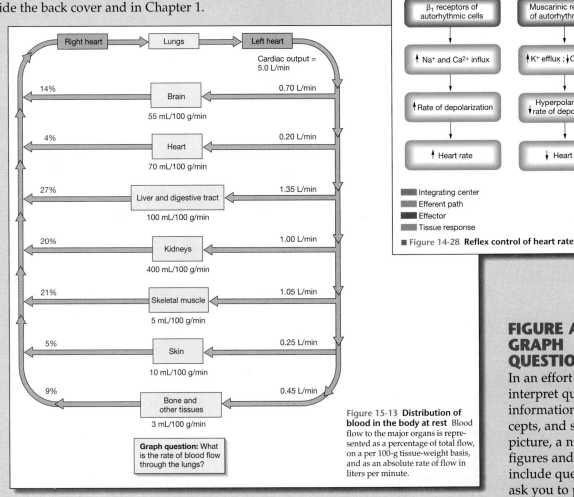

Figure 15-13 Distribution of blood in the body at rest Blood flow to the major organs is represented as a percentage of total flow, on a per 100-g tissue-weight basis, and as an absolute rate of flow in liters per minute.

■ Figure 14-28 **Reflex control of heart rate**

Graph question: What is the rate of blood flow through the lungs?

FIGURE AND GRAPH QUESTIONS

In an effort to help you interpret quantitative information, apply concepts, and see the big picture, a number of figures and graphs include questions that ask you to pause for a moment and consider what you really know about the information you just encountered. Answers to Graph and Figure Questions are in Appendix D.

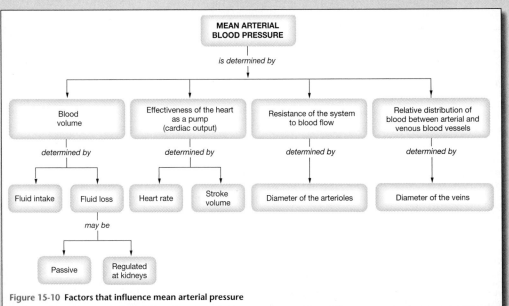

Figure 15-10 **Factors that influence mean arterial pressure**

Blood Substitutes Physiologists have been attempting to find a substitute for blood ever since 1878, when an intrepid scientist named T. Gaillard Thomas transfused whole milk in place of blood. Although milk seems an unlikely replacement for blood, it has two important properties: proteins to provide colloid osmotic pressure and molecules (emulsified lipids) capable of binding to oxygen. In the development of hemoglobin substitutes, oxygen transport is the most difficult property to mimic. A hemoglobin solution would seem to be the obvious answer, but hemoglobin that is not compartmentalized in red blood cells behaves differently than hemoglobin that is. First, the quaternary structure changes so that the four-subunit (tetramer) form is in equilibrium with a smaller two-subunit (dimer) form. The smaller version of hemoglobin is easily excreted by the kidneys, so almost half a dose of hemoglobin solution disappears from the blood in 2–4 hours. Second, hemoglobin found outside the red blood cells does not release oxygen as easily in peripheral tissues. Investigators are making progress by polymerizing hemoglobin into larger, more stable molecules and loading these hemoglobin polymers into phospholipid liposomes (∞ p. 109). By encapsulating hemoglobin into these artificial red blood cells, researchers hope to extend the life span of the blood substitute and make its properties approach those of real blood.

Shock *Shock* is a broad term that refers to generalized, severe circulatory failure. Shock can arise from multiple causes: failure of the heart to maintain normal cardiac output (*cardiogenic shock*), decreased circulating blood volume (*hypovolemic shock*), bacterial toxins (*septic shock*), and miscellaneous causes such as massive immune reactions that cause *anaphylactic shock*. No matter what the cause, the results are similar: low cardiac output and falling peripheral blood pressure. As tissue perfusion falls below the level needed to maintain adequate oxygen supply, the cells begin to sustain damage from inadequate oxygen and the buildup of metabolic wastes. Once this damage occurs, a positive feedback cycle begins. The shock becomes progressively worse until it becomes irreversible, and the patient dies. The management of shock includes administration of oxygen, fluids, and norepinephrine to stimulate vasoconstriction and increase cardiac output. If the shock arises from a cause that is treatable, such as a bacterial infection, measures must also be taken to remove the precipitating cause.

Rapid Action of Aldosterone For years, physiologists divided hormones into two groups: those with intracellular receptors that act by initiating new protein synthesis (known as genomic effects because the hormone acts on DNA) and those whose activity depends on plasma membrane receptors and second messenger systems that modify existing proteins (nongenomic effects). One of the most interesting discoveries in recent years has been the blurring of the distinction between these two groups. There is now good evidence that aldosterone, a steroid hormone, has both types of receptors: a cytoplasmic receptor that initiates a slow response and a cell membrane receptor that allows aldosterone to affect cell function within 1–2 minutes. (The aldosterone membrane receptor has not yet been described in Na^+-reabsorbing epithelia such as the kidney tubule.) In addition, the cytoplasmic receptor also seems to control a two-phase response. In the early phase, the apical Na^+ channels increase their open time under the influence of an as-yet-unidentified signal molecule. As intracellular Na^+ levels rise, the Na^+/K^+-ATPase speeds up, transporting cytoplasmic Na^+ out into the extracellular fluid. The net result is a rapid increase in Na^+ reabsorption that does not require the synthesis of new channel or ATPase proteins. In the later phase of aldosterone action, new channels and pumps are inserted into the epithelial cell membranes.

FOCUS BOXES
You will find three kinds of short focus boxes in this book, each one designated by a unique icon. All of these boxes were developed with the goal of helping you understand the role of physiology in science and medicine today.

 Biotechnology boxes discuss physiology-related applications and laboratory techniques from the fast-moving world of biotechnology.

 Clinical boxes focus on clinical applications and pathologies that clarify normal function. In addition to being interesting, these boxes will help you understand the normal function of the human body.

 Cellular and Molecular Physiology boxes offer additional insight concerning physiological events at the cellular and molecular levels. Most physiological research today is being done at the cellular and molecular levels, so these boxes will help you integrate current research with the physiology you are learning.

middle of the seventeenth century. Then, European medicine was still heavily influenced by the ancient belief that the cardiovascular system distributed both blood and air. Blood was thought to be made in the liver and distributed throughout the body in the veins. Air went from the lungs to the heart, where it was "digested" and picked up "vital spirits." From the heart, air was distributed to the tissues through vessels called *arteries*. Anomalies such as the fact that a cut artery squirted blood rather than air were ingeniously explained by unseen links between arteries and veins that opened with injury.

According to this model of the circulatory system, the tissues consumed all blood delivered to them, so the liver continuously synthesized new blood. It took the calculations of William Harvey (1578–1657), court physician to King Charles I of England, to show that the weight of the blood pumped by the heart in a single hour is greater than the weight of the entire body! Once it became obvious that the liver could not make blood as rapidly as the heart pumped it, Harvey looked for an anatomical route that would allow the blood to recirculate rather than be consumed in the tissues. He showed that valves in the heart and veins created a one-way flow of blood and that veins carried blood back to the heart, not out to the limbs. He also showed that blood returning to the right side of the heart from had to go to the lungs before it could go to the the heart. The results of these studies creat among Harvey's contemporaries, leading Ha in a huff that no one under the age of 40 cou stand his conclusions. Ultimately, Harve became the foundation of modern cardiovasc iology. Today, we understand the structure diovascular system at microscopic and molec that Harvey never dreamed existed. Yet so have not changed. Even now, with our sop technology, we are searching for "spirits" in although we call them by names such as "horr "cytokine."

OVERVIEW OF THE CARDIOVASCULAR SYSTEM

In the simplest terms, the cardiovascular syste tem of tubes (the blood vessels) filled with flu and connected to a pump (the heart). Pressure in the heart propels blood through the syster ously. The blood picks up oxygen at the lungs ents in the intestine, delivering them to c simultaneously removing cellular wastes for In addition, the cardiovascular system plays tant role in cell-to-cell communication and in the body against foreign invaders. This chap on an overview of the cardiovascular system a heart as a pump. The properties of the blood v

Problem

Myocardial Infarction

At 9:06 A.M., the blood clot that had been silently growing in Walter Parker's left coronary artery made its sinister presence known. The 53-year-old advertising executive had arrived at the Dallas Convention Center feeling fine, but suddenly a dull ache started in the center of his chest and he became nauseated. At first he brushed it off as aftereffects of the convention banquet the night before. However, when it persisted, he made his way to the Aid Station. "I'm not feeling very well," he told the medic. "I think it may be indigestion." The medic, on hearing Walter's symptoms and seeing his pale, sweaty face, immediately thought of a heart attack. "Let's get you over to the hospital and get this checked out."

continued on page 390

the homeostatic controls that regulate blood flow and blood pressure are described in Chapter 15.

The Cardiovascular System Transports Material throughout the Body

through a vessel; vasodilation will increase blood flow through a vessel.

In summary, by combining equations 1 and 2 above, we get the following equation:

(6) Flow $\propto \Delta P/R$

The flow of blood within the circulatory system is directly proportional to the pressure gradient in the system and inversely proportional to the resistance of the system to flow. If the driving pressure remains constant, then flow will vary inversely with resistance.

Velocity of Flow Depends on the Flow Rate and the Cross-Sectional Area

The word *flow* is often used imprecisely in cardiovascular physiology, leading to confusion. Flow usually means **flow rate,** the volume of blood that passes one point in the system per unit time. In the circulation, flow is expressed in liters per minute (L/min) or milliliters per minute (mL/min). For instance, blood flow through the aorta in a 70-kg man at rest is about 5 L/min.

Flow rate should not be confused with **velocity of flow,** the distance a fixed volume of blood will travel in a given period of time. Velocity of flow is a measure of how fast blood is flowing; flow rate measures the volume of blood that flows past a point in a given period of time. In a tube of fixed size, velocity of flow is directly related to flow rate. In a tube of variable diameter, blood velocity will also vary inversely with the diameter. Thus, velocity will be faster through the narrow sections and slower in the wider sections. The relationship between flow rate (Q), cross-sectional area of the tube (A), and velocity of flow (v) is expressed by the following equation:

(7) Velocity = flow rate/cross-section area

Figure 14-6 ■ shows how the velocity of flow will vary according to the cross-sectional area of the tube. The vessel in the figure has two widths: narrow, with a small cross-sectional area of 1 cm², and wide, with a large cross-

continued from page 384

The clot in Walter's coronary artery has restricted blood flow to his heart muscle, and heart muscle cells are beginning to die from lack of oxygen. In the next few hours, various interventions are critical to save Walter's life and prevent further damage. While waiting for the ambulance to arrive, the medic gives Walter oxygen, hooks him to a heart monitor, and starts an intravenous (IV) injection of normal (isotonic) saline. With an intravenous injection line in place, other drugs can be given rapidly if Walter's condition should suddenly get worse.

Question 1: Why is the medic giving Walter oxygen? What effect will the injection of isotonic saline have on Walter's extracellular fluid volume? On his intracellular fluid volume? On his total body osmolarity?

✓ The two identical tubes below have the pressures shown at each end. Which tube has the greater flow? Defend your choice.

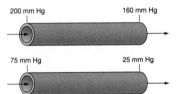

✓ All four tubes below have the same driving pressure. Which tube has the greatest flow? Which has the least flow? Defend your choices.

sectional area of 12 cm². The flow rate is identical in both parts of the vessel: 12 cm³ per minute.* This flow rate means that in one minute, 12 cm³ of fluid flows past point X in the narrow section, and 12 cm³ flows past point Y in the wide section. But *how fast* does the fluid need to flow to accomplish that rate? According to equation 7, the velocity of flow at point X is 12 cm/min, but at point Y it is only 1 cm/min. Thus, fluid flows more rapidly through narrow sections of a tube than through wide sections.

To see this principle in action, watch a leaf as it floats down a stream or gutter. Where the stream is narrow, the leaf moves rapidly, carried by the fast velocity of the water. In sections where the stream widens into a pool, the velocity of the water drops and the leaf meanders more slowly.

Table 14-2 summarizes the factors that influence blood flow. In the next chapter, you will learn more about blood flow through the cardiovascular system. In the remainder of this chapter, we explore the heart and how it creates the driving pressure for blood flow.

✓ Two canals in Amsterdam are identical in size, but the water flows faster through one than through the other. Which canal has the highest flow rate?

* 1 cm³ = 1 mL

RUNNING PROBLEMS
Each chapter includes a real-life problem that appears in segments throughout the chapter.

After the opening scenario, additional segments present material related to the problem and the chapter, then ask questions that prompt you to utilize information you have learned.

PROBLEM CONCLUSION

At the end of the chapter, you can check your problem-solving strategy against the **Problem Conclusion**, which lists the questions from each segment, the facts relevant to each question, and the integration and analysis steps that were required to successfully handle the running problem.

Problem Conclusion

In this running problem, you learned about the current treatments for heart attack. You also learned that many of these treatments depend on speed to work effectively.

Further check your understanding of this running problem by checking your answers against those in the summary table.

Question	Facts	Integration and Analysis
1a Why is the medic giving Walter oxygen?	The medic suspects that Walter has had a heart attack. A blood clot may be blocking blood flow and oxygen supply to the heart muscle.	If the heart is not pumping effectively, the brain may not receive adequate oxygen. Administration of oxygen will raise the amount of oxygen that reaches both the heart and the brain.
1b What effect will the injection of isotonic saline have on Walter's extracellular fluid volume? On his intracellular fluid volume? On his total body osmolarity?	An isotonic solution is one that does not change cell volume (∞ p. 134). Isotonic saline is isosmotic to the body.	The extracellular volume will increase because all of the saline administered will remain in that compartment. The intracellular volume and total body osmolarity will not change.
2a Some of the enzymes that are used as markers for heart attacks are also found in related forms in skeletal muscle. What are related forms of an enzyme called?	Related forms of an enzyme are called isozymes (∞ p. 80).	Although isozymes are variants of the same enzymes, their activity may vary with different conditions, and their structures are slightly different. Cardiac and skeletal muscle isozymes can be distinguished by their different structures.
2b What is troponin, and why would elevated blood levels of troponin indicate heart damage?	Troponin is the regulatory protein bound to tropomyosin (∞ p. 334). When Ca^{2+} binds to troponin, tropomyosin shifts and uncovers the myosin-binding site of actin.	Troponin is part of the contractile apparatus of the muscle cell. If troponin escapes from the cell and enters the blood, this is an indication that the cell has been damaged or is dead.
3a How do electrical signals move from cell to cell in the myocardium?	Electrical signals pass directly from cell to cell through gap junctions in the intercalated disks (∞ p. 147).	The cell-to-cell conduction of electrical signals in the heart allows rapid and coordinated contraction of the myocardium.
3b What happens to contraction in a myocardial contractile cell if a wave of depolarization passing through the heart bypasses it?	Depolarization in a muscle cell is the signal for contraction.	If a myocardial cell is not depolarized, it will not contract. Failure to contract creates a nonfunctioning region of heart muscle and impairs the pumping function of the heart.
4 If the ventricles of the heart are damaged, in which wave or waves of the electrocardiogram would you predict abnormal changes to occur?	The waves of the ECG are the P wave, QRS complex, and T wave. The P wave represents atrial depolarization; the QRS complex and T wave represent ventricular depolarization and repolarization, respectively.	The QRS complex and the T wave are mostly likely to show changes after a heart attack. Changes indicative of myocardial damage include enlargement of the Q wave, shifting of the S-T segment off of the baseline (elevated or depressed), and inversion of the T wave.
5 If Walter's heart attack has damaged the muscle of his left ventricle, what do you predict will happen to his left cardiac output?	Cardiac output equals stroke volume (the amount of blood pumped by one ventricle per contraction) times heart rate.	If the ventricular myocardium has been weakened, the stroke volume may decrease. Decrease in stroke volume in turn would decrease the cardiac output.

QUESTION
To find out more about how to answer a question (or to check your own solution), examine the *Facts* and *Integration and Analysis* columns to the right of each question.

FACTS
As you study the chapter, identify relevant facts from this chapter and from earlier chapters. Use the facts listed here to check your assessment.

INTEGRATION AND ANALYSIS
As you work through the chapter, try to analyze the information you've learned to arrive at an answer. Using the explanation under *Integration and Analysis*, check to see if your reasoning is sound.

Brief Contents

Contents

8
The Nervous System   201

9
The Central Nervous System   235

10

Sensory Physiology   263

11
Efferent Peripheral Nervous System: The Autonomic and Somatic Motor Divisions   307

12
Muscles   324

13
Integrative Physiology I: Control of Body Movement   362

14
Cardiovascular Physiology   383

15
Blood Flow and the Control of Blood Pressure   423

16
Blood   453

17
Respiratory Physiology   474

18
The Kidneys   518

19
Integrative Physiology II: Fluid and Electrolyte Balance   543

23

Integrative Physiology III: Exercise   684

24

Reproduction and Development   698

Appendices

Preface

There has never been a better time to apprentice oneself into the study of human physiology. We are the beneficiaries of centuries of study by physiologists who have constructed a foundation of knowledge about how the human body functions. That work has been supplemented since the 1970s by rapid advances in the fields of cellular and molecular biology, advances so spectacular that in comparison, students sometimes see physiology as a "dead" field, one in which everything is known. Indeed, we once thought that by sequencing the human genome, we would have the key to the secret of life. However, this deconstructionist view of biology has proved to have its limitations. Living organisms are much more than the simple sum of their parts, and sequencing genes and proteins in the molecular biology laboratory does not always tell us the significance of these substances in the living animal. As we enter the twenty-first century, molecular biologists are turning to physiologists to help them understand the function of molecules in the intact organism. The integration of function across all levels, from the molecules to the living body, is the special skill of a physiologist.

Our knowledge base in physiology and all of biology has been expanding so rapidly that it is impossible for any one individual to keep up with every detail. As students study human physiology using this book, their goal should be to learn the language of physiology and the basic concepts that give the language meaning. Using that language and those concepts, they can develop the problem-solving skills that they need to succeed in a career in physiology, medicine, and other challenging fields.

This textbook has some special features that are designed to make the study of physiology easier. Please take a few minutes to look at the Owner's Manual on page v. That section explains some of the symbols, boxes, and features that make this book unique.

LEARNING PHYSIOLOGY

Learning physiology should have more to do with applying information than with memorizing what will be on the test. Physiologists, medical professionals, and others whose work depends upon a background in physiology do not rely upon the facts they once memorized for an exam. Instead, these people are problem solvers, with the ability to pose questions and create answers. But how did these people gain this skill? People are not born problem-solvers; they learn how to approach problems and how to reason their way to an answer. And any skill that is learned can be taught.

One essential step in learning to problem solve is using higher level cognitive skills. Benjamin S. Bloom* and his associates divide cognitive skills into six levels: knowledge, comprehension, application, analysis, synthesis, and evaluation. At the lowest levels, Bloom places knowledge and comprehension, skills that require the student to recognize or recall information, or think on a low level so that the knowledge can be reproduced or communicated without a verbatim repetition. This is the kind of thinking that allows a student to memorize his or her class notes and do well on a test. Moving to the higher cognitive levels requires additional skills. Application of information requires that the student solve or explain a problem by applying what he/she has learned to new situations. Analysis requires that the student solve a problem through the systematic examination of facts or information, and synthesis requires that the student find a solution to a problem through the use of original, creative thinking. Application, analysis, and synthesis are the skills that one uses to solve problems. The highest cognitive level in Bloom's scheme is evaluation, which requires that the student assess the value of information according to some standards.

How can students develop higher-level cognitive skills? One of the most important ways is to actively build models of the information they are learning. These models may be mental models or a concrete representation of a visual image or concept. Creating maps is one excellent way to organize material and build models.

Another way to improve skill is to practice problem solving. This type of active learning can take place in the classroom, in the laboratory, and outside of formal instructional settings when students work individually,

---

*Benjamin S. Bloom, ed. *A Taxonomy of Educational Objectives: The Classification of Educational Goals, Handbook I: Cognitive Domain.* (New York: McKay Publishers, 1956).

in pairs, or in small groups. Learning to solve problems is an extension of the traditional "memorize and recite" paradigm. It does not mean that students will not memorize material. On the contrary, in order to solve problems, one must have an extensive mental file of information on which to draw. That information must be logically organized, and there must be some reproducible method for accessing it.

KEY THEMES OF THIS BOOK

This book has four key themes that grew directly out of my classroom teaching experience. These themes give this book its particular flavor and approach to physiology.

1. A Focus on Problem Solving

Problem solving requires more thought and effort than memorizing strings of facts. Students who merely memorize facts are seldom able to do anything with this information after the brief time that it resides in the short-term memory. But those who connect facts to concepts and apply concepts to real-world situations remember the information longer and are able to use it to think critically and solve new problems. The information becomes valuable and permanent because it has context and meaning. With this in mind, I have utilized several devices and strategies for helping students "do something" with the information in this book. These include the Running Problems in each chapter and higher-level questions in both the Concept Check and the end-of-chapter questions.

2. An Emphasis on Integration

The physiology discipline has become so large and so complex that it is easy to lose sight of the "big picture" in all of the details. Much like dots of color in a painting by the French impressionist Claude Monet, the individual details of physiology are interesting by themselves but gain additional significance when viewed as a whole. The theme of integration in physiology is evident in many of the illustrations in the book; it can be seen in certain analogies and narrative discussions within each chapter. End-of-chapter exercises ask students to create their own maps and to use maps as a tool to step back and consider the big picture. Additionally, three chapters in the book (Chapters 13, 19, and 23) highlight the integrative nature of physiology and the cooperative function of multiple systems in the human body when homeostasis is disturbed.

3. Cellular and Molecular Physiology

Most physiological research today is being done at the cellular and molecular level. Every chapter of the text has been developed with the goal of integrating physiology with cellular and molecular themes in order to enhance student understanding of the big picture. An emphasis on cellular and molecular elements also helps students to develop problem-solving skills as they see how physiologists analyze and apply new research to what is already known about how the human body works.

4. Physiology as a Dynamic Field

In the book, I have tried to present physiology as a dynamic discipline with numerous unanswered (even unasked) questions that merit further investigation and research. It is essential that students appreciate that many of the "facts" they are learning are really only our current theories. Thus, memorizing details for the next test should not be as important as learning the major themes and concepts that underlie the normal function of the body. Once students have a firm foundation and an understanding of the basic concepts and themes in physiology, they will be able to modify their models of physiological function as they incorporate new information.

SPECIAL FEATURES

The book has a number of special features designed to make it easy and fun to use.

Art

Hand-crafted illustrations by William Ober, M.D. Bill is an award-winning medical illustrator who served as the art coordinator for this book. He has the ability to help students visualize concepts through his clear, easy-to-understand illustrations, and his anatomical art is a highlight of the book.

Anatomy Summaries Perhaps the best example of Bill Ober's work are these unique anatomy summary figures that show the anatomy of a physiological system, from a macro- to micro-perspective, in one large spread. "Zoom" arrows are used to "explode" structures into their component parts, all the way down to the tissue level. These summaries allow students to see all of the essential features of each system in one figure.

Reflex Pathway and Concept Maps A unique pedagogical tool in this book is the use of color-keyed reflex pathways as the basis for all control loops. The steps of a reflex (stimulus, receptor, afferent path, integrating center, efferent path, effector, response) can be applied to all extrinsic control loops, from the simplest to the most complex. Students find it easier to remember complex pathways when they have a logical scheme by which they can classify the steps. In addition, a series of structure/function maps organize material in a logical, relational format to carefully walk students through the organization of physiological systems.

Figure and Graphs Using graphs and figures is a skill that most students need to practice. Graphs throughout the book have special graph questions that ask students to extrapolate from or interpret the information represented by the graph. Figure questions ask students to consider some aspect of the illustration, or to consider the figure in the context of material presented previously. Answers to the graph and figure questions are included in Appendix D.

Content

Running Problems Each chapter includes a real-life problem that appears runs in segments throughout the chapter. After the opening scenario, additional segments present material related to the problem and the chapter, then ask questions that prompt students to utilize information they have learned. At the end of the chapter, students can check their problem-solving strategy against the Problem Conclusion, which lists the questions from each segment, the facts relevant to each question, and the integration and analysis steps that were required to successfully handle the running problem. The goal of the running problem is to help students develop problem-solving skills in the context of real-world scenarios. Each running problem has been carefully developed with the assistance of the book's clinical consultant, Andrew Silverthorn, M.D.

Integration Chapters The subtitle of this book is "An Integrated Approach," and integration is a theme that appears in several forms. Units II, III, and IV end with an integration chapter that illustrates how the body systems coordinate their function. Thus, at the end of Unit II covering homeostasis and control systems, you will find a chapter (13) that focuses on the integration of muscle control. Likewise, Unit III, which discusses the cardiovascular, respiratory and renal systems, ends with an integration chapter (19) on fluid and acid base balance. Finally, Unit IV, which addresses digestion, endocrine control of metabolism, and the immune system, has an integration chapter (23) that discusses the coordinated response of body systems when challenged by exercise.

Focus Features on Organs There are several special focus features highlighting the following organs: skin, liver, thymus, pineal gland, bone marrow, and spleen. Too often, the anatomy and physiology of these organs are overlooked in physiology texts, so in an effort to show students the big picture, I have added these special focus features.

Focus Boxes Three kinds of small boxes are found in this book, each one designated by a unique icon. All of these boxes were developed with the goal of enhancing the presentation of normal physiology.

 Biotechnology boxes discuss physiology-related applications and laboratory techniques from the fast-moving world of biotechnology.

 Clinical boxes focus on clinical applications and pathologies that clarify normal function.

 Cellular and Molecular Physiology boxes offer additional insight concerning physiological events at the cellular and molecular levels.

Appendices *Appendix A: Physics and Math* This appendix reviews some of the basic principles of biophysics such as force and energy. It includes sections on bioelectric principles, osmotic principles, the Nernst and Goldman equations, and laws governing the behaviors of gases and liquids.

Appendix B: Genetics This appendix reviews the structure and function of DNA. It includes information on DNA replication, mitosis, mutations, and oncogenes.

Appendix C: Anatomical Positions of the Body

Appendix D: Answers to Concept Check and Graph Questions

Combined Glossary and Index Students like the accessibility of definitions in a glossary, but sometimes they want additional information from the text itself. Usually this means flipping from the Glossary to the Index. Our combined Glossary/Index meets both those needs, with definitions for selected words as well as numbers for the text pages where terms are discussed, all in one convenient location.

Pedagogy

Chapter Outline Each chapter begins with an outline listing the major headings for that chapter. Next to each heading is a page reference. This format serves as an advance organizer and also functions as a helpful review and reference tool.

Background Basics This is a handy tool found at the beginning of each chapter that shows students which important concepts they should have mastered prior to beginning the chapter. Page references allow students to revisit those topics that they need to review.

Concept Links These blue "chain-links" with page references are an extension of the Background Basics, linking students to concepts discussed earlier in the text. This allows them to review material that might be helpful for understanding new concepts.

Figure Reference Locators The red squares found next to every figure callout in the text function as place markers, making it easier for students to toggle between an illustration and the narrative associated with that illustration.

Concept Checks   The red check marks placed at intervals throughout the chapters act as stopping points where students can check their understanding of what they just read. Some of these questions are factual, whereas others ask students to think about or apply what they have learned. Answers to the Concept Check questions are in Appendix D for instant feedback.

Figure and Graph Questions   In an effort to help students interpret quantitative information, apply concepts, and see the big picture, a number of figures and graphs throughout the book include questions that ask students to pause for a moment and consider what they really know about the information they just encountered. Answers to Graph and Figure Questions are in Appendix D.

Chapter Summary   This outline with enumerated points and page references provides the student with a quick summary of every major concept in the chapter.

End-of-Chapter Questions   The end-of-chapter questions have been organized into three levels of difficulty. Level One questions review basic facts and terms—the language of physiology. Level Two questions test the student's ability to comprehend and integrate the chapter's key concepts. Level Three exercises are real-world scenarios designed to develop a student's problem-solving skills. Answers are available in the Instructor's Resource Guide and on the Web site for the book.

MORE THAN A TEXTBOOK

Technology is a wonderful thing. In the 1970s, it allowed publishers to make full-color textbooks affordable. In the 1980s, technology made computer test item files and student tutorials possible. And now, in the latter half of this decade, multimedia and access to the Internet have given us an altogether different dimension to explore. The book itself continues to be the single most important physical component of any physiology course, but we have also acknowledged that it takes more than a textbook to improve a student's ability to learn. With this in mind, we have assembled an array of innovative supplements for both students and instructors.

For Students

The Silverthorn Physiology Web Site   The companion web site for this text can be found at www.prenhall.com/silverthorn
  This site contains a study guide and a selection of frequently asked student questions. It also contains the answers to all end-of-chapter questions in the book.

Additionally, there are links to other interesting physiology-related sites. Because the World Wide Web is such a fluid medium, you never know what new information you might find on this site, so I encourage you to be a regular visitor.

The Physiology Workbook   This useful supplement provides an additional opportunity to review basic concepts and further develop your problem-solving skills. Written in a casual, conversational style, each chapter presents questions, exercises, and problems that reinforce the material and key themes from the text. [ISBN: 0-13-267543-9]

Themes of the Times   is a program sponsored jointly by Prentice Hall and The New York Times. It is designed to enhance student access to current, relevant information. Physiology-related articles have been compiled into a free supplement that helps students make the connection between the classroom and the outside world. Ask your instructor to order this free supplement directly from Prentice Hall.

The Prentice Hall Anatomy and Physiology Video Tutor   gives students the opportunity to see the dynamics of the most difficult physiological processes from the comfort of a favorite couch or chair. This two-hour video focuses on concepts that instructors across the country have consistently identified as the most challenging. Physiological concepts are highlighted using three-dimensional animations and video footage. On-camera narration and built-in review questions ensure that these concepts come to life for the viewer. [ISBN: 013-7518439]

For the Instructor

Instructor's Resource Guide   Designed to be useful to the new instructor and the veteran alike, this supplement contains an annotated chapter outline with additional detail that instructors will find useful in writing their lectures, references from the research and clinical literature, and demonstrations and activities designed to make any classroom an arena of active learning. [ISBN: 013-262536-9]

Test/Quiz Management System   A printed bank of over 2,000 questions is available with this text. Written in the same configuration as the end-of-chapter questions in the text, the test bank is a valuable complement to an instructor's own test/quiz files. Available in Macintosh and Microsoft Windows formats, this powerful software includes easy-to-use Wizards (the Wizard asks the user questions and offers prompts to simplify test creation), an editing function, and a grade book program. This software also includes a test item analysis program that generates helpful statistics on class performance. [ISBN: print version 013-262536-9; Mac version 013-262569-5; Windows version 013-262551-2]

Transparency Acetates   Prentice Hall has prepared approximately 250 full-color acetates containing key illustrations from the text and packaged them in a handy three-ring binder. [ISBN: 013-262577-6]

Presentational CD-ROM for the Lab or Classroom
Think of this supplement as a bank of images from the text along with a series of physiological animations. The proprietary software program embedded on the CD allows the instructor to preview, select, and then save images and animations in the order to be presented in the lab or classroom. [ISBN: 013-262585-7]

Custom Lab Exercises for Physiology   A selection of over 150 (and always growing) class-tested laboratory exercises authored by instructors from around the world and edited by Dee Silverthorn, Alice Mills, and Bruce Johnson. Using a special ordering menu, Prentice Hall will allow instructors to select and customize a laboratory manual that reflects their needs and inventory of equipment. For instructors who have grown weary of using a small percentage of traditional lab manuals, the Prentice Hall Custom Lab Program for Physiology is definitely worth considering. For more information, contact a Prentice Hall sales representative or visit the Prentice Hall Web site at www.prenhall.com

College Newslink   Let Prentice Hall's College Newslink bring each day's physiology news right to your computer. Subscribers to this program receive a personal e-mail each morning with news stories from the ever-changing world of physiology. Such highly respected sources as The New York Times, The Boston Globe, The San Francisco Chronicle, The London Times, Newsweek, and others can help instructors give physiology a real-world focus every day of the semester.

For more information contact a Prentice Hall sales representative or visit our web site at www.ssnewslink.com

EVOLUTION OF THIS TEXTBOOK

Textbooks are for students, and one of the unique aspects of this book is the role that students have played in its development. Creating a new textbook is a detailed process that involves many steps and many people. Despite the tremendous support team I have taken advantage of, I take ultimate responsibility for any errors, omissions, ambiguities or other problems that make this book less than what it could be.

Market Surveys

The process began with a series of market surveys that allowed us to test some ideas that we wanted to implement in the book. I was pleased to see that numerous physiology instructors across North America support active learning approaches in the classroom. It was also evident that almost all instructors would embrace a

book that could help their students become better problem solvers.

Peer Input

For the past several years, I have sought out opportunities to talk to other people about what they are doing in their classrooms and laboratories. I have learned a great deal from my colleagues who took the time to talk to me about their active learning techniques and educational philosophies. The book is better because of these valuable exchanges.

Focus Groups

The book was developed with the input from two focus groups, one at the beginning of my first draft and one as we moved into the second draft. This gave me another opportunity to talk to instructors about my approach, their preferences, and the contents of the ideal physiology text. We discussed illustrations, we debated features, and we dissected whole chapters of manuscript. My deepest thanks go to the following faculty members who took time from their busy schedules to meet and share their thoughts and ideas.

Patricia J. Berger, University of Utah

Timothy Bradley, University of California—Irvine

Alan P. Brockway, University of Colorado at Denver

Kim Cooper, Arizona State University

Gordon Garrett, Utah Valley State College

Richard W. Heninger, Brigham Young University

Bruce Johnson, Cornell University

Alice C. Mills, Middle Tennessee State University

Patricia L. Munn, Longview Community College

Mary Anne Rokitka, State University of New York (SUNY) at Buffalo

Norm Scott, University of Waterloo

Judy Sullivan, Antelope Valley College

Richard L. Walker, University of Calgary

David A. Woodman, University of Nebraska—Lincoln

Reviews

Numerous instructors, content specialists, and students reviewed each draft of my manuscript, including concept sketches for line art. Through this process, they provided valuable insights and expertise in a field where it is impossible for any one individual to be an expert in everything. They also kept me on track with respect to clarity, accuracy, organization, and level of detail, and saved me from some hilarious typos that the spell-checker missed, such as the "congenial defect" and the "fatty steak" of atherosclerosis. I am very grateful for their thoughtful input and comments, and take full responsibility for any errors.

Brenda Alston-Mills, North Carolina State University

Roland M. Bagby, University of Tennessee

Patricia J. Berger, University of Utah

Ronald Beumer, Armstrong State College

Sunny K. Boyd, University of Notre Dame

Timothy Bradley, University of California—Irvine

Alan P. Brockway, University of Colorado at Denver

David Byman, Pennsylvania State University Worthington-Scranton Campus

Robert G. Carroll, East Carolina University School of Medicine

Kim Cooper, Arizona State University

Charles L. Costa, Eastern Illinois University

Gordon Garrett, Utah Valley State College

Margaret M. Gould, Georgia State University

Janet L. Haynes, Hinds Community College

Richard W. Heninger, Brigham Young University

Carol Hoffman, San Jose City College

Bruce Johnson, Cornell University

William C. Kleinelp, Middlesex County College

James Larimer, University of Texas

Daniel L. Mark, Penn Valley Community College

Paul Matsudaira, Massachusetts Institute of Technology

Wayne Meyer, Austin College

Joel Michael, Rush Medical College

Alice C. Mills, Middle Tennessee State University

Ron Mobley, Wake Technical Community College

Harold Modell, National Resource for Computers in Life Science Education

John G. Moner, University of Massachusetts at Amherst

Patricia L. Munn, Longview Community College

Dell Redding, Evergreen Valley College

Mary Anne Rokitka, State University of New York (SUNY) at Buffalo

Allen Rovick, Rush Medical College

Evelyn Schlenker, University of South Dakota

Norm Scott, University of Waterloo

Brian R. Shmaefsky, Kingwood College

Sabyasachi Sircar, Delhi, India (IUPS Teaching Fellow, 1995-6)

Judy Sullivan, Antelope Valley College

Deborah Taylor, Kansas City Kansas Community College

Richard L. Walker, University of Calgary

Donald Whitmore, University of Texas at Arlington

Ronald L. Wiley, Miami University of Ohio

Jack Wilmore, University of Texas

David A. Woodman, University of Nebraska—Lincoln

I also took advantage of a small army of my students who spent hours examining manuscript, page proofs, and illustrations. I owe a tremendous debt of gratitude to my 1996-7 students who read and answered Running Problems. I would particularly like to thank those who volunteered their free time to proofread pages. The following students came in regularly to pore over the manuscript and figures; Shelly Spiller, who has missed her calling as an editor, merits special recognition. Shelly spent over 90 hours working on the pages.

John Brehm, Michael Cimo, James Dopson, Darlene Esper, Joel Grant, Manisha Gupta, Steve Hewitt, Damian Hill, Loc Ho, Andrea Kretzschmar, Kabel Morgan, Islam Mossaad, Manny Moustakakis, Seth Roberts, Michael Rubin, Vishal Shah, Shelly Spiller, Cathay Wang

Other students who helped proof the book include:

Sam Afshar, Pratap Agusala, Faisal Ahmad, Maneesha Ahluwalia, Jennifer Bang, Cynthia Barrett, Sital Bhavsar, Delbert Brod, Margaret Brown, Jeff Chen, Moya Chikiamco, Jennifer Cozart, Krista Day, Sarang Desai, Katie DeVaul, Rick Farnam, Kery Feferman, Kristina Gross, Christopher Guerin, Steve Hammond, Avital Harari, John Harrell, Razi Hussani, Nadia Islam, Omeed Khodaparast, Cindy Lee, Dana Ly, Laura Maltz, Jennifer Navarro, Harriett Ng, Lan Nguyen, Robin Oxley, Monika Parekh, John Schelter, Turner Slicho, Bill Stigall, Vivian Tsang, Russell Ward

Illustrations

A picture can be worth a thousand words, but getting an illustration just right is sometimes a challenge. The process began with my rough sketches. These were reviewed, discussed, and refined as they passed through a committee of instructors, editors, and students. Eventually my sketches became an artist's rough sketches, and these too were passed through the committee. Once the artist's roughs were approved, the final pieces were crafted and labels were carefully applied. Next, the captions were fine-tuned, and the art was joined to the copy-edited manuscript in a form that publishers call "pages."

Pages

In the old days, publishers pasted waxed columns of galley proof and myriad pieces of line art onto paper forms which represented the design and dimensions of the pages of a book. Today it is all done electronically, and I had the luxury of seeing three rounds of computer-generated pages. Each round of pages was reviewed and proofread. It was at this stage of the process that small groups of my students were of immeasurable help, gathering in my lab afternoons, evenings, and even occasional weekends to review and proof "the book." They all had been frustrated by typos and errors in their textbooks at one time or another, so they were eager to help and quickly learned that it was okay to question what they read. Their eagle-eyes spotted mistakes that even the professional proofreaders had missed, and they questioned some inconsistencies that had slipped by numerous other readers. The book is much improved from their input, and I cannot imagine ever doing a book again without this kind of help.

ACKNOWLEDGMENTS

Many other people devoted a lot of time and energy to making this book a reality, and I would like to make an attempt at thanking them. I apologize in advance to anyone whose name has been left out.

My Students

This book was written with students as well as for students. My graduate teaching assistants have played a huge role in my teaching ever since I arrived at the University of Texas, and their input has helped shape how I teach. Many of them are now faculty members themselves. I would particularly like to thank:

Eric Bauer	Lawrence Brewer, Ph.D.	Peter English
Chris Godell	Carol Cutler Linder, Ph.D.	Jan Melton-Machart
Alice Mills, Ph.D.	Tonya Thompson, M.D.	Kurt Venator
Kira Wennstrom		

During the early stages of the book, many students provided input about different topics. They are too numerous to list, but I would particularly like to thank:

Donell Baird-Oliver	Daniel Biller	Arthur Cortez
Arezu Daftarian	Marvin Garcia	Christopher Lord
Jason Nelson		

Prentice Hall

I am convinced that I am writing for the best college textbook publisher in the business. From the very beginning, these smart, hard-working people helped me to formulate a viable vision for this book, and they have been with me ever since. All of these folks worked well beyond their job descriptions to make this book happen. I would especially like to thank David K. Brake, my executive editor, whose creativity, vision, and belief in me ensured that this book will reach its

full potential. David saw that the considerable resources of Prentice Hall were available whenever we needed them during the development and production process. Special thanks also go to Shana Ederer, my development editor, who put a tremendous amount of energy and enthusiasm into the development of this book, and to ̄ ̇ a Wechsler, my production editor. Debra's meticulous attention to detail, stamina, and ability to coach a new author and keep us on schedule are without equal.

Additional Thanks

Some extraordinary people helped above and beyond the call of duty. At the top of that list is Bruce Johnson, Cornell University, Department of Neurobiology and Behavior, who guided me through the development of the chapters on the central nervous system and motor control. Pat Munn at Longview Community College in Kansas helped write the end-of-chapter and test bank questions. Lawrence Brewer and Damian Hill worked with me in developing the Instructor's Resource Guide and Physiology Workbook. Damian Hill and Daniel Biller wrote the physics and genetics appendices. Seth Roberts and Coco Kishi assisted with special art and video projects; Michael Cimo developed the concept for the inside front cover.

A special thanks goes to my colleagues at the University of Texas who encouraged me and who allowed me to bounce ideas off them. They include Ruth Buskirk, Judy Edmiston, Jeanne Lagowski, and Jim Larimer. Other support at the University came from Sue Speer and Linda Davis in the Zoology Department, Marilla Svinicki and staff in the Center for Teaching Effectiveness, and Nancy Elder and staff in the Life Sciences Library.

A special thank you goes to Ann Daniel, who as a Prentice Hall field editor first suggested that I write a book. Some day I may find a way to get even.

I also must thank my family, friends, and colleagues for their understanding whenever I said, "I'm sorry, I can't; I'm working on the book." And the biggest thank you goes to my husband Andy, who was always encouraging and supportive, even when it meant eating "Dinner at Your Doorstep" when schedules got tight.

A WORK IN PROGRESS

Because most textbooks are revised every three or four years, they are always works in progress. I invite you to contact me or my publisher with any suggestions, corrections, or comments about this first edition. I can be reached through E-mail at silverth@utxvms.cc.utexas.edu You can reach my editor at the following address:

David Brake
Executive Editor
Prentice Hall Publishing
1208 E. Broadway
Tempe, AZ 85282
Fax: 602/921-9319
E-mail: david_brake@prenhall.com

—Dee Silverthorn
University of Texas
Austin, Texas
July 1997

To Walter George Unglaub, M.D.,
who started me on this path,
John and Winona Vernberg,
who shaped my view of physiology,
Harold Hempling,
who hired a crustacean physiologist,
Dorothea Bennett,
for her support and encouragement,
and Saidee, Mr. Pepys, and Andy,
for being there.

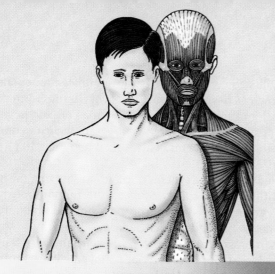

1

Introduction to Physiology

For most of recorded history, humans have been interested in how their bodies work. Early Egyptian, Indian, and Chinese writings describe attempts by physicians to treat various diseases and to restore health. Camel dung and powdered sheep horn may seem to us today to be bizarre therapies, but we must view them in light of what was known about the human body in those times. In order to appropriately treat disease and injury, we must first understand the human body in its healthy state. The study of the normal functioning of a living organism and its component parts, including all chemical and physical processes, is the field of **physiology.**

The term *physiology* literally means "knowledge of nature." Aristotle (384–322 B.C.) used the word in this broad sense to describe the functioning of all living organisms, not just of the human body. However, Hippocrates (ca. 460–377 B.C.), considered the father of medicine, used the word *physiology* to mean "the healing power of nature," and the field became closely associated with medicine. By the sixteenth century in Europe, physiology had been formalized as the study of the vital functions of the human body, although today the term is again used to refer to the study of the functions of all animals and plants. In contrast, **anatomy** is the study of structure, with minimum emphasis on function. Despite this distinction, anatomy and physiology cannot truly be separated. The function of a tissue or organ is closely tied to its structure, and the structure of an organism presumably evolved to provide an efficient physical base for its function.

Physiology is an integrative science that examines function at many levels of organismic complexity, from molecules and cells to intact organisms to populations of animals adapted to a particular set of environmental conditions. In this chapter, we examine the fundamental principles of physiology and explore how we know what we know about the way the human body functions.

LEVELS OF ORGANIZATION

To understand how the human body functions as a coordinated entity, we must first examine its component parts. Figure 1-1 ■ shows the different **levels of organization** of living organisms, ranging from the parts of a single organism to groups of the same species (*populations*) and populations of different species living together in *ecosystems* and the *biosphere*. The different disciplines of chemistry and biology related to the study of each level are shown on the right side of the figure. There is considerable overlap between the different fields, and these artificial divisions will vary according to who is defining them. One distinguishing feature of physiology is that it is a discipline that encompasses many levels of organization.

At a fundamental level, atoms of elements link together to form molecules. One of the great, and possibly unsolvable, mysteries of science is how interactions between certain groups of molecules result in the unique properties of a living organism. The smallest unit of structure capable of carrying out all life processes is the **cell.** Cells are collections of molecules separated from the external environment by a cell membrane. Simple organisms are composed of only one cell, but complex organisms have many cells with different structural and functional specializations.

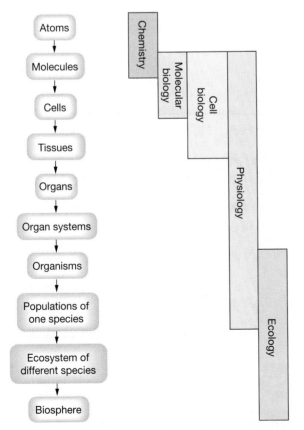

■ Figure 1-1 **Levels of organization and related fields of study**

Collections of cells that carry out related functions are known as **tissues** [*texere*, to weave]. Tissues form structural and functional units known as **organs** [*organon*, tool], and groups of organs integrate their functions to create **organ systems.** The human body has 10 physiological organ systems (Table 1-1).

Figure 1-2 ■ is a schematic diagram showing the interrelationship of the systems in the human body. The **integumentary system,** composed of the skin, forms a protective boundary that separates the body's internal environment from the external environment of the outside world. Support and body movement are provided by the **musculoskeletal system.**

Four systems exchange material between the external and internal environments: The **respiratory system** exchanges gases, the **digestive system** takes up nutrients and water and eliminates wastes, the **urinary system** removes excess water and waste material, and the **reproductive system** produces eggs or sperm. The major organs of these four systems are hollow, and their interior spaces, or **lumens** [*lumin*, window], are essentially extensions of the external environment. Material that enters the lumens of these organs is not truly part of the internal environment until it crosses the tissue wall of the organ.

An interesting illustration of the difference between the internal and external environment involves the bacterium *Escherichia coli*. This organism normally lives and reproduces in the lumen of the large intestine, an interior space that is continuous with the external environment. In this location, the organism does not harm the host. However, if the intestinal wall is accidentally punctured by disease or accident and *E. coli* enters the internal environment of the body, a serious infection can result.

The **immune system** of the body is responsible for protecting the internal environment from foreign invaders such as the *E. coli* bacterium. To carry out that function, the cells and tissues of the immune system are positioned to intercept material that may enter through the exchange surfaces or through a break in the skin. In addition, immune tissues are closely associated with the **circulatory system,** which distributes material throughout the body.

Coordination of body function is the responsibility of the **nervous** and **endocrine systems,** shown in the figure as a continuum rather than as two distinct systems. As we learn more about the integrative nature of physiological function, the lines between these systems blur. How do we classify nerve cells that secrete hormones as endocrine cells do? How should we classify responses that begin in the nervous system and end with the endocrine system? One of the current challenges in physiology is integrating information from the different body systems into a cohesive picture of the living human body. One way that physiologists do this is with the use of maps or flow charts, schematic representations of structure and function (see Fig. 1-3 ■, Focus on Mapping).

TABLE 1.1 Organ Systems of the Human Body

System Name	Representative Organs or Tissues	Functions
Circulatory	Heart, blood vessels, blood	Transport of materials between all cells of the body
Digestive	Stomach, intestines, liver, pancreas	Conversion of food into particles that can be transported into the body; elimination of some wastes
Endocrine	Thyroid gland, adrenal gland	Coordination of body function through synthesis and release of regulatory molecules
Immune	Thymus, spleen, lymph nodes	Defense against foreign invaders
Integumentary	Skin	Protection from external environment
Musculoskeletal	Skeletal muscles, bones	Support and movement
Nervous	Brain, spinal cord	Coordination of body function through electrical signals and release of regulatory molecules
Reproductive	Ovaries and uterus, testes	Perpetuation of the species
Respiratory	Lungs, airways	Exchange of oxygen and carbon dioxide between the internal and external environments
Urinary	Kidneys, bladder	Maintenance of water and solutes in the internal environment; waste removal

PHYSIOLOGY IS AN INTEGRATIVE SCIENCE

Physiology is a discipline that cuts across many levels of organization. At the systems level, there are relatively few unanswered questions about how the human body works. Most of these questions involve the nervous control of function, such as the nervous control mechanisms that govern breathing. The bulk of research in physiology today is focused on the cellular and molecular level. It is here that many questions about how the human body works remain unanswered. Nevertheless, explaining what happens in isolated cells cannot answer questions about how the intact body works. We must also figure out how events in a single cell influence neighboring cells, tissues, organs, and systems throughout the

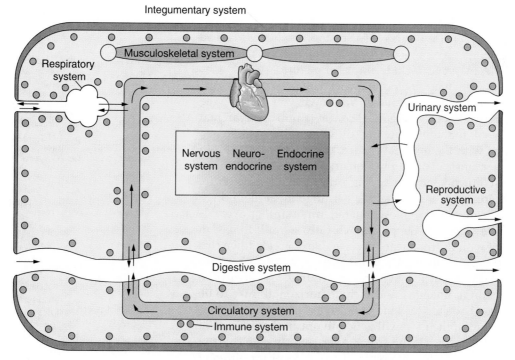

■ Figure 1-2 **The systems of the body** This schematic figure shows the relationships between the 10 systems of the human body. The lumens of the hollow organs of the respiratory, digestive, reproductive, and urinary systems open to the external environment. The immune system is positioned to guard against invaders entering the body from the external environment. The circulatory system transports material through the body. The integumentary and musculoskeletal systems provide support and protection. The nervous and endocrine systems, shown here as a continuum, integrate information and provide coordinated control of body functions.

Focus on   **Mapping**

■ Figure 1-3   **Maps for physiology**

(a) A map showing structure/function relationships

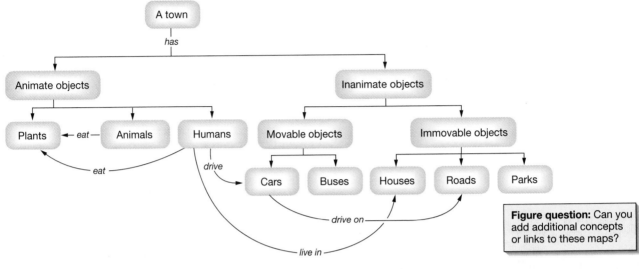

Figure question: Can you add additional concepts or links to these maps?

Mapping is a nonlinear way of organizing material. It is a useful study tool because mapping requires active processing of information rather than memorization. Maps in physiology usually focus on the relationships between anatomical structures and physiological processes (structure/function), normal homeostatic control processes (flow charts), or response pathways to abnormal (pathophysiological) events. A map can take a variety of forms but often consists of terms linked by explanatory arrows. The connecting arrows can be labeled with explanatory phrases (for example, tissues "are composed of" cells). A map may include graphs or pictures.

The real benefit from mapping occurs when you prepare the maps yourself. By organizing the material on your own, you examine the relationships between terms, organize concepts into a hierarchial structure, and look for similarities and differences between items. Such interaction with the material ensures that you process it into long-term memory instead of simply memorizing it and forgetting it.

Often the most difficult part of mapping is deciding where to begin. Select the concepts to map and mentally arrange them in an organized fashion. Begin with the most general, important, or overriding concept from which all the others naturally stem. If this is an event map, start with the first event to occur. Next, break down this one idea into progressively more specific parts using the other concepts, or follow the event through its time course. Use arrows to point the direction of linkages, and include horizontal links to tie concepts together. Labeling the kind of linkage is useful. The downward development of the map will generally mean the passage of time or an increase in complexity. Color is very effective on concept maps. You can use colors for different types of links or for different sections.

These maps are examples of structure/function maps and flow charts.

(b) A process map, or flow chart

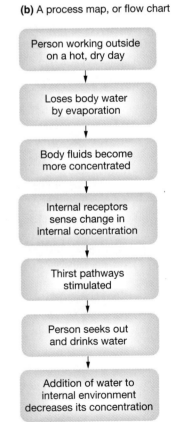

body. In this respect, physiologists are like children throwing rocks into a pond. They want to see what kind of splash the rocks make where they enter the water, but they also want to know where the ripples go and what happens when they encounter other ripples.

The integration of function among systems is a special focus of physiology. Traditionally, students have studied the cardiovascular system and the regulation of blood pressure in one unit, then studied the kidneys and the control of body fluid volume in a separate unit. But body fluid volume is a key factor in the determination of blood pressure, so how can the two units be completely separated? Creating an integrated view of physiology that links together the different systems is one of the most interesting and most challenging aspects of this field.

PROCESS AND FUNCTION

One important skill for you to acquire as you study physiology is the ability to distinguish between process and function. The **function** of a physiological system or event is the "why" of the system or event: Why does the system exist? This way of thinking about a subject is called the **teleological approach** to science. For example, the teleological answer to the question of why red blood cells transport oxygen is "because red blood cells bring oxygen to the cells that need it." This answer explains the reason red blood cells transport oxygen but says nothing about *how* the cells transport oxygen.

In contrast, the **mechanistic approach** to physiology examines the **process** by which events occur, or the "how" of a system. The mechanistic answer to the question of why red blood cells transport oxygen is "because there are hemoglobin molecules that combine reversibly with oxygen molecules." This very concrete answer explains exactly how oxygen transport occurs but says nothing about the significance of oxygen transport to the intact animal.

Although function and process seem to be two sides of the same coin, it is possible to study processes, particularly at the cellular and subcellular level, without understanding the function of those processes in the life of the organism. As biological knowledge becomes more complex, and as research focuses on the subcellular and molecular levels, scientists sometimes become so involved in the processes they study that they fail to step back and look at what those processes mean to cells, organ systems, or the intact animal. One role of physiology is to integrate function and process into a cohesive picture.

THE EVOLUTION
OF PHYSIOLOGICAL SYSTEMS

To understand why the systems of the human body work the way they do, note that human beings are large, mobile, terrestrial animals. Our physiological systems and processes are adapted to allow us to survive in a dry, highly variable external environment, a fact that becomes more obvious when we examine the adaptations of other animals to their habitats.

Life probably originated in tropical seas, a stable environment where salinity, oxygen content, and pH vary little and where light and temperature cycle in predictable ways. The earliest living organisms had an internal environment almost identical to that of seawater, just as today's marine invertebrates do. If external conditions changed, conditions inside the primitive organisms changed as well. Even today, marine invertebrates cannot tolerate significant changes in salinity and pH, as you know if you have ever maintained a saltwater aquarium. In both ancient and modern times, most marine organisms have relied on the constancy of their external environment to keep their internal environment in balance.

As animals migrated from the ancient seas into estuaries, where fresh and sea water mix, then into freshwater environments and onto the land, their external environment was no longer constant. Salinity varies when rains send freshwater into the estuaries or as the sun evaporates tidal pools among the rocks. Temperature fluctuates widely from day to night and from season to season. Animals that live in freshwater must cope with the influx of water into their body fluids, while animals that live on land constantly lose internal water to the dry air around them. Coping with the demands of variable external environments required the evolution of a variety of physiological, structural, and behavioral mechanisms. For terrestrial animals, these mechanisms include anatomical, metabolic, and behavioral adaptations to reduce water loss (impermeable body surfaces, internalized surfaces for gas exchange, and highly efficient kidneys that conserve water). Fertilization and fetal development, two processes that require aqueous environments, often take place internally.

Although the human body as a whole is adapted to cope with a variable external environment, most of the individual cells of the body are much less tolerant of change. Only a small minority of cells in a multicellular organism are actually indirect contact with the external environment (Fig. 1-4 ■). The vast majority of the cells are sheltered from the outside world by the buffer zone of the **extracellular fluid** [*extra-*, outside of], the body fluid that surrounds the cells. This internal environment serves as the interface between the external environment and the cells. When conditions outside the body change, the changes are reflected in the composition of the extracellular fluid, which in turn affects the cells. But our cells are not very tolerant of changes in their surroundings. As a result, a variety of mechanisms have evolved that maintain the composition of the extracellular fluid within a narrow range of values. Figure 1-5 ■ summarizes the ideal situation, in which any change in the internal environment is met by a

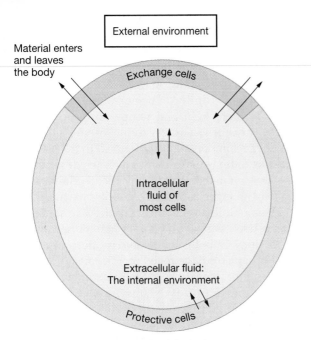

■ Figure 1-4 **The internal and external environments** Only a few cells in the body are able to exchange material with the external environment. Most cells are in contact with the internal environment, composed of the extracellular fluid.

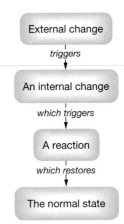

■ Figure 1-5 **Environmental variability and homeostasis**

response that restores the normal condition. The body's coordinated response in order to maintain internal stability is the process known as **homeostasis** [*homeo-*, similar + *-stasis*, condition]. Homeostasis and the regulation of the internal environment are central precepts of physiology and create an underlying theme of each chapter in this book. Failure to maintain homeostasis disrupts normal function.

THEMES IN PHYSIOLOGY

It has been said that "physiology is not a science or a profession but a point of view."* Physiologists pride themselves on relating the processes they study to the functioning of the animal as a whole. One key to critical thinking in physiology is the ability to recognize patterns. The amount of information known about science can seem overwhelming at times, but once you realize that many specific processes are variations on a few basic themes, learning physiology is easier.

One key theme of human physiology is homeostasis. A related theme is the integration of body systems, the ability of the different organ systems to coordinate with each other and work together to maintain homeostasis and carry out the functions of the body. Integration and homeostasis both require that the cells of the body communicate with one another rapidly and efficiently.

A variety of sensing mechanisms monitor changes in the internal and external environments. These sensors, which range from single cells to highly developed special sense organs such as the eyes, nose, and ears, must relay information about environmental changes to other cells. Most cell-to-cell communication uses chemical signals, although the nervous system adds speed by means of electrical signals. Information is sent between distant cells through the movement of blood in the circulatory system or along the specialized cells of the nervous system. Integration and coordination of responses take place in the brain and spinal cord and in endocrine and immune cells.

Communication between the internal environment of the cells and the extracellular fluid requires information transfer across the cell membrane, another key theme. You will learn how some molecules can move across the cell membranes and how others remain outside the cell and simply pass information through the membrane.

Living processes require the continuous input of energy. Where does this energy come from, and how is it stored? As we answer those questions, you will learn how energy is used to do work. In the body, energy is used for synthesis and breakdown of molecules, to transport molecules across cell membranes, and to create movement.

Another common theme and important principle of physiology is the **law of mass balance.** This law says that if the amount of a substance in the body remains constant, any gain must be offset by an equal loss (Fig. 1-6 ■). For example, in order to maintain constant body temperature, heat gain from the external environment and from metabolism must be offset by heat loss back to the external environment. Most substances enter the body from the outside environment, although some, like heat, can be produced internally. The major routes for loss from the body are the urinary and digestive systems (in the urine and feces), the respiratory system, and the integumentary system (skin). Some aspects of mass balance are monitored by sensors (receptors) that detect changes in the internal and external environments. When mass balance

*Ralph W. Gerard, *Mirror to Physiology: A Self Survey of Physiological Science* (Washington, D.C.: American Physiology Society, 1958).

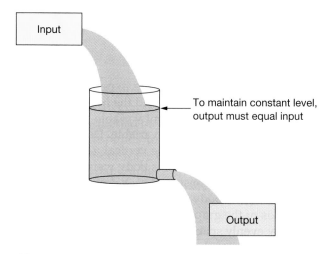

Input

To maintain constant level, output must equal input

Output

■ Figure 1-6 **Mass balance in the body**

is disturbed, physiological and behavioral homeostatic reflexes bring the body back into balance. The law of mass balance is summarized by the following equation:

Total amount of substance x in the body =
$$\text{intake} + \text{production} - \text{output}$$

Mass balance in the body is usually discussed as a function of time. We talk about sodium intake per day or carbon dioxide output per minute. Thus, the *rate* of intake, production, or output is an important consideration. This is expressed as **mass flow,** where

Mass flow (amount/min) =
concentration (amount/vol) $\times$ volume flow (vol/min)

For example, a person is given an intravenous infusion of glucose containing 5 grams of glucose per liter. If the infusion is given at a rate of 2 milliliters per minute, the mass flow for glucose would be

50 g glucose/1000 mL solution $\times$ 2 mL solution/min = 0.1 g glucose/min

Substances whose concentrations are maintained through mass balance include oxygen and carbon dioxide, water, salts, and hydrogen ions. In addition, our bodies have regulatory mechanisms to keep temperature and energy stores within an acceptable range.

THE SCIENCE OF PHYSIOLOGY

How do we know what we know about the physiology of the human body? Scientific knowledge is generated through observation and experimentation, the key elements of **scientific inquiry.** An investigator observes an event or phenomenon and, using prior knowledge, generates a **hypothesis** [*hypotithenai,* to assume], or logical guess, about how the event takes place. The next step is to test the hypothesis by designing an experiment in which some aspect of the phenomenon is manipulated by the investigator. A common type of experiment

removes or alters some element that the investigator thinks is an essential part of the observed phenomenon. The element altered in the experiment is the **independent variable.** For example, a biologist notices that birds at a feeder seem to eat more in the winter than in the summer. She generates a hypothesis that cold temperatures cause birds to increase their food intake. To test her hypothesis, she designs an experiment in which she will keep birds at different temperatures and monitor how much food they eat. In her experiment, temperature, the manipulated element, is the independent variable. Food intake, which is hypothesized to be dependent on temperature, becomes the **dependent variable.**

An essential feature of any experiment is a **control.** A control group is usually a duplicate of the experimental group in every respect except that the manipulated variable is not changed from its normal value. For example, in the bird-feeding experiment, the control group would be a set of birds maintained at warm temperatures but otherwise treated exactly like the birds held at cold temperatures. The purpose of the control is to ensure that any observed changes are due to the experimental manipulation and not to an outside factor. For example, if the investigator in the bird experiment changed to a different food, and food intake increased, she could not determine if the increased food intake was due to temperature or to more palatable food unless she had a control group that was also fed the different food.

During an experiment, the investigator carefully collects facts, or **data** [plural; singular *datum,* a thing given], about the effect that the manipulated (independent) variable has on the observed (dependent) variable. Once the investigator feels that she has sufficient information to draw a conclusion, she begins to analyze the data. Analysis can take many forms and usually includes statistical analysis to determine if apparent differences are statistically significant. A common format for presenting data is a graph (see Fig. 1-7 ■, Focus on Graphs).

If the hypothesis that cold causes birds to eat more is supported by one experiment, the experiment should be repeated to ensure that the results were not an unusual one-time event. This step is called **replication.** When a hypothesis is supported by data on multiple occasions, it may become a scientific **theory.** Most of the facts presented in textbooks like this one are really theoretical models that scientists have developed on the basis of the best experimental evidence available. On occasion, new experimental evidence that does not support a current theory or model is presented. In that case, the model must be revised to fit the available evidence. Thus, although you may learn a physiological "fact" while using this textbook, in 10 years that fact may be inaccurate because of what scientists have learned in the interval.

Where do our scientific models come from? Much of what we know about human physiology has been

Problem

Chromium Supplements

"Lose weight while gaining muscle," the ads promise. "Prevent heart disease." "Stabilize blood sugar." What is this miracle substance? It's chromium picolinate, a nutritional supplement being marketed to consumers looking for a quick fix. But does it work? Some athletes, like Stan—the star forward on the college football team—swear by it. Stan takes 500 micrograms of chromium picolinate daily. Many researchers, however, are skeptical and feel that the necessity for and safety of chromium supplements has not been established.

continued on page 16

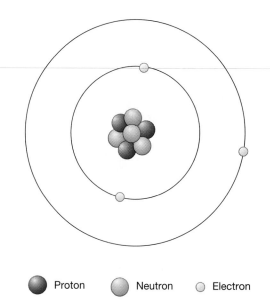

Proton   Neutron   Electron

■ **Figure 2-1 Atomic structure** An atom has a central nucleus composed of protons and neutrons. The nucleus is surrounded by rapidly moving, lightweight electrons, one for each proton in the nucleus. This lithium atom has three protons, four neutrons, and three electrons.

ELEMENTS AND ATOMS

The simplest kinds of matter are called **elements.** Some are substances that you encounter in everyday life: the oxygen that you breathe, the carbon in the soot of a barbecue grill. Elements are the building blocks of all matter, including the human body. Three of the elements—oxygen, carbon, and hydrogen—make up more than 90% of its mass. These elements and others bind together into more complex substances called molecules. To understand what distinguishes one element from another and how they combine into molecules, first we must look at atoms.

Atoms Are Composed of Protons, Neutrons, and Electrons

All elements are composed of small particles called **atoms,** which once were thought to be the smallest particles of matter [*atomos,* indivisible]. We now know, however, that an atom itself is composed of three types of even smaller particles: protons, neutrons, and electrons. These subatomic particles are distinguished from each other by their different electrical charges and masses. A **proton** has one positive charge (written +), an **electron** has one negative charge (written −), and a **neutron** has no charge, or zero charge—that is, it is neutral. Like the poles of a magnet, oppositely charged atomic particles (+ and −) attract, while those with like charge (+ and +, or − and −) repel. An atom usually has an equal number of protons and electrons, giving it an overall electrical charge of zero.

Protons and neutrons have nearly equal masses of about 1 atomic mass unit or amu (1 amu = 1.6605×10^{-27} kg). Electrons are much smaller: It takes the mass of 1836 electrons to equal the mass of one neutron. The arrangement of protons, neutrons, and electrons in an atom is always the same. The protons and neutrons of the atom, which account for almost all of its mass, are clustered into a dense body called the **nucleus** [*nuculeus,* little nut; plural, nuclei]. The space around the nucleus, almost all of the atom's volume, is occupied by the

rapidly moving, lightweight, negatively charged electrons, held in their orbit by the positively charged protons (Fig. 2-1 ■).

Atoms are very tiny, having diameters in the range of 1 angstrom to 5 angstroms (1 angstrom = 1 Å = 10^{-10} m). Most of their volume is empty space. To get a feel for how little of an atom is occupied by the subatomic particles, imagine that the atom has a diameter equal to the length of a football field. The nucleus would then be the size of a large apple in the center of the field, and tiny electrons, smaller than a pea, would move through the remainder of the space.

If all atoms have protons, neutrons, and electrons, what makes an atom of one element different from that of another? For example, why is carbon so unlike oxygen? The next three sections will answer those questions by showing what happens to an atom when it gains or loses one of its subatomic particles.

An Element Is Distinguished by the Unique Number of Protons in Its Nucleus

There are more than 100 different elements in the world, about 90 of which occur naturally; the remainder are manmade. All of the elements that we know of are listed in a **periodic table of the elements,** as shown in Figure 2-2 ■. Each box in the table contains the name of an element and its one- or two-letter symbol.

Only about one-fourth of the elements are commonly encountered in biological systems. Those that are necessary for life are called **essential elements.** Eleven **major essential elements** are found in the human body in fairly large amounts. Carbon (C), hydrogen (H), oxygen (O),

Periodic Table of the Elements

Key:
- Atomic number = number of protons — 6
- Name — Carbon
- Symbol — C
- Atomic mass — 12.0

Legend:
- Major essential elements
- Minor essential elements
- Not believed essential for life

Transitional metals

Group	Period	Element (No., Symbol, Name, Mass)

Group 1
- 1 H Hydrogen 1.0
- 3 Li Lithium 6.9
- 11 Na Sodium 23.0
- 19 K Potassium 39.1
- 37 Rb Rubidium 85.5
- 55 Cs Cesium 132.9
- 87 Fr Francium (223)

Group 2
- 4 Be Beryllium 9.0
- 12 Mg Magnesium 24.3
- 20 Ca Calcium 40.1
- 38 Sr Strontium 87.6
- 56 Ba Barium 137.3
- 88 Ra Radium 226.0

Group 3
- 21 Sc Scandium 45.0
- 39 Y Yttrium 88.9
- 57* La Lanthanum 138.9
- 89** Ac Actinium (227)

Group 4
- 22 Ti Titanium 47.9
- 40 Zr Zirconium 91.2
- 72 Hf Hafnium 178.5
- 104 Db Dubnium (261)

Group 5
- 23 V Vanadium 50.9
- 41 Nb Niobium 92.9
- 73 Ta Tantalum 181.0
- 105 Jl Joliotium (262)

Group 6
- 24 Cr Chromium 52.0
- 42 Mo Molybdenum 95.9
- 74 W Tungsten 183.9
- 106 Rf Rutherfordium (263)

Group 7
- 25 Mn Manganese 54.9
- 43 Tc Technetium (98)
- 75 Re Rhenium 186.2
- 107 Bh Bohrium (262)

Group 8
- 26 Fe Iron 55.8
- 44 Ru Ruthenium 101.1
- 76 Os Osmium 190.2
- 108 Hn Hahnium (265)

Group 9
- 27 Co Cobalt 58.9
- 45 Rh Rhodium 102.9
- 77 Ir Iridium 192.2
- 109 Mt Meitnerium (268)

Group 10
- 28 Ni Nickel 58.7
- 46 Pd Palladium 106.4
- 78 Pt Platinum 195.1
- 110 (269)

Group 11
- 29 Cu Copper 63.5
- 47 Ag Silver 107.9
- 79 Au Gold 197.0
- 111 (272)

Group 12
- 30 Zn Zinc 65.4
- 48 Cd Cadmium 112.4
- 80 Hg Mercury 200.6
- 112 (277)

Group 13
- 5 B Boron 10.8
- 13 Al Aluminum 27.0
- 31 Ga Gallium 69.7
- 49 In Indium 114.8
- 81 Tl Thallium 204.4

Group 14
- 6 C Carbon 12.0
- 14 Si Silicon 28.1
- 32 Ge Germanium 72.6
- 50 Sn Tin 118.7
- 82 Pb Lead 207.2

Group 15
- 7 N Nitrogen 14.0
- 15 P Phosphorus 31.0
- 33 As Arsenic 74.9
- 51 Sb Antimony 121.8
- 83 Bi Bismuth 209.0

Group 16
- 8 O Oxygen 16.0
- 16 S Sulfur 32.1
- 34 Se Selenium 79.0
- 52 Te Tellurium 127.6
- 84 Po Polonium (209)

Group 17
- 9 F Fluorine 19.0
- 17 Cl Chlorine 35.5
- 35 Br Bromine 79.9
- 53 I Iodine 126.9
- 85 At Astatine (210)

Group 18
- 2 He Helium 4.0
- 10 Ne Neon 20.2
- 18 Ar Argon 39.9
- 36 Kr Krypton 83.8
- 54 Xe Xenon 131.3
- 86 Rn Radon (222)

Lanthanide series (*)
- 58 Ce Cerium 140.1
- 59 Pr Praseodymium 140.1
- 60 Nd Neodymium 144.2
- 61 Pm Promethium (145)
- 62 Sm Samarium 150.4
- 63 Eu Europium 152.0
- 64 Gd Gadolinium 157.3
- 65 Tb Terbium 158.9
- 66 Dy Dysprosium 162.5
- 67 Ho Holmium 164.9
- 68 Er Erbium 167.3
- 69 Tm Thulium 168.9
- 70 Yb Ytterbium 173.0
- 71 Lu Lutetium 175.0

Actinide series (**)
- 90 Th Thorium 232.0
- 91 Pa Protactinium 231.0
- 92 U Uranium 238.0
- 93 Np Neptunium 237.0
- 94 Pu Plutonium (244)
- 95 Am Americium (243)
- 96 Cm Curium (247)
- 97 Bk Berkelium (247)
- 98 Cf Californium (251)
- 99 Es Einsteinium (252)
- 100 Fm Fermium (257)
- 101 Md Mendelevium (258)
- 102 No Nobelium (259)
- 103 Lr Lawrencium (260)

Modern name	Latin name	Symbol
Copper	Cuprium	Cu
Iron	Ferrum	Fe
Potassium	Kalium	K
Sodium	Natrium	Na

Note: Numbers in parentheses are mass numbers (the total number of protons and neutrons in the nucleus) of most stable or best-known isotope of radioactive elements.

■ Figure 2-2 **Periodic table of the elements**

and nitrogen (N) are by far the most common essential elements because they are the main constituents of most biological molecules. The essential elements that are found in very low concentrations in living tissues are called **trace elements,** or minor essential elements. All of the trace elements are essential in some way to cell function, and the list of such elements continues to change as scientists discover more about how cells work. The need for trace elements is evident only when they are absent from a person's diet or present in insufficient quantities. One trace element that has received a lot of attention in recent years is chromium. A deficiency of chromium in the body has been postulated (but not proven) to play a role in the development of the metabolic disease *diabetes mellitus* (sugar diabetes) in some people.

The atoms of different elements are distinguished from one another by the fact that each element has a unique number of protons in its nucleus. For example, the lightest element, hydrogen, has only one proton in its nucleus. The next element, helium (He), has two protons, followed by lithium (Li) with three protons, and so on. The number of protons in one atom of an element is called the **atomic number** of the element. The periodic table is arranged by atomic number, starting at hydrogen, with an atomic number of 1, and ending with meitnerium (Mt), with an atomic number of 109.

The average atomic mass of an element is shown below the symbol of the element. **Atomic mass** is calculated by adding the mass of the protons and neutrons in one atom of the element. For example, one atom of carbon has six protons and six neutrons. Because the mass of each proton or neutron is one atomic mass unit, carbon's atomic mass is 12 amu. However, in the periodic table, not all average atomic masses are whole numbers. For example, chlorine (Cl), with 17 protons in its

continued from page 14

What is chromium picolinate? Chromium (Cr) is an essential element that is linked to normal glucose metabolism. In the diet, chromium is found in brewer's yeast, broccoli, mushrooms, and apples. Chromium in food and in chromium chloride supplements is poorly absorbed from the digestive tract, so a scientist developed the compound chromium picolinate. Picolinate, derived from amino acids, enhances chromium uptake at the intestine. The recommended daily allowance (RDA) of chromium is 50 micrograms. As we've seen, Stan takes several times this amount.

Question 1: Locate chromium on the periodic table of the elements. What is its atomic number? Atomic weight? How many electrons does one atom of chromium have? Which elements close to chromium are also essential elements?

nucleus, has an average atomic mass of 35.5. This is because chlorine comes in two forms, one with 18 neutrons and one with 20. In the next section we will discuss these different forms of the elements, which are known as isotopes.

Isotopes of an Element Are Atoms with Different Numbers of Neutrons

Different numbers of protons are what distinguishes one element from the next. But, though the number of protons in an atom of any given element is constant, the number of neutrons may differ (Fig. 2-3 ■). Atoms of one element that have different numbers of neutrons are known as **isotopes** of the element [*iso-,* same + *topos,* place (that is, the same place in the periodic table)]. The

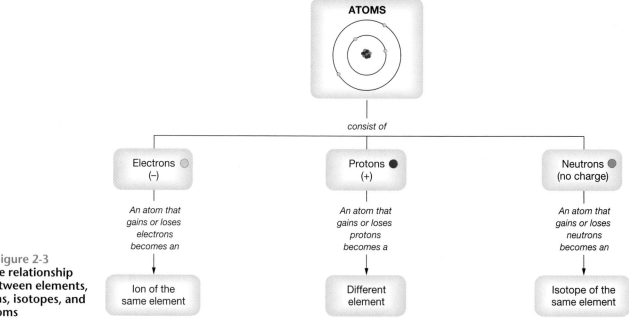

■ Figure 2-3
The relationship between elements, ions, isotopes, and atoms

symbol for an isotope shows the atomic mass number written to the upper left of the chemical letter symbol. For example, ^1H, hydrogen-1, is the most common naturally occurring isotope of hydrogen. It has a single proton in its nucleus (Fig. 2-4 ■). But hydrogen has two other isotopes, hydrogen-2, or *deuterium* (^2H), and hydrogen-3, or *tritium* (^3H), which are formed by the addition of one or two neutrons to the nucleus.

Isotopes are important in biology and medicine, because some are unstable and emit energy. These unstable isotopes are called **radioisotopes,** and the energy they emit is called **radiation.** Every element has at least one radioisotope, most of which are produced artificially by using high-energy particle accelerators to "shoot" extra neutrons into a stable nucleus. A small number of elements, such as carbon and uranium, have naturally occurring radioisotopes. Both artificial and natural radioisotopes have medical uses. For example, radioisotopes such as thallium-201 (^{201}Tl) produce images of the inside of the body for *diagnostic procedures* (those that identify disease). Radioisotopes are also used to treat cancer, an example of *therapeutic use.* The use of radioisotopes in the diagnosis and treatment of disease is a specialty known as *nuclear medicine.*

Electrons Form Bonds between Atoms and Capture Energy

Electrons, the third subatomic particle, play a significant role in physiology. The arrangement of electrons around the nucleus determines the ability of an atom to bind with other atoms and form molecules. The gain or loss of electrons changes the electrical charge on an atom and plays an essential role in communication throughout the body. Finally, electrons on certain atoms can capture energy from their environment and transfer that energy to other molecules, so that it can be used for movement and other life processes.

 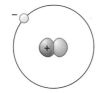 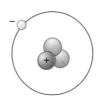

(a) Hydrogen-1, H   **(b)** Hydrogen-2, deuterium, ^2H **(c)** Hydrogen-3, tritium, ^3H

■ Figure 2-4 **Hydrogen atoms** (a) A typical hydrogen atom has one proton in the nucleus and one electron orbiting the nucleus, but no neutrons. Thus, its atomic mass is 1 amu. There are an equal number of protons and electrons, so the overall charge on the atom is zero. (b) A neutron has been added to the nucleus, increasing its atomic mass to 2 amu. This isotope is hydrogen-2, or ^2H, also known as deuterium. (c) A second neutron has been added to the nucleus, yielding a new isotope known as hydrogen-3, or ^3H, or tritium; its atomic mass is 3 amu. All three isotopes remain electrically neutral. Isotopes have the same chemical properties, because such traits are affected by the electron configuration and not the number of neutrons.

The electrons in an atom do not move around the nucleus at random. Instead, they are arranged in a series of energy levels, or **shells.** The lowest energy level, designated as level 1, is closest to the nucleus. As successive levels increase in energy, their distance from the nucleus also increases. Each energy level is divided into sublevels called, for historical reasons, *s, p, d,* and *f.* Each of these sublevels may be further subdivided into a certain number of **orbitals,** each of which holds a maximum of two electrons (Fig. 2-5 ■).

For example, the first energy level or shell, 1, has one sublevel, 1*s*, with one orbital. Thus, the first energy level can hold a total of two electrons. The second shell has two sublevels (2*s* and 2*p*) with a total of four orbitals. It can hold a total of eight electrons. If an energy sublevel has more than one orbital, each orbital is assigned one electron before any orbital gets two (Fig. 2-5c ■). The nitrogen atom is a good example of this principle: The highest energy level, 2*p*, has one electron in each of its three orbitals. The most stable electron configuration for an element is one in which all orbitals in the outer shell are filled with electrons. Neon, a completely unreactive element, is an example of an element whose outer electron shell is totally filled (Fig. 2-5b ■).

When an orbital is filled with two electrons, the electrons are said to be paired. Paired electrons are stable. An atom having an unpaired electron has a higher probability of losing the unpaired electron to another atom or gaining an unpaired electron from another atom. In some cases, two atoms with unpaired electrons share the

 Radiation and Radioisotopes Of the 2000 known isotopes of the elements, unstable isotopes that emit radiation outnumber the stable isotopes four to one. Some of these isotopes occur naturally, but many have been made artificially, including all those with atomic numbers of 93 or higher. There are three types of radiation: alpha (α), beta (β), and gamma (γ). Alpha and beta radiation are composed of fast-moving particles ejected from the nucleus of an unstable atom. Alpha particles consist of two protons and two neutrons. Beta particles are electrons created when a neutron in the nucleus converts into a proton and an electron. Because the nucleus of the unstable atom loses two protons when emitting alpha radiation or gains one proton when emitting beta radiation, the atom is converted to a different element. For example, unstable iodine ^{131}I (atomic number 53) converts a neutron to a proton and electron, becoming xenon, ^{131}Xe, with an atomic number of 54. Gamma radiation is composed of high-energy waves rather than particles. Gamma radiation is able to penetrate matter much more deeply than α or β particles. X-rays are similar to gamma radiation and are used to create images of the body's interior. Because gamma radiation is not composed of nuclear particles, the atom emitting energy does not change into a different element as it emits the radiation. Instead, it becomes a more stable atom of the same element.

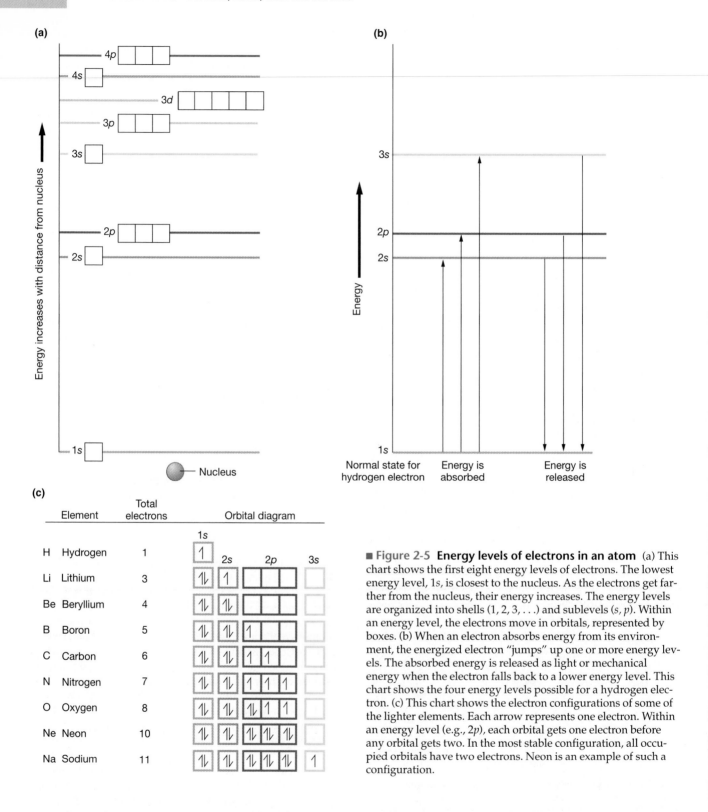

(a)

Energy increases with distance from nucleus

4p
4s
3d
3p
3s
2p
2s
1s

Nucleus

(b)

Energy

3s

2p
2s

1s

| Normal state for hydrogen electron | Energy is absorbed | Energy is released |

(c)

Element	Total electrons	Orbital diagram
		1s
H Hydrogen	1	↑
		1s 2s 2p 3s
Li Lithium	3	↑↓ ↑
Be Beryllium	4	↑↓ ↑↓
B Boron	5	↑↓ ↑↓ ↑
C Carbon	6	↑↓ ↑↓ ↑ ↑
N Nitrogen	7	↑↓ ↑↓ ↑ ↑ ↑
O Oxygen	8	↑↓ ↑↓ ↑↓ ↑ ↑
Ne Neon	10	↑↓ ↑↓ ↑↓ ↑↓ ↑↓
Na Sodium	11	↑↓ ↑↓ ↑↓ ↑↓ ↑↓ ↑

■ **Figure 2-5 Energy levels of electrons in an atom** (a) This chart shows the first eight energy levels of electrons. The lowest energy level, 1s, is closest to the nucleus. As the electrons get farther from the nucleus, their energy increases. The energy levels are organized into shells (1, 2, 3, . . .) and sublevels (s, p). Within an energy level, the electrons move in orbitals, represented by boxes. (b) When an electron absorbs energy from its environment, the energized electron "jumps" up one or more energy levels. The absorbed energy is released as light or mechanical energy when the electron falls back to a lower energy level. This chart shows the four energy levels possible for a hydrogen electron. (c) This chart shows the electron configurations of some of the lighter elements. Each arrow represents one electron. Within an energy level (e.g., 2p), each orbital gets one electron before any orbital gets two. In the most stable configuration, all occupied orbitals have two electrons. Neon is an example of such a configuration.

electrons. When electrons are transferred or shared in that manner, each participating atom ends up with a stable orbital of paired electrons.

Electrons are able to absorb energy from their environment and "jump" up one or more energy levels to a higher orbital (Fig. 2-5b ■). High-energy electrons are not terribly stable, so they eventually fall back to their stable orbital position, emitting energy as they do so. The released energy can be captured by other atoms or may be emitted as radiation. The bioluminescence of fireflies is visible light emitted by high-energy electrons returning to their normal low-energy state. When biologists speak of "high-energy electrons" that capture and transfer energy in the cell, they mean a single hydrogen electron that moves out of its 1s orbital into an unoccupied higher orbital.

Use the periodic table of the elements in Figure 2-2 to answer the following questions:

✓ Which element has 30 protons?

✓ How many electrons does one atom of sodium (Na) have?

✓ How many electrons does one atom of calcium (Ca) have?

✓ What is the atomic mass of nitrogen?

MOLECULES AND BONDS

The transfer or sharing of electrons is a critical part of forming **bonds,** or links between atoms. When two or more atoms link by sharing electrons, they make units known as **molecules.** Some molecules, such as those of oxygen (O_2) and nitrogen (N_2) gas, are made of two atoms of the same element. But most molecules contain more than one element and are known as **compounds.** For example, although oxygen gas (O_2) is a molecule, it is not a compound because it has only oxygen atoms. In contrast, glucose, a molecule composed of carbon, hydrogen, and oxygen, is a compound.

What Does the Chemical Formula of a Molecule Tell Us?

The **chemical formula** of a molecule is a form of notation that describes the type and number of atoms in the molecule. Each element in the molecule is represented by its chemical symbol followed by a numeric subscript showing the number of atoms of the element. Thus water, which has two hydrogen atoms and one oxygen atom, is written as H_2O. Glucose is written $C_6H_{12}O_6$, meaning it has six atoms each of carbon and oxygen, and twelve atoms of hydrogen.

From the chemical formula of a molecule, we can calculate its **molecular weight.** This is the weight of one molecule, expressed in atomic mass units or, more often, in daltons (1 amu = 1 Da).* To calculate molecular weight, multiply the atomic mass of each element in the molecule by the number of atoms of the element. For glucose, $C_6H_{12}O_6$, it would be:

Element	Number of Atoms in the Molecule	×	Atomic Mass of the Element	
Carbon	6		12.0 amu	= 72.0
Hydrogen	12		1.0 amu	= 12.0
Oxygen	6		16.0 amu	= 96.0
			SUM	180.0 amu
			or	180 Da

---

*In recognition of John Dalton, who proposed the modern theory of the atom around 1803.

 Imaging in Medicine X-rays, high-energy emissions from unstable atoms, are the oldest method of imaging used in medicine. Discovered in 1895 by Wilhelm Roentgen, X-rays were our only means of visualizing the interior structures of the body until the 1950s, when ultrasound was introduced. Ultrasound is a technique similar to sonar in the ocean, in which an image is created from sound waves that are reflected as they bounce off of tissues and fluid. In the 1970s, computerized tomography (CT scans), positron emission tomography (PET scans), and magnetic resonance imaging (MRI) came into use [*tomos,* a section]. CT scans are series of X-rays that create images of very thin slices of the body. CT images are assembled by computer rather than projected onto a photographic plate as X-rays are, so they can be used to visualize the three-dimensional structure of an organ. PET scans are made using radioisotopes that emit *positrons:* tiny positively charged particles released when a proton converts into a neutron. Because the body does not normally contain positron-emitting chemicals, these radioisotopes must be administered to a subject before a PET scan be taken. For MRI scans, the subject is placed inside a strong magnetic field, and radiation emitted by the nuclei of hydrogen atoms is captured to create an image of internal structure.

Although the chemical formula gives the number of atoms in a molecule, it doesn't tell you which atom is linked to which. To understand that aspect of molecular structure, you need to know the different types of bonds that atoms can form with each other.

There are four common bond types, two strong and two weak. (1) Covalent bonds are strong bonds that result when two atoms share a pair of electrons, one from each atom. Covalent bonds usually require the input of energy to break them apart. (2) Ionic bonds result when an atom has such a strong attraction for electrons that it pulls one or more electrons completely away from another atom. (3) Hydrogen bonds are weak attractive forces between hydrogens and certain other atoms. (4) Van der Waals forces are another type of weak attractive force that acts between all molecules. Each type of bond is discussed further in the sections below.

Covalent Bonds Are Formed When Adjacent Atoms Share Electrons

Covalent bonds are formed when atoms share one or more pairs of electrons. Each atom contributes one electron to each pair. Most of the bonds that link atoms together into molecules are covalent bonds.

It is possible to predict how many covalent bonds an atom will form by knowing how many unpaired electrons the atom has in its outer shell. Figure 2-6 ■ shows the electron configurations of the three most plentiful elements in the body: hydrogen, carbon, and oxygen.

Hydrogen, the lightest element, has one proton in its nucleus and one unpaired electron in its outer shell

(a) Hydrogen

H •

1 proton

(b) Carbon

•C•

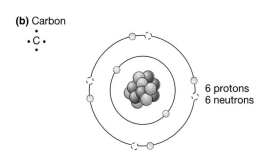

6 protons
6 neutrons

(c) Oxygen

:O:

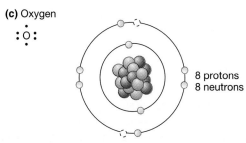

8 protons
8 neutrons

■ **Figure 2-6 Electron configuration of the three most common elements in the body** An electron-dot model shows the number of electrons found in the outer shell of an atom: Each electron is represented by a single dot. The first electron shell closest to the nucleus can hold a maximum of two electrons. The second shell can hold a maximum of eight. Each shell is most stable when completely filled with electrons. (a) Hydrogen, the lightest element, has only one electron, which occupies the lowest energy level. (b) Carbon has six electrons, two in the first shell and four occupying half of the eight possible places in the second shell. (c) Oxygen has eight electrons, two in the first shell and six in the second. This leaves two unpaired electrons in the second shell.

(Fig. 2-6a ■). Because it has only one electron to share, it always forms a single covalent bond. Chemists represent the single electron by a single dot in the *electron-dot shorthand*.

Carbon, with an atomic number of 6, has six protons in the nucleus and therefore six electrons surrounding the nucleus. Two electrons fill the first electron shell, while the outermost second shell has an additional four electrons (Fig. 2-6b ■). Because there is space for eight electrons in the second shell, there are four unoccupied places. An atom is most stable when its outer electron shell is filled. This means that carbon, with four of eight possible outer electrons, binds to other atoms to find the four electrons it needs to complete its outer shell. When

carbon is sharing all four of its outermost electrons with other elements, it forms four covalent bonds.

Oxygen has eight electrons: two electrons completely fill the first shell and six are found in the eight-place second shell (Fig. 2-6c ■). Because oxygen needs only two additional electrons to complete its outer shell, it can get them by sharing two of its outer electrons in covalent bonds. A water molecule is a simple example of covalent bonds formed by oxygen. In each molecule of water, two hydrogen atoms share their electrons with two electrons in the outer shell of the oxygen atom.

We can use electron-dot shorthand to show how atoms form covalent bonds in molecules. In the next example, one carbon combines with four hydrogens to form a methane molecule. Each hydrogen fills its two-electron outer shell by sharing its one electron. At the same time, the carbon atom fills its eight-place outer shell with the four shared hydrogen electrons. When the outer electron shells are filled, even if it is with shared electrons, the atoms are more stable than if the shells are unfilled.

$$4 \text{ H} \bullet + 1 \text{ }\bullet\overset{\bullet}{\underset{\bullet}{\text{C}}}\bullet \longrightarrow \text{H}\overset{\bullet\bullet}{\underset{\bullet\bullet}{\text{:}}}\overset{}{\text{C}}\overset{}{\text{:}}\text{H} \quad \text{or} \quad \text{H}-\overset{\text{H}}{\underset{\text{H}}{\text{C}}}-\text{H}$$

Methane CH_4

Another form of chemical shorthand shows the paired electrons of a covalent bond as a single line drawn between the chemical symbols for the elements of the molecule. Sometimes adjacent atoms share two pairs of electrons rather than just one. In this case the bond is called a *double bond* and is represented by a double line (=). The following example shows two carbon atoms sharing two pairs of electrons:

$$4 \text{ H} \bullet + 2 \bullet\overset{\bullet}{\text{C}}\bullet \longrightarrow \overset{\text{H}}{\underset{\text{H}}{}}\text{C}::\text{C}\overset{\text{H}}{\underset{\text{H}}{}} \quad \text{or} \quad \overset{\text{H}}{\underset{\text{H}}{}}\text{C}=\text{C}\overset{\text{H}}{\underset{\text{H}}{}}$$

Ethylene C_2H_4

Because two of the four carbon electrons are shared in the carbon-carbon bond, this leaves only two electrons per carbon to bond with hydrogen. But if you count the bonds made by the carbons, once again each carbon forms four bonds and each hydrogen forms one. The arrangement of bonds is different but the number of electrons being shared has not changed. The chemical formula for this molecule, called ethylene, is C_2H_4.

✓ You know that the number of electrons present in the outer shell determines the number of covalent bonds an atom can form. Use this fact to decide whether a hydrogen atom can form a double bond. Why or why not?

Functional groups   There are several types of partial molecules, or functional groups, that are commonly found in biological molecules. The atoms in functional groups tend to move from molecule to molecule as a single unit. For example, hydroxyl groups, —OH, common on many biological molecules, are added and removed as a group rather than as single hydrogen or oxygen atoms. Functional groups usually attach to molecules by single covalent bonds. The most common groups are listed in Table 2-1.

Molecular shape   Molecules can have many shapes. Molecular shape is important because it is closely related to molecular function. Figure 2-7 ■ shows how atoms are linked in two biological molecules. Notice the carbons linked in a straight chain in the leucine molecule, represented in Figure 2-7a ■. In the glucose molecule (Fig. 2-7b ■), the atoms join together to form a closed circle, or **ring**.

The three-dimensional structure of molecules is difficult to show on paper. The angles of the bonds between atoms may cause molecules to take on characteristic shapes. Two well-known examples are the double helix of the DNA molecule and the alpha helix of proteins. These molecules are shown in Figures 2-16 ■ and 2-18 ■.

Polar and nonpolar molecules   Covalent bonds are not always the result of an even sharing of electron pairs between linked atoms. In some molecules, the nucleus of one atom has a stronger attraction for the shared electrons than the other nucleus does. The electron pair spends more time near the nucleus with the stronger attraction, giving that atom a slightly more negative charge. Meanwhile the atom that is farther from the electrons

(a) Leucine: $C_6H_{13}NO_2$

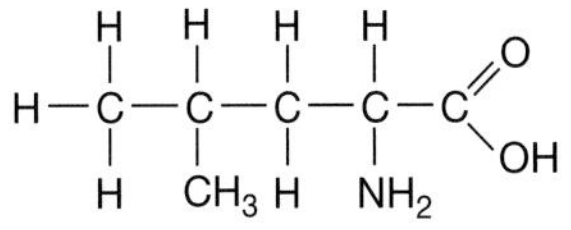

(b) Glucose: $C_6H_{12}O_6$

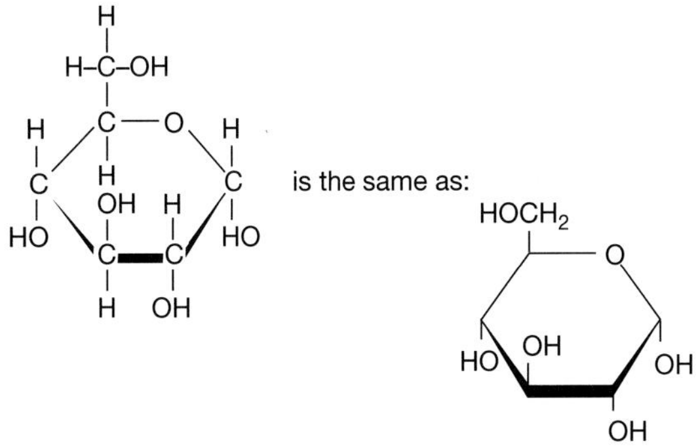

is the same as:

■ **Figure 2-7 Chemical structures and formulas of some biological molecules** (a) Leucine, an amino acid, is a straight chain of carbon atoms with hydrogens and other functional groups attached. (b) Glucose has a ring of five carbon atoms and one oxygen atom. On the left, the glucose molecule is drawn with all of its carbons and hydrogens shown. In the shorthand version on the right, the carbons at each corner of the ring are omitted, as are the hydrogens attached to the carbons. If the corner of the ring is occupied by an atom other than carbon, the chemical symbol of the atom must be included. You can figure out how many hydrogens have been omitted by remembering that each carbon makes four covalent bonds. If a carbon appears to have only three bonds (two to the carbons on each side of it in the ring and one to a hydroxyl group, for example), then the fourth bond (not shown) is to a hydrogen.

develops a slightly more positive charge. For both atoms, the change in charge is only a fraction of the charge carried by a single electron or proton. Molecules that develop regions of partial positive and negative charge are called **polar molecules** because they can be said to have positive and negative poles. Certain elements, particularly nitrogen and oxygen, have a strong attraction for electrons and are often found in polar molecules.

A good example of a polar molecule is water (H_2O). When the hydrogen atoms form covalent bonds with oxygen, their electrons are pulled away from the hydrogen nuclei toward the larger and stronger nucleus of the oxygen atom. This leaves the two hydrogen atoms of the molecule with a partial positive charge, and the single oxygen atom with a partial negative charge from the borrowed electrons (see Fig. 2-8 ■). These partial charges are represented by the lowercase Greek letter delta: δ^+ indicates a positive pole of the molecule, and δ^- indicates a negative pole. Note that the net charge for the entire water molecule is zero.

TABLE 2-1	Common Functional Groups	

Notice that oxygen, with two electrons to share, may bind to another atom with a double bond.

	Shorthand	Bond Structure
Carboxyl (acid)	—COOH	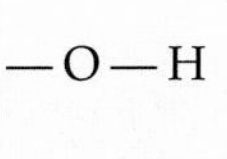 —C(=O)OH
Hydroxyl	—OH	—O—H
Amino	—NH₂	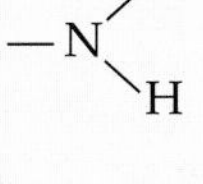 —N(H)(H)
Phosphate	—H₂PO₄	—O—P(=O)(OH)(OH)

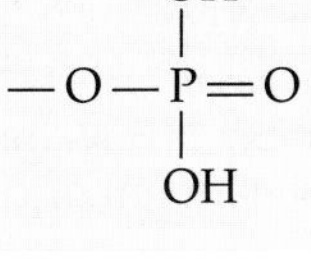

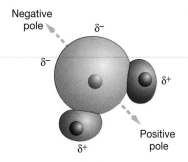

Negative pole

δ−

δ−

δ+

δ+

Positive pole

Water molecule

■ **Figure 2-8 Water is a polar molecule** A water molecule develops separate regions of partial positive and partial negative charge when the oxygen nucleus pulls the hydrogen electrons toward itself. The electron pairs are no longer shared evenly between the atoms, causing the oxygen to have a partial negative charge and the hydrogens to have a slight positive charge.

A **nonpolar molecule** is one whose electrons are distributed so evenly that there are no regions of partial positive or negative charge. Molecules composed mostly of carbon and hydrogen tend to form nonpolar bonds because carbon does not pull electrons away from hydrogen the way oxygen does. The polarity of a molecule is important in determining whether it will dissolve in water. Polar molecules generally dissolve easily; nonpolar molecules do not.

Ionic Bonds Form When Atoms Gain or Lose Electrons

When an atom or molecule completely gains or loses one or more electrons, it forms an **ion.** Generally, the electrons that are gained have been released by another atom or molecule. Table 2-2 lists the most common ions encountered in physiology. Notice that some of the ions are single atoms, and others are functional groups.

A molecule that gains electrons acquires one negative charge (-1) for each electron added. By convention, ions are indicated by a superscript notation following the symbol of the element, with the number of charges first and the sign of the charge second. Positively charged ions are called **cations;** negatively charged ions are called **anions.** For example, a chloride ion has one extra electron (Cl^-), and a phosphate ion has two (HPO_4^{2-}). Any molecule

that loses electrons has protons without matching electrons, leaving it with one positive charge ($+1$) for each electron lost. Hydrogen loses its only electron and ends up as a single proton, H^+; the ion formed by a calcium atom that loses two electrons is abbreviated as Ca^{2+}.

A chemical bond between a cation and an anion is called an **ionic bond.** Figure 2-9 ■ shows how sodium and chlorine atoms form ions, and how those ions form ionic bonds. A sodium (Na) atom has only one electron in its eight-place outer electron shell; a chlorine (Cl) atom has seven of the eight electrons needed to fill its outer shell. The less stable sodium atom gives up its electron to chlorine, resulting in a positively charged sodium ion (Na^+) and a negatively charged chloride ion (Cl^-). In table salt, the solid form of NaCl, the opposite charges attract, creating a neatly ordered crystal of alternating Na^+ and Cl^- ions held together by ionic bonds.

Hydrogen Bonds Are Weak Bonds between Molecules or Regions of a Molecule

A **hydrogen bond** is the weak attraction of a hydrogen atom to a nearby oxygen, nitrogen, or fluorine atom. Hydrogen bonds may occur between atoms in neighboring molecules or between atoms in different parts of the same molecule. For example, hydrogen bonds cause large biological molecules such as proteins to fold back on themselves, creating a three-dimensional shape that is essential for the function of the protein.

The hydrogen bonds of water are a good example of bonds between adjacent molecules. The positive hydrogen pole of one water molecule is attracted to the negative oxygen poles of as many as four other water molecules. As a result, the molecules line up with their neighbors in a somewhat ordered fashion (Fig. 2-10 ■). The weak interaction between the positive hydrogen regions and the negative oxygen regions is the hydrogen bond. Hydrogen bonding in water is responsible for the **surface tension** of water, the attractive force between water molecules that causes water to form spherical droplets when falling or to bead up when spilled onto a nonabsorbent surface (Fig. 2-10b ■). These weak forces of attraction also make it difficult to separate water mol-

TABLE 2-2	Important Ions of the Body		
Cations		*Anions*	
Na^+	Sodium	Cl^-	Chloride
K^+	Potassium	HCO_3^-	Bicarbonate
Ca^{2+}	Calcium	HPO_4^{2-}	Phosphate
H^+	Hydrogen	SO_4^{2-}	Sulfate
Mg^{2+}	Magnesium		

continued from page 16

Chromium is found in several ionic forms. The chromium usually found in biological systems and in dietary chromium supplements is the trivalent ion, represented by the symbol Cr^{3-}. A hexavalent ion, Cr^{6-}, is used in industries such as the manufacturing of stainless steel and the chrome plating of metal parts.

Question 2: How many electrons have been lost from the hexavalent ion of chromium? From the trivalent ion?

(a)

Step 1

Sodium atom   Chlorine atom

Step 2

Sodium ion (Na⁺)   Chloride ion (Cl⁻)

Step 3

Sodium chloride (NaCl) molecule

(b)

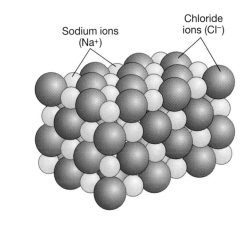

Sodium ions (Na⁺)   Chloride ions (Cl⁻)

Sodium chloride (NaCl) crystal

■ **Figure 2-9 Formation of ionic bonds and ions** (a) *Step 1:* Chlorine has seven electrons in its eight-place outer shell; sodium has only one electron of a possible eight. Sodium therefore gives up its one weakly held electron to chlorine. *Step 2:* This exchange of electrons fills chlorine's outer shell and leaves the sodium ion with a stable, filled second shell. *Step 3:* The sodium and chloride ions are attracted to each other because of their opposite charges. (b) In the solid state, the charged ions form a NaCl crystal.

ecules, as you may have noticed in trying to pick up a wet glass that has become "stuck" to a slick table top by a thin film of water.

Van der Waals Forces Are Weak Attractions between All Molecules

Van der Waals forces are even weaker bonds than hydrogen bonds. They are nonspecific attractions between the nucleus of any atom and the electrons of nearby atoms. Two atoms that are weakly attracted to

each other by van der Waals forces move together until they are so close that their electrons begin to repulse each other. Consequently, van der Waals forces allow atoms to pack closely together and occupy a minimum amount of space. A single van der Waals attraction between atoms is very weak, but multiple van der Waals forces supplement the hydrogen bonds that hold proteins in their three-dimensional shapes.

✓ Write the chemical formula for this molecule:

Citric acid

✓ Why does table salt (NaCl) dissolve in water?

✓ Why does carbon always form four bonds to other molecules?

✓ What do dishwashing detergents do that keeps water from beading up on glasses?

continued from page 22

The hexavalent form of chromium used in industry is known to be toxic to humans. In 1992, officials at California's Hazard Evaluation System and Information Service warned that inhaling chromium dust, mist, or fumes placed chrome and stainless steel workers at increased risk for lung cancer. Officials found no risk to the public from normal contact with chrome surfaces or stainless steel. The first link between the biological trivalent form of chromium and cancer was presented in a controversial 1995 study. In these experiments, isolated cells exposed to moderate levels of Cr^{3-} developed cancerous changes.

Question 3: From this information, can you conclude that hexavalent and trivalent chromium are equally toxic?

(a)

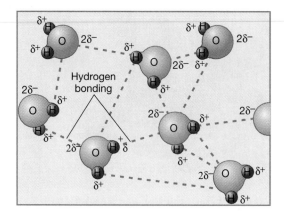

(b)

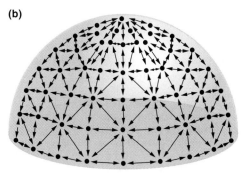

Drop of water

■ **Figure 2-10 Hydrogen bonds of water** (a) The polar regions of water molecules allow them to form hydrogen bonds with one another. The negative regions of one water molecule are attracted to the positive regions of another. (b) Surface tension is created by hydrogen bonds in a drop of water. The intermolecular forces hold the droplet together, which keeps it from spreading out.

SOLUTIONS AND SOLUTES

Substances that dissolve in liquid are collectively known as **solutes.** The liquids into which they dissolve are known as **solvents.** In biological solutions, water is the universal solvent. The combination of solutes dissolved in solvent is called a **solution.** This section examines some of the properties of biological solutions.

Solubility in Aqueous Solutions

Life as we know it is established on water-based, or *aqueous,* solutions that resemble dilute seawater in their ionic composition. Na^+, K^+, and Cl^- are the main ions; other ions make up a lesser proportion. All molecules and cell components are either dissolved or suspended in these saline solutions. It is useful therefore to classify molecules on the basis of their ability to dissolve in water.

The degree to which a molecule is able to dissolve in a solvent is its **solubility:** the more easily it dissolves, the higher its solubility. Molecules that dissolve readily in water are **hydrophilic** [*hydro-*, water + *-philic*, loving].

Most are polar or ionic molecules whose positive and negative regions can form hydrogen bonds with water molecules (Fig. 2-11 ■). Molecules that do not dissolve readily in water are **hydrophobic** [*-phobic*, hating]. They tend to be nonpolar molecules that are unable to make hydrogen bonds with water molecules.

The lipids (fats and oils) are the most hydrophobic group of biological molecules. Placed in an aqueous solution, they cannot interact with the water molecules and tend to form separate layers, like oil floating on vinegar in a bottle of salad dressing. In order for hydrophobic molecules such as the lipids to dissolve in body fluids, they must be combined with a hydrophilic molecule that can effectively carry them into solution. For example, cholesterol, a common animal fat, is a hydrophobic molecule. Fat from a piece of meat dropped into a glass of warm water will float to the top, undissolved. Similarly, cholesterol in the blood will not dissolve unless it is combined with water-soluble molecules called lipoproteins.

In addition to its role as medium for dissolved substances, water has other properties that affect the way our bodies work. The surface tension of water is important in lung function, as you will see in Chapter 17. Water also has the ability to capture and store large amounts of heat. The heat-retaining property of water is significant in heat loss from the body through sweating. Temperature control in the body is discussed in Chapter 21.

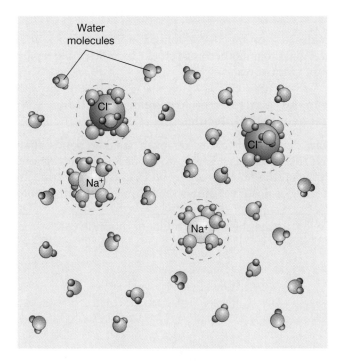

Sodium chloride in solution

■ **Figure 2-11 Sodium chloride dissolves in water** Sodium chloride dissolves into Na^+ and Cl^- in water. The positive charge on the sodium ion attracts the negative poles of many water molecules, whereas the negative charge on chloride ions attracts the positive poles of other water molecules. The interaction of the ions with the water molecules breaks the ionic bonds that held the NaCl crystal together.

The Concentration of Solutions

The **concentration** of a solution is the amount of solute per unit volume of solution. There are many different ways to express concentration in physiology. The amount of solute can be described as the weight of the solute before it dissolves or as the number of ions or molecules of solute in the solution. Weight is usually given in grams (g) or milligrams (mg). Numbers of solute molecules are given in moles (mol), and numbers of solute ions are given in equivalents (eq). Volume is usually expressed as liters (L) or milliliters (mL).

Moles and molarity A **mole** [*moles*, a mass] is defined as 6.02×10^{23} atoms, ions, or molecules of a substance. Thus, one mole of a substance has the same number of particles as one mole of any other substance. However, the weight of a mole depends on the weight of the molecules. The weight of a mole is equal to the atomic mass or molecular weight of the substance, expressed in grams. This value is called the **gram molecular weight.** For example, the molecular weight of glucose is 180 daltons. Therefore one mole of glucose weighs 180 grams. One mole of NaCl, with the same number of molecules, only weighs 58.5 grams.

Solution concentrations in biology are frequently given as **molarity,** the number of moles of solute in a liter of solution. Molarity is abbreviated as mol/L or, simply, M. A 1 molar solution of glucose (1 mol/L, 1 M) contains 6.02×10^{23} molecules of glucose per liter. It is made by dissolving 180 grams (one mole) of glucose in enough water to make one liter of solution.

Typical biological solutions are so dilute that solute concentrations are usually expressed as **millimoles** per liter. A millimole is 1/1000th of a mole [*milli-*, 1/1000].

Equivalents Concentrations of ions are sometimes expressed in equivalents per liter instead of moles per liter. An **equivalent** (eq) is equal to the molarity of the ion times the number of charges the ion carries. A sodium ion with a charge of +1 has one equivalent per mole. A calcium ion, Ca^{2+}, has two equivalents per mole. The same rule is true for negatively charged ions. One mole of chloride ions (Cl^-) has one equivalent per mole, one mole of phosphate ions (HPO_4^{2-}) has two equivalents. A **milliequivalent** (meq) is 1/1000th of an equivalent. Milliequivalents are useful for expressing the concentration of dilute biological solutions. Many clinical measurements, such as the concentration of sodium ions in the blood, are reported in meq/L.

Weight/volume and percent solutions In the laboratory or pharmacy, scientists cannot measure out solutes by the mole. Instead, they use the more conventional measurement of weight. The solute concentration of these solutions may then be expressed as **percent solution,** or a percentage of the total solution. A 10% solution means that there are 10 parts of solute per 100 parts *of total solution.* If the solute is normally a solid at room temperature, the percent solution is expressed as weight/volume. A 5% glucose solution would be made by weighing out 5 grams of glucose and adding water to make a final volume of 100 mL. If the solute is a liquid, such as concentrated hydrochloric acid, the solution is made using volume/volume measurements. To make 0.1% HCl, add 0.1 mL of concentrated acid to enough water to give a final volume of 100 mL.

Another convention is common in medicine. The concentrations of drugs and other chemicals in the body are often expressed as milligrams per deciliter, mg/dL. A **deciliter** (dL) is 1/10 of a liter, or 100 mL. A second way of expressing this, no longer commonly used, is mg%, where % means per 100 parts or 100 mL. A solution of 20 mg/dL could also be expressed as 20 mg%.

The Concentration of Hydrogen Ions in the Body Is Expressed in pH Units

One of the important ionic solutes in the body is the hydrogen ion, H^+. The concentration of H^+ in body fluids determines the body's acidity, a physiological parameter that is closely regulated because H^+ in solution can interfere with hydrogen bonding and van der Waals forces. Because these bonds are responsible for the proper function of many important molecules, their disruption can have serious consequences.

Where do hydrogen ions in body fluids come from? Some of them come from the separation of water (H_2O) into H^+ and OH^- ions. Others come from ionized molecules that release H^+ ions when they dissolve in water. A molecule that contributes H^+ to a solution is called an **acid.** Carbonic acid is an acid made in the body from CO_2 (carbon dioxide) and water. In solution, carbonic acid separates into a bicarbonate ion and a hydrogen ion:

$$CO_2 + H_2O \rightleftarrows H_2CO_3 \text{ (carbonic acid)} \rightleftarrows H^+ + HCO_3^-$$

Note that while the H^+ is part of the intact carbonic acid molecule, it does not contribute to acidity. It is only free H^+ that affects H^+ concentration.

Some molecules decrease the H^+ concentration of a solution by combining with free H^+. These molecules are called **bases.** Molecules that produce hydroxide ions, OH^-, in solution are bases because the hydroxide combines with H^+ to form water:

$$H^+ + OH^- \rightleftarrows H_2O$$

Another molecule that acts as a base is ammonia, NH_3. It reacts with free H^+ to form an ammonium ion:

$$NH_3 + H^+ \rightleftarrows NH_4^+$$

The concentration of H^+ in body fluids is measured in terms of **pH.** The expression pH stands for "power of hydrogen" and is calculated from the negative log of the hydrogen ion concentration: $pH = -\log [H^+]$. The square brackets around the symbol for a solute are the short-hand notation for "concentration": [H⁺] = hydrogen ion concentration.*

Another way to write the pH equation is pH = log(1/[H⁺]). This equation states that pH is inversely related to the H⁺ concentration. In other words, as the H⁺ concentration goes up, the pH goes down.

The pH of a solution is measured on a numeric scale between 0 and 14. The H⁺ concentration in pure water is 1×10^{-7} M, and so pure water has a pH value of 7.0. It is considered neutral. Solutions that have gained H⁺ from an acid have higher H⁺ concentrations. Because [H⁺] is inversely proportional to pH, **acidic** solutions have a pH less than 7. Solutions that have lost H⁺ to a base have a lower H⁺ concentration than pure water; they are called **alkaline.** They have a pH value greater than 7. Remember, because pH is proportional to the reciprocal of the [H⁺], as the hydrogen ion concentration *increases*, the pH number *decreases*.

The pH scale is logarithmic, so that a change in pH value of 1 unit indicates a tenfold increase or decrease in the hydrogen ion concentration. For example, if the pH of a solution is altered from a pH of 8 to a pH of 6, there has been a 100-fold (10 × 10) increase in the H⁺ concentration.

Figure 2-12 ■ shows a pH scale and the pH values of various solutions. The normal pH of blood in the human body is 7.40, slightly alkaline. Regulation of the body's pH within a narrow range is critical because a blood pH more acidic than 7.00 or more alkaline than 7.70 is incompatible with life.

Buffers are a key factor in the body's ability to maintain normal pH. A buffer is any molecule that helps prevent large changes in pH. Many buffers are anions that have a strong affinity for H⁺ molecules. When free H⁺ is added to a solution containing buffers, the buffers bond to the H⁺, thereby preventing a change in pH. The bicarbonate ion, HCO_3^-, is an important buffer in the human body. Sodium bicarbonate (baking soda) is a buffer used to adjust the pH of swimming pools and spas. The following equation shows how sodium bicarbonate buffers a solution to which hydrochloric acid (HCl) has been

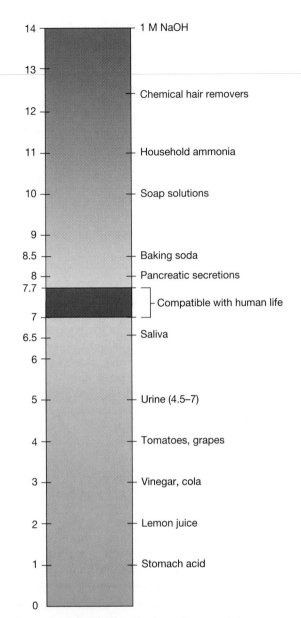

■ Figure 2-12 **pH scale** This logarithmic scale is a measure of the hydrogen ion concentration. At a pH of 14, the hydrogen ion concentration is 10^{-14} moles/liter, or mol/L. At the neutral pH of 7.0, the hydrogen ion concentration is 10^{-7} mol/L. At a pH of 2, the hydrogen ion concentration is 10^{-2} mol/L.

added. When placed in solution, HCl dissociates into H⁺ and Cl⁻, creating a high H⁺ concentration and low pH. If sodium bicarbonate is present, some of the bicarbonate ions combine with some of the H⁺ to form undissociated carbonic acid. The removal of free H⁺ lowers the H⁺ concentration and raises the pH:

$$H^+ + Cl^- + HCO_3^- + Na^+ \rightarrow H_2CO_3 + Cl^- + Na^+$$

BIOMOLECULES

Most of the molecules of concern in human physiology have three elements in common: carbon (C), hydrogen (H), and oxygen (O). Molecules that contain carbon are

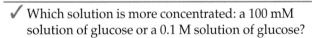

✓ Which solution is more concentrated: a 100 mM solution of glucose or a 0.1 M solution of glucose?

✓ In making a 5% solution of glucose, why wouldn't you weigh out 5 grams of glucose and add it to 100 mL of water?

✓ When the body becomes more acidic, does the pH increase or decrease?

✓ How can urine, stomach acid, and saliva all have pH values outside the pH range that is compatible with life?

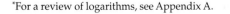

*For a review of logarithms, see Appendix A.

continued from page 23

In the human body, chromium helps move glucose—the simple sugar that cells use to fuel all their activities—from the bloodstream into cells. A small amount of chromium in the diet is therefore necessary for the proper functioning of cells.

Question 4: If people have a chromium deficiency, would you predict that their blood glucose level would be lower or higher than normal?

known as **organic molecules,** because it was once thought that they existed in or were derived from only plants and animals. Organic molecules associated with living organisms are also called **biomolecules.** There are four kinds of biomolecules: carbohydrates, lipids, proteins, and nucleotides. The first three groups are used by the body for energy and as the building blocks for cellular components. The fourth group, the nucleotides, includes DNA and RNA, the structural components of genetic material. Compounds that carry energy, such as ATP, or regulate metabolism, such as cyclic AMP, are also nucleotides. Each group of biomolecules has a characteristic composition and molecular structure, which are described briefly in the following sections.

Carbohydrates Are the Most Abundant Type of Biomolecule

Carbohydrates get their name from their structure: literally, carbons with water. The basic formula for a carbohydrate is $(CH_2O)_n$, showing that for each carbon there are two hydrogens and one oxygen, the same H:O ratio found in water. The n after the parentheses represents the number of repetitions of the CH_2O unit. For example, glucose, $C_6H_{12}O_6$, has an n of 6. Glucose is an example of a simple sugar, the smallest type of carbohydrate. The names of all the simple sugars end with the suffix *-ose.* There are two kinds of simple sugars: the **monosaccharides** [*mono-*, one + *sakcharon*, sugar] and the **disaccharides** [*di-*, two]. The monosaccharides are the building blocks of complex carbohydrates and have either five carbons, like **ribose,** or six carbons, like **glucose (dextrose), fructose,** and **galactose.** When two monosaccharides bond to each other, they form a disaccharide molecule. Typical disaccharides are **maltose, lactose** (milk sugar), and **sucrose** (table sugar). Figure 2-13 ■ shows the composition of most of these simple sugars.

When many glucose molecules join together, they form very large molecules. These complex carbohydrate molecules are known as **polysaccharides** [*poly-*, many]. A large molecule made up of repeating units is called a **polymer.** Thus, the complex carbohydrates are all glucose polymers (Fig. 2-13 ■). Because these polymers have only one type of molecule, glucose, the differences in the polysaccharides come from the manner in which

the glucose molecules are linked. All living cells store glucose for energy in the form of a polysaccharide, and some cells produce polysaccharides for structural purposes as well. Yeasts and bacteria make a glucose storage polymer called *dextran.* Animal cells make a storage polysaccharide called **glycogen** that is found in all tissues of the body. Many invertebrate animals make a structural polysaccharide called *chitin.* Plants make two types of polysaccharides: the storage molecule **starch,** which humans can digest, and the structural molecule **cellulose,** which humans cannot digest. It is unfortunate that we are unable to digest cellulose and obtain its energy, because it is the most abundant organic molecule on earth.

Lipids Are Made Mainly of Hydrogen and Carbon

Lipids are biomolecules made up of carbon, hydrogen, and oxygen, like the carbohydrates, but as a rule they contain much less oxygen. An important characteristic of this group is that they are not very soluble in water, because of their nonpolar structure. Technically, lipids are called fats if they are solid at room temperature and oils if they are liquid at room temperature. Most lipids derived from animal sources, like lard and butter, are fats, whereas most plant lipids are oils.

Lipids are the most diverse group of biomolecules. In addition to the true lipids, this category includes three lipid-related substances: phospholipids, steroids, and eicosanoids (Fig. 2-14 ■).

The true lipids and phospholipids are most similar to each other in structure. Both groups contain a simple 3-carbon molecule known as **glycerol** and long carbon-chain molecules known as **fatty acids.** Fatty acids are long chains of carbon atoms bound to hydrogens, with a carboxyl (—COOH) group at one end of the chain. Fatty acids are described as **saturated** if there are no double bonds between carbons; **monounsaturated** if there is one double bond in the molecule; or **polyunsaturated** if there are two or more double bonds in the molecule (Fig. 2-14 ■). You have probably heard these terms in commercials for margarine and cooking oil: Manufacturers like to boast that their products are "High in polyunsaturates!!" A lot of research is being done trying to relate the degree of fat saturation in our diets to the development of heart disease. Saturated fats, such as the fat that marbles a piece of choice beef, are generally considered to contribute to the development of artery-clogging fatty deposits. For a time, it was thought that the more unsaturated fats were the least likely to be harmful, but recent studies suggest that monounsaturated fats such as olive oil may be the most healthful.

Fatty acids in the body link to glycerol to form mono-, di-, or triglycerides. Triglycerides (more correctly known as **triacylglycerol**) are the most important form of lipid in the body; more than 90% of our lipids are in this form. Triglyceride levels in the blood are also important as a

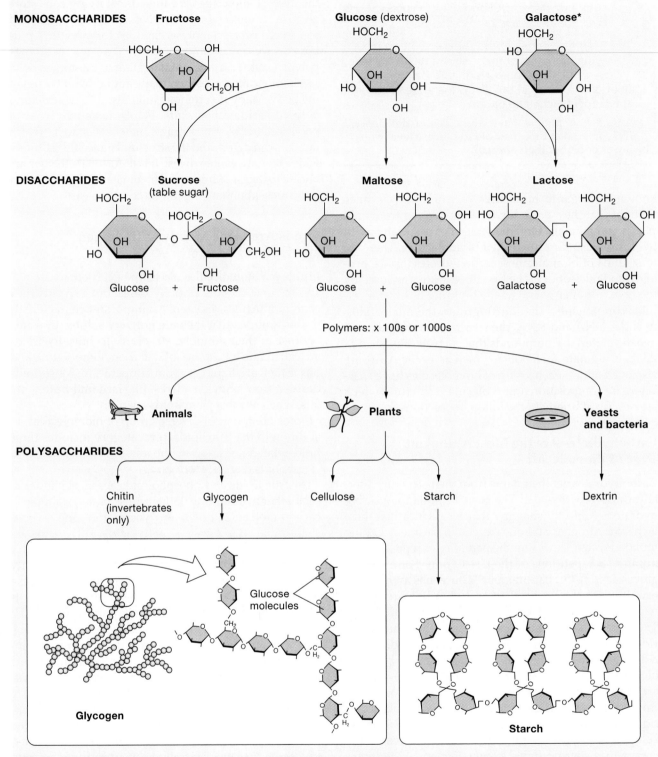

MONOSACCHARIDES   Fructose   Glucose (dextrose)   Galactose*

DISACCHARIDES   Sucrose
(table sugar)

Glucose + Fructose

Maltose

Glucose + Glucose

Lactose

Galactose + Glucose

Polymers: x 100s or 1000s

Animals   **Plants**   **Yeasts
and bacteria**

POLYSACCHARIDES

Chitin
(invertebrates
only)   Glycogen   Cellulose   Starch   Dextrin

Glucose
molecules

Glycogen

Starch

* Notice that the only difference between glucose and galactose is the spatial arrangement of the hydroxyl groups.

■ Figure 2-13 **Map of carbohydrates**

predictor of artery disease; an elevated fasting triglyceride level is linked to a higher risk of disease.

The lipid-related molecules include phospholipids, steroids, and eicosanoids. Most **phospholipids** are diglycerides with a phosphate group attached to the single carbon that lacks a fatty acid. Phospholipids are important components of cell membranes. **Steroids** are lipid-related molecules whose structure includes four linked carbon rings (Fig. 2-14 ■). Cholesterol is the source of steroids in the human body. It is also an impor-

LIPIDS

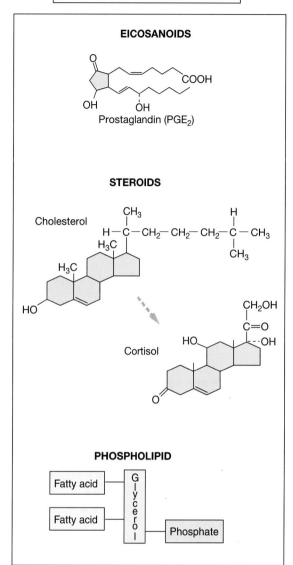

LIPID-RELATED MOLECULES

Palmitate, a saturated fatty acid

Glycerol

Oleate, an unsaturated fatty acid

FATTY ACIDS

Glycerol | Fatty acid

MONOGLYCERIDE

DIGLYCERIDE

TRIGLYCERIDE

EICOSANOIDS

Prostaglandin (PGE₂)

STEROIDS

Cholesterol

Cortisol

PHOSPHOLIPID

■ **Figure 2-14 Lipids and lipid-related molecules** The fatty acid palmitate is shown with all bonds and atoms. It has no double bonds between carbons and thus is a saturated fatty acid. The fatty acid oleate is shown without symbols for the carbons and hydrogens of the molecule in the middle. Oleate has one double bond between carbons and thus is a monounsaturated fatty acid.

tant component of animal cell membranes. The **eicosanoids** are modified 20-carbon fatty acids that are found in animals [*eikosi*, twenty]. These molecules all contain one open five- or six-carbon ring that makes the long carbon chain appear to double back on itself (Fig. 2-14 ■). The main eicosanoids are thromboxanes, leukotrienes, and prostaglandins. The eicosanoids are regulators of various physiological functions.

Proteins Are the Most Versatile of the Biomolecules

Proteins are large molecules made from smaller building-block molecules called **amino acids.** Twenty different amino acids commonly occur in natural proteins, and these building blocks may be assembled in an

almost infinite number of combinations. All amino acids have a similar core structure: a central carbon atom is linked to a hydrogen atom, an amino group (—NH₂), a carboxyl group (—COOH), and a group of atoms designated "R" that is different in each of the amino acids (Fig. 2-15a ■). The R groups differ in their size, shape, and ability to form hydrogen bonds or ions. Because of the different R groups, each of the different amino acids reacts with other molecules in unique ways.

The human body can synthesize all but nine of the 20 amino acids. Those nine must come from dietary proteins and are called the **essential amino acids** (Table 2-3). Because of the nitrogen in the amino group, proteins are the major dietary source of nitrogen.

When two amino acids link together, the amino group of one is joined to the acid group of the other. By linking

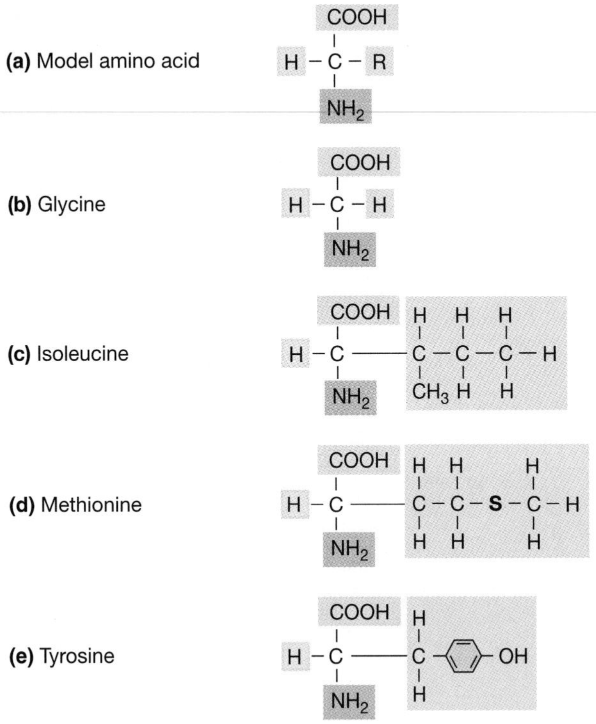

(a) Model amino acid

$$H - C - R$$
(with COOH above C and NH$_2$ below C)

(b) Glycine

$$H - C - H$$
(with COOH above C and NH$_2$ below C)

(c) Isoleucine

$$H - C \text{——} C - C - C - H$$
(with COOH above first C, NH$_2$ below first C; CH$_3$, H, H below the chain carbons, and H's above)

(d) Methionine

$$H - C \text{——} C - C - S - C - H$$
(with COOH above first C, NH$_2$ below first C; H's above and below the chain)

(e) Tyrosine

$$H - C \text{——} C - \text{(benzene ring)} - OH$$
(with COOH above first C, NH$_2$ below first C; H's on second C)

■ **Figure 2-15 Amino acid structure** All amino acids have an acid group (—COOH), an amino group (—NH$_2$), and a hydrogen attached to the carbon. The fourth carbon bond is to a variable "R" group. Notice the sulfur atom (—S—) in methionine.

many molecules in this way, long chains of up to hundreds of amino acids can be formed. If there are two to nine amino acids in the amino acid chain, the molecule is called a **peptide.** A chain of 10 to 100 amino acids is called a **polypeptide.** A chain of more than 100 amino acids is called a **protein.**

The sequence of amino acids in the chain is called the **primary structure** of the protein (Fig. 2-16 ■). For example, the primary structure of thyrotropin-releasing hormone, a peptide made in the brain, is glutamic acid-histidine-proline. The primary structure of a protein is genetically determined.

As the amino acid chain forms, it takes on its **secondary structure,** which is the spatial arrangement of the amino acids in the chain. Some amino acid chains curl into a spiral, or **α-helix,** whereas others create a **β-pleated sheet.** The secondary structure is determined by the angle of the bond between adjacent amino acids and is stabilized by hydrogen bonding between different parts of the molecule. Proteins that are destined for structural uses may be composed almost entirely of sheets, because this configuration is very stable. Other proteins combine sheets with helices.

The three-dimensional shape of an amino acid chain is its **tertiary structure.** Some long protein chains fold up into a ball or other compact shape when the R groups of amino acids in the alpha helix are attracted to each other. The attraction may be between adjacent amino acids or

TABLE 2-3 The Amino Acids of the Human Body

Each amino acid has both a three-letter abbreviation of its name and a single-letter abbreviation, used to spell out the sequence of amino acids in large proteins. The essential amino acids are marked with a star. These nine amino acids cannot be made by the body and must be obtained from the diet.

Amino Acid	Three-Letter Abbreviation	One-Letter Symbol
Alanine	Ala	A
Arginine	Arg	R
Asparagine	Asn	N
Asparagine or aspartic acid	Asx	B
Aspartic acid	Asp	D
Cysteine	Cys	C
Glutamic acid	Glu	E
Glutamine	Gln	Q
Glutamine or glutamic acid	Glx	Z
Glycine	Gly	G
*Histidine	His	H
*Isoleucine	Ile	I
*Leucine	Leu	L
*Lysine	Lys	K
*Methionine	Met	M
*Phenylalanine	Phe	F
Proline	Pro	P
Serine	Ser	S
*Threonine	Thr	T
*Tryptophan	Trp	W
Tyrosine	Tyr	Y
*Valine	Val	V

*Essential amino acids

between amino acids that are at different ends of the molecule. In some proteins, the tertiary structure is held in place by hydrogen bonds. In other proteins, sulfur atoms in two different *cysteine* molecules bond covalently to each other in a disulfide (S—S) bond.

Proteins are often classified as either globular or fibrous proteins based on their shape. **Globular proteins** have amino acid chains that are folded into ball-like shapes. **Fibrous proteins** are found as pleated sheets or in long chains that wind around each other.

A protein with more than one polypeptide chain is said to have **quaternary structure** (Fig. 2-16 ■). Hemoglobin, the oxygen-carrying pigment found in red blood cells, has a quaternary structure composed of four subunits: two polypeptide alpha chains of 141 amino acids each and two polypeptide beta chains of 146 amino acids each. The hemoglobin molecule is illustrated in Figure 16-7.

Proteins, peptides, and amino acids serve a number of important functions in the body. The fibrous proteins, which are insoluble in water, are important structural components of cells and tissues. Examples include collagen, a fibrous protein found in many types of connective tissue, and keratin, a fibrous protein found in hair and

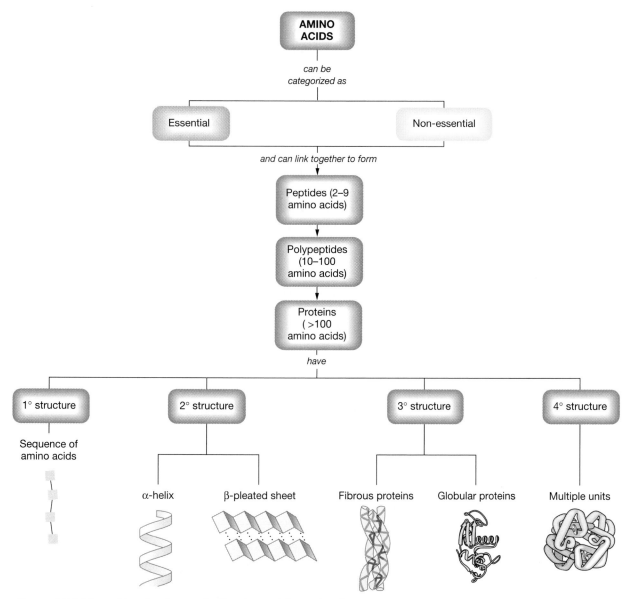

■ Figure 2-16 **Map of levels of organization in protein molecules**

nails. The globular proteins are soluble in water. They act as **carriers** for insoluble lipids in the blood, binding to them and making them soluble. They may also serve as **enzymes** that increase the rate of chemical reactions. Soluble proteins also act as cell-to-cell messengers in the form of hormones and neurotransmitters and as defense molecules to help fight foreign invaders.

Some Molecules Combine Carbohydrates, Proteins, and Lipids

Not all biomolecules are pure protein, pure carbohydrate, or pure lipid. **Conjugated proteins** are protein molecules combined with another kind of biomolecule. Proteins combine with lipids to form lipoproteins; proteins combine with carbohydrates to form glycoproteins. **Lipoproteins** are found mostly in cell membranes.

They also carry hydrophobic molecules such as cholesterol in the blood: The lipid section of the molecule interacts with insoluble cholesterol while the protein section makes the entire lipoprotein-cholesterol complex soluble in water.

A "glyco-" molecule is composed of a carbohydrate bonded to either a lipid or protein. Like lipoproteins, **glycoproteins** and **glycolipids** are important components of cell membranes (Chapter 5). The carbohydrate portions of these molecules are usually found on the external surface of the cell, where they can form hydrogen bonds with water and other molecules in body fluids. Glycolipids are believed to contribute to the stability of the cell membrane. Certain glycoproteins and glycolipids act as *cell membrane receptors* and *cell markers*. Membrane receptors bind to molecules outside the cell. The cell markers identify cells, like a label, so that the defense cells of the body can tell the difference between "self" and foreign invaders.

✓ Why must we include essential amino acids in our diet but not the other amino acids?

✓ What property of protein structure allows proteins to have more versatility than lipids or carbohydrates?

Nucleotides Are Responsible for the Transmission and Storage of Energy and Information

Nucleotides play an important role in many of the processes of the cell, including the storage and transmission of genetic information and the transfer of energy. A single **nucleotide** is a three-part molecule composed of (1) one or more phosphate groups, (2) a five-carbon sugar, and (3) a carbon-nitrogen ring structure called a **base** (Fig. 2-17 ■). Every nucleotide contains one of two

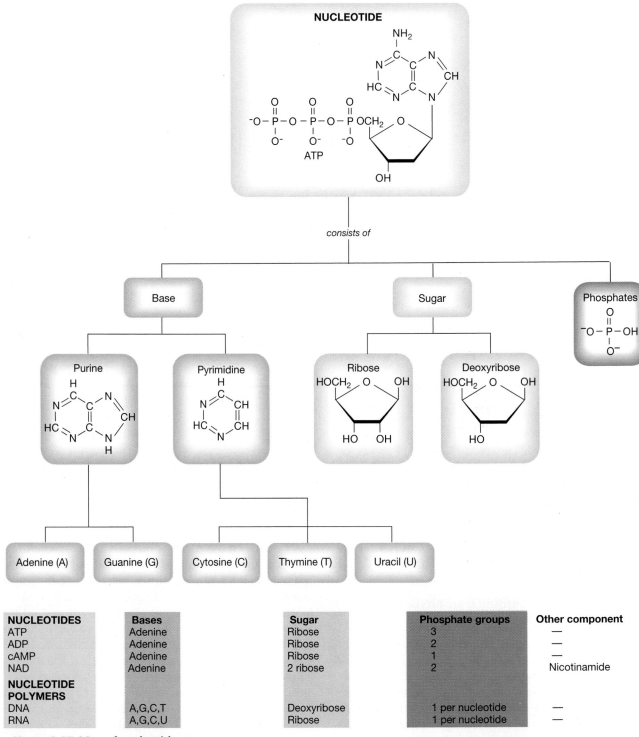

NUCLEOTIDES	Bases	Sugar	Phosphate groups	Other component
ATP	Adenine	Ribose	3	—
ADP	Adenine	Ribose	2	—
cAMP	Adenine	Ribose	1	—
NAD	Adenine	2 ribose	2	Nicotinamide
NUCLEOTIDE POLYMERS				
DNA	A,G,C,T	Deoxyribose	1 per nucleotide	—
RNA	A,G,C,U	Ribose	1 per nucleotide	—

■ Figure 2-17 **Map of nucleotide structure**

continued from page 27

In a few studies, chromium supplements were shown to lower blood glucose levels in some people with diabetes mellitus. Diabetes mellitus is a disease in which blood sugar levels are abnormally elevated.

Question 5: From the results of these studies, can you conclude that all people with diabetes suffer from chromium deficiency?

possible sugars: **ribose** or **deoxyribose** [*de-*, without; -*oxy-*, oxygen]. There are also two types of bases: the **purines** and the **pyrimidines**. The purines have a double ring structure; the pyrimidines have a single ring. There are two different purines: **adenine** and **guanine.** There are three different pyrimidines: **cytosine, thymine,** and **uracil.**

Different nucleotides are made up of sugars, phosphate groups, and various combinations of bases. The smallest nucleotides include the energy-transferring compounds **ATP** (adenosine triphosphate) and **ADP** (adenosine diphosphate). These molecules contain the base adenine, the sugar ribose, and two or three phosphate groups (Fig. 2-17 ■). A related nucleotide is **cyclic AMP** (cyclic adenosine monophosphate, or cAMP), a molecule important in the transfer of signals between the extracellular fluid and the cell. Other energy-transferring nucleotides include **NAD** (nicotinamide adenine dinucleotide) and **FAD** (flavin adenine dinucleotide). Molecules that take part in energy transfer and storage are discussed in more detail in Chapter 4.

The nucleotide polymers **DNA** and **RNA** are responsible for storing genetic information within the cell as well as transmitting it to future generations of cells. DNA stands for *deoxyribonucleic acid;* RNA is short for *ribonucleic acid.* In each case, the name describes the type of sugar found in the nucleotides of the polymer. DNA contains the bases adenine, guanine, cytosine, and thymine. RNA also contains adenine, guanine, and cytosine but has uracil instead of thymine. Both polymers are formed by linking

nucleotides into long chains, similar to the long amino acid chains of proteins. The sugar of one nucleotide links to the phosphate of the next, creating a chain or backbone of alternating sugar-phosphate-sugar-phosphate groups. The bases extend freely to the side (Fig. 2-18 ■).

The DNA molecule is composed of two linked nucleotide chains that form a double helix, a three-dimensional structure that was first described by J. D. Watson and F. H. C. Crick, in 1953. The double helix is formed when two DNA chains link through complementary **base pairing.** Because of space limitations, the larger purines (adenine or guanine) must always link to a smaller pyrimidine (cytosine or thymine). Adenine (A) always pairs with thymine (T), and guanine (G) always pairs with cytosine (C). These base pairs (A—T and G—C) hold the double strand of DNA in its spiral shape. Base pairing and DNA replication are reviewed in Appendix B.

RNA usually takes the form of a single chain, although it is temporarily bound to DNA while it is being formed. RNA plays an important role in protein synthesis, discussed in Chapter 4.

In this chapter we have looked at how atoms combine to form molecules, how molecules dissolve to form solutions, and how different molecular building blocks combine into biologically important compounds (see Figure 2-19 ■ for a summary). The next chapter will consider the role of biomolecules in the structure of cells and tissues.

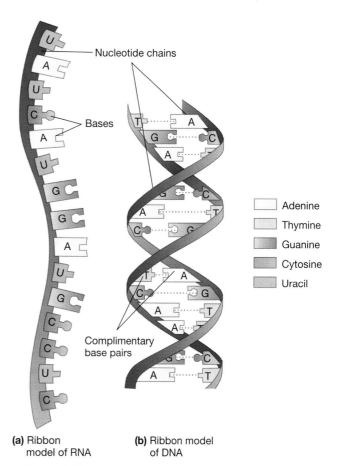

- Nucleotide chains
- Bases
- Complimentary base pairs

Adenine
Thymine
Guanine
Cytosine
Uracil

(a) Ribbon model of RNA

(b) Ribbon model of DNA

■ Figure 2-18 **RNA and DNA**

continued from above

To test the claim that chromium supplements enhance muscle development, a group of researchers gave high daily doses of chromium picolinate to football players during a two-month training period. Researchers found the players who took chromium supplements did not experience greater increase in muscle mass or strength than players who did not take the supplement.

Question 6: Based on this study, which did not show enhanced muscle development from chromium supplements, and on the study that suggested that chromium supplements might cause cancer, do you feel that Stan should continue taking chromium?

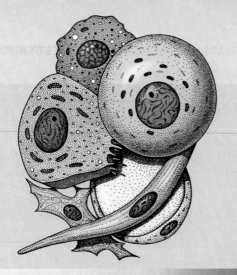

3

Cells and Tissues

BACKGROUND BASICS

Just as atoms combine to make molecules, a vast collection of molecules makes up the **cell,** the basic functional unit of most living organisms. Individual cells carry out the same life-sustaining functions as multicellular animals. They sense and respond to changes in their environment, and they communicate with neighboring cells by releasing chemicals or by generating electrical signals. Cells take in oxygen and nutrients, they extract energy from the nutrients for growth, repair, and reproduction, and they get rid of wastes such as carbon dioxide.

Much of what we know about cells comes from studies of simple organisms that consist of one cell. But humans are much more complex, with trillions of cells in their bodies. It has been estimated that there are more than 200 different types of cells within the human body, each cell with its own characteristic structure and function. In studying physiology, you must be aware both of how individual cells carry out their functions and of how they cooperate and communicate with each other in the human body.

Cytology [*kytos,* cell + *logia,* explaining] is the study of cells. In recent years, the word *cytology* has been used mostly to refer to the study of the structure of cells, whereas the more general term *cell biology* has referred mostly to the study of how cells work. The development of new research techniques has greatly expanded our understanding of how cells do what they do, and so we are better able to apply that knowledge to the study of tissues, organs, and the body as a whole.

The cooperation between cells is a central idea of human physiology. This chapter looks first at the structure and function of individual cells; it then examines how groups of cells with similar functions join together to form the tissues and organs of the body.

Problem

The Pap Smear

Dr. George Papanicolaou has saved the lives of millions of women by popularizing the Pap test, a screening method that detects the early signs of cervical cancer. In the 50 years the Pap test has been widely used, deaths from cervical cancer have dropped nearly 70%. If detected early, cervical cancer is one of the most treatable forms of cancer. Today, Jan Melton, who had an abnormal Pap test three months ago, returns to her family physician for a repeat test. The results will determine whether she should undergo treatment for cervical cancer.

continued on page 43

STUDYING CELLS AND TISSUES

The first recorded use of magnifying lenses to study the composition of biological material took place in the last half of the 1600s. In the Netherlands, Antoine van Leeuwenhoek discovered that pond water was teeming with "wee beasties" too small to be seen with the naked eye, while the English scientist Robert Hooke around 1665 used a simple light microscope to investigate the structure of cork, the dead tissue layer under the bark of trees. It was Hooke who named the holes in cork "cells," from the Latin word *cella*, meaning a small room. But Hooke's cork cells, like the rooms of a building, were only walls and did not contain living material. Not until the early part of the 1800s, with improved microscopes, did we begin to understand that living cells were filled with a semifluid substance, and that the cell was the fundamental structural unit of life.

The light microscope makes it possible to examine both living cells and preserved cells. But even at the highest magnification and with the aid of dyes that stain different cellular components, it is possible to see only the general shape and size of cells or the larger internal structures such as the nucleus. Because most internal components are too small to be seen with the light microscope, the fine details of cell structure went undescribed until the electron microscope came into widespread use in the 1950s. Figure 3-1 ■ shows a cell as seen under a light microscope, by transmission electron microscopy, and by scanning electron microscopy. The scanning electron microscope has greatly added to the understanding of cells and tissues by making it possible to view their fine structure in a realistic three-dimensional fashion rather than as thin flat slices that cut through internal structures. The next section of this chapter examines the details of cellular structure as seen with the aid of these microscopic techniques.

CELLULAR ANATOMY

Cells are not all alike. All animals begin life as a single cell, the **zygote,** or fertilized ovum (egg). As the zygote divides into two cells, then four cells, then eight, each daughter cell looks identical. But by a process known as **differentiation,** the cells take on more than 200 different shapes and functions as the embryo develops further. Each cell inherits the same genetic information in its DNA, but no one cell uses all of it. During differentiation, only selected genes are activated, transforming the cell into a specialized unit. In most cases, the final shape and size of a cell and its contents are a direct reflection of its function. Figure 3-2 ■ shows some of the different cells in the human body. Although these mature cells look very different from each other, they all started out alike

(a)
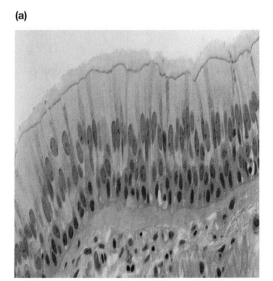

(b)
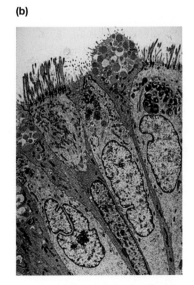

(c)

■ **Figure 3-1 One cell type viewed by light microscopy (a), transmission electron microscopy (b), and scanning electron microscopy (c)** In (a), note that the fingerlike microvilli appear as a fuzzy border; only the nucleus and cell membrane are easily seen.

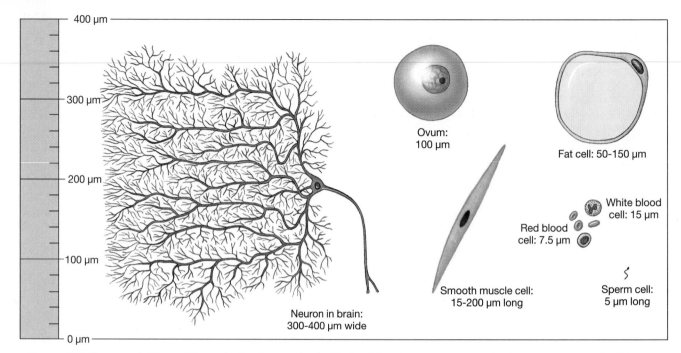

■ Figure 3-2 **Representative cell types in the human body** Each cell has the genetic potential to become any type of cell, but selective activation of genes during differentiation determines the ultimate size, shape, and function of the cell.

and they retain many features in common. Let us look at the anatomical organization that is shared by most cells.

You can compare the structural organization of a cell to that of a medieval walled city. The city is separated from the surrounding countryside by a high wall, with entry and exit strictly controlled through gates that can be opened and closed. The city inside the walls is a diverse collection of houses and shops with varied functions. Within the city, a ruler in the castle oversees the everyday comings and goings of the city's inhabitants. Because the city depends on food and raw material from outside the walls, the ruler negotiates with the farmers in the countryside. Foreign invaders are always a threat, so the city ruler communicates and cooperates with the rulers of neighboring cities.

In the cell, the outer boundary is the cell membrane. Like the city wall, it controls the movement of material between the cell interior and the outside by opening and closing "gates," made of protein. The inside of the cell is divided into compartments called organelles, rather than into shops and houses. Each of these compartments has a specific purpose that contributes to the function of the cell as a whole. In the cell, DNA in the nucleus is the "ruler in the castle," controlling both the internal workings of the cell and its interaction with other cells. Like the city, the cell depends on supplies from its external environment. It must also communicate and cooperate with other cells in order to keep the body functioning in a coordinated fashion.

Figure 3-3 ■ is a map of cell structure. The cells of the body are surrounded by a dilute salt solution known as

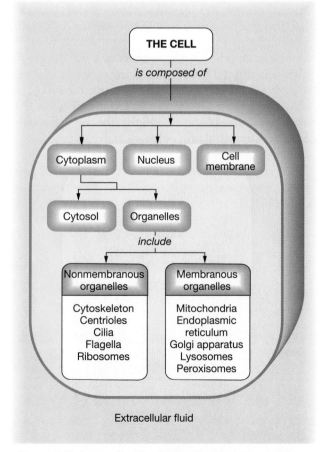

■ Figure 3-3 **A map for the study of cell structure** The cells of the body are surrounded by the extracellular fluid.

the **extracellular fluid (ECF)** [*extra-*, outside]. This is the external environment of the cell from which the cell gains its nutrients and into which the cell puts its wastes and the chemicals that it uses to communicate with other cells. The cell membrane separates the inside environment of the cell from the extracellular fluid.

Internally the cell is divided into the cytoplasm and the nucleus. The cytoplasm consists of a fluid portion, called the cytosol, and assorted structures, collectively known as organelles. The organelles work in an integrated manner, each organelle taking on one or more of the cell's functions. A typical cell from the lining of the small intestine is shown in Figure 3-4d ■; it has most of the structures found in animal cells. Now we will examine the structure and function of typical cell components.

The Cell Membrane

The **cell membrane** (also called the **plasma membrane**) separates the inside of the cell from the surrounding extracellular fluid. Structurally, the membrane is a bilayer (double layer) of phospholipid molecules, arranged so that the hydrophilic phosphate groups face the aqueous solutions inside and outside the cell; this leaves the hydrophobic lipid portions hidden in the center of the membrane. The membrane is studded with protein molecules, like raisins in a slice of bread (Fig. 3-5 ■, p. 46). Functionally, the cell membrane serves as both a gateway and a barrier. All substances that enter or leave the cell must cross the cell membrane, which regulates exchange between the cell and the extracellular fluid. Because of the importance of membranes in physiology, the details of membrane structure and function are covered separately in Chapter 5.

The Cytoplasm

The **cytoplasm** includes all material inside the cell membrane, except for the nucleus. The cytoplasm has two components:

1. **Cytosol,** or intracellular fluid: The cytosol is a semigelatinous substance, separated from the extracellular fluid by the cell membrane. The cytosol contains

---

continued from page 41

Cancer is a condition in which cells divide uncontrollably and fail to coordinate with the normal cells that surround them. Cancerous cells also fail to differentiate into specialized cell types, and can usually be recognized by the presence of a large nucleus surrounded by a relatively small amount of cytoplasm.

Question 1: Why is it necessary to remove or kill cancerous cells in order to treat cancer?

---

dissolved nutrients, ions, waste products, and particles of insoluble materials known collectively as **inclusions.** Among the most common inclusions are stored nutrients such as glycogen granules and lipid droplets. Also suspended in the cytosol are the:

2. **Organelles:** These "little organs" are like the organs of the body in that each has a specific role to play in the overall function of the cell. For example, mitochondria generate most of the cell's ATP, and lysosomes filled with enzymes act as the digestive system of the cell. Organelles are often classified according to whether they have a membrane around them. **Membranous organelles** are separated from the cytosol by one or more phospholipid membranes similar in structure to the cell membrane. Many membranous organelles have hollow interiors. The cavity of a hollow tube, organ, or organelle is known as its **lumen,** from the Latin word for light or window. Various anatomical structures have lumens. **Nonmembranous organelles** do not have boundary membranes and so are in direct contact with the cytosol. Movement of material between these organelles and the cytosol does not require transport across a membrane.

Nonmembranous Organelles

Nonmembranous organelles can be divided into two groups: ribosomes and the organelles made from insoluble protein fibers. The latter include the cytoskeleton, centrioles, cilia, and flagella.

Ribosomes participate in protein synthesis **Ribosomes** (Fig. 3-6 ■, p. 46) are small dense granules of RNA and protein that manufacture proteins under the direction of the cell's DNA (see Chapter 4). Most ribosomes are attached to the endoplasmic reticulum and called **fixed ribosomes.** Ribosomes also float freely in the cytoplasm, where some form groups of 10 to 20 known as **polyribosomes.** Ribosomes are most numerous in cells that synthesize proteins, such as protein hormones, for export out of the cell.

Protein fibers of the cytoplasm The protein fibers in the cytoplasm are classified by their diameter. The thinnest, **microfilaments,** are made of the protein **actin.** Somewhat larger **intermediate filaments** are made of several different types of protein, including **myosin** in muscle, **keratin** in hair and skin, and **neurofilament** in nerve cells. In muscle, intermediate myosin filaments combine to form **thick filaments.** The largest protein fiber organelles are the **microtubules.** Microtubules and other fibers link together to create the internal scaffolding of the cell, the cytoskeleton. In addition, microtubules combine to form the more complex structures of centrioles, cilia, and flagella.

Anatomy Summary   Levels of Organization— Cells to Systems

System	Organ	Tissue

■ Figure 3-4

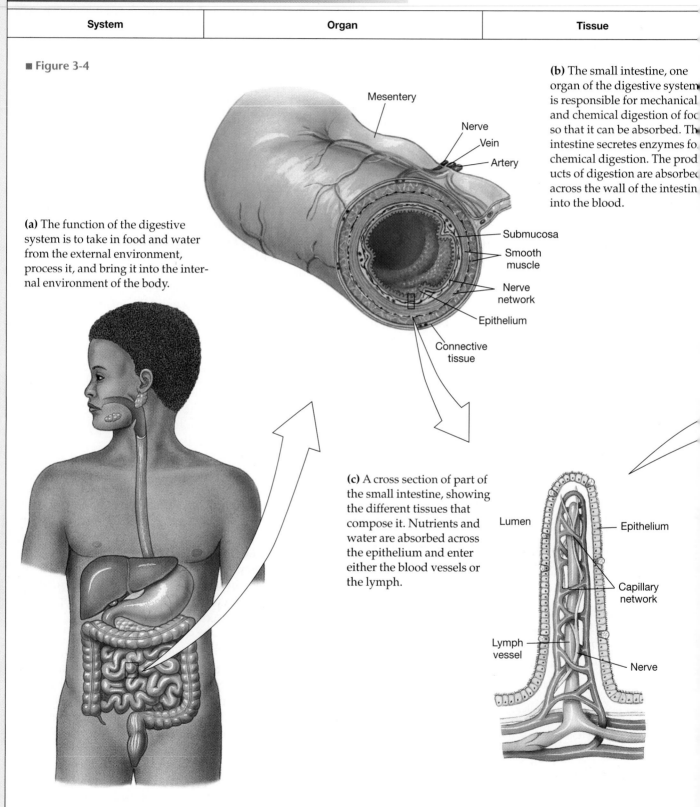

(b) The small intestine, one organ of the digestive system, is responsible for mechanical and chemical digestion of food so that it can be absorbed. The intestine secretes enzymes for chemical digestion. The products of digestion are absorbed across the wall of the intestine into the blood.

(a) The function of the digestive system is to take in food and water from the external environment, process it, and bring it into the internal environment of the body.

Labels in figure (b): Mesentery; Nerve; Vein; Artery; Submucosa; Smooth muscle; Nerve network; Epithelium; Connective tissue

(c) A cross section of part of the small intestine, showing the different tissues that compose it. Nutrients and water are absorbed across the epithelium and enter either the blood vessels or the lymph.

Labels in figure (c): Lumen; Epithelium; Capillary network; Lymph vessel; Nerve

(d) Each cell of the intestinal epithelium has structures that allow it to carry out the digestion and absorption of food. The microvilli on the side of the cell that faces the lumen of the intestine increase the surface area available for absorption. This epithelial cell also has most of the organelles of a typical cell.

Cell	Sub-cellular structures (organelles)	Characteristics	Function
	(e) Microvilli	Extensive folding of the cell membrane	Increase surface area for absorption
	Cell membrane	Bilayer of phospholipid molecules with inserted proteins	Acts as a gateway for and barrier to movement of material between the interior of the cell and the extracellular fluid
	Cytosol	Semi-gelatinous substance	Contains dissolved nutrients, ions, wastes, insoluble inclusions; suspends the organelles
	Centrioles	Bundles of microtubules	Direct movement of DNA during cell division
	Lysosomes and peroxisomes	Membrane-bound vesicles filled with digestive enzymes	Digest bacteria, old organelles; metabolize fatty acids
	Mitochondrion	Double wall with central matrix	Produces most of cell's ATP
	Golgi apparatus	Hollow membranous sacs	Modify and package proteins
	Nucleus	Central lumen with a two-membrane outer envelope with pores	Contains DNA to direct all functions of the cell
	Nucleolus	Region of DNA, RNA, and protein	Contains the genes that direct synthesis of ribosomal RNA
	Rough endoplasmic reticulum	Membrane tubules that are continuous with the outer nuclear membrane; edged with ribosomes	Site of protein synthesis
	Smooth endoplasmic reticulum	Same as rough ER without ribosomes	Synthesis of fatty acids, steroids, and lipids
	Ribosomes	Granules of RNA and protein	Assemble amino acids into proteins
	Microtubules and microfilaments	Protein fibers	Provide strength and support; enable cell motility; transport materials

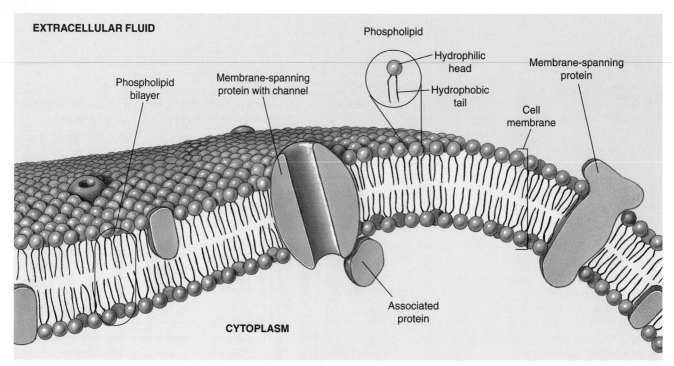

■ Figure 3-5 **The cell membrane** The cell membrane is a double layer of phospholipid molecules, arranged with the hydrophilic phosphate heads facing the aqueous solutions inside and outside the cell. The hydrophobic lipid tails are hidden in the middle of the membrane. Proteins of various kinds are inserted into and through the phospholipid bilayer.

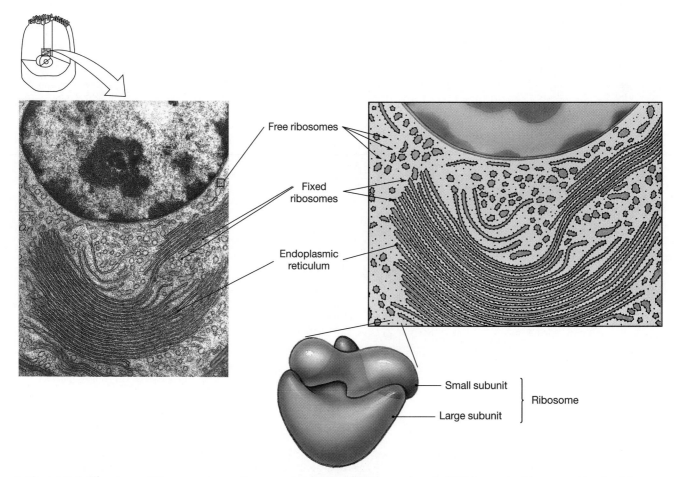

■ Figure 3-6 **Ribosomes** Ribosomes are small nonmembranous organelles composed of RNA organized into two subunits. Many ribosomes are attached to endoplasmic reticulum, but others are found free in the cytoplasm.

The Cytoskeleton   The **cytoskeleton** [Greek *kytos*, cell + skeleton] is a flexible, changeable three-dimensional scaffolding of actin microfilaments, intermediate filaments, and microtubules that extends throughout the cytoplasm. The cytoskeleton (1) provides mechanical strength to the cell, (2) stabilizes the positions of organelles, and (3) helps transport materials into the cell and within the cytoplasm. Fibers of the cytoskeleton may also connect with protein fibers in the extracellular fluid, (4) linking cells to each other and to support material outside the cells. Finally, the cytoskeleton is (5) responsible for various kinds of movement by certain cells of the human body. These mobile cells include white blood cells that squeeze out of blood vessels and travel to sites of infection, as well as growing nerve cells that send out long extensions as they elongate.

Some cytoskeleton fibers are permanent, but others can be synthesized or disassembled according to the cell's needs. Because of the cytoskeleton's changeable nature, its organizational details are not well understood. Figure 3-7a ■ is a diagram of a typical cell with organelles held in place by cytoskeletal fibers. In some cells, the cytoskeleton plays an important role in determining the shape of the cell. Figure 3-7b ■ shows how the cytoskeleton helps support **microvilli** [*micros*, small + *villus*, tuft of hair], fingerlike extensions of the cell membrane that increase the surface area for absorption of material.

Centrioles   **Centrioles** are bundles of microtubules that direct the movement of DNA strands during cell division (see Appendix B). Each centriole is a cylindrical bundle of 27 microtubules, arranged in nine triplets (Fig. 3-8a ■). In cell division, fibers of the cytoskeleton begin to grow close to the centrioles, then lengthen outward through the cytoplasm. Most cells contain a pair of centrioles arranged as shown in the typical cell of Figure 3-4d ■. Cells that have lost their ability to undergo cell division, such as mature nerve cells, lack centrioles.

Cilia and Flagella   **Cilia** [singular, *cilium*, Latin for eyelash] are short, hairlike structures projecting from the cell surface like the bristles of a brush. The surface of a cilium is a continuation of the cell membrane; its core contains nine pairs of microtubules surrounding a central pair (Fig. 3-8b ■). The microtubules terminate just inside the cell at the **basal body.** Cilia beat rhythmically back and forth when the microtubule pairs in their core

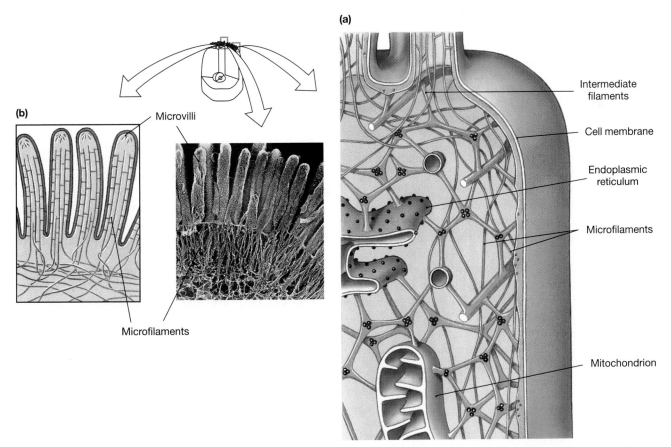

(a)

(b)

Microvilli

Microfilaments

Intermediate filaments

Cell membrane

Endoplasmic reticulum

Microfilaments

Mitochondrion

■ **Figure 3-7 The cytoskeleton and microfilaments** (a) The cytoskeleton of a cell is composed of microfilaments, intermediate filaments, and microtubules. The cytoskeleton provides support for the cell membrane and anchors cell organelles in position. (b) Microfilaments form a network just under the cell membrane. They also extend into microvilli that project above the membrane surface. Microvilli greatly increase the surface area of the cell.

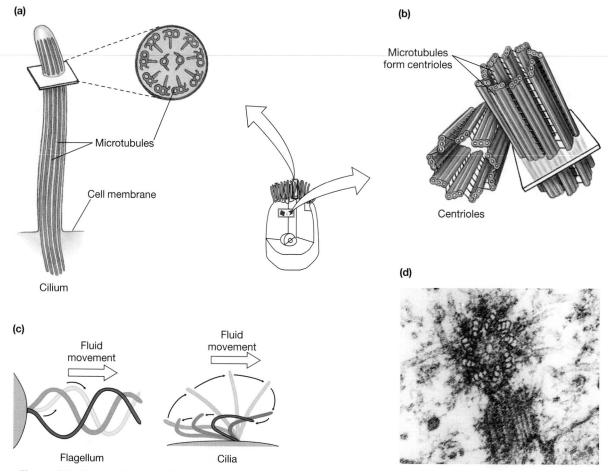

(a)

Microtubules

Cell membrane

Cilium

(b)

Microtubules form centrioles

Centrioles

(c)

Fluid movement

Fluid movement

Flagellum

Cilia

(d)

■ **Figure 3-8 Centrioles and cilia** (a) The outer surface of a cilium or flagellum is a continuation of the cell membrane. Internally, cilia and flagella are composed of nine pairs of microtubules surrounding a central pair. (b) Most cells contain a pair of centrioles that direct the movement of DNA during cell division. Each centriole is a cylindrical bundle of 27 microtubules, arranged in nine triplets. (c) The movement of cilia and flagella occurs when the microtubule pairs slide past each other. (d) A pair of centrioles.

slide past each other in a process that consumes energy. This movement creates currents that sweep fluids or secretions across the cell surface.

Flagella [singular, *flagellum*, Latin for whip] have the same microtubule arrangement as cilia but are considerably longer. In addition, a flagellated animal cell usually only has one or two flagella, whereas ciliated cells may be almost totally covered with cilia. Flagella are found on free-floating single cells; the only such cell in humans is the male sperm cell (see Fig. 3-2 ■). The function of a flagellum is to push the cell through fluid with wavelike movements of the flagellum, just as undulating contractions of a snake's body push it headfirst through its environment. Flagella bend and move by the same basic mechanism as cilia.

Membranous Organelles

The membranous organelles are separated from the cytosol by one or more phospholipid membranes similar to the cell membrane. This membrane barrier between the inside of the organelle and the cytosol allows the organelle to contain substances that might be harmful to the cell if they were free. It also allows the cell to separate different functions. For example, proteins are synthesized in one organelle, modified in another, and stored in a third. Figure 3-4e ■ shows five types of membranous organelles: mitochondria, the endoplasmic reticulum, the Golgi apparatus, lysosomes, and peroxisomes.

Mitochondria are the powerhouse of the cell

Mitochondria [singular, *mitochondrion; mitos*, thread + *chondros*, granule] are small spherical to elliptical organelles with an unusual double wall (Fig. 3-9 ■). The outer membrane of the wall gives the mitochondrion its shape; the inner membrane is folded into leaflets or tubules called **cristae.** In the center of the mitochondrion, inside the inner membrane, is the mitochondrial **matrix** that contains enzymes, ribosomes, granules, and strands of DNA. Between the outer and inner membranes is the **intermembrane space,** a region that plays an important role in the production of ATP by the mito-

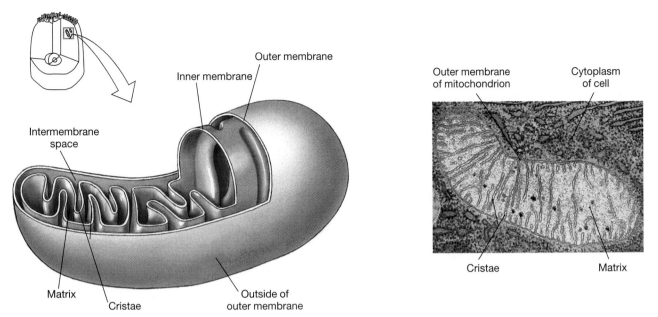

Outer membrane

Inner membrane

Intermembrane space

Outer membrane of mitochondrion

Cytoplasm of cell

Matrix

Cristae

Outside of outer membrane

Cristae

Matrix

■ **Figure 3-9 Mitochondria** Mitochondria have a double wall structure with a heavily folded inner membrane.

chondria. Mitochondria are where most ATP is generated, which is why they are nicknamed the cell's "powerhouse." The number of mitochondria in a particular cell depends on the cell's energy needs. Cells such as skeletal muscle cells, which use a lot of energy, have many more mitochondria than less active cells.

Mitochondria are unusual organelles in two ways. First of all, in their matrix they have their own unique DNA. This mitochondrial DNA, along with the ribosomes in the matrix, means that mitochondria can manufacture some of their own proteins. Why this is true of mitochondria and not other organelles has been the subject of intense scrutiny. According to one theory, mitochondria are the descendants of bacteria that invaded cells millions of years ago. The bacteria developed a mutually beneficial relationship with their hosts and soon became an integral part of the host cell. Supporting evidence for this theory is the fact that our mitochondria contain DNA, RNA, and enzymes similar to the type found in bacteria but unlike those found in our own nuclei.

The second unusual characteristic of mitochondria is their ability to replicate themselves even when the cell to which they belong is not undergoing cell division. This process is aided by the presence of mitochondrial DNA that allows the organelles to direct their own duplication. Mitochondrial replication takes place by budding, with small daughter mitochondria pinching off an enlarged parent. Cells such as exercising muscle cells, which are subjected to increased energy demands over a period of time, may meet the demand for more ATP by increasing the number of mitochondria in their cytoplasm.

The endoplasmic reticulum is the site of protein and lipid synthesis   The **endoplasmic reticulum,** or **ER,** is a network of interconnected membrane tubes (Fig. 3-10 ■) that are a continuation of the outer nuclear membrane. The name *reticulum* comes from the Latin word for *net* and refers to the netlike appearance of the tubules. Electron micrographs reveal that there are two forms of endoplasmic reticulum: **rough** (granular) **endoplasmic reticulum (RER)** and **smooth** (agranular) **endoplasmic reticulum (SER).** Rows of ribosomes dot the cytoplasmic surface of rough endoplasmic reticulum, giving it a granular or rough appearance. Smooth endoplasmic reticulum lacks the ribosomes and appears as smooth membrane tubes. Both types of endoplasmic reticulum have the same three major functions: synthesis, storage, and transport of biomolecules.

The smooth endoplasmic reticulum is the main site for the synthesis of fatty acids, steroids, and lipids (∞ p. 27). Phospholipids for the cell membrane are produced here, and cholesterol is modified into steroid hormones, such as the sex hormones estrogen and testosterone. The smooth endoplasmic reticulum of liver and kidney cells is responsible for the detoxification or inactivation of drugs. In skeletal muscle cells, a modified form of smooth endoplasmic reticulum stores calcium ions (Ca^{2+}) to be used in muscle contraction.

The rough endoplasmic reticulum is the main site for the synthesis of proteins. Proteins are assembled on the ribosomes attached to the cytoplasmic surface of the rough endoplasmic reticulum and are inserted into the lumen, where they may undergo chemical modification. Most of these proteins are packaged into spherical membrane-bound **transport vesicles** [*vesicula*, bladder] that

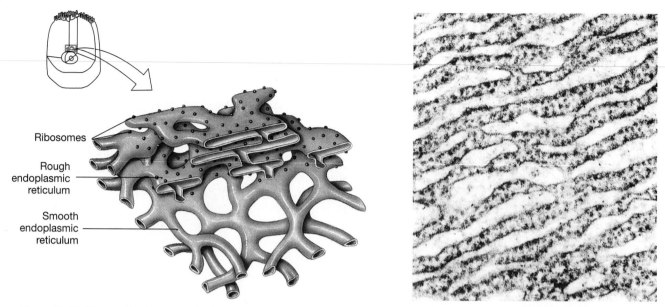

Ribosomes

Rough
endoplasmic
reticulum

Smooth
endoplasmic
reticulum

■ **Figure 3-10 The endoplasmic reticulum** The endoplasmic reticulum is a series of hollow tubules formed by a continuation of the outer nuclear membrane. The appearance of the rough endoplasmic reticulum is due to the fixed ribosomes attached to it.

pinch off from the tips of the endoplasmic reticulum. The transport vesicles shuttle their contents across the cytosol to the Golgi apparatus, the subject of the next section.

The Golgi apparatus packages proteins into membrane-bound vesicles   The **Golgi apparatus** (Fig. 3-11 ■) was first described by Camillo Golgi in 1898. For years, some investigators thought that this organelle was just a result of the fixation process needed to prepare tissues for viewing under the light microscope. However, we now know from electron microscope studies that the Golgi apparatus is indeed another membranous organelle. Its function is to modify proteins made by the rough endoplasmic reticulum and package them into membrane-bound vesicles.

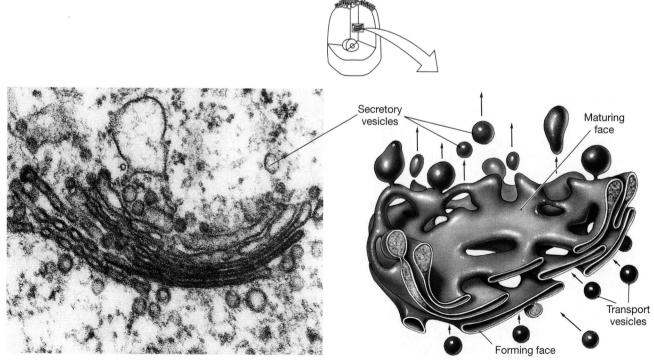

Secretory
vesicles

Maturing
face

Transport
vesicles

Forming face

■ **Figure 3-11 The Golgi apparatus** The Golgi apparatus is a series of hollow membranous sacs stacked on top of each other. Material arrives at the Golgi in transport vesicles from the endoplasmic reticulum and leaves the Golgi packaged in secretory vesicles.

The Golgi apparatus consists of five or six hollow curved sacs stacked on top of each other like a series of hot water bottles and connected so that they share a single lumen. The convex side of the stack faces the rough endoplasmic reticulum and receives the transport vesicles from it. These transport vesicles fuse with the membranes of the Golgi sac and discharge their contents into its lumen. As the proteins move through the sac, they may be modified by enzymes. Finally the proteins are enclosed in membrane-bound vesicles that pinch off from the concave face of the Golgi apparatus and move out into the cytosol. These vesicles are of two kinds: secretory vesicles and storage vesicles. **Secretory vesicles** contain proteins that will be exported to other parts of the body. (Secretion is the process by which a cell releases a substance into the extracellular space.) The contents of most **storage vesicles,** however, never leave the cytoplasm. Lysosomes and peroxisomes are the major storage vesicles of the cell.

Lysosomes are the intracellular digestive system

Lysosomes [*lysis*, dissolution + *soma*, body] are small spherical storage vesicles that appear as dark membrane-bound granules in the cytoplasm (Fig. 3-12 ■). They contain powerful enzymes that break down all types of biomolecules. Cells use the enzymes in lysosomes to destroy damaged organelles, to kill bacteria inside the cell, and occasionally to dissolve extracellular support material, such as the hard calcium carbonate

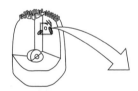

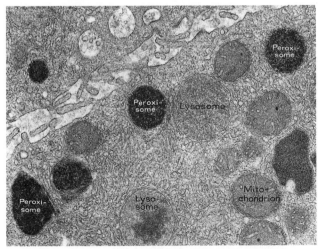

■ **Figure 3-12 Lysosomes and peroxisomes** Lysosomes and peroxisomes (stained dark in this electron micrograph) are membrane-bound vesicles filled with digestive enzymes. The enzymes are activated when a lysosome accumulates H⁺ in its lumen, creating an acid environment.

portion of bone. As many as 50 types of enzymes have been identified in lysosomes from different types of cells.

Because the enzymes of lysosomes are so powerful, one of the questions that puzzled researchers was why these enzymes do not normally destroy the cell that contains them. What they discovered was that lysosomal enzymes are activated only by very acid conditions, 1000 times more acid than the normal cytoplasm. When the lysosomes are first pinched off from the Golgi apparatus, their interior pH is about the same as that of the cytosol, 7.0–7.3. The enzymes are inactive at this pH. Their inactivity serves as a form of insurance: If the lysosome breaks or accidentally releases the enzyme, it will not harm the cell. However, as the lysosome sits in the cytoplasm, it accumulates H⁺ ions in a process that uses energy. The pH inside the vesicle drops to 4.8–5.0, and the enzymes are activated. Once this has occurred, the enzymes are capable of breaking down various biomolecules.

Lysosomes act as the digestive system of the cell. They take up bacteria or old organelles, such as mitochondria, and break them down into their component molecules. Those molecules that can be reused are reabsorbed into the cytosol, and the rest are dumped out of the cell. In the inherited conditions known as *lysosomal storage diseases*, lysosomes are not effective because they lack specific enzymes. As a result, harmful waste products accumulate, disrupting the normal function of cells, often with fatal results.

The digestive enzymes of lysosomes are not always kept isolated within the membranes of the organelle. Under some conditions, cells allow the enzymes of their lysosomes to come in contact with the cytoplasm, leading to self-digestion of all or part of the cell. When muscles *atrophy* (shrink) from lack of use or the uterus dimin-

Tay-Sachs Disease   One of the best-known lysosomal storage diseases is the inherited condition known as Tay-Sachs disease. Infants who inherit two copies of the gene for Tay-Sachs disease have lysosomes that lack enzymes that break down *gangliosides*, glycolipids that are an important component of nerve cells. As a result, gangliosides accumulate in the cells of the brain, causing the cells to swell and degenerate. The infants exhibit symptoms of nervous system dysfunction, including blindness and loss of coordination, and most die in early childhood. The Tay-Sachs gene is most commonly found among Ashkenazi Jews. One theory for the prevalence of the gene in this population notes that Ashkenazi Jews in previous centuries died from tuberculosis at a rate much lower than that of the population as a whole. The theory suggests that a person with one copy of the Tay-Sachs gene may have increased resistance to the tuberculosis bacterium. Today, it is estimated that one in thirty American Jews of eastern European origin is a carrier of the gene.

ishes in size after pregnancy, the loss of cell mass is due to the action of lysosomes. Damaged cells will digest themselves if their lysosomes rupture. The inappropriate release of lysosomal enzymes has been implicated in certain disease states: for example, the inflammation and destruction of joint tissue in *rheumatoid arthritis*.

Peroxisomes contain enzymes that neutralize toxins
Peroxisomes are storage vesicles that are even smaller than lysosomes. For years, they were thought to be a kind of lysosome, but we now know that they contain a different set of enzymes. Their main function appears to be to degrade long-chain fatty acids and potentially toxic foreign molecules. Peroxisomes get their name from the fact that the breakdown of fatty acids generates hydrogen peroxide (H_2O_2), a toxic molecule. The peroxisomes rapidly convert this peroxide to oxygen and water by using the enzyme *catalase*. Peroxisomal disorders disrupt the normal processing of lipids and can severely disrupt the normal function of the nervous system by altering the structure of nerve cell membranes.

The Nucleus

The **nucleus** of the cell contains DNA, the genetic material that ultimately controls all cell processes. How the nucleus receives information about conditions in the cell or elsewhere in the body and responds with fine-tuned control of the cell's protein synthesis is one of the most active areas of biological research. Figure 3-13 ■ illustrates the structure of a typical nucleus. The boundary, or **nuclear envelope,** is a two-membrane structure that separates the nucleus from the cytoplasmic compartment. The outer membrane of the envelope is connected with the endoplasmic reticulum, and both membranes of the envelope are pierced here and there by round holes, or **pores.**

Communication between the nucleus and cytosol occurs through the pores in the nuclear envelope.

> **Vaults, Newly Discovered Organelles** Although scientists have been looking at the fine structure of cells for nearly 50 years, they are still finding new organelles as they develop new staining techniques. One of these new organelles, first described in 1986, is a eight-part particle of ribonucleoprotein called a **vault.** Although they are found in all eukaryotic cells from yeasts to mammals, the function of vaults is still unclear. They were found localized within the nuclear pore in one study and found to be necessary for the most rapid cell growth in another. Their association with the nucleus, with RNA, and with cell growth leads scientists to suspect that they may in some way be linked to protein synthesis. Look for more research on vaults in the years to come!

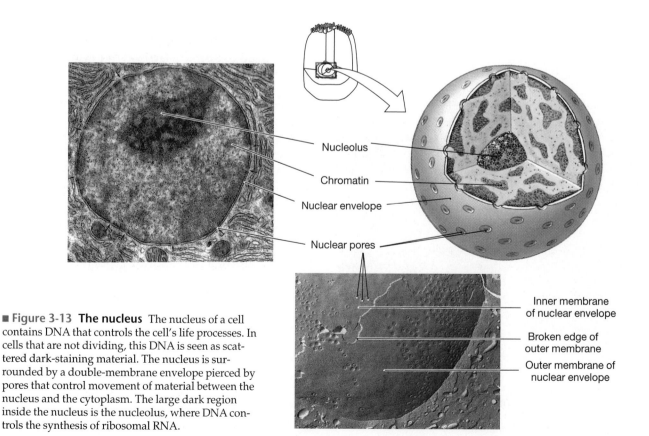

Nucleolus

Chromatin

Nuclear envelope

Nuclear pores

Inner membrane of nuclear envelope

Broken edge of outer membrane

Outer membrane of nuclear envelope

■ **Figure 3-13 The nucleus** The nucleus of a cell contains DNA that controls the cell's life processes. In cells that are not dividing, this DNA is seen as scattered dark-staining material. The nucleus is surrounded by a double-membrane envelope pierced by pores that control movement of material between the nucleus and the cytoplasm. The large dark region inside the nucleus is the nucleolus, where DNA controls the synthesis of ribosomal RNA.

Nuclear pores are not simply holes: They are protein complexes with a central opening. Experiments have shown that ions and small molecules can move freely through this pore complex, but proteins and RNA must be transported by a process that uses energy. This requirement allows the cell to restrict large molecules such as DNA to the nucleus and various enzymes to either the cytoplasm or the nucleus.

In electron micrographs of cells that are not dividing, most DNA in the nucleus appears as randomly scattered granular material. A nucleus usually also contains from one to four larger dark-staining bodies of DNA, RNA, and protein called **nucleoli** [singular, *nucleolus,* little nucleus]. Nucleoli contain the genes and proteins that control the synthesis of RNA for ribosomes.

✓ Microscopic examination of a cell reveals many mitochondria. What does this observation imply about the cell's energy requirements?

✓ You are examining tissue from a previously unknown species of fish. You discover a tissue with large amounts of smooth endoplasmic reticulum in its cells. What is one possible function of these cells?

✓ How would the absence of a flagellum affect a sperm cell?

✓ Which would have more rough endoplasmic reticulum: pancreatic cells that manufacture the protein hormone insulin, or adrenal cortex cells that synthesize the steroid hormone cortisol?

TISSUES OF THE BODY

Despite the amazing variety of intracellular structures, no single cell can carry out all the processes of the mature human body. Instead, cells assemble into larger units called **tissues,** collections of cells held together by specialized connections called cell junctions and by other support structures. Tissues range in complexity from simple tissues containing only one cell type to complex tissues containing many cell types and extensive extracellular material. The cells of most tissues work together to achieve a common purpose.

The study of tissue structure and function is known as **histology** [*histos,* tissue]. Histologists describe tissues by their physical features: (1) the shape and size of the cells, (2) how the cells are arranged in the tissue (in layers, scattered, and so on), (3) how the cells are connected to each other, and (4) the amount of extracellular material that is present in the tissue. There are four **primary tissue types** in the human body: epithelial, connective, muscle, and neural, or nerve (Table 3-1). Before we con-

continued from page 43

During a Pap test for cervical cancer, cell samples are swabbed from the cervix (neck) of the uterus. The cells are then smeared onto a glass slide and sent to a laboratory for examination by a trained cytologist. The cytologist looks for *dysplasia,* a change in the size and shape of the cells that is indicative of cancerous changes [*dys-,* abnormal + *-plasia,* growth or cell multiplication]. Jan's first Pap test showed all the hallmarks of dysplasia. A number of the cells scraped from her cervix were abnormally shaped and displayed a small amount of cytoplasm in comparison to the size of the nucleus.

Question 2: What property of cancer cells explains the large size of their nucleus and the relatively small amount of cytoplasm?

sider each tissue type specifically, let us look at how cells link together to form tissues.

Extracellular Matrix Helps Support Tissues

Matrix is extracellular material that is synthesized and secreted by the cells of a tissue. Its main roles are to hold cells together and provide a base upon which tissues can grow. The amount of matrix varies with different types of tissues. It is almost nonexistent in nerve and muscle tissue but is extensive in cartilage, bone, and blood. The composition of matrix is also variable. In many tissues, matrix is composed of complex glycoprotein molecules mixed with insoluble protein fibers. The consistency varies, ranging from the watery matrix of blood and lymph to the rigid matrix of bone. The variability of matrix is discussed further with the four tissue types.

Although for many years matrix was considered to be a passive support structure, recent experimental evidence suggests that it also plays a role in physiological function. When matrix proteins attach to proteins in the cell membrane, they provide a means of communication between the cell and its external environment. Molecules in the matrix play an important role during development in guiding the growth of neurons and the migration of other cells. Current research efforts are looking at the role of matrix in the spread of cancer cells, because it appears that tumor cells destroy matrix as they invade new tissues.

Cell Junctions Hold Cells Together to Form Tissues

Individual cells within tissues are linked to each other by various types of cell-to-cell connections, or **cell junctions.** There are three types of cell junctions: adhesive junctions, tight junctions, and gap junctions. In both adhesive and tight junctions, part of the connection is a layer of matrix "cement."

TABLE 3-1	**Characteristics of the Four Tissue Types**			
	Epithelial	*Connective*	*Muscle*	*Nerve*
Matrix amount	Minimal	Extensive	Absent	Absent
Matrix type	Basement membrane	Varied—protein fibers in ground substance that ranges from liquid to gelatinous to firm to calcified	NA	NA
Unique features	No direct blood supply	Cartilage has no blood supply	Able to generate electrical signals, force, and movement	Able to generate electrical signals
Surface features of cells	Microvilli, cilia	NA	NA	NA
Locations	Covers body surface; lines cavities and hollow organs and tubes. Secretory glands	Supports skin and other organs. Cartilage, bone, and blood	Makes up skeletal muscles, hollow organs, and tubes	Throughout body. Concentrated in brain and spinal cord
Cell arrangement and shapes	Variable number of layers, from one to many; cells flattened, cuboidal, or columnar	Cells not in layers; usually randomly scattered in matrix; cell shape irregular to round	Cells linked in sheets or elongated bundles; cells shaped in elongated, thin cylinders. Heart muscle cells may be branched	Cells isolated or networked; cell appendages highly branched and/or elongated

Adhesive junctions   Adhesive junctions have been compared to buttons or zippers that link cells together and hold them in position within a tissue. In vertebrates, adhesive junctions take the form of either desmosomes or adherens junctions.

Desmosomes [*desmos*, band + *soma*, body] are recognizable in electron micrographs by the dense glycoprotein bodies, or *plaques*, that lie just inside the cell membranes in the region where the desmosome joins two cells (Fig. 3-14 ■). Extending from these plaques are intermediate filaments that anchor the desmosome to the cytoskeleton and also penetrate the membrane to connect with the filaments of the adjacent cell. Desmosomes may be small points of contact between two cells (spot desmosomes), bands or belts that encircle the entire cell (belt desmosomes), or *hemidesmosomes* [*hemi-*, half] that tie cells to protein fibers in a supporting matrix outside the cell.

The protein linkage of adhesive cell junctions is very strong, and it allows sheets of tissue to resist stretching and twisting. For that reason, desmosomes are especially abundant in skin and in the tissues that line the body cavities. Even the tough protein fibers of adhesive junctions can be broken, however. If you have shoes that rub against your skin, the stress can shear the proteins connecting the different skin layers. When fluid accumulates in the resulting space and the layers separate, a *blister* is the result.

Disappearance of cell junctions is a characteristic of cancer. Cells that become cancerous tend to lose their desmosomes. It is suspected that this releases the cancer cell from its moorings and probably contributes to the spread, or metastasis, of cancer throughout the body.

Tight junctions   **Tight junctions** are designed to prevent the movement of material past the cells they link. In tight junctions, the cell membranes of adjacent cells partly fuse together, thereby making a barrier (Fig. 3-14 ■). Tissues held together with tight junctions are like a solid brick wall: Very little can pass from one side of the wall to the other between the bricks. By comparison, tissues held together with adhesive junctions are like a picket fence. The pickets form a continuous surface but the spaces between the individual pickets allow material to pass from one side of the fence to the other.

In physiology, tight junctions in the intestinal tract and kidney prevent substances from moving freely between the external and internal environments and thus allow cells to regulate what enters and leaves the body. Tight junctions also create the so-called blood-brain barrier that prevents potentially harmful substances in the blood from reaching the extracellular fluid of the brain.

Gap junctions   **Gap junctions** link cells together so that the cytoplasms of the joined cells are connected (Fig. 3-14 ■). They may be compared to hollow rivets connecting two metal boxes. The rivets are actually interlocking cylindrical proteins with narrow channels through their centers that permit small molecules and

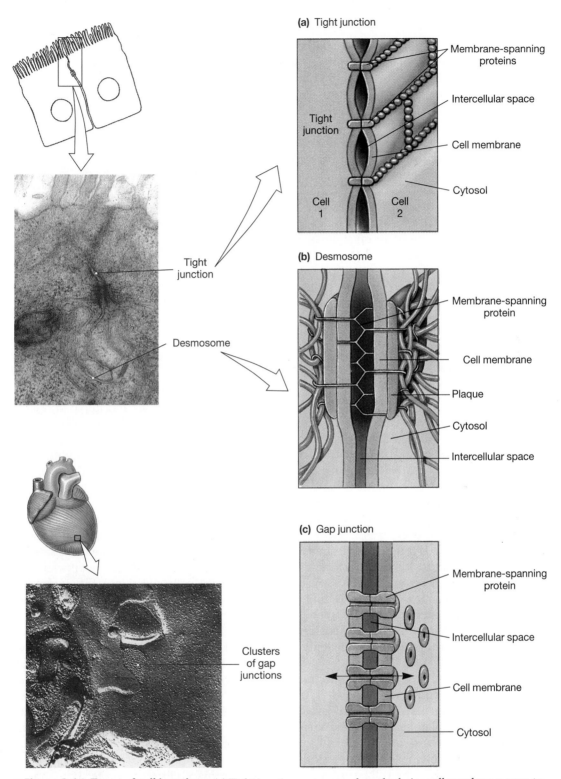

(a) Tight junction

Membrane-spanning proteins

Intercellular space

Tight junction

Cell membrane

Cytosol

Cell 1

Cell 2

Tight junction

Desmosome

(b) Desmosome

Membrane-spanning protein

Cell membrane

Plaque

Cytosol

Intercellular space

(c) Gap junction

Membrane-spanning protein

Intercellular space

Clusters of gap junctions

Cell membrane

Cytosol

■ Figure 3-14 **Types of cell junctions** (a) Tight junctions are areas where the facing cell membranes seem to fuse together, effectively preventing movement of material past the cells. (b) Desmosomes are smaller areas where cells link together by means of protein filaments and an intercellular matrix "cement." (c) Gap junctions are connected proteins that create a bridge between the cytoplasms of adjacent cells.

ions to pass from cell to cell. Gap junctions allow rapid transfer of chemical and electrical signals between cells. They are found primarily in cardiac muscle, smooth muscle, and certain nerve cells.

Now that you understand how cells are held together into tissues, let us look at the four different tissue types in the body: (1) epithelial, (2) connective, (3) muscle, and (4) nervous tissues.

Epithelia

The **epithelial tissues,** or **epithelia,** protect the internal environment of the body and regulate the exchange of material between the internal and external environments. They are found covering exposed surfaces, such as the skin, and lining internal passageways, such as the digestive tract. Some epithelia, such as those of the skin and mucous membranes of the mouth, act as a barrier to keep water in the body and invaders such as bacteria out. Other epithelia, such as those in the kidney and intestinal tract, control the movement of material between the external environment and the extracellular fluid of the body. **Any substance that enters or leaves the internal environment of the body must cross an epithelium.** Indeed, nutrients, gases, and wastes must often cross several different epithelia in their passage between cells and the outside world.

One type of epithelium is specialized to manufacture and secrete chemicals into the blood or to the external environment. Sweat and saliva are examples of substances secreted by epithelia into the environment, whereas hormones, chemicals used to maintain homeostasis, are secreted into the blood.

Characteristics of epithelia   The cells of most epithelial tissues retain the ability to divide and reproduce throughout their life span. This is necessary because many of them are subject to wear-and-tear from exposure to the outside environment and may live only a day or two. For example, it has been estimated that the intestinal epithelium lining the digestive tract is completely replaced with new cells every two to five days. Epithelial cells that are protected from the environment, such as those of the pancreas, may live for two months or more. The constant cell division in epithelia seems to make these cells prone to genetic mutations, leading to abnormal growth patterns such as cancer or benign tumors. It has been estimated that more than 90 percent of all cancer in adults over the age of 45 arises in epithelial tissues.

Epithelia typically have a thin layer of extracellular matrix that lies between the cells and their underlying tissues (Fig. 3-15 ■). This layer, called the **basal lamina** [*bassus,* low; *lamina,* a thin plate], or **basement membrane,** is composed of a network of fine protein filaments embedded in glycoprotein. The filaments hold the epithelial cells to the underlying cell layers, just as adhesive junctions hold the individual cells in the epithelium to each other.

The cell junctions in epithelia are variable. Physiologists classify epithelia either as "leaky" epithelia or "tight" epithelia. In a leaky epithelium, the cells are loosely connected to each other by adhesive junctions, leaving gaps or pores that allow molecules to pass across the epithelium between the cells. One leaky epithelium is the wall of the capillary blood vessels, where all dissolved mole-

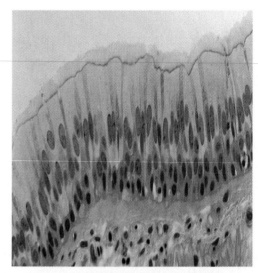

■ **Figure 3-15   Epithelial tissue**   Epithelial tissue is supported by a layer of extracellular matrix known as the basement membrane. This matrix is secreted by the cells of the epithelium.

cules except for large proteins can pass between the cells. In a tight epithelium, adjacent cells are bound to each other by tight junctions and infolded membranes that create a barrier to movement between the cells. In order to cross a tight epithelium, most substances must enter the cells and go *through* them. The tightness of an epithelium is directly related to how selective it is about what can move across it.

Types of epithelia   Epithelial tissues can be divided into two general types: (1) sheets of tissue that lie on the surface of the body or that line the inside of tubes and hollow organs and (2) secretory epithelia that synthesize and release substances into the extracellular space. Histologists classify sheet epithelia by the number of cell layers in the tissue and by the shape of the cells in the surface layer. This classification scheme recognizes two types of layering—**simple** (one cell thick) and **stratified** (multiple cell layers) [*stratum,* layer + *facere,* to make]—and three cell shapes—**squamous** [*squama,* flattened plate or scale], **cuboidal,** and **columnar.** However, physiologists are more concerned with the functions of these tissues, so we will divide epithelia into five groups according to their function.

There are five functional types of epithelia: exchange, transporting, ciliated, protective, and secretory. **Exchange epithelia** are designed for rapid exchange of material, in particular, gases. **Transporting epithelia** are selective about what can cross them; they are found primarily in the intestinal tract and the kidney. **Ciliated epithelia** are located in the airways of the respiratory system and in the female reproductive tract. **Protective epithelia** are found on the surface of the body and just inside the openings of body cavities. The **secretory epithelia** synthesize and release secretory products into

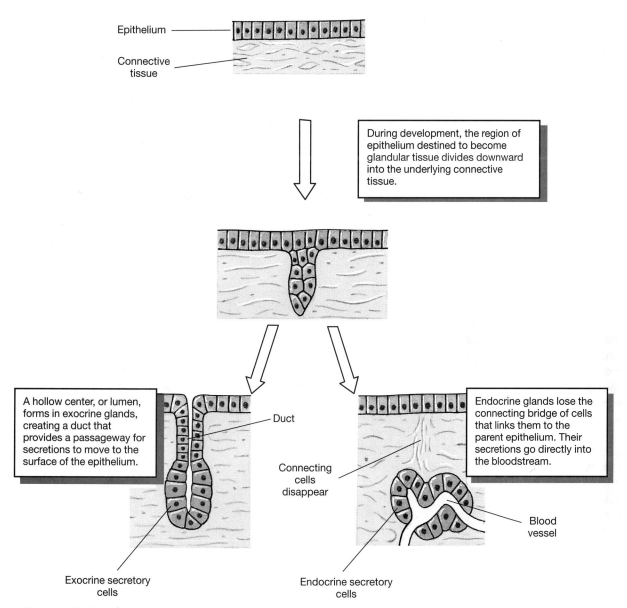

Epithelium

Connective tissue

During development, the region of epithelium destined to become glandular tissue divides downward into the underlying connective tissue.

A hollow center, or lumen, forms in exocrine glands, creating a duct that provides a passageway for secretions to move to the surface of the epithelium.

Duct

Connecting cells disappear

Endocrine glands lose the connecting bridge of cells that links them to the parent epithelium. Their secretions go directly into the bloodstream.

Blood vessel

Exocrine secretory cells

Endocrine secretory cells

■ Figure 3-20 **Development of endocrine and exocrine glands from epithelium**

Connective Tissues

Connective tissues, the second major tissue type, provide structural support and sometimes a physical barrier that, along with specialized cells, helps defend the body from foreign invaders such as bacteria. The distinguishing characteristic of connective tissues is the presence of extensive extracellular matrix containing widely scattered cells. The cells secrete and modify the matrix of the tissue. Connective tissues range from blood, to the support tissues for the skin and internal organs, to cartilage and bone.

Characteristics of connective tissue The matrix of connective tissue is a **ground substance** of glycoproteins and water in which insoluble fibrous protein fibers are

arranged, much like suspended pieces of fruit in a gelatin salad. The consistency of ground substance is highly variable, depending on the type of connective tissue. At one extreme is the watery matrix of blood, and at the other extreme is the hardened matrix of bone. In between are solutions of glycoproteins that vary in consistency from syrupy to gelatinous. The term *ground substance* is sometimes used interchangeably with *matrix*.

The actual cells of connective tissue lie embedded within the extracellular matrix. Connective tissue cells are described as *fixed* if they remain in one place, or *mobile* if they have the ability to move from place to place. **Fixed cells** are responsible for local maintenance, tissue repair, and energy storage, whereas the **mobile cells** of connective tissue are responsible mainly for

continued from page 53

Many kinds of cancer develop in epithelial cells that are subject to damage or trauma. The cervix consists of two types of epithelia. Secretory epithelium with mucus-secreting glands lines the inside of the cervix, while a protective epithelium covers the outside of the cervix. At the opening of the cervix, these two types of epithelia come together. As Jan lies on the examining table, her physician uses a small spatula and a little brush to take samples of cells from both inside and outside her cervix. She then smears the cells on a slide and sprays them with fixative. The slides will now be sent to a lab for evaluation.

Question 3: What kind of damage or trauma are cervical epithelial cells normally subjected to? Which of its two types of epithelia is more likely to be affected by trauma?

defense. The distinction between fixed and mobile cells is not absolute, because at least one cell type is found in both fixed and mobile forms.

Although matrix itself is nonliving, the connective tissue cells constantly modify it by the addition, deletion, or rearrangement of molecules. The suffix *-blast* [*blastos*, sprout] on a connective tissue cell name indicates a cell that is either growing or actively secreting extracellular matrix. Cells that are actively breaking down matrix are identified by the suffix *-clast* [*klastos*, broken]. Cells that are neither growing, secreting, nor breaking down matrix are identified by the suffix *-cyte*, meaning "cell." By remembering these suffixes you will be able to tell the functional difference between cells with similar names, such as the osteoblast, osteocyte, and osteoclast, three cell types found in bone.

In addition to secreting the glycoprotein ground substance, the connective tissue cells produce the fibers of the matrix. There are four types of fiber proteins found in the matrix of connective tissue, aggregated into insoluble protein fibers. **Collagen** [*kolla*, glue + *-genes*, produced] is the most abundant protein in the human body, almost a third of its dry weight. It is also the most diverse of the four protein types, with at least 12 different variations. Collagen is found almost everywhere connective tissue is found, from the skin to muscles and bones. Individual collagen molecules pack together to form collagen fibers, flexible but inelastic fibers whose strength per unit weight exceeds that of steel. The amount and arrangement of collagen fibers is one of the distinguishing characteristics of different types of connective tissue.

Three other protein fibers in connective tissue are elastin, fibrillin, and fibronectin. **Elastin** is a coiled, wavy protein that returns to its original length after being stretched; this property is known as *elasticity*. Elastin combines with the very thin, straight fibers of **fibrillin** to form filaments and sheets of elastic fibers. These two fibers are important in elastic tissues such as the lungs, blood vessels, and skin. **Fibronectin** forms a protein fiber that helps connect cells to their extracellular matrix. Fibronectins also play an important role in wound healing and in blood clotting.

Types of connective tissue   The different types of connective tissue are described in detail in Table 3-3. The most common types of connective tissue are loose and dense connective tissue, adipose tissue, cartilage, bone, and blood. **Loose connective tissues** (Fig. 3-21 ■) are the elastic tissues that underlie skin and provide support for small glands. **Dense connective tissues** are tissues whose primary function is strength or flexibility.

TABLE 3-3	**Types of Connective Tissue**			
Tissue Name	*Ground Substance*	*Fiber Type and Arrangement*	*Main Cell Types*	*Where Found*
Loose connective tissue	Gel; more ground than fibers and cells	Collagen, elastic, reticular; random	Fibroblasts	Skin, around blood vessels and organs, under epithelia
Dense, irregular connective tissue	More fibers than ground	Mostly collagen; random	Fibroblasts	Muscle and nerve sheaths
Dense, regular connective tissue	More fibers than ground	Collagen, parallel	Fibroblasts	Tendons and ligaments
Adipose	Very little	None	Brown fat, white fat	Depends on age and sex
Blood	Aqueous	None	Blood cells	In blood and lymph vessels
Cartilage	Firm but elastic; hyaluronic acid	Collagen	Chondroblasts	Joint surfaces, spine, ear, nose, larynx
Bone	Rigid due to calcium salts	Collagen	Osteoblasts and osteoclasts	Bones

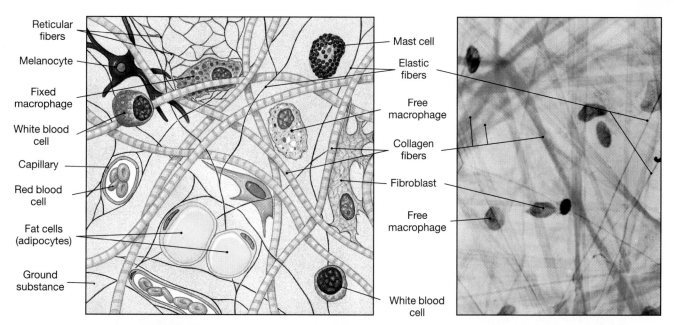

■ Figure 3-21 Cells and fibers of connective tissue This figure represents a typical loose connective tissue with extensive extracellular matrix. The cells labeled fibroblasts are the connective tissue cells that secrete the fibers and glycoproteins of the matrix. Macrophages and mast cells are two types of cells that defend against foreign invaders, such as bacteria.

Important examples of this tissue type are tendons, ligaments, and the sheaths that surround muscles and nerves. In these tissues, collagen fibers are the dominant type. **Tendons** (Fig. 3-22 ■) attach skeletal muscles to bones. **Ligaments** connect one bone to another. Because ligaments contain elastic fibers in addition to collagen fibers, they have a limited ability to stretch, whereas tendons cannot stretch.

Adipose tissue is made up of **adipocytes,** or fat cells. An adipocyte of **white fat** typically contains a single enormous lipid droplet that occupies most of the volume of the cell (Fig. 3-23a ■). This is the most common form of adipose tissue in adults. **Brown fat** is composed of adipose cells that contain multiple lipid droplets rather than a single large droplet (Fig. 3-23b ■). This type of fat is almost completely lacking in adults, but it plays an important role in temperature regulation in infants.

Blood is an unusual connective tissue that is characterized by its watery matrix, consisting of a dilute solution of ions and dissolved organic molecules. The matrix lacks insoluble protein fibers but contains a large variety of soluble proteins. These will be discussed in Chapter 16.

Cartilage and bone together are considered supporting connective tissues. These tissues have a dense ground substance that contains closely packed fibers. **Cartilage** is found in structures such as the nose, ears, knee, and windpipe. It is solid, flexible, and notable for its lack of blood supply. Without a blood supply, nutrients and oxygen must reach the cells of cartilage by diffusion. This is a slow process, which means that damaged cartilage heals slowly. The fibrous matrix of **bone** is said to be *calcified* because it contains mineral deposits, primarily calcium salts, such as calcium phosphate. These minerals give the bone strength and rigidity. The structure and formation of bone will be discussed along with calcium metabolism in Chapter 21. Figure 3-24 ■ is a study map showing the components of connective tissues.

Growing New Cartilage Replacing cartilage or bone that is lost because of injury or damaged because of disease is not often an option in surgery today. But researchers are trying to change that by growing *chondrocytes,* the cells that synthesize the matrix of cartilage, in the laboratory. They have discovered that chondrocytes grown in a single flat layer (monolayer) tend to become unspecialized and lose their ability to produce cartilage. But cells grown in a three-dimensional support system of agar gel or on a biodegradable polymer scaffold retain their special characteristics and continue to secrete the appropriate matrix. If the polymer scaffold can be shaped before the cells are grown on it, we may someday be able to grow new joints, noses, and tracheas to replace those lost or damaged.

Muscle and Nerve

Muscle and neural tissues are collectively called the *excitable tissues* because of their ability to generate electrical signals. Both of these tissue types have very small amounts of extracellular matrix.

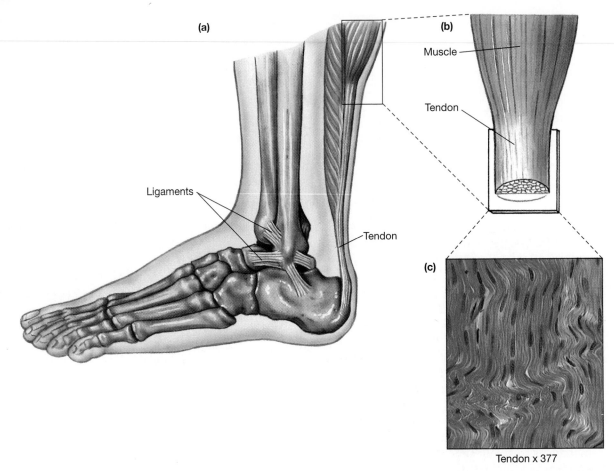

(a)

(b)

Muscle

Tendon

Ligaments

Tendon

(c)

Tendon x 377

■ **Figure 3-22 Tendons and ligaments** Tendons and ligaments are dense connective tissue. Tendons attach muscle to bone (a, b) whereas ligaments attach bone to bone (a). The collagen of the tendon is densely packed into parallel bundles and has a wavy appearance (c).

Muscle tissue has the ability to contract and produce force and movement. There are three types of muscle tissue in the body: cardiac muscle, in the heart; smooth muscle, which makes up most internal organs; and skeletal muscles. Most of the last type are attached to bone and are responsible for gross movement of the body. Muscle histology is discussed in detail in Chapter 12.

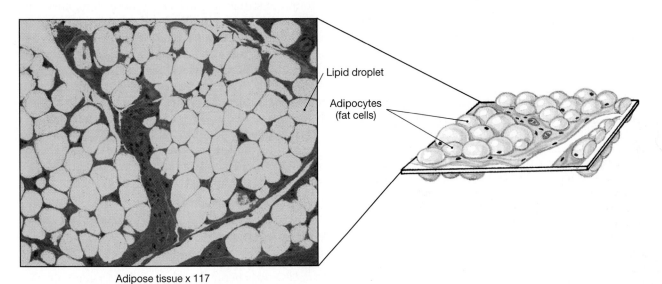

Lipid droplet

Adipocytes
(fat cells)

Adipose tissue x 117

■ **Figure 3-23 Adipose cells** Brown fat cells contain multiple lipid droplets in a single cell. The cytoplasm of a white fat adipose cell (not shown) is almost obscured by a single large lipid droplet.

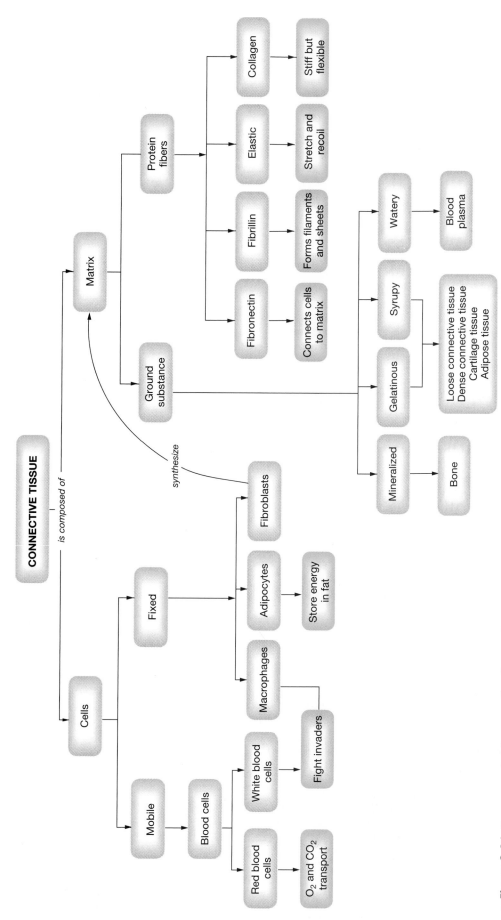

■ **Figure 3-24 Components of connective tissue** This map shows the relationship of cells and the components of matrix in connective tissue.

Focus on   The Skin

■ **Figure 3-25**

(a) The layers of the skin

The waterproof nature of the skin helps prevent body water from evaporating, and the tough outer layers keep bacteria and other pathogens out. The secretions of the sweat glands help cool the body in high temperatures, whereas hair and subcutaneous fat insulate against cold temperatures. Sensory receptors in the dermis constantly monitor external conditions and communicate with the rest of the body when conditions change.

Skin is constructed in three layers: an outer layer of epithelial cells called the **epidermis,** a middle layer of connective tissue called the **dermis,** and a variable inner layer called the **subcutis** or **hypodermis.** It is from these latter two names that we get the adjectives *subcutaneous* and *hypodermic*.

The epidermis consists of multiple layers of cells that serve as the protective layer of the skin. Derived from this epithelium are exocrine glands that contribute to the secretory function of the skin: sweat glands, hair follicles, sebaceous glands, apocrine glands, and the milk-producing ducts of the female breast.

The dermis is primarily composed of loose connective tissue interspersed with tiny blood vessels, nerve endings, glands, and muscles. Located here are the exocrine glands derived from the epidermal layer. **Hair follicles** secrete the nonliving keratin shaft of hair. Tiny muscles attached to the follicle pull the follicle into a vertical position when the muscle contracts. This is how you get "goosebumps" and how your hair "stands on end." **Sebaceous glands** open into the duct of the hair follicle or directly onto the surface of the skin. These glands are not very active until puberty, when, under hormonal influence, they begin to secrete a lipid mixture of waxes and triglycerides. These are the secretions that, in excess, cause oily skin. **Sweat glands** also open directly onto the skin's surface. The final type of exocrine gland is called an **apocrine gland.** These glands are located in the skin of the genitalia, anus, axillae [*axilla*, armpit], and eyelids. They release waxy or viscous milky secretions in response to fear or sexual excitement. The function of these secretions in

humans is not known, but, in other mammals, similar secretions are used to mark territories or for sexual attraction. The chemical *musk,* commonly used in perfumes, comes from the anal apocrine glands of musk deer.

The third and deepest layer of the skin is the **hypodermis.** This layer is also the most variable in thickness, because the amount of adipose tissue in it varies by age, sex, and location on the body. The hypodermis is loose connective and adipose tissue, and it contains the major blood vessels and nerves whose branches extend upwards into the dermis. The adipose tissue in the hypodermis insulates against cold and acts as a shock absorber.

(b) Epidermis

If we start at the surface of the skin and work inward, the outermost layer is composed of the fibrous protein keratin. The thickness of the keratin layer depends on the part of the body. Thick skin on the soles of the feet and palms of the hands has a much heavier layer of keratin than does the thin skin on the scalp. The keratin fibers are produced by **keratinocytes,** tied to each other by an extensive system of desmosomes. As these cells mature, they synthesize numerous keratin fibers and secrete a hydrophobic phospholipid matrix that acts as the skin's main waterproofing agent. By the time older keratinocytes are pushed to the surface of the epidermis by newer cells, their cytoplasm is thick with keratin fibers. At this point the cells die, and their nuclei and organelles disappear. The mats of keratin fibers remain behind, still linked to each other by the protein fibers of the desmosomes. It is this surface layer of keratin that forms the protective barrier between the deeper cells and the external environment.

(c) Basement membrane and the connection between epidermis and dermis

Between the epidermis and the underlying layer of connective tissue is an acellular basement membrane. The cells of the epidermis and fibers of the dermis are anchored to each other by their connection to the protein fibers that run throughout the basement membrane.

---

continued from page 62

During her examination, Jan learns that a group of researchers recently confirmed that 80% of cervical cancers worldwide are caused by infection with the *human papilloma virus* (HPV), a sexually transmitted disease that also causes genital warts. Researchers believe that HPV enters cervical cells and alters the genetic machinery contained in the nucleus, leading to uncontrolled cell growth. Jan also learns that the risk of developing cervical cancer is higher in women who begin sexual intercourse at an early age, who have more than one child, and/or who have multiple sexual partners.

Question 4: Which of the listed risk factors are directly associated with HPV infection? Why is beginning sexual intercourse at an early age associated with an increased risk of cervical cancer?

Neural tissue carries information from one part of the body to another. It is concentrated in the brain and spinal cord but also includes a network that extends to virtually every part of the body. The nerve cells that make up neural tissue have a unique structure that is closely linked to their ability to use electrical and chemical signals for communication throughout the body. Neural tissue will be discussed in Chapter 8.

✓ Blood is a connective tissue with two components: plasma and cells. What is the matrix in this connective tissue?

✓ Why does torn cartilage heal more slowly than an epithelial cut?

■ Figure 3-25

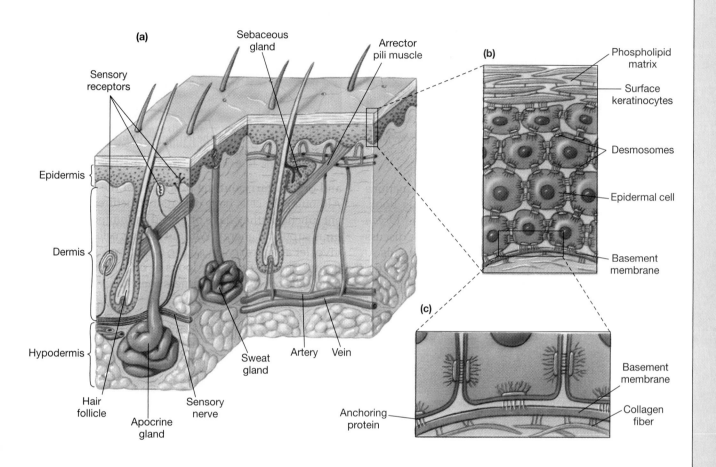

(a) Sebaceous gland · Arrector pili muscle · Sensory receptors · Epidermis · Dermis · Hypodermis · Hair follicle · Apocrine gland · Sensory nerve · Sweat gland · Artery · Vein

(b) Phospholipid matrix · Surface keratinocytes · Desmosomes · Epidermal cell · Basement membrane

(c) Anchoring protein · Basement membrane · Collagen fiber

ORGANS

Groups of tissues that carry out related functions may form structures known as organs. The organs of the body contain the four types of tissue in various combinations. The skin is an excellent example of an organ that incorporates all four types of tissue into an integrated whole. We think of skin as being a thin layer that covers the external surfaces of the body, but in reality it is the heaviest single organ, at about 16% of an adult's total body weight! If it were flattened out, it would cover a surface of between 1.2 and 2.3 square meters, about the size of a couple of card-table tops. Its size and weight make skin one of the most important organs of the body. The structure of this organ is highlighted in *Focus on the Skin*, Figure 3-25 ■.

continued from page 66

The day after Jan's visit, a cytologist stains her slide with dye and inserts it into PAPNET, a computerized analysis system that assists cytologists by identifying abnormal cells in cervical samples. The robotic arm of PAPNET picks up Jan's slide and places it under a camera. A computer rapidly scans the cells on the slide, looking for abnormal size or shape. The computer is programmed to record the 128 cells that appear most abnormal. When the computer finds an abnormal cell, it is projected on a large screen.

Question 5: Has Jan's dysplasia improved or worsened? What evidence do you have to support your answer (Fig. 3-26 ■)?

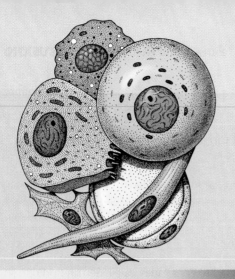

4

Cellular Metabolism

The structure of a cell is the framework for processes that are essential to a living thing: growth, movement, reproduction, and the maintenance of a relatively constant internal environment. Without these processes, the cell is a ghost town, filled with buildings but lacking the energy and organization brought to it by living people. Left to itself, a deserted town slowly crumbles into ruin, because there is no one to work on repairs. People provide the energy for growth and maintenance of a town, for building new buildings and repairing old ones. Similar energy-dependent processes take place within the cell. Raw materials are imported, new molecules are made, and aging parts are repaired or recycled. One of the most outstanding characteristics of a living cell is its ability to extract energy from its external environment and to use that energy to construct and maintain itself as an organized, functioning unit. In this chapter, we will look at the cell processes through which the human body obtains energy and maintains its ordered systems.

Problem

Tay-Sachs Disease

In many ultra-orthodox Jewish communities of the United States—in which arranged marriages are the norm—the rabbi is entrusted with an important, life-saving task. He keeps a confidential record of individuals known to carry the gene for Tay-Sachs disease, a fatal, inherited condition that strikes one in 3600 American Jews of Eastern European descent. Babies born with this disease rarely live beyond age four, and there is no cure. Based on the family trees he constructs, the rabbi can avoid pairing two individuals who carry the deadly gene.

Sarah and David, who met while working on their college newspaper, are not orthodox Jews. But both are aware that their Jewish ancestry might put any children they might have at risk for Tay-Sachs disease. Six months before their wedding, they decide to see a genetic counselor to determine whether they are carriers of the gene for Tay-Sachs disease.

continued on page 79

ENERGY IN BIOLOGICAL SYSTEMS

All cells obtain energy from their environment in order to grow, make new parts, and reproduce. Plants trap radiant energy from the sun and store it as chemical bond energy in the process of photosynthesis (Fig. 4-1 ■). They link together carbon and oxygen from carbon dioxide, nitrogen from the soil, and hydrogen and oxygen from water to form glucose, amino acids, and other biomolecules. Multiple glucose molecules are linked together into starch molecules for energy storage and into cellulose molecules, which act as structural elements (∞ p. 27). The energy from the sun trapped by photosynthesis is the ultimate energy source for all animals, including humans.

Animals cannot trap energy from the sun or obtain carbon and nitrogen from the air and soil to synthesize biomolecules. They must import energy and those atoms in the form of biomolecules synthesized by plants or contained in other animals, which ultimately did obtain them from plants. The energy animals extract from these biomolecules is used for growth, maintenance, move-

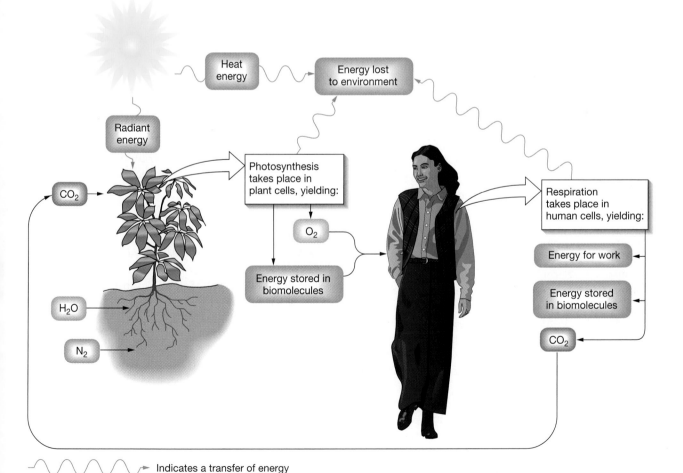

Indicates a transfer of energy

■ Figure 4-1 **Energy transfer in the environment** Radiant energy from the sun is trapped by plants and converted into the chemical energy of molecular bonds through photosynthesis. Animals eat the plants and convert the biomolecules to other forms of energy. Some of the energy is used for the work of growth, maintenance, reproduction, and movement. Some is lost to the environment as heat. The rest is stored as the chemical bond energy of glycogen and lipids.

ment, and reproduction. However, the extraction process is never totally efficient. As a result, some chemical bond energy is returned into the environment in the form of heat.

If animals ingest more energy than they need for immediate use, the excess energy is stored in chemical bonds, just as it is in plants. Glycogen (another glucose polymer) and lipid molecules are the main energy stores in animals (∞ p. 27). These storage molecules are available for use at times when energy from an animal's current intake of food does not meet its body's needs.

The cycling of energy in the environment and its use by living organisms is one of the fundamental concepts of biology. This section summarizes the ways that animals capture, store, and utilize energy. You will encounter many of these basic principles throughout the book as you study the physiology of the human body.

Energy Is Used to Perform Work

As the preceding discussion suggests, a cell must be able to obtain, store, or make use of energy to fuel its activities. **Energy** can be defined as the capacity to do work. But what is work? We use the word in everyday life to mean various things, from hammering a nail to sitting at a desk writing a paper. In biological systems, work has three basic forms: chemical work, transport work, and mechanical work.

Chemical work enables cells and organisms to grow, maintain a suitable internal environment, and store information needed for reproduction and normal activities. Protein synthesis for wound repair is an example of chemical work.

Transport work enables cells to move ions, molecules, and larger particles through the cell membrane and through membranes of organelles within the cell. Transport work is particularly useful for creating **concentration gradients,** in which the concentration of a molecule is higher on one side of a membrane than on the other. For example, certain types of endoplasmic reticulum use energy to import calcium ions from the cytosol. This creates a high calcium concentration inside the organelle and a low concentration in the cytosol. If calcium is then released back into the cytoplasm, it creates a "calcium signal" that tells the cell to perform an action such as muscle contraction.

Mechanical work in animals is used for movement. At the cellular level, this movement includes organelles that move around within the cell, as well as the beating of cilia and flagella (∞ p. 47). In whole animals, movement usually involves muscle contraction. Most mechanical work, whether in cells or entire animals, is mediated by the movement of intracellular fibers and filaments of the cytoskeleton (∞ p. 47).

The Two Major Categories of Energy Are Kinetic Energy and Potential Energy

Energy can be classified in various ways.* We often think of energy in terms that we deal with daily: thermal energy, electrical energy, mechanical energy. The introduction to this section discussed chemical bond energy. Each type of energy has its own characteristics. However, all share an ability to appear in two forms: as potential energy or as kinetic energy.

Kinetic energy is the energy of motion [*kinetikos,* motion]. A ball rolling down a hill, perfume spreading through the air, electric current flowing through power lines, heat warming a frying pan, and molecules moving across biological membranes are all examples of kinetic energy. **Potential energy** is stored energy. Potential energy may be stored in the position of a molecule with respect to its concentration gradient. In chemical bonds, potential energy is stored in the position of the electrons that form the bond (∞ p. 19). A key feature of all forms of energy is the ability of potential energy to become kinetic energy and vice-versa.

Energy Can Be Transformed from One Type to Another

Recall that a general definition of energy is the capacity to do work. Work always involves movement, and therefore always involves kinetic energy. Potential energy also can be used to perform work, but it must be converted to kinetic energy, and the conversion is never 100% efficient. A certain amount of energy is lost to the environment, usually in the form of heat. The amount of energy that is lost in the transformation depends on the efficiency of the process. Many physiological processes in the human body are not very efficient. For example, 70% of the energy used in physical exercise is lost in the form of heat rather than transformed into the work of muscle contraction.

Figure 4-2 ■ summarizes the relationship of kinetic energy, potential energy, and work. In Figure 4-2a ■, work is performed by pushing the ball up a ramp. Some of the kinetic energy of the moving ball is transformed into potential energy as it goes up the ramp. Figure 4-2b ■ shows potential energy, stored in the stationary ball at the top of the ramp. No work is being performed, but the capacity to do work is stored in the position of the ball. In Figure 4-2c ■, the potential energy of the ball becomes kinetic energy when the ball rolls down the ramp. Some of this kinetic energy is lost to the environment as heat resulting from friction between the ball and the air and ramp. In biological systems, potential energy is stored in concentration gradients and chemical bonds; it is transformed into

---

*To review the common forms of energy and their standard units, see Appendix A.

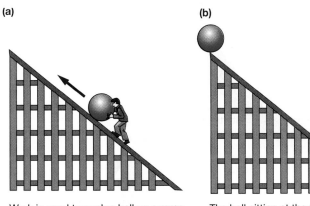

(a)

Work is used to push a ball up a ramp.

(b)

The ball sitting at the top of the ramp has potential energy, the potential to do work.

(c)

The ball rolling down the ramp is converting the potential energy to kinetic energy. However, the conversion is not totally efficient, and some energy is lost as heat due to friction between the ball, ramp, and air.

■ **Figure 4-2 Potential energy, kinetic energy, and work**

kinetic energy in order to do chemical, transport, or mechanical work.

Thermodynamics Is the Study of How Energy Is Converted to Work

Two basic rules govern the transfer of energy in biological systems and in the universe as a whole. The **first law of thermodynamics** states that the total amount of energy in the universe is constant. The universe is called a "closed system"—nothing enters and nothing leaves. Energy can be converted from one form to another, but the total amount of energy in a closed system never changes. The human body is not a closed system, however. As an open system, it exchanges material and energy with its surroundings. Our bodies cannot create energy, but must import it from outside in the form of food. By the same token, our bodies lose energy, especially in the form of heat, to the environment. Energy that stays within the body can be changed from one form to another or used to do work.

The **second law of thermodynamics** states that natural spontaneous processes move from a state of order (non-randomness) to a condition of randomness or disorder, also known as **entropy.** For an open system such as the body to create and maintain order, the input of energy is required. Disorder occurs when open systems only lose energy to their surroundings. The analogy of the deserted town illustrates the second law. When people put all their energy into activities outside the town, the town slowly falls into disrepair and becomes less organized. Without constant input of energy, a cell is unable to maintain its ordered internal environment. As the cell loses organization, its ability to carry out normal functions disappears, and it dies.

In the remainder of this chapter you will learn how cells obtain energy from and store energy in the chemical bonds of biomolecules. The chemical processes through which this takes place are collectively called metabolism. Using metabolism, cells transfer the potential energy of chemical bonds into kinetic energy for growth, maintenance, reproduction, and movement.

✓ How do animals store energy in their bodies?

✓ What is the difference between potential energy and kinetic energy?

CHEMICAL REACTIONS

Living organisms are characterized by their ability to extract energy from the environment and use it to support life processes. The study of energy flow through biological systems is a field known as **bioenergetics** [*bios*, life + *en-*, in + *ergon*, work]. In a biological system, chemical reactions are a critical means of transferring energy from one part of the system to another. This section will explain what a chemical reaction is, discuss different types of reactions, and present the general mechanisms for trapping, releasing, or transferring energy during reactions.

Energy Is Transferred between Molecules during Reactions

In a **chemical reaction,** a substance becomes a different substance by breaking covalent bonds and/or making new ones. A reaction begins with one or more molecules called **reactants** (or **substrates**) and ends with one or more **products** (Table 4-1). For the purpose of this discussion, we will consider a reaction that begins with two reactants and ends with two products:

$$A + B \rightarrow C + D$$

The speed with which the reaction takes place, the **reaction rate,** is the disappearance rate of the reactants (A and B) or the appearance rate of the products (C and D). Reaction rate is measured as change in concentration

Metabolism and coupled reactions   All the chemical reactions in the body are known collectively as **metabolism.** It is usually described as a series of pathways or a network of linked reactions. One of the basic features of metabolic pathways is the coupling of exergonic reactions that release energy to endergonic reactions that require energy. Without this coupling, many endergonic reactions would be unable to proceed. Some of the most familiar coupled reactions are those that drive an endergonic reaction with the energy released by breaking the high-energy bond of ATP:

$$ATP \qquad ADP + P_i$$

$$E + F \longrightarrow G + H$$

In this type of coupled reaction, the two reactions take place simultaneously and in the same location, so that the energy from ATP can be used immediately to drive the endergonic reaction of E and F. But it is not practical for all reactions to be directly coupled like this, so living cells have developed ways of trapping and saving the energy released by exergonic reactions. The most common method for trapping energy is in high-energy electrons carried on nucleotides (∞ p. 32). The nucleotide molecules NADH, FADH$_2$, and NADPH all have captured energy in the electrons of their hydrogen atoms. NADH and FADH$_2$ usually transfer most of that energy to ATP (Fig. 4-5 ■).

In a few cases, the energy released by an exergonic reaction is stored as the potential energy of a concentration gradient. Chapter 5 discusses how these differences in concentration can generate kinetic energy.

✓ If you mix baking soda and vinegar together in a bowl, the mixture reacts and foams up, releasing carbon dioxide gas. Name the reactant(s) and product(s) in this reaction.

✓ Do you think this reaction is endergonic or exergonic?

ENZYMES

Enzymes Lower the Activation Energy of Reactions

Enzymes are biological **catalysts,** molecules that speed up the rate of chemical reactions without themselves being changed. In the absence of enzymes, most chemical reactions in the cell would go so slowly that the cell would be unable to live. Because an enzyme is not permanently changed or used up in the reaction that it catalyzes, we might write it in a reaction equation this way:

$$A + B + enzyme \rightarrow C + D + enzyme$$

This way of writing shows that the enzyme participates in the reaction with reactants A and B but is unchanged at the end of the reaction. A more common shorthand for enzymatic reactions shows the name of the enzyme above the reaction arrow, like this:

$$A + B \xrightarrow{\text{enzyme}} C + D$$

In enzymatically catalyzed reactions, the reactants are called **substrates.**

How does an enzyme increase the rate of a reaction? In thermodynamic terms, it lowers the required activation energy, making it more likely that the reaction will start (Fig. 4-6 ■). Enzymes accomplish this by binding to the reactant molecules and bringing them together into the best position for reacting with each other. Without enzymes, the reaction would depend on the random collision of the molecules to bring them into alignment. Notice in the figure that neither the starting nor ending free energy content of the substrates and products has changed.

The rate of a reaction catalyzed by an enzyme is much more rapid than the same reaction taking place in the absence of the enzyme. For example, consider the enzyme carbonic anhydrase, which converts CO_2 and water to carbonic acid. A single molecule of carbonic anhydrase can catalyze the conversion of up to one million molecules of CO_2 and water to carbonic acid in one

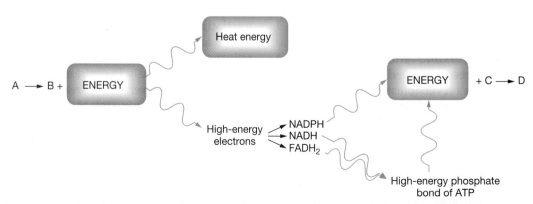

■ **Figure 4-5 Energy transfer and storage in biological reactions** Some of the energy released by exergonic reactions can be trapped in the high-energy electrons of NADH, FADH$_2$, or NADPH. NADH and FADH$_2$, in turn, transfer energy to the high-energy phosphate bond of ATP. The stored energy of ATP and NADPH can be used to run endergonic reactions. Energy that is not trapped is given off as heat.

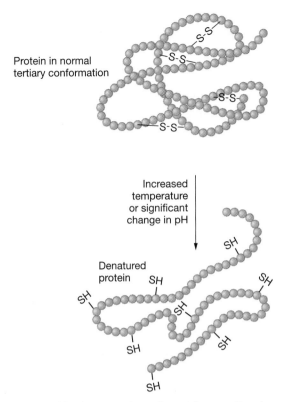

Protein in normal tertiary conformation

Increased temperature or significant change in pH

Denatured protein

■ **Figure 4-10 Denaturation of proteins results when intramolecular bonds are broken** Heat and acid can break the sulfur-sulfur or hydrogen bonds that hold a protein in its three-dimensional shape. When this happens to an enzyme, the active site is destroyed and the enzyme loses its activity.

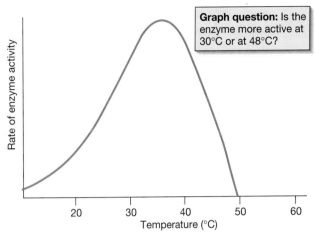

(a) Effect of changing temperature on enzyme activity.

Graph question: Is the enzyme more active at 30°C or at 48°C?

Rate of enzyme activity

Temperature (°C)

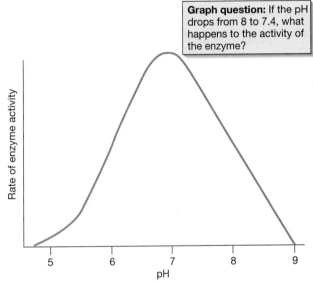

(b) Effect of changing pH on enzyme activity.

Graph question: If the pH drops from 8 to 7.4, what happens to the activity of the enzyme?

Rate of enzyme activity

pH

■ **Figure 4-11 Effect of pH and temperature on enzyme activity** These graphs show the typical results when an enzymatic reaction takes place at a range of temperatures or pH. (a) This enzyme denatures above temperatures of 50° C and activity slows to almost zero in the cold. (b) Extremes of pH also denature enzymes. Most enzymes in humans have optimal activity near the body's internal pH of 7.4.

once the changes exceed some critical point, the structure of the enzyme is altered so much that its activity is destroyed. When this occurs, the enzyme is *denatured*. In a few cases, activity can be restored if the modulator is removed. The protein then resumes its original shape as if nothing had happened. Usually, however, denaturation is a permanent loss of activity. There is certainly no way to unfry an egg or uncook a piece of fish. The potentially disastrous influence of temperature and pH on enzymes and other proteins is the reason that these factors are so closely regulated by the body.

Chemical Modulators Chemical modulators are molecules that bind to enzymes and alter their catalytic ability. **Competitive inhibitors** bind to the active site of the enzyme, blocking the site and thus preventing the enzyme from binding with the substrate (Fig. 4-12 ■). Some competitive inhibitors are alternative substrates that can also be acted upon by the enzyme. Others are simply molecules that bind to the enzyme and block the active site without being acted upon themselves. They are like the guy who slips into the front of the movie ticket line to chat with his girlfriend, the cashier. He has no interest in buying a ticket, but he prevents the people in line behind him from getting their tickets for the movie.

The next group, **allosteric modulators** [*allo-*, other + *stereos*, solid (as a shape)], bind to the enzyme away from the active site and change the shape of the active site (Fig. 4-13 ■). This type of modulation can either (1) increase the probability of enzyme-substrate binding and enhance enzyme activity or (2) decrease the affinity of the binding site for the substrate and inhibit enzyme activity. Allosteric modulation also occurs in proteins other than enzymes. For example, you will learn how the ability of the respiratory protein hemoglobin to bind oxygen changes with allosteric modulation by carbon dioxide, H^+, and several other factors.

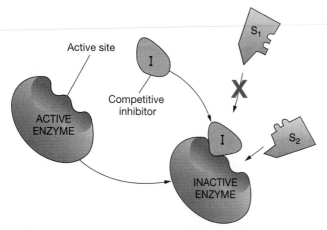

■ Figure 4-12 **Competitive inhibition** Competitive inhibitors bind to the active site of the enzyme, preventing the normal substrate from binding.

is the phosphate group. Many enzymes in the cell can be either activated or inactivated when a phosphate group forms a covalent bond with them. The ability to turn reactions on and off or up and down allows the cell to regulate the flow of biomolecules through synthetic and energy-producing pathways.

One of the best known chemical modulators is the antibiotic penicillin, discovered in 1928, when Alexander Fleming noticed that *Penicillium* mold inhibited bacterial growth. By 1938, researchers had extracted the active ingredient penicillin from the mold, and it was used to treat infections in humans. But it was not until 1965 that researchers figured out exactly how the antibiotic works. Penicillin binds to the active site of a key bacterial enzyme by mimicking the normal substrate. In doing so, it forms bonds that are unbreakable and irreversibly inhibits the enzyme. The enzyme is essential for the formation of a rigid cell wall in the bacterium; without it, the cell swells, ruptures, and dies.

The **covalent modulators** are atoms or functional groups that bind with covalent bonds to enzymes and affect their ability to decrease the activation energy of the reaction. Like allosteric modulators, covalent modulators may either increase or decrease the activity of the enzyme. One of the most common covalent modulators

Enzyme and Substrate Concentration Affect Reaction Rate

The rate of an enzymatic reaction is assessed by measuring either how fast the products are synthesized or how fast the substrates are consumed. There are several fac-

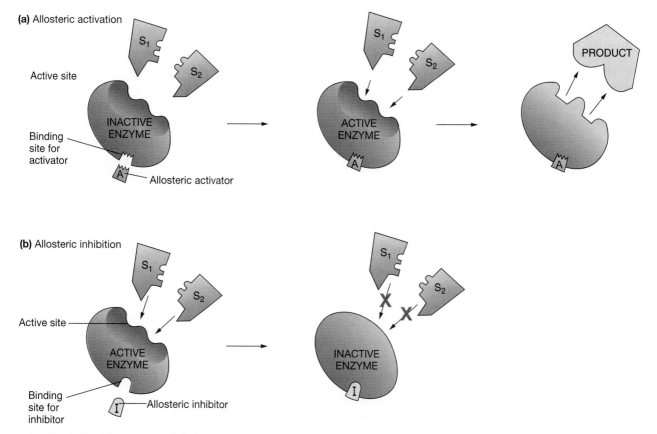

■ Figure 4-13 **Allosteric modulation** The modulator binds to the enzyme away from the active site and changes enzyme-substrate binding at the active site. (a) An allosteric activator enhances enzyme-substrate binding. Without modulation, the active site is poorly matched to the substrates. When the modulator binds to the enzyme, the active site is changed to a more suitable configuration. The enzymatically catalyzed reaction proceeds at a faster rate. (b) An allosteric inhibitor decreases enzyme-substrate binding by changing the active site so that it has a lower attraction for the substrates.

tors that affect the rate of a reaction. One is modulation of enzyme activity, as just discussed. Another is the amount of enzyme present in the cell. A third factor is the ratio of the concentrations of substrates and products. This section discusses how the concentration of enzyme, substrate, or product can alter the rate of a chemical reaction.

Reaction rate is directly related to the amount of enzyme present

A major determinant of the rate of an enzymatic reaction is the amount of enzyme present. As discussed previously, if there is no enzyme, most metabolic reactions go very slowly. If enzyme is present, the rate of the reaction will be proportional to the amount of enzyme. The graph in Figure 4-14 ■ shows the results of a typical experiment in which the amount of substrate remains constant while the amount of enzyme is varied. As the graph shows, an increase in the amount of enzyme present causes an increase in the rate of the reaction. This is like the checkout lines in a supermarket: Imagine that each cashier is an enzyme, the waiting customers are substrates, and those leaving the store with their purchases are products. One hundred people will get checked out faster when there are 25 lines open than when there are only 10 lines. Likewise, in a reaction, the presence of more enzyme molecules means that more active sites are available to interact with the substrate molecules. As a result, more reactants will be converted into products during any given period of time.

The relationship between enzyme concentration and reaction rate is an important way that cells regulate their physiological processes. Cells control the amount of an enzyme by regulating both its synthesis and its breakdown. If enzyme synthesis exceeds breakdown, enzyme accumulates and the reaction speeds up. If enzyme breakdown exceeds synthesis, the amount of enzyme decreases, as does the reaction rate. Even when the amount of enzyme is constant, there is still a steady turnover of enzyme molecules.

Reaction rate can reach a maximum

The amount of enzyme present is not the only factor that determines the rate of a reaction. If the concentration of the enzyme is constant, then the reaction rate will vary according to the concentration of substrate present. Figure 4-15 ■ shows the results of a typical experiment in which the enzyme concentration is constant, but the concentration of substrate varies. At low substrate concentrations, the reaction rate is directly proportional to the substrate concentration. But when the concentration of the substrate molecules increases beyond a certain point, the limited number of enzyme molecules have no more active sites free to bind substrate molecules. The enzyme is catalyzing reactions as fast as it can, and the rate of reaction reaches a maximum value. This condition is known as **saturation.**

An analogy to saturation appeared in the early days of television on the *I Love Lucy* show. Lucille Ball was working on the conveyor belt of a candy factory, loading chocolates into the little paper cups of a candy box. Initially, the belt was moving slowly, and she had no difficulty removing the candy and putting it into the box. Gradually, the belt brought candy to her more rapidly, and she had to increase her packing speed to keep up. Finally, the belt was bringing candy to her so fast that she couldn't pack it all in the boxes because she was working at her maximum rate. That was Lucy's saturation point. (Her solution was to stuff the candy into her mouth as well as into the box!)

Reversible reactions, reaction rate, and the Law of Mass Action

The study of reaction rates in cells is complicated by the fact that some reactions are **reversible.** This means that the forward reaction A + B → C + D can reverse, so that C + D → A + B. In the

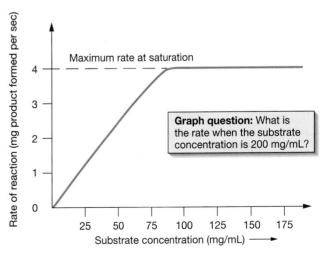

■ Figure 4-15 **Effect of changing substrate concentration on rate** In this experiment, the enzyme concentration remained constant while the substrate concentration was varied. The rate of the reaction increases as the substrate concentration increases, up to a maximum rate. The maximum rate represents the enzyme working at full capacity, with each of its active sites filled by substrate. This condition is known as saturation of the enzyme.

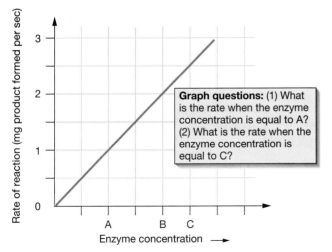

■ Figure 4-14 **Effect of changing enzyme concentration on rate** In this experiment, the substrate concentration remained constant while the enzyme concentration was varied.

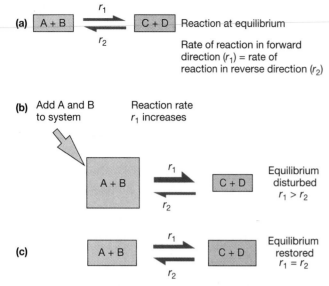

Figure 4-16 Law of Mass Action The Law of Mass Action states that when a reaction is at equilibrium, the ratio of the products and substrates will remain constant. Although this figure shows A + B equal to C + D at equilibrium, the actual ratio can vary, depending on the reaction. (a) Reaction at equilibrium. The rate of the forward reaction, r_1, is equal to the rate of the reverse reaction, r_2. As fast as substrates are converted into products, the products are converted back to substrates. (b) Substrate has been added to one side, sending the reaction out of equilibrium. Because reaction rate is proportional to substrate concentration, r_1 is now greater than r_2. Some of the new substrate is converted into product. (c) The reaction is returned to equilibrium when the substrate:product ratio returns to the constant value for this reaction.

reverse reaction, the products become the reactants and the reactants become the products.* What then determines in which direction the reaction is to go? The answer is that reversible reactions go to a state of **equilibrium,** where the rate of the reaction in the forward direction (A + B → C + D) is exactly equal to the rate of the reverse reaction (C + D → A + B), as shown in Figure 4-16a ■. At equilibrium, there is no net change in the amount of reactant or product. As fast as A and B convert to C and D, the reverse reaction takes place: C and D turn back into A and B at an equal rate.

But what happens if some factor disturbs this equilibrium by increasing the amount of A and B within the system? You learned in the previous section that reaction rate is proportional to substrate concentration. If the concentration of A and B in the system goes up, so does the rate of the forward reaction (Fig. 4-16b ■). More A and B is converted into products C and D. But since the reaction is reversible, as the product concentration increases, so does the rate of the reverse reaction. The system swings back and forth until finally it reaches equilibrium again.

Reversible reactions like this obey the **Law of Mass Action,** a simple relationship that holds for chemical reac-

*The naming of forward and reverse directions is arbitrary.

tions whether in a test tube or in the blood. In very general terms, the Law of Mass Action says that, when a reaction is at equilibrium, the ratio of the substrates to the products is always the same. If the concentration of a substrate (A or B) or product (C or D) changes, the equilibrium will be disturbed. The system then adjusts the substrate and product concentrations until the equilibrium ratio is restored. The Law of Mass Action is important in physiology because as the concentration of substrates changes, the change is reflected in the concentrations of the products.

Types of Enzymatic Reactions

Most reactions catalyzed by enzymes can be classified into one of three categories: oxidation-reduction reactions, hydrolysis-dehydration reactions, and exchange-addition-subtraction. These types of reactions are summarized in Table 4-4.

Oxidation-reduction reactions **Oxidation-reduction reactions** are the most important reactions in energy extraction and transfer within the cell. These reactions result in the transfer of electrons or protons (H^+) between atoms, ions, or molecules. A molecule that gains electrons or loses H^+ is said to be **reduced.** One way to think of this is to remember that the gain of negatively charged electrons *reduces* the electric charge on the molecule. Likewise, the loss of a positive proton leaves behind an electron with a negative charge. Conversely, molecules, atoms, or ions that lose electrons or gain H^+ are said to be **oxidized.**

Hydrolysis-dehydration reactions Hydrolysis and dehydration reactions are important in the breakdown and synthesis of large biomolecules. Generally, synthesis of polymers such as glycogen takes place when two molecules combine, yielding a water molecule in the process (see Table 4-4, 2b). One molecule contributes a hydroxyl group (–OH) to the water molecule formed during the reaction; the other molecule contributes a hydrogen. Because this process results in the removal of a water molecule and the synthesis of a new molecule, it is known as **dehydration synthesis.**

The breakdown of large polymers takes place by adding water in a **hydrolysis reaction** [*hydro,* water + *lysis,* to loosen or dissolve]. Water is split into a hydroxyl group (–OH) and a hydrogen (–H). The hydroxyl group binds to one product, and the hydrogen bonds to the other (Table 4-4, 2a). When an enzyme name consists of the substrate name with the suffix *-ase,* the enzyme causes a hydrolysis reaction. An example is *lipase,* an enzyme that breaks up large lipids into smaller lipids by hydrolysis.

Addition-subtraction-exchange reactions An **addition reaction** adds a functional group to one or more of the reactants; a **subtraction reaction** removes a func-

TABLE 4-4   Types of Enzymatic Reactions

Reaction Type	What Happens	Representative Enzymes
1. Oxidation-reduction	Add or subtract electrons or H^+	
a. Oxidized	Lose electrons	Oxidase
b. Reduced	Gain electrons	Reductase
	Remove electrons and H^+	Dehydrogenase
2. Hydrolysis-dehydration	Add or subtract a water	Hydrolase
a. Hydrolysis	Splits large molecules by adding water	Protease, lipase
b. Dehydration	Removes water to synthesize large molecules from several smaller ones	
3. Transfer chemical groups	Add, subtract, or exchange groups between molecules	
a. Exchange reaction	Phosphate	Kinase
	Amino group	Transaminase
b. Add a group	Phosphate	Phosphorylase
	Amino group	Aminase
c. Subtract a group	Phosphate	Phosphatase
	Amino group	Deaminase

tional group from one or more of the reactants. Likewise, functional groups are exchanged between or among reactants during an **exchange reaction.** For example, phosphate groups may be transferred between molecules during an addition, subtraction, or exchange reaction. Because transfer of phosphate groups is an important means of covalent modulation, several types of enzymes bind to phosphate groups and catalyze reactions that transfer them. The enzymes named *kinases* transfer a phosphate group from a substrate to an ADP molecule to create ATP, or from an ATP molecule to a substrate. For example, creatine kinase transfers a phosphate group from creatine phosphate to ADP, forming an ATP and leaving behind creatine.

The addition, subtraction, or exchange of amino groups (∞ p. 29) from an amino acid or peptide is also important in metabolism. Removal of an amino group is a **deamination** reaction, addition of an amino group is **amination,** and the transfer of an amino group from one molecule to another is called **transamination.**

✓ Name two ways an enzyme modulator can work to affect the rate of an enzymatic reaction.

✓ What happens to the rate of an enzymatic reaction as the amount of enzyme present increases?

METABOLISM

Metabolism refers to all chemical reactions that take place within an organism. These reactions extract energy from nutrient biomolecules (such as proteins, carbohydrates, and lipids) and synthesize or break down molecules. Metabolism is often divided into **catabolism,** reactions that release energy and result in the breakdown of large biomolecules, and **anabolism,** reactions that require a net input of energy and result in the synthesis

continued from page 79

Tay-Sachs disease is a recessive genetic disorder caused by a defect in the gene that directs the synthesis of hexosaminidase A. For a baby to be born with Tay-Sachs disease, it must inherit two defective genes, one from each parent. People with one Tay-Sachs gene and one normal gene are called carriers of the disease. Carriers will not develop the disease but can pass the defective gene on to their children. About 1 in 27 people of Eastern European Jewish descent in the United States carries the Tay-Sachs gene.

People who have two normal genes have normal amounts of hexosaminidase A in their blood. Carriers have lower-than-normal levels of the enzyme, but this amount is enough to prevent excessive accumulation of gangliosides in cells.

Question 2: How would you test Sarah and David for the presence of the Tay-Sachs gene?

of large biomolecules. Anabolic and catabolic reactions take place simultaneously in cells throughout the body, so that at any given moment, some biomolecules are being synthesized while others are being broken down. The energy released from or stored in the chemical bonds of biomolecules during metabolism is commonly measured in **kilocalories** (kcal). A kilocalorie is the amount of energy needed to raise the temperature of 1 liter of water by 1 degree Celsius. Much of the energy released during catabolism is trapped in the high-energy phosphate bonds of ATP or in the high-energy electrons of NADH, $FADH_2$, or NADPH (∞ p. 32). From these temporary carriers, energy is transferred to ATP or the covalent bonds of biomolecules through anabolic reactions.

Metabolism is a highly coordinated process in which activities at any given moment are matched to the needs of the cell. The enzymatic reactions of metabolism form a network of interconnected chemical reactions, or **pathways.** Each step of a pathway is a different enzymatic reaction, and the reactions of a pathway proceed in sequence. Starting substrate A is changed into product B, which then becomes the substrate for the next reaction, in which B is changed into C, and so forth:

$$A \longrightarrow B \longrightarrow C \longrightarrow D$$

Because the products of one reaction become the substrates for the next, we simply call the molecules of the pathway **intermediates.** You will sometimes hear metabolic pathways referred to as *intermediary metabolism.* Certain intermediates, called **key intermediates,** participate in more than one pathway and act as the branch points for channeling substrate in one direction or another.

In many ways, a network of metabolic pathways is similar to a detailed road map (Fig. 4-17 ■). Just as a map shows a network of roads that connect various cities and towns, metabolism can be thought of as a network of chemical reactions connecting various intermediate products. The greatest difference is that a molecule traveling a metabolic pathway is transformed into a new chemical intermediate at each "stop." But other similarities make this a useful image: think of an irreversible enzymatic reaction in a pathway as a one-way road, for example, or big cities with roads to several destinations as key intermediates. And keep in mind that there may be more than one way to get from one place—or one chemical intermediate—to another.

Next you will learn how cells regulate the flow of molecules through metabolic pathways. We will then examine the major metabolic pathways and trace the flow of chemical energy through animal cells.

Regulation of Metabolic Pathways

There are five basic ways that cells regulate the flow of molecules through their metabolic pathways:

1. By controlling the amount of enzyme
2. By producing allosteric and covalent modulators
3. By using two different enzymes to catalyze reversible reactions
4. By isolating enzymes within intracellular organelles
5. By maintaining an optimum ratio of ATP to ADP

The effect of changing enzyme amount was covered in an earlier section.

Enzyme modulation   Enzyme modulation, described on page 82, is frequently controlled by hormones and other signals coming from outside the cell. This type of outside regulation is a key element in the integrated control of the body's metabolism following a meal or during periods of fasting. Metabolic pathways may also have their own built-in form of modulation, called **feedback inhibition.** In this form of modulation, the end product of a pathway, shown as Z in Figure 4-18 ■, acts as an inhibitory modulator of the pathway. As the pathway proceeds and Z accumulates, the enzyme catalyzing the conversion of A to B is inhibited. This slows down production of Z until the cell can use it up. Once the levels of Z drop, the inhibition is removed and the pathway starts to run again. Because Z is the end product of the pathway, this type of inhibition is sometimes called *end-product inhibition.*

(a) Section of a road map

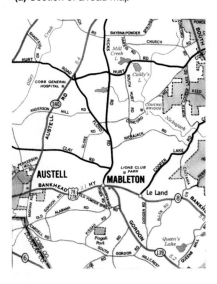

(b) Metabolic pathways drawn like a road map

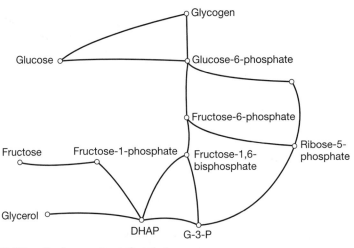

DHAP = dihydroxyacetone phosphate
G-3-P = glucose-3-phosphate

■ **Figure 4-17 A group of metabolic pathways resembles a road map** Cities on the map are equivalent to intermediates in metabolism. In metabolism, there may be more than one way to go from one intermediate to another, just as on the map there may be many ways to get from one city to another.

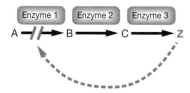

■ Figure 4-18 Feedback inhibition
The accumulation of product Z inhibits the first step of the pathway, usually by allosteric modulation. As product Z builds up, the pathway producing Z slows down. As the cell consumes Z in another metabolic reaction, the inhibition is removed and the pathway resumes production.

Enzymes and reversible reactions   The reversibility of a metabolic reaction can be used by cells to regulate the rate and direction of metabolism. If a single enzyme is able to run the reaction in either direction (Fig. 4-19a ■), the reaction will go to a state of equilibrium as determined by the Law of Mass Action. These reactions therefore cannot be closely regulated except by modulators and by control of the amount of enzyme. If a reaction requires two different enzymes, one for the forward and one for the reverse reaction, the cell is able to regulate the reaction more closely (Fig. 4-19b ■). If there is no enzyme for the reverse reaction, the reaction is irreversible (Fig. 4-19c ■).

Isolation of enzymes within the cell   Many of the enzymes of metabolism are localized in specific parts of the cell. Some, like the enzymes of carbohydrate metabolism, are found dissolved in the cytoplasm, whereas others are isolated within specific organelles. Mitochondria, endoplasmic reticulum, Golgi apparatus, and lysosomes all contain enzymes that are not found out in the cytoplasm. This separation of pathways allows the cell to control metabolism by regulating the movement of substrate between the different compartments. The isolation of enzymes within organelles is called **compartmentation.**

Ratio of ATP to ADP   The energy status of the cell is one final mechanism that can influence metabolic pathways. Through complex regulation, the ratio of ATP to ADP in the cell determines whether pathways that result in ATP synthesis are turned on or off. When ATP levels are high, production of ATP drops. When ATP levels are low, the cell sends substrates through pathways that result in more ATP synthesis. The next section will look further into the role of ATP in cellular metabolism.

ATP Transfers Energy between Reactions

The usefulness of metabolic pathways as suppliers of energy is often measured in terms of the net amount of adenosine triphosphate, or ATP, that the pathways can yield. ATP is a nucleotide molecule containing three phosphate groups (∞ p. 32). The third phosphate group is attached by a covalent bond that required net input of energy to form. The energy stored in a **high-energy phosphate bond** is released when the bond is broken during removal of the phosphate group. This relationship is shown by the following reaction

$$ADP + P_i + energy \longleftrightarrow ADP \sim P$$

in which the squiggle indicates a high-energy bond and P_i is the abbreviation for an inorganic phosphate group. Estimates of the amount of free energy released when a high-energy phosphate bond is broken range from 7 to 12 kcal per mole of ATP.

ATP is more important as a carrier of energy than as an energy storage molecule. For one thing, cells have a limited amount of ATP. A resting adult human requires 40 kg (88 pounds) of ATP to support one day's worth of metabolic activity, far more than our cells could store. Consequently, most of the body's daily energy requirement is met by energy stored in the chemical bonds of complex biomolecules. Metabolic pathways extract this energy through catabolism and convert it into the high-energy bonds of ATP.

The metabolic pathways that yield the most ATP molecules are those that require oxygen, the **aerobic,** or **oxidative,** pathways. **Anaerobic** (*an-*, without + *aer*, air] pathways that can proceed without oxygen also produce ATP molecules, but in much smaller quantities. The lower ATP yield of anaerobic pathways means that most animals (including humans) are unable to survive for extended periods of time on anaerobic metabolism alone. In the next section we will consider how the three

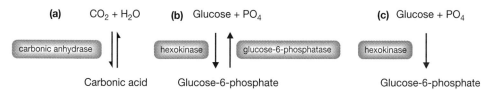

■ Figure 4-19 Reversible and irreversible metabolic reactions (a) Some enzymes catalyze a reaction in both directions. In this example, the reaction of carbon dioxide and water to form carbonic acid takes place in both directions with the enzyme carbonic anhydrase. (b) Other enzymes catalyze a reaction in only one direction. In this example, in order to make the reaction reversible, the cell must produce different enzymes for each direction. (c) If the cell has only one of the two enzymes, the reaction is irreversible. For example, most cells of the body can phosphorylate glucose to glucose-6-phosphate but cannot make glucose by dephosphorylation, because they lack the enzyme glucose-6-phosphatase.

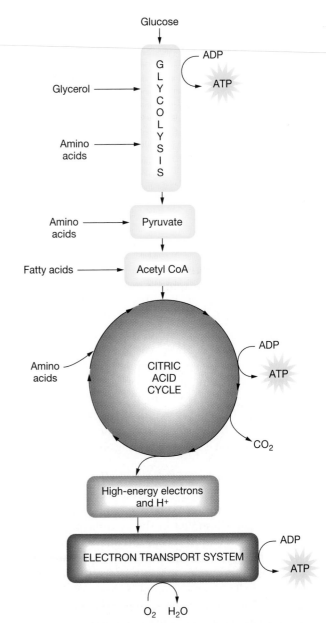

■ **Figure 4-20 Aerobic pathway for ATP production** ATP can be made through an aerobic pathway using the chemical bond energy from glucose, amino acids, or glycerol and fatty acids. Only glucose follows the entire pathway from glycolysis through the electron transport system. The other biomolecules enter the pathway at the points indicated.

major groups of biomolecules are metabolized in order to transfer energy to ATP.

ATP PRODUCTION

The catabolic pathways that produce ATP are summarized in Figure 4-20 ■. Aerobic production of ATP follows two common pathways: glycolysis and the citric acid (tricarboxylic acid) cycle. Both pathways produce small amounts of ATP directly, but their most important contribution to ATP synthesis is the production of high-energy electrons that are carried by NADH and $FADH_2$ to the electron transport system in the mitochondria. The electron transport system, in turn, transfers energy from the electrons to the high-energy phosphate bond of ATP. At various points, the process produces carbon dioxide and water. The water can be used by the cell, but carbon dioxide is a waste product and must be removed from the body.

All three types of biomolecules used for energy (carbohydrates, proteins, and lipids) use glycolysis, the citric acid cycle, and the electron transport system to produce ATP. Carbohydrates enter the pathways in the form of glucose. Lipids are broken down into fatty acids and glycerol (∞ p. 27), which enter the pathway at different points. Proteins are broken down into assorted amino acids, which also enter at various points. Because glucose is the only molecule that follows the entire pathway, we will look first at glucose catabolism. We will then see how protein and lipid catabolism differ from glucose catabolism.

The pathways for ATP production are a good example of compartmentation. The enzymes for glycolysis are located in the cytoplasm while the enzymes for the citric acid cycle are in the mitochondria.

Glycolysis Converts Glucose and Glycogen into Pyruvate

The main catabolic pathway of the cytoplasm is the series of reactions known as **glycolysis** [*glyco-*, sweet + *lysis*, dissolve]. During glycolysis, a molecule of glucose

is converted by a series of enzymatically catalyzed reactions into two pyruvate molecules, producing a net gain in energy. Thus, glycolysis is an exergonic pathway. A portion of the energy produced is used to phosphorylate ADP molecules, yielding ATP molecules. The events of glycolysis can be summarized as follows:

$$\text{Glucose} + 2\,\text{NAD}^+ + 2\,\text{ADP} + 2\,\text{P}_i \rightarrow$$
$$2\,\text{Pyruvate} + 2\,\text{ATP} + 2\,\text{NADH} + 2\,\text{H}^+ + 2\,\text{H}_2\text{O}$$

Notice that glycolysis does not require the presence of oxygen. It therefore serves as a common pathway for both aerobic and anaerobic catabolism of glucose.

In the first step of glycolysis, glucose is phosphorylated with the aid of ATP to glucose-6-phosphate.* Glucose-6-phosphate is a key intermediate that leads into glycolysis as well as to other pathways we shall not consider. By the time the six-carbon glucose molecule that entered the pathway has reached the end of glycolysis, it has been split into two molecules of **pyruvate,** each containing three carbons (Fig. 4-21 ■). Two steps of glycolysis are endothermic and require energy input from two ATP. Other steps are exothermic and produce four ATP and two NADH for each glucose that enters the pathway. The net energy yield from each glucose is therefore two ATP and two NADH. (See Table 4-5 for a summary of energy yield of various catabolic pathways.)

Pyruvate Is Converted into Lactate in Anaerobic Metabolism

Pyruvate is a branch point for metabolic pathways, like the hub cities on the road map. Depending on the cell's needs and condition, pyruvate can be shuttled off into one of two pathways (Fig. 4-22 ■). If the cell contains adequate oxygen, pyruvate continues on into the citric acid cycle. If the cell lacks sufficient oxygen for aerobic pathways (a condition that may be caused by a number of factors, including strenuous exercise), pyruvate is converted into lactate† with the assistance of the enzyme lactate dehydrogenase mentioned earlier:

$$\text{Pyruvate} \xrightarrow[\text{lactate dehydrogenase}]{\overset{\displaystyle \text{NADH} \quad \text{NAD}^+}{\curvearrowright}} \text{Lactate}$$

The conversion of pyruvate to lactate changes the NADH produced earlier in glycolysis back to NAD$^+$ when a hydrogen atom is transferred to the lactate molecule. As a result, the net energy yield for anaerobic metabolism of one glucose molecule is two ATP and no NADH.

Pyruvate Enters the Citric Acid Cycle in Aerobic Metabolism

If the cell has adequate oxygen for aerobic metabolism, then the pyruvate molecules formed from glucose are transported into the mitochondria (Fig. 4-22 ■). Once in the mitochondrial matrix, pyruvate is converted into another key intermediate called **acetyl coenzyme A,** or acetyl CoA for short. As its name suggests, this molecule has two parts: a two-carbon acetyl (acyl) unit, derived from pyruvate, and a coenzyme. Coenzyme A is made from the vitamin pantothenic acid. Like enzymes, it is

*The "6" in glucose-6-phosphate tells on which carbon of glucose the phosphate group has been attached.

†Lactate is the anion of lactic acid. At normal cell conditions, many of the organic acids such as pyruvic acid, lactic acid, and glutamic acid are ionized and are therefore named as the corresponding anion: pyruvate, lactate, and glutamate. The suffix *-ate* identifies the anions of organic and some inorganic acids.

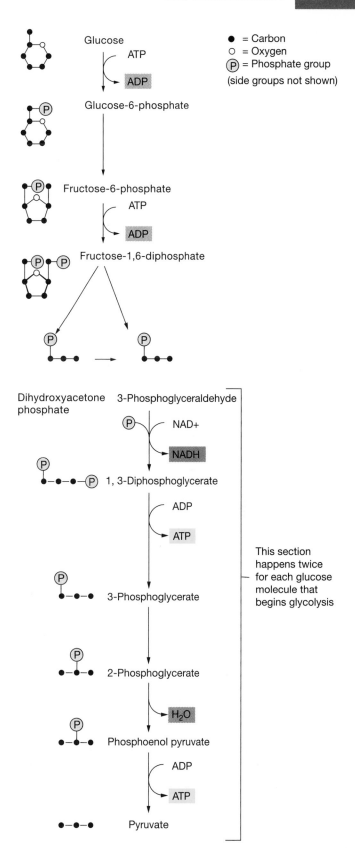

■ **Figure 4-21 Glycolysis** Glycolysis is the pathway that converts glucose to pyruvate. Glucose, a six-carbon sugar, is split into two molecules of 3-phosphoglyceraldehyde, which, in turn, is metabolized into two three-carbon pyruvate molecules. The steps of glycolysis consume two ATP and make four ATP and two NADH. Pyruvate is the branch point for aerobic and anaerobic metabolism of glucose.

TABLE 4-5 **Summary of Energy Yield from Catabolic Pathways**

Aerobic metabolism

Pathway	Uses	Makes	Net Energy Yield
Glycolysis glucose → 2 pyruvate	2 ATP	4 ATP 2 NADH	+2 ATP +2 NADH*
Aerobic metabolism 2 pyruvate → 2 acetyl CoA	—	2 NADH 2 CO_2	+2 NADH
2 acetyl CoA through citric acid cycle		2 ATP 6 NADH 2 $FADH_2$	+2 ATP +6 NADH +2 $FADH_2$
Total yield for 1 glucose through aerobic metabolism:			+4 ATP +2 NADH cytoplasm* +8 NADH mitochondria +2 $FADH_2$

Maximal ATP yield after mitochondrial oxidative phosphorylation:

4 ATP	4 ATP
2 cytoplasmic NADH × 2–3 ATP/NADH	4–6 ATP
8 mitochondrial NADH × 3 ATP/NADH	24 ATP
2 $FADH_2$ × 2 ATP/$FADH_2$	4 ATP
	= 36–38 ATP

Anaerobic metabolism

Pathway	Uses	Makes	Net Energy Yield
Glucose → 2 lactate	2 ATP 2 NADH	4 ATP 2 NADH	+2 ATP 0 NADH

*Cytoplasmic NADH molecules cannot enter mitochondria. Their ATP yield depends on whether their electrons go to NADH or to $FADH_2$ in the mitochondria.

not consumed during metabolic reactions and can be reused. The synthesis of acetyl CoA from pyruvate is an exergonic reaction that produces one NADH. Because pyruvate has three carbon atoms and acetyl CoA has only two, the extra carbon is combined with two oxygen atoms and turned into CO_2.

Acetyl CoA releases its two-carbon acetyl unit into a cyclical pathway known as the citric acid, or tricar-

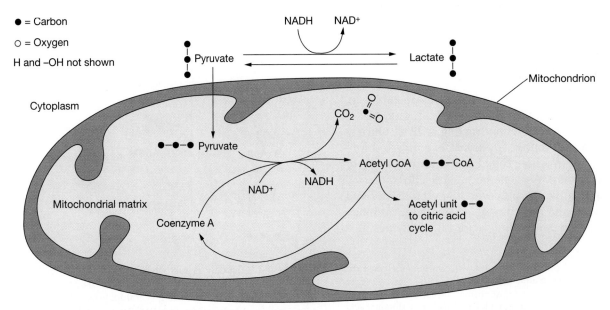

■ **Figure 4-22** **Pyruvate metabolism** Under anaerobic conditions, pyruvate is made into lactate (lactic acid). If adequate amounts of oxygen are present in the cell, pyruvate is transported into the mitochondrial matrix. There it is converted to acetyl CoA, with one carbon and two oxygens given off as carbon dioxide. The acetyl unit of acetyl CoA feeds into the citric acid cycle, while the coenzyme is released to react with another pyruvate.

boxylic acid, cycle (Fig. 4-23 ■). This pathway was first described by Sir Hans A. Krebs, so it is also known as the Krebs cycle. Because Sir Hans Krebs described other metabolic cycles, we shall avoid confusion by using the term citric acid cycle. The citric acid cycle makes a never-ending circle, bringing in an acetyl CoA with each turn of the cycle and producing ATP, high-energy electrons, and carbon dioxide.

The two-carbon acetyl unit released by acetyl CoA enters the cycle by combining with a four-carbon oxaloacetate molecule that is the last intermediate in the cycle. The resulting six-carbon citrate molecule then goes through a series of reactions until it completes the cycle as another oxaloacetate molecule. Most of the energy released by the enzymatic reactions of the citric acid cycle is captured as high-energy electrons on three NADH and one $FADH_2$. However, some energy is used to make the high-energy phosphate bond of one ATP, and the remainder is given off as heat. In two of the reactions, carbon and oxygen are removed in the form of carbon dioxide. By the end of the cycle, the four-carbon oxaloacetate molecule that remains is ready to begin the cycle again.

In the next step of aerobic metabolism, NADH and $FADH_2$ transfer two high-energy electrons to the electron transport system and return to the citric acid cycle as NAD^+ and FAD^+.

The Electron Transport System Transfers Energy from NADH and FADH₂ to ATP

The final step in aerobic ATP production is the transfer of energy from the electrons of NADH and $FADH_2$ to ATP; this transfer is made possible by a group of mitochondrial proteins known as the **electron transport system.** The synthesis of ATP using the electron transport system is called **oxidative phosphorylation,** because the electron transport system requires oxygen to act as the final acceptor of electrons and H^+.

The electron transport system couples transport work to chemical work: the work of moving H^+ from one mitochondrial compartment to another is linked to the chemical work of forming ATP bonds. This explanation has been named the **chemiosmotic theory** of oxidative phosphorylation. Mitochondrial NADH and $FADH_2$ give up two high-energy electrons and accompanying H^+ to the proteins of the electron transport system (Fig. 4-24 ■). Two at a time, the electrons pass through a series of protein complexes located in the inner mitochondrial membrane, losing energy with each transfer (∞ p. 48). During three of the transfers, enough energy is released to perform the work of pushing H^+ ions from the mitochondrial matrix into the space between the two mitochondrial membranes. During the other transfers, the

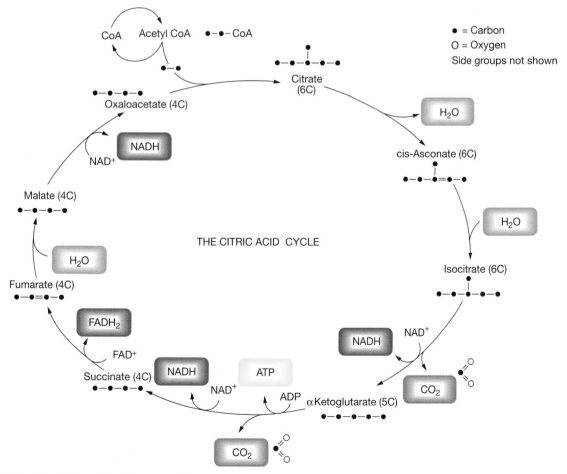

■ **Figure 4-23 The citric acid cycle** Two-carbon acetyl units from acetyl CoA combine with four-carbon oxaloacetate molecules to make citrate. Each full turn of the cycle releases two carbons as carbon dioxide and produces 1 ATP, 3 NADH, and 1 $FADH_2$.

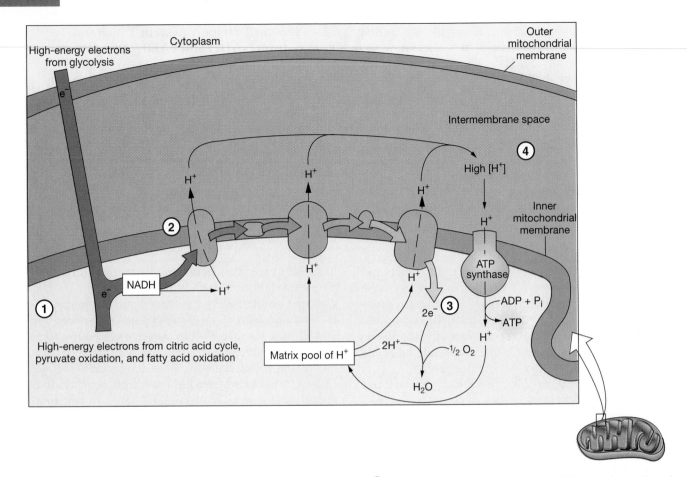

■ **Figure 4-24　The electron transport system and ATP synthesis**　① Some of the chemical energy released during glycolysis and the citric acid cycle is captured in the high-energy electrons of NADH. ② These high-energy electrons are transferred to the electron transport system, a series of protein complexes on the inner mitochondrial membrane. As the high-energy electrons are transferred between complexes, they give up their stored energy. Some of that energy is used to push H^+ from the matrix into the intermembrane space between the two mitochondrial membranes. ③ By the time electrons leave the electron transport system, they have given up their usable stored energy and are back to their normal energy state. They combine with H^+ and an oxygen atom to form water. ④ The H^+ concentrated in the intermembrane space store energy as potential energy. Some of this potential energy is converted to kinetic energy when H^+ return to the matrix by passing through a channel in an enzyme known as ATP synthase. Some of the kinetic energy of H^+ moving down their concentration gradient is captured in the high-energy bond of ATP.

released energy is given off as heat. By the end of the electron transport chain, the electrons have given up the usable portion of their stored energy. At that point, they combine with H^+ from the pool of ions in the matrix and an oxygen atom that entered the mitochondrion as an oxygen molecule (O_2). The result is a molecule of water, one for each two electrons that have passed through the system.

If oxygen is not present to act as the final electron acceptor, the electron transport system stops. This leaves NADH and $FADH_2$ unable to give up their high-energy electrons and return to the citric acid cycle as NAD^+ and FAD^+. The citric acid cycle in turn runs out of electron acceptors and the entire aerobic pathway comes to a halt.

The energy released by electrons moving through the electron transport system is stored temporarily in the H^+ ions that are concentrated in the intermembrane compartment. In the final step of oxidative phosphorylation, this energy is captured in high-energy bonds of ATP.

ATP Production Is Coupled to Movement of Hydrogen Ions across the Inner Mitochondrial Membrane

The potential energy stored by H^+ ions concentrated in the intermembrane space is converted into chemical energy when the ions move down their concentration gradient back into the mitochondrial matrix. Another mitochondrial protein known as the F_1F_0 ATPase, or **ATP synthase,** forms a pore in the inner mitochondrial membrane through which H^+ ions can move back into the matrix. As they do so, the protein captures the kinetic energy of H^+ ions moving down their concentration gradient and converts it into the stored energy of the high-energy phosphate bond of ATP. Because energy conversions are never completely efficient, a portion of the energy is released as heat. From studies of the isolated ATP synthase, scientists have determined that a maximum of one ATP is formed for each three H^+ that shuttle through the enzyme.

From studies on mitochondia, scientists have determined that for each NADH that sends electrons through the electron transport system, a maximum of three ATP will be produced. In contrast, $FADH_2$ electrons enter the electron transport system after the first protein complex, pumping fewer H^+ into the intermembrane space and resulting in a potential yield of only two ATP/$FADH_2$. Note that we say *potential* yield. This is because often the mitochondria do not work up to capacity. There are various reasons for this, including the fact that there is a certain amount of leakage of H^+ ions back into the matrix.

The Net Energy Yield of One Glucose Molecule Is 36 to 38 ATP

If we tally up the maximum potential energy yield for the catabolism of one glucose molecule through aerobic pathways, the total comes to either 36 or 38 ATP (Table 4-5). The variability in the maximum comes from the two cytoplasmic NADH molecules that are produced during glycolysis. These NADH molecules are unable to enter the mitochondria and directly donate their high-energy electrons to oxidative phosphorylation, so they must donate their electrons to carrier molecules that take them through the mitochondrial membrane. Two different molecules can act as recipients: mitochondrial NADH and mitochondrial $FADH_2$. If the two cytoplasmic NADH transfer their electrons to mitochondrial NADH, an additional six ATP molecules will be produced through oxidative phosphorylation. If they donate them to mitochondrial $FADH_2$, only four ATP molecules will be made.

Even when oxidative phosphorylation is not working at maximum efficiency, the ATP yield from aerobic metabolism of glucose far exceeds the puny yield of two ATP from the anaerobic metabolism of glucose to lactate. The low efficiency of anaerobic metabolism severely limits its usefulness in most vertebrate cells, because their metabolic energy demand is greater than can be met in this way. Some cells, such as exercising muscle, can tolerate anaerobic metabolism for a limited period of time, but they must then shift back to aerobic metabolism to get rid of the lactate that builds up.

Conversion of Large Biomolecules to ATP

Now that you have learned the common pathway for ATP production, let us examine the pathways that allow the large biomolecules such as glycogen, protein, and fat to enter glycolysis and the citric acid cycle.

Glycogen metabolism Glycogen, a glucose polymer, is the primary carbohydrate storage molecule in animals. Glycogen can be broken down into glycolysis pathway intermediates in one of two ways. About 10% is hydrolyzed back to plain glucose molecules (Fig. 4-25 ■). These molecules must then be phosphorylated to glucose-6-phosphate with the help of an ATP. The other 90% of glycogen molecules are converted directly to glucose-6-phosphate in a reaction that splits a glucose

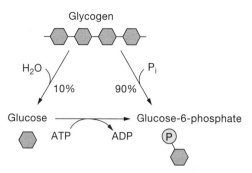

■ **Figure 4-25 Glycogen catabolism** When glycogen is broken down for ATP production, 90% of the glycogen molecules are converted directly to glucose-6-phosphate by combining with an inorganic phosphate group from the cytoplasm. The remaining 10% of the glycogen is broken down into glucose molecules that must then be phosphorylated with the help of ATP. The direct conversion of glycogen to glucose-6-phosphate saves one ATP.

molecule from the polymer with the help of an inorganic phosphate from the cytoplasm. Because the direct phosphorylation of glycogen to glucose-6-phosphate saves an ATP, this pathway has a net yield of one additional ATP molecule (maximum 37–39 ATP for aerobic metabolism).

Proteins can be catabolized to produce ATP The first step in protein catabolism is the digestion of proteins into smaller polypeptides by enzymes called **proteases.** Once that has been accomplished, other enzymes known as **peptidases** break the peptide bonds at the ends of the polypeptide, freeing individual amino acids. Most amino acids that are used for ATP synthesis come not from breaking down protein in cells, however, but from surplus amino acids in the diet.

Amino acids can be converted into intermediates for metabolism through either glycolysis or the citric acid cycle. The first step in this conversion is **deamination,** or removal of the amino group, creating an ammonia molecule and an organic acid (Fig. 4-26 ■). Some of the organic acids that are created by deamination include pyruvate, acetyl CoA, and several intermediates of the citric acid cycle. The former amino acids thus enter the pathways of aerobic metabolism and become part of the ATP production process previously described.

The amino group of the amino acid is removed as ammonia (NH_3) and then rapidly picks up a hydrogen ion to become ammonium ion (NH_4^+). Because both ammonia and ammonium are toxic, liver cells rapidly convert them into urea (CH_4N_2O). Urea is the main nitrogenous waste of the body and is excreted by the kidneys.

Lipids yield more energy per unit weight than glucose or proteins Lipids are the primary fuel storage molecule of the body, because they have a higher energy content than proteins or carbohydrates. In the reactions known as **lipolysis,** lipids are broken down by lipases into glycerol and fatty acids. Glycerol feeds into glycolysis about halfway through the pathway and goes through the same reactions as glucose from that point on (Fig. 4-27 ■).

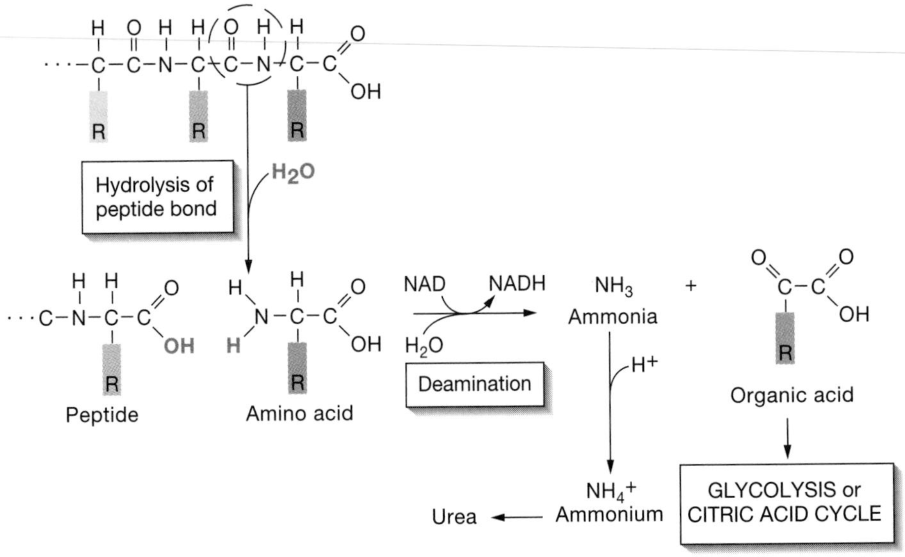

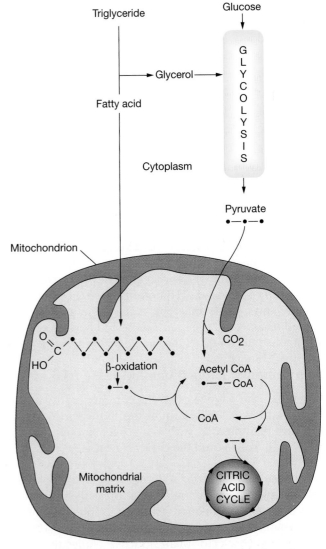

■ **Figure 4-26 Protein catabolism** Proteins are broken into amino acids by hydrolysis of their peptide bonds. Deamination, removal of the amino group (–NH$_2$) from the amino acid, is catalyzed by an enzyme known as a deaminase. The molecule remaining is an organic acid that becomes either part of glycolysis or the citric acid cycle. The amino group becomes an ammonium ion, then is detoxified by conversion to urea.

The long carbon chains of fatty acids are not as easy to turn into ATP as glucose or amino acids. A fatty acid must first be transported from the cytoplasm into the mitochondria. There it is slowly disassembled by chopping two-carbon units off the end of the chain, one unit at a time, in the process of **beta-oxidation.** In most cells, the two-carbon units from fatty acids are converted into acetyl CoA, which feeds directly into the citric acid cycle. Because of the many acetyl CoA molecules that can be produced from a single fatty acid, lipids contain nine kcal of stored energy per gram, compared with four kcal per gram for either proteins or carbohydrates.

The liver is one tissue that is unable to convert the two-carbon units into acetyl CoA and oxidize them to CO_2 and water. For this reason, liver cells must use alternative pathways to break down fatty acids. One result of these alternative pathways is the formation of **ketone bodies,** strong metabolic acids that can seriously disrupt the body's pH balance. The details of fatty acid metabolism will be discussed in Chapter 21.

SYNTHETIC PATHWAYS

The metabolic reactions discussed to this point have been catabolic processes that break up large molecules into smaller ones in order to trap energy in the high-energy phosphate bond of ATP. This section examines the anabolic reactions, those that synthesize polysaccharides, lipids, and proteins. In order to carry out synthetic reactions, the cell must have an adequate supply of organic nutrients, as well as energy in the form of ATP.

Glycogen Can Be Made from Glucose

Glycogen, the main storage form of glucose in the body, is found in all cells, but especially high concentrations are found in the liver and skeletal muscles. Glycogen in skeletal muscles provides a ready energy source for

■ **Figure 4-27 Lipolysis** Triglycerides are broken down into glycerol and long-chain fatty acids. Glycerol becomes a glycolysis substrate. Fatty acids lose two-carbon units, one at a time, in the process known as beta-oxidation. The two-carbon units become acetyl CoA, which can be used in the citric acid cycle.

muscle contraction, whereas glycogen in the liver acts as the main source of glucose for the body in periods between meals. It is estimated that the liver keeps about a four-hour supply of glucose stored as glycogen. Once that is exhausted by catabolism, the cells must turn to fats and proteins as their source of energy.

The synthesis of glycogen is essentially the reverse of glycogen breakdown (discussed in the previous section). Individual glucose molecules can be linked together into glycogen, or glucose-6-phosphate from glycolysis can be synthesized into the branching glucose chains of glycogen by removal of the phosphate group (see Fig. 4-25 ■).

Glucose Can Be Made from Glycerol or Amino Acids

Glucose is one of the key metabolic intermediates in the body for several reasons. As you have just learned, the aerobic metabolism of glucose is the most efficient way to make ATP, and so glucose is the primary substrate for energy production in all cells. More importantly, it is normally the only substrate for ATP synthesis in nervous tissue. If the brain is deprived of glucose for any period of time, its cells begin to die. For this reason, the body has multiple metabolic pathways that it can use to make glucose, ensuring a continuous supply for the brain. The simplest source of glucose is glycogen breakdown, or **glycogenolysis,** as previously discussed. But what happens when all the glycogen stores in the body are used up? In that case, cells can turn to fats and proteins for the intermediates needed to make glucose.

The production of glucose from nonglucose precursors such as proteins or glycerol is a process known as **gluconeogenesis,** or, literally, "birth of new glucose." The pathway is like glycolysis run in reverse, and it can be started with glycerol, various amino acids, or lactate (Fig. 4-28 ■). In all cells, the process can go as far as glucose-6-phosphate (G-6-P). However, only liver and kidney cells have the enzyme needed for the final reaction that removes the phosphate group from G-6-P to make glucose. During periods of fasting, the liver is the primary source of glucose synthesis, producing about 90% of the glucose that is made from the breakdown of fats and proteins.

Acetyl CoA Is an Important Precursor for Lipid Synthesis

Lipids are so diverse that it is difficult to generalize about their synthesis. Generally, lipids are synthesized by enzymes located in the smooth endoplasmic reticulum and in the cytosol. Fatty acids are made in the cytosol by an enzyme called fatty acid synthetase, which links the two-carbon acetyl groups of acetyl CoA together (Fig. 4-29 ■). The process also uses hydrogens and high-energy electrons from NADPH. Glycerol can be made from glucose or glycolysis intermediates. The combination of glycerol and fatty acids into triglycerides takes place in the endoplasmic reticulum. The phospho-

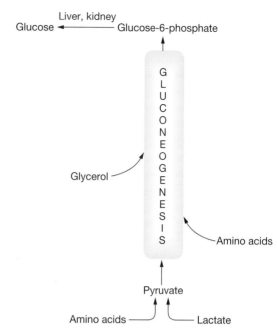

■ Figure 4-28 **Gluconeogenesis** The production of glucose from nonglucose precursors is known as gluconeogenesis. Glycerol, amino acids, and lactate can all be used to make glucose by a series of reactions that are the reverse of the glycolysis pathway. Only the liver and kidneys are able to convert glucose-6-phosphate to glucose.

rylation steps that turn triglycerides into phospholipids also take place in the endoplasmic reticulum.

Cholesterol is an important component of cell membranes, and it forms the skeleton of steroid hormones. Cholesterol is usually obtained from the diet, but it is such an important molecule for cells that the body will

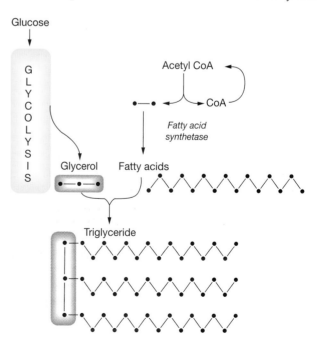

■ Figure 4-29 **Lipid synthesis** Lipids can be made from glucose through the glycolysis pathway. One glycolysis intermediate is used to make glycerol; two-carbon units from acetyl CoA are strung together by an enzyme called fatty acid synthetase to make long-chain fatty acids.

continued from page 87

In 1989, researchers discovered three genetic mutations responsible for Tay-Sachs disease. This discovery paved the way for a new carrier screening test that directly tests for the presence of the defective gene in blood cells, rather than testing for lower-than-normal hexosaminidase A levels. David and Sarah will undergo this new genetic test.

Question 3: Why is the new test for the Tay-Sachs gene more accurate than the old test, which detects decreased amounts of hexosaminidase?

synthesize it if the diet is deficient. Even vegetarians who eat no animal products (vegans) have substantial amounts of cholesterol in their cells. The entire structure of cholesterol can be made from acetyl CoA. From there, it is a simple matter to add or subtract functional groups to the carbon rings to change cholesterol into various hormones and other steroids. As you will see in Chapter 21, all the steroid hormones are very similar in structure.

Proteins Are Necessary to Many Cell Functions

Proteins are the molecules that run the cell from day to day. As enzymes, they control the synthesis and breakdown of carbohydrates, lipids, structural proteins, and signal molecules. Proteins in the cell membrane and organelle membranes regulate the movement of molecules into and out of the cell, or within the cell. Thus, protein synthesis is critical to cell function.

The power of proteins arises from their tremendous variability and specificity. Protein synthesis can be compared to creating a language with an alphabet of 20 letters and words up to hundreds of letters in length. The language is a rich one, with thousands of different words with different meanings. Like a language, a "spelling error" during the synthesis of a protein can change the function of the protein, just as changing the last letter turns the word "foot" into "food." In many inherited diseases, the substitution of a single amino acid alters the structure of a protein dramatically. The classic example is the disease sickle cell anemia. In this inherited disease, a change of one amino acid alters the shape of hemoglobin enough so that red blood cells take on a sickle shape; this causes them to get tangled up and block small blood vessels.

The protein "alphabet" begins with DNA One of the mysteries of biology until the 1960s was the question of how only four bases in the DNA molecule—adenine (A), guanine (G), cytosine (C), and thymine (T)—could code for over 20 different amino acids. If each base controlled the synthesis of one amino acid, a cell could only make four different amino acids. If pairs of bases represented different amino acids, it could make 4^2, or 16, possible combinations. Because we have 20 amino acids, this is still not satisfactory. But if triplets of bases were the

code for different molecules, DNA could create 4^3 or 64 different amino acids. These triplets, called **codons,** are indeed the way information is encoded in DNA (Fig. 4-30 ■). Of the 64 possible triplet combinations, one DNA triplet (TAC) acts the initiator or "start" codon that signifies the beginning of a coding sequence. Three codons are recognized as terminator, or "stop," codons that show where the sequence ends. The remaining 60 triplets all code for amino acids. Methionine and tryptophan have only one codon each, but the other amino acids have between two and six different codons each. It is from the information contained in these codons of DNA that the amino acid sequence of proteins is determined.

How is the sequence of codons found in DNA translated into a chain of linked amino acids? First, the information encoded in DNA is transferred to **messenger RNA** (mRNA). The mRNA leaves the nucleus and enters the cytoplasm, where it assembles amino acids in the order specified by the DNA. Two other forms of RNA, transfer RNA and ribosomal RNA, are needed to help mRNA complete the process of amino acid assembly. The conversion of the DNA code to mRNA is described in the next section.

During transcription, DNA guides the synthesis of a complementary mRNA molecule Because DNA is a very large molecule, it cannot pass through the nuclear envelope to the cytoplasm where protein synthesis takes place. Instead, a smaller single chain of mRNA is made by using DNA as a template during a process called **transcription** [*trans*, over + *scribe*, to write]. Messenger RNA is formed by base pairing in the same way that the double strand of DNA is (see Appendix B for a review of DNA synthesis). During transcription, part of the DNA chain unwinds and the complementary RNA base pairs come in to make a new RNA strand. The main difference

Second base of codon

		U	C	A	G	
First base of codon	U	UUU, UUC } Phe; UUA, UUG } Leu	UCU, UCC, UCA, UCG } Ser	UAU, UAC } Tyr; UAA, UAG } Stop	UGU, UGC } Cys; UGA Stop; UGG Trp	U C A G
	C	CUU, CUC, CUA, CUG } Leu	CCU, CCC, CCA, CCG } Pro	CAU, CAC } His; CAA, CAG } Gln	CGU, CGC, CGA, CGG } Arg	U C A G
	A	AUU, AUC, AUA } Ile; AUG Met Start	ACU, ACC, ACA, ACG } Thr	AAU, AAC } Asn; AAA, AAG } Lys	AGU, AGC } Ser; AGA, AGG } Arg	U C A G
	G	GUU, GUC, GUA, GUG } Val	GCU, GCC, GCA, GCG } Ala	GAU, GAC } Asp; GAA, GAG } Glu	GGU, GGC, GGA, GGG } Gly	U C A G

Third base of codon

■ Figure 4-30 **The genetic code** This table shows the genetic code as it appears in the codons of mRNA. The three-letter abbreviations to the right of the brackets indicate the amino acids that these codons represent. The start and stop codons are also marked.

CHAPTER REVIEW

Chapter Summary

Energy in Biological Systems

1. **Energy** is the capacity to do work. **Chemical work** enables cells and organisms to grow, reproduce, and carry out normal activities. **Transport work** enables cells to move molecules to create **concentration gradients.** **Mechanical work** is used for movement. (p. 74)

2. **Kinetic energy** is the energy of motion. **Potential energy** is stored energy. (p. 74)

Chemical Reactions

3. A **chemical reaction** begins with one or more **reactants** and ends with one or more **products. Reaction rate** is measured as the change in concentration of products with time. (p. 75)

4. The energy stored in the chemical bonds of a molecule and available to perform work is the **free energy** of the molecule. (p. 76)

5. **Activation energy** is the initial input of energy required to begin a reaction. (p. 76)

6. A reaction that results in net energy release is called an **exergonic reaction.** A reaction that requires net energy input is called an **endergonic reaction.** (p. 76)

7. A reaction that can proceed in both directions is called a **reversible reaction.** If a reaction can proceed in one direction but not the other, it is an **irreversible** reaction. The net free energy change of a reaction determines whether that reaction can be reversed. (p. 77)

8. All the chemical reactions in the body are known collectively as **metabolism.** Metabolic pathways couple exergonic reactions to endergonic reactions. (p. 78)

9. Energy for driving endergonic reactions is stored in ATP and in the potential energy of concentration gradients. (p. 78)

Enzymes

10. **Enzymes** are biological **catalysts** that speed up the rate of chemical reactions without themselves being changed. Enzymes bind to reactant molecules and bring them together into the best position for reacting with each other. In reactions catalyzed by enzymes, the reactants are called **substrates.** (p. 78)

11. The part of the enzyme that binds to the substrates is called the **active site,** or **binding site.** The **induced-fit model** of enzyme-substrate interaction says that the active site has an intermediate shape that can change to fit either the substrate or the product molecules. (p. 79)

12. **Specificity** is the ability of an enzyme to catalyze a certain reaction or a group of closely related reactions. (p. 79)

13. Some enzymes are produced as inactive precursors and must be activated. This may require the presence of a **cofactor.** Organic cofactors are called **coenzymes.** (p. 81)

14. Enzyme activity is altered by temperature, pH, and **modulator** molecules. A modulator either (1) changes ability of the substrate to bind to the active site or (2) changes ability of the enzyme to alter activation energy of the reaction. (p. 82)

15. **Competitive inhibitors** block the active site of the enzyme directly. **Allosteric modulators** bind to the enzyme away from the binding site and change the shape of the active site. **Covalent modulators** affect the ability of an enzyme to decrease the activation energy of the reaction. (p. 83)

16. Factors that affect the rate of enzymatic reactions include modulation, the amount of enzyme present, and the ratio of the concentrations of substrates and products. (p. 84)

17. When an enzyme is saturated with substrate, it works at its maximum possible rate. (p. 85)

18. Reversible reactions go to a state of **equilibrium,** where the rate of the reaction in the forward direction is exactly equal to the rate of the reverse reaction. Reversible reactions obey the **Law of Mass Action:** when a reaction is at equilibrium, the ratio of substrates to products is always the same. If the concentration of a substrate or product changes, the equilibrium will be disturbed. (p. 86)

19. Most reactions can be classified as either oxidation-reduction reactions, hydrolysis-dehydration reactions, or addition-subtraction-exchange reactions. (p. 86)

Metabolism

20. **Catabolic reactions** release energy and result in the breakdown of large biomolecules. **Anabolic reactions** require a net input of energy and result in the synthesis of large biomolecules. (p. 87)

21. Cells regulate the flow of molecules through their metabolic pathways by (1) controlling the amount of enzyme, (2) producing of allosteric and covalent modulators, (3) using different enzymes to catalyze reversible reactions, (4) isolating enzymes within intracellular organelles, or (5) maintaining an optimum ratio of ATP to ADP. (p. 88)

22. **Aerobic pathways** require oxygen and yield the most ATP. **Anaerobic pathways** can proceed without oxygen, but produce ATP in much smaller quantities. (p. 89)

ATP Production

23. During **glycolysis,** one molecule of glucose is converted into two **pyruvate** molecules, 2 ATP, 2 NADH, and 2 H$^+$. Glycolysis does not require the presence of oxygen. (p. 90)

24. In anaerobic metabolism, pyruvate is converted into lactate with a net yield of 2 ATP for each glucose molecule. (p. 91)

25. Aerobic metabolism of pyruvate through the citric acid cycle yields ATP, carbon dioxide, water, and high-energy electrons captured by NADH and FADH$_2$. (p. 91)

26. High-energy electrons from NADH and FADH$_2$ give up their energy as they pass through the electron transport system. Their energy is trapped in the high-energy bonds of ATP. (p. 93)

27. Glycogen and lipids are the primary energy storage molecules in animals. Lipids are broken down for ATP production in the process of **beta-oxidation.** (p. 96)

Synthetic Pathways

28. Protein synthesis is controlled by **genes** made of DNA in the nucleus. The code from a gene is transcribed into a matching code on **messenger RNA** (mRNA). The mRNA leaves the nucleus and, with the assistance of transfer and ribosomal RNA, assembles amino acids into the sequence designated by the gene. (p. 98)

29. Newly synthesized proteins may be modified in the rough endoplasmic reticulum or the Golgi apparatus. In the Golgi, they are packaged into membrane-bound vesicles that become lysosomes, storage vesicles, or secretory vesicles. (p. 102)

Questions

LEVEL ONE Reviewing Facts and Terms

1. List the three basic forms of work and give a physiological example of each.

2. Explain the difference between potential and kinetic energy.

3. State the two Laws of Thermodynamics in your own words.

4. The sum of all chemical processes through which cells obtain and store energy is called _____.

5. In the reaction $CO_2 + H_2O \rightarrow H_2CO_3$, water and carbon dioxide are the _____, and H_2CO_3 is the product. Because this reaction is catalyzed by an enzyme, it is also appropriate to call water and carbon dioxide _____. The speed at which this reaction occurs is called the reaction _____, often expressed as molarity/sec.

6. _____ are protein molecules that speed up chemical reactions by (increasing or decreasing?) the activation energy of the reaction.

7. Match these terms with their definitions (not all terms are used):
   1. exergonic
   2. endergonic
   3. activation energy
   4. reversible
   5. irreversible
   6. induced fit
   7. active site
   8. specificity
   9. free energy

   (a) A reaction that can run in either direction
   (b) The part of a protein molecule that actually binds the substrates
   (c) A reaction that releases energy
   (d) The ability of an enzyme to catalyze a certain reaction
   (e) Boost of energy needed to get a reaction started
   (f) The ability of a protein to flex its shape to fit more closely with that of the substrate

8. Since 1972, enzymes have been designated by adding the suffix _____ to their name.

9. An ion, such as Ca^{2+} or Mg^{2+}, that must be present in order for an enzyme to work is called a _____, whereas organic molecules that must be present for an enzyme to function are called _____. The precursors of these organic molecules come from _____ in our diet.

10. A protein whose tertiary structure is altered by extremes of heat or pH, so that its activity is destroyed, is said to be _____.

11. In an oxidation-reduction reaction, where electrons are moved between molecules, the molecule that gains an electron is said to be _____, and the one that loses an electron is _____.

12. The removal of H$_2$O from reacting molecules is called _____, and using H$_2$O to break down polymers, such as starch, is _____.

13. The removal of an amino group (NH$_2$) from a molecule (such as an amino acid) is called _____, while the movement of an amino group from one molecule to the carbon skeleton of another molecule (to form a different amino acid) is called _____. What is the fate of the amino group in the body after it has been removed from the amino acid?

14. Metabolism can be subdivided into _____ reactions, which release energy and result in the breakdown of large biomolecules, and _____ reactions, which require a net input of energy and result in the synthesis of large biomolecules. The energy of metabolism is measured in what units?

15. If the last product of a metabolic pathway (the end product) accumulates, allowing those molecules to react with an enzyme earlier in the pathway to slow or stop progress down that path, this results in a type of regulation called _____.

16. Explain how H$^+$ movement across the inner mitochondrial membrane is associated with ATP synthesis.

17. List the two carrier molecules that deliver electrons to the electron transport system.

18. The breakdown of lipids for energy is termed _____. The long chains of fatty acids are broken into two-carbon acetyl CoA molecules by a pathway called _____.

These acetyl CoA molecules then join which pathway in order to produce ATP?

19. How many kilocalories per gram are stored in carbohydrates, proteins, and fats, respectively?

20. Which cellular organelles are involved in lipid synthesis?

LEVEL TWO   Reviewing Concepts

21. Create a concept map to help you examine the factors that affect the rate of an enzymatic reaction.

22. When bonds are broken during a chemical reaction, what are the three possible fates for the potential energy found in those bonds?

23. Define, compare, and contrast the following terms: competitive inhibition, allosteric modulation, covalent modulation.

24. Explain why it is advantageous for a cell to store or secrete an enzyme in an inactive form.

25. Compare the energy yields from the aerobic breakdown of glucose to CO_2 and H_2O with anaerobic glycolysis end-ing with lactate. What are the advantages each pathway offers?

26. What is the advantage of converting glycogen to glucose-6-phosphate rather than to glucose?

27. Define the following terms and explain how they are related: gene, RNA polymerase, intron, exon, promoter, alternative splicing, sense strand, initiation factors.

28. Briefly describe the processes of transcription and translation. Which organelles are involved in these processes?

29. On what molecules does the anticodon appear? What class of biomolecule is also attached to this same molecule? Describe its location with respect to the anticodon.

LEVEL THREE   Problem Solving

30. Transporting epithelial cells in the intestine do not use glucose as their primary energy source. What molecules might they use instead?

31. Given the following strand of DNA, list the sequence of bases that would appear in mRNA. For the underlined triplets, underline the corresponding mRNA codon and list the appropriate anticodon.

DNA: CGCTACAAGTCAGGTACCGTAACG

mRNA:

Anticodons:

Problem Conclusion

In this running problem you have learned that Tay-Sachs disease is an incurable, recessive genetic disorder in which the enzyme that breaks down gangliosides in cells is missing. One in 27 Americans of Eastern European Jewish descent carries the gene for this disorder. You have also seen how a blood test can detect the presence of the deadly gene in a person's blood.

Further check your understanding of this running problem by checking your answers against those in the summary table.

Question	Facts	Integration and Analysis
1 What is another symptom of Tay-Sachs disease?	Hexosaminidase A breaks down gangliosides. In Tay-Sachs disease, the enzyme is absent and gangliosides accumulate in cells, including light-sensitive cells of the eye, and cause them to function abnormally.	Damage to light-sensitive cells of the eye could cause vision problems and even blindness.
2 How would you test for the presence of the defective gene?	Carriers of the gene have lower-than-normal levels of hexosaminidase A.	Run tests to determine the average enzyme levels in known carriers of the disease (i.e., people who have had children with Tay-Sachs disease) and in people who have little likelihood of being carriers. Compare the enzyme levels of suspected carriers to the averages for the known carriers and noncarriers.
3 Why is the new test for Tay-Sachs disease more accurate than the old test?	The new test detects the defective gene. The old test analyzed the levels of the enzyme produced by the gene.	The new test is a direct way to test if a person is a carrier. The old test was an indirect way. It is possible for factors other than a defective gene to alter a person's enzyme level. If enzyme levels are decreased due to some factor other than the Tay-Sachs gene, a false-positive test result will occur, and the subject may be erroneously told that he or she is a carrier. In addition, the variability in enzyme levels among the general population may be so broad that there is no distinct difference in the enzyme levels between carriers and noncarriers.
4 The Tay-Sachs gene is a recessive gene (t). What is the chance that any child of a carrier and a noncarrier will be a carrier?	Tt × TT ↓ TT Tt TT Tt	Each child has a 50% chance of being a carrier (Tt).

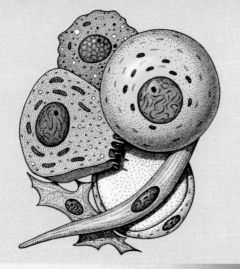

5

Membrane Dynamics

Up to this point, we have been concerned with the structures and processes of a single cell. But the human body is made of trillions of cells that work together. These cells must communicate and cooperate with each other while maintaining their separate natures. The cell membrane plays a key role in the process because it acts both as a barrier and as the means by which signals from outside the cell are communicated to the inside.

Membranes and their processes are increasingly the focus of physiological and medical research. Everything that enters and leaves the body or the cells must cross at least one membrane. As we have learned more about how membranes allow material to pass and how they relay signals from one area to another, we have begun to unravel many of the mysteries of physiology. Almost daily, you can pick up the newspaper and read about a new medical treatment that has arisen from our growing understanding of how membrane processes affect cells and tissues. Drugs for high blood pressure, depression, gout, stomach ulcers, and myriad other conditions have been developed as a result of research on membranes. You will be learning more about the action of these drugs in later chapters. This chapter introduces you to basic membrane structure and physiology.

The word *membrane* has come to have two different meanings in physiology. The simplest membrane, the plasma membrane or cell membrane, is the membrane that surrounds the cell. Cell membranes are a double layer (*bilayer*) of phospholipids (∞ p. 43) with protein molecules inserted in them. In addition to forming the outer boundary of the cell, phospholipid-protein membranes also enclose the membranous organelles within the cytoplasm (∞ p. 48). Confusion sometimes arises because the word *membrane* is also used to describe epithelial tissues that line a cavity or separate two compartments. Anatomically, we speak of mucous membranes in the mouth and vagina, the peritoneal membrane that lines the inside of the abdomen, the pleural membrane that covers the surface of the lungs, and the pericardial membrane that surrounds the heart. These visible membranes are actually tissues: thin, translucent layers of cells. To confuse matters further, epithelial membranes are often depicted in book illustrations as a single line, leading students to think of them as if they were similar in structure to the cell membrane (Fig. 5-1 ■). It seems a minor point until you realize that a molecule moving from one side of an epithelium to the other must pass through *two* cell membranes.

This chapter looks at both types of membranes and at how molecules move across them. The first section describes the structure and function of the phospholipid-protein cell membrane. The middle sections describe the fluid compartments of the body and the various mechanisms by which molecules cross membranes. The final section considers how the movement of molecules across cell and epithelial membranes affects the distribution of solutes and water in the body.

CELL MEMBRANES

Phospholipid membranes surround the contents of the cytoplasm and divide the interior of the cell into compartments such as the nucleus, mitochondria, endoplasmic reticulum, and Golgi complex. In this chapter we will be concerned with the external cell boundary that is called the cell membrane. (We will use the term *cell membrane* rather than *plasma membrane* to avoid confusion with blood plasma.) The general functions of the cell membrane include:

Problem

Cystic Fibrosis

Over a 100 years ago, midwives performed an unusual test on the infants they delivered: the midwife would lick the infant's forehead. A salty taste meant that the child was destined to die of a mysterious disease that withered the flesh and robbed the breath. Today, a similar "sweat test" will be performed in a major hospital—this time with state-of-the-art techniques–on Daniel Biller, a two-year old with a history of weight loss and respiratory problems. The name of the mysterious disease? Cystic fibrosis.

continued on page 113

1. *Physical isolation:* The cell membrane is a physical barrier that separates the inside of the cell (the cytosol or intracellular fluid) from the surrounding fluid.
2. *Regulation of exchange with the environment:* The cell membrane controls the entry of ions and nutrients, the elimination of wastes, and the release of secretory products.
3. *Communication between the cell and its environment:* The cell membrane is the part of the cell first affected by changes in the extracellular fluid (∞ p. 5). The membrane contains receptors that allow the cell to recognize and respond to specific molecules in its environment. Any alteration in the cell membrane may affect the cell's activities.
4. *Structural support:* Intracellular proteins of the cytoskeleton are linked to cell membrane proteins to maintain cell shape. Specialized connections between adjacent cells or between cells and the extracellular matrix stabilize the assembly of cells into tissues (∞ p. 53).

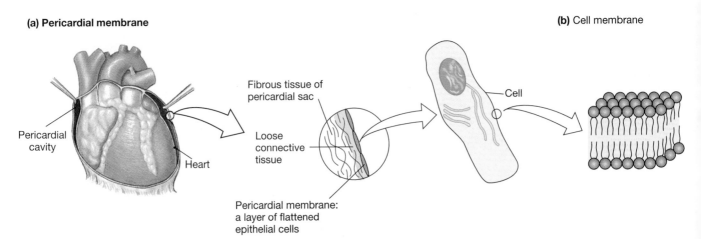

■ **Figure 5-1 Membranes in the body** (a) The pericardial membrane, depicted as a single line just like the cell membrane, and enlarged to show that it is really a layer of flattened cells. Substances crossing the pericardial membrane must cross two phospholipid bilayers. (b) The phospholipid bilayer of a cell membrane.

In order to understand how the cell membrane can carry out these diverse functions, let us first examine the current model of cell membrane structure.

Membranes Are Mostly Lipid and Protein

The structure of cell membranes has been a matter of great interest ever since scientists first observed cells with a microscope. By the 1890s, scientists had concluded that the outer membrane of cells was made of a thin layer of lipids that acted as a barrier between the aqueous interior and the watery environment outside the cell. But this simple and uniform structure did not account for the highly variable properties of membranes found in different types of cells. How could water cross the cell membrane to enter a red blood cell but not be able to enter certain cells of the kidney tubule? The answer had to lie in the structure of the cell membrane. To answer such questions, researchers in the first decades of the twentieth century ground up cells in order to isolate and analyze their membranes.

What they discovered was that the cell membrane and all membranes within the cell consist of a combination of lipids and proteins with a small amount of carbohydrate. The ratio of protein to lipid varies widely, depending on the source of the membrane (Table 5-1). Generally, the more metabolically active the cell membrane is, the more proteins it contains. For example, the inner mitochondrial membrane, which contains enzymes for oxidative phosphorylation (∞ p. 93), is three-quarters protein.

The analysis of the chemical composition of membranes was useful but still did not explain how lipids and proteins were structurally arranged in the membrane. Studies in the 1920s suggested that the amount of lipid contained in a given area of membrane created a double layer of lipid. This model was further modified in the 1930s to account for the presence of proteins.

By the early 1970s, freeze-fracture electron micrographs had revealed the three-dimensional arrangement of lipids and proteins within cell membranes. As a result of what scientists learned from looking at freeze-fractured membranes, S. J. Singer and G. L. Nicolson in 1972 proposed the **fluid mosaic model** of the membrane. Figure 5-2 ■ highlights the major features of this model of membrane structure. The phospholipids of the membrane are arranged in a bilayer, with a variety of proteins

inserted partly or completely through the bilayer. Carbohydrates on the extracellular surface are attached to both the proteins and the lipids, forming a "sugar coating." All cell membranes have similar components and are of uniform thickness, about 8 nm.

Membrane Lipids Form a Barrier between the Cytoplasm and Extracellular Fluid

There are two main types of lipids in the cell membrane: phospholipids and cholesterol. Phospholipids are made of a glycerol backbone (Fig. 5-3a ■) with two fatty acid chains extending to one side and a phosphate group extending to the other (∞ p. 29). The glycerol-phosphate region of the molecule is polar and thus hydrophilic, whereas the fatty acid region is nonpolar and thus hydrophobic (Fig. 5-3b ■). When placed in an aqueous solution, phospholipids will orient themselves so that the polar side of the molecule interacts with the water molecules while the nonpolar fatty acid tails hide by putting the polar heads between themselves and the water. This arrangement can be seen in three different structures: the **phospholipid bilayer,** the **micelle,** and the **liposome** (Fig. 5-3b ■). Micelles are small droplets of phospholipid, arranged so that the interior is filled with hydrophobic fatty acid tails. Micelles are important in the digestion and absorption of fats in the digestive tract. Liposomes are larger spherical structures with an exterior composed of a phospholipid bilayer. This arrangement of phospholipids leaves a hollow center with an aqueous core. This core can be loaded with water-soluble molecules. Liposomes are being tested as a medium for the delivery of drugs through the skin.

Although phospholipids are the major lipid of membranes, cholesterol is also a significant part of the cell membrane. Because cholesterol molecules do not have a polar region like that of phospholipids, they are hydrophobic, and so they insert themselves into the central lipid portion of the bilayer (Fig. 5-2 ■). Cholesterol helps make the membrane impermeable to small water-soluble molecules and keeps the membrane flexible over a wide range of temperatures.

The major role of the phospholipid-cholesterol bilayer is structural: to form a barrier across which only lipid-soluble molecules can migrate and to provide a framework for the membrane proteins. A few membrane lipids participate in communication between the extracellular environment and the cell.

Membrane Proteins May Be Tightly or Loosely Bound to the Membrane

Each cell has between 10 and 50 different *types* of proteins inserted into its phospholipid bilayer. Anatomically, membrane proteins are described as being either membrane-spanning or associated. The membrane-spanning proteins pass all the way through the membrane and may extend into the intracellular and extracellular fluid (Fig. 5-2 ■).

TABLE 5-1	Composition of Selected Membranes		
Membrane	*Protein*	*Lipid*	*Carbohydrate*
Red blood cell membrane	49%	43%	8%
Myelin membrane around nerve cells	18%	79%	3%
Inner mitochondrial membrane	76%	24%	0%

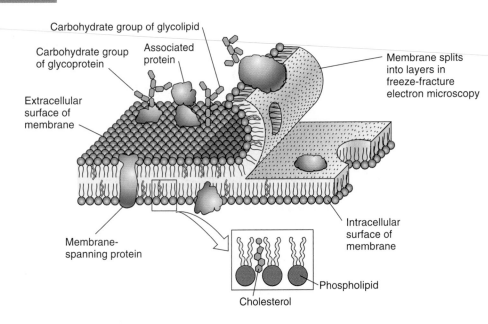

Carbohydrate group of glycolipid

Carbohydrate group of glycoprotein

Associated protein

Extracellular surface of membrane

Membrane splits into layers in freeze-fracture electron microscopy

Membrane-spanning protein

Intracellular surface of membrane

Phospholipid

Cholesterol

■ **Figure 5-2 The fluid mosaic model of the membrane** The rounded ends of the phospholipid molecules (purple) represent the polar glycerol-phosphate portion of the molecule, and the middle layer (yellow) is the nonpolar lipid tail. Cholesterol molecules insert themselves into the nonpolar lipid layer. Membrane-spanning proteins pass all the way through the membrane; associated proteins are found attached to either the intracellular or the extracellular surface. Both membrane proteins and lipids on the extracellular surface may have carbohydrates attached to them.

The associated proteins are attached less tightly to either the intracellular or extracellular side of the membrane.

Membrane-spanning proteins **Membrane-spanning proteins,** formerly called integral or intrinsic proteins, are tightly bound into the phospholipid bilayer.

The only way a membrane-spanning protein can be removed is by disrupting the membrane structure with detergents or by another harsh method that destroys the membrane integrity. Proteins that span the entire membrane are used both to transport molecules across the membrane and to relay signals from one side of the

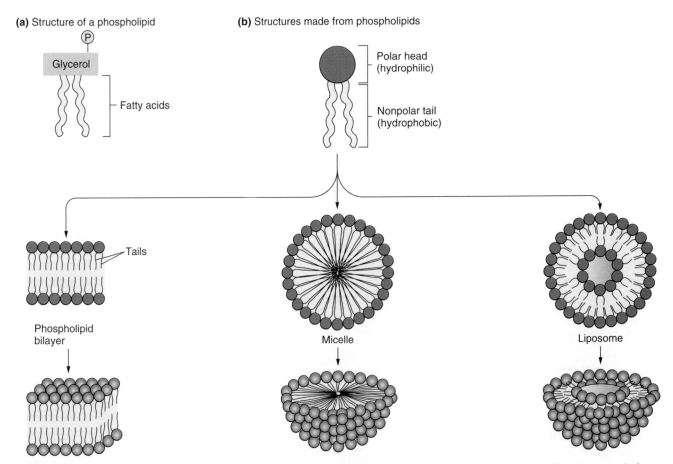

(a) Structure of a phospholipid

Ⓟ

Glycerol

Fatty acids

(b) Structures made from phospholipids

Polar head (hydrophilic)

Nonpolar tail (hydrophobic)

Tails

Phospholipid bilayer

Micelle

Liposome

■ **Figure 5-3 Membrane phospholipids** (a) Phospholipid molecule showing polar and nonpolar regions. (b) Structures made from phospholipids include the phospholipid bilayer, the micelle, and the liposome.

membrane to the other. These functions are discussed in more detail later in this chapter and in the next chapter.

Membrane-spanning proteins extend all the way across the membrane, with regions that protrude into the cytoplasm and/or the extracellular fluid (Fig. 5-4 ■). Some of the largest proteins span the membrane many times. The amino acid chains in between the membrane-spanning sections loop outward into the cytoplasm and the extracellular fluid. Carbohydrates attach to the extracellular loops; phosphate groups may attach to the intracellular loops.

Some membrane-spanning proteins move from location to location within the phospholipid bilayer, directed by fibers of the cytoskeleton that run just under the membrane surface. Immobile membrane-spanning proteins are held in position by cytoskeleton proteins (Fig. 5-5 ■).

The ability of the cytoskeleton to restrict the movement of some membrane proteins allows cells to develop *polarity*, in which different faces of the cell have different proteins and therefore different properties. This is particularly clear in the cells of the transporting epithelia.

Associated proteins   Associated proteins, formerly called peripheral or extrinsic proteins, attach themselves to membrane-spanning proteins or to the polar regions of the phospholipids. Because they are rather loosely bound to the membrane, they can be removed without disrupting the integrity of the membrane itself. Associated proteins include enzymes and some of the structural proteins that tie the cytoskeleton to the membrane (Fig. 5-5 ■).

Membrane Proteins Function as Structural Proteins, Enzymes, Receptors, and Transporters

For physiologists, it is more useful to classify membrane proteins by their function than by their structure. Cell membrane proteins function as (1) structural elements, (2) enzymes, (3) receptors, and (4) transporters that move molecules into and out of the cell. Many of the physiological processes you will be studying involve membrane enzymes, receptors, and transporters.

Structural proteins   The structural proteins have two major roles. The first is to link the membrane to the cytoskeleton in order to maintain the shape of the cell. The microvilli of transporting epithelia are a good example of membrane shaping by the cytoskeleton (∞ p. 47). The second role of structural membrane proteins is to form part of the cell-to-cell connections that hold tissues together. Membrane-spanning proteins link cytoskeleton fibers inside the cell to collagen and other fibers in the extracellular matrix (∞ Fig. 3-14, p. 55).

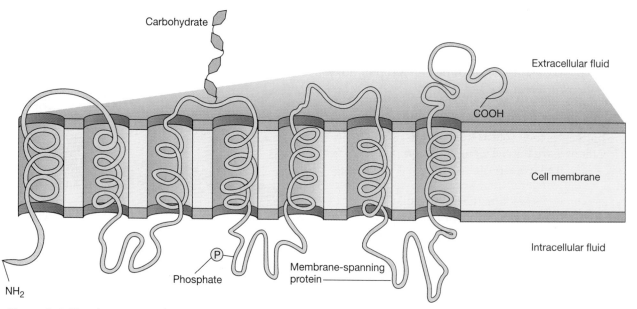

Carbohydrate

Extracellular fluid

COOH

Cell membrane

Intracellular fluid

P Phosphate

Membrane-spanning protein

NH₂

■ **Figure 5-4 Membrane-spanning proteins**   Membrane-spanning proteins cross the membrane several times with loops of peptide chains extending into both the cytoplasm and the extracellular fluid.

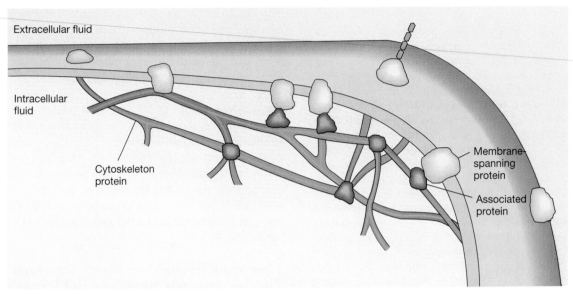

Extracellular fluid

Intracellular fluid

Cytoskeleton protein

Membrane-spanning protein

Associated protein

■ Figure 5-5 **Cytoskeleton proteins anchoring to proteins in the cell membrane**

Enzymes   Membrane-associated enzymes catalyze chemical reactions that take place on the cell's external surface or just inside the cytoplasm. For example, enzymes on the luminal surface of cells in the small intestine are responsible for the digestion of peptides and carbohydrates. Enzymes attached to the intracellular surface of many cell membranes play an important part in the transfer of signals from the extracellular environment to the cytoplasm of the cell.

Receptors   Receptor proteins on the outer surface of the cell are part of the body's chemical signaling system. Each receptor is specific for a certain molecule or family of related molecules—it will combine only with those molecules, just as enzymes recognize and bind to specific substrates. The molecule that binds to a receptor is called its **ligand** [*ligare*, to bind or tie]. The binding of receptor and ligand usually triggers another event at the membrane, such as activation of a membrane enzyme (Fig. 5-6 ■). An example of a ligand is the hormone insulin, which combines with an insulin receptor on a cell membrane in order to exert its effects.

Transporters   The fourth major function of membrane proteins is their role as transporters. Many molecules enter or leave the cell using transport proteins. We will classify transport proteins into two categories: channels and carriers. Channel proteins have water-filled passageways that link the intracellular and extracellular compartments. Carrier proteins never form a direct connection between the intracellular and extracellular fluid.

  Why do cells need both channels and carriers? The answer lies in the different properties of the two transporters. Channel proteins allow more rapid transport across the membrane but are not as selective about what

they transport. Carriers, while slower, are better at discriminating between closely related molecules.

Channel Proteins   The proteins that form channels are made of amino acid chains that zigzag back and forth across the membrane, creating a cluster of protein cylinders surrounding the water-filled channel (Fig. 5-7 ■). The diameter of these channels is narrow enough that movement through them is restricted to water, ions, and very small molecules such as urea. Ions and water are the primary substances to move through channel proteins. Because water-filled channels are open to both sides of the membrane at once, tens of millions of ions per second can whisk through them unimpeded.

  The selectivity of a channel may restrict molecules based on electrical charge. If the channel is lined with polar amino acids with a certain charge, ions with the same charge will be repelled while ions of opposite charge will be attracted. For example, there are cation channels that transport Na^+, K^+, lithium ions, and other ions of simi-

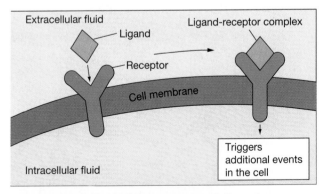

Extracellular fluid

Ligand

Ligand-receptor complex

Receptor

Cell membrane

Triggers additional events in the cell

Intracellular fluid

■ Figure 5-6 **Cell membrane receptor**   Cell membrane receptors combine with molecules known as ligands. The receptor-ligand complex triggers additional events within the cell.

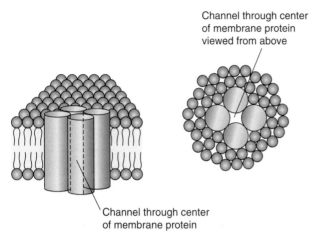

Channel through center of membrane protein viewed from above

Channel through center of membrane protein

■ Figure 5-7 **Structure of channel proteins** Many membrane channels are composed of multiple subunits that assemble in the membrane to form a unit. Hydrophilic amino acids facing the center of the channel create a water-filled passage that allows ions and very small molecules, such as urea, to pass through.

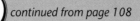

continued from page 108

Cystic fibrosis is a debilitating disease caused by a defect in a gated channel protein that normally transports chloride ions (Cl⁻). The channel protein—called the cystic fibrosis transmembrane regulator, or CFTR—is located in the cell membranes of epithelium lining the airways, sweat glands, and pancreas. The open state of the channel is regulated through the binding of nucleotides to specific regions of the channel. In people with cystic fibrosis, the CFTR channel is absent or defective. As a result, chloride transport (either secretion or reabsorption) across the epithelium is impaired.

Question 1: Is CFTR a chemically gated, voltage-gated, or mechanically gated channel protein?

lar size and electrical charge. These cation channels will not allow Cl⁻ or other negatively charged ions to pass.

Membrane channels have regions of the channel protein that act like "gates" that swing to open and close the channel. According to current models, many "gates" are part of the cytoplasmic side of the membrane protein (Fig. 5-8 ■). Such a gate can be envisioned as a ball on a chain that swings up and blocks the mouth of the channel. A few channels have a gate in the middle of the protein, and one type of channel in nerve cells has two gates.

Channels can be classified according to whether their gates are usually open or closed. **Open channels** spend most of their time in the open state, allowing ions to move back and forth across the membrane without regulation. Their gates may occasionally flicker closed, but for the most part, these channels behave as if they have no gates. Open channels are sometimes referred to as leak channels.

Gated channels spend most of their time in the closed state. This allows these channels to regulate the movement of molecules through them. If a gated channel is open, molecules diffuse through the channels just as they diffuse through open channels. If a gated channel is closed, which it may be much of the time, it allows no movement between the intracellular and extracellular fluid. The

opening and closing of gated channels is controlled by messenger molecules or ligands (**chemically gated channels**), by the electrical state of the cell (**voltage-gated channels**), or by a physical change such as increased temperature or a blow that stretches the membrane and pops the channel open (**mechanically gated channels**).

Carrier Proteins The second type of transporter protein is the **carrier protein.** These membrane transporters bind with specific molecules and carry them across the membrane by changing conformation, or shape. The molecules being transported across the membrane are referred to as substrates. The conformation change slows transport so that only 1000 to 1,000,000 molecules per second can be moved by each protein. Small organic molecules, such as glucose and amino acids, cross membranes only on carriers; ions, such as Na⁺ and K⁺, may be moved by carriers as well as through channels.

Carrier proteins differ from channel proteins in several ways. First, the molecule being moved binds temporarily with the carrier protein instead of simply traveling through a water-filled channel. Second, carrier proteins never create a continuous passage between the inside and outside of the cell. They do form channels, but the channels have two gates, one on the extracellular side and one on the intracellular side. One gate is always closed, preventing free exchange between inside and outside. In this respect, carrier proteins are like the

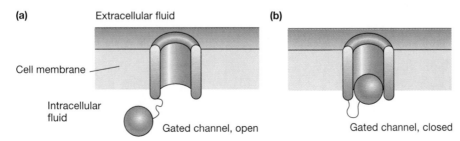

■ Figure 5-8 **Membrane-spanning proteins involved in membrane transport** (a) Open channels freely allow selected molecules through the cell membrane. (b) Gated channels open and shut in response to chemical or electrical signals.

Panama Canal (Fig. 5-9a ■). Picture the canal with only two gates, one on the Atlantic side and one on the Pacific side. Only one gate at a time is open. When the Pacific gate is closed, the canal opens into the Atlantic. A ship enters the canal from the Atlantic and the gate closes behind it. Now the canal is isolated from both oceans with the ship trapped in the middle. Then the Pacific gate opens, making the canal continuous with the Pacific Ocean. The ship sails out of the gate, having crossed the barrier of the land without the canal ever forming a direct connection between the two bodies of water.

Movement across the membrane through carrier proteins is similar (Fig. 5-9b ■). The substrate binds to the carrier on one side of the membrane. The binding changes the conformation of the protein so that one gate closes and the other one opens. As a result, the channel opens to the opposite side of the membrane. The carrier then releases the substrate into the other side, having brought it through the membrane without creating a direct connection between the extracellular fluid and the intracellular fluid.

The mechanisms by which molecules cross membranes, either on transporters or by moving through the phospholipid bilayer, is the subject of the section Movement Across Membranes.

Membrane Carbohydrates Attach to Both Lipids and Proteins

Most membrane carbohydrates are glucose polymers attached either to membrane proteins (*glycoproteins*) or to membrane lipids (*glycolipids*). They are found exclusively on the external surface of the cell, where they form a protective layer known as the **glycocalyx** [*glykys*, sweet + *kalyx*, husk or pod]. The glycoproteins on the surface of the cell play a key role in the body's immune response.

Figure 5-10 ■ is a summary map organizing the structure of the cell membrane. Figure 5-11 ■ is a map showing the structure and function of membrane proteins. In the next section, we will see how membranes divide the body into compartments, and how membranes regulate the movement of molecules between these compartments.

✓ Name two types of lipids found in cell membranes.

✓ Describe two ways that membrane proteins are physically associated with the cell membrane.

✓ Name four functions of membrane proteins.

BODY FLUID COMPARTMENTS

Cell membranes divide the inside of the body into different compartments, just as the membranes of organelles divide the inside of the cell into compartments. Most cells of the body are not in direct contact with the outside world. Instead they are surrounded by extracellular fluid (∞ Fig. 1-4, p. 6). If we think of all the cells of the body together as a single unit, we can divide the body into two main compartments: (1) the intracellular fluid within the cells and (2) the extracellular fluid outside the cells. The extracellular fluid can be further subdivided into two compartments. The fluid that directly bathes the cells is known as **interstitial fluid** [*inter-*, between + *stare*, to stand] and forms one compartment. **Plasma,** the fluid portion of the blood, constitutes the second extracellular compartment.

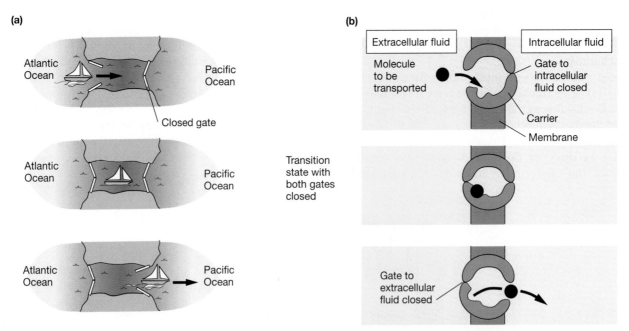

(a)

Atlantic Ocean — Pacific Ocean

Closed gate

Atlantic Ocean — Pacific Ocean

Transition state with both gates closed

Atlantic Ocean — Pacific Ocean

(b)

Extracellular fluid   Intracellular fluid

Molecule to be transported   Gate to intracellular fluid closed

Carrier

Membrane

Gate to extracellular fluid closed

■ Figure 5-9 **Facilitated diffusion by means of a carrier protein** The carrier protein has a channel that is alternately open to the extracellular and intracellular fluids. The carrier binds to the molecule to be transported on one side of the cell membrane, then changes configuration, opening the channel to the other side and releasing the molecule.

MOVEMENT ACROSS MEMBRANES

ents, ions, and other molecules dissolve in it because of its polar nature. But substances that are hydrophilic and dissolve in water are lipophobic as a rule; that is, they do not readily dissolve in lipids. Thus, the lipid layer between the cytoplasm and the cell's exterior acts as a boundary that prevents hydrophilic molecules from crossing.

Substances that are able to pass through the lipid center of the cell membrane move by diffusion. Diffusion directly across the phospholipid bilayer of a membrane is called **simple diffusion** and has the following properties in addition to the properties of diffusion listed earlier.

1. **The rate of diffusion depends on the ability of the molecule to dissolve in the lipid layer of the membrane.** Most molecules in solution can mingle with the polar phosphate-glycerol heads of the membrane's phospholipid bilayer, but only nonpolar molecules that are lipid-soluble (*lipophilic*) can traverse the central lipid core of the membrane. As a general rule, only lipids, steroids, and small lipophilic molecules can diffuse across the membrane by simple diffusion.

    Water, although a polar molecule, can diffuse across some phospholipid membranes. For years, it was thought that the polar nature of the water molecule excluded it from moving through the lipid center of the bilayer, but experiments done with artificial membranes have shown that the small size of the water molecule allows it to slip between the lipid tails in many membranes. How readily water passes through the lipid layer depends on the composition of the phospholipid bilayer. Membranes with high cholesterol content are less permeable to water than those with low cholesterol content, presumably because the cholesterol molecules fill the spaces between the fatty acid tails and thus exclude water. For example, the cell membranes of some sections of the kidney are essentially impermeable to water unless the cells insert special water channel proteins into the phospholipid bilayer.

2. **The rate of diffusion across a membrane is directly proportional to the surface area of the membrane.** In other words, the larger the surface area, the more molecules can diffuse across per unit time. This may seem obvious, but it has important implications in physiology. One striking example of how a change in surface area affects diffusion is the lung disease *emphysema*. As lung tissue is destroyed, the surface area for diffusion of oxygen decreases. Consequently, less oxygen can move into the body. In severe cases, the amount of oxygen that reaches the cells is not enough to sustain any muscular activity and the patient is confined to bed.

3. **The rate of diffusion across a membrane is inversely proportional to the thickness of the membrane.** The thicker the membrane, the slower diffusion will take place. This factor comes into play in certain lung con-

ditions in which the exchange epithelium of the lung is thickened with scar tissue. This slows down diffusion so that the amount of oxygen entering the body is not adequate to meet metabolic needs.

The rules for simple diffusion across membranes are summarized in Table 5-2. They can be combined mathematically into an equation known as Fick's Law of Diffusion. In an abbreviated form, Fick's Law says that:

rate of diffusion is proportional to

$$\frac{\text{Available surface area} \cdot \text{concentration difference}}{\text{Resistance of membrane} \cdot \text{thickness of membrane}}$$

The resistance of the membrane is the most complex of the four terms because it varies according to (1) the size and lipid-solubility of the molecular species that is diffusing and (2) the composition of the lipid layer across which it is diffusing. The principles of Fick's Law are illustrated in Figure 5-15 ■.

Lipophobic molecules cannot enter a cell by simple diffusion, and cells must make special provisions for their transport. Smaller lipophobic molecules and ions cross cell membranes through protein carriers. Ions also cross membranes through open water-filled channels. However, the diffusion of ions through these channels is affected by electric gradients as well as concentration gradients; therefore, ion movement cannot always be predicted by using Fick's Law and the diffusion rules in

---

TABLE 5-2 Rules for Simple Diffusion

1. Molecules diffuse from an area of higher concentration to an area of lower concentration.
2. Diffusion is a passive process that does not require an outside energy source. It uses only the kinetic energy of molecular movement.
3. There will be net movement of molecules until the concentration is equal. Molecular movement continues, however, so this is called a *dynamic equilibrium*.
4. Diffusion is rapid over short distances but much slower over long distances.
5. Diffusion is directly related to temperature. At higher temperatures, molecules move faster and diffusion is more rapid.
6. Diffusion is inversely related to molecular size. The larger a molecule is, the slower it will diffuse.
7. The rate of diffusion through a cell membrane depends on the ability of the molecule to dissolve in the lipid layer of the membrane. Lipid-soluble (lipophilic) molecules can diffuse across the membrane by simple diffusion.
8. The rate of diffusion through a membrane is directly proportional to the surface area of the membrane. The larger the surface area, the more molecules can diffuse across per unit time.
9. The rate of diffusion across a membrane is inversely proportional to the thickness of the membrane. The thicker the membrane, the slower diffusion will take place.

Vesicular Transport Across Membranes

Protein-mediated transfer of molecules across cell membranes is limited to small molecules that can interact with a protein transporter. Many macromolecules are too large to enter or leave cells through membrane transporters. Instead, they move in and out of the cell with the aid of bubblelike vesicles created from the cell membrane. There are two basic mechanisms by which large molecules are brought into the cell: phagocytosis and endocytosis. Phagocytosis formerly was considered a type of endocytosis, but as scientists learned more about the mechanisms behind the two processes, they decided that phagocytosis was fundamentally different. Material leaves cells by the process known as exocytosis, similar to endocytosis in reverse.

Phagocytosis If you studied *Amoeba* in your biology laboratory, you may have watched these one-cell creatures ingest their food by surrounding it and enclosing it into a vesicle that is brought into the cytoplasm. **Phagocytosis** is the process by which a cell engulfs a particle into a vesicle [*phagein,* to eat + *cyte,* cell + *-sis,* process]. In humans, this process is limited to certain types of white blood cells called **phagocytes** that specialize in "eating" bacteria and other foreign particles (Fig. 5-23 ■). When a phagocyte encounters a bacterium, the bacterium binds to the phagocyte membrane ① . The phagocyte then uses the fibers of its cytoskeleton to push its cell membrane out around the particle so that the bacterium becomes completely engulfed ②. Once the bacterium is surrounded, the arms of the membrane fuse, leaving the bacterium contained within a membrane-bound **vesicle** ③. The vesicle is pinched off from the cell membrane and moved to the interior ④, where it fuses with a lysosome (∞ p. 51), so that digestive enzymes in the lysosome can destroy the bacterium ⑤. Phagocytosis requires energy from ATP for the movement of the cytoskeleton and for the intracellular transport of the vesicles.

Endocytosis Another means by which molecules or particles can move into cells is called **endocytosis.** It differs from phagocytosis in that the membrane surface indents rather than pushes out, and the vesicles formed are much smaller. Endocytosis is an active process that requires energy input from ATP. It does not involve the cytoskeleton in the same way that phagocytosis does. Endocytosis can be unselective, allowing extracellular fluid to enter the cell (a process called **pinocytosis** [*pino-,* drink]) or it can be highly selective, allowing only specific molecules to enter the cell in **receptor-mediated endocytosis** (Fig. 5-24 ■).

In the first step of receptor-mediated endocytosis, membrane proteins recognize and bind to the substances that will be brought into the cell, the ligands ① .

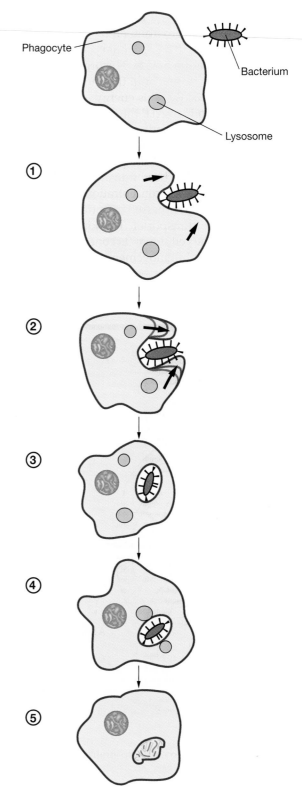

■ **Figure 5-23 Phagocytosis** (1) The phagocytic white blood cell encounters a bacterium that binds to the cell membrane. (2) The phagocyte uses its cytoskeleton to push its cell membrane around the bacterium, creating a large membrane-bound vesicle. (3) The vesicle containing the bacterium separates from the cell membrane and moves into the cytoplasm. (4) The vesicle fuses with lysosomes containing digestive enzymes. (5) The bacterium is killed and digested within the vesicle.

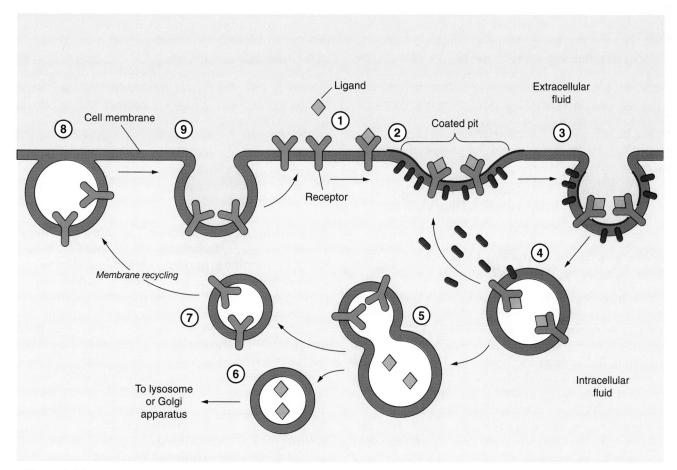

A receptor-ligand complex migrates along the cell surface until it encounters a **coated pit,** a region where the cytoplasmic side of the cell membrane has high concentrations of protein coating ②. The coated pits are the site of endocytosis. Once the receptor-ligand complex is in the coated pit, the membrane draws inward, or invaginates, and pinches off from the cell membrane as a vesicle ③. The protein coat is released and recycles back to the membrane ④. The vesicle moves into the cytoplasm and the ligand is released from the receptor ⑤. Once in the cytoplasm, the vesicle may fuse with a lysosome or transfer its contents to the Golgi apparatus ⑥. Sometimes the section of vesicle membrane with the receptors splits off and the receptors are inserted back into the cell membrane in the process known as **membrane recycling** ⑦-⑨.

Receptor-mediated endocytosis is used to transport a number of substances into the cell including protein hormones, growth factors, antibodies, and plasma proteins

that serve as carriers for iron, vitamins, and cholesterol. Receptor-mediated movement of cholesterol from the blood into cells plays an important part in total cholesterol metabolism.

Exocytosis   **Exocytosis** is the means by which cells secrete large impermeable molecules. In this process, intracellular vesicles move to the cell membrane, fuse with it (Figure 5-24 ■, step 8), and then open to expose their contents to the extracellular fluid (step 9). Exocytosis, like endocytosis, requires energy in the form of ATP.

Cells use exocytosis to export large lipophobic molecules, such as proteins synthesized within the cell, and to get rid of wastes left in lysosomes from intracellular digestion. Exocytosis takes place continually in some cells. For example, goblet cells (∞ p. 60) in the intestine continually release mucus by exocytosis, and fibroblasts

■ Figure 5-24 **Receptor-mediated endocytosis and exocytosis** (1) Protein receptors on the cell membrane bind to their specific ligand. (2) The receptor-ligand complex migrates along the surface of the cell membrane to a region known as a coated pit. The cytoplasmic side of the cell membrane in the coated pit is lined with a protein coating. (3) The coated pit invaginates to create a small membrane-bound vesicle. (4) The vesicle separates from the cell membrane. The protein coat molecules separate from the vesicle and return to the cell membrane. (5) The vesicles move into the cell and the receptors and ligands separate. Sometimes they form two distinct vesicles, one containing the ligands and one with the receptors still in the membrane. (6) The vesicles with the ligands move into the cytoplasm where they may be processed through lysosomes or the Golgi apparatus. (7) The secretory vesicles with the receptors move back to the cell membrane and fuse with it. (8) The fused section opens and the vesicle membrane again becomes part of the cell membrane (9). The fusion and opening of secretory vesicles with the cell membrane is known as exocytosis.

§ **LDL—The Lethal Lipoprotein**   Because cholesterol is an insoluble nonpolar molecule, it must be bound to a lipoprotein carrier molecule for transport in the blood. There are several forms of the carrier, the most common of which is **low-density lipoprotein** (LDL). Cholesterol-LDL is taken into cells by endocytosis when LDL receptors on the cell surface bind to the protein portion of the molecule. People who inherit a genetic defect that decreases the number of LDL receptors on their cell membranes cannot transport cholesterol normally into their cells. As a result, LDL-cholesterol remains in the plasma. Abnormally high blood levels of LDL-cholesterol predispose these people to the development of **atherosclerosis,** commonly known as hardening of the arteries [*atheroma-*, a tumor + *skleros*, hard + *-sis*, condition].

in connective tissue release collagen (∞ p. 62). In other cells, molecules are stored in secretory vesicles in the cytoplasm and released on demand. Exocytosis is also the means by which membrane proteins can be inserted into the cell membrane, as shown in Figure 5-24 ■.

Exocytosis in many cells is regulated by signals coming from outside the cell. One common signal is a ligand that increases the concentration of Ca^{2+} within the cell. You will encounter many examples of regulated exocytosis in your study of physiology.

✓ What is the difference between phagocytosis and endocytosis?

✓ If cholesterol is lipid-soluble, what keeps it from simply diffusing from the blood into the cells?

Molecules Move Across Epithelia Using Passive and Active Transport

The transport processes described in the previous sections all deal with the movement of molecules across a single membrane, that of the cell. But in the beginning of the chapter it was pointed out that all molecules entering and leaving the body across an epithelium must cross *two* cell membranes. Molecules cross the first membrane when they move into an epithelial cell from the external environment, and the second when they leave the epithelial cell to enter the extracellular fluid. Movement across epithelial cells uses a combination of active and passive transport.

The transporting epithelia of the intestine and kidney are specialized to transport molecules into and out of the body. Recall that epithelial cells are connected to each other by adhesive junctions and tight junctions (∞ p. 54). These act as a barrier to minimize the unregulated diffusion of material through the space between the cells. They also mark the division of the cell membrane into two regions. The surface of the epithelial cell that faces the lumen of an organ is called the **apical** (or mucosal)

surface (Fig. 5-25 ■); it is often folded into microvilli that increase its surface area. The sides of the cell below the tight junctions and the surface of the cell membrane in contact with the extracellular fluid are collectively referred to as the **basolateral** (or serosal) surface.

Transporting epithelial cells are said to be *polarized* because their apical and basolateral membranes have very different properties; this is due to the uneven distribution of membrane proteins. Certain proteins, such as the Na^+/K^+-ATPase, are found only on the basolateral membrane, whereas others, like the Na^+-glucose symporter, are found only on the apical membrane. This polarized distribution of transporters results in the one-way movement of certain molecules across the epithelium. The cells of transporting epithelia also have the ability to insert transport proteins into the membrane and withdraw them, selectively altering the permeability of the epithelium. Transporters that are pulled out of the membrane may be destroyed in lysosomes or they may be stored in vesicles inside the cell, ready to be reinserted into the membrane in response to a signal (another example of membrane recycling).

The transepithelial movement of molecules requires that the molecules first enter the epithelial cell and then leave it to enter the extracellular fluid. If the molecule can be transported through a protein channel or on transport carriers, then the two-step process usually has one uphill step that requires energy and one downhill step in which the molecule moves passively down its concentration gradient. Molecules that are too large to be moved by membrane proteins can be transported across the cell in vesicles.

Transepithelial transport using membrane proteins The movement of glucose from the lumen of the kidney tubule or intestine to the extracellular fluid is an important example of directional movement across a transporting epithelium. Transepithelial movement of glucose involves three different transport systems: the secondary active transport of glucose with Na^+ from the lumen into the cell at the apical membrane, followed by the active transport of Na^+ and the facilitated diffusion of glucose out of the cell and into the extracellular fluid on the basolateral side. Figure 5-26 ■ shows the process in detail. The glucose concentration in the transporting epithelium cell is higher than the glucose concentration in either the extracellular fluid or the lumen of the intestine. To move glucose from the lumen into the cell is an uphill process and requires the input of energy—in this case, the energy stored in the Na^+ concentration gradient. Sodium ions in the lumen bind to the symporter, as previously described, and bring glucose with them into the cell. The energy needed to move glucose against its concentration gradient comes from the kinetic energy of Na^+ moving down its gradient.

Once glucose is in the cell, it leaves by moving down its concentration gradient through the facilitated diffu-

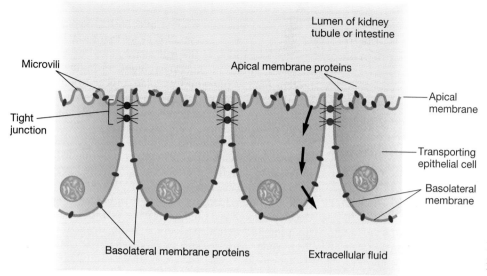

Microvilli

Tight junction

Lumen of kidney tubule or intestine

Apical membrane proteins

Apical membrane

Transporting epithelial cell

Basolateral membrane

Basolateral membrane proteins

Extracellular fluid

■ **Figure 5-25 Polarized cells of transporting epithelia** The cells of transporting epithelia are polarized into an apical membrane that faces the lumen of the organ lined by the epithelium and a basolateral membrane that faces the extracellular fluid. These two surfaces have different proteins, and consequently, they have very different transporting properties. The polarization of these cells allows directional transport of material from lumen to extracellular fluid or from the extracellular fluid into the lumen.

sion carrier in the basolateral membrane. The Na^+ is pumped out of the cell on the basolateral side by means of the Na^+/K^+-ATPase. The removal of Na^+ is an essential step if glucose is to continue to be absorbed, because the Na^+-glucose symporter depends on low intracellular concentrations of Na^+. If the basolateral Na^+/K^+-ATPase is poisoned, Na^+ that enters the cell will not be pumped out. The Na^+ concentration inside the cell then increases until it is equal to that outside the cell. Without a sodium gradient, there is no energy source to run the

Na^+-glucose symporter, so the movement of glucose across the epithelium stops.

Ions and amino acids can also be absorbed by secondary active transport, using a different set of transporters. For example, the apical membrane contains a $Na^+/K^+/2\,Cl^-$ symporter that brings K^+ into the cell from the lumen, *against* its concentration gradient, as Na^+ and Cl^- move in *down* their gradient. Because the K^+ concentration inside the cell is higher than in the extracellular fluid, the K^+ brought in by cotransport can

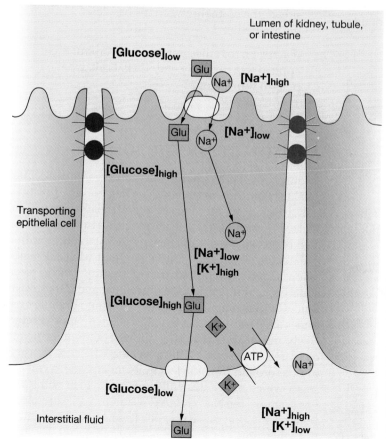

Lumen of kidney, tubule, or intestine

[Glucose]low   Glu   Na+   [Na+]high

Glu   Na+   [Na+]low

[Glucose]high

Na+

[Na+]low
[K+]high

[Glucose]high   Glu

K+

ATP   Na+

K+

[Glucose]low

Transporting epithelial cell

Interstitial fluid

Glu

[Na+]high
[K+]low

■ **Figure 5-26 Transport of glucose across an epithelial cell** The transport of glucose across an epithelial cell is an example of how multiple transport systems work together. Sodium and glucose enter the cell on the apical side using the Na^+/glucose symporter. The Na^+ is pumped out of the cell on the basolateral side by Na^+/K^+-ATPase using energy from ATP, while the glucose diffuses out on the basolateral side by a facilitated diffusion carrier. The apical uptake of glucose depends upon the energy stored in the Na^+ gradient. If the Na^+/K^+-ATPase ceases to function, the Na^+ entering the cell is not pumped out, the Na^+ gradient disappears, and uptake of glucose ceases.

move out of the cell on the basolateral side through open K^+ leak channels. The Na^+ leaves by the Na^+/K^+-ATPase. By this simple mechanism the body is able to absorb Na^+ and K^+ at the same time from the lumen of the intestine or the kidney.

Transcytosis and vesicular transport

Some molecules that are too large to cross membranes on protein carriers are brought into the body by **transcytosis,** a combination of endocytosis, vesicular transport across the cell, and exocytosis (Fig. 5-27 ■). In this process, the particle is absorbed on one side of the cell using pinocytosis, receptor-mediated endocytosis, or potocytosis (see Focus Feature). The vesicle that results attaches to microtubules in the cytoskeleton and is transported across the cell to the opposite side, where its contents are expelled into the extracellular fluid by exocytosis. The movement of the vesicles across the cell is known as *vesicular transport.*

Transcytosis makes it possible for large proteins to move across an epithelium and remain intact. It is the means by which infants absorb maternal antibodies in breast milk: The antibodies are absorbed on the apical surface of the infant's intestinal epithelium and then released into the extracellular fluid. Transcytosis also makes it possible for protein hormones to move across the capillary epithelium. Plasma containing the proteins is captured in vesicles and released into the interstitial fluid on the other side of the epithelium.

This section has examined how solutes are moved into the body and into and out of cells. The last section of this

Potocytosis and Caveolae Caveolae ("little caves") are small vesicles that form as invaginations of the cell membrane. These vesicles contain a variety of membrane proteins, including a unique protein named *caveolin.* Caveolae appear to have three functions: the concentration and internalization of small molecules and ions, transcytosis of macromolecules across capillary endothelium, and participation in cell-to-cell signaling. The concentration and uptake of small molecules by caveolae has been termed *potocytosis* to distinguish it from the receptor-mediated endocytosis in coated pits that is used for larger molecules. Caveolar transcytosis of macromolecules may turn out to play a significant role in the development of certain diseases. For example, cholesterol-lipoproteins are transported across blood vessel endothelium using caveolae. This transport may contribute to the development of atherosclerosis, the accumulation of cholesterol in blood vessels that blocks blood flow and causes heart attacks.

chapter considers how the distribution of these solutes in the various body compartments affects the volume and concentration of fluid in the body compartments.

✓ Ouabain is a poison that selectively binds to the Na^+/K^+-ATPase and inhibits it. Ouabain cannot pass through cell membranes. What would happen to the transepithelial glucose transport shown in Figure 5-26 ■ if ouabain were applied to the apical side of the epithelium? To the basolateral side of the epithelium?

✓ If a poison that disassembles microtubules is applied to a capillary endothelial cell, what happens to transcytosis?

DISTRIBUTION OF WATER AND SOLUTES IN THE BODY

Solutes in the body, such as ions and organic molecules, will distribute themselves according to whether they can cross cell membranes or the capillary epithelium. Molecules such as water that can diffuse through the cell membrane will do so until they reach equal concentration in the extracellular and intracellular fluid. Molecules such as ions that require protein transporters may never reach equal concentrations in the extracellular and intracellular fluid. Concentration differences in ions and organic molecules arise from the fact that the cell membrane is **selectively permeable;** that is, it allows some molecules to cross freely but restricts the movement of others. By varying the protein transporters in its membrane, a cell can select which molecules will enter and which will leave. For example, the Na^+/K^+-ATPase allows cells to concentrate K^+ in the cytoplasm while keeping Na^+ concentrations there low.

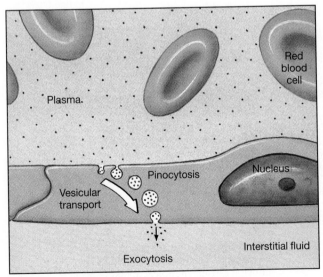

■ **Figure 5-27 Transcytosis across an epithelium**
Transcytosis is the means by which large molecules can be transported intact across an epithelium. In this example, plasma in a capillary is taken into the cell in vesicles by pinocytosis. The vesicles move across a cell in the capillary endothelium by using the cytoskeleton. Their contents are then released into the interstitial fluid by exocytosis.

TABLE 5-5	Comparing Osmolarities		
Solution A = 1 OsM Glucose	**Solution B = 2 OsM Glucose**	**Solution C = 1 OsM NaCl**	
A is hyposmotic to B	B is hyperosmotic to A	C is isosmotic to A	
A is isosmotic to C	B is hyperosmotic to C	C is hyposmotic to B	

well as their concentration. By nature of the particles, we mean whether the solute particles can cross the cell membrane. If the solute particle (ion or molecule) can enter the cell, we call it a **penetrating solute.** Particles that cannot cross the cell membrane are called **nonpenetrating solutes.** The most important nonpenetrating solute in physiology is NaCl. If a cell is placed in a solution of NaCl, the Na^+ and Cl^- will not cross the membrane into the cell. (In reality, a few Na^+ ions may leak across, but they are immediately transported back to the extracellular fluid by the Na^+/K^+-ATPase). NaCl is therefore considered a non-penetrating solute. If a saline solution of water and NaCl is put into the plasma, the Na^+ and Cl^- will stay in the extracellular fluid because they cannot enter the cells.

Tonicity always compares a cell and a solution, and it describes the behavior of the cell when it is placed in the solution. Tonicity has no units; you cannot say that a solution has a tonicity of X. You characterize the tonicity of the solution *relative to a particular cell.* Notice that tonicity SAYS NOTHING about the osmolarity of the solution. This is where tonicity becomes tricky. Suppose you know the composition and osmolarity of a solution. How can you figure out the tonicity of the solution without actually putting a cell in it? The key lies in knowing the relative concentrations of nonpenetrating solutes in the cell and in the solution.

To determine tonicity of a solution relative to a cell, you must consider *the relative concentrations of nonpenetrating solutes in the solution and in the cell.* If the cell has a higher concentration of nonpenetrating solutes than the solution, there will be net movement of water into the cell; the cell swells, and so the solution is hypotonic. If the cell has a lower concentration of nonpenetrating solutes, there will be net movement of water out of the cell; the cell shrinks, and so the solution is hypertonic. If the concentrations of nonpenetrating solutes are the same in the cell and the solution, there is no net movement of water, and so the solution is isotonic. The starting *osmolarities* of the cell and solution cannot be used to determine whether the solution is hypotonic, hypertonic, or isotonic. The presence of *penetrating solutes*

must be *ignored* when determining tonicity, because these solutes move freely into the cell as if the cell membrane did not exist. To understand why this is true, look at the following example.

In Figure 5-30a ■, the "cell" of compartment B contains six particles of nonpenetrating solute in one liter of solution. The solution in A also contains six particles of solute per liter: three nonpenetrating particles and three penetrating particles. Because A and B have the same concentrations (six particles per liter), they are isosmotic solutions.

To determine the tonicity of solution A, look at only the nonpenetrating solutes in the two compartments. Compartment A has three nonpenetrating particles per liter, whereas compartment B has six nonpenetrating particles per liter. By the rule stated above, this means that water will move from A into B. Since we have made B the "cell" in this example, the net movement of water into B will increase its volume, so solution A is hypotonic to cell B.

To understand why water moves into B as the system goes to equilibrium, ignore the penetrating solutes at first and focus on the net movement of water *based only on the concentration of nonpenetrating particles.* Because compartment B has twice as many nonpenetrating particles as compartment A, water will diffuse from A to B until the concentrations of the nonpenetrating solutes are equal in the two compartments (Fig. 5-30b ■). In this example, water will move until B has twice as much water as A, equalizing the concentrations of the nonpenetrating solutes. Once water has equilibrated according to the concentration of nonpenetrating solutes, mentally add back the penetrating solutes. The system is now out of *osmotic* balance, because compartment A has three additional penetrating solute particles. But penetrating solutes will cross the dividing membrane, obeying the laws of diffusion. Because compartment A has three penetrating particles and compartment B has none, a concentration gradient exists for the penetrating particles. They will diffuse from A into B until the concentration of penetrating solute particles in the compartments is equal. In this case, when two of the three penetrating particles

TABLE 5-6	Tonicity of Solutions	
Solution	Cell Behavior When Placed in the Solution	Description of the Solution Relative to the Cell
A	Cell swells	Solution A is hypotonic
B	Cell doesn't change size	Solution B is isotonic
C	Cell shrinks	Solution C is hypertonic

6

Communication, Integration, and Homeostasis

Complex living organisms are collections of molecules arranged into organelles, cells, tissues, and organs. The sheer size of the human body presents certain logistical problems because individual cells may be within micrometers of each other or meters apart. Diffusion is adequate for communication between neighboring cells, but as you learned in the last chapter, diffusion is ineffective over long distances. Multicellular organisms require additional means by which signals can move from one part of the body to another.

To meet the challenge of coordinating body functions, various anatomical and physiological mechanisms evolved. Blood flowing through our efficient circulatory system makes a complete circuit about once a minute, moving material from one part of the body to another. The nervous system takes care of more rapid communication, with electrical signals traveling as fast as 120 m/sec (268 mi/hr). The combination of simple diffusion across small distances, widespread distribution of materials through the circulatory system, and rapid, specific delivery of messages by the nervous system enables each cell in the body to communicate with most other cells.

The first section of this chapter examines the mechanisms of cell-to-cell communication. The second half of the chapter looks at why efficient cell-to-cell communication is important to the body as a whole. When change occurs as the result of either an external or an internal disturbance, the cells of the body must react in a unified fashion to return the body to its normal state. Without adequate communication, a coordinated response to change is impossible. The process of maintaining a relatively stable internal environment is called **homeostasis** (∞ p. 6). The control of homeostasis resides primarily in the nervous and endocrine systems, with their combination of chemical and electrical signals. When homeostatic mechanisms fail to maintain the normal internal environment, disease and illness result.

Problem

Diabetes Mellitus

It is 8:00 A.M., and Marvin Garcia, age 20, is hungry. As part of his routine physical examination before school starts, he is having a fasting blood glucose test. In this test, blood is drawn after an overnight fast, and the glucose concentration in the blood is measured. Marvin knows that he is in good physical condition, so he doesn't worry about the results. He is surprised, then, when the nurse practitioner at his family physician's office calls two days later. "Your fasting blood sugar is a bit elevated, Marvin. It is 150 milligrams per deciliter, and normal is 110 or less. Does anyone in your family have diabetes?" "Well, yeah—my dad has it. What exactly is diabetes?"

continued on page 151

CELL-TO-CELL COMMUNICATION

It has been estimated that the human body is composed of some 75 trillion cells. Those cells face a daunting task: to communicate with each other in a manner that is rapid and that conveys a tremendous amount of information. It is somewhat surprising then to find that there are only two basic types of physiological signals: chemical and electrical. Electrical signals are restricted to nerve and muscle cells and only rarely transfer directly from cell to cell. Chemical signals, on the other hand, are secreted by all cells, and these signals form the basis for most communication within the body.

There are three basic methods of cell-to-cell communication: (1) direct cytoplasmic transfer of electrical and chemical signals through gap junctions that connect adjacent cells; (2) local chemical communication through paracrines, autocrines, or neuromodulators; and (3) long-distance communication that uses a combination of electrical and chemical signals.

A given molecule can act as a signal through more than one method. For example, a molecule can be both a paracrine and a hormone or both a neurotransmitter and a hormone. The classification of a chemical signal depends on the location of the cells that respond to the signal. The cells that receive the signal are called **target cells,** or **targets** for short.

Gap Junctions Transfer Chemical and Electrical Signals Directly between Cells

The simplest form of cell-to-cell communication is the direct transfer of electrical and chemical signals through **gap junctions,** protein channels that connect to create cytoplasmic bridges between two adjacent cells (Fig. 6-1a ■) (∞ p. 54). A gap junction forms when opposing

membrane-spanning proteins called **connexins** on two adjacent cells unite. The connexins create a protein channel that is capable of opening and closing. In the open state, ions and small molecules such as amino acids, ATP, and cyclic AMP pass directly from the cytoplasm of one cell to the cytoplasm of the next. As with other membrane channels, larger molecules are excluded. When the channels are open, the connected cells function like a single cell with multiple nuclei (a *syncytium*).

In humans, gap junctions connect the muscle cells of the heart. Electrical signals pass rapidly from cell to cell through the gap junctions, triggering a wave of muscle contraction that becomes the heartbeat. Smooth muscle cells in the intestine are also connected by gap junctions, and the electrical signals that travel through them coordinate muscle contractions that sweep along the intestine. Because electrical signals cannot jump from cell to cell, the cytoplasmic bridges created by gap junctions are the only means by which electrical signals can pass *directly* between cells.

Paracrines and Autocrines Are Chemical Signals Distributed by Diffusion

Local communication is accomplished by autocrines and paracrines. A chemical that is secreted by a cell to act on cells in its immediate vicinity is called a **paracrine** [*para-,* beside + *krinein,* to secrete]. If the signal molecule acts on the cell that secreted it, the chemical is called an **autocrine** [*auto-,* self]. In some cases a chemical may act both as an autocrine and as a paracrine.

Paracrines and autocrines reach their target cells by diffusion through the interstitial fluid (Fig. 6-1b ■). Because distance is a limiting factor for diffusion, the effective range of the chemical signal is restricted to adjacent cells. A good example of a paracrine is **histamine,** a chemical released from damaged cells. When you scratch yourself with a pin, the red, raised *wheal* that results is due in part to the local release of histamine from the injured tissue. The histamine acts as a paracrine, diffusing over to the capillaries in the immediate area of the injury and making them more permeable. Fluid leaves the blood vessels and collects in the interstitial space, causing swelling around the area of injury.

There are several important groups of molecules that act as paracrines. **Neuromodulators** are paracrines and autocrines secreted by neurons. **Cytokines** are regulatory molecules that usually act close to the site where they are secreted. Cytokines are discussed further below.

Long-Distance Communication Is Carried Out by Electrical Signals, Hormones, and Neurohormones

All cells in the body can release paracrines, but long-distance communication between cells is the function of the nervous and endocrine systems. The endocrine system communicates with **hormones** [*hormon,* to excite], chemical signals that are secreted into the blood and distrib-

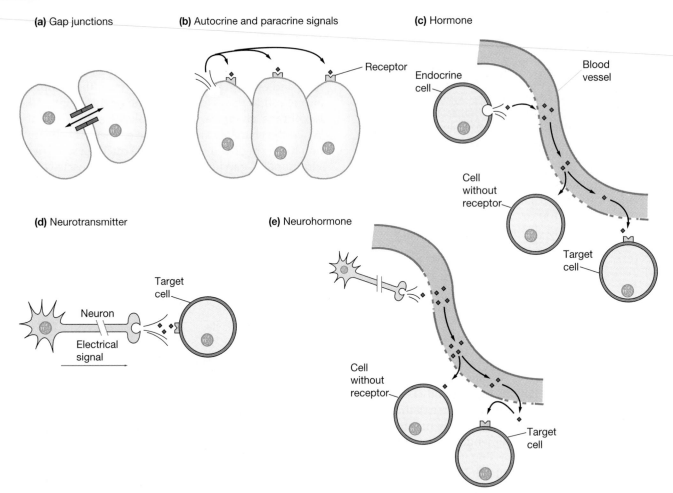

(a) Gap junctions

(b) Autocrine and paracrine signals

(c) Hormone

(d) Neurotransmitter

(e) Neurohormone

■ **Figure 6-1 Cell-to-cell communication** (a) Gap junctions form direct cytoplasmic connections between adjacent cells. (b) Paracrines are secreted by one cell and move by diffusion to act on adjacent cells. Autocrines act on the same cell that secreted them. (c) Hormones are secreted by endocrine glands or cells into the blood. Only cells with receptors for the hormone will be able to respond to the signal. (d) Electrical signals travel long distances along nerve cells. At the target, the nerve cell releases a chemical neurotransmitter that diffuses across a small gap to a receptor on the effector. (e) Neurohormones are chemicals released by a neuron into the blood. As with hormones, only cells with receptors will respond to the neurohormone.

uted all over the body by the circulation. Hormones come in contact with most cells of the body but only those cells that possess receptors for the hormone are target cells (Fig. 6-1c ■).

The nervous system uses a combination of chemical signals* and electrical signals to communicate. An electrical signal travels along a nerve cell (**neuron**) until it reaches the very end of the cell, where it is translated into a chemical signal. Chemical signals known as **neurotransmitters** diffuse from the neuron across a narrow extracellular space to the target cell (Fig. 6-1d ■). **Neurohormones** are released from nerve cells like neurotransmitters, but instead of going to an adjacent target, they are released into the blood like other hormones (Fig. 6-1e ■). The similarities between neurohormones and hormones secreted by the endocrine system blur the distinction between the nervous and endocrine systems and make them a contin-

uum rather than two distinct systems (see Fig. 6-20 ■, p. 167).

Cytokines Act as Both Local and Long-Distance Signals

One of the new topics in chemical communication is the group of chemicals known as **cytokines.** This term initially referred only to proteins that modulated immune responses, but in the past few years it has been broadened to include a variety of regulatory peptides. Cytokines are synthesized and secreted by all nucleated cells in response to a stimulus. They control cell development, differentiation, and the immune response. In development and differentiation, cytokines function as autocrines or paracrines. In stress and inflammation, cytokines act on targets relatively distant from the site of their production and are transported through the circulation as hormones are.

Cytokines differ from hormones in that they are not produced by specialized glands and they generally act on a broader spectrum of target cells. In addition, cytokines

---

*All molecules secreted by nerve cells (neuromodulators, neurotransmitters, and neurohormones) are known as **neurocrines.**

are made on demand, whereas most peptide hormones are made in advance and stored in the endocrine cell until needed. The distinction between cytokines and hormones is sometimes blurry. For example, erythropoietin, the molecule that controls synthesis of red blood cells, is considered to be both a hormone and a cytokine.

✓ Match the physiological signals on the right with the means by which they are transmitted.

1. autocrine
2. cytokine
3. gap junctions
4. hormone
5. neurohormone
6. neurotransmitter
7. paracrine

(a) electrical signals
(b) chemical signals
(c) both electrical and chemical signals

✓ Which methods of cell-to-cell communication on the list above involve transport through the circulatory system?

✓ A cat sees a mouse and pounces on it. Do you think the internal signal to pounce could have been transmitted by a paracrine? Give two reasons to explain why or why not.

RECEPTORS AND SIGNAL TRANSDUCTION

Chemical signals in the form of paracrines, autocrines, and hormones are released from cells into the extracellular compartment; this is not a very specific way to ensure that they find their target. Substances that travel through the blood reach nearly every cell in the body. And not all cells within diffusion distance of a secreting cell may be appropriate targets. So what determines the specificity of the response? Why do some cells respond to a given signal while others ignore it? How can one type of molecule trigger response A in tissue 1 and response B in tissue 2? The answer lies in the receptor proteins to which all chemical messengers bind in order to initiate a response.

Receptors as Well as Signal Molecules Determine the Cellular Response

The receptors for chemical signal molecules are found on the surface of the cell membrane or inside the cell. A cell will not respond to a chemical signal if it lacks the appropriate receptor for that signal. Different cells may have different receptors for the same molecule, giving rise to different responses when receptor and chemical signals bind. For example, the hormone *epinephrine* (*adrenaline*) will bind to both alpha (α) and beta (β) adrenergic receptors. (*Adrenergic* is the adjective relating to adrenaline). Blood vessels in the body may have either type of receptor. If epinephrine binds to an α-receptor, the blood vessel constricts (Fig. 6-2 ■). If it binds to a β-receptor, the blood vessel dilates. In this example, the response of the tissue to the messenger depends more on the *receptor* than it does on the chemical signal itself. For most signal molecules, *the cell's response to binding of the signal molecule is determined by the receptor, not by the ligand.* The signal molecule is needed to start the response, but the response varies according to the receptor. This is a very important concept to remember because it explains many phenomena that physiologists have observed but have not previously been able to explain.

Agonists and antagonists   Because activation of a signal pathway depends on the receptor and not on the ligand, all molecules that combine with a receptor and turn on its activity will elicit the same response. This property of receptors has led to the development of many drugs that have the same effect as a naturally occurring ligand. These "mimic molecules" are called **agonists.** Many agonist drugs have been developed that have the same effect as the normal signal molecule but are more resistant to

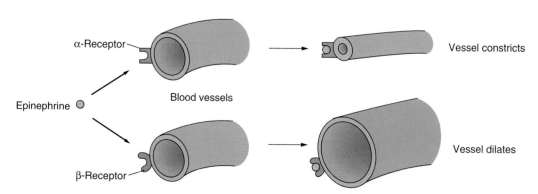

■ Figure 6-2 **Target response depends on the target receptor** The response of a cell depends on the receptor as much as on the signal ligand. In this example, one signal molecule, epinephrine, can either dilate or constrict blood vessels depending on the receptor found on the blood vessel.

enzyme breakdown and have a longer effect. One example is the family of estrogens (female sex hormones) that are found in birth control pills. These molecules have chemical groups added to protect the hormones from breakdown and extend the active life of a single dose.

On the other side of the coin, **antagonists** are molecules that bind to the receptor in such a way that they block the normal ligand from binding and turning the receptor on. As a result, the cell cannot respond. Antagonists have proven useful for treating many conditions. For example, antagonists to the sex hormone receptors for estrogens and androgens are used in the treatment of hormone-dependent cancers of the breast or prostate gland.

Up- and down-regulation of receptors enables cells to modulate cellular activity Because receptors are the key to cell-to-cell communication, it is not surprising to find that a single cell contains between 500 and 100,000 receptors on the surface of the cell membrane in addition to receptors found in the cytoplasm. In any given cell, the types of receptors may vary from one part of the cell membrane to another and also may change over time. Old receptors are withdrawn from the membrane by endocytosis and are broken down in lysosomes. New receptors are inserted into the membrane by exocytosis (∞ p. 127). Intracellular receptors are also made and broken down. This flexibility permits a cell to regulate its responses to chemical signals depending on the extracellular conditions and the internal needs of the cell.

If a signal molecule is present in the body in abnormally high concentrations for a sustained period of time and is creating an enhanced response, the target cell may attempt to bring its response back to normal by **down-regulation** of the receptors for the signal. Down-regulation takes two forms: Either the number of receptors decreases, or the binding affinity of the receptors for the ligand decreases. In both cases, the result is a lessened response of the target cell even though the concentration of the signal molecule remains high. Down-regulation is partially responsible for the development of *drug tolerance*, the condition in which the response to a given dose decreases despite continuous exposure to the drug. The development of tolerance to opiates such as morphine and codeine occurs when the receptors for these drugs down-regulate.

Up-regulation is the opposite of down-regulation. If the concentration of a ligand decreases, the target cell may insert more receptors into the cell membrane in an attempt to keep its response at a normal level.

Receptors and disease Many normal physiological events that we did not understand only a decade ago can now be explained in terms of receptors. Scientists are discovering that many disease processes are caused by changes in the number or structure of receptors (see Table 6-1 for some examples). Because receptors are proteins, a single change in the amino acid sequence can alter the shape of the receptor's binding site and destroy or modify its activity.

One example of a receptor-related disease is a rare form of *diabetes mellitus* ("sugar diabetes"). In this condition, the gene that controls a hormone receptor on the target cells mutates, leading to an abnormal receptor and abnormal cell response. In these patients, hormone levels in the blood may be normal but the target cell response is abnormal because of the receptors. In the most common form of diabetes, both the receptor and

---

TABLE 6-1 **Diseases or Conditions Caused by Changes in the Number or Structure of Receptors**

Genetically inherited abnormal receptors

Disease or Condition That Results	Receptor	Physiological Alteration
Congenital diabetes insipidus	Vasopressin receptor (X-linked defect)	Shortens half-life of the receptor
Familial hypercalcemia	Defective calcium-sensor in parathyroid gland	Fails to respond to increase in plasma Ca^{2+}
Retinitis pigmentosa	Rhodopsin receptor in retina of eye	Improper protein folding in rod cell discs

Drugs and toxins that affect signal pathways

Condition That Results	Drug or Toxin	Physiological Effect
Whooping cough	*Bordetella pertussis*	Blocks the inhibition of adenylate cyclase (i.e., keeps it active); link to symptoms unclear
Massive diarrhea	Cholera toxin	Blocks enzyme activity of G proteins; cell keeps making cAMP, and ions are secreted into lumen of intestine
Drug used to treat heart disease, respiratory illnesses, insomnia, convulsions	Forskolin (Indian herb, *Coleus forskohlii*)	Activates adenylate cyclase (converts ATP to cAMP)

The Role of Receptors in Diabetes Mellitus In normal people, the hormone insulin combines with insulin receptors on target cells to enhance glucose uptake and metabolism by the cells. In people with type II diabetes mellitus, also known as non-insulin-dependent diabetes, insulin binds to its receptor but the cell fails to respond normally to the signal. As a result, the enzymes and transporters involved in glucose metabolism become less active and the rate of glucose diffusion into the cell slows. Ultimately, glucose concentrations in the blood increase, a symptom characteristic of diabetes.

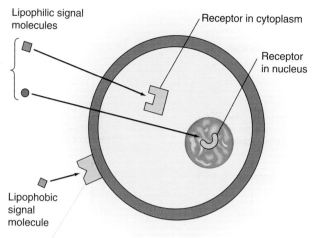

■ **Figure 6-3 Two classes of signal molecules** Lipophilic molecules can diffuse through the cell membrane and bind to receptors either in the cytoplasm or in the nucleus. However, lipophobic signal molecules cannot diffuse through the cell membrane. Their receptors are membrane proteins with binding sites on the extracellular surface.

the hormone level are normal but cellular response is still abnormal.

Signal Molecules Are the First Messengers in a Series of Events

The chemical signals released by cells are called the **first messengers** because they are responsible for starting the series of events that lead to a response. Chemically, signal molecules can be divided into two groups based on how they behave when they reach their target cells: (1) lipophilic signal molecules, which diffuse right through the cell membrane and enter the cytoplasm, and (2) lipophobic signal molecules, which are unable to cross the phospholipid bilayer of the cell membrane. Receptors for the two classes of signal molecules are located at different sites: Receptors for lipophilic molecules are found within the cell, whereas receptors for lipophobic molecules are found on the cell membrane surface.

Lipophilic messengers enter the cytoplasm and combine with receptors inside the cell Lipophilic messenger molecules such as the steroid hormones are able to diffuse through the cell membrane and reach the cytoplasm directly (Fig. 6-3 ■). As a result, the receptors for these molecules are usually found either in the cytoplasm or inside the nucleus. In a few cases, lipophilic messengers also have surface membrane receptors. The combination of steroid messenger and intracellular receptor initiates a response such as the production of new proteins.

Lipophobic messengers have surface membrane receptors The lipophobic messenger molecules are proteins and peptides, polar molecules that cannot enter the cell because they are unable to move through the lipid center of the cell membrane (Fig. 6-3 ■) (∞ p. 109). Because these signal molecules remain outside the cells, their receptors are membrane-spanning proteins with segments that extend out into the extracellular fluid. When the external signal ligand binds to the receptor, one of two events occurs: (1) If the receptor is a channel protein, binding of the ligand opens or closes the channel, or (2) information is transferred through the mem-

brane with the aid of membrane proteins, then transformed into an intracellular response. The transmission of information from one side of a membrane to the other using membrane proteins is the process known as **signal transduction**. A transducer is a device that converts a signal from one form into a different one [*trans*, across + *ducere*, to lead]. The transducer in a radio converts radio waves into sound waves (Fig. 6-4 ■). A biological transducer converts the message of an extracellular signal molecule into intracellular messages, which trigger a response. In biological systems, the signal is not only transformed but also amplified [*amplificare*, to make larger]. Peptide and protein hormones, paracrines, and cytokines act by binding to membrane receptors that carry out signal transduction.

continued from page 147

Later that day in the physician's office, the nurse practitioner explains diabetes to Marvin. Diabetes mellitus is a metabolic disease caused by a defect in one of the homeostatic pathways that regulate glucose metabolism in the body. Several forms of diabetes exist, and some can be inherited. One form, called insulin-dependent diabetes mellitus (IDDM), is caused by deficient production of insulin, a peptide hormone made in the pancreas. In another form, called non-insulin-dependent diabetes mellitus (NIDDM), insulin is present in normal or above normal levels. However, the insulin-sensitive cells of the body do not respond to the hormone.

Question 1: In which form of diabetes are the insulin receptors most likely to be up-regulated?

was lost, the original condition is restored. But if the fluid ingested is pure water without ions, the end result will be normal plasma volume but a drop in total plasma concentration. One problem was solved (volume), but another one was created (low plasma solute concentration) that now must be dealt with.

In situations in which the body cannot maintain parameters within their normal ranges, a disease state or **pathological** condition results [*pathos*, suffering]. Diseases generally can be divided into two groups according to their origin: those in which the problem arises from internal failure of some normal physiological process and those that originate from some outside source. Internal causes of disease include the abnormal growth of cells, which may cause cancer or benign tumors, the production of antibodies by the body against its own tissues (autoimmune diseases), and the premature death of cells or the failure of cell processes. Inherited disorders are also considered to have internal causes. External causes of disease include toxic chemicals, physical trauma, and foreign invaders such as viruses and bacteria. In both internally and externally caused diseases, when homeostasis is disturbed, the body attempts to compensate. If the compensation is successful, homeostasis is restored. If compensation fails, disease results (Fig. 6-11 ■). The functioning of the body in a disease state is known as **pathophysiology.** You will encounter examples of pathophysiology as we study the various systems of the body.

Homeostasis is a continuous process that involves monitoring and regulating multiple parameters and coordi-

continued from page 155

"Why is an elevated blood glucose concentration bad?" Marvin asks. "The elevated blood glucose itself is not bad," says the nurse practitioner. "But when it is high after an overnight fast, it suggests that there is something wrong with the way your body is regulating its glucose metabolism." Homeostasis of blood glucose levels is an essential function of the body because the brain and other nerve cells are unable to use any other kind of fuel. When a normal person absorbs a meal with carbohydrates, blood glucose levels increase and stimulate insulin release. Hours after the meal, when blood glucose levels fall as the cells metabolize glucose, another pancreatic hormone called glucagon is secreted. Glucagon increases blood glucose and helps keep the level within the homeostatic range.

Question 3: The homeostatic regulation of blood glucose levels by the hormones insulin and glucagon is an example of which of Cannon's postulates?

nating the responses in order to minimize the disturbance to the body's balance. The next section will examine the role of response loops and feedback loops in homeostasis.

CONTROL PATHWAYS: RESPONSE AND FEEDBACK LOOPS

Homeostasis May Be Maintained by Local or Long-Distance Pathways

The simplest control takes place strictly at the tissue or cell involved. In **local control,** a relatively isolated change occurs in the vicinity of a cell or tissue and evokes a paracrine or autocrine response (Fig. 6-12 ■). More complicated **reflex control pathways** respond to changes that are more widespread or systemic in nature. In a reflex control pathway, the decision that a response is needed is made away from the cell or tissue, and a chemical or electrical signal is sent to the cell or tissue to initiate a response (Fig. 6-12 ■). Long-distance reflex pathways are traditionally considered to involve two control systems, the nervous system and the endocrine system. However, cytokines are now considered to be the third control system. These proteins, secreted by all nucleated cells, can work by either local or reflex pathways depending on the circumstances. Acting as paracrines and autocrines, cytokines control the white blood cells that defend the body from outside invaders. During stress and systemic inflammatory responses, cytokines work together with the nervous and endocrine systems to integrate information from all over the body into coordinated reflex responses.

Local control   Paracrines and autocrines are responsible for the simplest control systems. In local control a cell or tissue senses a change in its immediate vicinity and responds. The response is restricted to the region

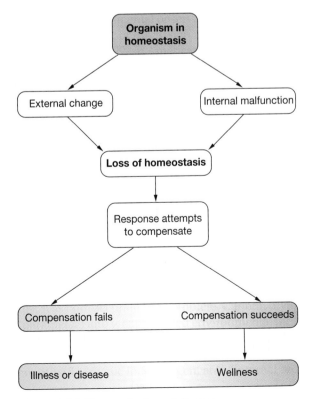

■ Figure 6-11 **Homeostasis and disease**

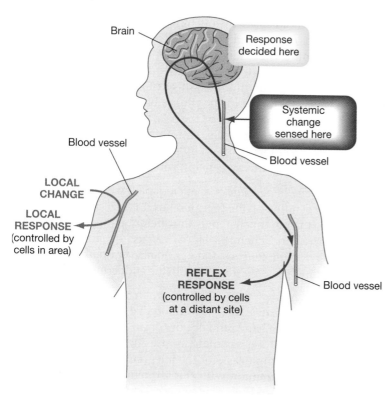

Brain

Response
decided here

Systemic
change
sensed here

Blood vessel

Blood vessel

**LOCAL
CHANGE**

**LOCAL
RESPONSE**
(controlled by
cells in area)

**REFLEX
RESPONSE**
(controlled by cells
at a distant site)

Blood vessel

■ Figure 6-12 **Comparison of local and reflex control** In local control, the response is initiated by cells in the vicinity of the change. In reflex control, a response is controlled by cells at a distant site. In the example shown, a change in systemic blood pressure is sensed in blood vessels in the neck. A signal is sent to the brain. The brain then sends a signal to create a response in blood vessels all over the body, represented here by a blood vessel in the arm.

where the change took place, hence the term *local control.* One example of this type of response occurs when oxygen concentration in a tissue drops. The cells lining the small blood vessels bringing blood to that area sense the drop in oxygen concentration and respond by secreting a paracrine. The paracrine relaxes muscles in the blood vessel wall, dilating the blood vessels and bringing more blood and oxygen to the area.

Some examples of paracrines involved in local control responses include carbon dioxide, metabolic products such as lactic acid, and histamine. One of the most recently discovered paracrines is **nitric oxide** (NO).

Reflex control   In a reflex pathway, control of the reaction lies outside the organ that carries out the response. In physiology, a reflex is any pathway that uses the nervous system or the endocrine system or both to receive input about a change, integrate the information, and react appropriately. Reflex pathways can be broken down into response loops and feedback loops. A **response loop** has three basic components: an input signal, integration of the signal, and an output signal. These three components can be broken down into the following sequence to form a pattern that is found with slight variations in all reflexes:

stimulus → receptor → afferent pathway →
integrating center →
efferent pathway → effector → response

The beginning of any homeostatic reflex pathway is (1) a **stimulus,** the disturbance or change that sets the path-

way in motion. The stimulus may be a change in temperature, oxygen content, blood pressure, or any one of myriad parameters. The stimulus is sensed by (2) a sensor, or **receptor,** that is continually monitoring its environment. When alerted to a change, the receptor sends out a signal. (3) The signal, or **afferent** (incoming) **pathway,** links the receptor to (4) an **integrating center.** The integrating center is the control center that evaluates the incoming signal, compares it with the **setpoint,** or desired value, and decides on an appropriate response. From the integrating center, (5) an outgoing signal, the **efferent** (outgoing) **pathway,** travels to an effector. (6) The **effector** is the cell or tissue that carries out (7) the appropriate **response** to bring the situation back to within normal limits.

Nitric Oxide   Nitric oxide, NO, is an uncharged soluble gas with a half-life of 2–30 seconds that is rapidly broken down in the body. This elusive signal molecule took years to identify because it does not linger once it initiates its second messenger cascade in the target cell. One important site of NO action is the blood vessels, where NO is produced in the endothelial lining of the vessel. The gas diffuses into adjacent smooth muscle cells, causing them to relax so that the vessel dilates. Nitric oxide also acts as a neurotransmitter in the nervous system. Large amounts of NO can be toxic to cells. Inappropriate production of NO has been suggested to play a role in certain conditions in which the body destroys its own cells.

Sensory Receptors The first step in a biological response loop is the receptor. NOTICE! This is a new and different use of the word *receptor*. Like many terms in physiology, *receptor* can have different meanings (Fig. 6-13 ■). The receptors of a biological reflex are not membrane proteins like those involved in signal transduction. Reflex receptors are specialized cells, parts of cells, or complex multicellular receptors such as the eye. There are many reflex receptors in the body, each located anatomically where it is in the best position to monitor the parameter it detects. The eyes, ears, and nose are specialized receptors that sense light, sound, motion, and odors. Your skin is covered with less complex receptors that sense touch, temperature, vibration, and pain. Other receptors are internal: receptors in the joints of the skeleton, pressure receptors in blood vessels, osmolarity receptors, and receptors for oxygen and carbon dioxide. Biological reflex receptors are divided into **central receptors,** located in or closely linked to the brain, and **peripheral receptors** that reside elsewhere in the body. All receptors have a **threshold,** a minimum stimulus that must be achieved in order to set the reflex response in motion. If a stimulus is below the threshold, no response loop will be initiated. You can demonstrate

threshold easily by touching the back of your hand with a sharp, pointed object such as a pin. If you touch the point to your skin lightly enough, you can see the contact between the point and your skin even though you do not feel anything. In this case, the stimulus (pressure from the point of the pin) is below threshold and the pressure receptors of the skin are not responding. As you press harder, the stimulus reaches threshold, and the receptors respond by sending a signal through the afferent pathway, causing you to feel the pin.

Afferent Pathway The afferent pathway in a reflex varies depending on the type of reflex. In a nervous reflex, such as the pin touch above, the afferent pathway is the electrical and chemical signal carried by a nerve cell. In endocrine reflexes, there is no afferent path because the stimulus comes directly into the endocrine cell.

Integrating Center The integrating center is the key to a reflex because it is the location that receives information about the change and decides on an appropriate response. If information is coming from a single stimulus, it is a relatively simple task for the integrating center to compare that information with the setpoint. Integrating centers really "earn their pay" on occasions when two or more

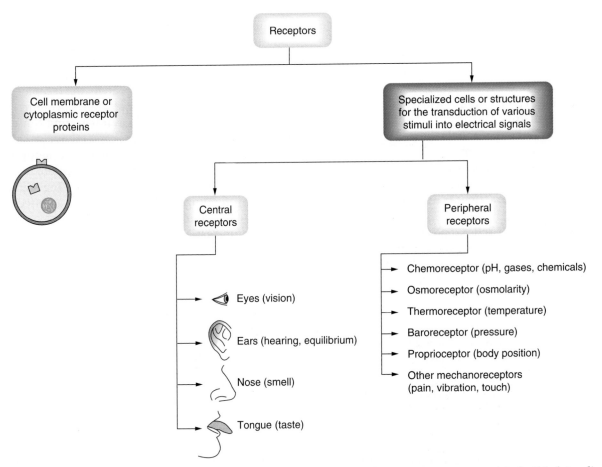

■ Figure 6-13 **Multiple meanings of the word *receptor*** In biology, the word *receptor* may mean a protein that binds to a ligand. It may also refer to sensory receptors, the specialized cells or structures for transduction of stimuli into electrical signals. Sensory receptors can be divided into peripheral receptors scattered throughout the body and central receptors located in or close to the brain.

conflicting signals come in from different receptors. The center must evaluate each signal on the basis of its strength and importance, and must come up with an appropriate response that integrates information from all the receptors. This is similar to the kind of decision making you must do when in one evening your parents want to take you to dinner, your friends are having a party, there is a television program you want to watch, *and* you have a major physiology test in three days. It is up to you to rank those items in order of importance and decide how you will act upon them.

In endocrine reflexes, the integrating center is the endocrine cell. In nervous reflexes, the integrating center lies within the *central nervous system*, which is composed of the brain and the spinal cord.

Efferent Pathway   Efferent pathways are relatively simple. In the nervous system, the efferent path is always the electrical and chemical signal transmitted by an efferent neuron. Because electrical signals traveling through the nervous system are identical, the distinguishing characteristic of the signal is the anatomical route taken by the nerve cell through which the signal goes. For example, the vagus nerve carries a nervous signal to the heart, and the phrenic nerve carries one to the diaphragm. Because the *nature* of the electrical message is always the same and there are relatively few types of neurotransmitters, nervous system efferent pathways are named using the anatomical name of the nerve that carries the signal.

In the case of a hormonal signal, the *anatomical* routing is always the same because all hormones travel in the blood to get to their target. Hormonal efferent pathways are distinguished by the chemical *nature* of the signal and are therefore named for the hormone that carries the message. For example, the efferent path for a reflex integrated through the pancreas will be either insulin or glucagon depending on the stimulus and the appropriate response.

Effectors   The **effectors** of reflex pathways are the cells or tissues that carry out the response. The targets of nervous reflexes are muscles, glands, and some adipose tissue. The targets of endocrine pathways are any cells that have the proper receptor for the hormone.

Responses   There are two levels of response that can be given for any reflex. One is the very specific cell or tissue response that results from the combination of the signal ligand with the receptor: opening a channel, initiating protein synthesis, or modifying enzyme activity. The more general systemic response describes what those specific events mean to the tissue or the organism as a whole. For example, when the hormone epinephrine combines with β-adrenergic receptors on the walls of certain blood vessels, the cellular response is relaxation of the smooth muscle. The systemic response to relaxation of the blood vessel wall is increased blood flow through the vessel.

Response Loops Begin with a Stimulus and End with a Response

Now that you have learned the parts of a reflex, we will apply that scheme to nonbiological and biological examples. A simple nonbiological analogy to a homeostatic reflex pathway is an aquarium heater, set to maintain water temperature at 30° C in a room whose temperature is 25° C (Fig. 6-14 ■). The temperature at which the heater is set is the setpoint for the parameter.

The components of the aquarium are listed in the left column below, with the corresponding part of the response loop in the right column.

Component	*Biological Reflex Equivalent*
1. thermometer	1. receptor
2. communication wire	2. afferent pathway
3. control box	3. integrating center
4. communication wire	4. efferent pathway
5. heater	5. effector

Assume that the aquarium is initially at room temperature and water temperature is 25° C. When you turn the heater on, you set the **response loop** in motion. The thermometer (receptor) in the water senses a temperature of 25° C. It sends information about the current temperature via a wire (afferent path) to the control box (integrating center). The control box evaluates the incoming temperature signal, compares it with the setpoint for the system (30°), and decides that a response is needed to bring the water temperature up to the setpoint. The control box sends a message via another wire (efferent path) to the heater (effector), which turns on and starts heating the water (the response). This sequence from stimulus to response is the response loop.

In physiological systems, the setpoint, or normal value for any given parameter, can vary from person to person, or even for the same individual over a period of time. Factors that influence an individual's setpoint for a given parameter include inheritance as well as the conditions to which the person has become accustomed. The adaptation of physiological processes to a given set of environmental conditions is known as *acclimatization* if it occurs naturally or *acclimation* if the process is induced artificially in a laboratory setting. Each winter, northerners go south in February, hoping to escape the bitter subzero temperatures and snows of the northern climes. As they walk around in 40° F weather in short-sleeve shirts, the southerners think they are crazy: The weather is cold! The locals are bundled up in coats and gloves. The difference in behavior is due to different temperature acclimatization, a difference in the setpoint for body temperature regulation that is a result of prior conditioning.

Physiological setpoints also vary within individuals in response to external cues such as the daily light-dark

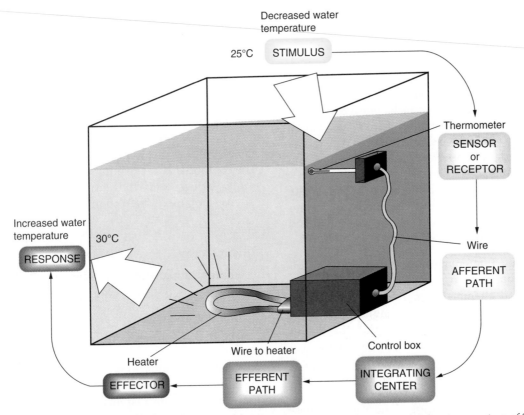

■ **Figure 6-14 Response and feedback loops** The control box of the aquarium is set to maintain a temperature of 30° ± 1° C in the water. The sensor (or receptor) sends a signal via the afferent path (a wire) to the control box that acts as the integrating center. The control box sends a message over the efferent path (another wire) to the effector, the heater. The heater turns on, completing the response loop. As the water heats, information about the water temperature is sent continuously to the control box. When the water temperature reaches the upper limit of the desired range, the control box turns off the heater. The increase in temperature that causes the response to cease is called negative feedback.

cycles and the seasons. These changing setpoints cause certain parameters to vary in predictable ways over a period of time, forming patterns of change known as *biorhythms* (see Biological Rhythms).

The response loop begins with a stimulus and ends with the response of the target cell. But it is only half of a reflex. The reflex is completed when the response becomes part of the stimulus and feeds back into the system, creating a feedback loop.

Feedback Loops Modulate the Response Loop

In the aquarium example above, the sensor sends a signal to the control box, telling it that the water is too cold. The response of the control box is to turn on the heater and warm up the water. But once the response starts, what keeps the heater from sending the temperature up to 50° C? The answer is a **feedback loop.** The response of the system, turning on the heater, alters the temperature of the water and changes the stimulus. The receptor monitoring the water sends a continuous signal back to the control box so that it receives constant input or feedback about the effect the heater is having. When the temperature warms up to the maximum acceptable, the control box shuts off the heater, thus completing the loop.

Negative feedback loops are homeostatic For most reflexes, feedback loops are homeostatic, that is, designed to keep the system at or near a setpoint so that the parameter will be relatively stable. How well an integrating center succeeds in maintaining stability depends on the sensitivity of the system, the range of normal values. In the case of our aquarium, the control box is programmed to have a **sensitivity** of plus or minus one degree Celsius. If the water temperature drops from 30° to 29.5°, it is still within the acceptable range and no

continued from page 158

Marvin is fascinated by the ability of the body to keep track of glucose. "How does the pancreas know which hormone to secrete?" Special cells in the pancreas called beta cells sense an increase in blood glucose concentrations after a meal, and they release insulin in response. Insulin then acts on many tissues of the body so that they take up and utilize glucose.

Question 4: In the insulin reflex that regulates blood glucose levels, what is the stimulus? the receptor? the integrating center? the efferent pathway? the effector(s)? the response(s)?

response is triggered. If the water temperature drops below 29° (30° − 1°), the control box turns the heater on (Fig. 6-15 ■). As the water heats up, the control box constantly receives information about the water temperature from the sensor. When the water reaches 31° (30° + 1°), the upper limit for the acceptable range, the feedback loop causes the control box to turn the heater off. The water then gradually cools off until the cycle starts all over again. The end result is a physiological variable that *oscillates* around the setpoint.

In biological systems, some receptors are more sensitive than others. For example, the sensors for osmolarity trigger reflexes to conserve water when blood concentration increases only 3% above normal, but the sensors for low oxygen in the blood will not respond until oxygen has decreased by 40%.

A pathway where the response opposes or removes the signal is known as **negative feedback.** Negative feedback loops stabilize the physiological variable that is being regulated. In the aquarium example, the heater warms up the water (the response) and removes the stimulus (low water temperature). With the loss of the stimulus for the pathway, the response loop shuts off. All homeostatic reflexes are controlled by negative feedback so that the parameter being controlled will stay within a normal range. Negative feedback loops can restore the normal state but cannot prevent the initial disturbance out of the normal range.

Positive feedback loops are not homeostatic
There are a few reflexes that are not homeostatic. In a **positive feedback loop,** the response *reinforces* the stimulus rather than decreasing it or removing it. In positive feedback, the response destabilizes the variable, triggering a vicious cycle of ever-increasing response and sending the system temporarily out of control (Fig. 6-16 ■). Because the response enhances the loop in positive feedback, these reflexes require some intervention or event outside the loop to stop them.

One example of a positive feedback loop involves digestive reflexes in the stomach (Fig. 6-17 ■). The presence of food in the stomach causes cells in the stomach wall to secrete hydrochloric acid (HCl) and an inactive enzyme called pepsinogen (the suffix -*ogen* indicates an inactive precursor of the root word). The HCl cleaves off part of the pepsinogen to make the molecule into the active enzyme **pepsin.** A positive feedback cycle is begun because the presence of pepsin in the stomach triggers more pepsinogen release, which HCl turns into more pepsin, and so on. The cycle will go on indefinitely unless another factor shuts it off. In this case, the outside factor is the removal of hydrochloric acid. As food moves out of the stomach into the small intestine, acid secretion stops. In the absence of HCl, pepsinogen is no longer converted to pepsin, and the vicious cycle of the positive feedback loop has been broken. Because positive feedback is not homeostatic, there are fewer examples of it than there are of negative feedback.

Feedforward Control Allows the Body to Anticipate Change and Maintain Stability

Negative feedback loops can stabilize a function and maintain it within a normal range but are unable to prevent the change that triggered the reflex in the first place. A few reflexes have evolved that allow the body to *predict* that a change is about to occur and start the response loop in anticipation of the change. These anticipatory responses are called **feedforward control.**

An easily understood physiological example of feedforward control is the reflex of salivation. The sight, smell, or even the thought of food is enough to start our mouths watering. The saliva is present in expectation of the food that will soon be eaten. This reflex extends even

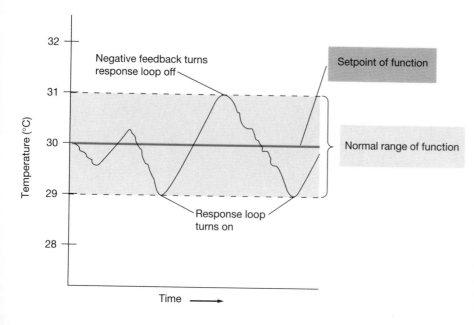

■ Figure 6-15 **Oscillation around the setpoint** Most functions that are controlled homeostatically have a setpoint, or normal value for that function. The reflexes that control the function will operate when the function gets outside a predetermined range. As a result, the function is not exactly at the setpoint all the time; instead, it oscillates around the setpoint.

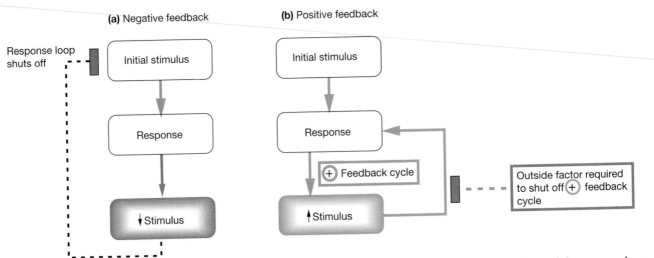

(a) Negative feedback

(b) Positive feedback

■ Figure 6-16 **Negative and positive feedback** (a) In negative feedback, the response offsets the stimulus and the response loop shuts off. (b) In positive feedback, the response sends a message that reinforces the response loop, sending the parameter under control farther away from the setpoint. The response will continue to increase until some outside intervention stops the reflex loop.

further, because the same stimuli can start the secretion of hydrochloric acid as the stomach anticipates food on the way. One of the most complex feedforward reflexes appears to be the body's response to exercise, to be discussed in Chapter 23.

Biological Rhythms Result from Changes in the Setpoint

In many reflex pathways, the triggers or stimuli are obviously related to the function of the reflex. In the aquarium example, a change in temperature is the trigger to ensure that temperature is maintained within the desired range. But this is not true of all reflexes. Many hormones, for example, are secreted continuously, with levels that rise and fall throughout the day. Most examples of these apparently spontaneous reflexes occur in a predictable manner and are often timed to coincide with a predictable environmental change such as light-dark cycles or the seasons. All animals exhibit some form of daily biological rhythm, called a **circadian rhythm** [*circa*, about + *dies*, a day]. Humans have circadian rhythms for many body functions, including blood pressure, body temperature, and metabolic processes. Body temperature peaks in the late afternoon and declines dramatically in the wee hours of the morning (Fig. 6-18a ■).

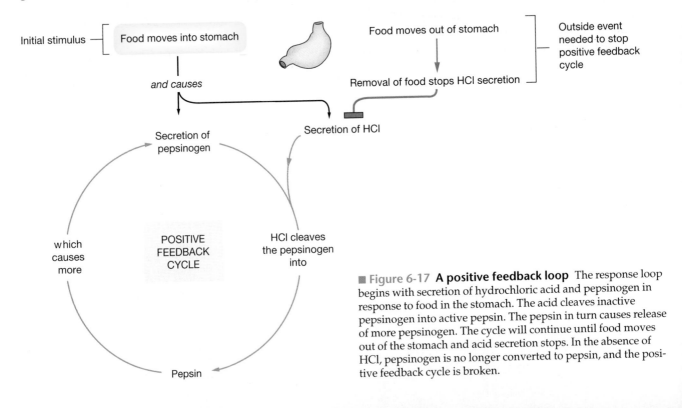

■ Figure 6-17 **A positive feedback loop** The response loop begins with secretion of hydrochloric acid and pepsinogen in response to food in the stomach. The acid cleaves inactive pepsinogen into active pepsin. The pepsin in turn causes release of more pepsinogen. The cycle will continue until food moves out of the stomach and acid secretion stops. In the absence of HCl, pepsinogen is no longer converted to pepsin, and the positive feedback cycle is broken.

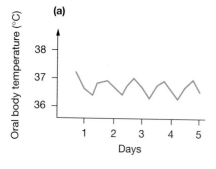

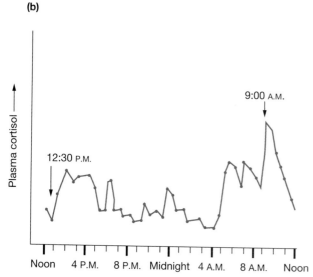

■ **Figure 6-18 Circadian rhythms** (a) Body temperature rises and falls in a regular pattern on a daily basis. (b) Blood levels of the hormone cortisol change during a 24-hour period. Cortisol levels in the blood vary significantly between 9:00 A.M. and 12:30 P.M.; therefore, analysis of hormone levels must take into account the time of day that a blood sample was drawn.

Have you ever been studying late at night and noticed that you feel cold? This is not because of a drop in environmental temperature but because your thermoregulatory reflex has turned down your internal thermostat.

Many hormones in humans are secreted so that their concentration in the blood fluctuates predictably through a 24-hour cycle as their setpoints change. Cortisol, growth hormone, and the sex hormones are among the most noted examples. If an abnormality in hormone secretion is suspected, it is important to know the time of day that the test for hormone level is made. A level that is normal in a morning sample may be abnormally high or low in an afternoon sample (Fig. 6-18b ■). One method used to avoid this error involves collecting all urine for a 24-hour period and getting an average value for the hormone metabolites in the sample.

What is the adaptive significance of having functions vary with a circadian rhythm? Our best answer is that biological rhythms create an anticipatory response to a predictable environmental variable. There are seasonal rhythms of reproduction in many nonmammalian vertebrates and invertebrates, timed so that the offspring

have food and other favorable conditions to maximize survival. Circadian rhythms cued by the light-dark cycle correspond to our rest-activity cycles. These rhythms allow our bodies to anticipate behavior and coordinate body processes accordingly. You may hear someone who is accustomed to eating dinner at 6 P.M. say that he cannot digest his food if he waits until 10 P.M. to eat because his digestive system has "shut down" in anticipation of going to bed.

One of the interesting correlations between circadian rhythms and behavior involves body temperature. Researchers found that self-described "morning people" have temperature rhythms that begin to climb before they awake in the morning, so that they get out of bed prepared to face the world. On the other hand, "night people" may be forced by school and work schedules to get out of bed while their body temperatures are still at their lowest point, before their bodies are prepared for activity. These same night people are still going strong and working productively into the wee hours of the morning when the morning people's body temperature is dropping and they are ready for bed.

Circadian rhythms arise from special groups of cells in the brain and are reinforced by information about the light-dark cycle that comes in through the eyes. The cellular basis for circadian rhythms will be discussed further in Chapter 21.

Now that we have discussed response loops and feedback loops, we will look at the various patterns of biological reflexes.

Control Systems Vary in Their Speed and Specificity

Biological reflexes are mediated by the nervous system, the endocrine system, or a combination of the two (Fig. 6-19 ■). A reflex mediated solely by the nervous system or solely by the endocrine system is relatively simple, but combination reflexes can be quite complex. In the most complex reflex, signals pass through three different integrating centers before finally reaching the target tissue. There is so much overlap between reflexes controlled by the nervous and endocrine systems that they should be considered as a continuum rather than two discrete systems (Fig. 6-20 ■).

Why does the body need two different types of control systems? To answer that question, let us compare endocrine control with nervous control and see what the differences are. These differences are summarized in Table 6-2.

Specificity   Nervous control is very specific because each nerve cell has a specific target cell or cells to which it sends its message. Anatomically, we can isolate a neuron and trace it from its origin to where it terminates on its target cell(s). Endocrine control is more general because the chemical messenger is released into the

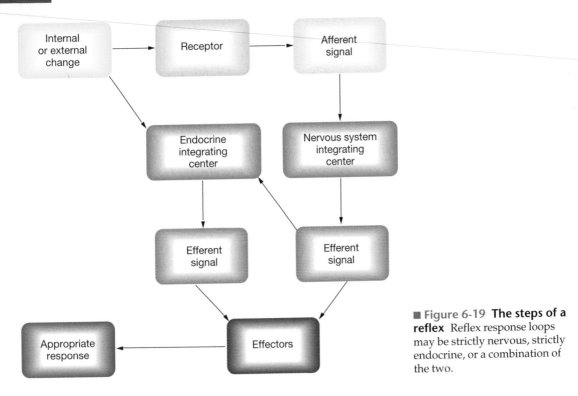

■ **Figure 6-19 The steps of a reflex** Reflex response loops may be strictly nervous, strictly endocrine, or a combination of the two.

blood and can reach virtually every cell in the body. As you learned in the first half of this chapter, the body's response to a specific hormone will depend on which cells have receptors for that hormone. Multiple tissues in the body can respond to a hormone simultaneously.

Nature of the signal The nervous system uses both electrical and chemical signals to send information throughout the body. Electrical signals travel long distances through nerve cells while chemical signals, the neurotransmitters, diffuse across the small gap between nerve cells or between nerve cell and target (Fig. 6-20 ■, pattern 1). In a limited number of instances, electrical signals pass directly from cell to cell through gap junctions.

The endocrine system uses only chemical signals: hormones secreted by endocrine glands or cells into the blood (∞ p. 60). Neurohormones represent a hybrid of the two systems. A nerve cell creates an electrical signal, but the chemical released in response to the electrical signal goes into the blood for general distribution rather than to a specific target (Fig. 6-20 ■, patterns 6 and 2).

Speed Nervous reflexes are much faster than endocrine reflexes. The electrical signals of the nervous system cover great distances very rapidly, with speeds of up to 120 m/sec. Neurotransmitters also create very rapid responses, on the order of milliseconds.

Hormones are much slower than nervous reflexes. Their distribution through the circulatory system and diffusion from the capillary to their receptor takes considerably longer than signals through nerve cells. In addition, hormones have a slower onset of action. In target tissues, the response may take minutes to hours before it can be measured.

Why do we need the speedy reflexes of the nervous system? Consider this example. A mouse ventures out of his hole and sees a cat ready to pounce on him and eat him. A signal must go from the eyes and brain of the mouse down to his feet, telling him to run back into the hole. If his brain and feet were only 5 micrometers (μm; 1/200 millimeter) apart, it would take a chemical signal 20 milliseconds (msec) to diffuse across the space. The mouse could escape. If the cells were 50 μm (1/20 millimeter) apart, diffusion would take two seconds: The mouse might get caught. But because the head and tail of a mouse are centimeters apart, at this rate of diffusion, it would take a chemical signal *three weeks* to diffuse from the head to the feet of the mouse. Poor mouse! Even if the distribution of the chemical were speeded by help from the circulatory system, the chemical message would still take 10 seconds to get to the feet, and the mouse would become cat food. The moral of this tale is that reflexes requiring a speedy response are mediated by the nervous system because it is so much more rapid.

Duration of action Nervous control is of shorter duration than endocrine control. The neurotransmitter released by a nerve cell combines with a receptor on the target cell and initiates a response. But the response is usually very brief because the neurotransmitter is rapidly removed from the vicinity of the receptor by various mechanisms. In order to get a sustained response, multiple repeating signals must be sent through the nerve cell.

Endocrine reflexes are slower to start but they are of longer duration. This means that most of the ongoing, long-term functions of the body such as metabolism and reproduction fall under the control of the endocrine system.

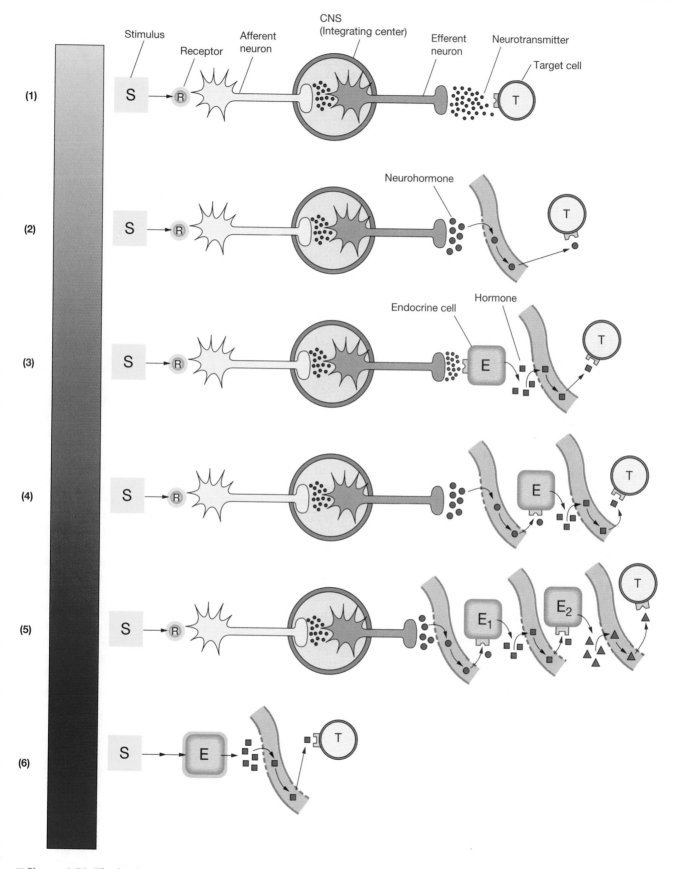

■ Figure 6-20 **The basic patterns of nervous, endocrine, and neuroendocrine control pathways** The colored bar along the left shows that there is no clear distinction between adjoining pathways; the nervous and endocrine systems are linked in a continuum.

TABLE 6-2 Comparison of Nervous and Endocrine Control

Property	Nervous Reflex	Endocrine Reflex
Specificity	Each neuron terminates on a single target cell or on a limited number of adjacent target cells.	Most cells of the body are exposed to a hormone. The response depends on which cells have receptors for the hormone.
Nature of the signal	Electrical signal through neuron with chemical neurotransmitters to pass the signal from cell to cell. In a few cases, cell-to-cell communication takes place through gap junctions.	Chemical signals that are secreted in the blood for distribution throughout the body.
Speed	Very rapid.	Distribution of the signal and onset of action are much slower than in nervous responses.
Duration of action	Usually very rapid. Slower responses are mediated by neuromodulators.	Duration of action is usually much longer.
Coding for stimulus intensity	Each signal is identical in strength. Stimulus intensity reflected by increased frequency of signaling.	Stimulus intensity reflected by amount of hormone secreted.

Coding for stimulus intensity As a stimulus increases in intensity, control systems must have a mechanism for conveying this information to the integrating center. The signal from any one neuron is constant in magnitude and therefore cannot reflect stimulus intensity. Instead, the frequency of signaling through the afferent neuron increases. In the endocrine system, stimulus intensity is reflected by the amount of hormone that is released: the stronger the stimulus, the more hormone.

✓ In Figure 6-19 ■, which boxes represent the following specific parts of a reflex: afferent neuron, brain and spinal cord, central and peripheral sense organs, efferent neuron, hormone, neurohormone? In the figure, add a dotted line and arrow connecting boxes to show a negative feedback loop to complete the reflex.

Pathways for Nervous, Endocrine, and Neuroendocrine Reflexes May Be Complex, with Several Integrating Centers

Figure 6-20 ■ is a summary figure of the different variations of endocrine, neuroendocrine, nervous, and mixed reflexes. The simplest reflexes are the pure nervous reflex (pattern 1), the pure endocrine reflex (pattern 6), and the simple neuroendocrine reflex (pattern 2).

In an endocrine reflex (pattern 6), the endocrine cell acts both as receptor and integrating center; there is no afferent pathway. Pure endocrine reflexes usually do not have a distinct sensory receptor in the same way that neural and neuroendocrine reflexes do. Instead, the endocrine cell itself acts as the receptor, monitoring the stimulus parameter directly. The efferent pathway is the hormone, and the target is any cell with the appropriate receptor.

An example of an endocrine reflex is the secretion of the hormone insulin in response to changes in blood glucose level. The endocrine cells that secrete insulin are able to directly monitor blood glucose concentrations.

As blood glucose increases above the threshold level, the endocrine cells sense the change and respond by secreting insulin into the blood. Any cell in the body with insulin receptors will respond to the hormone and initiate processes that take glucose out of the blood. The removal of the stimulus acts in a negative feedback manner, and the response loop shuts off as blood glucose levels drop below a certain concentration.

In a pure nervous reflex, all the steps of the reflex are present, from receptor to target. This reflex is represented in its simplest form by the knee jerk reflex (see Fig. 13-9, p. 372). A blow to the knee (the stimulus) activates a stretch receptor. An electrical and chemical signal travels through an afferent neuron to the spinal cord (the integrating center). If the blow is strong enough, a signal travels from the spinal cord through an efferent neuron to the muscles of the thigh. In response, the muscles contract, causing the lower leg to kick outward (the knee jerk).

The simple neuroendocrine reflex (pattern 2) is identical to the nervous reflex except that the chemical released by the neuron travels in the blood to its target, just like a hormone. A simple neuroendocrine reflex is the release of breast milk in response to a baby's suckling. The baby's mouth on the nipple stimulates sensory receptors that travel to the brain (integrating center). An electrical signal in the efferent neuron triggers the release of the neurohormone oxytocin from the brain into the circulation. Oxytocin is carried to the breast, where it causes contraction of smooth muscles in the breast (effectors) with resultant ejection of milk.

In complex pathways, there may be more than one integrating center. Patterns 3, 4, and 5 represent combinations of the three basic pathways. Pattern 3 stacks an endocrine reflex on top of a nervous reflex. The target of the initial nervous reflex is an endocrine cell that releases a hormone. An example of this can be found in the control of insulin release by the nervous system. The endocrine cells of the pancreas have both stimulatory and inhibitory neu-

TABLE 6-3 Comparison of Nervous, Neuroendocrine, and Endocrine Reflexes

	Nervous	Neuroendocrine	Endocrine
Receptor	Special and somatic sensory receptors	Special and somatic sensory receptors	Endocrine cell
Afferent path	Afferent sensory neuron	Afferent sensory neuron	None
Integrating center	Brain or spinal cord	Brain or spinal cord	Endocrine cell
Efferent path	Efferent neuron (electrical signal and neurotransmitter)	Efferent neuron (electrical signal and neurohormone)	Hormone (via the circulation)
Effector(s)	Muscles and glands	Most cells of the body	Most cells of the body
Response	Contraction and secretion	Change in enzymatic reactions, membrane transport, or cell proteins	Change in enzymatic reactions, membrane transport, or cell proteins

rons terminating on them, and they integrate the information coming in through the nervous system with their direct detection of blood glucose levels.

Pattern 4 combines a neurosecretory reflex with an endocrine reflex. The secretion of growth hormone is an example of this pathway. Pattern 5 is the most complex pathway, starting with a neuroendocrine reflex and continuing with two sequential endocrine reflexes. This pattern is typical of most of the hormones released by the anterior pituitary, an endocrine gland located just below the brain (see Chapter 7 for details). In describing all these complex pathways, it is simplest to mention only one receptor and afferent path, those that start the reflex. The brain is the first integrating center and the neurohormone is the first efferent path. The endocrine gland that is the target of the neurohormone becomes the second integrating center, and its hormone becomes the second efferent path. Finally, the endocrine target of efferent hormone 2 is the third integrating center, and its hormone is the third efferent path. The target of the last signal in the sequence is the effector.

Table 6-3 compares the various steps of reflexes in the nervous, neuroendocrine, and endocrine systems. There are some other combinations of endocrine and nervous reflexes that you may encounter in physiology. We will be using these general patterns throughout the book as an aid to simplifying endocrine pathways and as a way to classify them.

This chapter has described how the cells of the body communicate with each other in order to maintain a state of homeostasis. The next chapter looks in detail at the endocrine system and the role it plays in these control processes.

continued from page 162

"OK, just one more question," says Marvin. "You said that people with diabetes have high blood glucose levels. If glucose is so high, why can't it just seep into the cells?"

Question 5: Why can't glucose diffuse through the cell membrane if the blood glucose concentration is higher than the intracellular glucose concentration?

Question 6: What do you think happens to the rate of insulin secretion when blood glucose levels fall? What kind of feedback pathway is taking place?

CHAPTER REVIEW

Chapter Summary
Cell-to-Cell Communication

1. There are two basic physiological signals: chemical and electrical. Chemical signals are the basis for most communication within the body. (p. 147)

2. There are three methods of cell-to-cell communication: (1) direct cytoplasmic transfer through gap junctions, (2) local chemical communication, and (3) long-distance communication. (p. 147)

3. **Gap junctions** are protein channels that connect two adjacent cells. When they are open, ions and small molecules pass directly from one cell to the next. (p. 147)

4. Local communication is accomplished by **paracrines,** chemicals that act on cells in the immediate vicinity of the cell that secreted the paracrine. A chemical that acts on the cell that secreted it is called an **autocrine.** The action of paracrines and autocrines is limited by diffusion distance. **Neuromodulators** and cytokines are paracrines. (p. 147)

5. Long-distance communication is accomplished by chemical and electrical signals in the nervous system and by **hormones** in the endocrine system. Only cells that possess receptors for the hormone will be **target cells.** (p. 147)

6. **Cytokines** are regulatory molecules that control cell development, differentiation, and the immune response. They function as both local and long-distance signals. (p. 148)

Receptors and Signal Transduction

7. The response of a cell to a signal molecule is determined by the cell's receptor for the signal. (p. 149)

8. **Agonists** mimic the action of a signal molecule, whereas **antagonists** block the signal pathway. (p. 149)

9. Cells exposed to abnormally high concentrations of a signal for a sustained period of time attempt to bring their response back to normal by decreasing the number of receptors or decreasing the binding affinity of the receptors. This is known as **down-regulation** of the receptors. **Up-regulation** is the opposite of down-regulation. (p. 150)

10. Chemical signals released by cells are called **first messengers.** First messengers are classified into lipophilic signal molecules that enter the cell and combine with cytoplasmic or nuclear receptors and lipophobic signal molecules that are unable to cross the cell membrane and must combine with a membrane receptor. (p. 151)

11. Binding of a signal ligand to a membrane receptor opens or closes an ion channel or is transformed into an intracellular response through signal transduction. **Signal transduction** pathways use membrane proteins and second messengers, intracellular molecules that translate the signal from the first messenger into an intracellular response. (p. 152)

12. Signal transduction pathways create intracellular **cascades** that amplify the original signal. There are two general types of membrane proteins that act as transducers: **tyrosine kinases** and **G proteins.** Many G proteins are linked to amplifier enzymes that activate second messengers and start cascades. (p. 153)

Homeostasis Is the Maintenance of a Constant Internal Environment

13. There are four basic postulates of homeostasis. (1) The nervous system plays an important role in maintaining homeostasis. (2) Some parameters are under **tonic** control, which allows the parameter to be increased or decreased by a single signal. (3) Other parameters are under antagonistic control, in which one hormone or neuron increases the parameter while another decreases it. (4) Chemical signals can have different effects in different tissues of the body, depending on the type of receptor present at the target cell. (p. 155)

Control Pathways: Response and Feedback Loops

14. The simplest homeostatic control takes place at the tissue or cell level and is known as **local control.** (p. 158)

15. In **reflex control pathways,** the decision that a response is needed is made away from the cell or tissue. A chemical or electrical signal is sent to the cell or tissue to initiate a response. Long-distance reflex pathways involve the nervous system, the endocrine system, and cytokines. (p. 159)

16. Reflex pathways can be broken down into response loops and feedback loops. A **response loop** has an input signal, integration of the signal, and an output signal. A stimulus begins the reflex when it is sensed by a receptor. The receptor is linked by an afferent pathway to an integrating center that decides on an appropriate response. An efferent pathway travels from the integrating center to an effector that carries out the appropriate response. These steps create a response loop. (p. 161)

17. A homeostatic response is turned off by a **feedback loop** when the response of the system opposes or removes the original stimulus. This is known as **negative feedback.** (p. 162)

18. In **positive feedback loops,** the response reinforces the stimulus rather than decreasing it or removing it. This destabilizes the system until some intervention or event outside the feedback loop stops the response. (p. 163)

19. **Feedforward control** allows the body to predict that a change is about to occur and start the response loop in anticipation of the change. (p. 163)

20. Apparently spontaneous reflexes that occur in a predictable manner are called biological rhythms. Those that coincide with light-dark cycles are called **circadian rhythms.** (p. 164)

21. Nervous control is faster and more specific than endocrine control but is usually of shorter duration. Endocrine control is less specific and slower to start but is longer lasting and is usually **amplified.** (p. 166)

22. Many reflex pathways are combinations of nervous and endocrine control mechanisms. (p. 168)

Questions

LEVEL ONE Reviewing Facts and Terms

1. What are the two routes for long-distance signal delivery in the body?

2. Which two body systems are charged with maintaining homeostasis by responding to changes in the environment?

3. What two types of physiological signals are used to send messages through the body? Of these two types, which is available to all cells?

4. The process of maintaining a relatively stable internal environment is called _____.

5. List at least three parameters maintained by homeostasis.

6. Distinguish between the "target" and the "receptor" in physiological systems.

7. Activating G proteins results either in _____ ion channels or in activating an _____ on the inner surface of the cell's membrane.

8. The three main amplifier enzymes are (a) _____, which forms cAMP; (b) _____, which forms cGMP, and (c) _____, which converts a phospholipid from the cell's membrane into two different second messenger molecules.

9. An enzyme known as protein kinase adds the functional group _____ to its substrate, by transferring it from a(n) _____ molecule.

10. Put these parts of a reflex in the correct order for a physiological response loop: efferent path, afferent path, effector, stimulus, response, integrating center.

11. Distinguish between central and peripheral receptors.

12. Match each term with its description:
    (a) threshold   1. the desired target value for a
    (b) setpoint   parameter
    (c) effector   2. the distance allowed away from the
    (d) oscillates   setpoint before a response starts
    (e) sensitivity 3. the minimum stimulus to trigger a
                        response

4. the organ or gland that performs the change
5. moves back and forth, or higher and lower

13. Diseases or _____ conditions result when homeostasis cannot be maintained. _____ is the study of failed compensation mechanisms as the body tries to continue to function despite the disease.

14. The name for the daily fluctuations of body functions, including blood pressure, temperature, and metabolic processes, is _____ _____. These cycles arise in special cells in the _____.

15. Down-regulation results in a(n) (increased or decreased?) sensitivity to a prolonged signal, resulting in a condition called _____ _____, often seen when a drug is administered over a long term.

16. List the two ways that down-regulation may be achieved.

17. In a negative feedback loop, the effector moves the system in the (same/opposite) direction as the stimulus.

LEVEL TWO Reviewing Concepts

18. Explain the relationship among the terms in each of the following sets of terms. Give a physiological example or location if applicable.
    a. gap junctions, connexins, syncytium
    b. autocrine, paracrine, cytokine, neurocrine
    c. agonist, antagonist
    d. transduction, amplification, cascade

19. List and compare the two different mechanisms used by lipophobic first messengers to send their signal across the cell's membrane. Give an example of each.

20. Who was Walter Cannon? Restate his four postulates in your own words.

21. Briefly define each term in question 10 and give an anatomical example where applicable.

22. Explain the differences among positive feedback, negative feedback, and feedforward mechanisms. Under what circumstances would each one be advantageous?

23. Compare and contrast the advantages and disadvantages of nervous versus endocrine control mechanisms.

24. Label each system below as positive or negative feedback.
    a. glucagon secretion in response to declining blood glucose
    b. increasing milk letdown and secretion in response to more suckling
    c. urgency in emptying one's urinary bladder
    d. sweating in response to rising body temperature

25. Go back to the preceding question and identify the effector organ for each example.

26. Now go back to question 24 again and identify the integrating center for each example, as closely as you can.

LEVEL THREE Problem Solving

27. In each of the following situations, identify the components of the reflex.
    a. You are sitting quietly at your desk, studying, aware of the bitterly cold winds blowing about at 30 mph, and you begin to feel a little chilly. You start to turn up the thermostat, remember last month's bill, and reach for an afghan to pull around you instead. Pretty soon you are toasty warm again.
    b. You are strolling through the shopping district and the aroma of cinnamon sticky buns reaches you. You inhale appreciatively, but reassure yourself you're not hungry, since you just had lunch an hour ago. You go about your business, but 20 minutes later you're back at the bakery, sticky bun in hand, ravenously devouring its sweetness, saliva moistening your mouth.

Problem Conclusion

Marvin Garcia underwent further tests and was diagnosed with NIDDM. With careful attention to his diet and with a regular exercise program, he has managed to keep his blood glucose levels under control.

In this running problem, you have learned about how glucose homeostasis is maintained by insulin and glucagon, and how the disease diabetes mellitus is an indication that glucose homeostasis has been disrupted.

Further check your understanding of this running problem by comparing your answers to the questions against those in the summary table.

Question	Facts	Integration and Analysis
1 In which form of diabetes are the insulin receptors most likely to be up-regulated?	Up-regulation of receptors usually occurs if a signal molecule is present in unusually low concentrations (∞ p. 150). In IDDM, insulin is not secreted by the pancreas.	In IDDM, insulin levels are low. In NIDDM, insulin levels are normal to above-normal. Therefore, IDDM is most likely to cause up-regulation of the insulin receptors.
2 In which type of diabetes is the signal transduction mechanism for insulin more likely to be defective?	Because insulin is a peptide hormone and therefore lipophobic, it uses membrane receptors linked to second messenger systems to transmit its signal to cells (∞ p. 151). People with IDDM lack insulin; people with NIDDM have normal to above-normal levels of insulin.	Normal or high insulin levels in NIDDM suggest that the problem is not with amount of insulin but with the action of the insulin at the cell. The problem could be a defective signal transduction mechanism.
3 The homeostatic regulation of blood glucose levels by the hormones insulin and glucagon is an example of which of Cannon's postulates?	Cannon's postulates describe the role of the nervous system in maintaining homeostasis, and the concepts of tonic activity, antagonistic control, and different effects of signals in different tissues.	Insulin decreases blood glucose levels while glucagon increases them. Therefore, the two hormones are an example of an antagonistic control.
4 In the insulin reflex that regulates blood glucose levels, what is the stimulus? the receptor? the integrating center? the efferent pathway? the effector(s)? the response(s)?	See the steps of reflex pathways (∞ p. 158).	Stimulus: increase in blood glucose levels Receptor: beta cells of the pancreas that sense the change. Integrating center: beta cells. They evaluate the change and act upon it by releasing insulin. Efferent pathway (the signal): insulin Effectors: any tissues of the body that respond to insulin Responses: cellular uptake and use of glucose
5 Why can't glucose diffuse through the cell membrane if the blood glucose concentration is higher than the intracellular glucose concentration?	Glucose is lipophobic. In diffusion, molecules move from areas of high concentration to areas of low concentration.	Because glucose is lipophobic, it must cross the membrane by some form of mediated transport. It is too large to diffuse through an open channel; therefore it must be moving into the cell by facilitated diffusion on a carrier.
6 What do you think happens to the rate of insulin secretion when blood glucose levels fall? What kind of feedback pathway is taking place?	The stimulus for insulin release is an increase in blood glucose levels. In negative feedback, the response offsets the stimulus. In positive feedback, the response enhances the stimulus.	An increase in blood glucose concentration stimulates insulin release; therefore a decrease in blood glucose should decrease insulin release. In this example, the response (lower blood glucose) offsets the stimulus (increased blood glucose), so the pathway exhibits negative feedback.

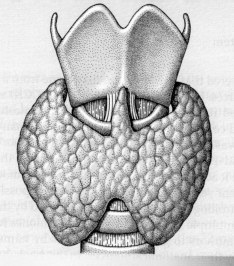

7

Introduction to the Endocrine System

BACKGROUND BASICS

Receptors (p. 149)
Peptides and proteins (p. 29)
Comparison of endocrine and nervous systems (p. 165)
Second messenger systems (p. 152)
Steroids (p. 28)
Specificity (p. 120)

David was seven years old when the symptoms first appeared. His appetite at meals increased, and he always seemed to be in the kitchen looking for food. But despite eating more, he was losing weight. When he started asking for water instead of soft drinks, David's mother became concerned. And when he wet the bed three nights in a row, she knew something was wrong. The doctor confirmed the suspected diagnosis after running tests to determine the concentration of glucose in David's blood and urine. David had diabetes mellitus. In David's case, the disease was due to lack of insulin, a hormone produced by the pancreas. David was placed on insulin injections, a treatment that he would continue for the rest of his life.

One hundred years ago, David would have died not long after the onset of symptoms. The field of **endocrinology,** the study of hormones, was at that time in its infancy. Most hormones had not been discovered, and the functions of known hormones were not well understood. There was no treatment for diabetes, no birth control pill for contraception. Babies born with inadequate secretion of thyroid hormone did not grow or develop normally. Today, all that has changed. We have identified a long and growing list of hormones. The endocrine diseases that once killed or maimed are under control with synthetic hormones and sophisticated medical procedures. This chapter provides an introduction to the basic principles of hormone structure and function. You will learn more about individual hormones as you encounter them in your study of the various systems.

Anatomy Summary   The Endocrine System

■ Figure 7-1 The endocrine system

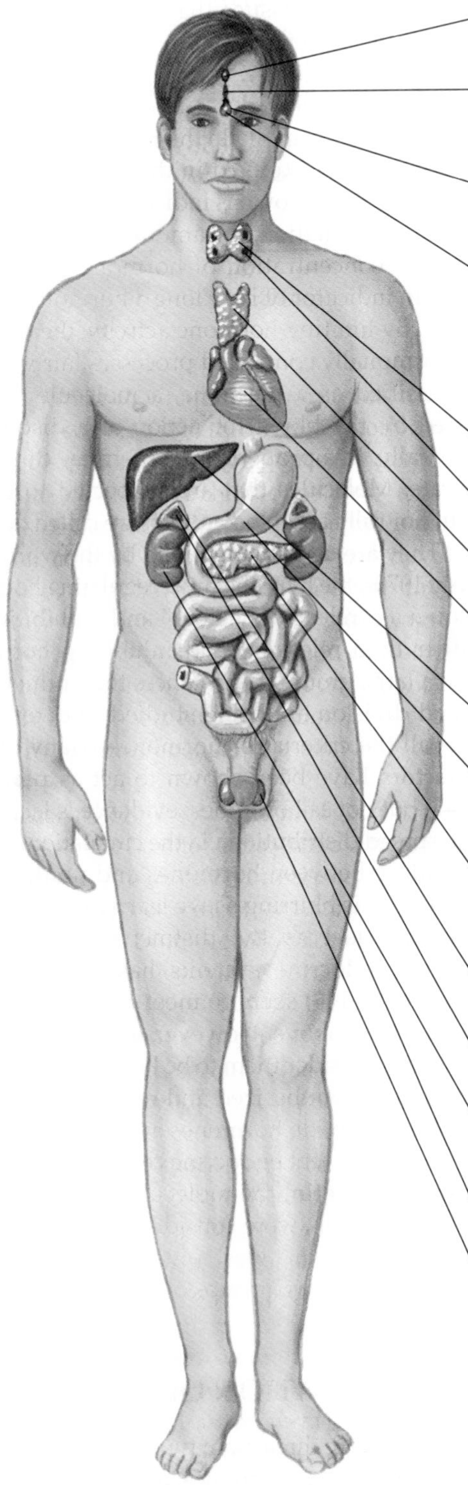

Location	Gland or Cell?	Chemical Class
Pineal gland	Gland	Amine
Hypothalamus	Clusters of neurons	Peptide Peptide
Posterior pituitary	Axon terminals from hypothalamic neurons	Peptide Peptide
Anterior pituitary	Gland	Peptide Peptide Peptide Peptide Peptide Peptide
Thyroid	Gland	Iodinated ami Peptide
Parathyroid	Gland	Peptide
Thymus	Gland	Peptide
Heart	Cells	Peptide
Liver	Cells	Peptide
Stomach and small intestine	Cells	Peptide
Pancreas	Gland	Peptide
Adrenal cortex	Gland	Steroid Steroid Steroid
Adrenal medulla	Gland	Amine
Kidney	Cells	Peptide Peptide Steroid
Skin	Cells	Steroid
Gonads	Glands	Steroids Steroid Peptide
Placenta (pregnant females only)	Gland	Steroids Peptide

rmone	Target	Main Effect
latonin	Unclear in humans	Circadian rhythms?
phic hormones (see Fig. 7-15) e posterior pituitary	Anterior pituitary	Release or inhibit pituitary hormones
ytocin	Breast and uterus	Milk ejection and delivery hormones
sopressin (ADH)	Kidney	Water reabsorption
lactin owth hormone (somatotropin) rticotropin (ACTH) yrotropin (TSH) llicle stimulating hormone (FSH) teinizing hormone (LH)	Breast Many tissues Adrenal cortex Thyroid gland Gonads Gonads	Milk production Growth and metabolism Cortisol release Thyroid hormone synthesis and release Egg or sperm production; sex hormone production Sex hormone production; egg or sperm production
iodothyronine and thyroxine lcitonin	Many tissues Bone	Metabolism, growth and development Plasma calcium levels (minimal effect in humans)
rathyroid hormone (PTH)	Bone, kidney	Plasma calcium and phosphate levels
ymosin, thymopoietin	Lymphocytes	Lymphocyte development
rial natriuretic peptide	Kidneys	Increase sodium excretion
giotensinogen matomedins	Adrenal cortex, blood vessels Many tissues	Aldosterone, blood pressure Growth
strin, cholecystokinin, cretin, and others	Digestive tract and pancreas	Assist digestion and absorption of nutrients
ulin, glucagon, somatostatin, ancreatic polypeptide	Many tissues	Metabolism of glucose and other nutrients
dosterone rtisol drogens	Kidney Many tissues Many tissues	Electrolyte balance Stress response Sex drive in females
inephrine, norepinephrine	Many tissues	Fight-or-flight response
ythropoietin nin 5 dihydroxy-vitamin D_3 ,25 dihydroxycholecalciferol)	Bone marrow Angiotensinogen in blood Intestine	Red blood cell production Angiotensin production Increase calcium absorption
tamin D_3	Intermediate form of hormone	Precursor of 1,25 dihydroxy vitamin D_3
drogens in males trogens and progesterone in females hibin	Many tissues Many tissues Anterior pituitary	Sperm production, secondary sex characteristics Egg production, secondary sex characteristics Decrease FSH secretion
trogens and progesterone; horionic gonadotropin	Many tissues	Fetal and maternal development

■ **Figure 7-2 Endocrine disorders in ancient art** This pre-Colombian stone carving of a woman shows a mass at her neck. This mass is an enlarged thyroid gland, a condition known as goiter. It was considered a sign of beauty among the people who lived high in the Andes mountains.

or more amino acids. The **steroid hormones** are all derived from cholesterol (∞ p. 28). The third group of hormones, the **amine hormones,** are derivatives of single amino acids. Peptide and steroid hormones differ in most respects, from their synthesis in the parent endocrine cell to their reception and action at their target. The amine hormones behave in some instances like peptide hormones and in other cases like steroids.

Most Hormones in the Body Are Peptides or Proteins

The peptide hormones range from small peptides of only three amino acids to proteins with 200 or more amino acids. Despite the wide variation in size among hormones in this group, for simplicity, they are all referred to as peptide hormones. You can remember which hormones are peptides most easily by exclusion: If the hormone is not a steroid and not an amine, then it must be a peptide.

Peptide hormone synthesis, storage, and release

The synthesis and packaging of peptide hormones into membrane-bound secretory vesicles is similar to that of other proteins (∞ p. 102). The initial peptide that comes off the rough endoplasmic reticulum is a large inactive protein known as a **preprohormone** (Fig. 7-3 ■). Preprohormones have one or more copies of a peptide hormone, a signal sequence, and other peptide sequences that may or may not have biological activity. As the inactive peptide moves through the endoplasmic reticulum and Golgi apparatus, the signal sequence is removed,

TABLE 7-1	Comparison of Peptide, Steroid, and Amine Hormones			
Property of Hormones	*Peptide Hormones*	*Steroid Hormones*	*Catecholamines*	*Thyroid Hormones*
Synthesis and storage	Made in advance; stored in secretory vesicles	Synthesized on demand from precursors	Made in advance; stored in secretory vesicles	Made in advance; precursor stored in secretory vesicles
Release from parent cell	By exocytosis	Simple diffusion	By exocytosis	Simple diffusion
Transport in blood	Freely dissolve in plasma	Bound to carrier proteins	Freely dissolve in plasma	Bound to carrier proteins
Half-life	Short	Long	Short	Long
Location of receptor	On cell membrane	In cytoplasm or nucleus; some may have membrane receptors also	On cell membrane	In nucleus
Response to receptor-ligand binding	Activate second messenger systems	Activate genes for transcription and translation	Activate second messenger systems	Activate genes for transcription and translation
General target response	Modification of existing proteins	Induction of new protein synthesis	Modification of existing proteins	Induction of new protein synthesis
Examples	Insulin, parathyroid hormone	Estrogen, androgens, cortisol	Epinephrine, norepinephrine	Thyroxine (T_4), triiodothyronine (T_3)

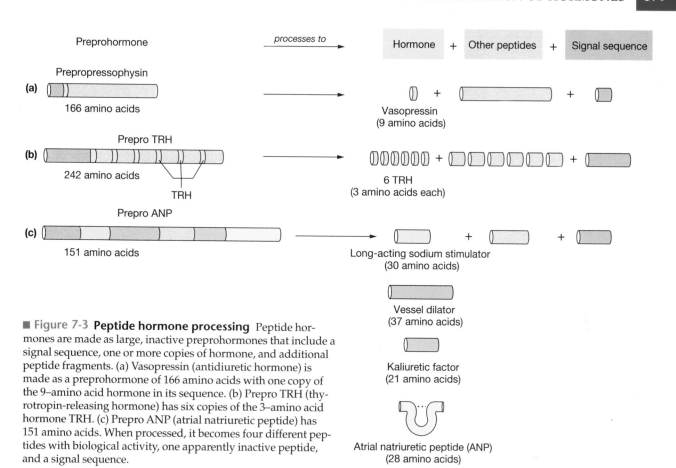

■ **Figure 7-3 Peptide hormone processing** Peptide hormones are made as large, inactive preprohormones that include a signal sequence, one or more copies of hormone, and additional peptide fragments. (a) Vasopressin (antidiuretic hormone) is made as a preprohormone of 166 amino acids with one copy of the 9–amino acid hormone in its sequence. (b) Prepro TRH (thyrotropin-releasing hormone) has six copies of the 3–amino acid hormone TRH. (c) Prepro ANP (atrial natriuretic peptide) has 151 amino acids. When processed, it becomes four different peptides with biological activity, one apparently inactive peptide, and a signal sequence.

creating a smaller, still inactive molecule called a **prohormone**. In the Golgi apparatus, the prohormone is packaged into secretory vesicles along with proteolytic [*proteo-*, protein + *lysis*, rupture] enzymes that chop the prohormone into active hormone and other fragments (Fig. 7-4 ■). The secretory vesicles are stored in the cytoplasm of the parent endocrine cell until the cell receives a signal for secretion. At that time, the vesicles move to the cell membrane and release their contents by calcium-dependent exocytosis (∞ p. 127). All of the peptide fragments created from the prohormone are released together into the extracellular fluid, in a process known as *co-secretion*.

Transport in the blood and half-life of peptide hormones   Peptide hormones are water-soluble and therefore dissolve easily in the extracellular fluid for transport throughout the body. The half-life for peptide hormones is usually quite short, in the range of several minutes. If the response to a peptide hormone must be sustained for an extended period of time, the hormone must continue to be secreted.

Cellular mechanism of action of peptide hormones
Because peptide hormones are lipophobic, they usually are unable to enter the target cell. Instead, they bind to surface membrane receptors. The hormone-receptor

complex initiates the cellular response by means of a signal transduction system (Fig. 7-5 ■). Many peptide hormones work through cAMP second messenger systems (∞ p. 152). A few peptide hormone receptors have tyrosine kinase activity or work through other signal transduction pathways.

The response of cells to peptide hormones is fairly rapid because second messenger systems modify existing proteins. Some of the changes triggered by peptide hormones include opening or closing membrane channels and modulating metabolic enzymes or transport proteins.

The Mystery of Insulin's Second Messenger System   Insulin, a peptide hormone from the pancreas that helps regulate blood glucose concentration, is one hormone whose second messenger system is still unknown. The insulin receptor is a membrane protein composed of two subunits: an α subunit on the extracellular fluid side that binds to insulin and a β subunit with tyrosine kinase activity on the cytoplasmic side. The signal transduction pathway that links the activated tyrosine kinase to altered metabolic enzymes is still unclear. Some novel second messenger pathways have been proposed, but no clear answer has emerged from the research. Many investigators are looking at this problem since insulin plays such a key role in diabetes mellitus.

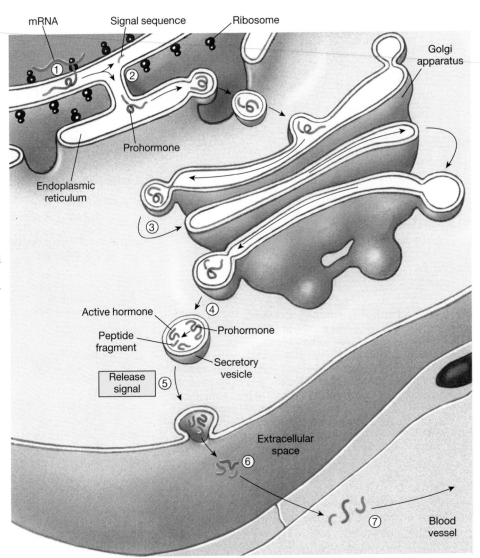

■ Figure 7-4 **Peptide hormone synthesis, packaging, and release** ① Messenger RNA on the ribosomes of the endoplasmic reticulum (ER) binds amino acids into a peptide chain that is directed into the lumen of the ER by an initial signal sequence of amino acids. This peptide chain is the preprohormone. ② Enzymes in the ER chop off the signal sequence, leaving behind an inactive prohormone. ③ The prohormone passes from the ER through the Golgi apparatus to a secretory vesicle. ④ Enzymes in the secretory vesicle chop the prohormone into one or more active peptides plus additional fragments that may or may not have biological activity. ⑤ When the signal for release occurs, the secretory vesicle releases its contents by exocytosis into the extracellular space. ⑥ The hormone, fragments, and enzymes in the vesicle are co-secreted. ⑦ The hormone moves into the circulation for transport to its target.

Steroid Hormones Are Derived from Cholesterol

Steroid hormones have similar chemical structures because they are all derived from cholesterol (Fig. 7-6 ■). Unlike peptide hormones that are made in tissues all over the body, steroid hormones are made in only a few organs. Three types of steroid hormones are made in the **adrenal cortex,** the outer portion of the adrenal glands. One adrenal gland sits on top of each kidney [*ad-,* upon + *renal,* kidney; *cortex,* bark]. The gonads produce the sex steroids (estrogens, progesterone, and androgens). In pregnant women, the placenta is also a source of steroid hormones.

Steroid hormone synthesis and release   Cells that secrete steroid hormones have unusually large amounts of smooth endoplasmic reticulum, the organelle where steroids are synthesized. Steroids are lipophilic, so they diffuse easily across membranes, both out of their par-

ent cell and into their target cell. This property also means that steroid-secreting cells cannot store hormones in secretory vesicles. Instead, they synthesize their hormones as hormone is needed. When a stimulus activates the endocrine cell, precursors in the cytoplasm are rapidly converted to active hormone. The hormone concentration in the cytoplasm rises, and the hormones move out of the cell by simple diffusion.

Transport in the blood and half-life of steroid hormones   Steroid hormones, like cholesterol, are not very soluble in plasma and other body fluids. For this reason, most of the steroid hormone molecules found in the blood are bound to protein carrier molecules (∞ p. 31). Some hormones have specific carriers, such as corticosteroid-binding globulin; others simply bind to general plasma proteins such as albumin. The binding of steroid hormone to a carrier protein protects the hormone from enzymatic degradation and results in an extended half-

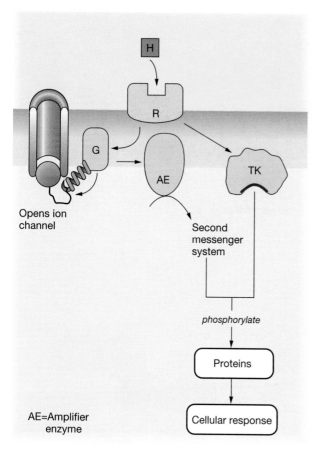

■ Figure 7-5 Cell surface receptor and second messenger systems for peptide hormones Because peptide hormones (H) cannot enter their target cells, they combine with membrane receptors (R) that activate second messenger systems. Many hormones use a cAMP second messenger, but some use tyrosine kinase (TK) or G protein–linked second messengers such as cGMP.

is the nucleus, where the complex acts as a *transcription factor*, binding to DNA and turning on one or more genes (Fig. 7-7 ■). The activated genes create mRNA that directs the synthesis of new proteins.

Because steroid hormones control the production of new proteins rather than the modification of existing proteins, there is usually a lag time between hormone-receptor binding and the first measurable biological effects. This lag can be as much as 90 minutes. Consequently, reflex pathways that require rapid responses are not mediated by steroid hormones.

A recent discovery suggests that there are exceptions to the "slow response" rule. Several steroid hormones, including aldosterone and estrogen, have receptors located on the surface of the target cell as well as within it. These receptors enable those hormones to initiate rapid responses in addition to their slower response. With the discovery of membrane receptors for steroid hormones, the difference in location of steroid and peptide hormone receptors seems to be blurring.

Amine Hormones Are Derived from Single Amino Acids

There are three groups of amine hormones in which the hormone molecule is created from a single amino acid. Melatonin is derived from the amino acid tryptophan (see Focus on the Pineal Gland, Fig. 7-21 ■). The other amine hormones, the catecholamines and thyroid hormones, are derived from the amino acid tyrosine, notable for its ring structure (Fig. 7-8 ■, p. 184). Despite their common precursor, these amines have little in common. The catecholamines (epinephrine, norepinephrine, and dopamine) behave like typical peptide hormones. The thyroid hormones behave more like steroid hormones. Thyroid hormones will be discussed in Chapter 21.

life. For example, cortisol, a hormone produced by the adrenal cortex, has a half-life of 60 to 90 minutes. (Compare this with epinephrine, an amine hormone whose half-life is measured in seconds.)

Although binding steroid hormones to protein carriers extends their half-life, it also blocks their entry into target cells. The carrier-steroid complex remains outside the cell since the carrier proteins are lipophobic and cannot diffuse through the membrane. Only unbound hormone molecules can diffuse into the target cell (Fig. 7-7 ■). Fortunately, hormones are active in minute concentrations, so a tiny amount of unbound steroid is enough to produce a response. As unbound hormone leaves the blood and enters cells, the carriers release bound steroid so that a small fraction of hormone remaining in the blood is always unbound.

Cellular mechanism of action of steroid hormones
Steroid receptors are found in the intracellular compartment, either in the cytoplasm or in the nucleus. The ultimate destination of steroid receptor-hormone complexes

continued from page 174

Shaped like a butterfly, the thyroid gland straddles the trachea just below the Adam's apple. Responding to hormonal signals from the hypothalamus and anterior pituitary, the thyroid gland concentrates iodine, an element found in food (most notably as an added ingredient to salt) and combines it with the amino acid tyrosine to make two thyroid hormones, thyroxine and triiodothyronine. These thyroid hormones perform many important functions in the body, including the regulation of growth and development, oxygen consumption, and the maintenance of body temperature.

Question 1

a. **To which of the three classes of hormones do the thyroid hormones belong?**

b. **If a person's diet is low in iodine, predict what happens to thyroxine production.**

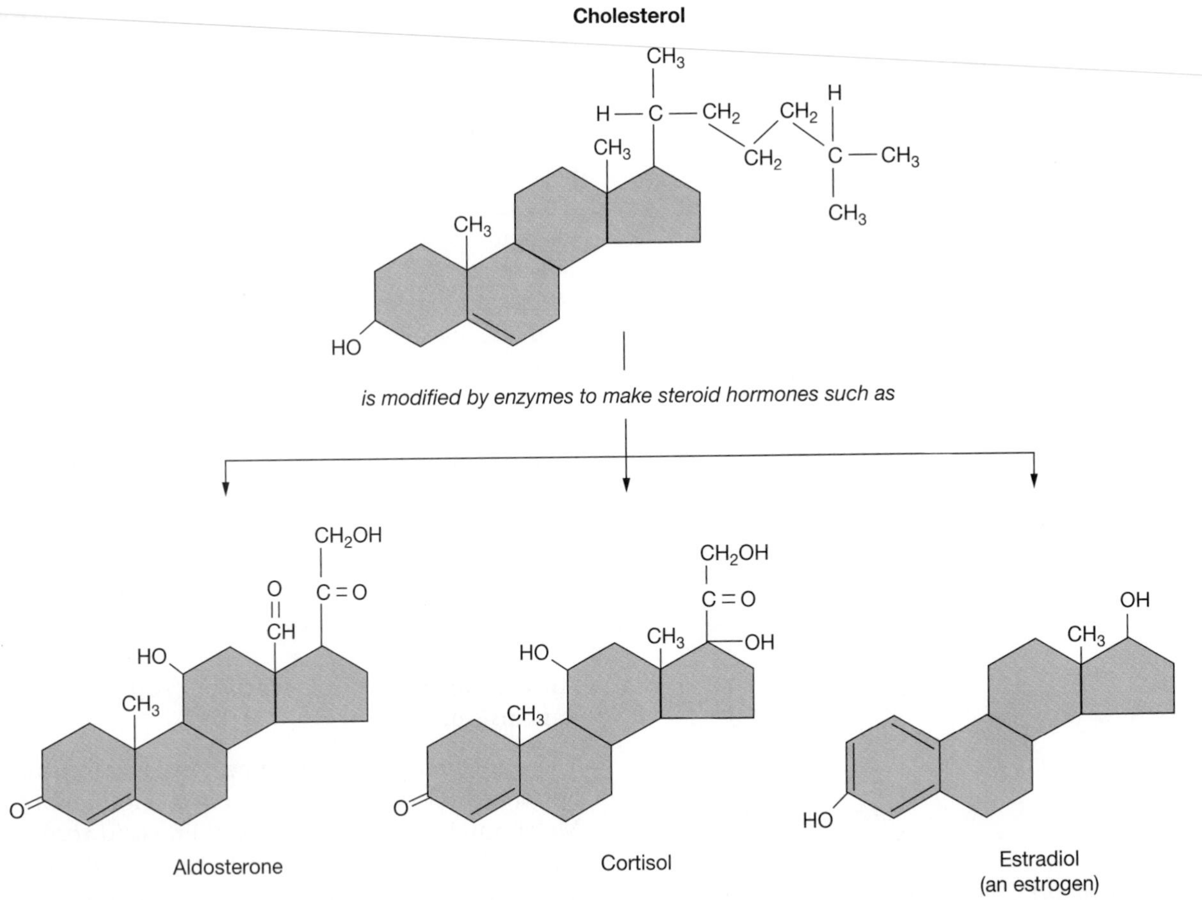

Cholesterol

is modified by enzymes to make steroid hormones such as

Aldosterone

Cortisol

Estradiol
(an estrogen)

■ **Figure 7-6   Steroid hormones** Cholesterol is the parent compound for all steroid hormones. Its basic ring structure remains unchanged while functional groups are added and subtracted. This figure shows three typical steroid hormones: aldosterone and cortisol secreted by the adrenal cortex, and estradiol, an estrogen produced by the ovaries and adrenal cortex.

This section has set the stage for your study of hormone action. The next section looks at the various patterns for endocrine reflexes.

✔ On the basis of what you know about the organelles involved in protein and steroid synthesis (∞ p. 49), what would be the major differences between the organelle composition of a steroid-producing cell and a protein-producing cell?

✔ The steroid hormone aldosterone has a short half-life for a steroid hormone, only about 20 minutes. What would you predict about the degree to which aldosterone is bound to blood proteins?

✔ Why do steroid hormones take so much longer to act than peptide hormones?

CONTROL OF HORMONE RELEASE

You learned in Chapter 6 that homeostasis is maintained by local and reflex control pathways. All reflex pathways have similar components: an input signal, integra-

tion of the signal, and an output signal. In endocrine and neuroendocrine reflexes, a hormone or neurosecretory hormone is the output signal. Some complex endocrine pathways involve as many as three different hormones.

The Target of Most Trophic Hormones Is Another Endocrine Gland or Cell

Complex reflex control pathways have multiple integrating centers and involve more than one hormone (∞ p. 168). In these complex pathways, any hormone that controls the secretion of another hormone is known as a **trophic (tropic) hormone.** The adjective *trophic* comes from the Greek word *trophikós,* which means "pertaining to food or nourishment" and refers to the manner in which the trophic hormone "nourishes" the target cell. Trophic hormones that act on endocrine cells may be recognized by the modifiers "releasing hormone" or "inhibiting hormone." Their names often end with the suffix-*tropin,* as in *gonadotropin.** The root word

---

*A few hormones whose names end in -*tropin* do not have endocrine cells as their targets. For example, melanotropin acts on pigment-containing cells in many animals.

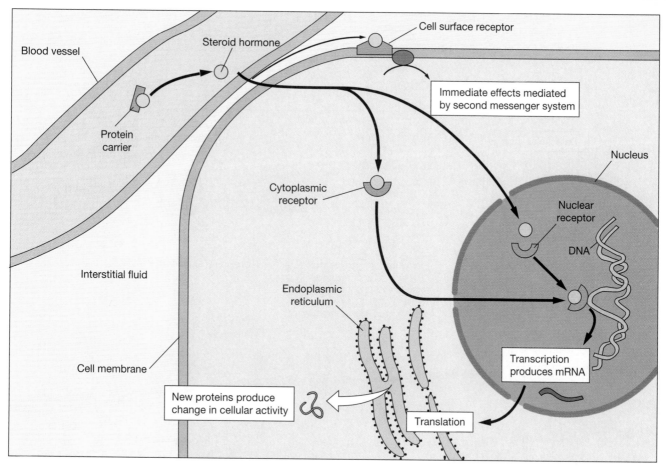

■ **Figure 7-7 Steroid hormone transport and cellular mechanism of action** Steroids are hydrophobic and must be bound to protein molecules in the blood. These same hormone-protein complexes are unable to cross cell membranes, and only unbound hormone can diffuse into the target cell. As unbound hormone leaves the extracellular fluid, more hormone unbinds from the carrier, obeying the Law of Mass Action. The primary receptors for steroid hormones are located in the cytoplasm or nucleus of the target cell. The receptor-hormone complex binds to DNA and activates one or more genes to produce mRNA. The resultant mRNA leaves the nucleus and directs the synthesis of new proteins in the cytoplasm. Recent evidence indicates that some steroid hormones also have surface receptors that activate second messenger systems and create rapid cellular responses.

to which the suffix is attached is the target tissue: The gonadotropins are hormones that are trophic to the gonads.

Negative Feedback Turns Off Hormone Reflexes

In many reflex pathways, the response of the pathway is also a negative feedback signal that turns off the reflex. For example, an increase in blood glucose concentration initiates an endocrine reflex to decrease blood glucose levels. Decreased blood glucose concentration then acts as a negative feedback signal to turn off the pathway and stop hormone secretion.

But in reflex control pathways involving multiple integrating centers, the hormones themselves act as negative feedback signals, as shown in Figure 7-9 ■. Each hormone in the pathway may feed back to suppress hormone secretion by integrating centers earlier in the reflex. This suppression decreases the output of trophic hormones and, in turn, means that less of the

final hormone is produced. With this system of negative feedback, the system stays within the appropriate response range.

Hormones Can Be Classified by Endocrine Reflex Pathways

Endocrine reflex pathways are a convenient method by which you can classify hormones and simplify learning the control pathways that regulate their secretion. In the sections that follow, we will discuss the major reflex control pathways for hormones. This discussion is not all-inclusive, and you will encounter a few hormones in later chapters that do not fit into these pathways.

Classic hormones The classic hormone has the simplest reflex control pathway in the endocrine system. The endocrine cell secreting the hormone acts as both receptor and integrating center, directly sensing the stimulus and responding by secreting its hormone

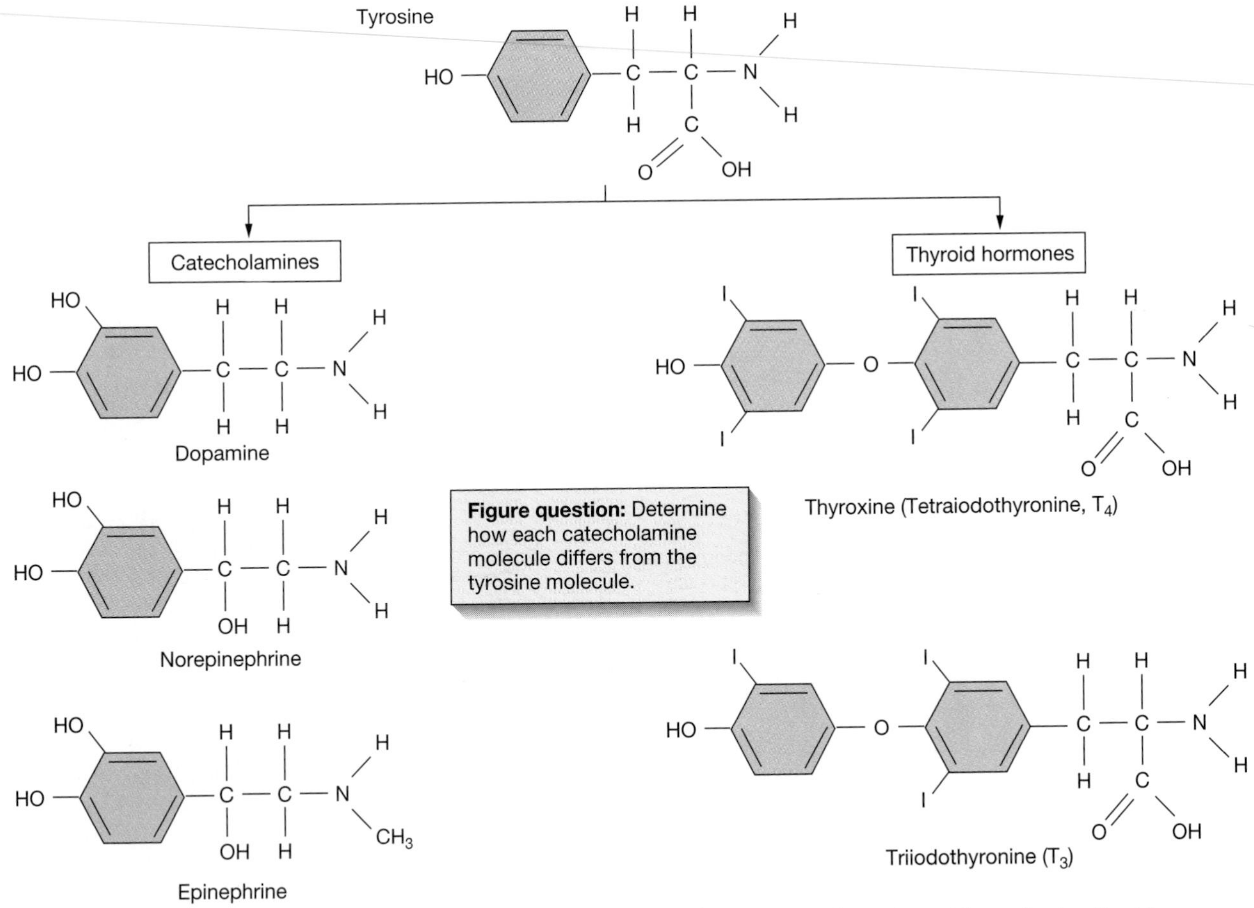

Figure question: Determine how each catecholamine molecule differs from the tyrosine molecule.

■ **Figure 7-8 Structure of tyrosine, catecholamines, and thyroid hormones** The amino acid tyrosine has its side groups modified to create the catecholamines. Thyroid hormones are synthesized from two tyrosines and iodine (I) atoms.

(∞ Fig. 6-20, pattern 6, p. 167). The hormone then travels through the blood to act on its target issues.

An example of a classic hormone is parathyroid hormone. The endocrine cells of the parathyroid glands monitor the concentration of Ca^{2+} in the blood (Fig. 7-10 ■). When the concentration of Ca^{2+} in the plasma falls below a certain point, the parathyroid cells secrete parathyroid hormone (PTH) in response. Parathyroid hormone travels through the blood to act on its target tissues, initiating responses that increase the concentration of calcium in the plasma. The increase in plasma calcium is a negative feedback signal that turns off the reflex, ending release of parathyroid hormone.

Classic hormones with multiple controls The secretion of some classic hormones is controlled by multiple stimuli. For example, insulin secretion can be triggered by an increase in the concentration of glucose in the blood, by stimulation from the nervous system, or by a hormone secreted from the digestive tract. The first two of these reflex pathways are diagrammed in Figure 7-11 ■, p. 187. No matter what the stimulus, insulin travels through the blood to act on its target tissues, whose responses ultimately remove glucose from the blood.

The resultant decrease in blood glucose concentration acts as a negative feedback signal and turns off the reflex, ending release of insulin.

Neurohormones There are three major groups of neurohormones: (1) the posterior pituitary hormones that are synthesized in the hypothalamus of the brain, (2) hormones released from the hypothalamus to control hormone release by the anterior pituitary, and (3) catecholamines made by the modified neurons that make up the adrenal medulla. Neurohormones are released when the nerve cell that synthesizes them receives a signal from the nervous system (∞ Fig. 6-20, pattern 2). Many neurohormones are associated with the pituitary gland, so we will first describe that important endocrine structure.

The Pituitary Gland The pituitary gland has been called the master gland of the body because its hormones control so many vital functions. The pituitary is a structure the size of a lima bean that extends downward from the brain, connected to it by a thin stalk and cradled in a protective pocket of bone (Fig. 7-12 ■, p. 188). The first accurate description of the function of the pituitary

down. If the pituitary remains suppressed and the adrenal cortex is deprived of ACTH long enough, the cells of both glands shrink and lose their ability to manufacture ACTH and cortisol. This condition is known as **atrophy** [*a-*, without + *trophikós*, nourishment]. If the cells of an endocrine gland atrophy because of exogenous hormone administration, they may be very slow or totally unable to regain normal function when the treatment with exogenous hormone is stopped. As you may know, steroid hormones are often used to treat poison ivy and severe allergies. However, when treatment is complete, the dosage must be tapered off gradually to allow the pituitary and adrenal gland to work back up to normal hormone production. As a result, packages of steroid pills direct patients ending treatment to take six pills one day, five the day after that, and so on.

The Effects of a Hormone Are Diminished or Absent with Hyposecretion

Symptoms of hormone deficiency occur when too little hormone is secreted **(hyposecretion)**. Hyposecretion may occur anywhere along the endocrine control axis, in the hypothalamus, pituitary, or other endocrine glands. The most common cause of hyposecretion syndromes is atrophy of the gland due to some disease process. Hyposecretion of thyroid hormone may occur if there is insufficient dietary iodine for the thyroid gland to manufacture the iodinated hormone.

Negative feedback pathways are affected in hyposecretion, but in the opposite direction from hypersecretion. The absence of feedback causes trophic hormone levels to rise in an effort to make the defective gland increase its production of hormone. For example, if the adrenal cortex atrophies as a result of tuberculosis, cortisol production diminishes. The hypothalamus and anterior pituitary sense that cortisol levels are below normal, so they increase secretion of CRH and ACTH, respectively, in an attempt to stimulate the adrenal gland into making more cortisol.

Abnormal Tissue Responsiveness Can Be Due to Problems with Hormone Receptors or Second Messenger Pathways

Endocrine diseases do not always arise from problems with endocrine glands. They may also be triggered by changes in the responsiveness of the target tissues to the hormones. In these situations, the target tissues show abnormal responsiveness even though the hormone levels are within normal range. Changes in the target tissue response are usually caused by abnormal interaction between the hormone and its receptor or by alterations in signal transduction pathways. These concepts have been covered for signal molecules in general in Chapter 6 (p. 146), so we will restrict this discussion to some typical examples of abnormal tissue responsiveness in the endocrine system.

Up- and down-regulation   If hormone secretion is abnormally high for an extended period of time, target cells will down-regulate (decrease the number of) their receptors in an effort to diminish their responsiveness to excess hormone. Hyperinsulinemia [*hyper-*, elevated + insulin + *-emia*, in the blood], in which sustained high levels of insulin cause target cells to remove insulin receptors from the cell membrane, is a classic example of down-regulation in the endocrine system.

Receptor and signal transduction abnormalities Many forms of inherited endocrine pathologies can be traced to problems with the hormone receptor in the target cell. If a mutation alters the protein structure of the receptor, the cellular response to receptor-hormone binding may be altered. Or the receptors may be absent or completely nonfunctional. For example, in testicular feminizing syndrome, the gene responsible for the synthesis of androgen receptors is absent in the male fetus. As a result, androgens produced by the developing fetus are unable to influence development of the genitalia, resulting in a child who appears to be female in every way except that she does not have functioning ovaries.

Genetic alterations in signal transduction pathways can lead to symptoms of hormone excess or deficiency. In the disease called *pseudohypoparathyroidism* [*pseudo-*, false + *hypo-*, decreased + parathyroid + *-ism*, condition or state of being], patients show signs of low parathyroid hormone even through blood levels of the hormone are normal to elevated. These patients have inherited a defect in the G protein that links the hormone receptor to the cAMP amplifier enzyme, adenylate cyclase. Because the signal transduction pathway does not function, the cells are unable to respond to parathyroid hormone, and symptoms of hormone deficiency appear.

Diagnosis of Endocrine Pathologies Depends on the Complexity of the Reflex

Diagnosis of endocrine pathologies may be simple or quite complex, depending on the complexity of the reflex. In the simplest endocrine reflex such as that for parathyroid hormone (see Fig. 7-10 ■), if there is too much or too little hormone, there is only one location where the problem can arise, the sole integrating center in the reflex. However, with the complex hypothalamic-pituitary-endocrine gland reflexes, the diagnosis can be much more difficult.

If a pathology (deficiency or excess) arises in the last endocrine gland in a reflex, the problem is considered to be a **primary pathology.** For example, if a tumor in the adrenal cortex begins to produce excessive amounts of cortisol, the resulting condition is called *primary hypersecretion*. If a dysfunction occurs in one of the tissues producing trophic hormones, the problem is a **sec-**

ondary pathology. For example, if the pituitary is damaged because of head trauma and ACTH secretion diminishes, the resulting cortisol deficiency is considered to be *secondary hyposecretion* of cortisol.

The diagnosis of pathologies in complex endocrine pathways depends on understanding negative feedback in the control axis. Figure 7-20 ■ shows three possible causes of excess cortisol secretion. To determine which is the correct *etiology* (cause) of the disease in a particular patient, the clinician must assess the levels of the three hormones in the control axis. If the problem is overproduction of CRH by the hypothalamus (Fig. 7-20a ■), CRH levels will be higher than normal. High CRH in

turn causes high ACTH, which in turn causes high cortisol. This is therefore secondary hypersecretion arising from a problem in the hypothalamus. Figure 7-20b ■ shows a secondary hypersecretion of cortisol due to an ACTH-secreting tumor of the pituitary. Once again, the high levels of ACTH cause high cortisol production. But in this example, high cortisol has a negative feedback effect on the hypothalamus, decreasing production of CRH. The combination of low CRH and high ACTH isolates the problem to the pituitary.

If cortisol levels are high but both trophic hormones are low, the problem must be a primary disorder (Fig. 7-20c ■). There are two possible explanations: endoge-

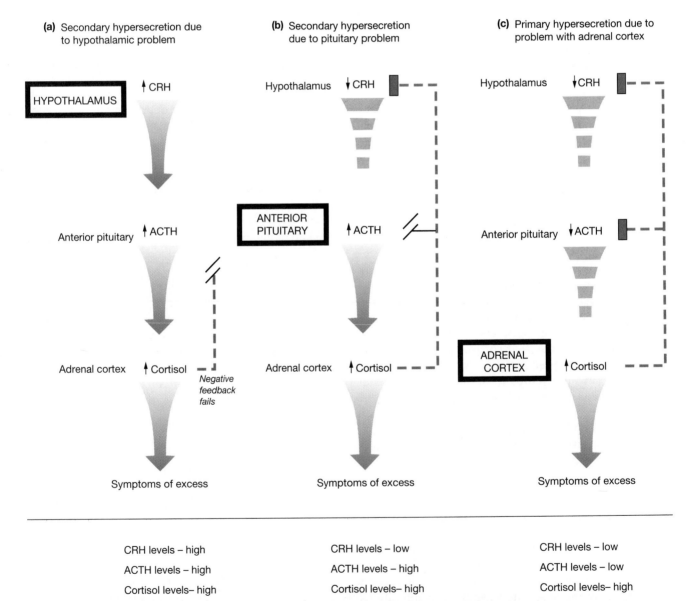

(a) Secondary hypersecretion due to hypothalamic problem	(b) Secondary hypersecretion due to pituitary problem	(c) Primary hypersecretion due to problem with adrenal cortex
CRH levels – high	CRH levels – low	CRH levels – low
ACTH levels – high	ACTH levels – high	ACTH levels – low
Cortisol levels– high	Cortisol levels– high	Cortisol levels– high

■ **Figure 7-20 Primary and secondary hypersecretion of cortisol** (a) Excessive CRH secretion at the hypothalamus causes secondary hypersecretion of cortisol. Levels of all three hormones in the axis are elevated and negative feedback fails. (b) A tumor in the pituitary might cause excessive secretion of ACTH, resulting in secondary hypersecretion of cortisol. The high levels of cortisol have a negative feedback effect on the hypothalamus. (c) In primary hypersecretion of cortisol, the problem arises in the adrenal gland itself. The high cortisol levels feed back and shut down secretion of CRH and ACTH.

nous cortisol hypersecretion or the exogenous administration of cortisol for therapeutic reasons (see Fig. 7-19 ■). In either case, high levels of cortisol act as a negative feedback signal that shuts off production of CRH and ACTH. The pattern of high cortisol with low trophic hormone levels points to a primary disorder. When the problem is an adrenal tumor that is churning out cortisol in an unregulated fashion, the normal control and feedback pathways are totally ineffective. The tumor must be removed or suppressed to correct the imbalance.

✓ Table 7-2 shows the three possible etiologies for hyposecretion of cortisol. Using your understanding of negative feedback in the hypothalamic-pituitary control axis, decide if the levels of CRH, ACTH, and cortisol will be high or low in each instance.

---

continued from page 192

Graves' disease causes the thyroid gland to produce too much thyroxine. People with Graves' disease therefore have elevated thyroxine levels in the blood. Their TSH levels are very low.

Question 4: If levels of TSH are low, but thyroxine levels are high, is Graves' disease a secondary disorder (arising as a result of a problem with the anterior pituitary or hypothalamus)? Explain your answer.

way out" in humans. Calcitonin is a good example of such a hormone. Although it plays a major role in calcium metabolism in fish, calcitonin apparently has no significant influence on daily calcium balance in humans: A deficiency is not associated with any pathological condition or symptom. Consequently, the results of animal experiments on calcitonin should be cautiously applied to human physiology.

HORMONE EVOLUTION

One point that should be raised before ending this discussion of hormones is the fact that hormone structure and function in many cases have changed very little from the most primitive vertebrates through the mammals. This situation is termed the *evolutionary conservation of hormone activity*. When Best and Banting discovered insulin in 1921 and the first diabetic patients were treated with the hormone, they were treated with insulin from cows, pigs, or sheep. Until recently, slaughterhouses were the major source of insulin for the medical profession. Now, with genetic engineering, the human gene for insulin has been inserted into bacteria, which then synthesize the hormone, providing us with a plentiful source of human insulin.

Many of our physiological models are based on research studies carried out on nonhuman vertebrates ranging from fish to frogs and rats. Although many hormones have the same function in most vertebrates, a few hormones that play a significant role in the physiology of lower vertebrates seem to be evolutionarily "on their

Do Humans Secrete Pheromones?   Pheromones [*pherein*, to bring] are hormones secreted into the external environment by an organism in order to affect other organisms of the same species. Pheromones, like other signal molecules, are used to communicate information. For example, sea anemones secrete alarm pheromones when danger threatens; ants release trail pheromones to attract fellow workers to food sources. Pheromones are also used to attract members of the opposite sex for mating purposes. Sex pheromones can be found throughout the animal kingdom, in animals from fruit flies to dogs. But do humans have pheromones? One perhaps relevant observation is that when women live together for an extended period of time, their menstrual cycles begin to synchronize (unless they are taking birth control pills). However, to date, experiments have not supported the existence of a pheromone that controls this phenomenon. Despite the lack of evidence for human pheromones, many people believe they exist, a fact that was counted on by the perfume manufacturer that named its fragrance simply "Pheromone."

TABLE 7-2   Patterns of Hormone Secretion in Hypocortisolism

For each condition, decide what will happen to the level of each hormone and circle the correct response (increased, decreased, no change).

	Secondary Problem Arising from Damage to the Pituitary				Primary Problem Arising from Atrophy of the Adrenal Cortex				Secondary Problem Arising from Damage to the Hypothalamus			
Hypothalamus	CRH	↑	↓	N/C	CRH	↑	↓	N/C	CRH	↑	↓	N/C
Anterior pituitary	ACTH	↑	↓	N/C	ACTH	↑	↓	N/C	ACTH	↑	↓	N/C
Adrenal cortex	Cortisol	↑	↓	N/C	Cortisol	↑	↓	N/C	Cortisol	↑	↓	N/C

Focus on **The Pineal Gland**

Theories about the pineal gland, a pea-sized structure buried deep in the brain of humans (Fig. 7-21 ■), have alternated between giving it a prominent role in human physiology and relegating it to obscurity. Nearly 2000 years ago, the pineal was thought to act as a valve that regulated the flow of vital spirits and knowledge into the brain, so it came to be known as "the seat of the soul." But, by 1950, scientists had decided that the human pineal was an evolutionary remnant with no known function. Then one of the wonderful coincidences of scientific research occurred. An investigator heard about a factor in beef pineal glands that could lighten the skin of amphibians. Using the classical methodology of endocrinology, he obtained pineal glands from a slaughterhouse and started making extracts. His biological assay consisted of dropping pineal extracts into bowls of live tadpoles to see if their skin color blanched. Several years and hundreds of thousands of pineal glands later, he had isolated a small amount of **melatonin,** an amine molecule synthesized by the pineal gland. This was the first evidence that the pineal gland produced a hormone. But what was its function?

Forty years later, we still do not know the role of the pineal gland in humans. We do know that melatonin is the "darkness hormone," secreted at night as we sleep. We know that in many seasonally breeding mammals, melatonin helps set the body clock that controls the onset of sexual activity. But the best we can say of the pineal gland in humans is that it transmits information about light-dark cycles to the center in the brain that governs the body's biological clock, using melatonin as the chemical message. Scientists and the popular press have tried to link melatonin to sexual function, the onset of puberty, seasonal affective depressive disorder (SADD) in the darker winter months, and sleep-wake cycles. To date, the only human function for melatonin that has been supported with scientific evidence is the hormone's ability to help shift the timing of the body's internal clock, making it useful in overcoming jet lag. As interest in melatonin grows, the pineal gland has come full cycle. No longer considered a vestigial structure

of unknown function, the pineal may turn out to hold the key to the biological rhythms that provide an underlying pattern to our lives.

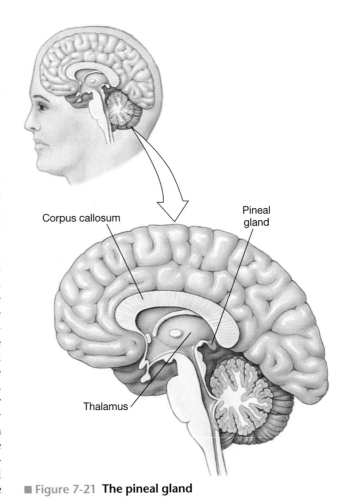

■ Figure 7-21 **The pineal gland**

Some endocrine structures that are important in lower vertebrates are vestigial [*vestigium*, footprint] in humans. For example, melanocyte-stimulating hormone (MSH) from the intermediate lobe of the pituitary controls pigmentation in reptiles and amphibians. But adult humans have only a rudimentary intermediate lobe and normally do not have measurable levels of MSH in their blood.

There are certainly many instances in which we have been able to use endocrinology that we have learned from other animals in our quest to understand the human body. The function of the pineal gland in humans is a good example (see Focus on the Pineal Gland, Fig. 7-21 ■).

This chapter has examined how the endocrine system with its hormones helps regulate the slower processes in the body. In the next chapter, you will learn how the nervous system can take care of the more rapid responses needed to maintain homeostasis.

continued from page 195

Researchers have learned that Graves' disease is a disorder involving the body's system of defense, the immune system. In this condition, the body produces proteins that mimic TSH and fool the thyroid gland into producing thyroid hormone. More women than men are diagnosed with Graves' disease, perhaps because of the influence of female hormones on thyroid function. Stress has also been implicated in hyperthyroidism. In fact, political pundits and physicians cite the stress of the Gulf War, the sluggish U.S. economy, and the impending presidential election as possible triggers for the Graves' disease that afflicted the Bushes in 1991.

Question 5: In Graves' disease, why doesn't negative feedback shut off thyroid hormone production before it becomes excessive?

CHAPTER REVIEW

Chapter Summary

Hormones

1. The **specificity** of a hormone depends on the location and structure of its receptors. (p. 174)

2. A **hormone** is a chemical, secreted by a cell or group of cells into the blood, that acts on a distant target and is effective at very low concentrations. (p. 174)

3. The rate of hormone breakdown is indicated by a hormone's **half-life.** (p. 175)

The Classification of Hormones

4. There are three types of hormones: **peptide hormones** composed of three or more amino acids, **steroid hormones** derived from cholesterol, and **amine hormones** derived from single amino acids. (p. 175)

5. Peptide hormones are made as inactive **preprohormones** and processed to **prohormones.** Prohormones are chopped into active hormone and peptide fragments that are co-secreted. (p. 178)

6. Peptide hormones dissolve in the plasma and have a short half-life. They bind to surface receptors and initiate cellular responses through second messenger systems. As a result, cells respond rapidly to peptide hormones. (p. 179)

7. Steroid hormones are synthesized as they are needed. They are hydrophobic, and most steroid hormones in the blood are bound to protein carriers. Steroids have an extended half-life. (p. 180)

8. The receptors for steroid hormones are inside the target cell. Steroid hormones enter the nucleus, turn on genes, and direct the synthesis of new proteins. Cell response, as a result, is slower than with peptide hormones. (p. 181)

9. Amine hormones are derived from a single amino acid. They may behave like typical peptide hormones or like a combination of steroid and peptide hormone. (p. 181)

Control of Hormone Release

10. **Trophic hormones** control the secretion of other hormones. (p. 182)

11. In many endocrine reflex pathways, the hormones of the pathway act as negative feedback signals. (p. 183)

12. Classic endocrine cells act as both receptor and integrating center in the reflex pathway. (p. 183)

13. Some classic hormones are controlled by multiple stimuli. (p. 184)

14. The pituitary gland has two parts. The **anterior pituitary** is a true endocrine gland, and the **posterior pituitary** is an extension of the brain. (p. 184)

15. The posterior pituitary releases two neurohormones, **oxytocin** and **vasopressin,** that are made in the hypothalamus. (p. 185)

16. The hormones of the anterior pituitary are controlled by releasing and inhibiting hormones from the hypothalamus. (p. 185)

17. The hypothalamic trophic hormones reach the pituitary through the **hypothalamic-hypophyseal portal system.** (p. 185)

18. There are six anterior pituitary hormones: **prolactin, growth hormone, follicle-stimulating hormone, luteinizing hormone, thyroid-stimulating hormone,** and **adrenocorticotrophic hormone.** (p. 186)

19. Hormones interact at their target cells. If the combination of two or more hormones yields a result that is more than additive, the interaction is **synergism.** If one hormone cannot exert its effects fully unless a second hormone is present, the second hormone is said to be **permissive** to the first. If one hormone opposes the action of another, they are **antagonistic.** (p. 187)

Endocrine Pathologics

20. Diseases of hormone excess are usually due to **hypersecretion.** Symptoms of hormone deficiency occur when too little hormone is secreted **(hyposecretion).** Changes in target tissue response may also result from abnormal interaction between the hormone and its receptor or from alterations in signal transduction pathways. (p. 192)

21. **Primary pathologies** arise in the last endocrine gland in a reflex. A **secondary pathology** is a problem with one of the tissues producing trophic hormones. (p. 193)

22. Many hormones that are active in humans are also found in other vertebrate animals. (p. 195)

Questions

1. The study of hormones is called _____.

2. List the three basic ways hormones act on their target cells.

3. List five endocrine glands and give one example of a hormone each secretes. Give one effect of each hormone you listed.

4. Match the researchers with their experiments:
   - (a) Lower
   - (b) Berthold
   - (c) Guillemin and Shalley
   - (d) Brown-Séquard
   - (e) Banting and Best

   1. isolated trophic hormones from the hypothalami of pigs and sheep
   2. claimed sexual rejuvenation after he injected himself with testicular extracts
   3. isolated insulin
   4. accurately described the function of the pituitary gland
   5. studied comb development in roosters that had been castrated

5. Put these steps for identifying an endocrine gland in order:
   a. Purify the extracts and separate the active substances.
   b. Perform replacement therapy with the gland or its extracts and see if the abnormalities disappear.
   c. Implant the gland or administer the extract from the gland to a normal animal and see if symptoms characteristic of hormone excess appear.
   d. Put the subject into a state of hormone deficiency by removing the suspected gland. Monitor the development of abnormalities.

6. In order for a chemical to be defined as a hormone, it must be secreted into the _____ for transport to a _____ and take effect at _____ concentrations.

7. What is meant by the term *half-life* in connection with the activity of hormone molecules?

8. Metabolites are inactivated hormone molecules, broken down by enzymes found primarily in the _____ and _____, to be excreted in the _____ and _____, respectively.

9. Candidate hormones often have the word _____ as part of their name.

10. List and define the three chemical classes of hormones; give an example of one hormone in each class.

11. Decide if each statement below applies best to peptide hormones, steroid hormones, amine hormones, all of these, or none of these.
    a. all are lipophobic, so cannot enter the target cell and must use a signal transduction system
    b. short half-life, measured in minutes
    c. usually trigger the synthesis of new proteins, so a lag time of 90 minutes is not uncommon
    d. water-soluble, therefore easily dissolve in the extracellular fluid for transport
    e. most hormones belong to this class
    f. all hormones in this group are derived from cholesterol
    g. consist of three or more amino acids linked together
    h. released into the blood to travel to a distant target organ
    i. transported in the blood bound to protein carrier molecules
    j. all are lipophilic, so diffuse easily across membranes

12. Give one reason why steroid hormones are transported through the blood attached to a carrier protein.

13. The ultimate target of steroid hormones is the _____ of the cell, where the hormone-receptor complex acts as a/an _____ factor, binds to DNA, and activates one or more _____, which create mRNA to direct the synthesis of new _____.

14. Researchers have discovered that some cells have additional steroid hormone receptors on their _____, enabling a faster response.

15. The amino acids that give rise to the amine hormones are _____, the basis for melatonin, and _____, from which the catecholamines and thyroid hormones are made.

16. A hormone that controls the secretion of another hormone is known as a _____ hormone.

17. In reflex control pathways involving trophic hormones and multiple integrating centers, the hormones themselves act as _____ _____ signals, suppressing trophic hormone secretion earlier in the reflex.

18. What characteristic defines neurohormones?

19. List the two hormones secreted by the posterior pituitary gland. To what chemical class do they belong?

20. What is the hypothalamic-hypophyseal portal system? Why is it important?

21. List the six hormones of the anterior pituitary gland; give an action of each. Which one(s) are considered to be trophic hormones?

22. How do long- and short-loop negative feedback differ? Give an example.

23. When two hormones work together to create a result that is greater than additive, that interaction is called _____. When two hormones must both be present to achieve full expression of an effect, that interaction is called _____. When hormone activities oppose each other, that effect is called _____.

LEVEL TWO Reviewing Concepts

24. Map the major endocrine glands of the body, their secretions, their targets, and the effects of the hormones on those target organs. Start with a big sheet of paper, and leave room to expand these ideas as you study the hormones in subsequent chapters. You may even enjoy working ahead now on some of the specific hormones that interest you.

25. Compare and contrast the following sets of terms:
    a. paracrine, hormone, cytokine
    b. primary and secondary endocrine pathologies
    c. hypersecretion and hyposecretion
    d. anterior and posterior pituitary glands

26. Use these terms in a short paragraph that describes peptide hormone synthesis and release: endoplasmic reticulum, Golgi apparatus, exocytosis, preprohormone, prohormone, co-secretion, vesicle.

27. Compare and contrast the cellular mechanism of action of the three chemical classes of hormones.

LEVEL THREE Problem Solving

28. You have encountered the terms *specificity, receptors, up-regulation,* and *down-regulation* in other chapters in this text. Have their meanings changed? What chemical and physical characteristics do hormones, enzymes, and transport proteins have in common that makes specificity important? What makes it possible? Compare these molecules with the structure of receptors.

29. Dexamethasone is a drug that is used to suppress the secretion of adrenocorticotrophic hormone (ACTH) from the anterior pituitary (see Fig. 7-17). Two patients with hypersecretion of cortisol are given dexamethasone. Patient A's cortisol secretion level falls to normal levels as a result, but patient B's cortisol secretion remains elevated. Which patient has primary hypercortisolism, also known as Cushing's syndrome? Explain your reasoning.

30. Some birth control pills consist of estrogen and progesterone in various levels. Devise a possible feedback loop to explain how administering additional amounts of these hormones and thus raising their levels in the blood would affect gonadotropin-releasing hormone (GnRH) and lead to birth control. Is this a long-loop or a short-loop feedback circuit? Include each particular organ or important structure involved. (Hint: See Fig. 7-15.)

Problem Conclusion

In this running problem, you have learned that the thyroid gland concentrates iodine to use for synthesis of thyroid hormones. You have also learned that TRH and TSH control the secretion of thyroid hormones. In Graves' disease, thyroid hormone levels are high due to a immune system protein that mimics TSH.

Further check your understanding of this running problem by checking your answers against those in the summary table.

Question	Facts	Integration and Analysis
1a To which of the three classes of hormones do the thyroid hormones belong?	The three classes of hormones are peptides, steroids, and amines.	Thyroid hormones are made from the amino acid tyrosine; therefore they are amines.
1b If a person's diet is low in iodine, predict what happens to thyroxine production.	In order to make thyroid hormones, the thyroid gland concentrates iodine and combines it with the amino acid tyrosine.	Iodine is an essential part of thyroid hormones; therefore, if it is lacking in the diet, a person will be unable to make thyroid hormone.
2 If thyroid hormones are needed for the regulation of oxygen consumption and the maintenance of body temperature, predict possible symptoms of hyperthyroidism.	Oxygen consumption is a result of oxidative phosphorylation in the mitochondria. In this process, some of the energy released from nutrients such as glucose is released as heat while the rest is trapped in the high-energy bond of ATP (∞ p. 33).	People who are hyperthyroid will have an increased rate of oxygen consumption. Because oxidative phosphorylation requires nutrients as substrates, these people will have increased appetite and/or will lose weight. In addition, the extra heat they generate will tend to make them intolerant of hot environmental conditions.
3 Why is radioactive iodine rather than some other radioactive element such as cobalt used to destroy thyroid tissue?	In order to make thyroid hormones, the thyroid gland concentrates iodine.	Radioactive iodine will be concentrated in the thyroid gland and therefore selectively destroys that tissue. Other radioactive elements would distribute more widely throughout the body and might harm normal tissues.
4 If levels of TSH are low, but thyroxine levels are high, is Graves' disease a secondary disorder (arising as a result of a problem with the anterior pituitary or hypothalamus)? Explain your answer.	In secondary hypersecretion disorders, you would expect the levels of the hypothalamic and/or anterior pituitary trophic hormone levels to be elevated.	In Graves' disease, the level of the anterior pituitary trophic hormone TSH is very low. Therefore, the hypersecretion of thyroid hormones is not the result of hypersecretion of TSH. This means that Graves' disease is not a secondary disorder.
5 In Graves' disease, why doesn't negative feedback shut off thyroid hormone production before it becomes excessive?	In the normal negative feedback loop, as thyroid hormone levels increase, they shut off TSH. Consequently, without TSH stimulation, the thyroid stops producing thyroid hormone.	In Graves' disease, the high levels of thyroid hormone have shut off TSH production. However, the thyroid gland is producing hormone in response to the abnormal immune protein, not in response to TSH. Therefore, thyroid hormone production continues despite negative feedback.

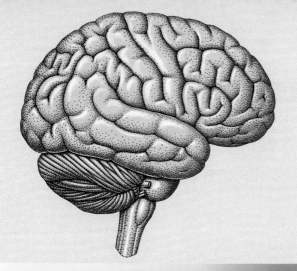

8
The Nervous System

I n an eerie scene from a science fiction movie, white-coated technicians move quietly through a room filled with bubbling cylindrical fish tanks. As the camera zooms in on one tank, no fish are seen darting through aquatic plants. The lone occupant of the tank is a gray mass with a convoluted surface like a walnut and a long tail that appears to be edged with beads. Floating off the beads are hundreds of fine fibers, waving softly as the oxygen bubbles weave through them on their way to the surface. This is no sea creature . . . it is a brain and spinal cord, removed from its original owner and awaiting transplantation into another body. Can this be real? Is this scenario possible? Or is it just the creation of an imaginative movie screenwriter?

The brain is regarded as the seat of the soul, the mysterious source of those traits that we think of as setting humans apart from animals. The brain and spinal cord are also integrating centers for homeostasis, movement, and many other body functions. They are the control center of the **nervous system,** a network of billions or trillions of nerve cells linked together in a highly organized manner to form the rapid control system of the body.

Nerve cells, or **neurons,** are designed to carry electrical signals rapidly over long distances. They are uniquely shaped cells, and most have long, thin extensions, or *processes,* that can extend up to a meter in length. Neurons link together to form reflex pathways. In most pathways, neurons communicate with chemical signals called **neurotransmitters** across a small gap known as a *synapse*. In a few pathways, neurons are linked by gap junctions (∞ p. 147), allowing electrical signals to pass directly from cell to cell.

Reflex pathways in the nervous system do not necessarily follow a straight line from one neuron to the next. One neuron may influence multiple neurons, or many neurons may affect the function of a single neuron. Because of the complexity of the nervous system, we still do not fully understand how it functions. As a result, neurophysiology is one of the most active research areas in physiology today.

Before you begin your study of the nervous system, you may want to become familiar with some specialized language used by neuroscientists. Table 8-1 lists some terms used in this book along with some common synonyms.

TABLE 8-1   Synonyms in Neuroscience

Term Used in this Book	Synonym
Action potential	Spike, nerve impulse, conduction signal
Axonal transport	Axoplasmic flow
Axon terminal	Synaptic knob, synaptic bouton, presynaptic terminal
Axoplasm	Cytoplasm of the axon
Cell body	Cell soma
Cell membrane of the axon	Axolemma
Glial cells	Neuroglia, glia
Rough endoplasmic reticulum	Nissl substance
Sensory neuron	Afferent neuron, afferent

Problem

Mysterious Paralysis

"Like a polio ward from the 1950s" is how Guy McKhann, M.D., a neurology specialist at the Johns Hopkins School of Medicine, describes a section of Beijing Hospital that he visited on a trip to China in 1986. Dozens of paralyzed children—some attached to respirators to assist their breathing—filled the ward to overflowing. The Chinese doctors were convinced that the children had Guillain-Barré syndrome (GBS), an usually rare paralytic condition. But Dr. McKhann immediately sensed that the diagnosis was wrong. There were simply too many stricken children for the illness to be the rare Guillain-Barré syndrome. Was it polio—as some of the Beijing staff feared? Or was it another illness, perhaps one that had not yet been discovered?

continued on page 204

ORGANIZATION OF THE NERVOUS SYSTEM

Information flow through the nervous system follows the basic pattern of a reflex described in Chapter 6 (∞ p. 158). Sensory receptors throughout the body continuously monitor conditions in both the internal and external environments (Fig. 8-1 ■). These receptors send information along **afferent,** or **sensory, neurons** to the central nervous system. (Afferent pathways carry incoming information.) The **central nervous system,** or **CNS,** is composed of the **brain** and **spinal cord.** The neurons of the central nervous system integrate incoming information and determine if a response is needed. Signals directing an appropriate response (if any) are sent back to the effector cells of the body through **efferent neurons.** The afferent and efferent neurons together form the **peripheral nervous system.**

The efferent neurons of the peripheral nervous system are subdivided into the **somatic motor division,*** which controls skeletal muscles, and the **autonomic division,** which controls smooth and cardiac muscles, exocrine and some endocrine glands, and some types of adipose tissue.

The autonomic division is also called the *visceral nervous system* because it controls contraction and secretion in the various internal organs [*viscera,* internal organs]. Autonomic neurons are further divided into **sympathetic** and **parasympathetic divisions,** which can be distinguished by their anatomical organization and by the chemicals that they use to communicate with their target cells. Many internal organs receive innervation from both types of autonomic neurons, and it is a common pattern to find that the two divisions exert antagonistic control over a single target (∞ p. 157).

In recent years, a third division of the nervous system has been receiving attention. The **enteric nervous system** is a network of neurons in the walls of the digestive tract. It is frequently controlled by the autonomic division of the nervous system, but it is also able to function autonomously. The enteric nervous system will be discussed further in Chapter 20 on the digestive system.

CELLS OF THE NERVOUS SYSTEM

The nervous system is composed primarily of two cell types: support cells known as glial cells (or glia or neuroglia) and nerve cells, or neurons, the basic signaling units of the nervous system.

---

*According to some neuroscience texts, all efferent neurons can be called motor neurons. However, clinically, the term *motor neuron* (or *motoneuron*) is used to describe a somatic motor neuron that controls skeletal muscles. For that reason, this book will restrict the use of *motor* to somatic motor neurons.

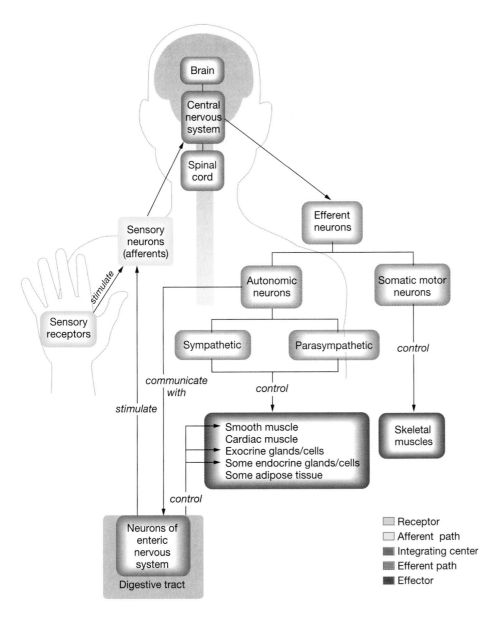

■ Figure 8-1 **Organization of the nervous system** The central nervous system (CNS) consists of the brain and spinal cord. The peripheral nervous system sends information to the CNS through afferent (sensory) neurons and takes information from the CNS to the target cells via efferent neurons. The neurons of the enteric nervous system can act autonomously or can be controlled by the central nervous system through the autonomic division.

Neurons Are Excitable Cells That Generate and Carry Electrical Signals

The neuron is the functional unit of the nervous system. A *functional unit* is the smallest structure that can carry out all of the functions of the system. Neurons are uniquely shaped cells with long appendages that extend outward from the cell body. These appendages, or processes, are usually classified as either dendrites or axons. The shape, number, and length of axons and dendrites vary from one neuron to the next, but they are an essential feature that allows neurons to communicate with each other and with other cells. Many neurons are similar to the model neurons shown in Figure 8-2 ■. In some neurons, processes shown in the model may be missing or modified. However, if you understand these model neurons, you will understand how all neurons work.

Neurons in the nervous system may be classified either structurally or functionally. Structurally, neurons are classified by the number of processes that originate from the cell body. But, because physiology is concerned chiefly with function, we will classify neurons according to their function: (afferent) sensory neurons, interneurons, and efferent neurons.

Sensory neurons carry information about temperature, pressure, light, and similar stimuli from sensory receptors to the central nervous system. Structurally, they vary from the model neuron. For example, somatic sensory neurons have cell bodies that are found close to the central nervous system, with very long processes that extend to receptors in the limbs. In these neurons, the cell body is out of the direct path of signals originating at the dendrites (Fig. 8-3a ■). In contrast, sensory neurons for smell and vision have two long extensions, one an axon and one a dendrite (Fig. 8-3b ■). Signals

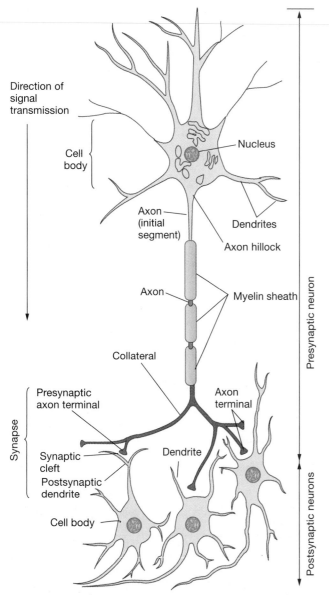

Direction of signal transmission

Cell body

Nucleus

Axon (initial segment)

Dendrites

Axon hillock

Axon

Myelin sheath

Presynaptic neuron

Collateral

Presynaptic axon terminal

Axon terminal

Synapse

Synaptic cleft

Postsynaptic dendrite

Dendrite

Cell body

Postsynaptic neurons

■ **Figure 8-2 Model neurons** The neurons illustrated are typical neurons, consisting of a cell body and numerous extensions. Dendrites receive information from other neurons or from sensory receptors. Axons carry information from the integrating center in the cell body to the axon terminals. The region where axon terminals communicate with their target cells is known as a synapse.

that begin at the dendrites travel through the cell body to the axon.

Neurons that lie entirely within the central nervous system are known as **interneurons** (short for *interconnecting neurons*). They come in a variety of forms but often have quite complex branching processes that allow them to form synapses with many other neurons (Fig. 8-3c ■).

The **efferent neurons** of the somatic motor and autonomic divisions are generally very similar to the model neuron in Figure 8-2 ■. In the autonomic division, some neurons have enlarged regions along the axon called *varicosities*. The varicosities store and release neurotransmitter.

continued from page 202

Guillain-Barré syndrome is a rare paralytic condition that strikes after a viral infection or immunization. There is no cure, but the illness usually resolves spontaneously. In Guillain-Barré, patients can neither feel sensations nor move their muscles.

Question 1: Which division(s) of the nervous system may be involved in Guillain-Barré syndrome?

The cell body is the control center of the neuron

The nerve **cell body** resembles the typical cell described in Chapter 3, with a nucleus and all organelles needed to direct cellular activity (∞ p. 45). An extensive cytoskeleton extends outward into the axon and dendrites. The position of the cell body varies in different types of neurons, but in most neurons the cell body is small, generally making up one-tenth or less of the total cell volume. Despite its small size, the cell body with its nucleus is essential to the well-being of the cell. If a neuron is cut apart, any sections separated from the cell body are likely to degenerate slowly and die. With injured somatic motor neurons, the degeneration of the distal (distant) portions can cause permanent paralysis of the muscles that are *innervated* by the neuron. (The term *innervate* means "linked to a neuron.") If the damaged neuron is a sensory neuron, the person may experience numbness or tingling in the region innervated by the neuron.

Dendrites receive incoming signals

Dendrites [*dendron*, tree] are thin, branched processes whose main function is to receive incoming signals. Dendrites increase the surface area of the neuron so that it can communicate with multiple other neurons. Simple neurons may have only a single dendrite, whereas cells in the brain may have multiple dendrites with incredibly complex branching (Fig. 8-3c ■). Signals reaching the dendrites are transferred to an integrating region within the neuron.

Axons carry outgoing signals to the target

Most neurons have a single **axon**, a projection of the cell body that originates from a special region of the cell body called the **axon hillock** (Fig. 8-2 ■). Axons vary from over a meter in length to only a few micrometers long. (A **nerve** is a collection of axons that carry information between the central nervous system and receptors or target cells.) Functionally, axons transmit electrical signals from the integrating center of the neuron to the end of the axon. At the distal end of the axon, the electrical signal is usually translated into a chemical message with the secretion of a *neurocrine* molecule (∞ p. 148).

Axons may branch sparsely along their length, forming *collaterals* [*col-*, with + *lateral*, something on the side]. In our model neuron, each collateral ends in a swelling

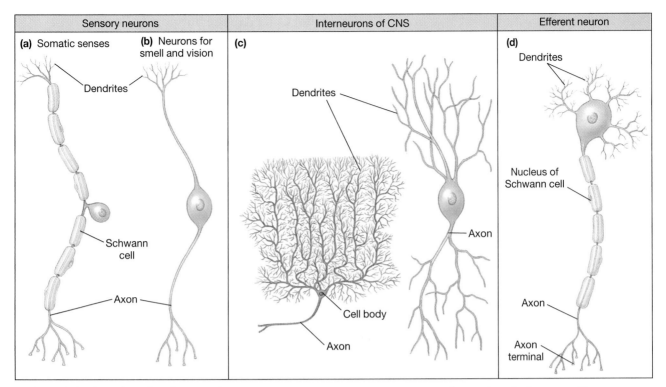

Sensory neurons	Interneurons of CNS	Efferent neuron

(a) Somatic senses   **(b)** Neurons for smell and vision

Dendrites

Schwann cell

Axon

(c)

Dendrites

Cell body

Axon

Axon

(d)

Dendrites

Nucleus of Schwann cell

Axon

Axon terminal

■ **Figure 8-3** **Three functional types of neurons** (a) Many sensory neurons appear to have only a single process. However, during development, the axon and dendrite fused, creating a single process called an axon. (b) Some sensory neurons have two relatively equal fibers extending off the central cell body. These sensory neurons are found in the nose and eyes. (c) The interneurons of the CNS are highly branched but without long extensions. (d) The typical efferent neuron has five to seven dendrites coming off the cell body, each branching four to six times. The single long axon may branch several times. Somatic motor neurons such as the one illustrated terminate at swollen regions called axon terminals or synaptic boutons.

called an **axon terminal.** The axon terminal contains mitochondria and membrane-bound vesicles filled with neurocrine molecules. Some axon terminals are found near blood vessels so that neurohormones can enter the circulation. Neurons that secrete neurotransmitters and neuromodulators end near their target cells, which are other neurons, muscles, or glands. The region where an axon terminal meets its target cell is called a **synapse** [*syn-*, together + *hapsis*, to join]. The neuron that delivers the signal to the synapse is known as the **presynaptic cell,** and the cell that receives the signal is called the **postsynaptic cell.**

The cytoplasm of axons is filled with many types of fibers and filaments but lacks ribosomes and endoplasmic reticulum. Therefore, any proteins destined for the axon, even the distant axon terminal, must be synthesized on the rough endoplasmic reticulum in the cell body. The proteins are then packaged into vesicles and moved down the axon by a process known as **axonal transport.**

Axonal transport has two components: slow movement through **axoplasmic** (cytoplasmic) **flow** and rapid movement that uses energy to "walk" vesicles and mitochondria along stationary microtubules. **Slow axonal transport** moves material from the cell body to the axon terminal at a rate of 0.2–2.5 mm/day. It carries components that are not consumed rapidly by the cell, such as

enzymes and cytoskeleton proteins. **Fast axonal transport** (Fig. 8-4 ■) goes in two directions. Forward, or anterograde, transport takes synaptic and secretory vesicles and mitochondria from the cell body to the axon terminal. Backward, or retrograde, transport returns old membrane components from the axon terminal to the cell body for recycling. There is evidence that nerve growth factors and some viruses also reach the cell body by fast retrograde transport. The energy-requiring process of fast axonal transport moves organelles at rates of up to 400 mm (about 15.75 in.) per day.

Foot Proteins for Fast Axonal Transport   Fast axonal transport takes place along stationary microtubules that run the length of the axon from cell body to terminal. The microtubules act as tracks along which transported particles "walk" with the aid of attached footlike proteins. These foot proteins, called **kinesins** [*kinesis*, motion], alternately bind and unbind to the microtubules with the help of ATP, stepping their attached organelles down the axon in a stop-and-go fashion. The role of proteins in axonal transport is similar to their role in muscle contraction and in the movement of chromosomes during cell division.

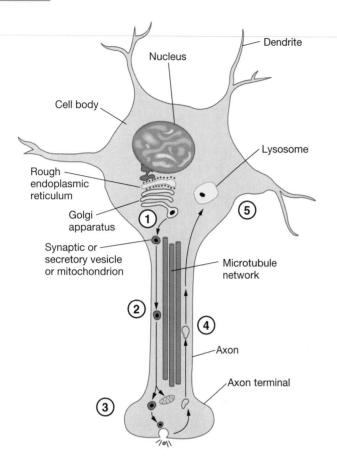

■ **Figure 8-4 Axonal transport of membranous organelles**
① Peptides are synthesized in the cell body and packaged into secretory vesicles. ② These vesicles, along with mitochondria and synaptic vesicles filled with neurotransmitter, are transported down the axon by fast axonal transport. The vesicles are "walked" along a network of microtubules by protein "feet." ③ In the axon terminal, the vesicles are stored or their contents are released. ④ Old membrane components are transported back to the cell body by backward (retrograde) fast axonal transport. ⑤ Back in the cell body, they are digested in lysosomes.

Glial Cells Are the Support Cells of the Nervous System

Glial cells [*glia,* glue] are the "unsung heroes" of the nervous system, outnumbering neurons in the nervous system by 10–50:1 (Fig. 8-5 ■). Although these neural support cells do not participate directly in the transmission of electrical signals over long distances, they do communicate with neurons and with each other using electrical and chemical signals. Glial cells provide physical support for neurons since neural tissue has very little extracellular matrix (∞ p. 53). They also direct the growth of neurons during repair and development. **Schwann cells** are the major type of glial cell associated with the peripheral nervous system, whereas the central nervous system has several types of support cells. **Oligodendrocytes** wrap around axons to support and insulate them. **Astrocytes** are in contact with both neurons and blood vessels and may transfer nutrients between the two. In addition, they take up potassium ions from the extracellular fluid. **Microglia** are specialized immune cells rather than true support cells; they remove damaged cells and foreign invaders.

One important function of glial cells is the formation of **myelin,** multiple concentric layers of phospholipid membrane that form an insulating sheath around portions of neurons. Myelin is created when Schwann cells or oligodendrocytes wrap around an axon, squeezing out the cytoplasm of the glial cell so that each wrap becomes two membrane layers (Fig. 8-6c ■). As an analogy, think of wrapping a deflated balloon tightly around a pencil. Some neurons have as many as 150 wraps (300 membrane layers) in the myelin sheath that surrounds their axons.

Myelin formation is different in the central and peripheral divisions of the nervous system. In the central nervous system, one oligodendrocyte forms myelin around

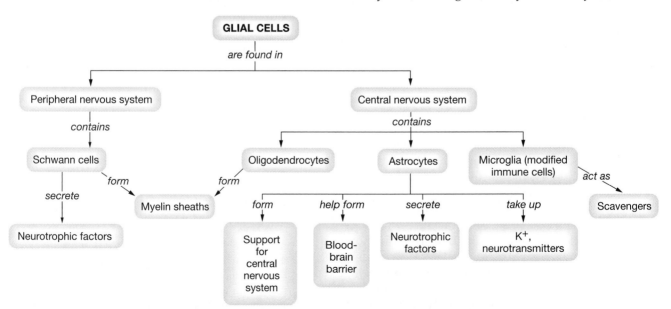

■ **Figure 8-5 Map of types and functions of glial cells**

(a)

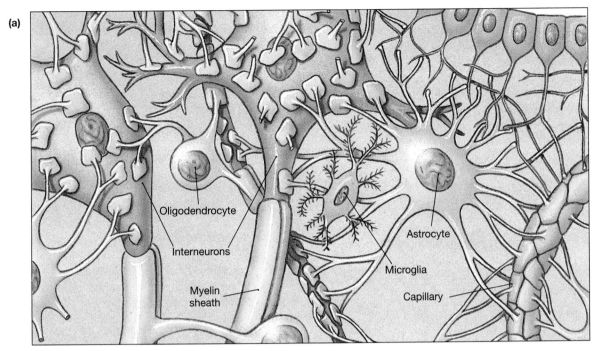

(b)

(c)

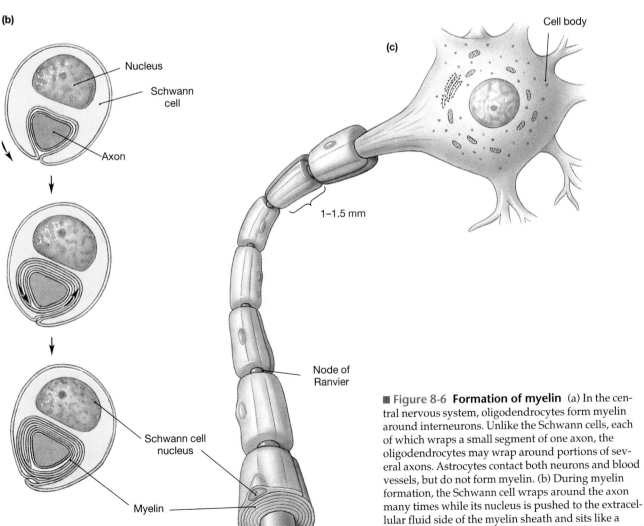

■ **Figure 8-6** **Formation of myelin** (a) In the central nervous system, oligodendrocytes form myelin around interneurons. Unlike the Schwann cells, each of which wraps a small segment of one axon, the oligodendrocytes may wrap around portions of several axons. Astrocytes contact both neurons and blood vessels, but do not form myelin. (b) During myelin formation, the Schwann cell wraps around the axon many times while its nucleus is pushed to the extracellular fluid side of the myelin sheath and sits like a bulge on the outer surface of the axon. (c) The axons of peripheral neurons are wrapped in myelin formed from Schwann cells.

portions of several axons (Fig. 8-6a ■). In the peripheral division, each Schwann cell associates with a single axon (Fig. 8-6b ■). One axon may have as many as 500 Schwann cells, each wrapped around a 1–1.5 mm segment of the axon. Between the insulated areas, a tiny region of axon membrane remains in direct contact with the extracellular fluid. These gaps, called the **nodes of Ranvier,** play an important role in the transmission of electrical signals along the axon.

ELECTRICAL SIGNALS IN NEURONS

The unique property of nerve and muscle cells that characterizes them as *excitable tissues* is their ability to generate and propagate electrical signals. This section of the chapter looks at the nature of these electrical signals and how they are created. Two basic types of electrical signals travel through neurons: graded potentials, which travel over short distances, and action potentials, which travel very rapidly over longer distances.

Changes in the Membrane Potential Create Electrical Signals

You should recall from Chapter 5 that all living cells have a resting membrane potential difference. The membrane potential, as it is called for short, is the electrical disequilibrium that results from the uneven distribution of ions across the cell membrane (∞ p. 139).

Two factors influence the membrane potential: (1) the concentration gradients of different ions across the membrane and (2) the permeability of the membrane to those ions. Sodium ions (Na^+) are more concentrated in the extracellular fluid than inside the cells; potassium ions (K^+) are more concentrated in cells than in extracellular fluid. At the same time, the resting cell membrane is much more permeable to K^+ than to Na^+. This means that, although strong concentration gradients favor movement of both ions across the cell membrane, the resting cell admits very little Na^+. Consequently, K^+, which can leak out of the cell down its concentration gradient, is the major ion contributing to the average resting membrane potential of −70 mV (inside the cell relative to outside).

Membrane potential (V_m) is measured experimentally by inserting one electrode into a cell and placing a second electrode in the extracellular fluid (Fig. 8-7 ■). By convention, the extracellular electrode serves as the ground and is set to 0 mV. The relative charge inside the cell is then measured, using a voltmeter. If a chart recorder is connected to the voltmeter, a recording of the membrane potential difference versus time is obtained. In the diagrams in this chapter, recordings of membrane potential will often be accompanied by figures showing an electrode inserted into a neuron at the location from which the recording was made.

Figure 8-8 ■ is a recording of membrane potential plotted against time. The membrane potential (V_m) begins at a steady resting value of −70 mV. If the tracing moves upward, the potential *difference* between the inside of the cell (−70 mV) and the outside (0 mV) *decreases*, and the cell is said to have **depolarized.** If the tracing moves downward, the membrane potential becomes more negative, the potential *difference* has *increased*, and the cell has **hyperpolarized.** A return to the resting membrane potential from either direction is termed **repolarization.**

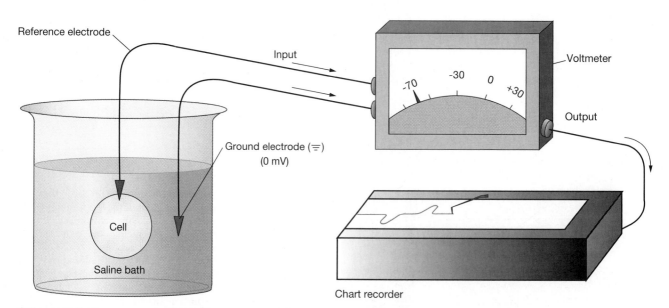

■ **Figure 8-7 Measurement of membrane potential** The membrane potential of a living cell is measured by placing a recording electrode inside the cell and measuring the electrical potential difference between it and the reference or ground electrode that is placed in the extracellular fluid. The voltmeter artificially assigns a value of zero millivolts to the reference electrode. A chart recorder hooked into the voltmeter shows changes in membrane potential over time.

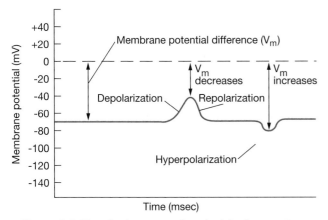

■ Figure 8-8 Terminology associated with changes in membrane potential If the cell becomes less negative, the membrane potential difference decreases and the cell depolarizes. If the cell becomes more negative, the membrane potential difference increases and the cell hyperpolarizes.

A major point of confusion that arises in neurophysiology is the use of the phrases "the membrane potential decreased" or "the membrane potential increased." Normally, in English, we associate "increase" with becoming more positive and "decrease" with becoming more negative—the opposite of what is happening in this case. One way to avoid confusion is to add the word "difference" after "membrane potential." If the membrane potential *difference* is increasing, the V_m must be moving away from the ground value of zero and becoming more negative. If the membrane potential *difference* is decreasing, the V_m is moving closer to the ground value of 0 mV and is becoming less negative.

Another way to keep the terminology straight is to relate an *increase* in membrane potential difference to *hyper*polarization, because the prefix *hyper-* usually means an increase in a value or parameter. Similarly, the prefix *de-* implies a diminution in a value, so *de*polarization can be linked to a *decrease* in membrane potential difference.

✓ Ouabain is a poison that inhibits the sodium-potassium pump. What happens to the resting membrane potential of a neuron over time if it is treated with ouabain?

✓ Most neurons are permeable to Cl^- so the ion distributes itself passively across the cell membrane until the concentration gradient of Cl^- is equal and opposite to the electrical gradient for Cl^-. A few cells contain an active pump for Cl^- that moves the ion out of the cell against its concentration gradient. If the ratio of Cl^- in the extracellular fluid to intracellular fluid is 11:1 for cells without the active pump, what might it be for a cell that has the chloride pump?

Ion Movement across the Cell Membrane Creates Electrical Signals

Although Na^+ contributes minimally to the resting membrane potential, it plays a key role in generating electrical signals in excitable tissues. At rest, the cell membrane of a neuron is almost impermeable to Na^+. However, if the membrane suddenly increases its Na^+ permeability, Na^+ will cross the membrane. With Na^+ more concentrated in the extracellular fluid, the concentration gradient favors Na^+ diffusion into the cell. In addition, the membrane potential with its net negative charge inside the cell attracts the positive Na^+. The addition of positive ions to the intracellular fluid depolarizes the cell membrane and creates an electrical signal.

The movement of ions across the membrane can also hyperpolarize a cell. If the cell membrane suddenly becomes more permeable to K^+, positive charge is lost from inside the cell and the cell becomes more negative (hyperpolarizes). A cell may also hyperpolarize if negatively charged ions such as Cl^- enter the cell from the extracellular fluid.

A significant change in membrane potential occurs with the movement of very few ions. For example, to change the membrane potential by 100 mV, only one out of every 100,000 K^+ must enter or leave the cell. This is such a tiny fraction of the total number of K^+ in the cell that the concentration of K^+ remains essentially unchanged. An equivalent situation is picking up one grain of sand (one K^+) from a beach: There are so many grains of sand on the beach that the loss of one grain is not significant.

Gated Ion Channels Control the Ion Permeability of the Neuron

For a cell to change its ion permeability, existing ion channels in the membrane must open or close or the cell must insert new channels into the membrane. Neurons contain a variety of gated ion channels that alternate between open and closed states, depending on the intracellular and extracellular conditions (∞ p. 113).

Ion channels are usually named according to the primary ion(s) that they allow to pass through them. For example, Na^+ channels permit Na^+ to move across the membrane. Other channels are less specific, such as the nonspecific monovalent cation channels through which both Na^+ and K^+ pass. The ratio of Na^+ to K^+ moving through such a channel depends on the relative electrochemical gradients for the two ions. There are four major types of ion channels in the neuron: (1) Na^+ channels, (2) K^+ channels, (3) Ca^{2+} channels, and (4) Cl^- channels.

Sodium permeability in neurons increases when Na^+ channels in the membrane open in response to external or internal signals. Mechanically gated Na^+ channels are found in sensory neurons and open in response to physical forces such as pressure or light. Chemically gated Na^+ channels in neurons respond to a variety of ligands found in the internal and external environments and to neurotransmitters and neuromodulators. Voltage-gated Na^+ channels play an important role in the conduction of electrical signals along the axon.

Sodium is not the only ion to cross neuronal membranes through gated ion channels. The voltage-gated K^+ channels of the axon open in response to depolarization of the membrane, allowing K^+ to flow from the cytoplasm to the extracellular fluid, down its concentration gradient. Voltage-gated Ca^{2+} channels in the membrane of the axon terminal also open in response to depolarization. When these channels open, Ca^{2+} enters the cell from the extracellular fluid, moving down its electrochemical gradient. Calcium ion entry serves as a signal to initiate exocytosis of neurotransmitter into the synapse (∞ p. 127). Chemically gated Cl^- channels are opened by a variety of neurotransmitters. When open, Cl^- channels allow Cl^- to move into the cell down its concentration gradient, hyperpolarizing the cell.

To summarize: Electrical signaling in neurons results from changes in the cell membrane potential. Ion channels in the cell membrane open in response to a variety of stimuli, including mechanical force, the binding of a chemical ligand such as a neurotransmitter, or a change in the cell membrane potential. When ion channels open, ions move into or out of the cell, their direction of movement depending on the concentration and electrical gradients for the ion. Potassium ions usually move out of the cell; Na^+, Cl^-, and Ca^{2+} usually flow into the cell. The net movement of electrical charge across the membrane depolarizes or hyperpolarizes the cell, creating an electrical signal.

There are two electrical signals in neurons: graded potentials and action potentials. Graded potentials are variable-strength signals that lose strength as they travel through the cell. Action potentials are signals that can travel for long distances through the neuron without losing strength. The relationship between these two types of signals is discussed in the sections that follow.

✓ Would the cell membrane depolarize or hyperpolarize if a small amount of Na^+ leaked into the cell?

Graded Potentials Reflect the Strength of the Stimulus That Initiates Them

Graded potentials are depolarizations or hyperpolarizations that occur in the dendrites, cell body, or less frequently, near the axon terminals. They are called "graded" because the *amplitude* (strength) of a graded potential is directly proportional to the strength of the triggering event [*amplitudo*, large]. A large stimulus will cause a strong graded potential, and a small stimulus will result in a weak graded potential.

Graded potentials begin on the cell membrane at the point where ions enter from the extracellular fluid (Fig. 8-9 ■). For example, a neurotransmitter combines with membrane receptors on the dendrite, opening Na^+ channels. Sodium ions move into the neuron down their electrochemical gradient, bringing in electrical energy. The positive charge carried by the Na^+ spreads as a wave of depolarization through the cytoplasm, just as a stone thrown into water creates ripples or waves that spread outward from the point of entry. The wave of depolarization that moves through the cell is known as **local current flow.** (By convention, current flow in biological systems is the net movement of positive electrical charge. Local current flow in action potentials is discussed later.)

The strength of the initial depolarization in a graded potential is determined by how much charge enters the cell, just as the size of waves in the water is determined by the size of the stone. If more Na^+ channels open, more Na^+ enters and the graded potential has a higher initial amplitude. The stronger the initial amplitude, the farther the graded potential can spread through the neuron before it dies out.

Graded potentials may be hyperpolarizing as well as depolarizing. For example, opening Cl^- or K^+ channels in the neuron membrane will create a hyperpolarizing graded potential.

Graded potentials typically occur in the cell body and dendrites of a neuron. In neurons of the central and efferent nervous systems, graded potentials occur when chemical signals from other neurons open chemically gated ion channels.

Graded potentials travel through neurons until they reach the region known as the **trigger zone.** In efferent neurons, the trigger zone is at the axon hillock and in the very first part of the axon, a region known as the **initial segment** (see Fig. 8-2 ■). In sensory neurons, the trigger zone is immediately adjacent to the receptor, where the dendrites join the axon.

The trigger zone is the integrating center of the neuron. If graded potentials reaching the trigger zone depolarize the membrane to a minimum level known as the **threshold** voltage, an *action potential* is initiated. If the depolarization does not reach threshold, no action potential is begun, and the graded potential simply dies out. Because depolarization is necessary to *excite* the cell to fire an action potential, a depolarizing graded potential is known as an **excitatory postsynaptic potential (EPSP).** A hyperpolarizing graded potential moves the membrane potential farther from the threshold value and makes the neuron less likely to fire an action potential. Consequently, hyperpolarizing graded potentials are called **inhibitory postsynaptic potentials (IPSPs).**

Figure 8-10 ■ shows a neuron that has several recording electrodes placed at intervals along the cell body and trigger zone. In the first example (Fig. 8-10a ■), a single stimulus triggers a *subthreshold* excitatory postsynaptic potential, one that is below threshold by the time it reaches the trigger zone. Although the cell is depolarized to -40 mV at the site where the graded potential begins, the current decreases as it travels through the cell body. As a result, the graded potential is below threshold by the time it reaches the axon hillock. (For the average mammalian neuron, threshold is about -55

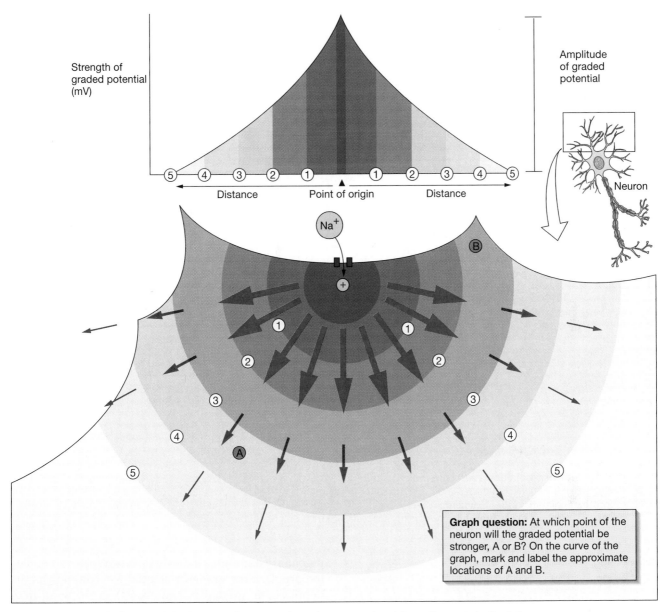

Graph question: At which point of the neuron will the graded potential be stronger, A or B? On the curve of the graph, mark and label the approximate locations of A and B.

■ Figure 8-9 **Graded potentials decrease in strength as they spread out from the point of origin**

mV.) The stimulus is not strong enough to depolarize the cell to threshold at the trigger zone, so the graded potential dies out without triggering an action potential.

The second example (Fig. 8-10b ■) shows a stronger initial stimulus that initiates a stronger EPSP and ultimately causes an action potential. Although this excitatory graded potential also diminishes with distance through the neuron, its higher initial strength ensures that it is above threshold at the axon hillock. In this example, an action potential is triggered.

Multiple Graded Potentials Are Integrated in the Trigger Zone

What happens if more than one stimulus arrives at the dendrites or cell body simultaneously? The likelihood of this event is quite high, because a single postsynaptic neu-

ron may form synapses with as many as 10,000 presynaptic neurons (Fig. 8-11 ■). If multiple stimuli arrive simultaneously, their graded potentials will be additive. For example, several subthreshold excitatory graded potentials can add to create a suprathreshold potential [*supra-*, above] at the trigger zone. The initiation of an action potential from several simultaneous subthreshold graded potentials is known as **spatial summation**. The word *spatial* refers to the fact that the graded potentials originate simultaneously at different locations on the neuron.

Figure 8-12a ■ shows one neuron receiving input from three sources simultaneously. Any one of the three stimuli is too weak to trigger an action potential by itself, but the sum of the three stimuli is above threshold, so an action potential is initiated.

Not all summation results in an action potential. Figure 8-12b ■ shows one inhibitory and two excitatory graded

(a)   **(b)**

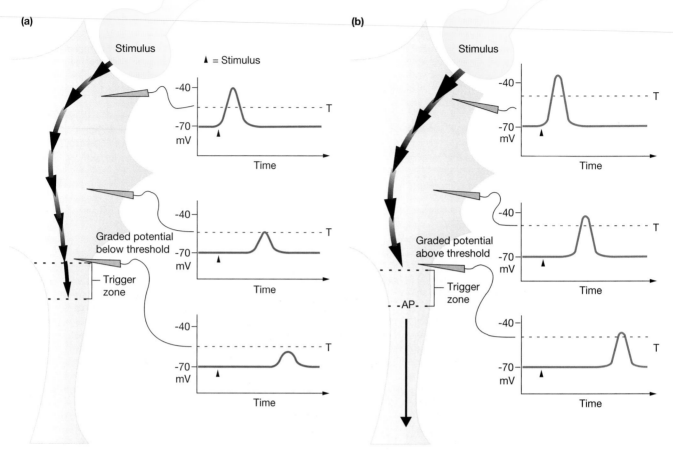

■ **Figure 8-10 Subthreshold and suprathreshold graded potentials in a neuron** (a) A graded potential starts above threshold at the point where it is initiated but decreases in strength as it travels through the cell body. At the trigger zone it is below threshold (T) and therefore does not initiate an action potential. (b) A stronger stimulus at the same point on the cell body creates a graded potential that is still above threshold by the time it reaches the trigger zone, so an action potential results.

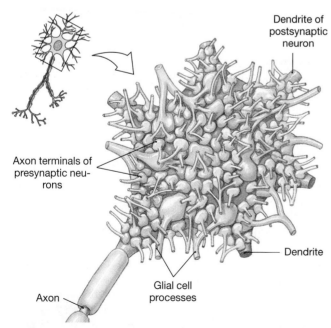

■ **Figure 8-11 Locations of synapses on a postsynaptic neuron** The cell body and dendrites of a somatic motor neuron are nearly covered in hundreds of axon terminals from other neurons.

potentials summing as they reach the trigger zone. The hyperpolarizing graded potential (−) counteracts the depolarizing graded potentials (+), creating an integrated signal that is below threshold. As a result, no action potential leaves the trigger zone.

Summation of graded potentials can be achieved by one other method. Two subthreshold graded potentials that follow each other sequentially can be summed if they arrive at the trigger zone close enough together in time. Figure 8-13 ■ shows how this can happen. Figure 8-13a ■ shows recordings from an electrode placed in the trigger zone of a neuron. A subthreshold graded potential starts on the cell body at point X at time A_1. The graded potential reaches the trigger zone and depolarizes it as shown on the graph. A second stimulus occurs at time A_2, and its subthreshold graded potential reaches the trigger zone a short time after the first. The interval between the two stimuli is so long that the two graded potentials do not overlap. Neither potential by itself is above threshold, so no action potential is triggered.

In Figure 8-13b ■, the two stimuli are given closer together in time. As a result, the two subthreshold graded potentials arrive at the trigger zone at almost the

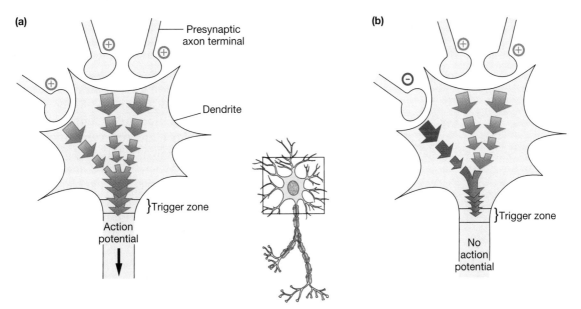

■ **Figure 8-12 Spatial summation** (a) Three graded potentials occur at the same time. Any one of them would be below threshold at the trigger zone, but they arrive there simultaneously and sum to create an above-threshold graded potential, resulting in an action potential. (b) Two stimuli that, summed, would be above threshold are diminished by a hyperpolarizing inhibitory graded potential. As a result, the sum of the three graded potentials is below threshold, and no action potential results.

same time. The second graded potential adds its depolarization to that of the first, allowing the trigger zone to depolarize to threshold. Summation that occurs from graded potentials overlapping in time is called **temporal summation** [*tempus*, time].

Summation of graded potentials demonstrates a key property of neurons: *postsynaptic integration*. When multiple, sometimes conflicting, signals reach a neuron, postsynaptic integration allows the neuron to evaluate the strength and duration of the signals. If the resultant integrated signal is above threshold, the neuron fires an action potential.

Action Potentials Travel Long Distances without Losing Strength

Action potentials are rapid electrical signals that pass along the axon to the axon terminal. Action potentials differ from graded potentials in several ways: (1) Action potentials are identical to one another, and (2) they do not diminish in strength as they travel through the cell. The mechanism through which actions potentials are generated and conducted along the axon allows them to stay constant: An action potential measured at the distal end of an axon is identical to the action potential that started at the trigger zone. This property is essential for the transmission of signals over long distances, such as from a fingertip to the spinal cord.

The strength of the graded potential that initiates an action potential has no influence on the action potential, as long as the graded potential is above threshold. Recordings of action potentials show them to be identical depolarizations of about 100 mV amplitude. Because

continued from page 204

In Guillain-Barré syndrome, the disease affects both sensory and somatic motor neurons. Dr. McKhann observed that, although the Beijing children could not move their muscles, they could feel a pin prick.

Question 2: Do you think that the paralysis found in the Chinese children affected both sensory and somatic motor neurons? Why or why not?

action potentials either occur as a maximal depolarization (if a stimulus reaches threshold) or do not occur at all (if a stimulus is below threshold), they are sometimes called **all-or-none** phenomena. Table 8-2 compares the key properties of action potentials and graded potentials.

Action potentials represent movement of Na$^+$ and K$^+$ across the membrane Action potentials are changes in membrane potential that occur when voltage-gated ion channels open, altering membrane permeability to Na$^+$ and K$^+$. Figure 8-14 ■ shows the voltage and ion permeability changes that take place during an action potential. The graph can be divided into three phases: the rising phase of the action potential, the falling phase, and the after-hyperpolarization phase. Before and after the action potential (① and ⑨), the neuron is at its resting membrane potential of −70 mV.

The rising phase of the action potential is due to a temporary increase in the cell permeability to Na$^+$. Action potentials begin when a graded potential reaching the trigger zone ② depolarizes the membrane to threshold

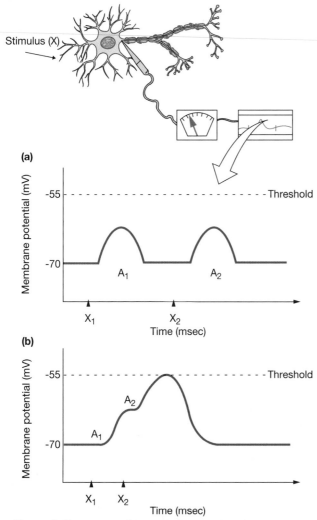

(a)

(b)

■ **Figure 8-13 Temporal summation** (a) Two graded potentials initiated by the same neuron will not cause an action potential if they are far apart in time. (b) If the same two subthreshold potentials arrive at the trigger zone within a short period of time, they may sum and create an action potential.

(-55 mV) ③. As the cell depolarizes, voltage-gated Na^+ channels open and the membrane suddenly becomes much more permeable to Na^+. Because Na^+ is more concentrated outside the cell, and because the negative membrane potential inside the cell attracts these positively charged ions, Na^+ flows into the cell. The addition of positive charge to the intracellular fluid depolarizes the cell membrane, making it progressively more positive ④. On the graph, this is shown by the steep rising phase of the action potential. When the cell membrane potential becomes positive, there is no longer an electrical driving force moving Na^+ into the cell because like charges repel. However, the Na^+ concentration gradient remains, so Na^+ continues to diffuse in. In the top third of the rising phase, the membrane potential has reversed polarity; that is, the inside of the cell has become more positive than the outside. This reversal is represented on the graph by the **overshoot,** that portion of the action potential above 0 mV. Sodium ions continue to diffuse into the neuron, sending the membrane potential toward the Na^+ equilibrium potential (E_{Na}) of $+60$ mV. (The E_{Na} is the membrane potential at which the tendency of Na^+ to move into the cell down its concentration gradient is exactly opposed by the positive membrane potential repelling Na^+; ∞ p. 140.) However, before the E_{Na} is reached, the Na^+ channels close. Sodium permeability decreases dramatically, and the action potential peaks at $+30$ mV ⑤.

The falling phase of the action potential is due primarily to an increase in K^+ permeability. Voltage-gated K^+ channels, like Na^+ channels, open in response to depolarization. But the K^+ channel gates are slower to open, and peak K^+ permeability occurs later than peak Na^+ permeability (Fig. 8-14 ■). By the time the K^+ channels are open, the membrane potential of the cell has reached $+30$ mV because of Na^+ influx. When the Na^+ channels close, the concentration and electrical gradients for K^+ favor move-

TABLE 8-2	Comparison of Graded Potential and Action Potential	
	Graded Potential	*Action Potential*
Type of signal	Input signal	Conduction signal
Where occurs	Usually dendrites and cell body	Trigger zone through axon
Types of gated ion channels involved	Mechanically or chemically gated channels	Voltage-gated channels
Ions involved	Usually Na^+ and Cl^-	Na^+ and K^+
Type of signal	Depolarizing (e.g., Na^+) or hyperpolarizing (e.g., Cl^-)	Depolarizing
Strength of signal	Depends on initial stimulus; can be summed.	Is always the same as long as graded potential is above threshold (all-or-none); cannot be summed
What initiates the signal	Entry of ions through chemically or mechanically gated ion channels	Above-threshold graded potential arrives at the integration zone
Unique characteristics	No minimum level required to initiate a graded potential	Threshold stimulus required to initiate action potential
	Two signals coming close together in time will sum	Refractory period: two signals too close together in time cannot sum
		Initial stimulus strength is indicated by frequency of a series of action potentials

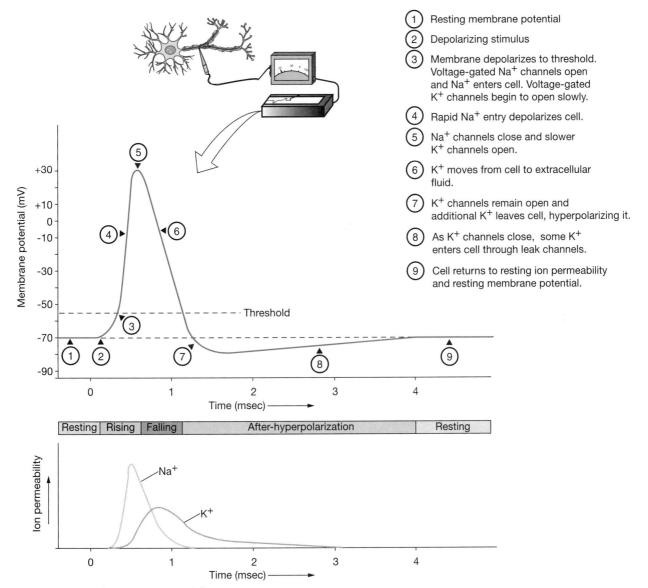

1 Resting membrane potential

2 Depolarizing stimulus

3 Membrane depolarizes to threshold.
   Voltage-gated Na$^+$ channels open
   and Na$^+$ enters cell. Voltage-gated
   K$^+$ channels begin to open slowly.

4 Rapid Na$^+$ entry depolarizes cell.

5 Na$^+$ channels close and slower
   K$^+$ channels open.

6 K$^+$ moves from cell to extracellular
   fluid.

7 K$^+$ channels remain open and
   additional K$^+$ leaves cell, hyperpolarizing it.

8 As K$^+$ channels close, some K$^+$
   enters cell through leak channels.

9 Cell returns to resting ion permeability
   and resting membrane potential.

■ Figure 8-14 **The action potential**

ment of K$^+$ *out* of the cell. As K$^+$ diffuses out of the cytoplasm, the membrane potential becomes more negative,
creating the *falling phase of the action potential* and sending
the cell toward its resting potential ⑥. The K$^+$ channels
close slowly, so instead of repolarizing to exactly -70 mV,
the cell hyperpolarizes, approaching the E_K of -90 mV.
The additional loss of K$^+$ creates an **after-hyperpolariza-**

tion (also called the *undershoot*) ⑦. Once the K$^+$ channels
close and normal K$^+$ permeability is restored, a few potassium ions move back into the cell ⑧, restoring the membrane potential to its normal resting value of -70 mV ⑨.

The phases of the action potential and the contributions of ions to each phase are outlined in Table 8-3. To
summarize, the action potential is a change in mem-

TABLE 8-3	**Comparison of Phases of the Action Potential in a Typical Neuron**			
	Neuron at Rest	*Rising Phase*	*Falling Phase*	*After-Hyperpolarization*
V_m	-70 mV	-70 to $+30$ mV	$+30$ to -80 mV	-80 to -70 mV
Na$^+$ permeability	Low	High	Low	Low
Na$^+$ movement	Leak in balanced by being pumped out	Na$^+$ into neuron	Minimal leak in	Leak in balanced by being pumped out
K$^+$ permeability	Moderate	Moderate	High	Moderate
K$^+$ movement	Leak out balanced by being pumped in	Leak out balanced by being pumped in	High K$^+$ out of neuron	Net leak back in to restore V_m

brane potential that occurs when gated ion channels in the membrane open, increasing the cell's permeability first to Na$^+$ and then to K$^+$. The influx (movement into the cell) of Na$^+$ depolarizes the cell. This depolarization is followed by K$^+$ efflux (movement out of the cell), which restores the cell to the resting membrane potential.

Na$^+$ channels in the axon have two gates One question that puzzled scientists for many years was how the voltage-gated Na$^+$ channels could close when the cell was depolarized, because depolarization was the stimulus for Na$^+$ channel *opening*. After many years of study, they found the answer. The voltage-gated Na$^+$ channel has *two* gates to regulate ion movement rather than a single gate. These two gates, known as the *activation* and *inactivation gates,* flip-flop back and forth to open and shut the Na$^+$ channel.

When the neuron is at its resting membrane potential, the **activation gate** is closed and no Na$^+$ moves through the channel (Fig. 8-15a ■). At the same time, the **inactivation gate,** apparently a ball-and-chain-like amino acid sequence on the cytoplasmic side of the channel protein, is open. When the cell membrane close to the channel depolarizes, the activation gate opens (Fig. 8-15b ■). The Na$^+$ channel opens and allows Na$^+$ to move into the cell down its electrochemical gradient (Fig. 8-15c ■). The addition of positive charge further depolarizes the inside of the cell and causes a *positive feedback loop* (∞ p. 163) (Fig. 8-16 ■). More Na$^+$ channels open and more Na$^+$ enters, further depolarizing the cell. As long as the cell remains depolarized, the activation gates remain open.

As happens in all positive feedback loops, outside intervention is needed to stop the escalating depolarization of the cell. This outside intervention is the role of the second gate. The movement of the inactivation gate is coupled to the movement of the activation gate, but its response time is slower. As the activation gate opens, the signal is passed along the channel protein to the inactivation gate. During the 0.5-msec delay before the inactivation gate responds, the Na$^+$ channel is open and Na$^+$ can move into the cell. Only a tiny number of sodium ions need to move to create a 100-mV change in the membrane potential, so the short open time of the channel is sufficient to create the rising phase of the action potential. Then, after the rising phase of the action potential has begun, the inactivation gate of the Na$^+$ channel closes (Fig. 8-15d ■). At that point, Na$^+$ permeability decreases, Na$^+$ influx stops, and the action potential peaks. As the neuron repolarizes during K$^+$ efflux, the Na$^+$ channel gates reset to their original positions so that they can respond to the next depolarization (Fig. 8-15e ■). Thus, the voltage-gated Na$^+$ channel uses a two-step process for opening and closing rather than a single gate that swings back and forth. This is an important property that allows electrical signals along the axon to be conducted in only one direction, as you will see in the next section.

(a) At the resting membrane potential, the activation gate closes the channel.

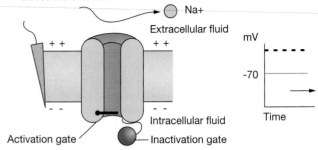

(b) Depolarizing stimulus arrives at the channel.

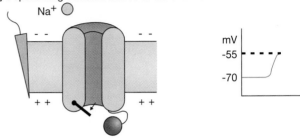

(c) With activation gate open, Na$^+$ enters the cell.

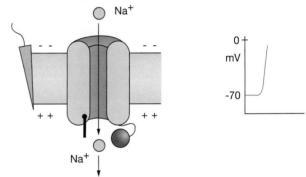

(d) Inactivation gate closes and Na$^+$ entry stops.

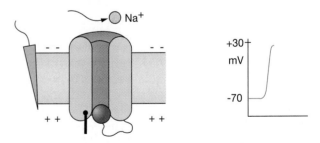

(e) During repolarization caused by K$^+$ leaving the cell, the two gates reset to their original positions.

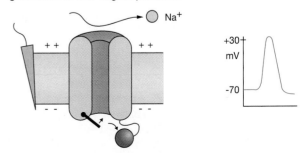

■ **Figure 8-15 Model of the voltage-gated Na$^+$ channel**

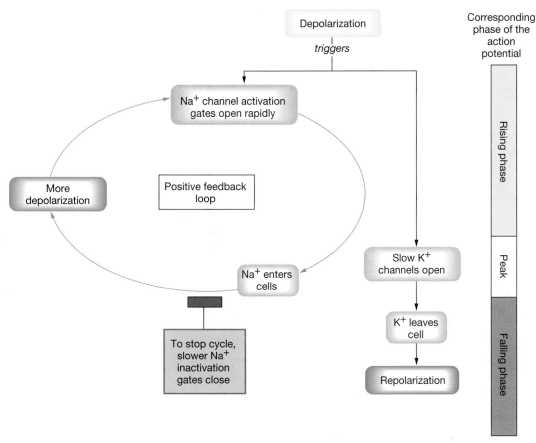

■ Figure 8-16 **Ion movements during the action potential** Entry of Na$^+$ creates a positive feedback loop that stops when the Na$^+$ channel inactivation gates close.

Action Potentials Will Not Fire during the Refractory Period

The double gating of the Na$^+$ channel plays a major role in the phenomenon known as the **refractory period.** The adjective *refractory* comes from a Latin word meaning "stubborn." The "stubbornness" of the neuron refers to the fact that once an action potential has begun, for about 1 msec, a second action potential cannot be triggered, no matter how large the stimulus. This period is called the **absolute refractory period** (Fig. 8-17 ■). The absolute refractory period ensures that a second action potential will not occur before the first has finished. *Action potentials cannot overlap because of their refractory periods, so they cannot be summed the way that graded potentials can.*

After the Na$^+$ channel gates have reset to their original positions, but before the membrane is restored to its resting potential, a higher-than-normal graded potential can start another action potential. During this time, the neuron is said to be in its **relative refractory period.** A threshold-level depolarization during the relative refractory period can open those Na$^+$ channels that have returned to their resting position. However, Na$^+$ entry through these channels is offset by K$^+$ loss through still-open K$^+$ channels. The opposing flow of charge balances out, and the membrane potential does not reach threshold. For that reason, a stronger-than-normal depolarizing graded potential is needed to bring the cell up to the threshold for an action potential.

The refractory period is a key characteristic that distinguishes action potentials from graded potentials. If two stimuli reach the dendrites of a neuron within a short time, the successive graded potentials can be added to each other by temporal summation (see Fig. 8-13 ■). But if two suprathreshold graded potentials reach the action potential trigger zone within too short a time, the second will be ignored because the Na$^+$ channels are inactivated and are incapable of being stimulated again so soon. Refractory periods limit the rate at which signals can be transmitted down a neuron. The absolute refractory period also assures one-way travel of an action potential from cell body to axon terminal by preventing backward conduction of the action potential.

Stimulus Intensity Is Coded by the Frequency of Action Potentials

One distinguishing characteristic of action potentials is that every action potential in a given neuron will be identical to every other action potential in that neuron. But if this is so, then how does the neuron transmit information about the strength and duration of the stim-

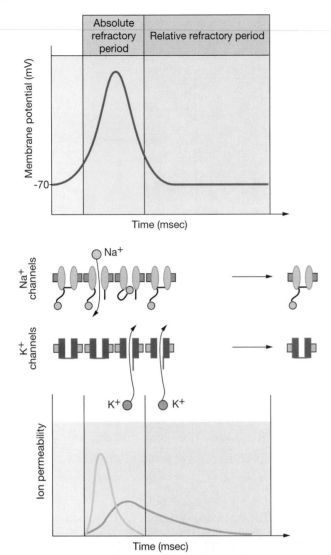

■ **Figure 8-17 Refractory periods** During the absolute refractory period, no depolarizing stimulus can trigger another action potential. During the relative refractory period, a larger-than-normal stimulus can initiate a new action potential. During this period, the Na^+ channel gates are resetting to their resting positions, and the K^+ channels are open. Potassium ion efflux hyperpolarizes the neuron and offsets a depolarizing stimulus.

increase in signal strength will increase the neurotransmitter output, which in turn changes the magnitude of the graded potential in the postsynaptic cell.

The Na^+/K^+-Pump Plays No Direct Role in the Action Potential

As you have just learned, the action potential results from ion movements: First, Na^+ moves into the cell and then K^+ moves out. However, it is important that you understand that very few ions move across the membrane in a single action potential: The relative Na^+ and K^+ concentrations remain *essentially unchanged*. If a second action potential is triggered immediately after the first, it will be identical to the first one. The tiny amounts of Na^+ and K^+ that cross the membrane during action potentials do not disrupt the normal concentration gradients until a thousand or more action potentials have been fired.

The concentration gradients for Na^+ and K^+ in all cells are maintained over time through the action of the Na^+/K^+-ATPase (Na^+/K^+-pump) in the cell membrane. The pump removes Na^+ that enters the cell and recaptures K^+ that has left it. But *the Na^+/K^+-ATPase plays no direct role in the action potential*. It treats the Na^+ that enters the cell during an action potential like any other Na^+ that leaks in, pumping it back out to the extracellular fluid. Since the number of Na^+ that enters the cell with each action potential is very small, a neuron poisoned with the pump inhibitor *ouabain* will continue to fire action potentials for an extended time before a change in the Na^+ concentration gradient can be noticed.

✓ If you could disable the inactivation gates of the Na^+ channels so that the channels remained open, what would happen to the membrane potential? Explain your answer.

✓ If you place an electrode in the middle of an axon and artificially depolarize the cell above threshold, in which direction will an action potential travel: to the axon terminal, to the cell body, or to both? Explain your answer.

Action Potentials Are Conducted from the Trigger Zone to the Axon Terminal

The movement of an action potential through the axon at high speed is called **conduction**. Conduction represents the flow of electrical energy from one part of the cell to another in a process that ensures that any energy lost to friction or leakage out of the cell is immediately replenished. Thus, an action potential does not lose strength over distance as a graded potential does. To understand this, let us examine conduction at the cellular level to see why the action potential that reaches the

ulus that started the action potential? The answer lies not in the amplitude of the action potential, but in the *frequency* of action potential propagation. A stimulus in the form of a graded potential reaching the trigger zone does not usually trigger a single action potential. Instead, even a minimal graded potential that is above threshold will trigger a burst of action potentials (Fig. 8-18a ■). If the graded potential increases in strength (amplitude), the frequency of action potentials fired (in action potentials per second) increases (Fig. 8-18b ■). The amount of neurotransmitter released at the axon terminal is directly related to the total number of action potentials that arrive at the terminal per unit time. An

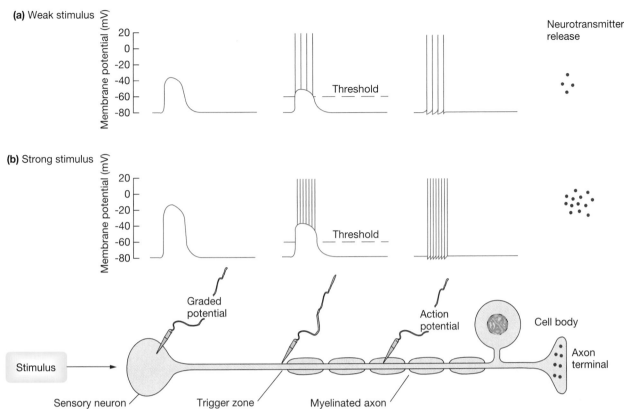

(a) Weak stimulus

Membrane potential (mV)

Threshold

Neurotransmitter release

(b) Strong stimulus

Membrane potential (mV)

Threshold

Graded potential

Action potential

Cell body

Axon terminal

Stimulus

Sensory neuron

Trigger zone

Myelinated axon

■ **Figure 8-18 Coding for stimulus intensity** Stimuli of different intensity change the frequency of action potential firing along the axon. Since all action potentials in a neuron are identical, the strength of the stimulus is indicated by the frequency of action potential firing. (a) A graded potential that is barely above threshold causes a series of action potentials to pass down the axon and release neurotransmitter. (b) A stronger graded potential increases the frequency of action potential firing in the axon. The higher frequency releases more neurotransmitter and creates a stronger response in the neuron's target cell.

end of the axon is identical to the action potential that started at the trigger zone.

When we talk about the conduction of an action potential, it is important to realize that there is no single action potential that moves through the cell. The action potential that occurs at the trigger zone is like the movement in the first domino of a series of dominos standing on end. As the first domino falls, it strikes the next, passing on its kinetic energy. As the second domino falls, it passes kinetic energy to the third domino, and so on. If you could take a snapshot of the line of falling dominos, you would see that as the first domino is coming to rest in the fallen position, the next one is almost down, the third one most of the way down, etc., until you reach the domino that has just been hit and is starting to fall (Fig. 8-19a ■). In the same way, *a wave of action potentials moves down the axon.* An action potential is simply a representation of the membrane potential in a given segment of cell membrane at any given moment in time. As the electrical energy of the action potential passes from one part of the axon to the next, the energy state is reflected in the membrane potential of that region. If we were to insert a series of recording electrodes along the length of an axon and start an electrical signal at the trigger zone, we would see a series of overlapping action potentials at different parts of the wave form, just like the dominos that are frozen in different positions (Fig. 8-19b ■).

Once a graded potential reaches threshold in the trigger zone, voltage-gated Na^+ channels open and the cell depolarizes (Fig. 8-20a ■). The positive current from Na^+ entry spreads through the cytoplasm in all directions, just as the depolarization of a graded potential spreads through the cell by local current flow (Fig. 8-20b ■). The current flow backward into the cell body can be ignored because there are few voltage-gated ion channels in that region. The current flow in the forward direction, down into the initial portions of the axon, is critical for conduction. The axon membrane is lined by voltage-sensitive Na^+ channels, just as the membrane of the trigger zone is. Positive charge from the depolarization of the trigger zone spreads into adjacent sections of membrane, attracted by the negative charge of the resting membrane potential. When the wave of depolarization reaches the voltage-sensitive Na^+ channels adjacent to the trigger zone, they open, allowing Na^+ into the cell (Fig. 8-20c ■). The membrane in the next segment of the axon is depolarized, and the positive feedback loop of depolarization begins: Na^+ channels open, Na^+ enters,

(a)

Dominos falling

(b)

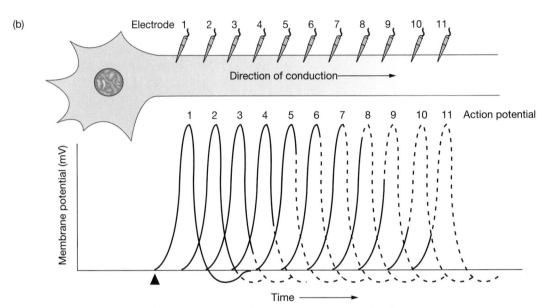

■ **Figure 8-19 Action potentials along an axon** There are many action potentials taking place along an axon at any point in time. (a) The transmission of action potentials can be compared to a "snapshot" of dominos falling, where each domino is in a different position. (b) Similarly, simultaneous recordings of membrane potentials along an axon will show that each section of membrane is experiencing a different phase of the action potential. Twelve electrodes have been placed along the axon; the recordings from them are shown below the axon.

causing depolarization. The result is always an action potential in the next segment of membrane. Meanwhile, the first segment of axon is in the falling phase of the action potential. The Na$^+$ channels are inactivated and K$^+$ channels are open, allowing K$^+$ to leave the cytoplasm. Although positive charge from the distal, depolarizing segment of membrane may flow back into this section by local current flow, the membrane is in its absolute refractory period because of inactivation of the Na$^+$ channels, so nothing happens; the propagation of the action potential will proceed in the forward direction, toward the axon terminal.

Local current flow into the distal section of the axon will continue conduction of the electrical signal down the axon. As each region depolarizes, the Na$^+$ ions entering the cell transfer their energy to the next segment of membrane, opening more Na$^+$ channels. Because of the continuous entry of Na$^+$ along the axon, the strength of the signal does not diminish. Compare this process with the graded potential, in which Na$^+$ entered only at the point of stimulus, resulting in a membrane potential change that died out over distance.

In this section, you learned that the regeneration of the action potential occurs by continuous Na$^+$ entry into the

axon as the action potential moves along its length. In the next section, we examine the factors that influence the speed with which an action potential is conducted.

The Diameter and Resistance of the Neuron Influence the Speed of Conduction

Two key physical parameters influence the speed of action potential conduction in mammalian neurons: (1) the diameter of the neurons and (2) the resistance of the neuron membrane to current leak out of the cell. The larger the diameter of the axon, the faster an action potential will move. To understand this relation, think of a water pipe with water flowing through it. The water that touches the walls of the pipe encounters resistance from friction between the flowing water molecules and the stationary walls. The water in the center of the pipe meets no direct resistance from the walls and therefore flows faster. In the same way, current flowing inside an axon meets resistance from the membrane. Thus, the larger the diameter of the axon, the lower its resistance to current flow.

The connection between axon diameter and speed of conduction is especially evident in the nerve cells of invertebrate animals such as the squid and the earthworm. The giant axons of these organisms may be up to 1 mm in diameter and have been very important to the development of our understanding of electrical signaling (Fig. 8-21a ■). The evolution of larger-diameter

axons for rapid conduction works only in animals with simple nervous systems, however. If you compare a cross section of a squid giant axon with a cross section of a mammalian nerve, you will find that the mammalian nerve contains about 400 fibers in the same cross-sectional area (Fig. 8-21b ■). If each of those 400 axons were as large as a squid giant axon, the nerve would have to be 1.6 cm in diameter, much too large for our bodies to have many of them! Because vertebrates cannot rely on large axons for rapid transmission, another mechanism evolved to increase conduction: wrapping the axons in the insulating membranes of the myelin sheath.

The multiple layers of membrane in myelin create a high-resistance sheath that prevents current flow between the cytoplasm of the neuron and the extracellular fluid. The membranes are analogous to heavy coats of plastic surrounding electrical wires, increasing the effective thickness of the axon membrane by as much as 100-fold. In regions where the axon lacks the myelin sheath, the cell membrane has low resistance to current flow. In those areas, current leaks out of the cell in addition to moving forward into the next section of axon. Consequently, forward current flow is slower in regions of the axon that lack myelin.

When an action potential starts at the trigger zone, current flows rapidly down the myelinated axon to the first unmyelinated node of Ranvier (Fig. 8-22a ■). Each node has a high concentration of voltage-gated ion channels. When depolarization reaches the node, Na^+ channels open and Na^+ enters the axon, as described for the action potential. The Na^+ entry reinforces the depolarization, keeping the amplitude of the action potential

(a)

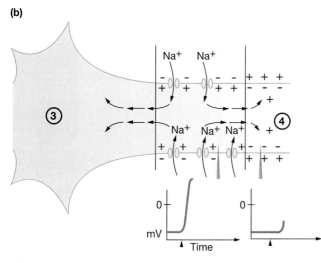

(b)

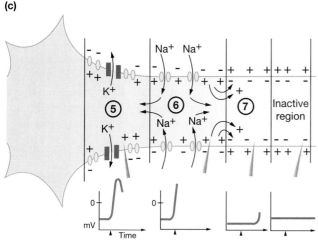

(c)

■ **Figure 8-20 Conduction of action potentials** Action potentials are conducted undiminished along an axon as Na^+ channels open and replenish the depolarization because of Na^+ influx. (a) A graded potential at or above threshold reaches the trigger zone ①. The depolarization by the graded potential opens voltage-gated Na^+ channels and Na^+ enters the axon, moving down its electrochemical gradient ②. (b) Sodium ion entry further depolarizes the membrane ③, opening additional Na^+ channels. The membrane potential becomes more positive inside. ④ Positive charge from the trigger zone moves into adjacent sections of the axon by local current flow. The next section of membrane becomes depolarized, opening Na^+ channels in that region of the axon. Some current flows back into the cell body but has no effect because the cell body does not have voltage-gated channels. (c) In the trigger zone ⑤, K^+ gates have opened and Na^+ gates have closed. Potassium ions move out of the axon along the electrochemical gradient. The loss of positive ions causes the membrane to repolarize. ⑥ The next section of axon is in the rising phase of the action potential as Na^+ flows into the cell. Once again, positive charge flows to adjacent sections of axon by local current flow. The backward movement of positive charge toward the cell body has no effect because the Na^+ channels in the trigger zone are inactivated. Farther from the cell body, however, ⑦ the depolarization opens Na^+ channels and the action potential begins in that section.

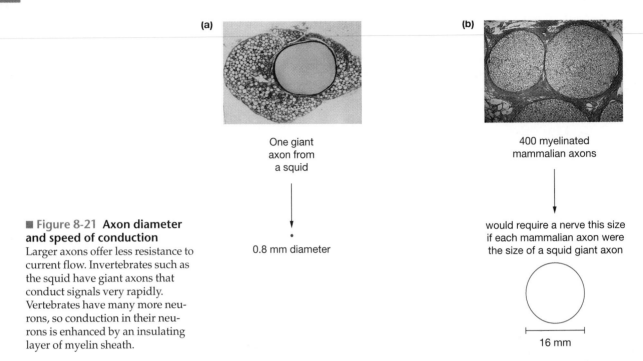

(a)

One giant
axon from
a squid

0.8 mm diameter

(b)

400 myelinated
mammalian axons

would require a nerve this size
if each mammalian axon were
the size of a squid giant axon

16 mm

■ **Figure 8-21 Axon diameter
and speed of conduction**
Larger axons offer less resistance to
current flow. Invertebrates such as
the squid have giant axons that
conduct signals very rapidly.
Vertebrates have many more neurons, so conduction in their neurons is enhanced by an insulating
layer of myelin sheath.

constant, but it also slows the current flow. When local
current flow at the node reaches the next myelinated
region, flow speeds up, zipping along until slowed at
the next node of Ranvier. The apparent leapfrogging of
the action potential from node to node has given this
mode of current flow the name **saltatory conduction,**
from the Latin word *saltare*, meaning "to leap." Saltatory

conduction is an effective alternative to large-diameter
axons and allows rapid action potentials through small
axons. A myelinated frog axon 10 μm in diameter con-
ducts action potentials at the same speed as an unmyelinated 500 μm squid axon. Thus, electrical current flows
through different axons at different rates, depending on
the two parameters of axon diameter and myelination.

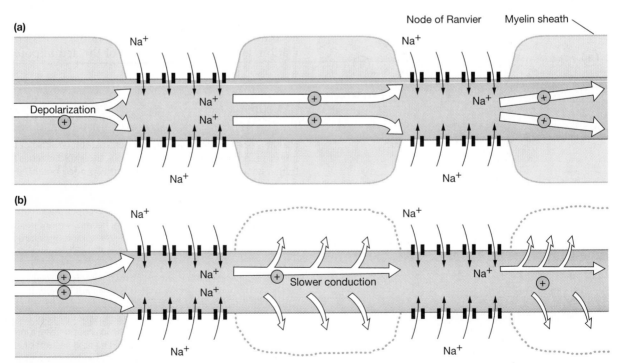

■ **Figure 8-22 Saltatory conduction** (a) Action potentials in myelinated axons appear to "leapfrog" from one node of
Ranvier to the next. Voltage-gated Na$^+$ channels are scarce between the nodes, and action potentials occur only at the nodes,
where the cytoplasm is in contact with the extracellular fluid. The local current flow moves rapidly through the myelinated
sections between the nodes. (b) In demyelinating diseases, conduction slows down as current leaks out of the previously
insulated regions between the nodes. Since there are few Na$^+$ channels in those regions, the action potentials are not conducted as strongly between the nodes.

In *demyelinating diseases,* the loss of myelin from vertebrate neurons can have devastating effects on neural signaling. In the central and peripheral nervous systems, the loss of myelin slows the conduction of action potentials. The strength of the action potentials decreases when current leaks out of the uninsulated regions of membrane between the channel-rich nodes of Ranvier (Fig. 8-22b ■). Multiple sclerosis is the most common and best-known demyelinating disease. It is characterized by a variety of neurological complaints, including fatigue, muscle weakness, difficulty walking, and loss of vision. At this time, we can treat the symptoms but not the causes of demyelinating diseases. Currently, researchers are using recombinant DNA technology to study similar demyelinating disorders in mice.

Electrical Activity in the Nervous System Can Be Altered by a Variety of Chemical Factors

A large variety of chemicals alter the conduction of action potentials by binding to Na^+ or K^+ channels in the membrane of the axon (Table 8-4). For example, chemicals called *neurotoxins* bind to and inactivate Na^+ channels; *local anesthetics,* which block sensation, function the same way. If the Na^+ channel is inactive, Na^+ cannot enter the axon. Thus, a depolarization that begins at the trigger zone loses strength as it moves down the axon, much like a normal graded potential. If the wave of depolarization manages to reach the axon terminal, it is too weak to release neurotransmitter. As a result, the message of the presynaptic neuron is not passed on to the postsynaptic cell.

Alterations in the extracellular fluid concentrations of K^+, Na^+, and Ca^{2+} are also associated with abnormal electrical activity in the nervous system. The relationship between extracellular fluid K^+ levels and the conduction of action potentials is the most straightforward and easiest to understand.

The concentration of K^+ in the blood and interstitial fluid is the major determinant of the resting potential of all cells. If K^+ levels in the blood move out of the nor-

Mutant Mouse Models   One technique that scientists use to study human diseases is to identify animals with a similar disease. The afflicted animal is then used as a model to explore the cause of the disease and to test potential treatments. Sometimes, natural mutations produce diseases that resemble human diseases. Two examples of such mutants are the twitcher mouse, in which normal myelin degenerates owing to an inherited metabolic problem, and the wobbler mouse, in which somatic motor neurons controlling the limbs die. In other cases, scientists have used biotechnology techniques to create mice that lack specific genes ("knock-out mice") or to breed mice that contain extra genes that were inserted artificially (transgenic mice). By comparing the genetically engineered mice with normal mice, we have expanded our understanding of how the nervous system responds during normal development and to disease.

mal range of 3.5–5 millimoles/liter (mmol/L), the result will be a change in the resting membrane potential of the cells. This change is not important to most cells, but it can have serious consequences to the body as a whole because of the relationship between resting potential and the excitability of nervous and muscle tissue. An increase in blood K^+, **hyperkalemia** [*hyper-,* above + *kalium,* potassium + *-emia,* in the blood], will shift the resting membrane potential of a neuron closer to threshold and cause the cells to fire action potentials in response to smaller graded potentials (Fig. 8-23c ■). If blood K^+ levels drop too low **(hypokalemia),** the resting membrane potential of the cells hyperpolarizes, moving farther from threshold and requiring a larger-than-normal stimulus to fire an action potential (Fig. 8-23d ■). This condition shows up as muscle weakness because the neurons that control skeletal muscles are not firing normally. Hypokalemia and its resultant muscle weakness are one reason that sport drinks supplemented with Na^+ and K^+, such as Gatorade®, were developed. When people sweat excessively, they lose

---

TABLE 8-4 Chemicals That Affect Electrical Signals in Neurons

Name of Chemical	Channel Blocked	Comments
Tetrodotoxin (TTX)	Voltage-gated Na^+	Made in ovaries and liver of Japanese puffer fish
Saxitoxin	Voltage-gated Na^+	Made by marine organism that causes "red tide"
Procaine	Voltage-gated Na^+	Local anesthetic
Tetraethylammonium chloride (TEA)	Voltage-gated K^+	Amine derivative
Ethanol	Ca^{2+} and assorted receptor-operated channels	Depresses action potentials and depolarizes the membrane; inhibitory in higher doses and excitatory in low doses

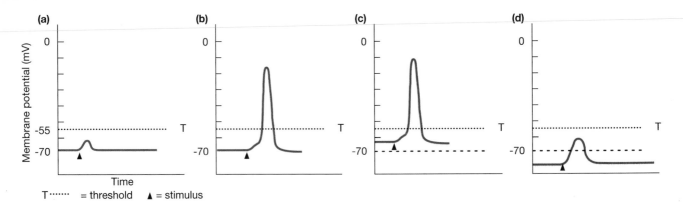

T⋯⋯ = threshold   ▲ = stimulus

■ **Figure 8-23 Effect of changing extracellular potassium concentration on the excitability of neurons** Potassium is the major ion contributing to the resting membrane potential. If the extracellular concentration of K^+ changes, the resting membrane potential changes and the neuron becomes more or less likely to fire an action potential. (a) A subthreshold graded potential does not fire an action potential when blood K^+ concentration is in the normal range (normokalemia). (b) An above-threshold (suprathreshold) stimulus will fire an action potential when K^+ concentration is normal. (c) Hyperkalemia, increased blood K^+ concentration, brings the membrane closer to threshold. Now a subthreshold stimulus can trigger an action potential. (d) Hypokalemia, decreased blood K^+ concentration, hyperpolarizes the membrane and makes the neuron less likely to fire an action potential in response to a stimulus that would normally be above threshold.

both salts and water. If they replace this fluid loss with pure water, the K^+ remaining in the blood will be diluted, causing hypokalemia. By replacing sweat loss with a dilute salt solution, they can prevent potentially dangerous drops in blood K^+ levels.

Because of the importance of K^+ to normal function of the nervous system, the body regulates blood K^+ levels within a narrow range. The important role of the kidneys in maintaining ion balance is discussed in Chapter 19.

CELL-TO-CELL COMMUNICATION IN THE NERVOUS SYSTEM

Information flow through the nervous system uses both electrical and chemical signals to send information from cell to cell. Unlike hormones secreted into the blood and distributed throughout the body, neurotransmitters are secreted very close to the target cells with their receptors. Much of the specificity of neural communication depends on these connections, which occur in regions known as synapses.

 continued from page 213

Like multiple sclerosis, Guillain-Barré syndrome is an illness in which the myelin that insulates axons is destroyed. One way that multiple sclerosis and other demyelinating illnesses are diagnosed is through the use of a nerve conduction test. This test measures the strength of action potentials and the rate at which they are conducted as they travel down neurons.

Information Passes from Cell to Cell at the Synapse

A synapse is the point at which a neuron meets its target cell (another neuron or a non-neuronal cell). Each synapse has three parts: the axon terminal of the presynaptic cell; the synaptic cleft, or space between the cells; and the membrane of the postsynaptic cell (see Fig. 8-2 ■). In a neural reflex, information moves from presynaptic cell to postsynaptic cell. In most synapses, the presynaptic cell ends next to the dendrites or cell body of the postsynaptic cell. In general, cells with many dendrites will also have many synapses. A moderate number of synapses is 10,000, but some cells in the brain are estimated to have 150,000 or more synapses! Synapses can also occur on the axon and even at the axon terminal. Synapses are classified as electrical or chemical depending on the type of signal that passes between cells.

Electrical synapses   **Electrical synapses** pass an electrical signal or current directly from the cytoplasm of one cell to another through gap junctions (∞ p. 147). Electrical synapses are uncommon and occur mainly in the central nervous system. Electrical synapses are also found in some glial cells and in cardiac and smooth muscle. The primary advantage of electrical synapses is rapid conduction of signals from cell to cell.

Chemical synapses   The vast majority of synapses in the nervous system are **chemical synapses** that use neurotransmitters to carry information from one cell to the next. Much of what we know about chemical synapses comes from the study of the *neuromuscular junction,* formed by the synapse of a somatic motor neuron on a skeletal muscle fiber.

When the structure of a chemical synapse is examined with an electron microscope, many small **synaptic vesicles** and large mitochondria are found in the axon terminal of the presynaptic cell (Fig. 8-24 ■). Some vesicles line up along the membrane closest to the synaptic cleft and appear to be "docked" at the membrane, waiting for a signal to release their contents. Other vesicles act as a reserve pool, clustering close to the docking sites. Each vesicle contains a certain amount of neurotransmitter that is released on demand.

The Nervous System Uses a Variety of Neurotransmitters

The neurotransmitters found in chemical synapses can be roughly grouped into five classes according to their structure: (1) amino acids, (2) amino acid–derived amines, (3) purines, (4) polypeptides, and (5) acetylcholine.

Acetylcholine (ACh), in a class by itself, is made from choline and acetyl coenzyme A (acetyl CoA). Choline is a small molecule also found in membrane phospholipids, and acetyl CoA is the metabolic intermediate that links glycolysis to the citric acid cycle (∞ p. 91). The synthesis of ACh from these two precursors is a simple pathway that takes place in the axon terminal itself (Fig. 8-25 ■). Once in the synaptic cleft, ACh is rapidly broken down by the enzyme acetylcholinesterase in the extracellular matrix and in the membrane of the postsynaptic cell. Choline from degraded ACh is actively transported back into the presynaptic axon terminal and used to make new acetylcholine to refill recycled synaptic vesicles.

Some amino acid, amine, and purine neurotransmitters are also synthesized in the axon terminals. The enzymes for their synthesis are brought from the cell body to the axon terminal by slow axonal transport. Once released into the synaptic cleft, these neurotrans-

continued from page 224

Dr. McKhann decided to perform nerve conduction tests on some of the paralyzed children. He found that, although the rate of conduction along the children's neurons was normal, the strength of the action potentials was greatly diminished.

Question 4: Is the paralytic illness that affected the Chinese children a demyelinating condition? Why or why not?

mitters are not broken down by enzymes as acetylcholine is. Instead they diffuse into the circulation to be carried away or they are actively transported back into the axon terminal to be metabolized or used again.

Polypeptide neurotransmitters, like the axon terminal enzymes, must be made in the cell body because axon terminals lack the rough endoplasmic reticulum needed for protein synthesis. These neurotransmitters are consumed more rapidly than the enzymes, however, so they are moved to the axon terminals by fast axonal transport.

One of the most interesting neurotransmitters is **nitric oxide** (NO), an unstable gas synthesized from oxygen and the amino acid arginine. Nitric oxide is an unusual neurotransmitter that diffuses freely into the target cell rather than binding to a membrane receptor. Once inside the target cell, nitric oxide binds to proteins or nucleic acids. With a half-life of only 2–30 seconds, nitric oxide is an elusive and difficult-to-study molecule. It is also released from cells other than neurons and often acts as a paracrine. For example, nitric oxide from the endothelial lining of blood vessels causes relaxation of the smooth muscle in the blood vessel wall.

The array of neurotransmitters in the body is truly staggering. The largest variety of neurotransmitters is

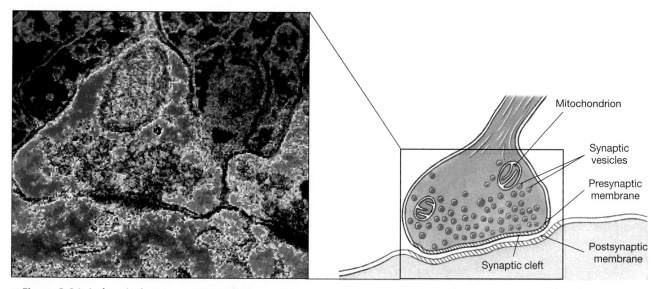

■ **Figure 8-24 A chemical synapse** This figure shows the axon terminal, synaptic cleft, and postsynaptic cell of an axon terminal. Notice the synaptic vesicles clustered close to the membrane.

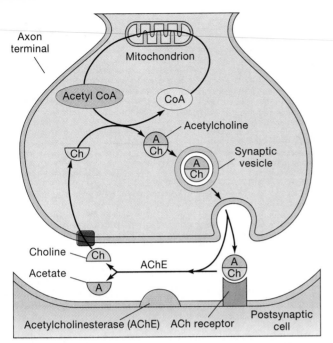

ters except nitric oxide have one or more receptors with which they bind. Receptors will be discussed later when we look at the neurotransmitters in more detail.

Calcium Is the Signal for Neurotransmitter Release at the Synapse

We now turn to the mechanism that allows cells within the nervous system to communicate with each other and with their non-neural targets: the release of neurotransmitters into the synapse. When the depolarization wave of the action potential reaches the axon terminal, it sets off a sequence of events (Fig. 8-26 ■). The membrane of the axon terminal has voltage-gated Ca^{2+} channels that open in response to depolarization. Calcium ions are more concentrated outside the cell than in the cytoplasm, so they diffuse into the cell. The entry of Ca^{2+} serves as a signal for neurotransmitter contained in synaptic vesicles to be released by exocytosis (∞ p. 127). Once the

Docking Proteins and Synaptic Vesicle Release The release of neurotransmitter from synaptic vesicles occurs only in active zones of the axon terminal membrane. Although the details are still not fully understood, the process involves at least eight different proteins with names such as synaptophysin, VAMP, and SNAP. Synaptic vesicles stored in the axon terminal are delivered to the active zone along fibers of the cytoskeleton. Once here, the vesicles fuse with the axonal membrane when vesicle proteins interact with cell membrane proteins. One vesicle protein, synaptotagmin I, binds Ca^{2+} and appears to be the Ca^{2+}-sensing mechanism that initiates exocytosis. It now appears that neurotoxins that block neurotransmitter release, including tetanus and botulinum toxins, exert their action by inhibiting specific proteins of the axon's exocytosis apparatus.

■ **Figure 8-25 Synthesis and recycling of acetylcholine at the synapse** Acetylcholine (ACh) is made from choline and acetyl CoA in the axon terminal. It is packaged into synaptic vesicles and released by exocytosis in the synaptic cleft. Acetylcholine combines with receptors on the postsynaptic cell membrane to initiate a response. It is rapidly broken down by acetylcholinesterase, an enzyme found in the matrix and on the postsynaptic membrane. The choline portion of the acetylcholine molecule is transported back into the axon terminal and is used to make more neurotransmitter.

found within the central nervous system (Table 8-5). These neurotransmitters include many polypeptides known primarily for their hormonal activity. In contrast, three primary neurotransmitters are used within the peripheral nervous system: acetylcholine (ACh), norepinephrine (NE), and epinephrine (E). All neurotransmit-

TABLE 8-5	Major Neurotransmitters in Mammals		
Neurotransmitter	*Chemical Structure*		*Comments*
Peripheral nervous system			
Acetylcholine (ACh)	Unique structure		Autonomic and somatic motor neurons
Norepinephrine (NE)	Amine		Autonomic neurons
Central nervous system			
Dopamine	Amine		Brain
Norepinephrine, epinephrine (E)	Amines		Brain, spinal cord; also act as hormones
Serotonin (5-hydroxytryptamine, or 5-HT)	Amine; related to preceding three transmitters		Brain
Histamine	Amine		Parts of brain; more common as paracrine
Glutamate, aspartate	Amino acids		Excitatory
Glycine, GABA (gamma-aminobutyric acid)	Amino acids		Inhibitory
Adenosine, ATP	Purines		Often co-secreted with other neurotransmitters
Endorphins, enkephlins, dynorphins	Opioid peptides		Analgesics
Substance P	Polypeptide		Transmitter in pain pathways
Neuropeptide Y	Peptide		Autonomic neurons

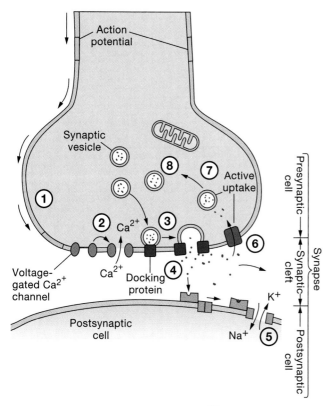

Figure 8-26 Events at the synapse ① An action potential coming down the axon depolarizes the membrane of the axon terminal. ② The depolarization opens voltage-gated Ca^{2+} channels in the membrane so Ca^{2+} enters the cell. ③ Calcium entry signals the synaptic vesicles to release their contents by exocytosis. ④ The neurotransmitter released from the axon diffuses across the synaptic cleft and binds with receptors on the postsynaptic cell. ⑤ The receptor-transmitter complex opens ion channels in the postsynaptic membrane, creating a rapid response. ⑥ The response is terminated by neurotransmitter uptake back into the presynaptic cell. Some neurotransmitter diffuses away from the synapse. Acetylcholine is enzymatically degraded by enzymes in the synaptic cleft (not shown). ⑦ Synaptic vesicles reload with neurotransmitter by ion-driven active uptake. ⑧ Vesicles store neurotransmitter until the next signal arrives for release.

neurotransmitter molecules are free in the extracellular fluid of the synaptic cleft, they diffuse across the tiny distance to bind with receptors on the membrane of the postsynaptic cell. When the neurotransmitter molecules act as ligands, binding to their receptors, a response is initiated in the postsynaptic cell.

continued from page 225

Dr. McKhann then asked to see autopsy reports on some of the children who had died of their paralysis at Beijing Hospital. In the reports, pathologists noted that the patients had normal myelin but damaged axons. In some cases, the axon had been completely destroyed, leaving only a hollow shell of myelin.

Question 5: Do the results of Dr. McKhann's investigation suggest that the Chinese children had Guillain-Barré syndrome? Why or why not?

The short duration of a neural signal is assured by the rapid removal or inactivation of the neurotransmitter in the synaptic cleft. Some neurotransmitter molecules simply diffuse away from the synapse, separating them from their receptors. Other neurotransmitters are inactivated by enzymatic breakdown or transported back into the presynaptic cell, or by both.

Not All Postsynaptic Responses Are Rapid and of Short Duration

The combination of a neurotransmitter with its receptor sets in motion a series of responses in the postsynaptic cell (Fig. 8-27 ■). In the simplest response, the neurotransmitter opens a chemically gated ion channel, leading to ion movement between the cell and the extracellular fluid. The resulting change in membrane potential is called a **fast synaptic potential** because it begins quickly and lasts only a few milliseconds. These synaptic potentials are graded potentials, the excitatory (depolarizing) and inhibitory (hyperpolarizing) postsynaptic potentials (EPSPs and IPSPs) previously described.

Fast responses always open ion channels, but slow responses can close ion channels as well as open them. In slow responses, neurotransmitters bind to receptors linked to G proteins and second messenger systems (∞ p. 152). The second messengers act from the cytoplasmic side of the cell membrane to open or close ion channels. Membrane potentials resulting from this process are called **slow synaptic potentials** because the second messenger method takes longer to create a response. In addition, the response itself lasts longer, from seconds or minutes.

Slow responses are not limited to altering the open state of ion channels. Neurotransmitter activation of second messenger systems may also modify existing cell proteins or regulate the production of new cell proteins. This type of slow response has been linked to the growth and development of neurons and to the mechanisms underlying long-term memory.

✓ In an experiment on synaptic transmission, a synapse was bathed in a Ca^{2+}-free medium that was otherwise equivalent to extracellular fluid. An action potential was triggered in the presynaptic neuron. Although the action potential reached the axon terminal at the synapse, the usual response of the postsynaptic cell did not occur. What conclusion did the researchers make on the basis of these results?

✓ Why are axon terminals sometimes called "biological transducers"?

Some Signals Move from the Postsynaptic to the Presynaptic Cell

The traditional view of chemical synapses was that they were the sites of one-way communication, with messages moving only from the presynaptic cell to the post-

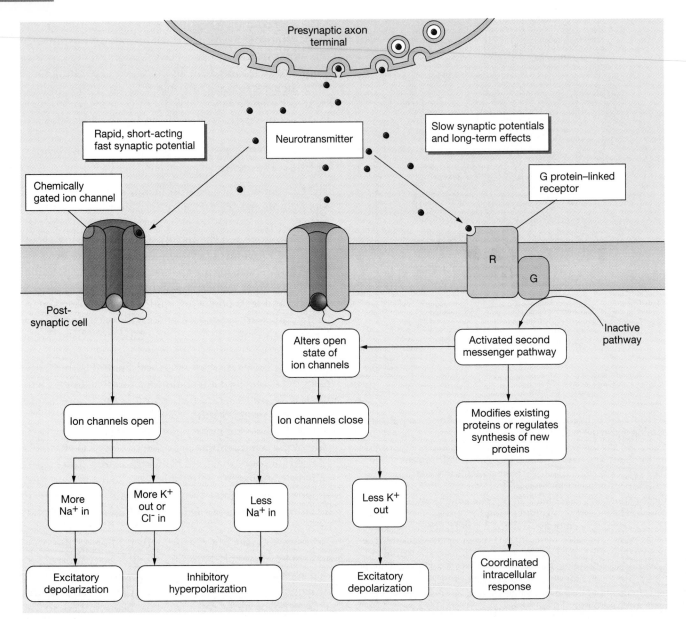

■ Figure 8-27 **Fast and slow responses in postsynaptic cells** Many neurotransmitters create rapid, short-acting responses by opening ion channels. Some neurotransmitters create slower, longer-lasting responses by activating second messenger systems. These responses include closing ion channels as well as opening them. In addition, slow responses may alter existing proteins or cause the synthesis of new proteins.

synaptic cell. We now know that this is not always the case. In the brain, there are some synapses where the cells on *both* sides of the cleft release a neurotransmitter that acts on the opposite cell. Perhaps more important, we have learned that many postsynaptic cells "talk back" to their presynaptic neurons, sending neuromodulators to influence how neurons assemble during development or to modulate the interactions of connecting neurons. For example, **nerve growth factor** is a protein secreted by the targets of certain neurons during embryonic development. It is taken up at the axon terminal and transported to the nucleus in cell body. There, the nerve growth factor acts as a gene control signal that directs the neuron to maintain its synaptic connection to

the postsynaptic target cell. Nitric oxide is another molecule that acts as a neuromodulator on presynaptic cells.

Disorders of Synaptic Transmission Are Responsible for Many Diseases

Synaptic transmission is the most vulnerable step in the process of signaling through the nervous system. It is the point at which many things go wrong, leading to disruption of normal function. Yet, at the same time, the receptors at synapses are exposed to the extracellular fluid, making them more accessible to drugs than intracellular receptors. In recent years, we have discovered that a variety of nervous system disorders are related to

Myasthenia Gravis Myasthenia gravis is an autoimmune disease in which the body fails to recognize the acetylcholine (ACh) receptors on skeletal muscle as part of "self" and therefore produces molecules to attack the receptors. The molecules bind to the receptors at a site that is distinct from the ACh binding site. In doing so, they change the receptor protein in some way that causes the muscle cell to speed up turnover of the receptor. The bound receptors are withdrawn from the membrane by endocytosis and destroyed inside lysosomes. Their destruction leaves the muscle with fewer ACh receptors in the membrane and therefore a diminished response to ACh from the controlling somatic motor neuron. Assume for the moment that you are a researcher trying to help people who suffer from myasthenia gravis. On the basis of what you know about membrane receptors and synapses, think of possible methods for treating this disease.

---

problems with synaptic transmission. These include Parkinson's disease, schizophrenia, and depression. The best understood diseases of the synapse are those that involve the neuromuscular junction. Diseases resulting from synaptic transmission problems within the central nervous system have proved more difficult to study because, anatomically, they are more difficult to isolate.

Drugs that act on synaptic activity, particularly synapses in the central nervous system, are the oldest known and most widely used of all pharmacological agents. Caffeine, nicotine, and alcohol are common drugs in many cultures. Medically, some drugs that we use to treat conditions such as schizophrenia, anxiety, and epilepsy act by influencing events at the synapse. In many disorders arising in the central nervous system, we do not yet fully understand either the cause of the disorder or the mechanism of action of the drug. This subject is one major area of pharmacological research, and new classes of drugs are being formulated and approved every year.

Development of the Nervous System Depends on Chemical Signals

In a system as complex as the nervous system, how can more than 100 billion neurons in the brain find their correct targets and make synapses among more than 10

continued from page 227

Currently, Dr. McKhann believes that the disease afflicting the Chinese children—which he has named acute motor axonal polyneuropathy (AMAN)—may be caused by a bacterial infection. He also believes that the disease initiates its damage of axons at the neuromuscular junctions.

Question 6: Based on information provided in this chapter, can you name another disease that alters synaptic transmission?

times that many glial cells? How can a somatic motor neuron in the spinal cord find the correct pathway to form a synapse with its target muscle in the big toe? The answer is found in the chemical signals used by the developing embryo. Embryonic nerve cells send out axons that can grow until they find their target cell. In experiments in which the target cells are moved to an unusual location in the embryo, the axons in many instances are still able to find their targets by "sniffing out" their chemical scent. Once the axon reaches the target cell, a synapse forms. But synapse formation must be followed by electrical and chemical activity, or the synapse will disappear. This "use it or lose it" scenario is most dramatically reflected by the fact that the infant brain is only about one-fourth the size of the adult brain. Further brain growth is due not to an increase in cell number but to an increase in size and number of axons, dendrites, and synapses. Development is dependent on electrical activity in the brain, on action potentials moving through sensor pathways and interneurons. Babies who are neglected or deprived of sensory input may have delayed development because of the lack of stimulation of the nervous system. On the other hand, there is no evidence that extra stimulation in infancy will enhance intellectual development, despite a popular movement to expose babies to art, music, and foreign languages before they can even walk. Once synapses form, they are not fixed for life. Variations in electrical activity can cause rearrangement of the connections. This process, known as the *plasticity* of synapses, continues throughout life and is one reason that older adults are urged to keep learning new skills and information.

When Neurons Are Injured, Segments Separated from the Cell Body Die

The responses of mature neurons to injury are similar in many ways to the growth of neurons during development, relying on a combination of chemical and electrical signals. This is another area of active research because spinal cord injuries and brain damage from illnesses and accidents disable many people each year. The loss of one neuron from a reflex can have drastic consequences on an entire reflex pathway.

When a neuron is injured, if the cell body dies, the entire neuron dies. If the cell body is intact and only the axon is severed, most of the neuron will survive. At the site of injury, the cytoplasm leaks out until membrane is recruited to seal the opening. The proximal segment (closest to the cell body) swells as organelles and filaments carried into the axon by axonal transport accumulate. Chemical factors produced by Schwann cells near the injury site move by retrograde transport to the cell body, telling it that an injury has occurred. The distal segment of axon, deprived of its source of protein, begins to degenerate slowly (Fig. 8-28 ■). The death of

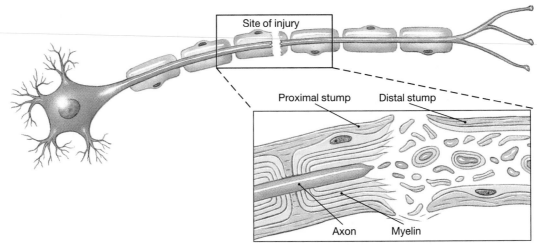

■ Figure 8-28 **Injury to neurons** When an axon is cut, the section attached to the cell body continues to live, but the section of the axon distal to the cut will begin to disintegrate and die. Under some circumstances, the stump of the axon may regrow through the existing sheath of Schwann cells and reform a synapse with the proper target.

this part of the neuron may take a month or longer, although synaptic transmission ceases almost immediately. The myelin sheath around the distal axon begins to unravel and the axon itself collapses. The fragments are cleared away by scavenger microglia or phagocytes that ingest and digest the debris.

Under some conditions, axons in the peripheral nervous system can regenerate and reestablish their synaptic connections. The Schwann cells of the damaged neuron secrete certain **neurotrophic factors** that keep the cell body alive and that stimulate regrowth of the axon. The growing tip of a regenerating axon behaves much like that of a developing axon, following chemical signals in the extracellular matrix along its former path until the axon rejoins its target cell. But sometimes the loss of the distal axon is permanent, and the pathway is destroyed. Many scientists are studying these mechanisms of axon growth and inhibition in the hopes of finding treatments that will restore function to victims of spinal cord injury and degenerative neurological disorders.

CHAPTER REVIEW

Chapter Summary

1. The **nervous system** is a complex network of **neurons** that form the rapid control system of the body. (p. 201)

Organization of the Nervous System

2. The nervous system is divided into the **central nervous system (CNS),** composed of the **brain** and **spinal cord,** and the **peripheral nervous system.** (p. 202)

3. The peripheral nervous system has **afferent (sensory) neurons** that bring information into the CNS and **efferent neurons** that carry information away from the CNS back to various parts of the body. (p.202)

4. The efferent neurons include **somatic motor neurons** that control skeletal muscles and **autonomic neurons** that control smooth and cardiac muscles, glands, and some adipose tissue. (p. 202)

5. Autonomic neurons are subdivided into **sympathetic** and **parasympathetic neurons.** (p. 202)

Cells of the Nervous System

6. Neurons have a **cell body** with a nucleus and organelles to direct cellular activity, **dendrites** to receive incoming signals, and an **axon** to transmit electrical signals from the cell body to the **axon terminal.** (p. 203)

7. **Interneurons** are neurons that lie entirely within the central nervous system. (p. 204)

8. The region where an axon terminal meets its target cell is called a **synapse.** The target cell is called the **postsynaptic cell,** and the neuron that releases the chemical signal is known as the **presynaptic cell.** (p. 205)

9. Material is transported between the cell body and axon terminal by **axonal transport.** (p. 205)

10. **Glial cells** provide physical support and direct the growth of neurons during repair and development. **Schwann cells** are glial cells associated with the peripheral nervous system; **oligodendrocytes** and **astrocytes** are found in the CNS. **Microglia** are modified immune cells that act as scavengers. (p. 206)

11. Schwann cells and oligodendrocytes form insulating **myelin** sheaths around neurons. The **nodes of Ranvier** are sections of uninsulated membrane between Schwann cells. (p. 206)

Electrical Signals in Neurons

12. Membrane potential is influenced by the concentration gradients of ions across the membrane and by the permeability of the membrane to those ions. Potassium is the major ion contributing to the average resting membrane potential. (p. 208)

13. When the membrane potential becomes less negative, the cell is **depolarized.** Cells usually depolarize as a result of addition of cations to the cytoplasm. They may also depolarize if the cell permeability to K^+ decreases. (p. 208)

14. If the membrane potential becomes more negative, the cell is **hyperpolarized.** Cells will hyperpolarize if anions enter the cell from the extracellular fluid or if cell permeability to K^+ increases. (p. 208)

15. The permeability of a cell to ions changes when ion channels in the membrane open and close, or when the cell inserts or withdraws channels in the membrane. (p. 209)

16. Gated Na^+ channels in neurons open in response to chemical or mechanical signals or in response to depolarization of the cell membrane (p. 209)

17. Voltage-gated Na^+ channels along the axon open in response to depolarization, allowing Na^+ to enter the cell. (p. 210)

18. Voltage-gated Ca^{2+} channels in the axon terminal open in response to depolarization. Calcium entry is the signal that initiates exocytosis of synaptic vesicles containing neurotransmitter. (p. 210)

19. Chemically gated Cl^- channels allow Cl^- to enter the cell, hyperpolarizing it. (p. 210)

20. **Graded potentials** are depolarizations or hyperpolarizations whose strength is directly proportional to the strength of the triggering event. Graded potentials lose strength as they move through the cell. (p. 210)

21. The wave of depolarization that moves through the cell with a graded potential is known as **local current flow.** (p. 210)

22. **Action potentials** are rapid electrical signals that travel undiminished in amplitude from the cell body to the axon terminals. (p. 210)

23. Action potentials begin in the **trigger zone** if a single graded potential or the sum of multiple graded potentials exceeds a minimum depolarization known as the **threshold.** (p. 210)

24. The summation of simultaneous graded potentials is known as **spatial summation.** The summation of graded potentials that follow each other sequentially is called **temporal summation.** (p. 211)

25. The rising phase of the action potential occurs when voltage-gated Na^+ channels open, allowing Na^+ to enter the cell. The action potential peaks when the Na^+ channels close. (p. 213)

26. The falling phase of the action potential occurs when voltage-gated K^+ channels open, allowing K^+ to leave the cell. (p. 214)

27. The voltage-gated Na^+ channels have two gates: a fast activation gate that opens with depolarization and a slower inactivation gate that closes shortly after the activation gate opens. (p. 216)

28. Once an action potential has begun, there is a brief period of time known as the **absolute refractory period** during which a second action potential cannot be triggered, no matter how large the stimulus. Because action potentials cannot overlap owing to their refractory periods, they cannot be summed the way that graded potentials can. (p. 217)

29. Immediately after the absolute refractory period, there follows another period of time during which a higher-than-normal graded potential is required to trigger an action potential. This period is known as the **relative refractory period.** (p. 217)

30. Information about the strength and duration of a stimulus is conveyed by the frequency of action potential propagation. (p. 217)

31. The Na^+/K^+-ATPase plays no direct role in the action potential because very few ions cross the membrane during an action potential. (p. 218)

32. The movement of an action potential through the axon at high speed is called **conduction.** It results from the opening of voltage-gated Na^+ channels along the axon that creates continuous entry of sodium along the axon. (p. 218)

33. The two parameters that influence the speed of action potential conduction are the diameter of the neuron and the resistance of the neuron membrane to leakage of current out of the cell. The myelin sheath around axons speeds up conduction by preventing current from leaking out of the membrane underneath it. (p. 220)

34. The apparent leapfrogging of action potentials from node to node of the axon is called **saltatory conduction.** (p. 222)

35. Changes in blood K^+ concentration affect resting membrane potential and the conduction of action potentials. (p. 223)

Cell-to-Cell Communication in the Nervous System

36. In electrical synapses, an electrical signal passes directly from the cytoplasm of one cell to another through gap junctions. **Chemical synapses** use **neurotransmitters** to carry information from one cell to the next. (p. 224)

37. Neurotransmitters are synthesized in the cell body or in the axon terminal. They are stored in **synaptic vesicles** and are released by exocytosis when an action potential reaches the axon terminal. Once in the synapse, the neurotransmitter combines with a receptor on the target cell. (p. 225)

38. Neurotransmitter action is rapidly terminated by re-uptake into the axon terminal, diffusion away from the synapse, or, in the case of **acetylcholine (ACh),** enzymatic degradation in the synapse. (p. 225)

39. When neurotransmitters open ion channels they create **fast synaptic potentials,** graded potentials that begin quickly and last only a few milliseconds. (p. 226)

40. If neurotransmitters bind to G protein–linked receptors, they activate second messenger systems and create **slow synaptic potentials** by opening or closing ion channels. These neurotransmitters may also modify existing cell proteins or regulate the production of new cell proteins. (p. 227)

41. Developing neurons find their way to their targets by using chemical signals. (p. 229)

Questions

LEVEL ONE **Reviewing Facts and Terms**

1. Name the divisions of the nervous system and describe in as much detail as possible.

2. Match the term with its description:
   (a) axon
   (b) dendrite
   (c) afferents
   (d) efferents
   (e) axon hillock
   1. process of a neuron that receives incoming signals
   2. sensory neurons, transmit information to central nervous system
   3. long process that transmits signal to the target cell
   4. region of neuron where action potential begins
   5. neurons that transmit information about a response from central nervous system to the rest of the body

3. Somatic motor neurons control _____; _____ neurons control smooth and cardiac muscles, exocrine and some endocrine glands, and some types of adipose tissue.

4. Autonomic neurons are classified as either _____ or _____ neurons.

5. Name the two primary cell types found in the nervous system.

6. Which part of the neuron lacks ribosomes and endoplasmic reticulum? What is the significance of this lack?

7. Axonal transport refers to:
   a. the release of neurotransmitter molecules into the synaptic cleft
   b. the use of microtubules to send secretions from the cell body to the axon terminal
   c. the transport of vesicles containing proteins down the length of the axon
   d. the movement of the axon terminal to synapse with a new postsynaptic cell
   e. none of these

8. A typical neuron synapses with how many other neurons?
   a. 1   b. fewer than 10
   c. 1000   d. 10,000
   e. more than 100,000

9. If a cell (gains/loses) K^+, the cell hyperpolarizes. A cell will also hyperpolarize if Cl^- (enters/exits) the cell.

10. List the four major types of ion channels found in neurons. For each type of channel, tell what factor(s) open(s) and/or close(s) the channel.

11. Arrange these events in the proper sequence:
    a. Efferent neuron reaches threshold and fires an action potential.
    b. Afferent neuron reaches threshold and fires an action potential.
    c. Effector organ responds by performing output.
    d. Integrating center reaches decision about response.
    e. Sensory organ detects change in the environment.

12. Where are interneurons found?

13. An action potential is (circle all correct answers):
    a. a reversal of the concentrations of Na^+ and K^+ inside and outside of the neuron
    b. the same size and shape at the beginning and end of the axon
    c. the event following depolarization to threshold
    d. a signal that is transmitted to the distal end of a neuron and causes the release of neurotransmitter

In questions 14–17, choose the correct ion to fill in the blanks: Na^+, K^+, Ca^{2+}, Cl^-

14. The resting cell membrane is more permeable to _____ ions than to _____ ions. Although _____ ions contribute little to the resting membrane potential, they play a key role in generating electrical signals in excitable tissues.

15. _____ is 12 times more concentrated outside the cell than inside.

16. _____ is 30 times more concentrated inside the cell than outside.

17. An action potential is due to _____ flooding into the cell.

18. The resting membrane potential is due to _____ being trapped in the cell.

19. What is the function of the myelin sheath?

20. Define excitability as it pertains to nervous tissue.

21. List two factors that will enhance conduction speed.

22. Match the neurotransmitter with its method for removal from the synapse:
    (a) enzymatic degradation   1. acetylcholine
    (b) reabsorption   2. norepinephrine
    (c) diffusion

23. Draw an action potential and draw the positioning of the K$^+$ and Na$^+$ channel gates during each phase.

LEVEL TWO   Reviewing Concepts

24. Sketch a typical efferent neuron, showing the cell body, axon, dendrites, nucleus, trigger zone, axon hillock, and axon terminal. Use each component as a starting point for a concept map, and include the functions and/or activities of each area.

25. List the three functional classes of neurons, and explain how they differ from one another.

26. List four different types of glial cells, and briefly explain the role of each type.

27. Arrange the following terms to describe the sequence of events that happens after a neurotransmitter binds to a receptor on a postsynaptic neuron. Terms may be used more than once or not at all.
    a. action potential fires at axon hillock
    b. axon hillock reaches threshold
    c. cell depolarizes
    d. exocytosis
    e. graded potential occurs
    f. ligand-gated ion channel opens
    g. local current flow
    h. saltatory conduction
    i. voltage-gated Ca^{2+} channels open
    j. voltage-gated K$^+$ channels open
    k. voltage-gated Na$^+$ channels open

28. Match the best term (hyperpolarize, depolarize, repolarize) to the events described. The cell in question has a resting membrane potential of -70 mV.
    a. membrane potential changes from -70 mV to -50 mV
    b. membrane potential changes from -70 mV to -90 mV
    c. membrane potential changes from $+20$ mV to -60 mV
    d. membrane potential changes from -80 mV to -70 mV

29. Use the best term (hyperpolarized, depolarized) to describe the changes in a neuron if the resting membrane potential is at -70 mV and each event below occurs. More than one answer may apply; list all those that are correct. What would happen if
    a. Na$^+$ enters the cell   b. K$^+$ leaves the cell
    c. Cl$^-$ enters the cell   d. Ca^{2+} enters the cell

30. Define, compare, and contrast the following concepts:
    a. threshold, subthreshold, spatial summation, temporal summation, all-or-none, overshoot, undershoot
    b. graded potential, EPSP, IPSP, absolute refractory period, relative refractory period
    c. afferent neuron, efferent neuron, interneuron
    d. sensory, somatic motor, sympathetic, autonomic, parasympathetic neuron
    e. fast and slow synaptic potentials
    f. temporal and spatial summation

31. If all action potentials within a given neuron are identical, how does the neuron transmit information about the strength and duration of the stimulus?

32. The presence of myelin allows an axon to
    a. produce more frequent action potentials
    b. conduct impulses more rapidly
    c. produce action potentials of larger amplitude
    d. produce action potentials of longer duration

LEVEL THREE   Problem Solving

33. If human babies' muscles and neurons are fully developed and functional at birth, why can't they focus their eyes, sit up, or learn to crawl within hours of being born? (Hint: Muscle strength is not the problem.)

34. The voltage-gated Na$^+$ channels of the neuron open when the neuron depolarizes. If depolarization opens the channels, why do they close when the neuron is maximally depolarized?

35. In each scenario below, will an action potential be produced? The postsynaptic neuron has a resting membrane potential of -70 mV.
    a. Fifteen neurons synapse on one postsynaptic neuron. At the trigger zone, 12 of the neurons produce EPSPs of 2 mV each, and the other three produce IPSPs of 3 mV each. The threshold for the postsynaptic cell is -50 mV.
    b. Fourteen neurons synapse on one postsynaptic neuron. At the trigger zone, 11 of the neurons produce EPSPs of 2 mV each, and the other three produce IPSPs of 3 mV each. The threshold for the postsynaptic cell is -60 mV.
    c. Fifteen neurons synapse on one postsynaptic neuron. At the trigger zone, 14 of the neurons produce EPSPs of 2 mV each, and the other one produces an IPSP of 9 mV. The threshold for the postsynaptic cell is -50 mV.

36. Jim has a kidney problem, with the end result that the electrolyte levels in his blood are abnormal. His K$^+$ is elevated. What is this condition called? What effect, if any, does this condition have on the resting membrane potential of his neurons? Do you expect an effect on the action potentials of his neurons?

37. Match the example to the proper type of stimulus:
    (a) mechanical   1. bath water at 106°F
    (b) chemical   2. acetylcholine
                              3. a hint of perfume
                              4. epinephrine
                              5. bright light
                              6. lemon juice
                              7. a punch on the arm

Problem Conclusion

In this running problem, you learned about a baffling paralytic illness that appears to be a new disease. Although it bears some resemblance to Guillain-Barré syndrome, upon closer investigation, it is actually quite different.

Further check your understanding of this running problem by checking your answers against those in the summary table.

Question	Facts	Integration and Analysis
1 Which division(s) of the nervous system may be involved in Guillain-Barré syndrome (GBS)?	The nervous system is divided into peripheral divisions (sensory and efferent), and the central nervous system. The efferent division is composed of somatic motor neurons that control skeletal muscles, and autonomic neurons that control smooth and cardiac muscles, as well as glands.	Patients with GBS can neither feel sensations nor move their muscles. This suggests a problem with the sensory and somatic motor divisions. However, there could also be a problem within the central nervous system, which integrates information flow between the sensory and efferent divisions. You do not have enough information to determine which division is affected.
2 Do you think that the paralysis found in the Chinese children affected both sensory and efferent neurons? Why or why not?	The Chinese children can feel a pin prick but cannot move their muscles.	Because the children can feel the pin prick, their sensory neurons are normal. Inability to move their muscles indicates a problem with the motor neurons controlling the muscles, with the portions of the brain controlling movement, or with the muscles themselves.
3 In Guillain-Barré syndrome, what would you expect the results of a nerve conduction test to be?	Nerve conduction tests measure the strength and conduction speed of action potentials. GBS is a condition in which myelin around the axons is destroyed.	Without myelin, electrical current flowing down the axon leaks out through the axon membrane, and saltatory conduction cannot occur. Thus, demyelinated axons conduct action potentials with reduced strength and slower speed.
4 Is the paralytic illness that affected the Chinese children a demyelinating condition? Why or why not?	Nerve conduction tests showed a normal speed of action potential conduction but a decrease in the strength of the action potentials.	Loss of myelin should decrease both strength and conduction speed of action potentials. Because the conduction speed was normal in these patients, their disease is probably not a demyelinating disease.
5 Do the results of Dr. McKhann's investigation suggest that the Chinese children had Guillain-Barré syndrome? Why or why not?	In addition to the nerve conduction tests, autopsy reports on children who died from the disease showed that they had damaged axons but undamaged myelin.	GBS is a demyelinating disease that affects both sensory and motor neurons. The Chinese children had normal sensory neurons, and nerve conduction tests and histological studies indicated normal myelin. Therefore, it is reasonable to conclude that the Chinese children did not have GBS.
6 Based on information provided in this chapter, can you name another disease that alters synaptic transmission?	Synaptic transmission can be altered by blocking neurotransmitter release, binding at the target cell, and removal from the synapse.	Parkinson's disease, depression, schizophrenia, and myasthenia gravis are related to problems with synaptic transmission (p. 227). Some neurotoxins also block neurotransmitter release (p. 226).

cerebrospinal fluid, one can get an indication of the chemical environment within the brain. This sampling procedure, known as a *spinal tap,* is generally done by withdrawing fluid from the subarachnoid space between vertebrae at the lower end of the spinal cord. The presence of proteins or blood cells in the cerebrospinal fluid is suggestive of infection.

✓ If the concentration of H^+ in the cerebrospinal fluid is higher than that in the blood, what can you say about the pH of the CSF?

✓ If a blood vessel running between the meninges is ruptured as a result of a blow to the cranium, what do you predict might happen?

✓ Is cerebrospinal fluid more like plasma or interstitial fluid? Defend your answer.

The Blood-Brain Barrier Protects the Brain from Harmful Substances in the Blood

Nervous tissue has special metabolic requirements. The only fuel source for neurons under normal circumstances is glucose, and, therefore, blood glucose concentrations are homeostatically regulated to ensure normal brain function. Neurons also exhibit high rates of oxygen consumption that reflect their need for ATP to transport ions against their concentration gradients (∞ p. 121).

To meet the oxygen and glucose needs of the neurons, about 15% of the blood pumped by the heart goes to the brain. Interruption of the blood supply to the brain can have devastating results, and brain death occurs after only a few minutes without oxygen. Progressive *hypoglycemia* (low blood glucose levels) leads to confusion, unconsciousness, and eventually death.

Because the brain is the main control center for the body, its cells must be protected from potentially harmful substances in the blood. Capillaries in the brain are much less permeable than most capillaries elsewhere in the body and they exclude many molecules that cross capillary walls elsewhere. These relatively impermeable capillaries form what is known as the **blood-brain barrier** (Fig. 9-4 ∎). The limited permeability of the blood-brain barrier is a protective mechanism that shelters the brain from fluctuations in the blood concentration of hormones, ions, and neuroactive substances such as neurotransmitters. However, lipid-soluble molecules can enter the interstitial fluid bathing the neurons by simply diffusing through the cell membranes (∞ p. 118). This is one reason some antihistamines make you sleepy but others do not. The older antihistamines were lipid-soluble amines that readily crossed the blood-brain barrier and acted on brain centers controlling alertness. The newer drugs are much less lipid-soluble and as a result do not have the same sedative effect.

The blood-brain barrier, like the choroid plexus, is able to transport nutrients and other useful material from the blood into the brain interstitial fluid by using specific

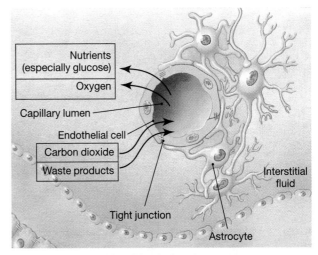

∎ **Figure 9-4 The blood-brain barrier** Tight junctions in cerebral capillaries create the blood-brain barrier. Paracrines secreted by astrocytes stimulate the development of the tight junctions. The movement of material across the blood-brain barrier is regulated in order to protect the neurons from potentially harmful substances in the blood.

membrane carriers. One interesting illustration of how the blood-brain barrier works is seen in Parkinson's disease, a neurological disorder in which brain levels of the neurotransmitter dopamine are too low. Dopamine administered therapeutically is ineffective because it is unable to cross the blood-brain barrier. But the dopamine precursor, L-DOPA, is transported across the cells of the blood-brain barrier on an amino acid transporter (∞ p. 129). Once in the cerebrospinal fluid, L-DOPA is metabolized to dopamine, thereby correcting the deficiency.

Why are capillaries in the brain so much less leaky than capillaries elsewhere in the body? The endothelial cells that form the walls of the brain capillaries are the same as those in other capillaries. However, the cells are connected to each other with tight junctions rather than leaky junctions and pores (∞ p. 54). The formation of these tight junctions is induced by paracrines released by adjacent astrocytes. Thus, it is the brain tissue itself that creates the blood-brain barrier.

In a few areas of the brain, the blood-brain barrier is missing. In these areas, the function of adjacent neurons depends in some way on direct contact with the blood. For instance, the posterior pituitary releases neurosecretory hormones that must pass into the capillaries for distribution to the body (∞ p. 185). Another example is the vomiting center in the medulla (Fig. 9-1h ∎). These neurons sense the presence of possibly toxic substances in the blood, such as drugs. They respond by inducing vomiting, thereby removing the contents of the digestive system in case the substance was ingested.

Now that we have described the non-neural elements associated with the central nervous system, let us look at the general anatomical characteristics of the brain and spinal cord.

continued from page 236

A stroke is a form of cerebrovascular disease in which a blood vessel to the brain ruptures or becomes blocked by a blood clot, or by the buildup of a fatty substance called plaque. The loss of blood flow to an area of brain tissue (a process called "ischemia") deprives brain cells of oxygen, and they die. Often, strokes occur without warning. Sometimes, however, a transient ischemic attack (TIA)—or a series of them—heralds the onset of a stroke. In fact, a recent study has found that TIAs occurred in 30% of people who later experienced full-fledged strokes.

Question 1: In computerized tomography (CT) scans of stroke victims, the tissue supplied by the affected blood vessel looks less dense than usual. Why is this? (Note: Computerized tomography is a specialized X-ray technique that makes thousands of sequential X-rays of a section of the body. The X-rays are then compiled into a three-dimensional composite of images.)

Neurons of the Central Nervous System Are Grouped into Nuclei and Tracts

The central nervous system, like the peripheral nervous system, is composed of neurons and supportive glial cells. Interneurons do not extend outside the central nervous system; sensory and efferent neurons link interneurons in the central nervous system to peripheral receptors and effectors.

When viewed on a macroscopic level, the tissues of the central nervous system are described by their color. The unmyelinated **gray matter** consists of nerve cell bodies, dendrites, and axon terminals (see Fig. 9-1e, i ■). The cell bodies are assembled in an organized fashion within both the brain and the spinal cord. They form layers in some parts of the brain or cluster into groups of neurons with a similar function. These clusters in the brain and spinal cord are known as **nuclei,** and they are usually identified by specific names.

White matter is made up mostly of axons with very few cell bodies. Its pale color comes from the myelin sheaths that surround the axons. Long projections of white matter that connect different regions of the central nervous system are known as **tracts. Ascending tracts** mainly carry sensory information from the cord to the brain; **descending tracts** carry mostly efferent signals from the brain to the cord. **Propriospinal tracts** remain within the cord. Tracts in the central nervous system are equivalent to nerves in the peripheral nervous system.

THE SPINAL CORD

The spinal cord is the major pathway of information flowing back and forth between the brain and the skin, joints, and muscles of the body. In addition, the spinal cord contains neural networks responsible for locomotion. If the spinal cord is severed, there is loss of sensation from the skin and muscles as well as loss of the ability to voluntarily control muscles (paralysis).

The interneurons of the spinal cord route sensory information from peripheral receptors to the brain and commands from the brain to the muscles and glands of the body. In many cases, the interneurons also modify information as it passes through. Spinal interneurons play a critical role in the coordination of movement (see Chapter 13).

The spinal cord is divided into four regions (*cervical, thoracic, lumbar,* and *sacral*), named to correspond to the adjacent vertebrae (Fig. 9-1a ■). Each region is further subdivided into segments, and each segment gives rise to a bilateral pair of spinal nerves. Just before a spinal nerve joins the spinal cord, it divides into two branches called **roots** (Fig. 9-1d ■). The **dorsal root** of each spinal nerve is specialized to carry incoming sensory information, and the **ventral root** carries information from the central nervous system to the muscles and glands.

In cross section, the spinal cord has a butterfly- or H-shaped core of gray matter and a surrounding rim of white matter (Fig. 9-5a ■). The color of the gray matter comes from the cell bodies of interneurons and efferent neurons. The cell bodies of sensory neurons are contained in the **dorsal root ganglia,** swellings found on the dorsal roots just before they enter the cord. Sensory fibers from the dorsal roots enter the **dorsal horns** of the gray matter and synapse with other neurons in sensory nuclei. Efferent fibers leave the gray matter via the **ventral horns.** The cell bodies of somatic motor neurons are found in nuclei in the ventral horns, and those of autonomic neurons are in nuclei in the lateral regions of the gray matter.

The white matter of the spinal cord is the biological equivalent of the fiber-optic cables that phone companies use to carry our communications systems. The white matter can be divided into a number of **columns** composed of tracts of axons that transfer information up and down the cord. The dorsal column of the white matter is a pathway dedicated to carrying somatic sensory information up the spinal cord to the brain (Fig. 9-5b ■). The lateral and ventral columns of white matter carry both sensory information to the brain and descending signals from the brain back to the spinal cord. Not all information from peripheral receptors must go to the brain. The spinal cord itself functions as a self-contained integrating center for the simplest reflexes.

✓ What are the differences between horns, roots, columns, and tracts of the spinal cord?

✓ If a dorsal root of the spinal cord is cut, what function will be disrupted?

(a) Gray matter

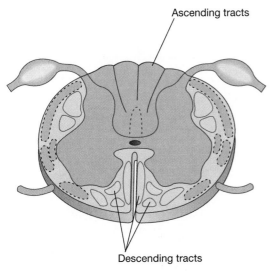

Visceral sensory nuclei

Somatic sensory nuclei

Incoming sensory information from dorsal root

Dorsal root ganglion

Dorsal horn

Ventral horn

Ventral root

Autonomic efferent nuclei

Somatic motor efferent nuclei

Efferent signals to muscles and glands via ventral root

(b) White matter

Ascending tracts

Descending tracts

■ **Figure 9-5 Specialization in the spinal cord** The slight differences in the two figures are due to the fact that they are representations of different regions of the spinal cord. (a) Neurons in the dorsal horn of the gray matter receive sensory information from peripheral receptors. The cell bodies of neurons receiving sensory information are organized into two distinct nuclei, one for somatic information and one for visceral information. Similarly, the cell bodies of neurons sending efferent signals to muscles and glands are organized into somatic motor and autonomic nuclei. The same areas exist on both sides of the spinal cord but are shown here on only one side. (b) The white matter consists of tracts of axons that carry information up and down the spinal cord. Ascending tracts take sensory information to the brain; they occupy the dorsal and external lateral portions of the spinal cord. Descending tracts that carry commands to effector organs occupy the ventral and interior lateral portions of the white matter.

THE BRAIN

Thousands of years ago, Aristotle declared that the heart was the seat of the soul. However, most people now agree that the brain is the organ that gives the human species its unique attributes. The adult human brain weighs about 1400 g and contains an estimated 10^{12} neurons. When you consider that each one of these millions of neurons may receive as many as 200,000 synapses, the number of possible neuronal connections is mind-boggling. Yet, despite this tremendous potential for variability, the central nervous system is remarkably orderly. The human brain is compartmentalized so that different regions have different functions. It also has built-in backup mechanisms in case of failure. A single function may be carried out in more than one region (a feature known as *parallel processing*), and there is a certain degree of plasticity in the networks so that the functions of a damaged region can sometimes be taken over by other sections.

Microscopically, the brain contains two neural elements: (1) neuron cell bodies, some of which are specialized for neurohormone secretion, and (2) nerve **fibers** (axons) in bundles.

When viewed intact in profile, the brain of an adult can be grossly divided into three areas: the brain stem, the cerebellum, and the cerebrum (Fig. 9-1g ■). The cerebrum [*cerebrum,* brain; adjective *cerebral*] is the largest and most obvious structure we see when looking at a human brain. In the early embryo, the area that will become the cerebrum is not much larger than the other regions of the brain (Fig. 9-6a ■). But, as development proceeds, the growth of the cerebrum greatly outpaces that of the other regions (Fig. 9-6b ■). When viewed in sagittal section, the adult cerebrum surrounds the hollow ventricles and many of the structures of the brain stem and diencephalon (Fig. 9-1h ■).

continued from page 242

Dorothy was surprised when her doctor prescribed a daily dose of half an aspirin. Acetylsalicylic acid, the active ingredient in aspirin, is a platelet anti-aggregate. "In other words," he said, "it helps keep blood cells from sticking together, even at this very low dose." He also asked Dorothy about her risk factors for stroke: high blood pressure, high cholesterol levels, heavy alcohol consumption, and smoking. Dorothy did not smoke nor drink, but she did have high blood pressure and elevated cholesterol levels. She was started on treatment to lower her blood pressure and given instructions for a cholesterol-reducing diet.

Question 2: Explain the rationale for taking aspirin to prevent stroke.

(a)

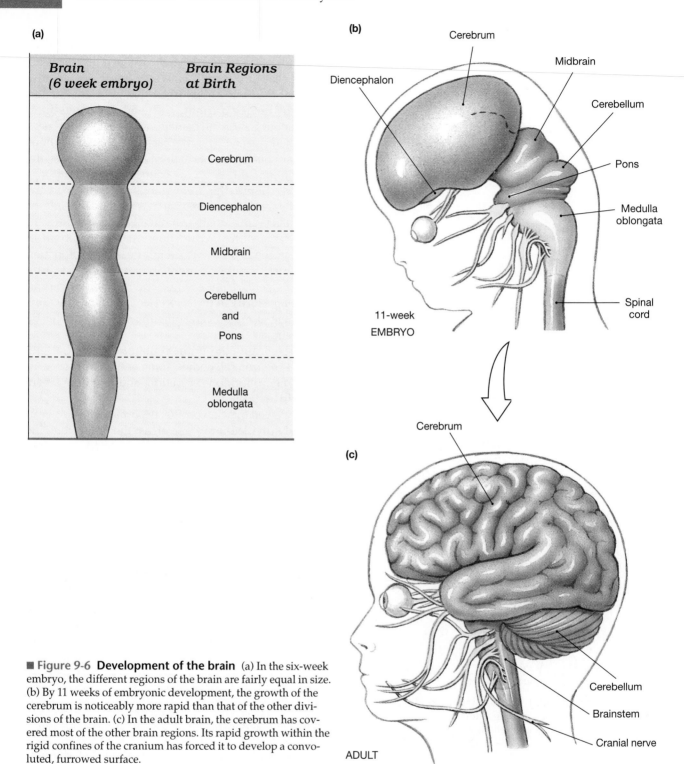

Brain (6 week embryo)	Brain Regions at Birth
	Cerebrum
	Diencephalon
	Midbrain
	Cerebellum and Pons
	Medulla oblongata

(b)

Cerebrum

Midbrain

Diencephalon

Cerebellum

Pons

Medulla oblongata

Spinal cord

11-week EMBRYO

(c)

Cerebrum

Cerebellum

Brainstem

Cranial nerve

ADULT

■ **Figure 9-6 Development of the brain** (a) In the six-week embryo, the different regions of the brain are fairly equal in size. (b) By 11 weeks of embryonic development, the growth of the cerebrum is noticeably more rapid than that of the other divisions of the brain. (c) In the adult brain, the cerebrum has covered most of the other brain regions. Its rapid growth within the rigid confines of the cranium has forced it to develop a convoluted, furrowed surface.

The Brain Stem

The brain stem is an extension of the spinal cord and is divided into three main sections: the medulla oblongata, the pons, and the midbrain, or mesencephalon (Fig. 9-6b ■). The brain stem also contains the third and fourth ventricles. Emerging from the brain stem are nine cranial nerves. (Cranial nerve I, the olfactory nerve,

enters the cerebrum.) The cranial nerves include both sensory and efferent fibers (Table 9-1).

The **medulla oblongata,** frequently just called the *medulla* [*medulla*, marrow; adjective *medullary*], is the transition from the spinal cord into the brain proper. Anatomically, it contains the *corticospinal fiber tracts* that convey information between the cerebrum and the spinal cord. Many of the corticospinal tracts cross the

TABLE 9-1   The Cranial Nerves

Number	Name	Type	Function
I	Olfactory	Sensory	Olfactory (smell) information from nose
II	Optic	Sensory	Visual information from eyes
III	Oculomotor	Motor	Eye movement
IV	Trochlear	Motor	Eye movement
V	Trigeminal	Mixed	Sensory information from face; motor signals for chewing
VI	Abducens	Motor	Eye movement
VII	Facial	Mixed	Sensory for taste; efferent signals for tear and salivary glands, facial expression
VIII	Vestibulocochlear	Sensory	Hearing and equilibrium
IX	Glossopharyngeal	Mixed	Sensory from oral cavity, baro- and chemoreceptors in blood vessels; efferent for swallowing, parotid salivary gland
X	Vagus	Mixed	Sensory and efferents to many internal organs, muscles, and glands
XI	Accessory	Motor	Muscles of oral cavity
XII	Hypoglossal	Motor	Tongue muscles

midline to the opposite side of the body in a region of the medulla known as the **pyramids.** As a result of this crossover, each side of the brain controls the opposite side of the body. The medulla also contains control centers for many involuntary functions such as blood pressure, breathing, swallowing, and vomiting.

The **pons** [*pons,* bridge; adjective *pontine*] is a bulbous protrusion on the ventral side of the brain stem below the midbrain. Because its primary function is to act as a relay station for information transfer between the cerebellum and cerebrum, the pons is often grouped with the cerebellum. The pons also coordinates the control of breathing along with centers in the medulla.

The **midbrain,** or **mesencephalon** [*mesos,* middle], is a relatively small area between the lower brain stem and the diencephalon. Its primary function is the control of eye movement, but it also relays signals for auditory and visual reflexes.

Running throughout the brain stem is the **reticular formation** (also known as the *reticular activating system*), diffuse groups of neurons that probably constitute the oldest part of the brain phylogenetically. The name *reticular* literally means "network" and comes from the redundant crisscrossed axons that branch profusely into superior sections of the brain and into the spinal cord. The reticular formation is best known for its role in arousal and sleep, but in recent years it has also been shown to be involved in muscle tone and stretch reflexes, coordination of breathing, blood pressure regulation, and modulation of pain.

The Cerebellum

The name **cerebellum** [adjective *cerebellar*] means "little brain," and, indeed, most of the nerve cells in the brain are in the cerebellum. The specialized function of the cerebellum is to process sensory information and coordinate the execution of movement. Sensory input into the cerebellum comes from somatic receptors in the periphery of the body and from receptors for equilibrium and balance located in the inner ear. The cerebellum also receives motor input from neurons in the cerebral cortex.

The Diencephalon

The **diencephalon,** or "between-brain," lies between the brain stem and the cerebrum (Fig. 9-7 ■). It is composed of two main sections, the thalamus and the hypothalamus. The diencephalon also contains the **pineal gland,** a small endocrine gland that secretes the hormone **melatonin** (∞ p. 196). Most of the diencephalon is occupied by many small nuclei that make up the **thalamus** [*thalamus,* bedroom; adjective *thalamic*]. This region is often described as a relay station because almost all sensory information is transmitted from lower parts of the central nervous system through the thalamus on its way to the cerebral cortex. But, like the spinal cord, the thalamus can shape the information passing through it, making it an integrating center as well as a relay station. The thalamus receives sensory fibers from the optic tract, ears, and spinal cord, and it projects fibers to the cerebrum.

The **hypothalamus** lies beneath the thalamus. The stalk of the pituitary gland and the neuroendocrine tissue of the posterior pituitary are downgrowths of this region of the diencephalon (∞ p. 184). The hypothalamus contains various centers for behavioral drives, such as hunger and thirst, and plays a key role in homeostasis. Its neural output controls many functions of the autonomic division of the nervous system as well as a variety of endocrine functions (Table 9-2). For example, neuroendocrine cells in the hypothalamus control the

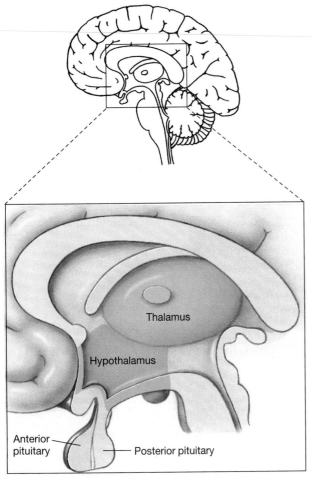

■ **Figure 9-7 The diencephalon** The diencephalon lies between the brain stem and the cerebrum. It is composed of the thalamus, a relay station for information going to and from the cerebrum, and the hypothalamus, a center for many homeostatic reflexes. The pituitary appears to be an extension of the hypothalamus, but the anterior pituitary is actually an endocrine gland that developed from the same epithelium as the roof of the mouth.

TABLE 9-2	Functions of the Hypothalamus

1. Activation of sympathetic nervous system
   - Control of catecholamine release from adrenal medulla (as in fight-or-flight reaction)
   - Helps maintain blood glucose concentrations through effects on endocrine pancreas
2. Maintenance of body temperature
   - Shivering and sweating
3. Control of body osmolarity
   - Thirst and drinking behavior
   - Secretion of vasopressin
4. Control of reproductive functions
   - Secretion of oxytocin (uterine contractions and milk release)
   - Trophic hormone control of anterior pituitary hormones FSH and LH
5. Control of food intake
   - Satiety center
   - Feeding center
6. Interacts with limbic system to influence behavior and emotions
7. Influences cardiovascular control center in medulla oblongata
8. Secretes trophic hormones that control release of hormones from anterior pituitary gland

release of anterior pituitary hormones by the secretion of neurohormones.

The hypothalamus receives input from multiple sources including the cerebral cortex, the reticular formation, and various sensory receptors. Its output goes to the thalamus and eventually to multiple effector pathways. Most pathways in the hypothalamus are bidirectional with two exceptions: A descending pathway takes neurohormones from the hypothalamus to the posterior pituitary, and an ascending pathway leads from the retina to the **suprachiasmatic nucleus**. This region of the hypothalamus appears to be the center for the biological clock that creates circadian rhythms in the body (∞ p. 164).

The Cerebrum

The **cerebrum** (Fig. 9-8 ■) fills most of the cranial cavity. It is composed of two hemispheres connected at the **corpus callosum** (see Fig. 9-12 ■). This connection ensures that the two hemispheres communicate and cooperate with each other. Each cerebral hemisphere is divided into four lobes, named for the bones of the skull under which they are located: frontal, parietal, temporal, and occipital.

The surface of the cerebrum of humans and other primates has a furrowed walnutlike appearance. During

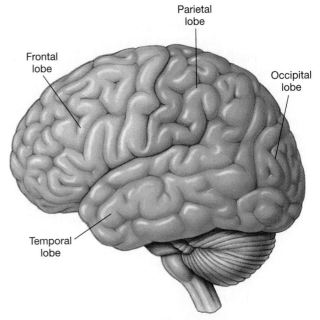

■ **Figure 9-8 The cerebrum** The cerebrum has four lobes in each half, or hemisphere. The lobes are named for the bones of the cranium under which they lie.

development, the cerebrum grows faster than the surrounding cranium, causing the tissue to fold back on itself in order to fit into a smaller volume. The degree of folding is directly related to the level of processing of which the brain is capable; less advanced mammals, such as rodents, have brains with a relatively smooth surface. The human brain is so convoluted that if it were inflated enough to smooth the surfaces, it would be three times as large and would need a head the size of a beachball.

The interior of the cerebrum contains three clusters of nuclei: the basal ganglia, the amygdala, and the hippocampus. The **basal ganglia** (also known as the *corpus striatum*) are involved in the control of movement. The amygdala and the hippocampus are part of the **limbic system,** a region of cerebrum that surrounds the brain stem (Fig. 9-9 ■). The limbic system probably represents the most primitive region of the cerebrum. It acts as the link between higher cognitive functions such as reasoning and more primitive emotional responses such as fear. The **amygdala** is linked to emotion and memory; the **hippocampus,** to learning and memory.

Organization of the cerebral cortex   The outer layer of neurons in the cerebrum, only a few millimeters thick, is called the **cerebral cortex** [*cortex*, bark or rind, adjective *cortical*]. It is within this layer that our higher brain functions such as reasoning arise. The neurons of the cortex are arranged in anatomically distinct horizontal layers and functionally distinct vertical columns. If you

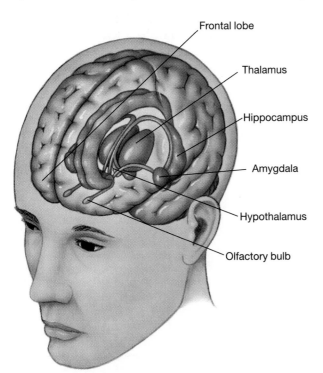

■ **Figure 9-9 The limbic system** The limbic system is a section of the interior of the cerebrum that surrounds the brain stem. Structures in the limbic system have been linked to learning, memory, and emotion.

look at a microscopic section of the cortex, you can see that the neurons are arranged into six layers parallel to the surface (Fig. 9-1i ■). The functional columns of the brain are not as obvious morphologically but have been identified by electrophysiological recordings. Information reaching a single fiber is transferred vertically as each neuron synapses on the next layer. The lateral spread of information is limited to a few millimeters, thus creating vertical columns of information transfer. Some of the best descriptions of cortical column function come from studies of the visual system.

The cerebral cortex contains three functional specializations: sensory areas (*fields*) that direct perception, motor areas that direct movement, and regions known as *association areas* (association cortices) that integrate information and direct voluntary behaviors (Fig. 9-10 ■). These functional areas do not necessarily correspond to the anatomical lobes but have been identified through experimental studies and observations of patients with various brain lesions.

Sensory areas include regions for perception of sensations from peripheral receptors as well as regions devoted to vision, hearing, and olfaction (smell). The **primary somatic sensory cortex** in the parietal lobe receives sensory information from the skin, musculoskeletal system, viscera, and taste buds. Damage to this part of the brain leads to reduced sensitivity of the skin on the opposite side of the body because sensory fibers cross to the opposite side of the midline as they ascend through the medulla. The **visual cortex,** located in the occipital lobe, receives information from the eyes, and the **auditory cortex,** located in the temporal lobe, receives information from the ears. The **olfactory cortex,** a small region in the temporal lobe, receives input from chemoreceptors in the nose.

The **motor areas,** or **primary motor cortices,** in the Frontal lobes process information about skeletal muscle movement and play a key role in voluntary movements. If part of one motor area is damaged by a stroke, paralysis on the opposite side of the body can result. The left side of the brain controls movement on the right side of the body and vice versa. Descending pathways from one motor cortex cross to the opposite side of the body as they pass through the medulla.

Association areas integrate sensory information such as somatic, visual, and auditory stimuli into perception. **Perception** is the brain's interpretation of sensory stimuli, and the perceived stimulus may be very different from the actual stimulus. For instance, the photoreceptors of the eye receive light waves of different frequencies, but we perceive the different wave energies as different colors. Similarly, the brain translates pressure waves hitting the ear into sound or interprets chemicals binding to chemoreceptors as taste or smell. One interesting aspect of perception is the way that our brain fills in missing information to create a complete picture or translates a two-dimensional drawing into a three-

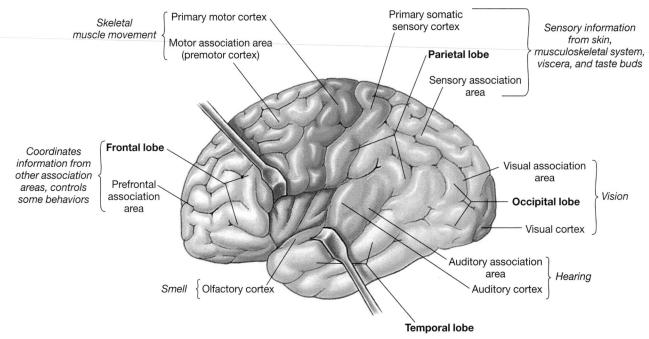

Skeletal muscle movement { Primary motor cortex

Motor association area (premotor cortex)

Primary somatic sensory cortex

Parietal lobe

Sensory information from skin, musculoskeletal system, viscera, and taste buds

Sensory association area

Coordinates information from other association areas, controls some behaviors { **Frontal lobe**

Prefrontal association area

Visual association area

Occipital lobe } Vision

Visual cortex

Auditory association area } Hearing

Smell { Olfactory cortex

Auditory cortex

Temporal lobe

■ Figure 9-10 **Functional areas of the cerebral cortex** The cerebral cortex is specialized into sensory areas for perception, motor areas that direct movement, and association areas that integrate information.

dimensional shape (Fig. 9-11 ■). Our perceptual translation of sensory stimuli allows the information to be acted upon and used in voluntary motor control or complex cognitive functions such as language.

The distribution of different areas of functional specialization across the cerebral cortex is not symmetrical: Each lobe has developed special functions not shared by the other lobe. This **cerebral lateralization** of function is sometimes referred to as cerebral dominance, more popularly known as left brain–right brain dominance. Language and verbal skills tend to be concentrated on the left side of the brain, the dominant hemisphere for right-handed people; spatial skills are concentrated on the right side (Fig. 9-12 ■).

Neural connections in the cerebrum, like those in other parts of the nervous system, exhibit a certain degree of plasticity, an ability to change neuronal connections on the basis of experience. For example, if a person loses a finger, the regions of motor and sensory cortex previously

devoted to control of the finger do not go dormant. Instead, adjacent regions of the cortex extend their functional fields and take over the parts of the cortex that are no longer used by the absent finger. Similarly, skills normally associated with one side of the cerebral cortex can be developed in the other hemisphere, as when a right-handed person with a broken hand learns to write with the left hand.

BRAIN FUNCTION

The brain is an information processor, much like a computer. It receives informational input from the internal and external environments, integrates and processes the information, and, if appropriate, creates a response. What is amazing is that the brain is capable of generating infor-

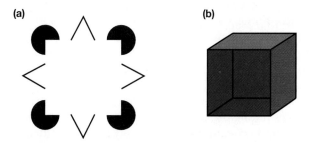

(a)

(b)

■ Figure 9-11 **Perception** The brain has the ability to interpret sensory information to create the perception of shapes (a) or a three-dimensional object (b).

continued from page 243

Dorothy took aspirin for about three months. Suddenly one morning, she became dizzy and lost consciousness. When she awoke, the left side of her body was numb, and she was unable to move her left arm and leg. At the hospital, tests showed that she had a blood clot causing a stroke that affected her cerebral cortex. To further test the extent of damage caused by the ischemia, the doctor drew a tree and asked her to copy it. Dorothy drew a cluster of lines that did not even resemble a tree.

Question 3: Based on her symptoms, do you think Dorothy's stroke occurred in the right or left hemisphere?

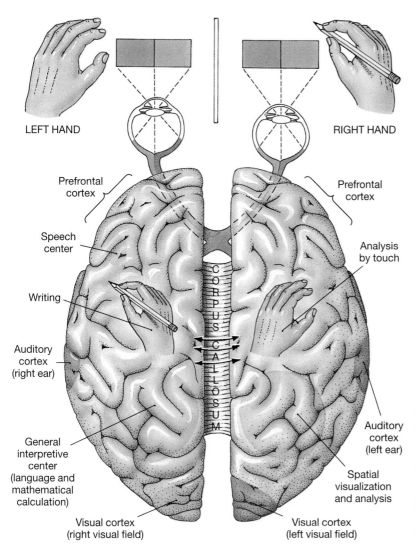

LEFT HAND

RIGHT HAND

Prefrontal cortex

Prefrontal cortex

Speech center

Analysis by touch

Writing

Auditory cortex (right ear)

CORPUS CALLOSUM

General interpretive center (language and mathematical calculation)

Auditory cortex (left ear)

Spatial visualization and analysis

Visual cortex (right visual field)

Visual cortex (left visual field)

■ Figure 9-12 **Cerebral lateralization** The functional areas in the two cerebral hemispheres are not symmetrical. Language and verbal skills are concentrated on the left side of the brain, and spatial skills are concentrated on the right. Information is exchanged between the two hemispheres through the fibers of the corpus callosum.

mation and output signals *in the absence of external input.* Another interesting aspect of higher brain function is the ability of this internally generated information to significantly affect tangible functions such as motor output or immune function. For example, you may be familiar with visual imaging, a technique sometimes used as an adjunct to traditional chemotherapy treatments for cancer. The patient is instructed to imagine the immune cells of the body attacking and destroying the malignant cells. According to some studies, the efficiency of the immune cells is improved by visual imaging. This field is currently receiving a lot of attention as scientists seek the physical links between the brain and the body that allow an internally derived mental image to enhance the effectiveness of the immune system.

For many years, studies of brain function were restricted to anatomical descriptions. As we developed techniques for monitoring electrical activity in neurons, research turned to physiological function. Studies on learning and memory can be performed using a variety of animal models, ranging from the mollusk *Aplysia* to rodents and primates. Brain functions dealing with per-

ception are the most difficult to study because they require communication between the subject and the investigator. Much of what we know today about human perception comes from the study of patients with inherited neurological defects, wounds from accidents or war, or surgical lesions that have been made in order to treat some medical condition such as epilepsy. These studies are enhanced by the fact that we can now visualize the human brain at work using noninvasive techniques such as positive emission tomography (PET) scans and magnetic resonance imaging (MRI) (Fig. 9-13 ■).

As a result of modern methods in neurobiology, we have discovered that some mental illnesses once thought to be untreatable have physiological origins. In many instances, they appear to result from abnormalities of neurotransmitter release or reception in different brain regions. For example, depression, bipolar disorder (manic-depressive disease), and schizophrenia are now treated as chemical imbalances in the brain. The search for a biological basis to disturbances of brain function has led to a whole battery of new drugs that work with varying degrees of effectiveness. Use of some of these

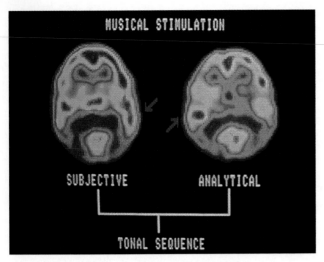

MUSICAL STIMULATION

SUBJECTIVE   ANALYTICAL

TONAL SEQUENCE

■ Figure 9-13 PET scan of the brain at work

drugs raises moral and ethical questions, however, because they may alter personality in the process of treating disease. One example of such a drug is the antidepressant drug Prozac®, a drug that prevents the normal reuptake of the neurotransmitter serotonin by presynaptic neurons. As a result of uptake inhibition, serotonin lingers in the synaptic cleft longer than usual, so serotonin-dependent activity within the central nervous system increases. Although Prozac is effective in alleviating depression, some patients stop taking it because they feel that it alters their personality so that they are no longer themselves.

Other research into brain function has become quite controversial, particularly that dealing with sexuality and the degree to which behavior in general is genetically determined in humans. We will not delve deeply into any of these subjects because they are complex and would require lengthy explanations to do them justice. Instead, we will look briefly at some of the recent models proposed to explain the mechanisms that are the basis for higher brain functions.

✓ Name the anatomical locations where neurons from one side of the body cross to the opposite side.

✓ Name the divisions of the brain in anatomical order, starting from the spinal cord.

Neurotransmitters and Neuromodulators Influence Communication in the Central Nervous System

In the last chapter, you learned how neurons communicate with each other at synapses (∞ p. 224). In a neural reflex, the central nervous system acts as the integrating center, receiving sensory information from various receptors, processing it through networks of interacting neurons, and sending commands to effectors. The meaning of signals in the nervous system depends on two factors: (1) where the

signals originate and terminate and (2) the physical and chemical connections between neurons in the pathway. Chemical communication has two components: the neurotransmitters and neuromodulators released by the presynaptic cells, and the target cell receptors for these molecules. Various combinations of receptor and ligand create a large number of possible communication signals.

Neurotransmitters are responsible for fast synaptic communication; neuromodulators act more slowly, often by using second messenger systems (∞ p. 152). A single chemical can be both a neurotransmitter and a modulator. For example, acetylcholine (ACh) acts as a neurotransmitter in the peripheral divisions of the nervous system but as a modulator in the central nervous system.

The neurotransmitters and neuromodulators in the central nervous system include amino acids, ACh, an assortment of peptides, and amines (Table 9-3). Two of the amino acids, gamma-aminobutyric acid (GABA) and glycine, are the most common inhibitory neurotransmitters of the central nervous system. They act by opening Cl^- channels, hyperpolarizing their postsynaptic targets, and making them less likely to fire an action potential. GABA acts primarily in the brain; glycine acts chiefly in the spinal cord. The amino acid glutamate is the primary excitatory neurotransmitter in the central nervous system; it depolarizes cells by opening channels to allow sodium ions to enter.

Many of the central nervous system neuromodulators are released from small but organized collections of neurons known as the **diffuse modulatory systems.** These neurons are clustered mostly in the brain stem and project their axons so that they affect large areas of the brain (Table 9-4). The diffuse modulatory systems influence such things as attention, motivation, wakefulness, memory, motor control, mood, and metabolic homeostasis. In medicine, the pharmacological use of neuromodulators to treat various disorders is a hot topic. Neuromodulatory drugs that enhance serotonin activity are being used to treat depression and other mood disorders, and the dopamine precursor L-DOPA is administered to alleviate the symptoms of Parkinson's disease.

TABLE 9-3   Neurotransmitters in the CNS
Acetylcholine (ACh)
Amino Acids
Gamma-aminobutyric acid (GABA)
Glycine
Glutamate
Amines
Dopamine
Norepinephrine
Epinephrine
Serotonin
Histamine

TABLE 9-4   The Diffuse Modulatory Systems

System (Neuromodulator)	Neurons Originate	Neurons Innervate	Functions Modulated by the System
Noradrenergic (norepinephrine)	Locus coeruleus of the pons	Cerebral cortex, thalamus, hypothalamus, olfactory bulb, cerebellum, midbrain, spinal cord	Attention, arousal, sleep-wake cycles, learning, memory, anxiety, pain, and mood
Serotonergic (serotonin)	Raphe nuclei along brain stem midline	Lower nuclei project to spinal cord	Pain; locomotion
		Upper nuclei project to most of brain	Sleep-wake cycle, mood and emotional behaviors such as aggression and depression
Dopaminergic (dopamine)	Substantia nigra in midbrain	Cortex	Motor control
	Ventral tegumentum in midbrain	Cortex and parts of limbic system	"Reward" centers linked to addictive behaviors
Cholinergic (acetylcholine)	Base of cerebrum; pons and midbrain	Cerebrum, hippocampus, thalamus	Sleep-wake cycles, arousal, learning, memory, sensory information passing through thalamus

States of Arousal and the Reticular Formation

One aspect of brain function that is difficult to define is consciousness: a state of arousal or awareness of self and environment. What distinguishes the awake state from various stages of sleep, sleep from coma, and coma from death? One way to define arousal states is by the pattern of electrical activity created by the cortical neurons. In awake states, many neurons are firing but not in a coordinated fashion. In the deepest stages of sleep, waves of synchronous depolarization sweep across the cerebral cortex. The desynchronization of electrical activity in waking states is produced by ascending signals from the reticular formation, the diffuse network of neurons that run through the brain stem. If connections between the reticular formation and the cortex are disrupted surgically, the animal becomes comatose for long periods of time. Other evidence for the importance of the reticular formation in states of arousal comes from studies showing that general anesthetics depress synaptic transmission in that region of the brain. Presumably, blocking the ascending pathways between the reticular formation and the cortex creates a state of unconsciousness.

The measurement of brain activity is recorded by a procedure known as **electroencephalography.** Surface electrodes placed on the scalp pick up depolarizations of the cortical neurons in the region just under the electrode. In the awake, aroused state, an electroencephalogram, or **EEG,** shows a rapid, irregular pattern with no dominant waves. In awake-resting states, sleep, or coma, electrical activity of the neurons synchronizes into series of waves with characteristic patterns. The more synchronous the firing of cortical neurons, the larger the amplitude of the waves. As the person's state of arousal lessens, the frequency of the waves slows down. Thus, deep sleep is marked by large-amplitude, slow-frequency waves known as **delta waves;** the awake-resting state is characterized by small waves with high frequency known as **alpha waves.** The significance of EEG wave patterns and why we need to sleep is one of the unsolved mysteries in neurophysiology.

Sleep   All organisms, even plants, have alternating daily patterns of rest and activity. When an organism is placed in conditions of constant light or dark, these activity rhythms persist, apparently cued by an internal clock. In mammals, the clock appears to be located in the suprachiasmatic nucleus of the hypothalamus, where it receives sensory information about light cycles from the eyes. Sleep-wake rhythms, like other biological cycles, generally follow a 24-hour light-dark cycle like our day and are known as **circadian rhythms** [*circa*, about + *dies*, a day].

In humans, our major rest period is marked by a behavior known as sleep. Sleep is an easily reversible state of inactivity, unlike coma. It is characterized by lack of interaction with the external environment. During some periods of sleep, sensory input to and motor output from the brain are interrupted. Until the 1960s, sleep was thought to be a passive state that resulted from withdrawal of stimuli to the brain. Then, experiments showed that neuronal activity in ascending tracts from the brain stem to the cortex was *required* for sleep. As a result, we now consider sleep to be an active state. Each stage of sleep is marked by identifiable, predictable events associated with characteristic EEG patterns and somatic changes (Fig. 9-14 ■).

Sleep is divided into two major phases known as **REM (rapid eye movement) sleep** and **slow-wave sleep** (non-REM sleep). During a sleep cycle, a person alternates between these two phases, spending variable periods of time in each phase, with transition stages in between. Slow-wave sleep is marked on the EEG by the presence

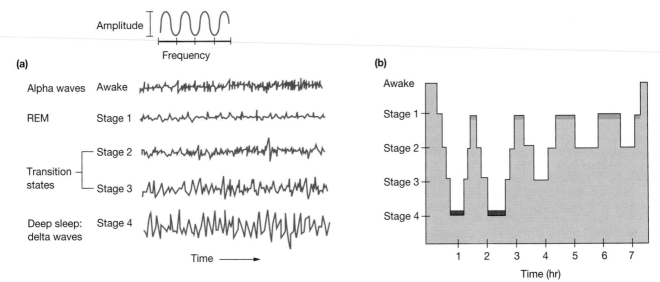

■ Figure 9-14 **Electroencephalograms (EEGs) and the sleep cycle** (a) Recordings of electrical activity in the brain during waking and sleep periods show characteristic patterns. Neuronal depolarizations in the awake state are unsynchronized and appear as low-amplitude, high-frequency (many waves per second) waves. (b) During sleep periods, a person cycles between slow-wave sleep, characterized by high-amplitude, low-frequency waves, and REM sleep, characterized by low-amplitude, high-frequency waves. The transition states between slow-wave sleep and REM sleep show waves with intermediate characteristics.

of delta waves, low-frequency waves of long duration. During this phase of the sleep cycle, sleepers adjust body position without conscious commands from the brain to do so.

In contrast, REM sleep is marked by an EEG pattern that resembles that of an awake person, with rapid, low-amplitude waves. During REM sleep, brain activity inhibits motor neurons to skeletal muscles, thus paralyzing them. Exceptions to this pattern are the muscles that move the eyes and those that control breathing. The control of homeostatic functions is depressed during REM sleep, and body temperature falls toward ambient temperature. REM sleep is the period of time during which most dreaming takes place. The eyes move behind closed lids, as if following the action of the dream. Sleepers are most likely to wake up spontaneously from periods of REM sleep.

If sleep is a neurologically active process, what is it that makes us sleepy? The possibility of a sleep-inducing factor was first proposed back in 1913, when scientists found that cerebrospinal fluid from sleep-deprived dogs could induce sleep in normal animals. Since then, a variety of sleep-inducing factors have been identified. Curiously, many of them are also substances that enhance the immune response, such as interleukin-1, interferon, serotonin, and tumor necrosis factor. As a result of this finding, some investigators have suggested that one answer to the puzzle of the biological reason for sleep is that we need to sleep to enhance our immune response. Whether or not that is a reason for why we sleep, the link between the immune system and sleep induction may help explain why we tend to sleep more when we are sick.

Sleep disorders are relatively common, as you can tell by looking at the variety of sleep-promoting

agents available over the counter in drugstores. Among the more common sleep disorders are *insomnia*, the inability to go to sleep or remain asleep long enough to awake refreshed, and *sleep apnea* [*apnoos*, breathless], a condition in which sleepers awake when their airway muscles relax to the point of obstructing normal breathing.

The Hypothalamus Is the Primary Integrating Center for Many Homeostatic Reflexes

Although the hypothalamus occupies less than 1% of the total brain volume, it is the center for homeostasis. Output signals from this region of the brain influence endocrine and autonomic reflexes, as well as behaviors such as eating and drinking. In coordinating homeostasis, the hypothalamus receives sensory input from cen-

Sleepwalking Sleepwalking, or somnambulism [*somnus*, sleep + *ambulare*, to walk], is a sleep behavior disorder that for many years was thought to represent the acting out of dreams. However, most dreaming occurs during REM sleep, whereas sleepwalking takes place during deep sleep. During sleepwalking episodes, which may last from 30 seconds to 30 minutes, the subject's eyes are open and registering the surroundings. The subject is able to avoid bumping into objects, can negotiate stairs, and in some cases is reported to perform tasks such as preparing food or folding clothes. The subject usually has little if any conscious recall of the sleepwalking episode upon awakening. Sleepwalking is most common in children, and the frequency of episodes declines with age. There is also a genetic component, as the tendency to sleepwalk runs in families.

tral and peripheral receptors. In addition, the limbic system provides the hypothalamus with emotional input that may affect homeostatic responses.

Probably the best-known hypothalamic pathway is the fight-or-flight response, a generalized reaction that occurs with fear or anger. When something occurs to threaten the well-being of the body, the hypothalamus takes over, initiating the fight-or-flight response. All of the physical manifestations of fear, such as pounding heart, sweaty palms, and increased blood pressure, can be traced to increased sympathetic output directed by the hypothalamus. Sympathetic innervation of various organs is reinforced by sympathetically induced release of epinephrine (adrenaline) from the adrenal medulla.

In addition to regulation of the fight-or-flight response, the hypothalamus contains centers for temperature regulation, eating, and control of body osmolarity. The responses to stimulation of these centers may be neural or hormonal reflexes or a behavioral response. Stress, reproduction, and growth are also mediated by the hypothalamus by way of multiple hormones. You will encounter these homeostatic reflexes in later chapters as we discuss the various systems of the body.

Emotion and Motivation Are Complex Neural Pathways

Emotion and motivation are two aspects of brain function that are linked together through the hypothalamus, limbic system, and cerebral cortex. The pathways are complex and form closed circuits that cycle information between various parts of the brain.

Emotions are difficult to define. We know what they are and can name them, but in many ways they defy description. One characteristic of emotions is that they cannot be voluntarily turned on or off. The most commonly described emotions that arise in different parts of the brain are anger, aggression, sexuality, fear, pleasure, and contentment or happiness.

The limbic system, particularly the region known as the amygdala, is the center of emotion in the human brain. We learned about the role of this brain region through experiments in humans and animals. If the amygdala is artificially stimulated in humans, they report experiencing feelings of fear and anxiety. Experimental lesions that destroy the amygdala in animals cause the animals to become tamer and to display hypersexuality. As a result, neurobiologists feel that the amygdala is the center for basic instincts such as fear and for learned emotional states.

The pathways for emotions are complex (Fig. 9-15 ■). Sensory stimuli feeding into the cerebral cortex are constructed within the brain to create a representation (perception) of the world. After that information is integrated by the association areas, it is passed on to the limbic system. Feedback from the limbic system to the cortex creates awareness of the emotion, while descending pathways to the hypothalamus and brain stem initiate voluntary behaviors and unconscious responses mediated by autonomic, endocrine, immune, and somatic motor systems. The physical result of emotions can be as dramatic as the pounding heart of a fight-or-flight reaction or as insidious as the development of a gastric ulcer. The links between mind and body are difficult to study, and we may never completely understand them.

Motivation is defined as internal signals that shape voluntary behaviors. Some of these behaviors, such as eating, drinking, and sex, are related to survival; others, such as curiosity and sex (again), are linked to emotions. Motivational states are known as **drives** and generally have three properties in common: (1) They create an increased state of central nervous system arousal or alertness; (2) they create goal-oriented behavior; and (3) they are capable of coordinating disparate behaviors in order to achieve that goal.

Motivated behaviors often work in parallel with autonomic and endocrine responses in the body, as you might expect from behaviors originating in the hypothalamus. For example, if you eat salty popcorn, your body osmolarity increases. This stimulus acts on the thirst center of the hypothalamus, motivating you to seek something to drink. Increased osmolarity also acts on an endocrine center in the hypothalamus, releasing a hormone that increases water retention by the kidneys. Thus, one stimulus triggers both a motivated behavior and a homeostatic endocrine response.

Some motivation states can be activated by internal stimuli that may not even be obvious to the person in whom they are occurring. Many motivated behaviors stop when the person has reached a certain level of satisfaction, or **satiety.** Eating, curiosity, and sex drive are three examples of behaviors that have complex stimuli underlying their onset.

Pleasure is a motivational state that is being intensely studied because of its relationship to addictive behaviors such as drug use. Animal studies have shown that pleasure is a physiological state that is accompanied by increased activity of the neurotransmitter dopamine in certain parts of the brain. Drugs that are addictive, such as cocaine and possibly nicotine, act by enhancing the effectiveness of dopamine, thus increasing the pleasurable sensations perceived by the brain. As a result, use of these drugs rapidly becomes a learned behavior. Interestingly, not all behaviors that are addictive are pleasurable. For example, there are a variety of compulsive behaviors that involve self-mutilation, such as pulling out hair by the roots. Fortunately, many behaviors that are learned can also be unlearned, given motivation.

Learning and Memory Change Synaptic Connections in the Brain

For many years, motivation, learning, and memory were considered to be in the realm of psychology rather than biology. Neurobiologists in decades past were more con-

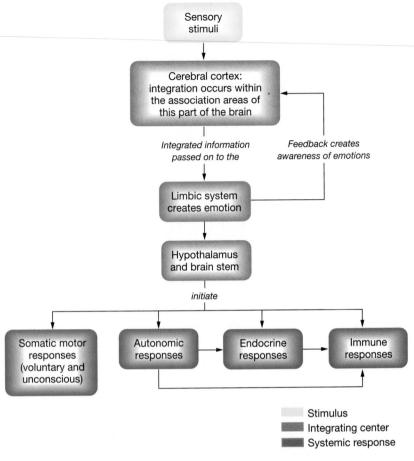

■ Figure 9-15 **The link between emotions and physiological functions** Sensory stimuli are perceived and integrated within the cerebral cortex, which passes the information on to the limbic system. The limbic system creates emotional reactions in response to the sensory information. It passes the emotions back to the cerebral cortex, where they are perceived, and on to the hypothalamus and brain stem. Physiological responses mediated through those parts of the brain can involve multiple systems, including skeletal muscle activity (somatic motor division), endocrine and autonomic responses, and even immune responses. The well-known link between stress and increased susceptibility to viruses is an example of an emotionally linked immune response.

cerned with the cellular aspects of neuronal function than with the integrated action of the central nervous system. However, in recent years, the two fields have overlapped more and more as scientists have discovered that the underlying basis for cognitive function seems to be explainable in terms of cellular events. One of the newest ideas is the concept of plasticity: that neuronal connections are not fixed and have the ability to change with experience. Two aspects of cognitive ability that probably involve plasticity are learning and memory.

Learning   Learning is the acquisition of knowledge about the world around us. Learning is demonstrated by behavioral changes, but behavioral changes are not required for learning to occur. Learning can be internalized and is not always reflected by overt behavior.

Learning can be classified into two broad types: associative and nonassociative. **Associative learning** occurs when two stimuli are associated with each other, such as Pavlov's classic experiment in which he simultaneously presented dogs with food and rang a bell. After a period of time, the dogs came to associate the sound of a bell with food and began to salivate in anticipation of food whenever the bell was rung. Another form of associative learning occurs when an animal associates a stimulus with a given behavior. An example would be a mouse that gets a shock each time it touches a certain

part of its cage. It soon associates that part of the cage with an unpleasant experience and avoids that area.

Nonassociative learning includes imitative behaviors such as learning a language. This type of learning includes habituation and sensitization, two adaptive behaviors that allow us to filter out and ignore background stimuli while responding more sensitively to potentially disruptive stimuli. In **habituation,** an animal shows a decreased response to a stimulus that is repeated over and over. For example, a sudden loud noise may startle you, but if the noise is repeated over and over again, your brain begins to ignore it. Habituated responses allow us to filter out stimuli that we have evaluated and found to be insignificant.

Sensitization is the opposite of habituation and helps to increase an organism's chances for survival. In sensitization learning, exposure to a noxious or intense stimulus causes an enhanced response upon subsequent exposure. For example, people who become ill while eating certain foods may find that they lose their desire to eat that food again. Sensitization is adaptive since it helps us to avoid potentially harmful stimuli.

Memory   In humans, the hippocampus seems to be an important structure in both learning and memory. For example, patients who had part of the hippocampus destroyed in order to relieve a certain type of epilepsy also had trouble remembering new information to which they were exposed. When given a list of words to repeat, they could remember the words as long as their attention stayed focused on the task. But if they were distracted, the memory of the words disappeared and they had to learn the list again. Information stored in long-term memory before the operation was not affected. This inability to remember newly acquired information is a defect known as **anterograde amnesia** [*amnesia*, oblivion].

Memory has multiple levels of storage, and our memory bank is constantly changing. When a stimulus comes into the central nervous system, it first goes into **short-term memory,** a limited store of information that can hold only about 7 to 12 pieces of information at a time. Items in short-term memory will disappear unless an effort is made to put them into a more permanent form (Fig. 9-16 ■). A special form of short-term memory is **working memory.** One region of the cerebral cortex is devoted to keeping track of bits of information long enough to put them to use in a task that takes place after the information has been acquired. Working memory in these regions is linked to long-term memory stores, so that newly acquired information can be integrated with stored information and acted upon. For example, you are trying to cross a busy road. You look to the right and see that there are no cars coming for several blocks. You then look left and see that there are no cars coming from that direction either. Working memory has stored the information that the road to the right is clear, so using stored knowledge about safety, you are able to conclude that there is no traffic from either direction and it is safe to cross the road. In people with damage to the prefrontal lobes of the brain, this task becomes more difficult because they are unable to recall whether the road is clear from the right once they have looked away to assess traffic coming from the left. Working memory allows us to collect a series of facts from short- and long-term memory and connect them together in logical order to solve problems or plan actions.

The processing of information that converts short-term memory into **long-term memory** is known as **consolidation.** Consolidation can take varying periods of time, from seconds to minutes. There are many intermediate levels of memory that information passes through during consolidation, and in each of these stages, information can be located and recalled.

Long-term memory has been divided into two types that are consolidated and stored using different neuronal pathways (Table 9-5). **Reflexive (implicit) memory** is automatic and does not require conscious processes for its creation or recall. Information stored in this kind of memory is acquired slowly through repetition. Motor skills fall into the category of reflexive memory, as do procedures and rules. For example, you do not need to think about putting a period at the end of each sentence or about how to pick up a fork. Reflexive mem-

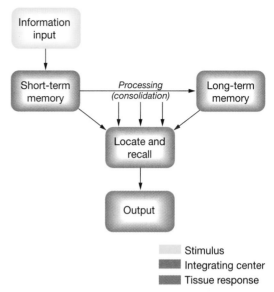

■ **Figure 9-16 Memory processing** New information goes first into short-term memory, from which it can be recalled and acted upon. The amount of information that can be stored in short-term memory is limited, however, and information will be lost unless it is processed through consolidation and stored in long-term memory.

ory has also been called procedural memory because it generally concerns how to do things. Reflexive memories can be acquired through either associative or nonassociative learning processes.

Declarative (explicit) memory, on the other hand, requires conscious attention for its recall. Its creation generally depends on the use of higher-level cognitive skills such as inference, comparison, and evaluation. Declarative memories deal with knowledge about ourselves and the world around us that can be reported or described verbally.

Sometimes information can be transferred from declarative memory to reflexive memory. The quarterback on a football team is a good example. When he learned to throw the football as a small boy, he had to pay close attention to gripping the ball and coordinating his muscles to throw the ball accurately. At that point of learning to throw the ball, the process was in declarative memory and required conscious effort as the boy analyzed his movements. With repetition, however, the mechanics of throwing the ball were transferred to reflexive memory: They became a reflex that could be executed without conscious thought. That transfer allowed the quarterback to use his conscious mind to analyze the path and timing of the pass while the mechanics of the pass became automatic. Athletes often refer to this automaticity of learned body movements as "muscle memory."

Memory is an individual thing. We process information on the basis of our experiences and perception of the world. Because people have widely different experiences throughout their lives, it follows that no two people will

TABLE 9-5 Types of Memory

Reflexive (Implicit) Memory	Declarative (Explicit) Memory
• Recall is automatic and does not require conscious attention	• Recall requires conscious attention
• Acquired slowly through repetition	• Depends on higher-level thinking skills such as inference, comparison, and evaluation
• Includes motor skills and rules and procedures	• Memories can be reported verbally
• Memories can be demonstrated	

process a piece of information in the same way. If you ask a group of people about what happened during a particular event such as a lecture, a robbery, or an accident, no two descriptions will be identical. The different people will have processed the event according to their own perceptions and experiences. Experiential processing is important to remember when studying in a group situation, because it is unlikely that all group members will learn or recall information using the same pathways.

Memory processing for reflexive and declarative memory appears to take place through different pathways. With noninvasive imaging techniques such as MRI and PET scans, researchers have been able to track brain activity as individuals learned to perform tasks. Memories are stored throughout the cerebral cortex in pathways known as **memory traces.** Declarative memories involve the temporal lobes of the brain; reflexive memory storage requires the amygdala and the cerebellum. Some components of memories are stored in the sensory cortices where they are processed. For example, pictures are stored in the visual cortex and sounds in the auditory cortex.

Learning or recall of a single task may involve multiple circuits in the brain that work in parallel. This parallel processing provides backup in case one of the circuits is damaged. It is also believed to be the means by which specific memories are generalized, allowing new information to be processed and matched to stored information. For example, a person who has never seen a volleyball will recognize it as a ball because the volleyball has the same general characteristics as all other balls the person has seen.

As scientists studied the consolidation of short-term memory into long-term memory, they discovered that the process involves changes in the synaptic connections of the circuits involved in learning. In some cases, new synapses form; in others, the effectiveness of synaptic transmission is altered. These changes are evidence of plasticity and show us that the brain is not "hard-wired," as we once had thought.

Plasticity and long-term potentiation Scientists who study learning and memory now look for evidence of plasticity in the neuronal circuits involved in a particular behavior. One of the most popular organisms for studying this phenomenon is *Aplysia,* a shell-less mollusk with stereotyped behaviors controlled by a limited number of neurons that are readily accessible for elec-

trophysiological studies. In vertebrates, the hippocampus is a prime study site for looking at plasticity.

One of the synaptic changes that occurs with learning is known as **long-term potentiation,** or **LTP.** Long-term potentiation is a form of prolonged facilitation, in which the response of the postsynaptic cell to a constant stimulus is enhanced for a period of time that ranges from hours to weeks. The initiation of LTP requires a stimulus that exceeds some critical threshold. As a result of this stimulus, potentiation causes physical changes in the synapse.

One such change may be an increase in membrane neurotransmitter receptors (up-regulation; ∞ p. 150). The addition of receptors to the membrane allows the postsynaptic cell to depolarize more in response to a given amount of neurotransmitter. After long-term potentiation, a stimulus that previously would not have activated the postsynaptic cell may now do so.

Long-term potentiation may also create new synaptic connections, a process thought to be involved in memory consolidation. There is evidence showing that the transfer of information from short-term into long-term memory involves new synaptic connections that create additional pathways for information transfer. For example, the number of synapses in the occipital cortex of rats increases 25% when the rats are placed in stimulating environments.

The cellular mechanisms by which long-term potentiation occurs are of great interest to neurobiologists because of the link to information processing and memory. Of all types of memory, the storage and recall of the

continued from page 248

Dorothy's doctor administered an intravenous solution containing a new drug called tissue plasminogen activator (t-PA). t-PA is a natural substance that the body produces to dissolve blood clots during the process of wound repair. The drug that Dorothy received is a genetically engineered version of t-PA. For maximum effect, it must be given within a few hours of the onset of a stroke.

Question 4: Explain the rationale for giving t-PA to people who have had a stroke. Why is it most effective when given shortly after the stroke occurs? Is there any type of stroke for which t-PA would not be appropriate? (Hint: What are the two causes of strokes?)

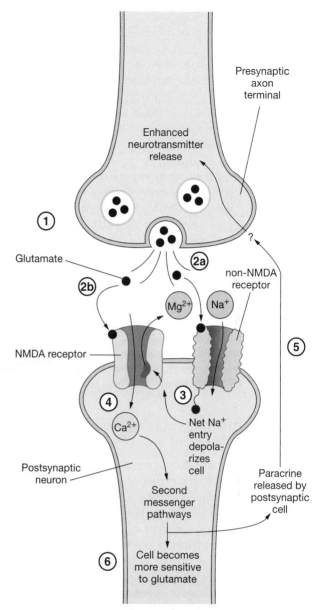

NMDA Receptors and Long-Term Potentiation
A key element in the process of long-term potentiation is the amino acid glutamate, the main excitatory neurotransmitter in the central nervous system. Postsynaptic neurons have two types of glutamate receptors: NMDA receptors, named for the glutamate agonist N-methyl-D-aspartate, and non-NMDA receptors. In order for long-term potentiation to take place, several pathways must be active simultaneously (Fig. 9-17 ■). When all conditions are right, the NMDA channel opens and allows Ca^{2+} into the postsynaptic cell. A cascade of second messenger events follows. The details of long-term potentiation from this point on are unclear, but it seems to involve changes within the postsynaptic cell as well as release of a paracrine. The paracrine diffuses back to the presynaptic cell and somehow enhances its neurotransmitter release in response to a normal action potential. The postsynaptic cell simultaneously becomes more sensitive to the neurotransmitter and may form new synapses.

symbols and sounds of language may represent the most complex use of our brain power.

Language Is the Most Elaborate Cognitive Behavior

One of the hallmarks of an advanced nervous system is the ability of a species to exchange complex information with other members of the same species. Although found predominantly in birds and mammals, this ability also occurs in certain insects that convey amazingly detailed information by means of sound (crickets), touch and sight (bees), and odor (pheromones, also found in many other animals). In humans, communication takes place primarily through spoken and written language. Because language is considered the most elaborate cognitive behavior, it has received considerable attention from neurobiologists.

Language skills require the input of sensory information (primarily from hearing and vision), processing in various centers in the cerebral cortex, and the coordination of motor output for vocalization and writing. Language ability is found primarily in the left hemisphere of the cerebrum, and even 70% of people who are left-handed or ambidextrous use their left brain for speech. The ability to communicate through speech has been divided into two processes: the combination of different sounds to form words (vocalization) and the combination of words into grammatically correct and meaningful sentences.

The integration of spoken language in the human brain involves two regions in the cortex: **Wernicke's area** in the temporal lobe and **Broca's area** in the frontal lobe close to the motor cortex (Fig. 9-18 ■). Most of what we know about these areas comes from studies of people with brain lesions because nonhuman animals are not capable of speech. Even primates that communicate on the level of a small child through sign language and other visual means do not have the physical ability to vocalize the sounds of human language.

■ **Figure 9-17 Long-term potentiation** The NMDA receptors for glutamate are linked to long-term potentiation and have an unusual property: Their channel is blocked by a chemically dependent gate and also by a magnesium ion (Mg^{2+}). ① Glutamate released from a presynaptic neuron acts as both a neuromodulator (NMDA receptor) and a neurotransmitter (non-NMDA receptor). ②a Glutamate binding to the non-NMDA receptor opens an ion channel that allows net Na^+ influx into the postsynaptic cell, depolarizing it. ②b The NMDA receptor has a gate that opens with glutamate binding. However, the channel of the NMDA receptor is blocked by a Mg^{2+} ion that is not displaced unless the cell depolarizes. ③ Net entry of Na^+ through the non-NMDA channel depolarizes the cell and displaces the Mg^{2+} that is blocking the channel. ④ Once the NMDA channel opens, Ca^{2+} enters the cytoplasm of the postsynaptic cell and acts as a signal to initiate second messenger pathways. ⑤ As a result of these pathways, the postsynaptic cell releases a paracrine that acts on the presynaptic cell to enhance neurotransmitter release. ⑥ The postsynaptic cell also becomes more sensitive to glutamate, possibly by inserting more glutamate receptors in the postsynaptic membrane.

Input into the language areas comes from either the visual cortex (reading) or the auditory cortex (listening). Sensory input goes first to Wernicke's area, then to Broca's area. After integration and processing, output to the motor cortex initiates a spoken or written response. If damage occurs to Wernicke's area, a person is unable to understand any spoken or visual information. The person's own speech as a result is nonsense, since the person is unaware of his or her own errors. This condition is known as **receptive aphasia** because the person is unable to understand sensory input [*a-*, not + *phatos*, spoken]. On the other hand, people with damage to Broca's area understand spoken and written language but are unable to express their response in normal syntax. Their response to a question often consists of appropriate words strung together in random order. These patients may have a difficult time dealing with their disability because they are aware of their mistakes but are powerless to correct them. Damage to Broca's area causes an **expressive aphasia.**

Mechanical forms of aphasia occur as a result of damage to the motor cortex. These patients find themselves unable to physically shape the sounds that make up words or unable to coordinate the muscles of their arm and hand to write.

Personality and Individuality Are a Combination of Experience and Inheritance

One of the most difficult aspects of brain function to translate from the abstract realm of philosophy into the physical circuits of neurobiology is the combination of attributes that we call personality. What is it that makes us individuals? The parents of more than one child will tell you that their offspring were different from birth, and even before that in the womb. If we all have the same brain structure, what makes us different?

This question is one that fascinates many people. The answer that is evolving as the result of neurobiology research is that we are a combination of our experiences and the genetic constraints that we inherit. What we learn or experience and what we store in memory create a unique pattern of neuronal connections in our brains. Sometimes these circuits malfunction, creating depression, schizophrenia, or any number of personality disturbances. Psychiatrists for many years attempted to treat these disorders as if they were due solely to events in the person's life, but now we know that there is a genetic component to many of these disorders. One of the biggest advances in treating altered mental states was the discovery of diffuse modulatory pathways and the neuromodulatory action of serotonin and dopamine. The development of drugs that enhance the effectiveness of information exchange at serotonergic and dopaminergic synapses has created a major breakthrough in the treatment of depression and other conditions.

On the other hand, we still have much to learn about repairing damage to the central nervous system. One of the biggest tragedies in life is the personality change that

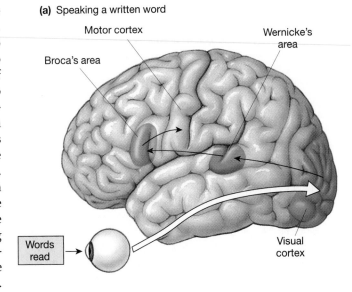

(a) Speaking a written word

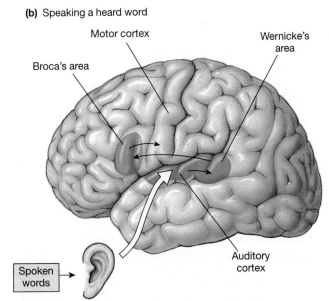

(b) Speaking a heard word

■ Figure 9-18 **Cerebral processing of spoken and visual language** Spoken and written language are processed through their respective sensory areas. The information is passed on to Wernicke's area, where it is interpreted, then to Broca's area for the coordination of speech and writing. People with damage to Wernicke's area do not understand spoken or written communication; those with damage to Broca's area understand, but are unable to respond appropriately.

sometimes accompanies head trauma. Physical damage to the delicate circuits of the brain can not only alter intelligence and memory storage but also can create a new personality. The person who exists after the injury may not be the same person who inhabited that body before the injury. Although the change may not be noticeable to the injured person, it can be devastating to the family and friends of the victim. Perhaps as we learn more about how neurons link to each other, we will be able to find a means of restoring damaged networks and preventing the lasting effects of head trauma.

CHAPTER REVIEW

Chapter Summary

Anatomy of the Central Nervous System

1. The brain and spinal cord are encased in the membranes of the **meninges** and the bones of the **cranium** and vertebrae. (p. 237)

2. The **choroid plexus** secretes **cerebrospinal fluid (CSF)** into the **ventricles** of the brain. Cerebrospinal fluid helps cushion the brain and spinal cord and also creates a chemically controlled environment. (p. 237)

3. Because the normal fuel source for neurons is glucose, the body closely regulates blood glucose concentrations. (p. 241)

4. Relatively impermeable capillaries in the brain create a **blood-brain barrier** that prevents possibly harmful substances in the blood from entering the interstitial fluid.

Paracrines released by astrocytes induce formation of tight junctions between capillary endothelial cells. (p. 241)

5. The **gray matter** of the CNS consists of nerve cell bodies, dendrites, and axon terminals. The cell bodies form layers in parts of the brain or cluster into groups of neurons known as **nuclei**. (p. 242)

6. **White matter** in the CNS is made up mostly of myelinated axons with very few cell bodies. In the spinal cord, **ascending tracts** of white matter carry sensory information to the brain, and **descending tracts** carry efferent signals from the brain. **Propriospinal tracts** remain within the cord. (p. 242)

The Spinal Cord

7. The **dorsal root** of each spinal nerve carries incoming sensory information, and the **ventral root** carries information from the central nervous system to the muscles and glands. (p. 242)

8. The nerve cell bodies of sensory neurons are found in the **dorsal root ganglia**. (p. 242)

The Brain

9. The brain can be divided into three areas: the brain stem, the cerebellum, and the cerebrum. (p. 243)

10. The brain stem is divided into the medulla oblongata, the pons, and the midbrain (mesencephalon). (p. 244)

11. The **medulla oblongata** contains the tracts that convey information between the cerebrum and the spinal cord or cranial nerves. Many of these tracts cross the midline in a region known as the **pyramids**. The medulla contains control centers for many involuntary functions. (p. 244)

12. The primary function of the **pons** is to act as a relay station for information transfer between the cerebellum and cerebrum. (p. 245)

13. The primary function of the **midbrain** is the control of eye movement. It also relays signals for auditory and visual reflexes. (p. 245)

14. The **reticular formation** is a diffuse group of neurons that play a role in arousal, sleep, muscle tone, stretch reflexes, coordination of breathing, blood pressure regulation, and modulation of pain. (p. 245)

15. The **cerebellum** processes sensory information and coordinates the execution of movement. (p. 245)

16. The **diencephalon** is composed of the thalamus and hypothalamus. The small nuclei that make up the **thalamus** serve as an integration and relay station for sensory information on its way to the cerebral cortex. (p. 245)

17. The **hypothalamus** contains centers for behavioral drives and plays a key role in homeostasis. It controls many functions of the autonomic division of the nervous system as well as a variety of endocrine functions. (p. 245)

18. The **cerebrum** is composed of two hemispheres connected at the **corpus callosum**. Each cerebral hemisphere is divided into frontal, parietal, temporal, and occipital lobes. (p. 246)

19. The interior of the cerebrum contains the **basal ganglia**, involved in the control of movement, and the **limbic system**, which acts as the link between higher cognitive functions and more primitive emotional responses. The **amygdala** of the limbic system is linked to emotion and memory, and the **hippocampus** is involved with learning and memory. (p. 247)

20. Our higher brain functions such as reasoning arise within the **cerebral cortex**, the outer layer of the cerebrum whose neurons are arranged in anatomically distinct horizontal layers and functionally distinct vertical columns. (p. 247)

21. The cerebral cortex contains three functional specializations: sensory areas, motor areas, and association areas. (p. 247)

22. Sensory areas direct perception of sensations. The **primary somatic sensory cortex** receives sensory information from the skin, musculoskeletal system, viscera, and taste buds. The **visual cortex, auditory cortex,** and **olfactory cortex** receive information about vision, sound, and odors, respectively. (p. 247)

23. **Motor areas** direct skeletal muscle movement. A **primary motor cortex** is located in each hemisphere of the cerebrum. (p. 247)

24. **Association areas** integrate sensory information into perception. **Perception** is the brain's interpretation of sensory stimuli. (p. 247)

25. Each hemisphere of the cerebrum has developed special functions not shared by the other hemisphere. This division of labor is known as **cerebral lateralization.** (p. 248)

26. Neural connections in the CNS exhibit **plasticity,** the ability to change neuronal connections on the basis of past experience. (p. 248)

Brain Function

27. A variety of neurotransmitters and neuromodulators in the brain create complex pathways for the transmission and storage of information. Gamma-aminobutyric acid (GABA) and glycine are the most common inhibitory neurotransmitters of the central nervous system. Glutamate is the primary excitatory neurotransmitter. (p. 250)

28. Many of the central nervous system neuromodulators are released from the **diffuse modulatory systems.** These neurons influence attention, motivation, wakefulness, memory, motor control, mood, and metabolic homeostasis. (p. 250)

29. Neural activity in the reticular formation is linked to states of arousal. Varying levels of arousal are marked by different patterns of electrical activity in the neurons. Brain activity is recorded by **electroencephalography.** (p. 251)

30. Sleep is an easily reversible state of inactivity with characteristic stages marked by identifiable, predictable events. The two major phases of sleep are **REM (rapid eye movement) sleep** and **slow-wave sleep** (non-REM sleep). (p. 251)

31. The physiological reason for sleep is unknown but may be linked to maintenance of the immune system. (p. 252)

32. The hypothalamus interacts with the limbic system in a way that causes emotional events to influence physiological functions in the body. (p. 253)

33. The limbic system is the center of emotion in the human brain. The pathways for emotions are complex and difficult to study. (p. 253)

34. **Motivation** is defined as internal signals that shape voluntary behaviors related to survival or emotions. Motivational **drives** share three properties: (1) an increased state of central nervous system arousal, (2) goal-oriented behavior, and (3) the ability to coordinate disparate behaviors. (p. 253)

35. **Learning** is the acquisition of knowledge about the world around us. **Associative learning** occurs when two stimuli are associated with each other. **Nonassociative learning** includes imitative behaviors such as learning a language. (p. 254)

36. In **habituation,** an animal shows a decreased response to a stimulus that is repeated over and over. In **sensitization,** exposure to a noxious or intense stimulus creates an enhanced response upon subsequent exposure. (p. 254)

37. Memory has multiple levels of storage and is constantly changing. Information is first stored in **short-term memory,** but will disappear unless consolidated into long-term memory. (p. 255)

38. **Long-term memory** includes **reflexive memory,** which does not require conscious processes for its creation or recall, and **declarative memory,** which uses higher-level cognitive skills for formation and requires conscious attention for its recall. (p. 255)

39. The **consolidation** of short-term memory into long-term memory appears to involve changes in the synaptic connections of the circuits involved in learning. **Long-term potentiation** is one mechanism by which neurons change the quality or quantity of their synaptic connections. (p. 256)

40. Language is considered the most elaborate cognitive behavior. The integration of spoken language in the human brain involves information processing in **Wernicke's area** and **Broca's area.** (p. 257)

Questions

LEVEL ONE   **Reviewing Facts and Terms**

1. A property that human brains have that most computer programs lack is _____, the ability to change circuit connections and function in response to sensory input and past experience.

2. Behaviors related to feeling and emotion are termed _____, whereas those related to thinking are _____.

3. The part of the brain called the _____ is what makes us human, allowing human reasoning and cognition.

4. In vertebrates, the central nervous system is protected by bony cases. The brain is found inside the _____, and the spinal cord is found in a canal inside the _____.

5. Name the meninges, beginning with the layer next to the bones.

6. The _____ is a salty fluid continuously secreted into hollow cavities of the brain called the _____ by the _____, a specialized tissue.

7. List and explain the purposes of cerebrospinal fluid (CSF).

8. Compare the concentrations of these substances in CSF and plasma.
   a. HCO_3^-
   b. Ca^{2+}
   c. glucose
   d. H^+
   e. Na^+
   f. K^+

9. The only fuel source for neurons under normal circumstances is _____. Low concentrations of this fuel in the blood is termed _____ and if not treated can lead to confusion, unconsciousness, and eventually death. In order to synthesize enough ATP to continually transport ions,

the neurons also exhibit high rates of _____ consumption. To supply these needs, about _____% of the blood pumped by the heart goes to the brain.

10. Match each area with its function.

(a) medulla oblongata	1. processes sensory information and coordinates execution of movement
(b) pons	2. composed of the thalamus and hypothalamus
(c) midbrain	
(d) reticular formation	3. composed of two hemispheres; this part fills most of the cranial cavity
(e) cerebellum	
(f) diencephalon	4. control centers for blood pressure, vomiting, breathing, and swallowing
(g) thalamus	
(h) hypothalamus	
(i) cerebrum	5. transmits and integrates sensory information between lower central nervous system and cerebral cortex

6. transfers information to cerebellum, coordinates breathing centers
7. contains centers for behavioral drives such as hunger and thirst; plays a key role in homeostasis
8. relays signals and visual reflexes, eye movement
9. arousal and sleep, muscle tone, and stretch reflexes

11. Relatively impermeable capillaries form the _____, sheltering the brain from what?

12. How are gray matter and white matter different, both anatomically and functionally?

13. Name the cerebral cortex areas that direct perception, direct movement, and integrate information and direct voluntary behaviors.

14. What does cerebral lateralization refer to? What functions tend to be centered in each hemisphere?

15. Name the two most common inhibitory neurotransmitters of the CNS. Where do they act? Which ion channels do they open?

16. List and define the two major phases of sleep. How are they different from each other?

17. The center for homeostasis is the _____, which occupies less than 1% of the total brain volume. List several reflexes and behaviors influenced by output from this region. What is the source of emotional input into this area?

18. The _____ region of the limbic system is believed to be the center for basic instincts such as fear and learned emotional states.

19. What are the broad categories of learning? Define habituation and sensitization. What anatomical structure (of the cerebrum) is important in both learning and memory?

20. What two regions of the cortex are involved in integrating spoken language?

LEVEL TWO   Reviewing Concepts

21. Map the functional areas of the brain. Start with the brain stem, diencephalon, cerebellum, and cerebrum at the top of your page (or give each one its own page).

22. Trace the pathway that the cerebrospinal fluid follows through the nervous system.

23. How do the capillaries of the brain regulate their "leakiness"?

24. Fill in the question marks in the chart below:

Cerebral Area	Lobe	Functions
Primary somatic sensory cortex	?	Receives sensory information from peripheral receptors
?	Occipital	Processes information from the eyes
Auditory cortex	Temporal	?
?	Temporal	Receives input from chemoreceptors in the nose
Motor cortices	?	?
Association areas	NA	?

25. How can a single molecule act as a neurotransmitter at one time and a neuromodulator at another?

26. Compare and contrast the following concepts:
   a. the diffuse modulatory systems, the reticular formation, and the limbic system
   b. the different forms of memory
   c. nuclei and ganglia
   d. tracts, nerves, horns, nerve fibers, and roots

27. Given the wave below, draw examples with (a) lower frequency, (b) higher amplitude, (c) higher frequency.

28. Define motivational states. What properties do they have in common?

29. What changes occur at synapses as memories are formed?

LEVEL THREE   Problem Solving

30. Mr. Andersen, a stroke patient, experiences expressive aphasia. His savvy therapist, Cheryl, teaches him to sing to communicate his needs. What symptoms did he exhibit before therapy? How do you know he did not have receptive aphasia? Using what you have learned about cerebral lateralization, hypothesize why singing worked for him.

31. A study was done in which typical citizens were taught about using seatbelts in their cars. At the end of the presentations, all participants scored at least 90% on a comprehensive test. However, the people were secretly videotaped entering and leaving the parking lot. Twenty subjects entered wearing their seatbelts, 22 left wearing them. Did learning occur? What is the relationship between learning and actually buckling the seatbelts?

32. In 1913, Henri Pieron took a group of dogs and kept them awake for several days. Before allowing them to sleep, he withdrew cerebrospinal fluid from the sleep-deprived animals. He then injected this CSF into normal, rested dogs. The recipient dogs promptly went to sleep for periods ranging from two to six hours. What conclusion can you draw about the possible source of a sleep-inducing factor? What controls should Pieron have run?

Problem Conclusion

In this running problem, you have learned how a stroke—a sudden blockage or rupture of a blood vessel that supplies oxygenated blood to the brain—can lead to loss of body function. You have also learned about the risk factors and current treatments for stroke.

Further check your understanding of this running problem by comparing your answers against those in the summary table.

Question	Facts	Integration and Analysis
1 Why does the tissue supplied by the affected blood vessel look less dense than usual in CT scans?	Strokes are caused by rupture or blockage in a blood vessel that supplies oxygenated blood to the brain. Without oxygen, brain cells die.	The less dense appearance of the area surrounding a blocked blood vessel in the brain is the result of brain cell death.
2 Explain the rationale for taking aspirin to prevent stroke.	Aspirin helps prevent platelets from sticking together and forming blood clots.	Strokes may be caused by a blood clot that blocks a blood vessel. Therefore, a drug that prevents blood clot formation may help prevent a stroke.
3 Did Dorothy's stroke occur in the right or left hemisphere?	Dorothy lost function on the left side of her body and lost the ability to draw a tree. Motor functions on the left side of the body are controlled by the right motor cortex of the brain. Spatial visualization is controlled by the right side of the brain.	Dorothy's loss of function is consistent with damage to her right cerebral cortex.
4a Explain the rationale for giving t-PA to people who have had a stroke.	t-PA is a natural substance produced by the body that dissolves blood clots.	If a stroke is caused by a blood cot, t-PA may help remove the blood clot and restore blood flow to the affected area of the brain.
4b Why is t-PA most effective when given shortly after the stroke occurs?	Brain cells have a high rate of oxygen consumption and die rapidly when deprived of oxygen.	Rapid administration of t-PA may help restore blood flow to ischemic areas of brain tissue and help minimize the damage caused by the stroke.
4c Is there any type of stroke for which t-PA would not be appropriate?	Strokes are caused by rupture or blockage in a blood vessel. If a blood vessel ruptures, it causes bleeding. The body's normal response to bleeding is clot formation.	In strokes caused by blood vessel rupture, bleeding can damage brain tissue due to pressure buildup within the rigid skull. t-PA would prevent the normal clot formation that would stop bleeding, so it would not be appropriate in this type of stroke.

middle ear from an infection, to diseases or trauma that impede vibration of the malleus, incus, or stapes. Correction of conductive hearing loss includes new microsurgical techniques in which the bones of the middle ear can be reconstructed. *Sensorineural hearing loss* arises from damage to the structures of the inner ear, including degeneration of hair cells as a result of loud noises. This form of hearing loss is more difficult to treat. *Central hearing loss* results from damage to the neural pathways between the ear and cerebral cortex or from damage to the cortex itself. This form of hearing loss is relatively uncommon.

Hearing is probably our most important social sense. Suicide rates are higher among deaf people than among those who have lost their sight. Hearing connects us to other people and the world around us more than any other sense.

✓ Why is somatosensory information projected to only one hemisphere of the brain but auditory information projected to both hemispheres?

✓ Would a cochlear implant help a person who suffers from nerve deafness? From conductive hearing loss?

THE EAR: EQUILIBRIUM

Equilibrium is a state of balance, whether the word is used to describe ion concentrations in body fluids or the positioning of the body in space. In body positioning, sensory information from the inner ear combines with that from proprioceptors in joints and muscles to tell our brain the location of different body parts in relation to each other and to the environment. Visual information also plays an important role in equilibrium, as you know if you have ever gone to one of the 360° movie theaters where the scene tilts suddenly to one side and the audience tilts with it!

Our sense of equilibrium is mediated through hair cells lining the fluid-filled vestibular apparatus and semicircular canals of the inner ear. These receptors respond to

continued from page 272

Anant reports to the otolaryngologist that his attacks of dizziness strike without warning and last from 10 minutes to an hour. They often cause him to vomit. He also reports that he has a persistent low buzzing sound in one ear and that he does not seem to hear low tones as well as he could before. The buzzing sound (tinnitus) often gets worse during his dizzy attacks.

Question 1: Subjective tinnitus occurs when an abnormality somewhere along the anatomical pathway for hearing causes the brain to perceive a sound that does not exist outside the auditory system. Starting from the ear canal, name the auditory structures in which problems may arise.

changes in rotational, vertical, and linear acceleration. The hair cells function just like those of the cochlea (see Sound Transduction through the Cochlea), but gravity and acceleration rather than sound waves provide the force that moves the stereocilia.

The Vestibular Apparatus Is Filled with Endolymph

The **vestibular apparatus** consists of two saclike **otolith organs,** the **utricle** and the **saccule,** along with three **semicircular canals** (Fig. 10-22a ■). The semicircular canals connect to the utricle at their bases. They are oriented at right angles to each other, like three planes that come together to form the corner of a box (Fig. 10-22b ■). At one end of each canal is an enlarged chamber, the **ampulla** [bottle], that contains a sensory receptor known as a **crista** [a crest; plural *cristae*]. The crista consists of a gelatinous mass, the **cupula** [small tub], that stretches from floor to ceiling of the ampulla, closing it off (Fig. 10-22c ■). Embedded in the cupula are the cilia of hair cells. The basal membranes of the hair cells synapse on the sensory neurons of the **vestibular nerve.**

The sensory receptors of the utricle and saccule, the **maculae,** are slightly different from those of the cristae (Fig. 10-22d ■). Each macula consists of a gelatinous mass known as the **otolith membrane** in which small crystals of calcium carbonate called **otoliths** are embedded [*oto,* ear + *lithos,* stone]. Underneath the otolith membrane are hair cells whose cilia are connected to the membrane.

The vestibular apparatus, like the cochlear duct, is filled with fluid endolymph secreted by the epithelial cells. Like cerebrospinal fluid, endolymph is secreted continuously and drains from the inner ear into the venous sinus in the dura mater of the brain. If endolymph production exceeds the drainage rate, buildup of fluid in the inner ear may cause *Ménière's disease.* This condition is marked by episodes of dizziness and nausea, apparently as a result of changes in fluid pressure within the vestibular apparatus. It may also cause hearing loss if the organ of Corti in the cochlear duct is damaged.

The Vestibular Apparatus Provides Information about Movement and Position in Space

The special sense of equilibrium has two components: a dynamic component that tells us about our movement through space and a static component that tells us if our head is displaced from its normal upright position. The three semicircular canals sense rotational acceleration in various directions, and the otolith organs tell us about linear acceleration and head position.

The receptor cells of both sets of sense organs, the hair cells, function just like the hair cells of the organ of Corti: When the cilia bend in one direction, they depolarize, and when they bend in the opposite direction, they hyperpolarize. Vestibular hair cells have one long cilium called a

Anatomy Summary **Vestibular Apparatus**

■ Figure 10-22

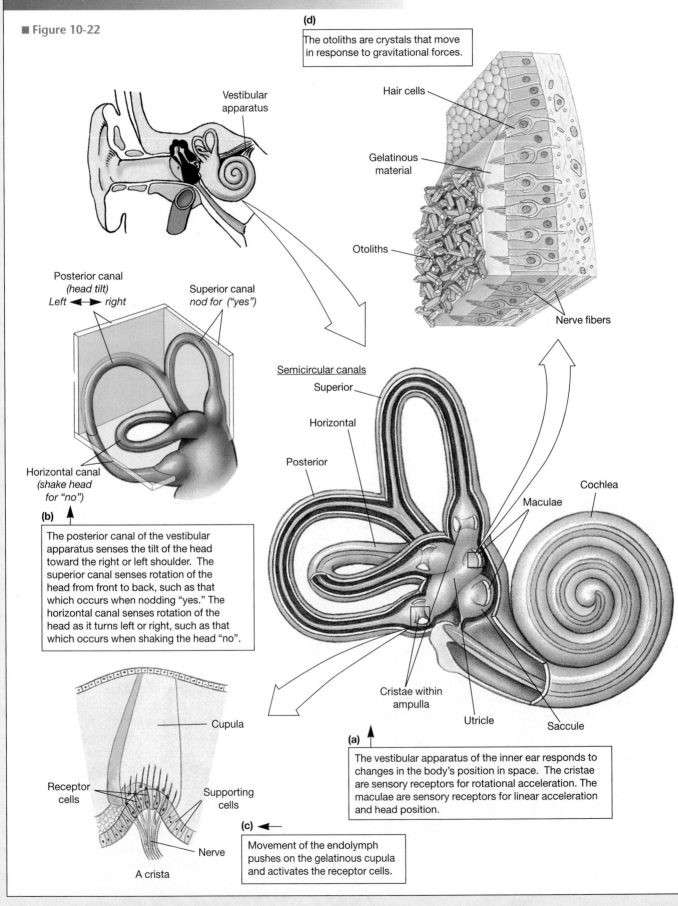

(d) The otoliths are crystals that move in response to gravitational forces.

Hair cells

Gelatinous material

Otoliths

Nerve fibers

Vestibular apparatus

Posterior canal
(head tilt)
Left ◀▶ *right*

Superior canal
nod for ("yes")

Horizontal canal
*(shake head
for "no")*

(b) The posterior canal of the vestibular apparatus senses the tilt of the head toward the right or left shoulder. The superior canal senses rotation of the head from front to back, such as that which occurs when nodding "yes." The horizontal canal senses rotation of the head as it turns left or right, such as that which occurs when shaking the head "no".

Semicircular canals
Superior

Horizontal

Posterior

Maculae

Cochlea

Cristae within
ampulla

Utricle

Saccule

(a) The vestibular apparatus of the inner ear responds to changes in the body's position in space. The cristae are sensory receptors for rotational acceleration. The maculae are sensory receptors for linear acceleration and head position.

Cupula

Receptor
cells

Supporting
cells

Nerve

A crista

(c) Movement of the endolymph pushes on the gelatinous cupula and activates the receptor cells.

kinocilium that is located at one side of the bundle, creating a reference point for the direction of bending.

Rotation is sensed in the following manner. As the head turns, the bony structure of the skull and the membranous labyrinth move, but the endolymph within the canals cannot keep up because of inertia. The drag of the fluid bends the cupula and its hair cells in the direction opposite to the way the head is turning. For an analogy, think of pulling a paintbrush (the crista attached to the wall of the canal) through sticky wet paint (the endolymph) on a board. If you pull the brush to the right, the drag of the paint on the bristles bends them back to the left (Fig. 10-23 ■). In the same way, the inertia of the fluid in the semicircular canal pulls the cupula and the cilia of the hair cells to the left when the head turns right. If the rotation continues, the rotation of the endolymph finally catches up. Then if the head rotation stops suddenly, the fluid has built up momentum and cannot stop as fast. The fluid continues to rotate in the direction of the head rotation, leaving the person with a turning sensation. If the sensation is strong enough, the person may throw his or her body in the direction opposite the rotation in a reflexive attempt to compensate for the apparent loss of equilibrium.

The otolith organs are arranged to sense linear forces. The maculae of the utricle are horizontal when the head is in its normal upright position. If the head tips back, the otoliths in the gelatinous membrane slide backward owing to gravity acting on the dense particles (Fig. 10-24 ■). The cilia of the hair cells bend and set off a signal. The maculae of the saccule are oriented vertically when the head is erect and are sensitive to vertical forces, such as dropping downward in an elevator. The brain analyzes the pattern of which hair cells are depolarized and which are hyperpolarized in order to compute direction of movement.

Equilibrium Pathways Project Primarily to the Cerebellum

The hair cells of the vestibular apparatus stimulate primary sensory neurons in the vestibular nerve. Those neurons either synapse in the vestibular nuclei of the

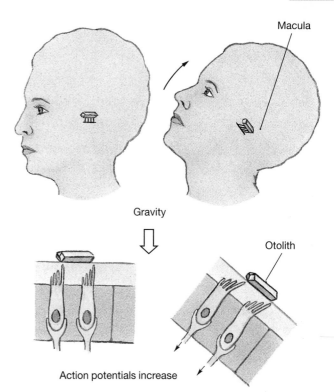

■ Figure 10-24 Otolith organs The crystalline otoliths are attached to gelatinous material in the maculae. When the head tilts, gravity causes the otoliths to slide, pulling the stereocilia of the hair cells out of their vertical position and increasing the action potentials in the sensory neurons.

medulla or run without synapsing to the cerebellum (Fig. 10-25 ■). Collateral pathways run from the medulla to the cerebellum or upward through the reticular formation and thalamus. There are some poorly defined pathways to the cerebral cortex, but most integration for equilibrium comes from the cerebellum. Descending pathways from the vestibular nuclei go to certain motor neurons involved in eye movement. These pathways help keep the eyes locked on an object as the head turns.

✓ Why does hearing decrease if you have an ear infection with fluid buildup in the middle ear?

✓ When dancers perform multiple turns, they try to keep their vision fixed on a single point ("spotting"). How does spotting help keep the dancer from getting dizzy?

THE EYE AND VISION

The eye is a sensory receptor that functions much like a camera. It focuses light on a light-sensitive surface (the retina) using a lens and an aperture or opening (the pupil), whose size can be adjusted to change the amount of entering light. **Vision,** the process through which light reflected from objects in our environment is translated into a mental image, can be divided into three

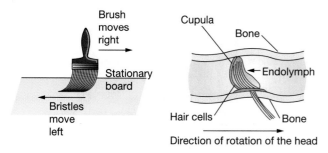

■ Figure 10-23 Transduction of rotational forces in the cristae When the head turns, inertia keeps the endolymph inside the ampulla from moving as rapidly as the surrounding cranium. When the head turns right, the endolymph pushes the cupula to the left, just as a brush pulled to the right results in bristles bent to the left. The force of the endolymph deforms the cupula and activates the hair cells embedded in it.

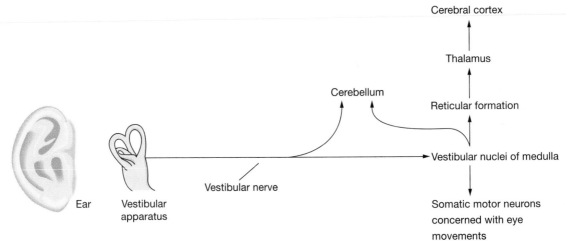

■ Figure 10-25 **Central nervous system pathways for equilibrium**

steps. In the first step, light enters the eye and is focused by a lens on the retina. In the second step, the photoreceptors of the retina transduce light energy into an electrical signal. The third step involves neural pathways and processing of these electrical signals.

The electromagnetic energy that we call **visible light** is composed of waves with a frequency of $4.0–7.5 \times 10^{14}$ waves per second (hertz) and a wavelength of 400–750 nanometers (nm). But these waves are only a small fraction of the electromagnetic spectrum that extends from high-energy, very short wavelength waves such as X-rays and gamma rays to low-energy, lower-frequency microwaves and radio waves (Fig. 10-26 ■). The waves of ultraviolet and infrared light border the ends of the visible light spectrum. Although our eyes are unable to perceive these wavelengths, some animals can see light in the infrared and ultraviolet range. For example, bees use ultraviolet "runways" on flowers to guide them to pollen and nectar.

The Optic Tract Extends from the Eye to the Visual Cortex

The anatomy of the eye and the neural pathways for vision are shown in the Anatomy Summary, Figure 10-27 ■. The eye itself is a hollow sphere divided into two compartments separated by a lens. The compartment in front of the lens and its attachment is filled with the **aqueous humor,** a low-protein plasmalike fluid that is secreted by the ciliary epithelium supporting the lens (Fig. 10-27b ■). Behind the lens is a much larger chamber filled mostly with the **vitreous humor,** a clear, gelatinous matrix. Light enters the eye at the **cornea,** a transparent disk of tissue on the anterior surface. After crossing the aqueous humor, light passes through the opening of the pupil and strikes the **lens,** a transparent structure with two convex surfaces. The cornea and lens together bend incoming light rays so that they focus on the **retina,** the

continued from page 287

Although many vestibular disorders can cause the symptoms Anant is experiencing, two of the most common are benign positional vertigo (BPV) and Ménière's disease. In benign positional vertigo, the calcium crystals normally embedded in the otolith membrane of the maculae become dislodged and float toward the semicircular canals. The primary symptom of BPV is brief episodes of severe dizziness brought on by a change in position. People with this condition often say that they cannot go to sleep because they feel dizzy when they turn over in bed.

Question 2: When a person with BPV changes position, the displaced crystals float toward the semicircular canals. Why would this cause dizziness?

Question 3: Compare the symptoms of BPV and Ménière's disease. On the basis of Anant's symptoms, which condition do you think he has?

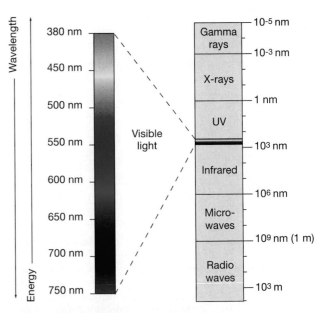

■ Figure 10-26 **The electromagnetic spectrum**

light-sensitive lining of the eye that contains the **photoreceptors.** When viewed through the pupil of the eye with an ophthalmoscope [*ophthalmos,* eye], the retina is crisscrossed with small arteries and veins that radiate out from one spot, the **optic disk** (Fig. 10-27c ■). The optic disk is the location where neurons from photoreceptors form the optic nerve and exit the eye (Fig. 10-27d ■). The optic nerves from each eye go to the **optic chiasm** in the brain, where some of the fibers cross to the opposite side. After synapsing in the **lateral geniculate body** (nucleus) of the thalamus, the neurons of the tract terminate in the occipital lobe at the visual cortex (Fig. 10-27a ■).

External structures associated with the eye include six extrinsic eye muscles that attach to the outer surface of the sphere and control eye movements, eyelids that close over the anterior surface of the eye, and the *lacrimal* apparatus, a system of glands and ducts that keep a continuous flow of tears washing across the cornea so that it remains moist and free of debris.

✓ What function(s) might the aqueous humor serve?

✓ What would happen to the eye if the normal drainage pathways for the aqueous humor became blocked?

The Lens Focuses Light on the Retina

Light striking the eye is modified two ways before it strikes the retina: (1) The amount of light that reaches photoreceptors is altered by changing the size of the pupil, and (2) the light waves are focused by changing the shape of the lens. This section looks at the principles of **optics,** the physical relationship between light and lenses.

Image focusing and the lens   Light waves passing from air into a medium of different density such as glass or water will bend, or **refract.** The angle of refraction depends on two factors: (1) the angle at which the light meets the face of the object into which it is passing and (2) the difference in density of the two objects. If the surface of the object is perpendicular to the light waves, the light will pass through without bending. But if the surface is not perpendicular, light waves striking the surface will bend. Parallel light waves striking a concave lens such as that shown in Figure 10-28a ■, p. 294, will be refracted into a wider beam; a light beam striking a spherical convex lens will be bent inward and focused into a point (Fig. 10-28b ■). The point where the waves converge is known as the **focal point,** and the distance from the center of the lens to the focal point is known as the **focal distance.** You may have used the ability of a convex lens to focus light if you ever took a magnifying glass and used it to focus sunlight onto a piece of paper or other surface.

When you focus your eyes on an object, the focal distance for that object depends on two factors: how far the object is from your eye and the shape of the lens of your eye. These principles are illustrated in Figure 10-29 ■, p. 294. For us to see an object in focus, the focal distance must be such that the focal point falls precisely on the retina of the eye. In Figure 10-29a ■, light rays from a distant source are parallel as they strike a lens, resulting in a short focal distance. If an object is closer to the lens (Fig. 10-29b ■), the light rays reflected from it are not parallel and therefore strike the lens at a more oblique angle. The focal distance is moved back from the lens. If an object moves closer to the eye without any adjustment on the part of the eye, the longer focal point will fall behind the retina and the object will go out of focus. In order to keep the object in focus, the lens must round in order to shorten the focal distance.

You can demonstrate this easily by closing one eye and holding your hand up about 8 inches in front of your open eye, fingers spread widely apart. Focus your eye on some object in the distance that is between your fingers. Notice that when you do so, your fingers are visible but out of focus. Your lens is flattened for distance vision, so the focal point for near objects falls behind the retina. Those objects appear out of focus. Now shift your gaze to your fingers and notice that they come into focus. The light waves coming off your fingers have not changed their angle, but your lens has become more rounded. The focal distance has been shortened so that the focal point falls on the retina (Fig. 10-29c ■). The process by which the eye adjusts the shape of the lens to keep objects in focus is known as **accommodation.**

How can the lens, which does not have any muscle fibers in it, change shape? The lens is attached to the ciliary muscle of the eye by inelastic fibers known as **zonulas** (Fig. 10-30 ■, p. 294). If no tension is placed on the lens by the zonulas, the lens assumes its natural rounded shape because of the elasticity of its capsule. If the zonulas pull on the lens, it flattens out and assumes the shape required for distance vision. The tension on the zonulas is controlled by the **ciliary muscle,** a ring of muscle that surrounds the lens. When the circular ciliary muscles contract, the muscle ring gets smaller, releasing tension on the zonulas so that the lens rounds. When the ciliary muscles are relaxed, the ring is more open and the lens is pulled into a flatter shape. Young people are able to focus on items as close as 8 cm, but as people age, the accommodation reflex diminishes. By age 40, it is only about half of what it was at age 10, and by age 60, many people lose the reflex completely. The loss of accommodation, **presbyopia,** is the reason that most people begin to wear reading glasses in their 40s.

Two other common vision problems, near-sightedness (**myopia**) and far-sightedness (**hyperopia**), occur when the focal point falls either in front of or behind the retina, respectively (Fig. 10-31 ■, p. 295). These conditions can be corrected by placement of a lens with the appropriate curvature in front of the eye to change the focal distance. A third common vision problem, **astigmatism,** is caused by a cornea that is not a perfectly shaped dome.

Anatomy Summary **The Eye**

■ **Figure 10-27**

(a) Neural pathway for vision

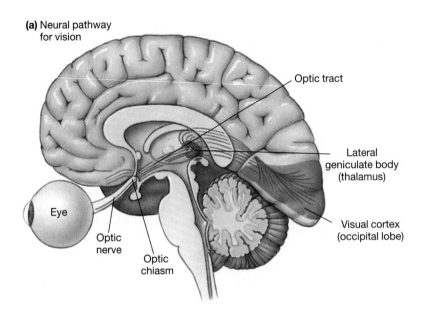

Optic tract

Lateral geniculate body (thalamus)

Visual cortex (occipital lobe)

Eye

Optic nerve

Optic chiasm

(b) Cross section of the eye

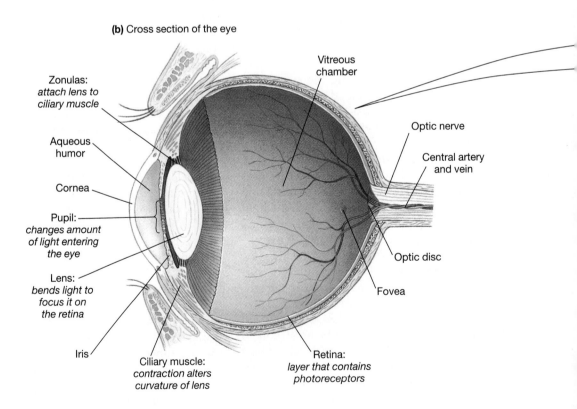

Vitreous chamber

Zonulas: *attach lens to ciliary muscle*

Aqueous humor

Cornea

Pupil: *changes amount of light entering the eye*

Lens: *bends light to focus it on the retina*

Iris

Ciliary muscle: *contraction alters curvature of lens*

Optic nerve

Central artery and vein

Optic disc

Fovea

Retina: *layer that contains photoreceptors*

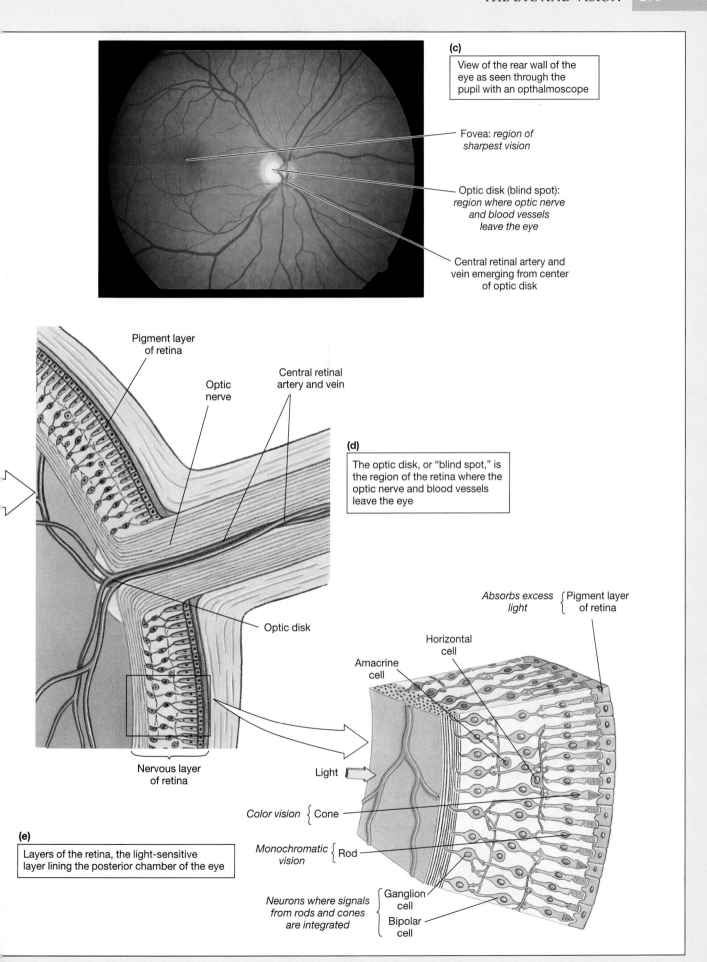

(c)

View of the rear wall of the eye as seen through the pupil with an opthalmoscope

Fovea: *region of sharpest vision*

Optic disk (blind spot): *region where optic nerve and blood vessels leave the eye*

Central retinal artery and vein emerging from center of optic disk

Pigment layer of retina

Optic nerve

Central retinal artery and vein

(d)

The optic disk, or "blind spot," is the region of the retina where the optic nerve and blood vessels leave the eye

Optic disk

Nervous layer of retina

Light

Absorbs excess light { Pigment layer of retina

Horizontal cell

Amacrine cell

Color vision { Cone

Monochromatic vision { Rod

Neurons where signals from rods and cones are integrated { Ganglion cell / Bipolar cell

(e)

Layers of the retina, the light-sensitive layer lining the posterior chamber of the eye

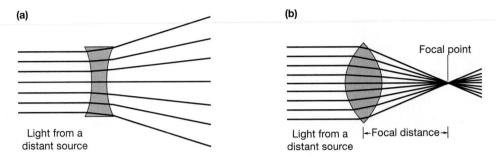

(a)

Light from a
distant source

(b)

Focal point

Light from a |←Focal distance→|
distant source

■ **Figure 10-28 Refraction of light** (a) Light is refracted and scattered by a concave lens. (b) A spherical convex lens focuses a beam of light to a single point called the focal point. The distance between the center of the lens and the focal point is called the focal distance (focal length).

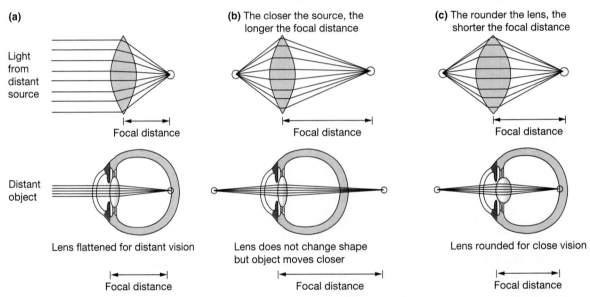

(a)

Light from distant source

Focal distance

Distant object

Lens flattened for distant vision

Focal distance

(b) The closer the source, the longer the focal distance

Focal distance

Lens does not change shape but object moves closer

Focal distance

(c) The rounder the lens, the shorter the focal distance

Focal distance

Lens rounded for close vision

Focal distance

■ **Figure 10-29 Optics** (a) Light reflecting off a distant object reaches the eye as nearly parallel rays. The lens is flattened so that the focal point falls on the retina. (b) As the object being viewed moves closer, the focal distance gets longer. Without any changes to the lens, the object goes out of focus since the light beam is not focused on the retina. (c) In order to keep an object in focus as it moves closer, the lens becomes more rounded. This adjustment is known as accommodation.

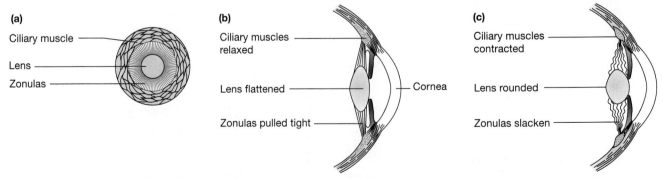

(a)

Ciliary muscle

Lens

Zonulas

(b)

Ciliary muscles relaxed

Lens flattened

Zonulas pulled tight

Cornea

(c)

Ciliary muscles contracted

Lens rounded

Zonulas slacken

■ **Figure 10-30 Accommodation** (a) The lens is attached to the circular ciliary muscles by inelastic fibers known as zonulas. (b) When the ciliary muscle is relaxed, the zonulas pull tight and keep the elastic lens flattened. (c) When the ciliary muscle contracts, it releases tension on the zonulas, which go slack. The elastic lens can then assume a more rounded shape.

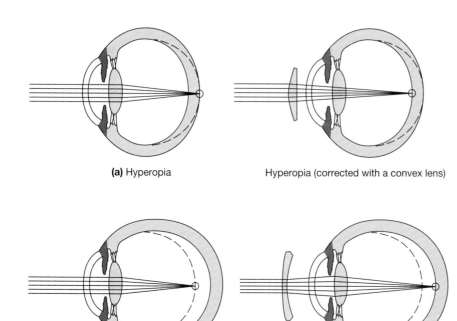

(a) Hyperopia   Hyperopia (corrected with a convex lens)

(b) Myopia   Myopia (corrected with a concave lens)

■ **Figure 10-31 Visual defects**
(a) Hyperopia, or far-sightedness, occurs when the length of the eyeball is too short and the focal point falls behind the retina. It is corrected with a convex lens. (b) Myopia, or near-sightedness, occurs when the eyeball is too long and the focal point falls in front of the retina. It is corrected with a concave lens.

Figure question: Explain (1) what the corrective lenses do to the refraction of light, and (2) why each lens type corrects the visual defect.

Light entering the eye and the pupil The human eye functions over a 100,000-fold range of light intensity. Most of this ability comes from the sensitivity of the photoreceptors, but the pupils of the eyes assist by regulating the amount of light that falls on the retina. In bright sunlight, the pupils constrict to about 1.5 mm; in the dark, the opening dilates to 8 mm, a 28-fold increase. In addition to regulating the amount of light that hits the retina, the pupils create what is known as **depth of field.** A simple example comes from photography. Imagine a picture of a puppy sitting in the foreground of a field of wildflowers. If only the puppy and the flowers immediately around her are in focus, the picture is said to have a shallow depth of field. But if the puppy and the wildflowers all the way back to the horizon are in focus, the picture has full depth of field. Depth of field is created by constricting the pupil (or shutter on the camera) so that only a narrow beam of light enters the eye. In this way, more of the depth of the image is focused on the retina.

The reaction of the pupil to light is known as the **pupillary reflex.** Light hitting the retina activates a parasympathetic pathway that causes constriction of the circular pupillary muscles. To open the pupil, a set of radial muscles that lie perpendicular to the circular muscles contract under the influence of sympathetic neurons.

✓ What kind of drug would be used to dilate the pupils before an eye exam?

✓ What kind of lens is needed to correct myopic vision? Explain.

Phototransduction Occurs at the Retina

The retina, the sensory organ of the eye, develops from the same embryonic tissue as the central nervous system, and, like the cortex of the brain, its neurons are

continued from page 290

The otolaryngologist strongly suspects that Anant has Ménière's disease, with excessive endolymph within the vestibular apparatus and cochlea. Many treatments are available, from simple dietary changes to surgery. For now, the physician suggests that Anant limit his salt intake and take diuretics, drugs that cause the kidneys to remove excess fluid from the body.

Question 4: Why is limiting salt (NaCl) intake suggested as a treatment for Ménière's disease? (Hint: Remember that water follows solute.)

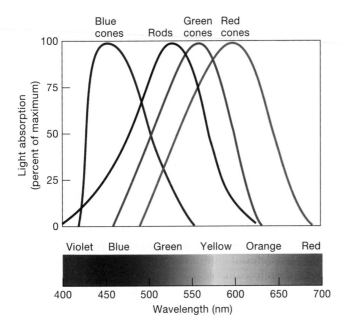

Graph question: Which pigment absorbs light over the broadest spectrum of wavelengths? Over the narrowest? Which pigment absorbs the most light at 500 nm?

■ Figure 10-35 **Light absorption of visual pigments**
There are three types of cone pigment, each with a characteristic light absorption spectrum.

Phototransduction   The process of phototransduction is similar for rhodopsin and the three color pigments. The visual pigment of the rods, rhodopsin, is composed of two sections: **opsin,** a protein embedded in the membrane of the rod disks, and **retinal,** a vitamin A derivative that is the light-absorbing portion of the pigment. In the absence of light, retinal binds snugly into a binding site on the opsin (Fig. 10-36a ■). When activated by as little as one photon of light, retinal changes shape to a new configuration. The activated retinal no longer binds to opsin and is released from the pigment in the process known as **bleaching.**

But how does rhodopsin bleaching lead to an action potential's traveling through the optical pathway? In order to understand the link, we must look at other properties of the rods. As you learned in Chapters 5 and 8, electrical signals in cells occur as a result of the movement of ions between the intracellular and extracellular compartments. Rods contain two types of ion channels that are of interest in this regard: a large number of cation channels that allow Na^+ to enter the rod and K^+ channels that allow K^+ to leak out of the cell. When the rods are in darkness and rhodopsin is not active, cyclic GMP levels in the rod are high and both sets of channels are open. Sodium ion influx is greater than K^+ efflux, so the rod stays depolarized to an average membrane

potential of −40 mV, instead of the more usual −70 mV. At this membrane potential, there is tonic (continuous) release of neurotransmitter from the synaptic portion of the rod onto the adjacent bipolar cell.

When light activates rhodopsin, a second messenger cascade is initiated through a G protein known as **transducin** (Fig. 10-36b ■). The second messenger cascade decreases the concentration of cGMP, and, in turn, the cation channels close. As a result, Na^+ influx slows or stops. Less Na^+ influx but continued K^+ efflux causes the inside of the cell to hyperpolarize. The amount of neurotransmitter released onto the bipolar neurons consequently decreases. Bright light will close all Na^+ channels and stop all neurotransmitter release; dimmer light will cause a response that is graded in proportion to the light intensity.

Since a change in neurotransmitter release by rods is mediated through a second messenger system, the response of rods to changes in light intensity is considerably slower than you might expect for a process associated with the nervous system. When you move from bright light into the dark, or vice versa, the time required for your eyes to adapt to the new light level is partially determined by the slow change in rod membrane potential as ion channels open or close. The recovery of rhodopsin from bleaching also can take some time. After activation, retinal diffuses out of the rod and is transported into the pigment epithelium. There, it is reduced to its inactive form before moving back into the rod and being reunited with opsin (Fig. 10-36c ■).

We now turn from the cellular mechanism of light transduction to the processing of light signals in the bipolar and ganglion cells of the retina.

Signal Processing in the Retina Occurs When Light Strikes Visual Fields

One of the hallmarks of signal processing in the retina is **convergence,** in which multiple neurons synapse onto a single postsynaptic cell (Fig. 10-37 ■). Most photoreceptors synapse onto bipolar neurons in groups of 15 to 45, although in the fovea, a few cones have a 1:1 relationship with their bipolar neurons. Multiple bipolar neurons in turn innervate a single ganglion cell, so that the information from hundreds of millions of photoreceptors is condensed down to a mere 1 million axons leaving the eye in each optic nerve.

The release of the neurotransmitter glutamate from the photoreceptors onto bipolar neurons is the first step in the signal pathway. Glutamate excites some bipolar neurons but inhibits others, depending on the type of glutamate receptor on the bipolar neuron. In this fashion, one stimulus (light) creates two responses with a single neurotransmitter.

The second synapse in the pathway occurs between the bipolar neurons and the ganglion cells. These connections can also be either excitatory or inhibitory. We

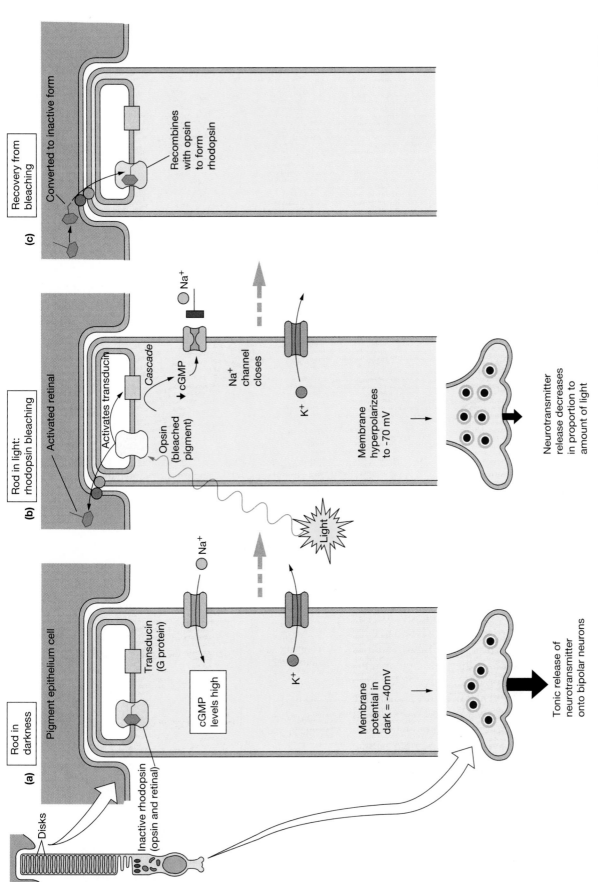

(a) Rod in darkness

Pigment epithelium cell

Disks

Inactive rhodopsin (opsin and retinal)

Transducin (G protein)

cGMP levels high

Na⁺

K⁺

Membrane potential in dark = –40mV

Tonic release of neurotransmitter onto bipolar neurons

(b) Rod in light: rhodopsin bleaching

Activated retinal

Activates transducin

Cascade

Opsin (bleached pigment)

↓cGMP

Na⁺

Na⁺ channel closes

K⁺

Light

Membrane hyperpolarizes to –70 mV

Neurotransmitter release decreases in proportion to amount of light

(c) Recovery from bleaching

Converted to inactive form

Recombines with opsin to form rhodopsin

■ **Figure 10-36 Phototransduction in rods** (a) In darkness, the visual pigment rhodopsin is in its inactive form. Cyclic GMP levels are high within the rod, and both Na⁺ and K⁺ channels are open. The net influx of Na⁺ depolarizes the cell to –40 mV. At this membrane potential, there is tonic release of neurotransmitter onto the bipolar neurons. (b) When a photon of light strikes the rhodopsin molecule, it activates the retinal molecule, causing it to unbind from the rhodopsin, a process known as bleaching. The resultant opsin molecule activates a second messenger cascade that decreases intracellular cGMP and closes the Na⁺ channels. With less Na⁺ influx, the cell hyperpolarizes and neurotransmitter release decreases. (c) In order for the rod to recover, the active retinal is reduced to its inactive form in the cells of the pigment layer. It then returns to the rod and recombines with the opsin to form rhodopsin.

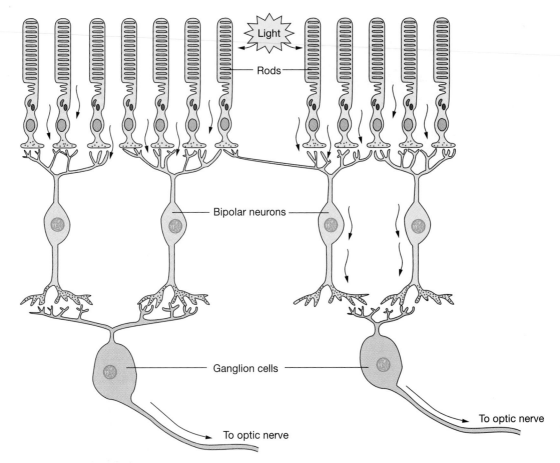

■ **Figure 10-37 Convergence in the retina** Multiple photoreceptors (rods or cones) can converge onto a single bipolar neuron, although rods and cones do not converge onto the same cell. A single receptor can provide input into more than one bipolar neuron. Multiple bipolar neurons can converge onto a single ganglion cell.

know more about ganglion cells because they lie on the surface of the retina, and their axons are the most accessible to researchers. Extensive studies have been done in which the retina was stimulated with carefully placed light and the response of the ganglion cells evaluated.

Each ganglion cell has a particular area of retina to which it is linked. These areas, known as **visual fields,** have two interesting properties. First of all, they are circular, unlike the somewhat irregular shape of receptive fields in the somatic sensory system. The visual field of a ganglion cell near the fovea is quite small. Only a few receptors are associated with each ganglion, so visual acuity is greatest in these areas. At the edge of the visual field, a single ganglion cell may have multiple receptors, so vision is not as sharp.

The second property of visual fields is that they are divided into two sections: a round center and its doughnut-shaped **surround** (Fig. 10-38 ■). There are two basic types of visual fields. In the "on-center/off-surround" field, the associated ganglion cell will respond most strongly with a series of action potentials when light is brightest in the center of the field. If light is brightest on the receptors in the "off" surround region of the field,

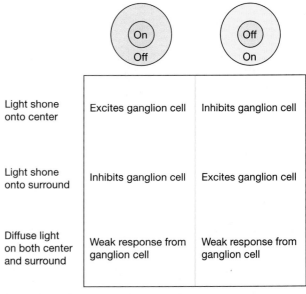

	On (center) / Off (surround)	Off (center) / On (surround)
Light shone onto center	Excites ganglion cell	Inhibits ganglion cell
Light shone onto surround	Inhibits ganglion cell	Excites ganglion cell
Diffuse light on both center and surround	Weak response from ganglion cell	Weak response from ganglion cell

■ **Figure 10-38 Ganglion cell receptive fields** The shape of the visual fields allows the retina to use contrast rather than absolute light intensity for better detection of weak stimuli.

the ganglion cell will be inhibited and stop firing action potentials. The reverse happens with off-center/on-surround visual fields. If light is uniform across the visual field, the ganglion cell responds weakly. Thus, the retina uses *contrast* rather than absolute light intensity to recognize objects in the environment. One advantage of using contrast is that it allows better detection of weak stimuli.

Scientists have now identified two types of ganglion cells in the retina. Large *magnocellular* ganglion cells, or **M cells,** carry information about movement, location, and depth perception. Smaller *parvocellular* ganglion cells, or **P cells,** transmit signals that pertain to color, form, and texture of objects in the visual field. Once action potentials leave these cells, they travel along the optic nerves to the central nervous system for further processing.

✓ Why is the difference in visual acuity between the fovea and the edge of the visual field similar to the difference in touch discrimination between the fingertips and the skin of the arm?

Visual Processing in the Central Nervous System Takes Place in the Visual Cortex

The optic nerves enter the brain in a region known as the **optic chiasm.** This is also the point at which some fibers from each eye cross to the contralateral side of the brain for processing. Figure 10-39 ■ shows how information from the right side of the visual field is processed on the left side of the brain, while information from the left side of the field is processed on the right. The central portion of the visual field, where left and right visual fields overlap, is the **binocular zone** (Fig. 10-40 ■). The two eyes have slightly different views of objects in this region, so the brain processes and integrates the two views to create three-dimensional representations of the objects. Our sense of depth perception—that is, whether one object is in front of or behind another—depends on binocular vision. Objects that fall within the visual field of only one eye are in the **monocular zone** and are viewed in two dimensions.

Once axons leave the optic chiasm, a few fibers project to the midbrain, where they participate in control of eye movement or coordinate with somatosensory and auditory information for balance and movement. Most axons, however, project to the **lateral geniculate body** of the thalamus, where the optic fibers synapse onto neurons leading to the visual cortex (see Fig. 10-27a ■). The lateral geniculate is organized in layers that correspond to the different parts of the visual field. Thus, information from adjacent objects is processed together. This **topographical organization** is maintained in the visual cortex, with the six layers of neurons grouped

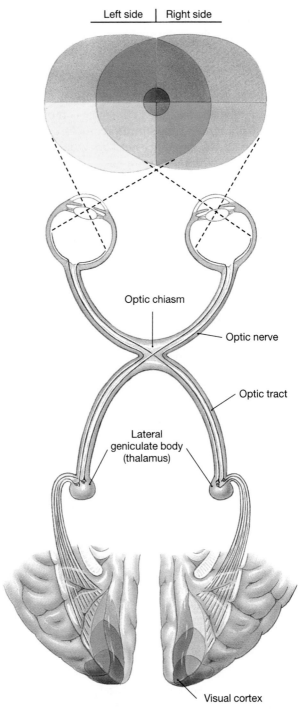

■ **Figure 10-39 Projection of visual fields** The left side of the visual field in each eye is projected to the right side of the brain, while the right side of the field projects to the left visual cortex. Visual fibers from each eye meet at the optic chiasm. From there, they project through the lateral geniculate nucleus to the visual cortex. The cells of the cortex reflect the spatial organization of the visual fields.

into vertical columns. Within each portion of the visual field, information is further sorted by form, color, and movement.

Visual processing and our perception of the world around us are exceedingly complex subjects whose details are beyond the scope of this book. It is sufficient to say that it is in the cortex that monocular information from the two eyes is merged to give us a binocular view of our surroundings. Information from on/off combinations of ganglion cells is translated into sensitivity to line orientation in the simplest pathways or into color, movement, and detailed structure in the most complex. Each of these attributes of visual stimuli is processed through a separate pathway, creating a network whose complexity we are just beginning to unravel.

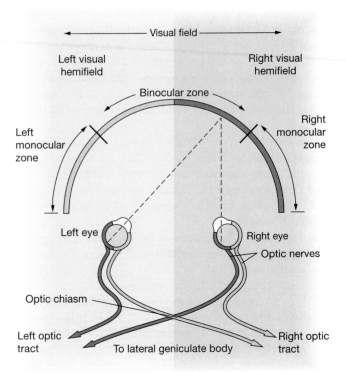

■ **Figure 10-40 Binocular vision** The central portion of the visual field, known as the binocular zone, is seen by both eyes. Objects in the binocular zone will be perceived in three dimensions. Objects falling outside the binocular zone will only be perceived in two dimensions.

---

continued from page 295

Anant's condition does not improve with the low-salt diet and diuretics, and he continues to suffer from disabling attacks of vertigo with vomiting. As a last resort, surgery is sometimes performed to treat severe cases of Ménière's disease. In one surgical procedure for the disease, the vestibular nerve is severed. This surgery is difficult to perform, as the vestibular nerve lies near many other important nerves, including facial nerves and the auditory nerve. Patients who undergo this procedure are advised that the surgery can result in deafness if the auditory nerve is inadvertently severed.

Question 5: Why would severing the vestibular nerve alleviate Ménière's disease?

CHAPTER REVIEW

Chapter Summary

1. Sensory stimuli are divided into the **special senses** of vision, hearing, taste, smell, and equilibrium, and the **somatic senses** of touch-pressure, temperature, pain, and proprioception. (p. 264)

General Properties of Sensory Systems

2. Sensory pathways begin with a stimulus that is converted by a receptor into an electrical potential. (p. 264)

3. If the stimulus is above threshold, action potentials pass along a sensory neuron to the central nervous system. We become aware of some stimuli but never conscious of others. (p. 264)

4. Sensory receptors vary from free nerve endings to encapsulated nerve endings to specialized receptor cells. (p. 264)

5. There are five types of sensory receptors, based on the stimulus to which they are most sensitive: **chemoreceptors, mechanoreceptors, thermoreceptors, photoreceptors,** and **nociceptors.** The specificity of a receptor for a particular type of stimulus is known as the **law of specific nerve energies.** (p. 264)

6. **Primary sensory neurons** synapse onto **secondary sensory neurons** in the central nervous system. (p. 264)

7. Sensory information from the spinal cord projects to the thalamus, then on to the sensory areas of the cortex. Only olfactory information is not routed through the thalamus. (p. 265)

8. The central nervous system is able to process information and change our level of awareness of sensory input. The **perceptual threshold** is the level of stimulus intensity necessary for us to be aware of a particular sensation. (p.265)

9. Multiple primary sensory neurons may converge on one secondary neuron and create a single large **receptive field.** (p. 266)

10. Each receptor type has an **adequate stimulus,** a particular form of energy to which it is most responsive. (p. 268)

11. The **modality** of a signal and its location are indicated by which sensory neurons are activated. (p. 268)

12. The association of a receptor with a specific sensation is called **labeled line coding.** (p. 268)

13. Localization of auditory and olfactory information depends on the timing of receptor activation in each ear or nostril. (p. 269)

14. **Lateral inhibition** enhances the contrast between the center and side of the receptive field so that a sensation is more easily localized. (p. 269)

15. Stimulus intensity is coded by the number of receptors that are activated and by the frequency of the action potentials coming from those receptors. (p. 270)

16. The duration of a stimulus is coded by the duration of action potentials in the sensory neuron if the receptor is a **tonic receptor. Phasic receptors** respond to a change in stimulus intensity, but turn off if the strength of the stimulus remains constant. (p. 270)

Somatic Senses

17. There are four somatosensory modalities: touch-pressure, proprioception, temperature, and nociception. The **somatosensory cortex** is the part of the brain that recognizes where ascending sensory tracts originate. (p. 272)

18. **Fast pain** is transmitted rapidly by small, myelinated fibers. **Slow pain** is carried on small, unmyelinated fibers. The receptors for pain are free nerve endings that respond to chemical, mechanical, or thermal stimuli. (p. 273)

19. Pain is a conscious sensation. However, not all irritant responses must go through the cortex in order to be acted upon. (p. 273)

20. Pain may be modulated by descending pathways from the brain or by **gating mechanisms** in the spinal cord. (p. 274)

21. **Referred pain** from internal organs occurs when multiple primary sensory neurons converge onto a single ascending tract. (p. 275)

Chemoreception: Smell and Taste

22. Chemoreception is divided into the special senses of smell (**olfaction**) and taste (**gustation**). (p. 276)

23. Olfactory receptor cells in the nasal cavity are bipolar neurons whose pathways project directly to the olfactory cortex. Olfactory neurons are the only neurons in the body that are continuously dividing. (p. 277)

24. Odorant molecules dissolve in mucus and bind to olfactory receptors with the aid of **olfactory binding proteins.** (p. 277)

25. Taste is a combination of five sensations: sweet, sour, salty, bitter, and umami. (p. 277)

26. The receptors for taste are the **taste buds.** The four primary taste sensations utilize different cellular mechanisms to increase intracellular Ca^{2+} levels and trigger exocytosis of neurotransmitter. (p. 277)

27. Bitter-tasting ligands activate a G protein named **transducin;** sweet ligands activate the G protein **gustducin.** (p. 279)

28. Salty and sour ligands bind directly to membrane ion channels. (p. 280)

The Ear: Hearing

29. Hearing is our perception of the energy carried by sound waves. Sound transduction turns air waves into mechanical vibrations, then fluid waves, chemical signals, and finally action potentials. (p. 280)

30. The **cochlea** of the inner ear contains three parallel, fluid-filled channels: the **cochlear duct,** the **vestibular duct,** and the **tympanic duct.** (p. 281)

31. The cochlear duct contains the **organ of Corti,** which lies along the **basilar membrane.** The organ of Corti is composed of **hair cell** receptors and is partially covered by the **tectorial membrane.** (p. 282)

32. Hair cells are topped by flexible cilia that open ion channels when they bend. When sound bends the cilia, the cells depolarize and release neurotransmitter onto the sensory neurons. (p. 285)

33. The initial processing for pitch, loudness, and duration of sound takes place in the cochlea. Localization of sound is a higher function that requires sensory input from both ears and sophisticated computation by the brain. (p. 286)

The Ear: Equilibrium

34. **Equilibrium** is mediated through hair cells in the **vestibular apparatus** and **semicircular canals** of the inner ear. Gravity and acceleration provide the force that moves the cilia. (p. 287)

The Eye and Vision

35. **Vision** is the translation of reflected light into a mental image. **Photoreceptors** of the **retina** transduce light energy into an electrical signal that passes to the visual cortex for processing. (p. 289)

36. Light entering the eye is modified two ways before it strikes the retina: (1) The amount of light that reaches photoreceptors is altered by changing the size of the pupil, and (2) the light waves are focused by changing the shape of the lens. (p. 290)

37. The eye adjusts the shape of the lens by contracting or relaxing the **ciliary muscle.** This movement changes the tension on fibers known as **zonulas** and creates tension that pulls on the lens. (p. 291)

38. Light is converted into electrical energy by the rods and cones of the retina. Signals pass through **bipolar neurons** to **ganglion cells,** whose axons form the optic nerve. (p. 296)

39. The **fovea** is the region of retina on which light is focused when you look at an object. It has the most acute vision because it has the smallest receptive fields and because light does not pass through multiple cell layers before striking the photoreceptors. (p. 296)

40. **Rods** are responsible for monochromatic nighttime vision; **cones** are responsible for high-acuity vision and color vision during the daytime. (p. 296)

41. In both types of photoreceptors, light-sensitive **visual pigments** convert light energy into a change in membrane potential. The visual pigment in rods is **rhodopsin;** cones have three related visual pigments. (p. 296)

42. Rhodopsin is composed of **opsin** and **retinal.** In the absence of light, retinal binds snugly to opsin. When activated by light, retinal changes shape and is released from the pigment in the process known as **bleaching.** (p. 298)

43. When light bleaches rhodopsin, a G protein known as **transducin** begins a second messenger cascade that decreases the concentration of cGMP and closes cation channels in the rod membrane. As a result, the rod hyperpolarizes and releases less neurotransmitter onto the bipolar neurons. (p. 298)

44. Signals pass from the photoreceptors through bipolar neurons to ganglion cells. These cells use contrast rather than absolute light intensity to recognize objects in the environment. (p. 298)

45. There are two types of ganglion cell. **M cells** carry information about movement, location, and depth perception; **P cells** transmit signals that pertain to color, form, and texture of objects in the visual field. (p. 301)

46. Information from the right side of the visual field is processed on the left side of the brain, and information from the left side of the field is processed on the right. (p. 301)

Questions

LEVEL ONE Reviewing Facts and Terms

1. What is the role of the afferent division of the nervous system?

2. Define proprioception.

3. What are the common elements of all sensory pathways?

4. List and briefly describe the five major types of somatic receptors based on the type of stimulus to which they are most sensitive.

5. Match the brain area with the sensory information processed there:
   (a) midbrain   1. sound
   (b) cerebellum   2. odors
   (c) cerebrum   3. visual information
   (d) medulla   4. taste
   (e) none of the above

6. The receptors of each primary sensory neuron pick up information from a specific area, known as the _____.

7. The conversion of stimulus energy into a change in membrane potential is called _____. The form of energy to which a receptor responds is called its _____. The minimum stimulus required to activate a receptor is known as the _____.

8. When a sensory receptor membrane depolarizes (or hyperpolarizes in a few cases), the change in membrane potential is called the _____ or _____ potential. Is this a graded potential or an all-or-none potential?

9. Explain what is meant by adequate stimulus to a receptor.

10. The organization of sensory regions in the _____ of the brain preserves the topographical organization of receptors on the skin, eye, or other regions, so that perceptions are localized to the area where the stimulus occurred.

However, there are exceptions to this rule; in what two senses does the brain rely on the timing of receptor activation to determine the location of the initial stimulus?

11. What is lateral inhibition?

12. Define tonic receptors and list some examples. Define phasic receptors and give some examples. Which type adapts? How is adaptation useful?

13. When cardiac ischemia is perceived as coming from the neck and down the left arm, this is an example of _____ pain, which often occurs when pain arises in visceral receptors.

14. What are the five basic tastes? Describe the hypothesis that attempts to account for the existence of each taste.

15. The unit of sound wave measurement is _____, which measures the frequency of the sound waves per second. The loudness or intensity of a sound is a function of the _____ of the sound waves and is measured in _____. The range of hearing for the average human ear is from _____ to _____ [units], with the most acute hearing in the range of _____ to _____ [units].

16. Which structure of the inner ear codes sound for pitch? Define spatial coding.

17. Loud noises cause action potentials to:
    a. fire more frequently
    b. have higher amplitudes
    c. have longer refractory periods

18. Once sound waves have been transformed into electrical signals in the cochlea, sensory neurons transfer information to the _____, from which collaterals then take the information to the _____ and _____. The main auditory

pathway synapses in the _____ and _____ before finally projecting to the _____ in the _____.

19. The parts of the vestibular apparatus that tell our brain about our movements through space are the _____, which sense rotation, and the _____, which responds to linear forces. Movement is the _____ component, whereas the otolith organs also function in telling us about head position while standing, the _____ component.

20. Put these structures in the correct order for a beam of light entering the eye: (a) pupil, (b) cornea, (c) retina, (d) lens, and (e) aqueous humor.

21. The three primary colors of vision are _____, _____, and _____. White light containing these colors stimulates photoreceptors called _____. Lack of the ability to distinguish some colors is called _____.

LEVEL TWO   Reviewing Concepts

22. **Concept map:** Map the special senses. In the first row, start with these stimuli: sound, standing on the deck of a rocking boat, light, a taste, an aroma. Follow each to the location of the receptor, then continue by detailing the structure or features of each receptor. Last, name the cranial nerve(s) that convey(s) each sensation to the brain. (∞ p. 245)

23. Compare and contrast the following:
    a. the special senses with the somatic senses
    b. different types of touch-pressure receptors with respect to structure, size, and location
    c. transmission of sharp localized pain with the transmission of dull and diffuse pain (include the particular fiber types involved as well as the presence or absence of myelin in your discussion)
    d. the forms of hearing loss

24. Describe the neural pathways that link pain with emotional distress, nausea, and vomiting.

25. Trace the neural pathways involved in olfaction. Define olfactory binding protein, vomeronasal organ, pheromones, and G_{olf}.

26. Explain the process of signal transduction in the taste buds.

27. Put these structures in the order in which a sound wave would encounter them: (a) pinna, (b) cochlear duct, (c) stapes, (d) ion channels, (e) oval window, (f) hair cells/stereocilia, (g) tympanic membrane, (h) incus, (i) vestibular duct, and (j) malleus.

28. Sketch the structures and receptors of the vestibular apparatus for equilibrium. Label the components. Briefly describe how they function to notify the brain of movement.

29. Explain how accommodation by the eye occurs. What is the loss of accommodation called?

30. List four common visual problems and explain how they occur.

31. Briefly define the following terms and tell how they are related: depth of field, ciliary muscle, blind spot, pupillary reflex, photo transduction, fovea, melanin, rods, cones, visual field.

LEVEL THREE   Problem Solving

32. You are prodding your blindfolded lab partner's arm with two needle probes (with her permission). Sometimes she can tell you are using two probes, but when you probe less-sensitive areas, she thinks there is just one probe. What sense are you testing? Which receptors are being stimulated?

33. Each of us makes varying kinds of physical contact with other people each day. What parts of our brain sort out these incidental experiences and "decide" how to react? For example: brushing up against a stranger in an elevator compared with taking a patient's pulse; responding to a hug from a loved one or responding to someone grabbing your arm in a threatening way.

34. Consuming alcohol depresses the nervous system and vestibular apparatus. In a sobriety check, police officers use this information to determine if an individual is inebriated. What kinds of tests can you suggest that would show evidence of this inhibition?

35. Often, children are brought to medical attention because of speech difficulties. If you were a clinician, which sense would you test first, and why?

36. Draw a diagram to explain the processes of phototransduction. Start by showing how bleaching occurs and continue with the release of the neurotransmitter from the rod. Explain how phototransduction differs in bright and low-level light. Continue to map the pathway to the visual cortex.

37. Explain how the intensity and duration of a stimulus are coded so that the stimulus can be interpreted by the brain. (Remember, action potentials are all-or-none phenomena.)

Problem Conclusion

In this running problem, you have learned that vestibular problems can result in severe dizziness. You have also learned about some treatments that are available to alleviate Ménière's disease.

Further check your understanding of this running problem by checking your answers against those in the summary table.

	Question	Facts	Integration and Analysis
1	Subjective tinnitus occurs when the brain perceives a sound that does not exist outside the auditory system. Starting from the ear canal, name the auditory structures in which problems may arise.	The middle ear consists of the three bones that vibrate with sound. The hearing portion of the inner ear consists of the hair cells in the fluid-filled cochlea. The neural portion of the auditory system is the auditory nerve that leads to the brain.	Subjective tinnitus could arise from a problem with any of the structures named. Abnormal bone growth can affect the bones of the middle ear. Excessive fluid accumulation in the inner ear in Ménière's disease will affect the hair cells. Neural defects may cause the auditory nerve to fire spontaneously, creating the perception of sound.
2	When a person with BPV changes position, the displaced crystals float through the semicircular canals. Why would this cause dizziness?	The ends of the semicircular canals contain the sensory cristae, each consisting of a cupula with embedded hair cells. Displacement of the cupula creates a sensation of rotational movement.	If the floating crystals displace the cupula of the cristae, the brain will perceive movement that is not matched to sensory information coming from the eyes. The result is vertigo, an illusion of movement.
3	Compare the symptoms of BPV and Ménière's disease. On the basis of Anant's symptoms, which condition do you think he has?	The primary symptom of BPV is brief dizziness following a change in position. Ménière's disease combines vertigo with tinnitus and hearing loss.	Anant complains of tinnitus and hearing loss. These are not typical of BPV. In addition, his dizzy attacks last up to an hour. BPV attacks are very brief because they are caused by a change in position. It is most likely that Anant has Ménière's disease.
4	Why is limiting salt (NaCl) intake suggested as a treatment for Ménière's disease? (Hint: Remember that water follows solute.)	Ménière's disease is a condition characterized by excessive endolymph in the inner ear. Endolymph is an extracellular fluid. However, its composition resembles intracellular fluid with high concentrations of K^+ and lower concentrations of Na^+.	Because water follows solute, reducing salt intake should also reduce the amount of fluid in the extracellular compartment. This reduction may result in less fluid accumulation in the inner ear. On the other hand, the primary cation in endolymph is K^+, not Na^+, so decreasing NaCl intake may not be effective. Some diuretics remove K^+ along with water.
5	Why would severing the vestibular nerve alleviate Ménière's disease?	The vestibular nerve transmits information about balance and rotational movement from the vestibular apparatus to the brain.	Severing the vestibular nerve prevents false information about body rotation from reaching the brain, thus alleviating the vertigo of Ménière's disease.

The axons of parasympathetic neurons in the brain stem leave the central nervous system in one of two mixed cranial nerves: the glossopharyngeal nerve (IX) or the vagus nerve (X) (∞ p. 245). The major parasympathetic tract is part of the **vagus nerve,** a large nerve carrying both afferent sensory neurons from and parasympathetic fibers to most of the internal organs (Fig. 11-5 ■). *Vagotomy,* a procedure in which the vagus nerve is surgically cut, was an experimental technique used in the nineteenth and early twentieth centuries to study the impact of the autonomic nervous system on various organs. For a time, vagotomy was the preferred treatment for stomach ulcers because removal of parasympathetic innervation to the stomach decreased the secretion of stomach acid. However, this procedure had many unwanted side effects and has been abandoned in favor of drug therapies that treat the problem more specifically.

The Adrenal Medulla Secretes Catecholamines

The **adrenal medulla,** part of the adrenal gland, is a neuroendocrine gland that is closely allied to the sympathetic branch of the autonomic division. During development, the neural tissue destined to secrete catecholamines (epinephrine, norepinephrine) splits into two functional entities: the sympathetic branch of the nervous system and the adrenal medulla.

The adrenal glands are paired glands that sit on top of the kidneys (Fig. 11-6a ■). Like the pituitary gland, the adrenals are two glands of different embryological origin that fused during development (Fig. 11-6b ■). The outer portion, the adrenal cortex, is a true endocrine gland of epidermal origin (∞ p. 60). It secretes steroid hormones that regulate ion balance and intermediary metabolism. The adrenal medulla [*ad-,* upon + *renal,* kidney; *medulla,* marrow] forms the small core of the gland and is a neuroendocrine gland. It is considered a modified sympathetic ganglion, with normal preganglionic input from the spinal cord but postganglionic neurons that lack axons projecting to target cells (Fig. 11-6c ■). Instead, the neurons of the adrenal medulla synthesize and secrete neurohormones directly into the blood. In response to alarm signals from the central nervous system, the adrenal medulla releases large amounts of catecholamines, primarily epinephrine, for general distribution throughout the body.

Acetylcholine and Norepinephrine Are the Primary Autonomic Neurotransmitters

The sympathetic and parasympathetic branches of the autonomic nervous system exert their control through a combination of neurotransmitters and receptors. The two most common neurotransmitters secreted by autonomic neurons are acetylcholine (ACh) and norepi-

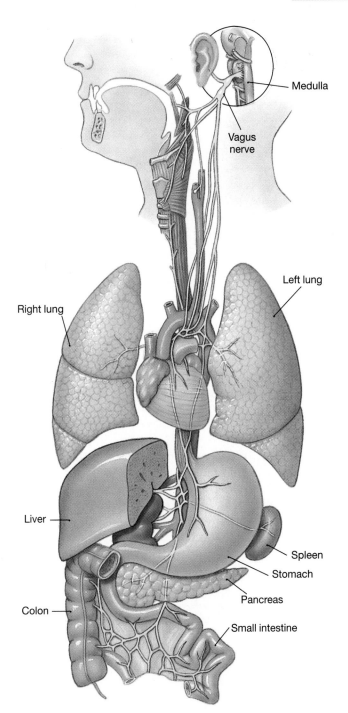

■ Figure 11-5 **The vagus nerve**

nephrine (NE). Neurons that secrete acetylcholine and receptors that bind ACh are described as **cholinergic** (Fig. 11-7a ■). Neurons that secrete norepinephrine are called **adrenergic neurons,** or, more properly, noradrenergic neurons (Fig. 11-7b ■). The adjective *adrenergic* does not have the same obvious link to its neurotransmitter as cholinergic does to acetylcholine. Instead, the adjective derives from the British name for epinephrine, *adrenaline.* In the early part of the twenti-

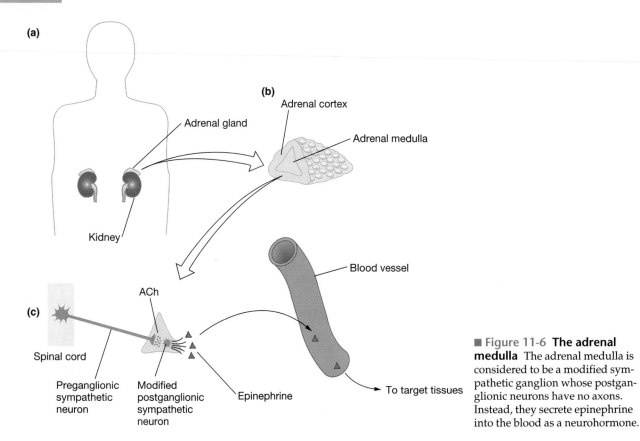

■ Figure 11-6 **The adrenal medulla** The adrenal medulla is considered to be a modified sympathetic ganglion whose postganglionic neurons have no axons. Instead, they secrete epinephrine into the blood as a neurohormone.

eth century, British researchers thought that sympathetic neurons secreted adrenaline (= epinephrine), hence the modifier *adrenergic*. Although our understanding changed, the name persisted. Whenever you see reference to "adrenergic control" of a function, you must make the connection to autonomic sympathetic neurons.

Chemical classification of autonomic neurons Autonomic neurons are often described according to the neurotransmitter they secrete.

1. *Adrenergic neurons.* Most sympathetic neurons release norepinephrine at their targets and are described as adrenergic neurons.

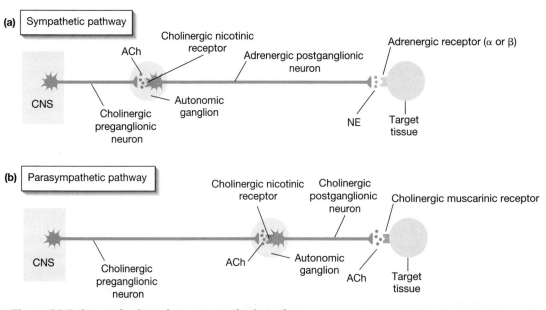

■ Figure 11-7 **Sympathetic and parasympathetic pathways** (a) Sympathetic pathways are distinguished by ganglia close to the central nervous system. They use norepinephrine (NE) as the postganglionic neurotransmitter. (b) Parasympathetic pathways generally have their ganglia closer to their target tissues. They use acetylcholine (ACh) as the postganglionic neurotransmitter.

continued from page 308

Nicotine is an addictive substance that binds to one subtype of receptor for the neurotransmitter acetylcholine (ACh). People who smoke have elevated nicotine levels in their blood.

Question 1: What is the usual response of cells that are chronically exposed to increased concentrations of a signal molecule? (Hint: This topic was discussed in Chapter 6, ∞ p. 150.)

2. *Cholinergic neurons.* All autonomic preganglionic neurons release acetylcholine onto postganglionic neurons. As a rule, parasympathetic neurons also release ACh at the neuroeffector synapse. A few sympathetic neurons, such as those that terminate on sweat glands, secrete ACh rather than norepinephrine. These neurons are therefore called *sympathetic cholinergic neurons.*

3. *Non-adrenergic, non-cholinergic neurons.* Some autonomic neurons secrete neither norepinephrine nor acetylcholine and are therefore known as *non-*

adrenergic, non-cholinergic neurons. The chemicals they use as neurotransmitters in place of or in addition to the catecholamines include substance P, somatostatin, vasoactive intestinal peptide (VIP), adenosine, nitric oxide, and ATP. These unusual neurotransmitters will be mentioned in later chapters when they play a significant role. The non-adrenergic, non-cholinergic neurons are assigned to either the sympathetic or parasympathetic branches according to their anatomical organization.

Neurotransmitter release Neurotransmitter synthesis, storage, and release were introduced in Chapter 8 (∞ p. 225). In the autonomic division, neurotransmitter synthesis takes place in the axon varicosities, and the neurotransmitter is packaged into synaptic vesicles for storage (Fig. 11-8 ■). When an action potential arrives at the varicosity, the synaptic vesicle contents are released by exocytosis.

The release of autonomic neurotransmitters is subject to modulation from a variety of sources. For example, the membranes of sympathetic varicosities contain receptors for hormones and for paracrines such as histamine. These modulators may either facilitate or inhibit

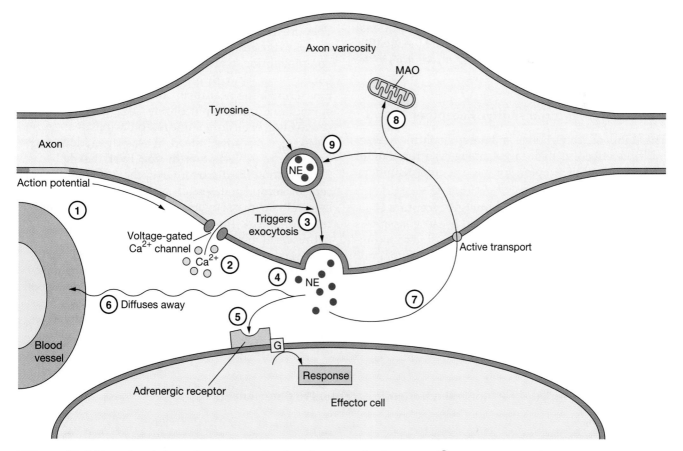

■ Figure 11-8 **Norepinephrine release at a varicosity of a sympathetic neuron** ① An action potential arriving at the varicosity opens voltage-gated Ca²⁺ channels ②, triggering exocytosis of synaptic vesicles ③. Norepinephrine (NE) in the synaptic cleft can ④ combine with an adrenergic receptor ⑤, diffuse away from the synapse ⑥, or be transported back into the axon ⑦. Once back in the axon, norepinephrine can be metabolized by monoamine oxidase (MAO) ⑧ or taken back into synaptic vesicles for re-release ⑨.

the release of norepinephrine from the varicosity. Some preganglionic neurons co-secrete polypeptides along with acetylcholine. The peptides act as neuromodulators, producing slow synaptic potentials that modify the activity of postganglionic neurons (∞ p. 227).

Termination of neurotransmitter activity   Once released in the synapse, autonomic neurotransmitters diffuse through the interstitial fluid until they encounter a receptor on the effector cell or diffuse away from the synapse. Neurotransmitter action at the target tissue is terminated when the neurotransmitter is either metabolized by enzymes in the extracellular fluid (e.g., ACh) or actively transported back into the presynaptic axon terminal (e.g., catecholamines). Once back in the axon terminal, the neurotransmitter is either repackaged into vesicles or broken down by intracellular enzymes such as *monoamine oxidase (MAO)*. Monoamine oxidase, found in mitochondria, is the main enzyme responsible for degradation of catecholamines.

The concentration of neurotransmitter in the synapse is a major factor in the control that an autonomic neuron exerts on its target: The more neurotransmitter, the longer or stronger the response. The concentration of neurotransmitter in turn depends both on its rate of release from the presynaptic neuron and its rate of breakdown or uptake from the synapse. Additional factors that influence autonomic control are the identity, affinity, and number of neurotransmitter receptors in the target tissues.

Autonomic Neurotransmitter Receptors

The ability of acetylcholine and norepinephrine to exert multiple effects at their target tissues is a function of multiple forms of receptors for these neurotransmitters.

Cholinergic receptors   **Cholinergic receptors** come in two main types: nicotinic, named because *nicotine* is an agonist, and muscarinic, for which *muscarine*, a compound found in some fungi, is an agonist. Autonomic cholinergic **nicotinic receptors** are found in autonomic ganglia, where preganglionic neurons release ACh onto postganglionic neurons. Combination of ACh with a nicotinic receptor directly opens cation channels through which both Na^+ and K^+ can pass. Sodium ion

entry into cells exceeds K^+ exit because the electrochemical gradient for Na^+ is stronger. As a result, net Na^+ entry depolarizes the postsynaptic cell.

Cholinergic **muscarinic receptors** are found at the neuroeffector junctions of the parasympathetic branch. Muscarinic receptors come in five related subtypes. They are all coupled to G proteins and are linked either to second messenger systems or to gated K^+ channels. The tissue response to activation of a muscarinic receptor varies with the receptor subtype.

Adrenergic receptors   **Adrenergic receptors,** like muscarinic cholinergic receptors, are linked to G proteins. The receptors come in two varieties: α (alpha) and β (beta). **Alpha receptors** are the most common sympathetic receptor found on target tissues of the sympathetic nervous system. They respond strongly to norepinephrine and only weakly to epinephrine (Table 11-1). Beta receptors have two main subtypes, β_1 and β_2. **β_1 receptors** are found on heart muscle and in the kidney; they respond equally strongly to norepinephrine and epinephrine. **β_2 receptors** are found in locations not innervated by sympathetic neurons. The sensitivity of β receptors is correlated to the likelihood that they will encounter different catecholamines. For example, β_2 receptors are not innervated by sympathetic neurons. They are therefore more sensitive to epinephrine, which travels to the receptors through the bloodstream, than to norepinephrine, which is primarily a neurotransmitter. On the other hand, β_1 receptors have sympathetic neurons terminating near them and are equally sensitive to epinephrine and norepinephrine (see Fig. 11-10 ■).

Activation of all adrenergic receptors initiates a second messenger cascade. Indeed, it was the action of epinephrine on β receptors in dog liver that led E. W. Sutherland to the discovery of cyclic AMP and the concept of second messenger systems as transducers of extracellular messengers. The two types of adrenergic receptors work through different second messenger pathways (Table 11-1).

Activation of alpha receptors increases intracellular Ca^{2+} levels by opening Ca^{2+} channels in the cell membrane, by releasing Ca^{2+} from intracellular stores, or by doing both. In general, activation of alpha receptors causes muscle contraction or secretion by exocytosis. Some alpha receptors in the digestive tract cause smooth muscle relaxation.

TABLE 11-1   **Sensitivity of Peripheral Adrenergic Receptors to Catecholamines**

Receptor	Found in	Sensitivity	Second Messenger
α	Most sympathetic target tissues	NE > E	Activates phospholipase C; inhibits cAMP
β_1	Heart muscle, kidney	NE = E	Activates cAMP
β_2	Certain blood vessels and smooth muscle of some organs	E > NE	Activates cAMP

NE = norepinephrine, E = epinephrine.

Catecholamine binding to beta receptors activates cyclic AMP and triggers the phosphorylation of intracellular proteins. Activation of β_1 receptors enhances cardiac muscle contraction and speeds up electrical conduction in cardiac muscle. Activation of β_2 receptors by the hormone epinephrine results in smooth muscle relaxation in many tissues.

For all adrenergic receptors, second messenger activity within the target tissue can persist for a longer time than is normally associated with the rapid action of the nervous system. These ongoing metabolic effects result from modification of existing proteins or from the synthesis of new proteins. The specific effects of catecholamines on various tissues will be discussed in subsequent chapters.

Agonists and antagonists   The study of autonomic neurotransmitters and receptors has been greatly simplified by two advances in molecular biology. The genes for many autonomic receptors and their subtypes have been cloned, allowing researchers to create mutant receptors and study their properties. In addition, researchers have discovered or synthesized a variety of agonists and antagonist molecules (Table 11-2; ∞ p. 149). Some agonists and antagonists combine with the target receptor for the neurotransmitter. Others act less directly by affecting secretion, reuptake, or degradation of the neurotransmitters. For example, cocaine blocks the reuptake of norepinephrine into adrenergic nerve terminals, thereby increasing its stimulatory effect on the target. Many new drugs have been developed from our studies of agonists and antagonists. The discovery of alpha and beta adrenergic receptors led to the development of drugs that block only one of the two receptor types. Antagonists of alpha receptors proved to have little clinical significance, but the drugs known as beta-blockers have given physicians a powerful tool for treating high blood pressure, one of the most common disorders in the United States today.

continued from page 313

Nicotinic cholinergic receptors are found on skeletal muscle fibers at the neuromuscular junction (∞ p. 319) and on the neurons of autonomic ganglia and the central nervous system. Researchers speculate that nicotinic receptors in certain regions of the brain may play a key role in nicotine addiction.

Question 2: Cholinergic receptors are classified as either nicotinic or muscarinic, on the basis of the agonist molecules that bind to them. What happens to the postsynaptic cell when nicotine rather than ACh binds to a nicotinic cholinergic receptor?

The Sympathetic and Parasympathetic Branches Maintain Homeostasis

Though often working against each other, the sympathetic and parasympathetic branches cooperate to fine-tune the internal functioning of the body. Together, the two divisions of the autonomic division display all of Walter Cannon's properties of homeostasis: (1) maintenance of the internal environment, (2) "up-down" regulation by varying tonic control, (3) antagonistic control, and (4) variable tissue responses that result from different membrane receptors (∞ p. 156).

Antagonistic control and variable response in target tissues depend on the two branches of the autonomic division and multiple receptor types for the neurotransmitters. For example, the beating of the heart is regulated by both autonomic divisions: Sympathetic innervation increases heart rate, and parasympathetic stimulation decreases it. Both branches of the autonomic system are tonically active, so heart rate can be changed by altering the relative proportion of sympathetic and parasympathetic control.

In other pathways, the membrane receptor determines the response of the target tissue. For instance, some blood vessels contain α adrenergic receptors that cause

TABLE 11-2	Agonists and Antagonists of Autonomic and Somatic Neurotransmitter Receptors		
Receptor	*Agonists*	*Antagonists*	*Other*
Cholinergic	Acetylcholine		*AChE* inhibitors:* neostigmine, parathion
			Inhibit ACh release: botulinus toxin
Muscarinic	Muscarine	Atropine, scopolamine	
Nicotinic	Nicotine	α-bungarotoxin (muscle only), tetraethylammonium (TEA) (ganglia only), curare	
Adrenergic	Norepinephrine, epinephrine		*Stimulate NE release:* ephedrine, amphetamines
			Prevent NE uptake: cocaine
Alpha	Phenylephrine	"Alpha-blockers"	
Beta	Isoproterenol	"Beta-blockers": propranolol (β_1 and β_2), metoprolol (β_1 only)	

*AChE = acetylcholinesterase.

TABLE 11-3 Comparison of Sympathetic and Parasympathetic Divisions		
	Sympathetic	*Parasympathetic*
Point of CNS origin	8th cervical, 1st thoracic to 4th lumbar segments	Midbrain, medulla, and 2nd–4th sacral segments
Location of peripheral ganglia	Primarily in paravertebral sympathetic chain; 3 outlying ganglia located alongside descending aorta	On or near target organs
Structure of region from which neurotransmitter is released	Varicosities	Boutons and varicosities
Neurotransmitter at target synapse	Norepinephrine (adrenergic neurons)	ACh (cholinergic neurons)
Inactivation of neurotransmitter at synapse	Uptake into varicosity, diffusion	Enzymatic breakdown, diffusion
Types of neurotransmitter receptors	Alpha and beta	Nicotinic and muscarinic
Ganglionic synapse	ACh on nicotinic receptor	ACh on nicotinic receptor
Neuron-target synapse	NE on α or β receptors	ACh on muscarinic receptor

Occasionally, patients suffer from primary autonomic failure when sympathetic neurons degenerate. In the face of continuing diminished sympathetic input, target tissues up-regulate (∞ p. 150), putting more receptors into the cell membrane to maximize the cell's response to available norepinephrine. This increase in receptor number leads to *denervation hypersensitivity*, a state in which the administration of exogenous adrenergic agonists causes a greater-than-expected response.

Understanding the autonomic and hormonal control of organ systems is the key to understanding maintenance of homeostasis in virtually every system of the body. We now turn to the somatic motor division, the division of the efferent nervous system that controls contraction in skeletal muscles.

THE SOMATIC MOTOR DIVISION

Somatic motor pathways differ from autonomic pathways both anatomically and functionally (Table 11-4). Somatic pathways have a single neuron that originates in the central nervous system and projects its axon to the target tissue, which is always a skeletal muscle. Unlike autonomic pathways, which may be either excitatory or inhibitory, somatic pathways are always excitatory.

Anatomy of the Somatic Division

The cell bodies of somatic motor neurons are located within the gray matter of the spinal cord or brain, with a long single axon projecting to the skeletal muscle target (Fig. 11-10 ■). These myelinated axons may be up to a meter or more in length, like the somatic motor neurons that innervate the muscles of the foot and hand. Somatic motor neurons branch close to their targets, each branch dividing into a cluster of enlarged axon terminals, or *boutons* [from the French word for "knob" or "button"], that lie on the surface of the skeletal muscle cells (Fig. 11-11 ■). This branching structure allows a single motor neuron to control many muscle fibers at one time.

The synapse of a somatic motor neuron on a muscle fiber is called the **neuromuscular junction** (Fig. 11-12 ■).

TABLE 11-4 Comparison of Somatic and Autonomic Divisions		
	Somatic	*Autonomic*
Number of neurons in efferent path	1	2
Neurotransmitter/receptor at neuron-target synapse	ACh (nicotinic)	ACh (muscarinic) or NE (α or β)
Target tissue	Skeletal muscle	Smooth and cardiac muscle; some endocrine and exocrine glands; some adipose tissue
Structure of axon terminal regions	Boutons	Boutons and varicosities
Effects on target tissue	Excitatory only: muscle contracts	Excitatory or inhibitory
Peripheral components found outside the CNS	Axons only	Preganglionic axons, ganglia, postganglionic neurons
Summary of function	Posture and movement	Visceral function, including movement in internal organs and secretion; control of metabolism

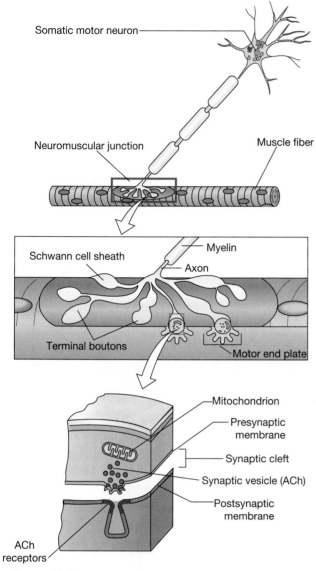

■ Figure 11-11 Anatomy of the neuromuscular junction
A somatic motor neuron branches at its distal end. Each axon terminal (bouton) lies opposite a special region of the skeletal muscle membrane, the motor end plate, that contains a high density of ACh receptors.

Schwann Cells May Do More Than Form Myelin Schwann cells are best known for creating the myelin wrapping that surrounds axons. However, the myelin sheath ends before the motor neuron branches, and extensions of the Schwann cell continue down to the axon terminals. These extensions form a thin layer covering the top of the terminal boutons (Fig. 11-11 ■). For years, it was thought that this cell layer simply provided insulation to speed up the conduction of the action potential. But it has now been discovered that these accessory cells contain neurotransmitter receptors. They respond to ACh release at the neuromuscular junction by transiently increasing their intracellular calcium concentrations. The purpose of this response in Schwann cells is still unclear, making this one of the "hot topics" in neuromuscular junction research.

The response of skeletal muscle to somatic innervation is always excitatory and results in contraction of the muscle. Relaxation in skeletal muscles results from inhibition of the motor neurons rather than inhibition of the muscle tissue itself.

✓ The Huaorani Indians of South America use blowguns to shoot darts poisoned with curare at monkeys. Curare is a plant toxin that binds to and inactivates nicotinic ACh receptors. What happens to the monkey if struck by these darts?

✓ Compare the ion channel in the motor end plate of a muscle with the ion channels along the axon of the somatic motor neuron.

✓ Patients with myasthenia gravis have a deficiency of ACh receptors on their skeletal muscles and have weak muscle function as a result. Why would administration of an anticholinesterase drug that inhibits acetylcholinesterase improve their muscle function?

The Neuromuscular Junction

As you learned in Chapter 8, action potentials arriving at the axon terminal open voltage-gated Ca^{2+} channels in the membrane. Calcium diffuses into the axon terminal down its electrochemical gradient, triggering the release of ACh-containing synaptic vesicles. Acetylcholine diffuses across the synaptic cleft and combines with nicotinic receptors on the skeletal muscle membrane (Fig. 11-12 ■). The nicotinic receptors of the neuromuscular junction are similar but not identical to the nicotinic ACh receptors found in ganglia (and those found in the central nervous system). This difference is evidenced by the fact that the snake toxin α-bungarotoxin binds to nicotinic muscle receptors but not to those in autonomic ganglia.

Nicotinic receptors are chemically gated ion channels with binding sites for ACh (Fig. 11-12b ■). When ACh

Like all synapses, this region has three components: the motor neuron's presynaptic axon terminal filled with synaptic vesicles and mitochondria, the synaptic cleft, and the postsynaptic membrane of the skeletal muscle fiber. Within the neuromuscular junction, the membrane of the muscle cell that lies opposite the axon terminal is modified into a **motor end plate,** a series of folds that look like shallow gutters. Along the upper edge of each gutter, ACh receptors cluster together into an active zone. Between the axon and the muscle, the synaptic cleft is filled with a fibrous matrix whose collagen fibers hold the axon terminal and the motor end plate in the proper alignment. The matrix also contains acetylcholinesterase (AChE), the enzyme that rapidly deactivates ACh by degrading it into acetyl and choline (∞ p. 225).

(a)

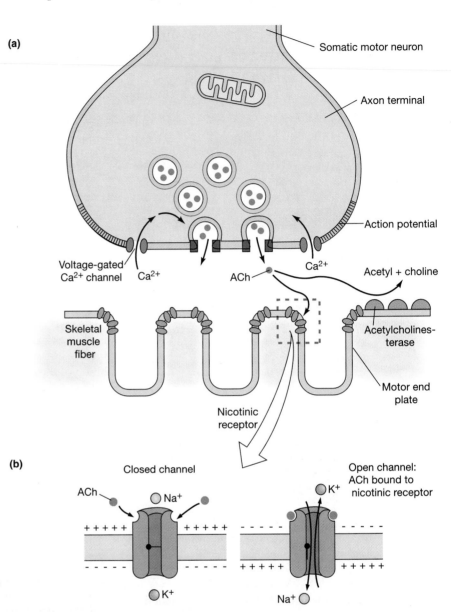

■ **Figure 11-12 Events at the neuromuscular junction** (a) An action potential opens voltage-gated Ca^{2+} channels in the axon terminal. Calcium ions enter the terminal, triggering exocytosis of synaptic vesicles. Acetylcholine (ACh) in the synaptic cleft can combine with a nicotinic receptor on the motor end plate or be metabolized by acetylcholinesterase (AChE) to acetyl and choline. (b) The nicotinic cholinergic receptor binds two ACh molecules, opening a nonspecific monovalent cation channel. Sodium ion influx exceeds K^+ efflux, and the muscle fiber depolarizes.

continued from page 317

One of the hallmarks of addiction is the withdrawal symptoms that occur when a person stops ingesting a substance. Nicotine withdrawal symptoms include lethargy, hunger, and irritability. To avoid withdrawal symptoms, people who are trying to stop smoking may use a nicotine patch. These adhesive patches deliver a small, continuous dose of nicotine that is usually enough to keep withdrawal symptoms in check. For Shanika, the first 24 hours without cigarettes have not been pleasant. Although she wears a nicotine patch, her coordination is off, and all she wants to do is sleep. She also notices that her heart is beating fast. Alarmed, she reads the nicotine package insert. The insert states that some people may experience an increased heart rate while wearing the nicotine patch. An overdose of nicotine could result in complete paralysis of the respiratory muscles that include the diaphragm and the skeletal muscles of the chest wall.

Question 4: Why might excessive levels of nicotine cause respiratory paralysis?

binds to the receptor, the channel gate opens and allows monovalent cations to flow through. Net Na^+ entry into the muscle fiber depolarizes it, triggering an action potential and resulting in contraction of the skeletal muscle cell.

"Use it or lose it" is a cliché that is very appropriate to the dynamics of muscle mass, so it is not surprising that the disruption of synaptic transmission at the neuromuscular junction has devastating effects on the entire body. Without communication between the motor neuron and the muscle, for whatever reason, the skeletal muscles for movement and posture weaken, as do the skeletal muscles used for breathing. In the severest cases, loss of respiratory function can be fatal unless the patient is placed on artificial ventilation. Myasthenia gravis, the subject of a Focus Feature in Chapter 8, is the most common disorder of the neuromuscular junction (∞ p. 229).

In the next chapter, you will learn how the electrical signals from efferent neurons initiate contraction in different types of muscle.

CHAPTER REVIEW

Chapter Summary

1. The efferent side of the peripheral nervous system is divided into **somatic motor neurons,** which control skeletal muscles, and **autonomic neurons,** which control smooth muscle, cardiac muscle, many glands, and some adipose tissue. (p. 307)

The Autonomic Division

2. The autonomic division is subdivided into a **sympathetic division** and a **parasympathetic division.** (p. 308)

3. The maintenance of homeostasis within the body is a balance between sympathetic and parasympathetic control. (p. 308)

4. An autonomic pathway is composed of a **preganglionic neuron** from the central nervous system that synapses with a **postganglionic neuron** in an **autonomic ganglion.** (p. 308)

5. A single preganglionic neuron typically synapses on eight or nine postganglionic neurons. This divergence of pathways allows rapid control over multiple targets. (p. 309)

6. The synapse between a postganglionic autonomic neuron and its target cells is called the **neuroeffector junction.** (p. 309)

7. Autonomic axons end with **varicosities** from which neurotransmitter is released. (p. 309)

8. Most sympathetic preganglionic neurons originate in the thoracic and lumbar regions of the spinal cord. Most sympathetic ganglia lie close to the spinal cord or along the descending aorta. (p. 309)

9. The cell bodies of parasympathetic preganglionic neurons are found either in the brain stem or in the sacral region of the spinal cord. Parasympathetic ganglia are located on or near their target organs. (p. 309)

10. Autonomic ganglia are capable of modulating information passing through them with the assistance of **intrinsic neurons** that lie completely within the ganglia. (p. 309)

11. The **adrenal medulla** is a neurosecretory endocrine gland that is controlled by sympathetic preganglionic neurons; it secretes catecholamines. (p. 310)

12. The two most common autonomic neurotransmitters are acetylcholine and norepinephrine. As a rule, sympathetic postganglionic neurons secrete norepinephrine, and postganglionic parasympathetic neurons secrete acetylcholine. All preganglionic neurons secrete ACh. (p. 310)

13. The level of nervous control that an autonomic neuron exerts on its target depends on the concentration of neurotransmitter in the synapse as well as the identity and number of neurotransmitter receptors in the target tissues. (p. 313)

14. **Cholinergic receptors** come in two main types. **Nicotinic receptors** are found in autonomic ganglia. Binding of ACh to a nicotinic receptor opens cation channels and depolarizes the postsynaptic cell. **Muscarinic receptors** are found at the neuroeffector junctions of the parasympathetic branch. They are coupled to G proteins, and the tissue response varies with the receptor subtype. (p. 314)

15. **Adrenergic receptors** are G protein–linked receptors whose activation initiates a second messenger cascade. **Alpha receptors** are found on most target tissues of the sympathetic nervous system. β_1 **receptors** are found on heart muscle and in the kidney. β_2 **receptors** are found in locations not innervated by sympathetic neurons. (p. 314)

16. The two autonomic divisions demonstrate all properties of homeostasis: maintenance of the internal environment, "up-down" regulation by varying tonic control, antagonistic control, and variable tissue responses that result from different membrane receptors. (p. 315)

17. Central nervous system control of the autonomic division is closely linked to centers in the hypothalamus, pons, and medulla of the brain. Autonomic output can be influenced by the cerebral cortex and limbic system. Some autonomic reflexes are spinal reflexes, capable of taking place without input from the brain, although many of these spinal reflexes are modulated by input from higher control centers. (p. 316)

The Somatic Motor Division

18. Somatic motor pathways have a single neuron that originates in the central nervous system and projects its axon to a skeletal muscle. Somatic motor pathways are always excitatory and result in contraction of the muscle. (p. 318)

19. Somatic motor neurons branch at their distal end so that a single motor neuron controls many muscle fibers at one time. (p. 318)

20. The synapse of a somatic motor neuron on a muscle fiber is called the **neuromuscular junction.** The membrane of the muscle cell at the synapse is modified into a **motor end plate** that contains a high concentration of ACh receptors. (p. 318)

21. Acetylcholine released from motor neurons combines with nicotinic receptors on the skeletal muscle membrane. Binding of ACh to the ACh receptor opens ion channels and allows net Na^+ entry into the muscle fiber, depolarizing it. Acetylcholine in the synapse is broken down by the enzyme acetylcholinesterase. (p. 319)

Questions

LEVEL ONE Reviewing Facts and Terms

1. What are the two divisions of the efferent part of the peripheral nervous system? What effectors does each control?

2. The autonomic nervous system is sometimes called the _____ nervous system. Why is this an appropriate name? List some of the functions controlled by the autonomic nervous system.

3. What are the two divisions of the autonomic nervous system? How are these divisions distinguished anatomically and physiologically?

4. Which neurosecretory endocrine gland is closely allied to the sympathetic branch of the autonomic division?

5. Neurons that secrete acetylcholine are described as _____, whereas those that secrete norepinephrine are called _____ or _____ neurons.

6. List four things that can happen to autonomic neurotransmitters after they are released into the synapse.

7. The main enzyme responsible for catecholamine degradation is _____, abbreviated _____.

8. Somatic motor pathways
   a. are excitatory or inhibitory?
   b. are composed of a single neuron or a preganglionic and a postganglionic neuron?
   c. synapse with glands or with smooth, cardiac, or skeletal muscle?

9. What is acetylcholinesterase? Describe its action.

10. What kind of receptors are found on the postsynaptic cell in the neuromuscular junction?

LEVEL TWO Reviewing Concepts

11. **Concept map:** Make a map comparing the somatic, sympathetic, and parasympathetic divisions. Use the following terms (a term may be used more than once): acetylcholine, adipose tissue, alpha receptor, autonomic division, beta receptor, ganglion, cardiac muscle, cholinergic receptor, efferent division, endocrine gland, exocrine gland, muscarinic receptor, nicotinic receptor, norepinephrine, one-neuron pathway, smooth muscle, parasympathetic division, skeletal muscle, somatic motor division, sympathetic division, two-neuron pathway.

12. What is the advantage of divergence of neural pathways in the autonomic nervous system?

13. Compare and contrast the following:
   a. the neuroeffector junctions of the autonomic nervous system with the neuromuscular junction of the somatic nervous system
   b. alpha, beta, muscarinic, and nicotinic receptors in the efferent side of the nervous system (describe where each is found, and the ligands that bind to them)

14. What are the autonomic ganglia? Why are they important?

15. How are the adrenal glands similar to the posterior pituitary gland?

16. What is the difference between a bouton and a varicosity?

LEVEL THREE Problem Solving

18. Little Danny was riding his tricycle in the driveway when a horrible thing happened. The parking brake on his Mom's car failed, and it rolled backward, flipping the tricycle and pinning Danny underneath, with the car on the handlebars, anchoring it down firmly. Sandy, his mother, was close by and was able to lift the car away so that she could get Danny out. Everybody was OK. Sandy weighs about 125 pounds. Is there a physiological explanation for what happened?

Problem Conclusion

Nicotine is an agonist for nicotinic ACh receptors, but with chronic exposure, nicotine inactivates the ACh receptors. With fewer active receptors, target cells up-regulate their receptor number. When smokers stop smoking, the removal of nicotine from the system allows all receptors to return to an active state. A larger than usual number of active receptors is probably responsible for some of the withdrawal symptoms noted by people who are trying to stop smoking. The nicotine patch allows the former smoker to gradually decrease nicotine levels in the body, preventing withdrawal symptoms during the time the cells are down-regulating their receptors back to the normal number.

In this running problem, you learned that nicotine is an addictive chemical that is an agonist of ACh. You also learned how cells respond to long-term exposure to nicotine.

Further check your understanding of this running problem by checking your answers against those in the summary table.

	Question	Facts	Integration and Analysis
1	What is the usual response of cells that are chronically exposed to increased concentrations of a signal molecule? (Hint: This topic was discussed in Chapter 6.)	A cell that is exposed to increased concentrations of a signal molecule will decrease its receptors for that molecule (∞ p. 150).	Down-regulation of receptors allows a cell to respond normally even if the concentration of ligand is elevated.
2	What happens to the post-synaptic cell when nicotine rather than ACh binds to a nicotinic cholinergic receptor?	Nicotine is an agonist of ACh. Agonists mimic the activity of a ligand.	Nicotine binding to a nicotinic cholinergic receptor will cause ion channels in the postsynaptic cell to open, and the cell will depolarize. This is the same effect that ACh binding creates (∞ p. 315).
3	Although nicotine has been shown in short-term studies to be an agonist at nicotinic receptors, chronic exposure to nicotine causes up-regulation of the receptors. Speculate on why up-regulation might occur.	Chronic exposure to an agonist usually causes down-regulation. Chronic exposure to an antagonist usually causes up-regulation.	Although nicotine is a short-term agonist, it appears to be acting as an antagonist during long-term exposure. Researchers have found that nicotinic ACh receptors have several functional states: unbound and inactive, bound to ligand and active, and bound to ligand but inactivated. Long-term exposure to nicotine apparently places many nicotinic ACh receptors into the third state.
4	Why might excessive levels of nicotine cause respiratory paralysis?	Nicotinic receptors are found at the neuro-muscular junction that controls skeletal muscle contraction. The diaphragm and chest wall muscles that regulate breathing are skeletal muscles.	The nicotinic receptors of the neuromuscular junction are not as sensitive to nicotine as are those of the CNS and autonomic ganglia. However, excessively high amounts of nicotine will activate the nicotinic cholinergic receptors of the motor end plate, causing the muscle fiber to depolarize and contract. The continued presence of nicotine keeps the ion channels open, and the muscle remains depolarized. In this state, it is unable to contract again, resulting in paralysis.

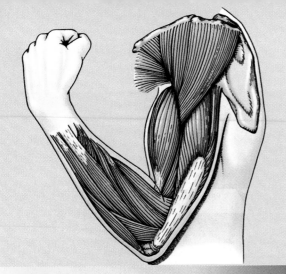

12

Muscles

It was his first time to be the starting pitcher. As he ran from the bullpen onto the field, his heart was pounding and his stomach felt as if it were tied in knots. He stepped onto the mound and gathered his thoughts before throwing the first practice pitch. Gradually, as he went through the familiar routine of throwing and catching the baseball, his heart slowed and his stomach relaxed. It was going to be a good game.

The pitcher's pounding heart, queasy stomach, and movements as he runs and throws the ball are all the results of muscle contraction. Our muscles have two common functions: (1) to generate motion and (2) to generate force. Our skeletal muscles also generate heat and contribute significantly to the homeostasis of body temperature. When homeostasis is threatened by cold conditions, our brain may direct our muscles to shiver, creating additional heat.

Three types of muscle tissue occur in the human body: skeletal muscle, cardiac muscle, and smooth muscle. Most **skeletal muscles** are attached to the bones of the skeleton, enabling them to control body movement. **Cardiac muscle** is found only in the heart. Skeletal and cardiac muscles are classified as **striated muscles** [*stria*, groove] because of their alternating light and dark bands under the light microscope (Fig. 12-1a, b ■).

Smooth muscle is the primary muscle of internal organs and tubes such as the stomach, urinary bladder, and blood

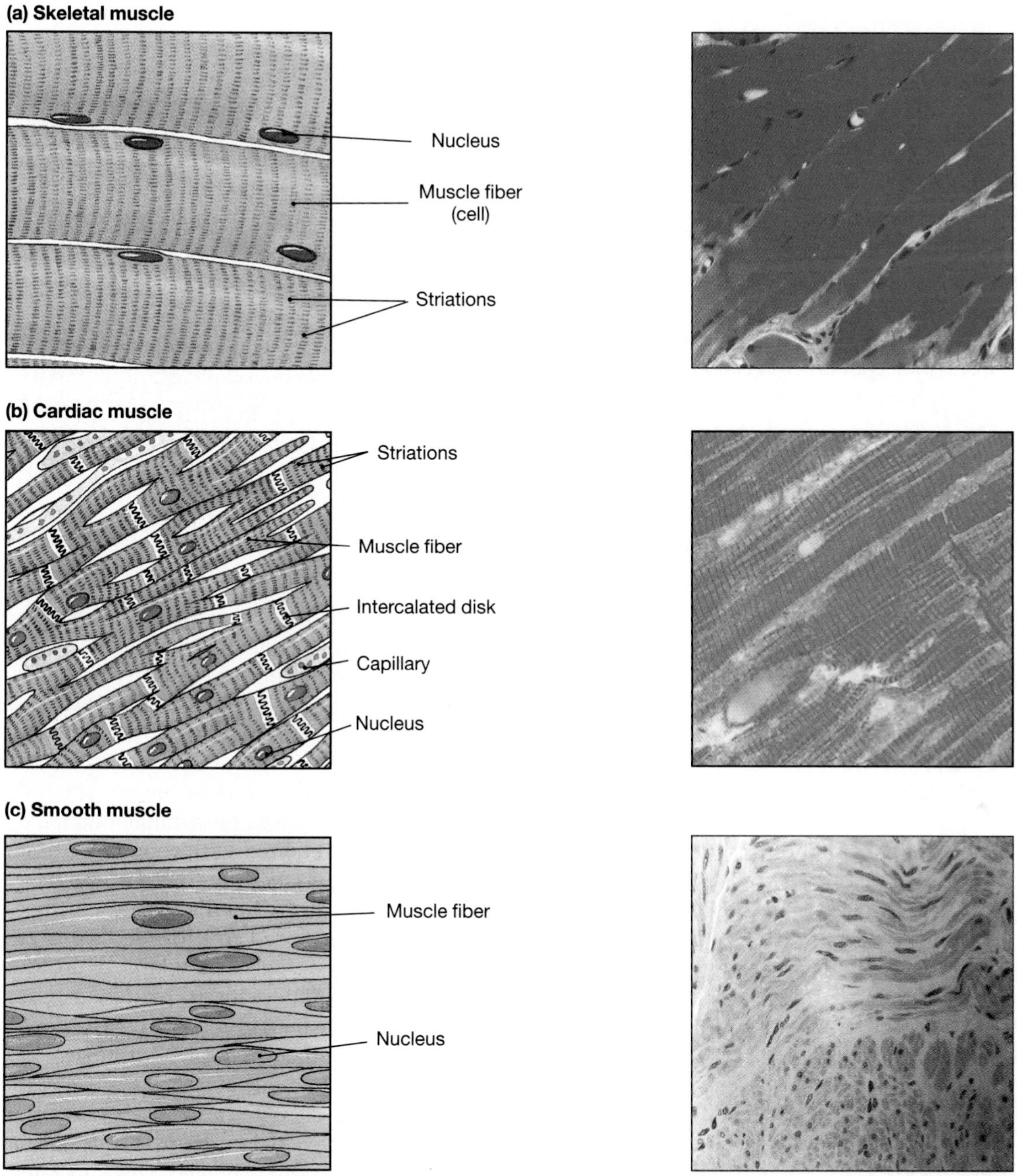

(a) Skeletal muscle

Nucleus

Muscle fiber
(cell)

Striations

(b) Cardiac muscle

Striations

Muscle fiber

Intercalated disk

Capillary

Nucleus

(c) Smooth muscle

Muscle fiber

Nucleus

■ Figure 12-1 **Three types of muscles** (a) Skeletal muscle fibers are large multinucleate cells that appear striated under the microscope. (b) Cardiac muscle fibers are striated uninucleate cells that are branched. They connect to each other through specialized junctions known as intercalated disks. (c) Smooth muscle fibers are smaller, uninucleate cells without obvious banding patterns.

vessels. Smooth muscle has a more homogeneous appearance under the microscope, without obvious bands (Fig. 12-1c ■). Its lack of banding results from the less organized arrangement of the contractile fibers within the muscle cells.

Skeletal muscles are often described as the voluntary muscles, and smooth and cardiac muscle are described as involuntary. Nevertheless, as we discussed in the previous chapter, this is not a precise classification. Skeletal muscles can contract without conscious direction, and we can learn a certain degree of conscious control over some smooth and cardiac muscle.

Skeletal muscles are unique in that they contract only in response to a signal from a somatic motor neuron. They cannot initiate their own contraction, nor is their contraction influenced directly by hormones. In contrast, cardiac and smooth muscle have multiple levels of control. Although their primary extrinsic control arises through autonomic innervation, some types of smooth and cardiac muscle can contract spontaneously, without input from the central nervous system. In addition, the activity of cardiac and some smooth muscle is subject to modulation by the endocrine system.

Problem

Periodic Paralysis

This morning, Paul Leong, six years old, gave his mother the fright of her life. One minute he was happily playing in the backyard with his new beagle puppy. The next minute, after sitting down to rest, he could not move his legs. In answer to his screams, his mother came running and found her little boy unable to walk. Panic-stricken, she scooped him up, brought him in the house, and dialed 9-1-1. But as she hung up the phone and prepared to wait for the paramedics, Paul got to his feet and walked over to her. "I'm OK now, Mom. I'm going outside."

continued on page 330

In this chapter, we first examine skeletal muscle in detail. Many properties of skeletal muscle are shared by smooth and cardiac muscle, whose descriptions conclude the chapter.

SKELETAL MUSCLE

Skeletal muscles make up the bulk of muscle in the body, about 40% of total body weight. Skeletal muscles are responsible for positioning and movement of the skeleton, as their name suggests. They are usually attached to bones by collagenous **tendons** (∞ p. 63). The **origin** of a muscle is the end of the muscle that is attached closest to the trunk or to the more stationary bone. The **insertion** of the muscle is the more distal or mobile attachment. When the bones attached to a muscle are connected by a flexible **joint,** contraction of the muscle results in movement of the skeleton. If the centers of the connected bones are brought closer together when the muscle contracts, the muscle is called a **flexor.** If the bones move away from each other when the muscle contracts, the muscle is called an **extensor.** Most joints in the body have both flexor and extensor muscles, since a contracting muscle can pull a bone in one

direction but cannot push it back. Flexor-extensor pairs are called **antagonistic muscle groups** because they exert opposite effects. Figure 12-2 ■ shows a pair of antagonistic muscles in the arm: the biceps muscle, which acts as the flexor, and the triceps muscle, which acts as the extensor. When the biceps contracts, the hand and forearm move toward the body. Contraction of the triceps extends the flexed arm. In each case, when one muscle contracts and shortens, the antagonistic muscle must relax and lengthen.

✓ Identify as many pairs of antagonistic muscle groups in the body as you can. If you cannot name them, point out the probable location of the flexor and extensor of each group.

Skeletal Muscles Are Composed of Muscle Fibers

Muscles function together as a unit. Just as a nerve is a collection of neurons, a skeletal **muscle** is a collection of muscle cells, or **muscle fibers** (see Anatomy Summary, Fig. 12-3 ■). A small muscle, such as those attached to the bones of the middle ear, may have several hundred muscle fibers. Larger muscles, such as the quadriceps of the thigh, may have several thousand fibers. The individual fibers of a skeletal muscle are arranged with their long axes in parallel and bound with connective tissue (Fig. 12-3a ■). Collagen, elastic fibers, nerves, and blood vessels are found between the bundles of muscle fibers. The entire muscle is enclosed in a connective tissue sheath that is continuous with the connective tissue around the muscle fibers and with the tendons holding the muscle to underlying bones.

Each skeletal muscle fiber is a long, cylindrical cell with up to several hundred nuclei on the surface of the fiber (Fig. 12-3b ■). Muscle fibers are the largest cells in the body, created by the fusion of many individual embryonic muscle cells. Muscle physiologists, like neurophysiologists, have adopted a specialized vocabulary

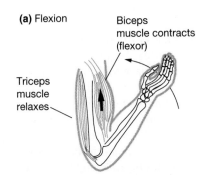

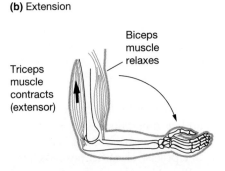

(a) Flexion — Biceps muscle contracts (flexor) — Triceps muscle relaxes

(b) Extension — Biceps muscle relaxes — Triceps muscle contracts (extensor)

■ **Figure 12-2 Antagonistic muscle groups** Muscles in the body are often arranged in pairs or groups to provide movement in several directions. A muscle can shorten and thereby pull on a bone, but cannot push a bone away.

(Table 12-1). The cell membrane of a muscle fiber is called the **sarcolemma** [*sarkos*, flesh + *lemma*, shell], and the cytoplasm is called the **sarcoplasm.** A muscle fiber contains little cytosol. The main intracellular structures are the **myofibrils,** bundles of contractile and elastic proteins that carry out the work of contraction.

Skeletal muscle fibers contain an extensive **sarcoplasmic reticulum** (SR), a form of modified endoplasmic reticulum that wraps around each myofibril like a piece of lace (Figs. 12-3b, 12-4 ■). The role of the sarcoplasmic reticulum is to concentrate and sequester calcium ions [*sequestrare*, to put in the hands of a trustee]. Closely associated with the sarcoplasmic reticulum is a branching network of **transverse tubules,** also known as **t-tubules.** The membranes of the t-tubules are a continuation of the surface membrane of the muscle fiber, making the lumen of the t-tubules continuous with the extracellular fluid. To understand how this network of tubules in the heart of the muscle fiber communicates with the outside, take a lump of soft clay and stick your finger into the middle of it several times. Notice how the outside surface of the clay (analogous to the cell membrane of the muscle fiber) is now continuous with the sides of the holes that you have poked in the clay (the membrane of the t-tubules). T-tubules allow action potentials that originate on the cell surface at the neuromuscular junction to move rapidly into the interior of the fiber. Without t-tubules, the action potential could reach the center of the fiber only by the diffusion of positive charge through the cytosol, a slower process that would delay the response time of the muscle fiber.

The cytosol between the myofibrils contains many glycogen granules and mitochondria. Glycogen, the storage form of glucose found in animals, is a reserve source of energy. Mitochondria provide much of the ATP for muscle contraction through oxidative phosphorylation of glucose and other biomolecules.

Myofibrils are the contractile structures of a muscle fiber

Each muscle fiber contains a thousand or more myofibrils that occupy most of the intracellular volume, leaving little space for cytosol and organelles (Fig. 12-3b ■). Each myofibril is composed of several types of proteins: the contractile proteins *myosin* and *actin*, the regulatory proteins *tropomyosin* and *troponin*, and the giant accessory proteins *titin* and *nebulin*.

Myosin [*myo-*, muscle] is the protein that makes up the thick filaments of the myofibril. Various *isoforms* [*iso-*,

equal], or different types, of myosin occur in different types of muscle. Each myosin molecule is composed of two heavy protein chains that intertwine to form a long coiled tail and a pair of tadpolelike heads (Fig. 12-3e ■). Two lightweight protein chains are associated with the heavy chain of each head. In skeletal muscle, about 250 myosin molecules join to create a **thick filament.** The thick filament is arranged so that the myosin heads are clustered at the ends, and the central region of the filament is a bundle of myosin tails. The rodlike portion of the thick filament is stiff, but the protruding myosin heads have an elastic hinge region where the heads join the rods. This hinge region allows the heads to swivel around their point of attachment.

Actin [*actum*, to do] is the protein that makes up the thin filaments of the muscle fiber. One actin molecule is a globular protein (*G-actin*), represented in Figure 12-3f ■ by a round ball. Usually, multiple globular actin molecules polymerize to form long chains or filaments (*F-actin*). In skeletal muscle, two F-actin polymers twist together like a double strand of beads, creating the **thin filaments** of the myofibril.

Most of the time, the parallel thick and thin filaments of the myofibril are connected by **crossbridges** that span the space between the filaments. The crossbridges are the myosin heads that loosely bind to the actin filaments. Each G-actin molecule has a single binding site for a myosin head.

Under a light microscope, the arrangement of thick and thin filaments in a myofibril creates a repeating pattern of alternating light and dark bands (Figs. 12-1a, 12-3c ■). One repeat of the pattern forms a **sarcomere** [*sarkos*, flesh + *-mere*, a unit or segment], which has the following elements:

- **Z disks:** These zigzag structures are made of proteins that serve as the attachment site for the thin filaments. One sarcomere is composed of two Z disks and the filaments found between them. The abbreviation *Z* comes from *zwischen*, the German word for "between."
- **I band:** These are the lightest color bands of the sarcomere and represent a region occupied only by thin filaments. The abbreviation *I* comes from *isotropic*, a description from early microscopists meaning that this region reflects light uniformly under a polarizing microscope. A Z disk runs through the middle of an I band, so each half of the I band belongs to a different sarcomere.
- **A band:** This is the darkest of the bands in a sarcomere and encompasses the entire length of a thick filament. At the outer edges of the A band, the thick and thin filaments overlap, whereas the center of the A band is occupied by thick filaments only. The abbreviation *A* comes from *anisotropic* [*an-*, not], meaning that the protein fibers in this region scatter light unevenly.

TABLE 12-1 Muscle Terminology

General Term	Muscle Equivalent
Muscle cell	Muscle fiber
Cell membrane	Sarcolemma
Cytoplasm	Sarcoplasm
Modified endoplasmic reticulum	Sarcoplasmic reticulum

Anatomy Summary **Skeletal Muscle**

■ Figure 12-3

(a)

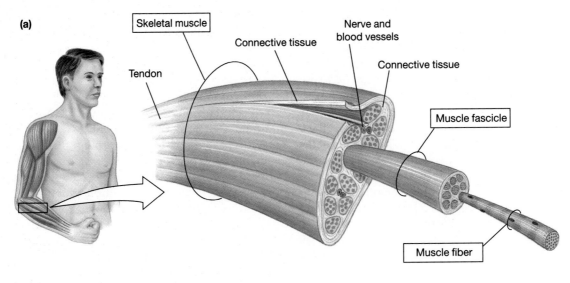

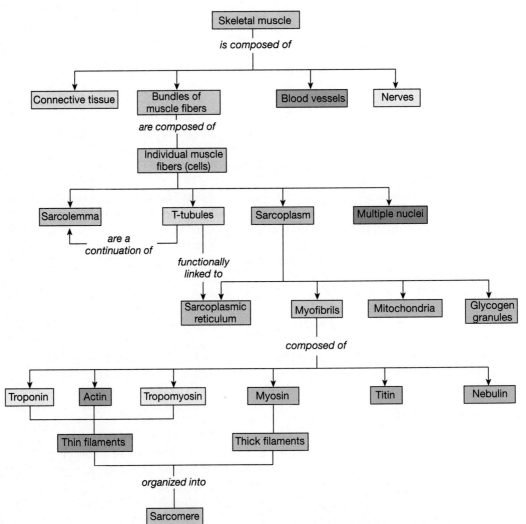

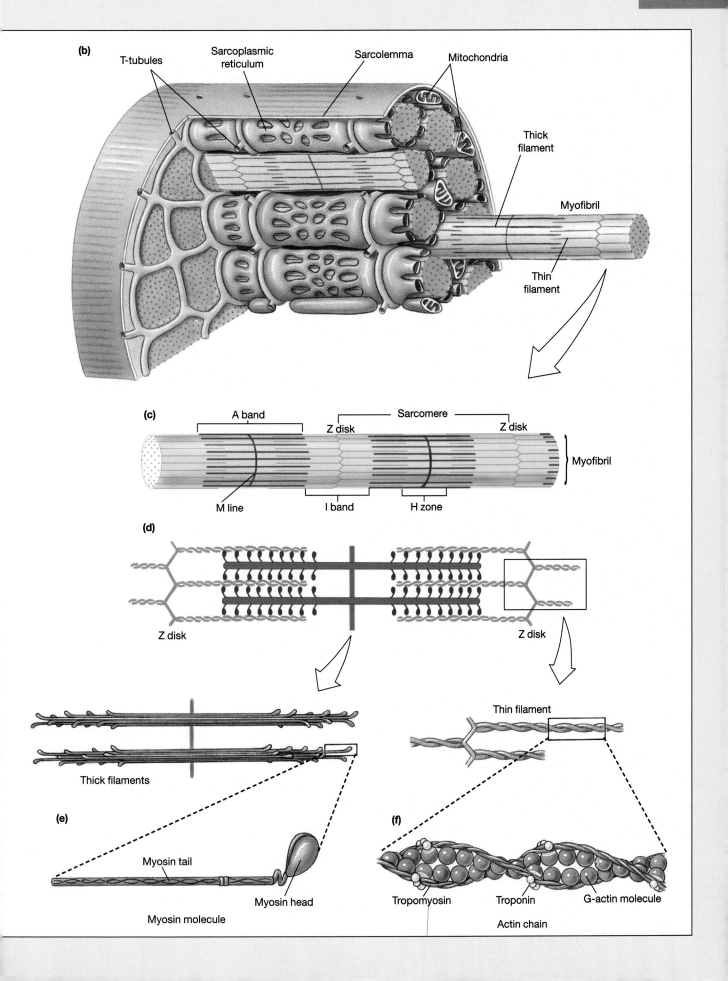

(b)

T-tubules
Sarcoplasmic reticulum
Sarcolemma
Mitochondria
Thick filament
Myofibril
Thin filament

(c)

A band
Sarcomere
Z disk
Z disk
Myofibril
M line
I band
H zone

(d)

Z disk
Z disk

Thick filaments

(e)

Myosin tail
Myosin head
Myosin molecule

Thin filament

(f)

Tropomyosin
Troponin
G-actin molecule
Actin chain

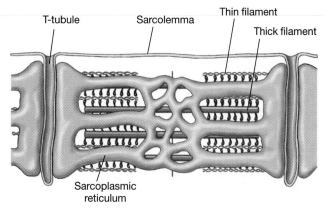

■ Figure 12-4 T-tubules and the sarcoplasmic reticulum The sarcoplasmic reticulum wraps around each myofibril. The t-tubule system is closely associated with the sarcoplasmic reticulum.

■ **H zone:** This central region of the A band is lighter than the outer edges of the A band because the H zone is occupied by thick filaments only. The *H* comes from *helles*, the German word for "clear."

■ **M line:** This band represents the attachment site for the thick filaments, equivalent to the Z disk for the thin filaments (see Fig. 12-5b ■). One A band is divided in half by an M line. *M* is the abbreviation for *mittel*, the German word for "middle."

In three-dimensional array, the actin and myosin molecules form a lattice of parallel, overlapping thin and thick filaments (Fig. 12-5 ■). When viewed end-on, each thin filament is surrounded by three thick filaments, and six thin filaments encircle each thick filament. The proper alignment of filaments within a sarcomere is ensured by two types of proteins, titin and nebulin (Fig. 12-6 ■). **Titin** is a huge elastic molecule; it is the largest known protein and has more than 25,000 amino acids. A single titin molecule stretches from one Z disk to the next M line. To get an idea of the immense size of titin, imagine that one titin molecule is an 8-foot-long piece of the very thick rope used to tie ships to a wharf. By comparison, a single actin molecule would be about the length and weight of a single eyelash.

Titin has two functions: It stabilizes the position of the contractile filaments, and its elasticity returns stretched muscles to their resting length. Titin is helped by **nebulin,** an inelastic giant protein that lies alongside thin filaments and attaches to the Z disk. Nebulin helps align the actin filaments of the sarcomere.

In the next section we will look at how this arrangement of contractile filaments results in muscle contraction.

✓ Why are the ends of the A band the darkest region of the sarcomere when viewed under the light microscope?

✓ What is the function of t-tubules?

✓ Why are skeletal muscles called striated muscle?

Muscles Shorten When They Contract

The contraction of skeletal muscle fibers is a remarkable process that allows us to create force to move or resist a load. In muscle physiology, the force created by the contracting muscle is called the muscle **tension;** the **load** is a weight or force that opposes contraction of a muscle. The creation of tension in a muscle is an active process that requires energy input from ATP.

In previous centuries, scientists observed that when muscles move a load, they shorten. This observation led to early theories of contraction, which proposed that muscles were made of molecules that curled up and shortened when active, then relaxed and stretched at rest, like elastic in reverse. The theory received support when myosin was found to be a helical molecule that shortened upon heating (the same reason meat shrinks when you cook it). However, in 1954, Sir Andrew Huxley and Rolf Niedeigerke discovered that the length of the A band of a myofibril remains constant during contraction. Because the A band represents the myosin filament, they realized that shortening of the myosin molecule could not be responsible for contraction. Subsequently, they proposed an alternative model, the **sliding filament theory of contraction.** In this model, overlapping muscle fibers of fixed length (the thick and thin filaments) slide past each other in an energy-requiring process, resulting in muscle contraction. The sliding filament theory takes into account the observation that when a muscle contracts, it does not always shorten. A muscle can contract and create force without creating movement. For example, if you push on a wall, you are creating tension in many muscles of your body without shortening the muscles and without moving the wall. According to the sliding filament theory, tension generated in a muscle fiber is directly proportional to interaction between the thick and thin filaments.

Sliding filament theory of contraction   If you examined a myofibril at its resting length, you would see that within each sarcomere, the ends of the thick and thin fil-

continued from page 326

Paul had experienced mild attacks of muscle weakness in his legs before, usually in the morning. Twice, the weakness had come on after exposure to cold. Each attack had disappeared within minutes, and Paul seemed to suffer no lasting effects. On the advice of Paul's family doctor, Mrs. Leong takes her son to see a specialist in muscle disorders who diagnoses a condition called periodic paralysis. This condition is caused by a genetic defect in the Na^+ channels in the membranes of his skeletal muscle fibers. Under certain conditions, the defective Na^+ channels remain open and continuously admit Na^+ into the interior of the muscle cells.

Question 1: What effect does the continuous influx of Na^+ have on the membrane potential of Paul's muscle fibers?

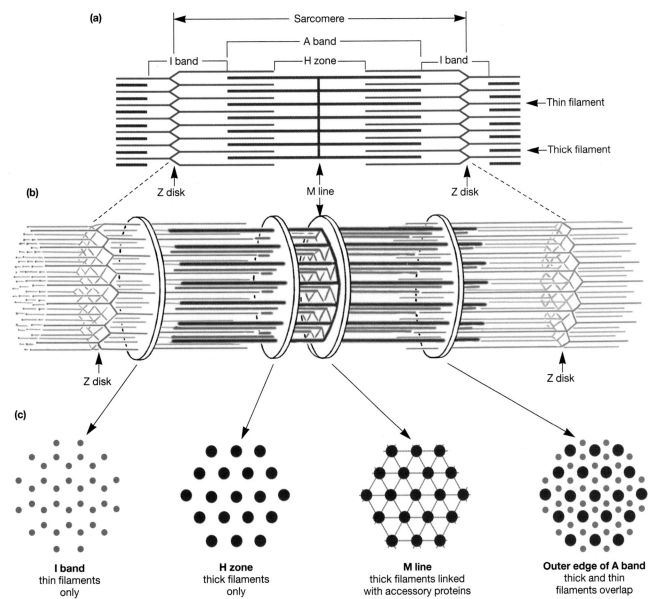

■ **Figure 12-5** **The two- and three-dimensional organization of a sarcomere** (a) A sarcomere runs from Z disk to Z disk. (b) Within a myofibril, the thick and thin filaments overlap in a three-dimensional array. (c) The thick and thin filaments are arranged so that when they overlap, each thick filament is surrounded by six thin filaments (as seen in the A band) and each thin filament is surrounded by three thick filaments. The M line is the region where the thick filaments are linked together. The Z disk (not shown) has a similar arrangement of accessory proteins that link the thin filaments.

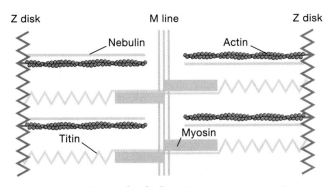

■ **Figure 12-6** **Titin and nebulin** Titin is a giant protein that spans the distance from one Z disk to the next M line. Titin has an elastic component that helps return a stretched sarcomere to its resting length. It also stabilizes the arrangement of the filaments within the sarcomere, aided by the nonelastic giant protein nebulin.

aments overlap slightly. As the muscle contracts, the thick and thin filaments slide past each other, moving the Z disks of the sarcomere closer together. This phenomenon can be seen in light micrographs of relaxed and contracted muscle (Fig. 12-7 ■). In the relaxed state, a sarcomere has a large I band (thin filaments only) and an A band whose length is the length of the thick filament. As contraction occurs, the sarcomere shortens. The two Z disks at each end move closer together while the I band and H zone, regions where actin and myosin do not overlap in resting muscle, almost disappear. Despite the shortening of the sarcomere, the length of the A band remains constant. These changes are consistent with the sliding of thin actin filaments along the thick myosin filaments as they move toward the M line in the center of

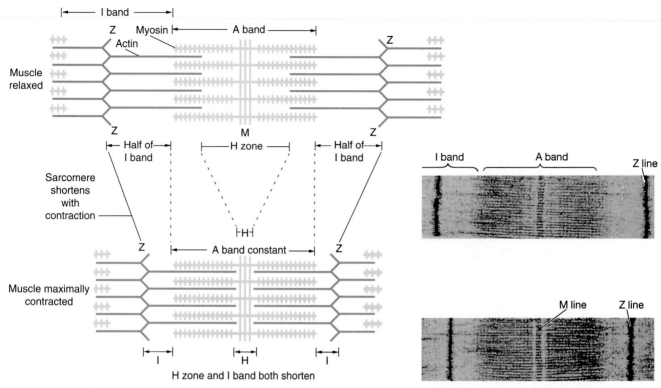

■ Figure 12-7 **Changes in a sarcomere during contraction** When a sarcomere contracts, the thick and thin filaments do not change in length. The thin actin filaments slide over the thick myosin filaments, moving toward the M line in the center of the sarcomere. The A band does not change in length, but both the H zone and the I band get shorter as the filaments overlap.

the sarcomere. It is from this process that the sliding filament theory of contraction derives its name.

The force that pushes the actin filament is the movement of myosin crossbridges that link actin and myosin. Myosin is a *motor protein* that converts the chemical bond energy of ATP into the mechanical energy of motion. Each myosin molecule acts as an ATPase, binding ATP and hydrolyzing it to ADP and inorganic phosphate. The energy released by ATP changes the angle between the head of the myosin molecule and the long axis of the myosin filament. This rotation of the myosin head on its flexible neck creates the **power stroke** that is the basis for muscle contraction.

The actin filaments serve as the "road" along which the myosin heads walk. Each myosin head has two binding sites on it: one for an ATP molecule and one that binds to actin. During the power stroke, the movement of the myosin head pushes the actin filaments toward the center of the sarcomere. At the end of a power stroke, the myosin releases the actin, swings back and binds to a new actin molecule, ready to start another contractile cycle. This process repeats many times as a muscle fiber contracts. The myosin heads repeatedly bind, swing, and release the actin molecules as they push the thin filaments toward the M line in the center of the sarcomere.

Figure 12-8 ■ shows the molecular events of a contractile cycle in detail. The starting point for this cycle is chosen arbitrarily.

1. Our contractile cycle starts with the myosin head of the crossbridge tightly bound to a G-actin molecule. In this state, no nucleotide (ATP or ADP) occupies the second binding site on the myosin head. This tight binding is known as the **rigor state** [*rigere*, to be stiff]. In living muscle, the rigor state occurs for only a very brief period.

The *In Vitro* Motility Assay One big step forward in understanding the power stroke of myosin was the development of the *in vitro* motility assay in the 1980s. In this assay, isolated myosin molecules are randomly bonded to a specially coated glass coverslip. A fluorescently labeled actin molecule is placed on top of the myosin molecules. With ATP as a source of energy, the myosin heads bind to the actin and move it across the coverslip, marked by a fluorescent trail as it goes. In even more ingenious experiments, developed in 1995, a single myosin molecule is bound to a tiny bead that elevates it above the surface of the cover slip. An actin molecule is placed on top of the myosin molecule, like the balancing pole of a tightrope walker. As the myosin "motor" moves the actin molecule, lasers measure the nanometer movements and piconewton forces created with each cycle of the myosin head. Because of this technique, researchers can now measure the mechanical work being done by a single myosin molecule!

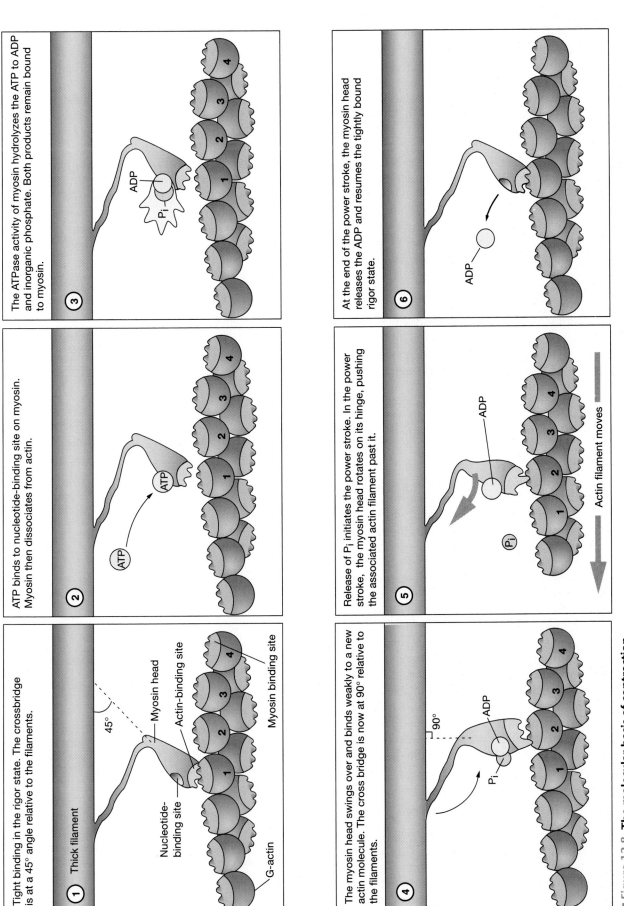

1. Tight binding in the rigor state. The crossbridge is at a 45° angle relative to the filaments.

2. ATP binds to nucleotide-binding site on myosin. Myosin then dissociates from actin.

3. The ATPase activity of myosin hydrolyzes the ATP to ADP and inorganic phosphate. Both products remain bound to myosin.

4. The myosin head swings over and binds weakly to a new actin molecule. The cross bridge is now at 90° relative to the filaments.

5. Release of Pi initiates the power stroke. In the power stroke, the myosin head rotates on its hinge, pushing the associated actin filament past it.

6. At the end of the power stroke, the myosin head releases the ADP and resumes the tightly bound rigor state.

■ Figure 12-8 **The molecular basis of contraction**

2. Next, an ATP molecule binds within the pocketlike nucleotide-binding site on the myosin head. As ATP binds, it changes the affinity of the actin-binding site so that the myosin head releases from actin.

3. The nucleotide-binding site on the myosin closes around the ATP and hydrolyses it to ADP and inorganic phosphate. Both products remain bound to myosin.

4. The free myosin head rotates and binds weakly to a new G-actin molecule, one or two positions away from the G-actin to which it was previously bound. At this point, the myosin head is cocked, like a stretched spring, and ready to execute the power stroke that will move the actin filament past it.

5. The power stroke of the myosin head begins when the inorganic phosphate is released from the myosin binding site. As the myosin head moves, it pushes the attached actin filament toward the center of the sarcomere. The power stroke is also called *crossbridge tilting* since the myosin head tilts from a 90° angle relative to the two filaments to a 45° angle.

6. In the last step of the contractile cycle, myosin releases ADP, the second product of ATP hydrolysis. At this point, the myosin head is again tightly bound to actin. The cycle is ready to begin again when a new ATP binds to the myosin.

Although our contractile cycle began with the rigor state in which no ATP or ADP was bound to myosin, relaxed muscle fibers remain mostly in stage 4 above. The rigor state in living muscle is normally brief because the muscle fiber has a sufficient supply of ATP that binds to myosin when ADP is released in step 6. After death, however, when metabolism stops and ATP supplies are exhausted, muscles are unable to bind more ATP, so they remain in the tightly bound state described in step 1. In the condition known as *rigor mortis*, the muscles "freeze" owing to immovable crossbridges. The tight binding of actin and myosin persists for a day or so after death, until enzymes within the decaying fiber begin to break down the muscle proteins.

Actin-myosin binding and cycling of the crossbridges explain how actin filaments slide past myosin filaments during contraction. But what regulates this process? If ATP is always available in the living muscle fiber, what keeps the filaments from continuously interacting? The answer lies with the two regulatory proteins tropomyosin and troponin.

Contraction Is Regulated by Troponin and Tropomyosin

The thin actin filaments of the myofibril are associated with two regulatory proteins that prevent myosin heads from completing their power stroke, just as the safety latch on a gun keeps the cocked trigger from being pulled. **Tropomyosin** [*tropos*, to turn] is an elongated protein polymer that wraps around the actin filament. In doing so, tropomyosin blocks part of the myosin-binding site on each actin molecule (Fig. 12-9a ■). When tropomyosin is in this blocking, or "off," position, weak actin-myosin binding can take place, but myosin is blocked from completing its power stroke. For contraction to occur, tropomyosin must be shifted to an "on" position that uncovers the remainder of the binding site and allows the power stroke to take place.

The "off-and-on" positioning of tropomyosin is regulated by troponin. **Troponin** is a complex of three proteins associated with tropomyosin. As contraction begins, one protein of the complex, **troponin C**, binds reversibly to Ca^{2+} (Fig. 12-9b ■). Calcium binding pulls tropomyosin toward the groove of the actin filament

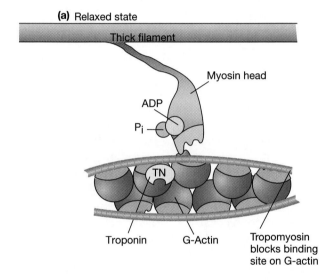

(a) Relaxed state

Thick filament

Myosin head

ADP

P_i

TN

Troponin   G-Actin   Tropomyosin blocks binding site on G-actin

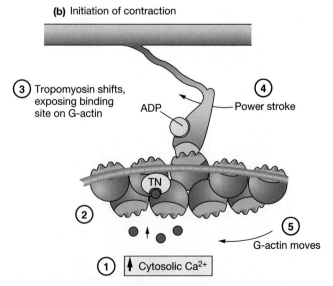

(b) Initiation of contraction

③ Tropomyosin shifts, exposing binding site on G-actin

④ Power stroke

ADP

TN

②

⑤ G-actin moves

① ↑ Cytosolic Ca^{2+}

■ **Figure 12-9 Regulatory role of tropomyosin and troponin** (a) In the relaxed state, troponin partially blocks the myosin-binding sites on actin. (b) Contraction is initiated when Ca^{2+} binds to troponin C. The binding of Ca^{2+} changes the shape of the associated tropomyosin molecule and uncovers the remainder of the myosin-binding site so that myosin can complete the power stroke.

and unblocks the myosin-binding sites. This "on" position enables the myosin heads to carry out their power strokes and move the actin filament. Contractile cycles repeat as long as the binding sites are uncovered.

For relaxation to occur, Ca^{2+} concentrations in the cytosol must decrease so that Ca^{2+} unbinds from troponin. Without Ca^{2+}, the troponin-tropomyosin complex returns to its "off" position, covering most of the myosin-binding site. During the portion of the relaxation phase when actin and myosin are not bound to each other, the filaments of the sarcomere slide back to their original positions with the aid of elastic connective tissue within the muscle.

Although the preceding discussion sounds as if we know everything there is to know about the molecular basis of muscle contraction, in reality these are simply our current models for the process. Studying the movement of molecules within a myofibril has proved very difficult. Many research techniques rely on crystallized molecules, electron microscopy, and other tools that cannot be used with living tissues. Often we can see filaments only at the beginning and end of contraction. This is like studying a baseball pitcher's delivery from two photos. The start of the pitch can be seen in a photo taken at the end of the windup, when the pitcher rears back, with his right arm cocked and left knee raised. The second photo shows the pitcher after the ball has been released: His right arm is extended forward toward home plate, his left foot is on the ground, and his right foot is in the air behind him. The dilemma is to reconstruct the motion that occurred between the two poses. This is one problem facing scientists who are trying to explain exactly how actin and myosin slide past each other. Progress is being made, however, and perhaps in the next decade you will see a "movie" of muscle contraction, constructed from photos of sliding filaments.

Now we will examine how a signal from a somatic motor neuron triggers the process of muscle contraction.

✓ In the sliding filament theory of contraction, what prevents the filaments from sliding back to their original position each time they release to bind to the next binding site? (Hint: What would happen if all crossbridges released simultaneously?)

Acetylcholine from Somatic Motor Neurons Initiates Excitation-Contraction Coupling

Signals for muscle contraction come from the central nervous system to skeletal muscles by way of somatic motor neurons. Acetylcholine from the somatic motor neuron initiates an action potential in the muscle fiber that in turn triggers contraction. This combination of electrical and mechanical events in a muscle fiber is called **excitation-contraction coupling.**

Acetylcholine released into the synapse at a neuromuscular junction binds to receptors on the motor end plate. These nicotinic cholinergic receptors are cation channels that allow Na^+ and K^+ to cross the sarcolemma. When the channels are open, Na^+ influx exceeds K^+ efflux because the electrochemical driving force is greater for Na^+. The addition of net positive charge to the muscle fiber depolarizes the membrane, creating an **end-plate potential (EPP).** Normally, end-plate potentials are always above threshold and always result in a muscle action potential. This action potential is conducted across the surface of the cell and into the t-tubules by the opening of voltage-gated Na^+ channels (Fig. 12-10 ■). The process is similar to the conduction of action potentials in axons, although action potentials in skeletal muscle are conducted more slowly than the action potentials of neurons (∞ p. 218).

The action potential that moves across the membrane and down the t-tubules is responsible for Ca^{2+} release from the sarcoplasmic reticulum. The t-tubule membrane contains voltage-sensing receptors (the **dihydropyridine,** or **DHP, receptors**) that are mechanically linked to Ca^{2+} channels in the membrane of the adjacent sarcoplasmic reticulum. When a wave of depolarization reaches a DHP receptor, its conformation changes. The conformation change opens Ca^{2+} channels in the sarcoplasmic reticulum, and Ca^{2+} diffuses into the cytosol. Free cytosolic Ca^{2+} levels in a resting muscle are normally quite low, but after an action potential, they increase about 100-fold. When cytosolic Ca^{2+} levels are high, Ca^{2+} binds to troponin, tropomyosin moves to the "on" position, and contraction occurs.

Relaxation of the muscle fiber occurs when the sarcoplasmic reticulum pumps Ca^{2+} back into its lumen using a Ca^{2+}-ATPase. As the concentration of cytosolic Ca^{2+} decreases, Ca^{2+} releases from troponin, tropomyosin slides back to block the myosin-binding site, and the fiber relaxes. The events of excitation-contraction coupling and contraction are summarized in Table 12-2.

The discovery that Ca^{2+}, not the action potential, is the signal for contraction was the first piece of evidence suggesting that calcium acts as a messenger inside cells. For years, it was thought that calcium signals occurred only in muscles, but we now know that calcium is an almost universal second messenger (∞ p. 152).

Figure 12-11 ■ shows the temporal sequence of events during excitation-contraction coupling. The somatic motor neuron action potential is followed by the skeletal muscle action potential, which in turn is followed by contraction. A single contraction-relaxation cycle in a skeletal muscle fiber is known as a **twitch.** Notice that there is a short delay, the **latent period,** between the muscle action potential and the beginning of muscle tension development. This interval represents the time required for excitation-contraction coupling to take place. Once contraction begins, muscle tension increases steadily to a maximum value. Tension then decreases in the **relaxation** phase of the twitch. During relaxation, the sarcoplasmic reticulum removes Ca^{2+} from the cyto-

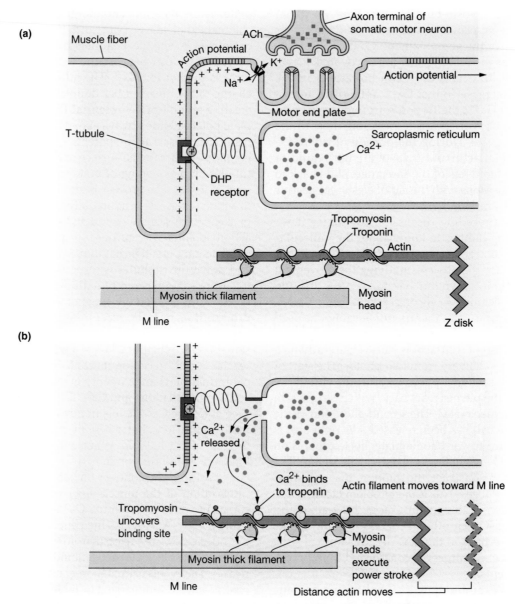

(a)

(b)

■ **Figure 12-10 Excitation-contraction coupling** (a) Acetylcholine is released from the somatic motor neuron. It opens ion channels in the motor end plate, leading to an action potential that moves across the fiber membrane and down into the t-tubules. (b) When the action potential reaches a dihydropyridine (DHP) receptor, the mechanical link between the receptor and the sarcoplasmic reticulum opens Ca^{2+} channels. Opening of the channels releases Ca^{2+} into the cytoplasm, where the ions can combine with troponin and initiate contraction.

plasm while the elastic elements of the muscle return the sarcomeres to their resting length.

A single action potential in a muscle fiber evokes a single twitch. But muscle twitches vary from fiber to fiber in the speed with which they develop tension (the rising slope of the twitch curve), the maximum tension they achieve (the height of the twitch curve), and the duration of the twitch (the width of the twitch curve). Factors that affect these parameters will be discussed in upcoming sections. First, in the next section, we look at how muscles produce the ATP that they require for contraction and relaxation.

Skeletal Muscle Contraction Depends on a Steady Supply of ATP

The muscle fiber's use of ATP is a key feature of muscle physiology. Muscles require energy constantly: during contraction for crossbridge movement and release, during relaxation to pump Ca^{2+} back into the sarcoplasmic reticulum, and after excitation-contraction coupling to restore Na^+ and K^+ to the extracellular and intracellular compartments, respectively. Where do muscles get the ATP they need for this work?

TABLE 12-2 Muscle Excitation and Contraction

1. Action potential in somatic motor neuron arrives at axon terminal.
2. Voltage-gated calcium channels open. Ca^{2+} entry triggers exocytosis of ACh-containing synaptic vesicles.
3. ACh diffuses into synaptic cleft and binds with nicotinic receptors on motor end plate of muscle.
4. ACh binding opens a nonspecific cation channel. Both Na^+ and K^+ move through the channel in response to their electrochemical gradients. Net influx of positive charge depolarizes the muscle membrane, creating an end-plate potential.
5. The end-plate potential is always above threshold and always results in a muscle action potential.
6. The action potential at the neuromuscular junction spreads along muscle fiber membrane, moving into the interior of the fiber via the t-tubules.
7. The action potential in the t-tubule activates dihydropyridine receptors. The DHP receptors open Ca^{2+} channels in the membrane of the sarcoplasmic reticulum.
8. Ca^{2+} diffuses out of the sarcoplasmic reticulum and binds to troponin, pulling tropomyosin away from the myosin-binding site. This action allows myosin to release inorganic phosphate from ATP hydrolysis and complete its power stroke.
9. At the end of the power stroke, the myosin crossbridge releases ADP and remains tightly bound to actin. Myosin must bind an ATP molecule in order to release from this rigor state.
10. The muscle fiber relaxes when Ca^{2+} releases from troponin, and tropomyosin again blocks the myosin-binding site. Calcium is transported back into the sarcoplasmic reticulum via a Ca^{2+}-ATPase.
11. The myosin ATPase hydrolyses ATP to ADP and inorganic phosphate, which both remain bound to the myosin head. The myosin swivels and binds to a new actin molecule, ready to execute its next power stroke.

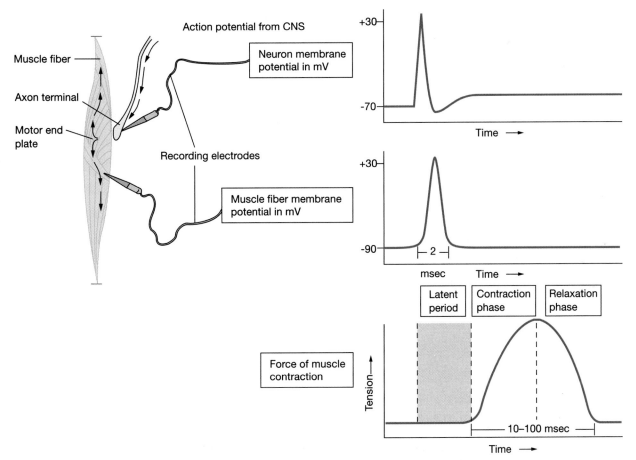

■ Figure 12-11 **Temporal sequence of electrical and mechanical events in muscle contraction** The tracings at right show action potentials in the axon terminal and the muscle fiber, followed by the tension curve of the muscle twitch. The latent period represents the time required for Ca^{2+} to be released from the sarcoplasmic reticulum and diffuse over to the filaments of the myofibril.

continued from page 330

Two forms of periodic paralysis exist. One form, called hyperkalemic periodic paralysis, is characterized by increased blood levels of K^+ during paralytic episodes. The other form, called hypokalemic periodic paralysis, is characterized by decreased blood levels of K^+ during episodes. Results of a blood sample reveal that Paul has the hyperkalemic form. It is not known why altered levels of K^+ trigger these attacks in people with the Na^+ channel defect.

Question 2: In people with hyperkalemic periodic paralysis, attacks tend to occur during a period of rest after exercise (i.e., repeated muscle contractions). Why? (Hint: What ion is responsible for the repolarization phase of muscle contraction, and in which direction does this ion move across the muscle fiber membrane?)

Question 3: Why does the Na^+ channel defect cause paralysis? (Hint: The gated channels that release Ca^{2+} from the sarcoplasmic reticulum close in a time-dependent manner, even if the cell remains depolarized.)

The amount of ATP within the muscle fiber at any one time is sufficient for only about eight twitches. As a backup energy source, muscles contain **phosphocreatine,** a molecule whose high-energy phosphate bonds are created from creatine and ATP when muscles are at rest (Fig. 12-12 ■). When muscles become active, such as during exercise, the high-energy phosphate group of phosphocreatine is transferred to ADP, creating ATP. The enzyme responsible for transferring the phosphate group is *creatine kinase (CK)*, also known as *creatine phosphokinase (CPK)*. Muscle cells contain large amounts of this enzyme. Consequently, elevated blood levels of creatine kinase usually indicate damage to skeletal or cardiac muscle. Because the two muscle types contain different isozymes (∞ p. 80), clinicians can distinguish tissue damage during a heart attack from skeletal muscle damage.

Energy stored in high-energy phosphate bonds is very limited, so muscle fibers must use metabolism to transfer energy from the chemical bonds of nutrients to ATP. Carbohydrates, particularly glucose, are the most rapid and efficient source of energy for ATP production. Glucose is metabolized through glycolysis to pyruvate (∞ p. 90). In the presence of adequate oxygen, pyruvate goes into the citric acid cycle, producing about 30 ATP for each molecule of glucose. When oxygen concentrations are too low to maintain aerobic metabolism, the muscle fiber shifts to anaerobic glycolysis. In this pathway, glucose is metabolized to lactic acid with a yield of only 2 ATP per glucose. Anaerobic metabolism of glucose is a quicker source of ATP but produces many fewer ATP per glucose. When muscle energy demands outpace the amount of ATP that can be produced through anaerobic metabolism of glucose, muscles can function for only a short period of time.

Muscle fibers also obtain energy from fatty acids, although this process always requires oxygen (∞ p. 95). During rest and light exercise, skeletal muscles burn fatty acids along with glucose, one reason that modest exercise programs of brisk walking are an effective way to reduce body fat. But beta-oxidation, the process by which fatty acids are converted to acetyl CoA, is a slow process and cannot produce ATP rapidly enough to meet the energy needs of muscle fibers during heavy exercise. Under these conditions, muscle fibers rely more on glucose.

Proteins are not normally a source of energy for muscle contraction since most of the amino acids found within muscle fibers are used to synthesize proteins rather than to produce ATP.

Do muscles ever run out of ATP? You might think so if you have ever exercised to the point of fatigue, where you feel that you cannot continue or your limbs refuse to obey the commands from your brain. But most studies show that even intense exercise uses only 30% of the ATP within a muscle fiber. The condition we call fatigue must come from other changes in the exercising muscle.

Muscle Fatigue Has Multiple Causes

The term **fatigue** is used to describe a condition in which a muscle is no longer able to generate or sustain the expected power output. Fatigue is highly variable and is influenced by the intensity and duration of the contractile activity, by whether the muscle fiber is using aerobic or anaerobic metabolism, by the composition of the muscle, and by the fitness level of the individual. Several factors are known to play a role in fatigue,

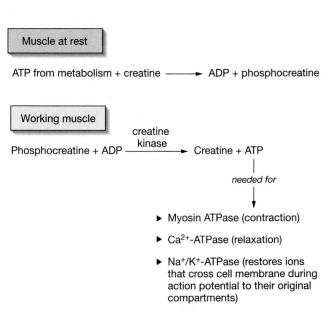

■ Figure 12-12 **Phosphocreatine** In resting muscle, ATP from metabolism is used to store energy in high-energy phosphate bonds of phosphocreatine. In working muscle, as ATP is consumed during contraction, phosphocreatine transfers its energy and phosphate back to ADP, producing more ATP.

including changes in the ionic composition of the muscle fiber after numerous contractions, depletion of muscle nutrients, and occasionally, diminished neurotransmitter output from the somatic motor neuron. Most experimental evidence suggests that muscle fatigue arises from excitation-contraction failure within the muscle fiber itself rather than from failure of control neurons or neuromuscular transmission.

Fatigue within the muscle fiber itself can occur in any one of several sites. In extended submaximal exertion, fatigue is associated with the depletion of glycogen stores within the muscle. Because most studies show that lack of ATP is not a limiting factor, glycogen depletion may be affecting some other aspect of contraction, such as the release of Ca^{2+} from the sarcoplasmic reticulum.

In short-duration maximal exertion, fatigue apparently results from the accumulation of H^+ from lactic acid, and increased production of inorganic phosphate from ATP breakdown. Both H^+ and phosphate have been shown to interfere with crossbridge function. Intracellular acidosis from elevated H^+ is also known to inhibit some enzymes of glycolysis and slow ATP production.

Another factor implicated in fatigue is K^+. During maximal exercise, K^+ leaves the cell with each action potential, and K^+ concentrations rise in the extracellular fluid of the t-tubules. The shift in K^+ alters the membrane potential of the muscle fiber and is believed to decrease Ca^{2+} release from the sarcoplasmic reticulum. However, there appears to be no single cause for excitation-contraction coupling failure in fatigue, so the study of this phenomenon is quite complex.

Neural causes of fatigue could arise either from communication failure at the neuromuscular junction or from failure of the central nervous system command neurons. If ACh is not synthesized in the axon terminal fast enough to keep up with the firing rate of the neuron, neurotransmitter release at the synapse will decrease. Consequently, the muscle end-plate potential may fail to reach the threshold value needed to trigger a muscle fiber action potential, resulting in contraction failure. This type of fatigue is associated with some neuromuscular diseases, but it is probably not a factor in normal exercise.

Central fatigue includes subjective feelings of tiredness and a desire to cease activity. Several studies have shown that this fatigue actually precedes physiological fatigue in the muscles and therefore may be a protective mechanism. Lactic acid production from anaerobic metabolism is often mentioned as a possible cause for fatigue. Some evidence suggests that acidosis caused by lactic acid dumped into the blood may influence the sensation of fatigue perceived by the brain. However, because the homeostatic mechanisms for pH balance maintain blood pH at normal levels until exertion is nearly maximal, pH is not a factor in central fatigue in most instances of submaximal exertion.

Skeletal Muscle Fibers Are Classified by Speed of Contraction and Resistance to Fatigue

Skeletal muscle fibers can be classified into three broad groups based on their speed of contraction and their resistance to fatigue with repeated stimulation. The three groups include **fast-twitch glycolytic fibers, fast-twitch oxidative fibers,** and **slow-twitch (oxidative) fibers.**

Fast-twitch muscle fibers develop tension two to three times as fast as slow-twitch fibers. The speed with which a muscle fiber contracts is determined by the isoform of myosin present in the fiber's thick filaments. Different myosin isoforms have differing ATPase activity. Fast-twitch fibers split ATP more rapidly and can therefore complete multiple contractile cycles more rapidly than slow-twitch fibers. This speed translates into faster tension development in the muscle fiber.

The duration of contraction also varies according to the fiber type. Twitch duration is determined largely by how fast the sarcoplasmic reticulum removes Ca^{2+} from the cytosol. As cytosolic Ca^{2+} concentrations fall, Ca^{2+} unbinds from troponin, allowing tropomyosin to move back and partially block the myosin-binding sites. With the power stroke inhibited, the muscle fiber relaxes. Fast-twitch fibers pump Ca^{2+} into their sarcoplasmic reticulum more rapidly than slow-twitch fibers, so fast-twitch fibers have quicker twitches. Fast-fiber twitches last only about 7.5 milliseconds (msec), making these muscles useful for fine, quick movements such as playing the piano. Contractions in slow-twitch muscle fibers may last more than 10 times as long. These muscles are used for strong, sustained movements such as lifting heavy loads. Fast fibers are used occasionally, but the slow-fiber types are used almost constantly for maintaining posture, standing, or walking.

Another major difference between fast-twitch glycolytic and slow-twitch fibers is their ability to resist fatigue. Fast glycolytic muscle fibers rely primarily on anaerobic glycolysis to produce ATP. However, the accumulation of lactic acid contributes to acidosis, a condition implicated in the development of fatigue. As a result, fast-twitch fibers fatigue more easily than do slow-twitch fibers, which do not produce large amounts of lactic acid.

Slow-twitch fibers, on the other hand, depend primarily on oxidative phosphorylation for production of ATP, hence their descriptive name as oxidative fibers. Slow-twitch fibers have more mitochondria, the site of enzymes for the citric acid cycle and oxidative phosphorylation. They also have more blood vessels in their connective tissue to bring oxygen to the muscle fibers (Fig. 12-13 ■).

The efficiency with which muscle fibers obtain oxygen is a factor in their preferred route of glucose metabolism. Slow-twitch fibers are described as "red muscle" owing to large amounts of **myoglobin,** a red oxygen-binding pigment that gives them their characteristic color. Oxygen brought from the lungs to the muscles must dif-

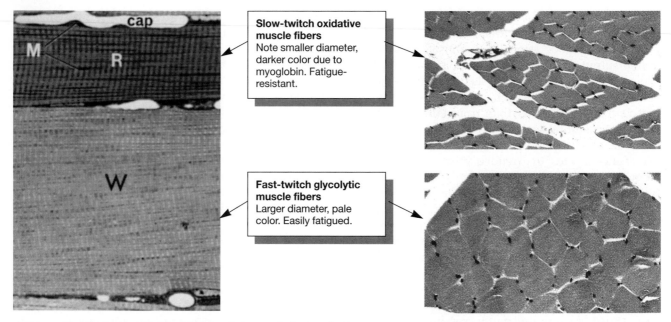

■ Figure 12-13 **Fast-twitch glycolytic and slow-twitch muscle fibers** Large amounts of red myoglobin, numerous mitochondria (M), and extensive capillary blood supply (cap) distinguish slow-twitch oxidative muscle (R) from fast-twitch glycolytic muscle (W).

fuse from the interstitial fluid into the interior of the fiber to reach the mitochondria. In muscles with myoglobin, this diffusion process is faster because myoglobin has a high binding affinity for oxygen. This affinity allows myoglobin to act as a transfer molecule, bringing oxygen more rapidly to the interior of the fiber. In addition, slow-twitch fibers have a smaller diameter, so the distance through which oxygen must pass before reaching the mitochondria is less. Because slow-twitch muscle fibers have more capillaries to bring blood to the cells, are smaller, and have a high myoglobin content, they maintain a better supply of oxygen and are more likely to use oxidative phosphorylation for ATP production.

Fast-twitch fibers, on the other hand, are described as "white muscle" because of their lower myoglobin content. These muscle fibers are also larger in diameter. The combination of larger size, less myoglobin, and fewer blood vessels means that fast-twitch fibers are more likely to run out of oxygen with repeated contractions. They are therefore forced to use anaerobic metabolism for ATP synthesis.

Fast-twitch fibers have been divided into two subcategories based on their relative diameters and fatigue resistance. Fast-twitch glycolytic fibers are the largest in diameter and rely primarily on anaerobic metabolism. They fatigue most rapidly of the three fiber types. Fast-twitch oxidative fibers are smaller, contain some myoglobin, and use a combination of oxidative and glycolytic metabolism to produce ATP. Because of their intermediate size and the use of oxidative phosphorylation for ATP synthesis, fast-twitch oxidative fibers are more fatigue-resistant than their fast glycolytic cousins. Characteristics of the three muscle fiber types are compared in Table 12-3.

Tension Developed by Individual Muscle Fibers Is a Function of Fiber Length

Within a single muscle fiber, the tension of a twitch is a direct reflection of the length of individual sarcomeres before contraction begins (Fig. 12-14 ■). Each sarcomere will contract with optimum force if it is an optimum length, neither too long nor too short, before the contraction begins. Fortunately, the normal resting length of skeletal muscles usually ensures that sarcomeres within the individual muscle fibers are at optimum length when they begin a contraction.

At the molecular level, sarcomere length reflects the overlap between the thick and thin filaments. The sliding filament theory predicts that the tension a muscle fiber can generate is directly proportional to the number of crossbridges formed between the thick and thin filaments. If the fibers start a contraction at a very long sarcomere length, the thick and thin filaments are barely overlapped, forming few crossbridges (Fig. 12-14e ■). This means that in the initial part of the contraction, the sliding filaments can interact only minimally and therefore cannot generate much force. At the optimum sarcomere length, the filaments begin contracting with more crossbridge linkage between the thick and thin filaments, allowing the fiber to generate optimum force in that twitch (Fig. 12-14c ■). If the sarcomere is shorter than optimum length at the beginning of the contraction, the thick and thin fibers were too well overlapped before the contraction began. Consequently, the thick filaments can move the thin filaments only a short distance before the thin actin filaments from opposite ends of the sarcomere start to overlap. This overlap prevents crossbridge formation. Finally, if the sarcomere shortens

TABLE 12-3 Characteristics of Muscle Fiber Types

	Slow-Twitch Oxidative; Red Muscle	*Fast-Twitch Oxidative; Red Muscle*	*Fast-Twitch Glycolytic; White Muscle*
Time to development of maximum tension	Slowest	Intermediate	Fastest
Myosin ATPase activity	Slow	Fast	Fast
Diameter	Small	Medium	Large
Contraction duration	Longest	Short	Short
Ca^{2+}-ATPase activity in SR	Moderate	High	High
Endurance	Fatigue-resistant	Fatigue-resistant	Easily fatigued
Use	Most used: standing, walking		Least used: jumping
Metabolism	Oxidative; aerobic; numerous large mitochondria	Glycolytic but become more oxidative with endurance training	Glycolytic; more anaerobic than fast-twitch oxidative type
Color	Dark red owing to myoglobin	Red	Pale

too much, the thick filaments run into the Z disks at the ends of the sarcomere. Once this happens, the thick filaments are unable to find new binding sites for crossbridge formation, so tension decreases rapidly (Fig. 12-14a ■). Thus, the development of single-twitch tension within a muscle fiber is a passive property that depends on filament overlap and sarcomere length.

Force of Contraction Increases with Summation of Muscle Twitches

Although we have just shown that single-twitch tension is determined by the length of the sarcomere, it is important to note that a single twitch does not represent the maximum force that the muscle fiber can develop. The force generated by the contraction of a single muscle fiber can be increased by increasing the rate at which muscle action potentials stimulate the muscle fiber. A typical muscle action potential lasts between 1 and 3 msec, whereas the muscle contraction itself may last 100 msec. If repeated action potentials are separated by long intervals of time, the muscle fibers have time to relax completely between stimuli (Fig. 12-15a ■). If the interval of time between action potentials is shortened, the muscle fiber will not have relaxed completely at the time of the second stimulus, resulting in a more forceful contraction (Fig. 12-15b ■). This process is known as **summation** and is similar to the summation that takes place in neurons (∞ p. 211).

If action potentials continue to stimulate the muscle fiber repeatedly at short intervals, relaxation between contractions diminishes until the muscle fiber achieves a state of maximal contraction known as **tetanus.** There are two types of tetanus. In *unfused tetanus,* the stimulation rate of the muscle fiber is slower, and the fiber relaxes slightly between stimuli (Fig. 12-15c ■). In *complete,* or *fused, tetanus,* the stimulation rate is fast enough that the muscle fiber does not have time to relax.

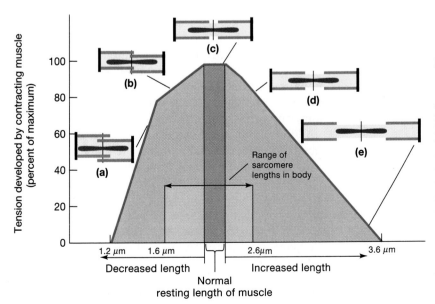

■ Figure 12-14 **Length-tension relationships in contracting muscle** The graph shows the amount of tension generated by a muscle compared with its resting length before contraction starts. The inserts of the sarcomeres show the amount of overlap between thick and thin filaments at each resting muscle length. If the muscle is too long, the filaments in the sarcomere barely overlap and cannot form as many crossbridge links (e). If the muscle begins its contraction at a very short length, the sarcomere cannot shorten very much before the myosin filaments run into the Z disks at each end (a).

(a) Single twitches

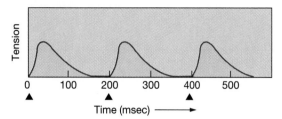

(b) Summation

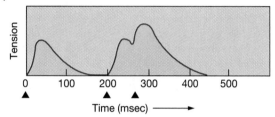

(c) Summation leading to unfused tetanus

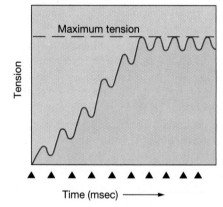

(d) Summation leading to complete tetanus

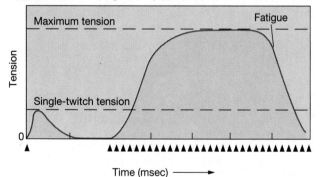

■ **Figure 12-15 Summation of contractions** A muscle fiber will respond to a stimulus (▲) with a twitch. (a) If the stimuli are far enough apart in time, the muscle will relax completely between twitches. (b) As the stimuli move closer and closer together in time, the muscle fiber does not have time to relax and the contractions sum, creating a contraction with greater tension. (c) If the stimuli come very rapidly, the muscle will reach its maximum tension. If the muscle has a chance to relax slightly between stimuli, it has reached unfused tetanus. (d) If the muscle reaches a steady tension, it is said to be in complete tetanus. Once the muscle fiber starts to fatigue, the tension it creates diminishes even though the stimuli continue.

Instead, it reaches maximum tension and remains there (Fig. 12-15d ■). Thus, it is possible to increase the tension developed in a single muscle fiber by changing the rate at which action potentials occur in the fiber. Muscle action potentials are initiated by acetylcholine released from somatic motor neurons, so we now turn to whole muscles and the neurons that control them.

✓ Summation in muscle fibers means that the _____ of the fiber increases with repeated action potentials. Temporal summation in neurons means that the _____ of the neuron increases when two depolarizing stimuli occur close together in time.

✓ Which type of runner would you expect to have more slow-twitch fibers, a sprinter or a marathoner?

One Somatic Motor Neuron and the Muscle Fibers It Innervates Form a Motor Unit

The basic unit of contraction in an intact skeletal muscle is a **motor unit,** composed of a group of muscle fibers and the somatic motor neuron that controls them (Fig. 12-16 ■). When the somatic motor neuron fires an action potential, all muscle fibers in the motor unit contract. Note that although one somatic motor neuron innervates multiple fibers, each muscle fiber is innervated by only a single neuron.

The number of muscle fibers in a motor unit varies. In muscles used for fine motor actions, such as the muscles that move the eyes or the muscles of the hand, a motor unit contains as few as three to five muscle fibers. If one motor unit is activated, only a few fibers contract, and the muscle response is quite small. If additional motor units are activated, the response increases by small increments since only a few more muscle fibers contract with the addition of each motor unit. This arrangement allows fine gradations of movement. In muscles used for gross motor actions, such as standing or walking, a single somatic motor neuron may innervate hundreds or even thousands of muscle fibers. The gastrocnemius muscle in the calf of the leg, for example, has about 2000 muscle fibers in each motor unit. Each time an additional motor unit is activated in these muscles, many more muscle fibers contract, and the muscle response jumps by correspondingly greater increments.

All muscle fibers in a single motor unit are of the same fiber type. Thus, there are fast-twitch motor units and slow-twitch motor units. The determination of which kind of fibers associates with a particular neuron appears to lie with the neuron itself. During embryological development, the somatic motor neuron secretes a growth factor that directs the differentiation of all muscle fibers in its motor unit so that they develop into the same fiber type.

Intuitively, it would seem that people who inherit a predominance of one fiber type over another would

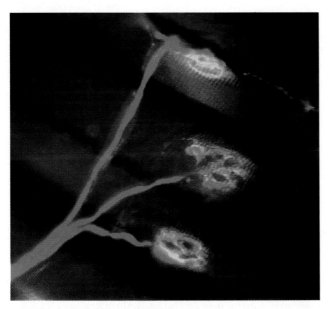

■ **Figure 12-16 A motor unit** One motor neuron branches to innervate several muscle fibers.

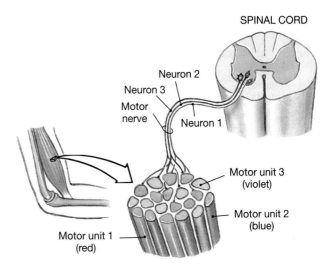

■ **Figure 12-17 A muscle is composed of multiple motor units** A muscle may have many motor units of different fiber types.

excel in certain sports. They do, to some extent. Endurance athletes such as distance runners and cross-country skiers have a predominance of aerobic slow-twitch fibers, whereas sprinters, ice hockey players, and weight lifters tend to have larger percentages of fast-twitch fibers.

But inheritance is not the only determining factor for fiber composition in the body. The metabolic characteristics of muscle fibers are plastic and can be changed to some extent. With endurance training, the aerobic capacity of some fast-twitch fibers can be enhanced until they are almost as fatigue-resistant as a slow-twitch fiber. Since the conversion occurs only in those muscles that are being trained, a neuromodulator chemical is probably involved. In addition, endurance training increases the number of capillaries and mitochondria in the muscle tissue, allowing more oxygen-carrying blood to reach the contracting muscle and contributing to the increased aerobic capacity of the muscle fibers.

Contraction in Intact Muscles Depends on the Types and Numbers of Motor Units in the Muscle

Within a skeletal muscle, each motor unit contracts in an all-or-none manner. How then can muscles create graded contractions of varying force and duration? The answer lies in the fact that intact muscles are composed of multiple motor units of different types (Fig. 12-17 ■). This diversity allows the muscle to vary contraction by (1) changing the types of motor units that are active or (2) changing the number of motor units that are responding at any one time.

The force of contraction within a skeletal muscle can be increased by recruiting additional motor units.

Recruitment is controlled by the nervous system and proceeds in a stereotyped fashion. A weak stimulus directed onto a pool of somatic motor neurons in the central nervous system activates only the neurons with the lowest thresholds (∞ p. 210). Studies have shown that these low-threshold neurons control fatigue-resistant slow-twitch fibers that generate minimal force. As the stimulus onto the motor neuron pool increases in strength, additional motor neurons with higher thresholds begin to fire. These neurons in turn stimulate motor units composed of fatigue-resistant oxidative fast-twitch fibers. Because more motor units (and thus, more muscle fibers) are participating in the contraction, greater force is generated. As the stimulus increases to even higher levels, somatic motor neurons with the highest thresholds begin to fire. These neurons stimulate motor units composed of glycolytic fast-twitch fibers. At this point, the muscle contraction is approaching its maximum force. Because of differences in myosin and cross-bridge formation, fast-twitch fibers can generate more force than slow-twitch fibers. However, because fast-twitch fibers fatigue more rapidly, it is impossible to hold a muscle contraction at maximum force for an extended period of time. You can demonstrate this by clenching your fist as hard as you can: How long can you hold it before some of the muscle fibers begin to fatigue?

Sustained contractions within a muscle require a continuous train of action potentials from the central nervous system to the muscle. But, as you learned earlier, increasing the stimulation rate of a muscle fiber results in summation of its contractions. If the muscle fiber is easily fatigued, summation will lead to fatigue and diminished tension. One way that the nervous system avoids fatigue in a sustained contraction is by **asynchronous recruitment** of motor units. The nervous system modulates the firing rates of the motor neurons so that different motor units take turns maintaining the muscle tension. The alternation of active motor units allows some of the motor units to rest between contractions, preventing fatigue.

Asynchronous recruitment will prevent fatigue only in submaximal contractions, however. In high-tension, sustained contractions, the individual motor units may reach a state of unfused tetanus, in which the muscle fibers cycle between contraction and partial relaxation. In general, we do not notice this cycling, because the different motor units in the muscle are contracting and relaxing at slightly different times. As a result, the contractions and relaxations of the motor units average out and appear to be one smooth contraction. But, as different motor units fatigue, we notice that we are unable to maintain the same amount of tension in the muscle, and the force of the contraction gradually decreases.

✓ What is the response of a muscle fiber to an increase in the firing rate of the somatic motor neuron?

✓ How does the nervous system increase the force of contraction in a muscle composed of many motor units?

MECHANICS OF BODY MOVEMENT

Because one main role of skeletal muscles is to move the body, we now turn to the mechanics of body movement. The term *mechanics* refers to how muscles move loads and how their anatomical relationship to the bones of the skeleton maximizes the work they can do.

Isotonic Contractions Move Loads but Isometric Contractions Create Force without Movement

When we described the function of muscles, we noted that they create force to generate movement and that they can also create force without generating movement. A contraction that creates force and moves a load is known as an **isotonic contraction** [*iso*, equal + *teinein*, to stretch]. A contraction that creates force without movement is known as an **isometric contraction** [*iso*, equal + *metric*, measurement]. These two types of contraction are illustrated in Figure 12-18 ■. To demonstrate an isotonic contraction experimentally, we hang a weight (the load) from the muscle and electrically stimulate the muscle to contract. The muscle contracts, lifting the weight. The graph on the right shows the development of force throughout the contraction.

To demonstrate an isometric contraction experimentally, we attach a heavier weight to the muscle. When the muscle is stimulated, it develops tension, but the force created is not enough to move the load. In isometric contractions, muscles create force without shortening significantly. For example, when your exercise instructor

continued from page 338

Paul's doctor explains to Mrs. Leong that the paralytic attacks associated with hyperkalemic periodic paralysis last only a few minutes to a few hours and generally involve only the muscles of the extremities. "Is there any treatment?" asks Mrs. Leong. The doctor replies that, although the condition is presently incurable, attacks can be prevented with several drugs. Diuretics, for example, increase the body's excretion of water and ions (including Na^+ and K^+), and they have been shown to prevent attacks of paralysis in people with this condition.

Question 4: Speculate on how diuretics prevent attacks in people with hyperkalemic periodic paralysis.

yells at you to "tuck in your tummy," your response is isometric contraction of abdominal muscles.

How can an isometric contraction create force if the length of the muscle does not change? The elastic elements of the muscle provide the answer. All muscles contain elastic fibers in the tendons and other connective tissues that attach muscles to bone and in the connective tissue between muscle fibers. Within the muscle fibers, elastic cytoskeletal proteins occur between the myofibrils and as part of the sarcomere. Finally, the contractile filaments themselves apparently can stretch even while generating force. All of these elastic components behave collectively as if they were connected in series (one after the other) to the contractile elements of the muscle. Consequently, they are often called the **series elastic elements** of the muscle (Fig. 12-19a ■). When the sarcomeres shorten in an isometric contraction, the elastic elements stretch. This stretching of the elastic elements allows the fibers to maintain constant length even though the sarcomeres are shortening and creating tension (Fig. 12-19c ■). Once the elastic elements have been stretched and the force generated by the sarcomeres equals the load, the muscle shortens in an isotonic contraction and lifts the load (Fig. 12-19d ■).

It is also possible for a muscle to lengthen while it contracts and creates tension. A lengthening contraction is known as an **eccentric action.** You can demonstrate eccentric action as well as isotonic and isometric contractions with a pair of heavy weights. If you pick up the weights and hold them out in front of you, the muscles of your arms are creating tension to overcome the load of the weights. Since the muscles are not shortening, the contraction is isometric. If you bend your arms at the elbows and bring the weights into your shoulders, you have performed a shortening, or isotonic, contraction. If you then slowly extend your arms, resisting the tendency of the weights to pull them down, you are performing an eccentric, or lengthening, contraction. Eccentric exercise is thought to contribute most to cellular damage after exercise and to lead to delayed muscle soreness.

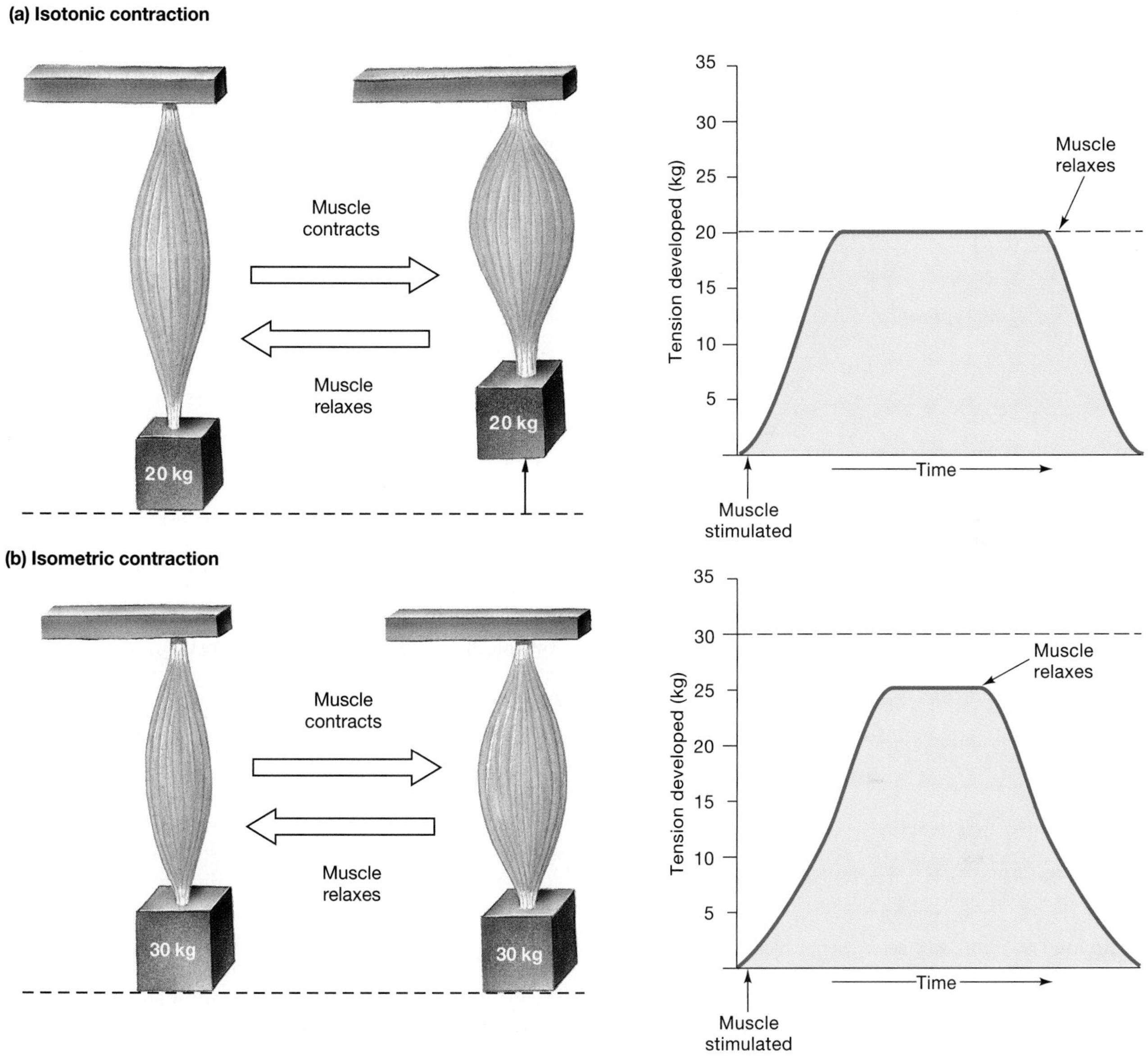

(a) Isotonic contraction

Muscle
contracts

Muscle
relaxes

20 kg

20 kg

Muscle
relaxes

Time

Muscle
stimulated

(b) Isometric contraction

Muscle
contracts

Muscle
relaxes

30 kg

30 kg

Muscle
relaxes

Time

Muscle
stimulated

■ **Figure 12-18 Isotonic and isometric contractions** (a) In an isotonic contraction, the muscle creates enough tension to move the load. (b) In an isometric contraction, the muscle is unable to create enough tension to move the load. The muscle contracts but does not shorten.

Bones and Muscles around Joints Form Levers and Fulcrums

The anatomical arrangement of muscles and bones in the body is directly related to how muscles work. The body uses its bones and joints as levers and fulcrums on which muscles exert force to move or resist a load. A lever is a rigid bar that pivots around a point known as the fulcrum. In the body, bones form levers, flexible joints form the fulcrums, and muscles attached to bones create force when they contract. The work being performed by the muscle can be expressed as

Work = force × distance

In a lever and fulcrum system, distance is measured from the fulcrum to the point along the lever at which force is being exerted (Fig. 12-20a ■). Most lever systems

in the body are similar to the one shown; the fulcrum is located at one end of the lever. This arrangement maximizes the distance and speed with which the lever can move the load but also requires that the muscles do more work. We will use the flexion of the forearm to illustrate how this lever system functions.

In the lever and fulcrum system of the forearm, the elbow joint acts as a fulcrum around which movement of the forearm, the lever, takes place (Fig. 12-20b ■). The biceps muscle is attached at its origin at the shoulder and inserts into the radius bone of the forearm (the lever) about 5 cm away from the elbow joint (the fulcrum). When the biceps contracts, it creates an upward force as it pulls on the bone.

If the biceps muscle is to hold the arm stationary and flexed at a 90° angle, it must do work that exactly

(a) Schematic of the series elastic elements

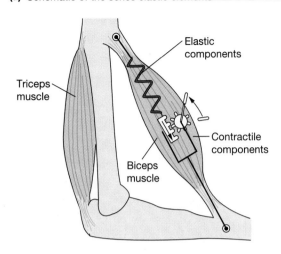

Triceps muscle

Elastic components

Contractile components

Biceps muscle

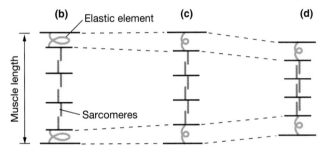

(b) Elastic element

Muscle length

Sarcomeres

(c)

(d)

(b) Muscle at rest

(c) Isometric contraction: Muscle has not shortened. Sarcomeres shorten, generating force, but elastic elements stretch, allowing muscle length to remain the same.

(d) Isotonic contraction: Sarcomeres shorten more but, since elastic elements are already stretched, the entire muscle must shorten.

■ **Figure 12-19** **Series elastic elements in muscle** (a) A muscle has both contractile components (sarcomeres, shown here as a gear and rachet) and elastic components (shown here as a spring). (b) When the muscle starts to contract, the sarcomeres shorten and create tension but their shortening is offset initially by the stretching of the elastic elements (c). In this way, the muscle can contract isometrically, creating force without a change in overall muscle length. Once the elastic elements are maximally stretched (d), any further contraction requires that the muscle shorten in an isotonic contraction.

opposes the downward pull of the weight of the fore-arm and hand. For example, suppose that the muscle and bone of the forearm weigh 2 kg and that the center of gravity of the forearm is 15 cm from the elbow (Fig. 12-20b ■). The work of the biceps muscle ($W_1 = F_1 \times D_1$) must equal the work created by the downward pull of the arm ($W_2 = F_2 \times D_2$):

$$W_1 = F_1 \times D_1$$
Work of biceps = (upward force of biceps muscle)
× (5 cm between fulcrum and insertion)

$$W_2 = F_2 \times D_2$$
Work of arm = (2 kg downward force of arm weight)
× (15 cm from fulcrum to center of gravity)

If $W_1 = W_2$, then $F_1 \times D_1 = F_2 \times D_2$

Therefore

(Upward force of biceps) × (5 cm)
= (2 kg downward force of arm) × (15 cm)

Upward force created by biceps muscle
$$= \frac{(2 \text{ kg downward force of arm}) \times (15 \text{ cm})}{(5 \text{ cm})}$$

$$= (2 \text{ kg}) \times (3)$$

$$= 6 \text{ kg}$$

Solving the equation shows us that the biceps must create 6 kg of force simply to hold the arm at a 90° angle. Since the muscle is not shortening, this is an isometric contraction.

Now what happens if a 7-kg weight is placed in the hand? Unless the biceps muscle creates additional upward force to offset the downward force created by the weight, the hand will drop. If the weight on the hand is 25 cm from the elbow, the equation for the force needed to offset the load becomes

(Upward force of biceps) × (5 cm)
= (7 kg force of weight) × (25 cm)

The biceps must create an additional 35 kg of force to keep the arm from dropping the 7-kg weight! As you can see from these calculations, genetic variability in the distance between the fulcrum and the muscle insertion could have a dramatic effect on the force required to move or resist a load. For example, if the biceps inserts 7 cm from the fulcrum instead of 5 cm, it need generate only 25 kg of force to offset the same 7-kg weight. Some studies have shown a correlation between the insertion points of muscles and success in certain athletic events.

In the example so far, we have assumed that the load is stationary and that the muscle is contracting isometrically. What happens if we want to flex the arm and lift the load? Then, the biceps muscle must exert a force that exceeds the force created by the load. The disadvantage of a lever system in which the fulcrum is positioned near one end of the lever is that the muscle is required to create large amounts of force to move or resist small loads.

But this type of lever system also maximizes speed and mobility. A small movement of the bone at the point where the muscle inserts becomes a much larger movement at the hand (Fig. 12-20d ■). In addition, the two movements occur in the same amount of time, so the speed of contraction at the insertion point is amplified at the end of the arm. For instance, if the biceps muscle in our example contracts and shortens 1 cm, the hand elevates 5 cm. If the contraction takes 1 second, the muscle shortens at a speed of 1 cm/sec, but the hand moves

(a)

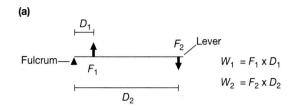

(b)

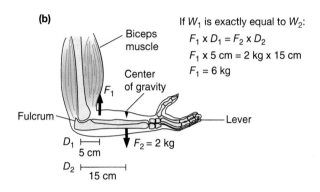

(c)

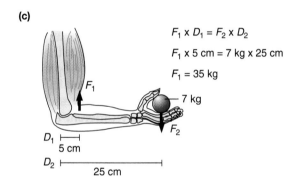

(d)

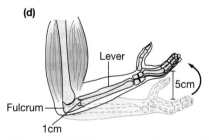

■ **Figure 12-20 The arm is a lever and fulcrum system**
(a) The amount of work required to move a load at the end of the lever is equal to the force created by the load (F_2) times the distance of the load from the fulcrum (D_2). If a muscle is attached to the lever at a distance D_1 from the fulcrum, it must create a force F_1 in order to move the load. (b) In order to hold the arm, which exerts a downward force of 2 kg at its center of mass, a distance of 15 cm from the elbow (the fulcrum), the biceps must pull upward with a force of 6 kg. (c) If a 7-kg load is added to the hand 25 cm from the elbow, the biceps muscle must exert an additional 35 kg of force to keep the hand from dropping. (In this example, we are disregarding the weight of the arm.) (d) Because the insertion of the biceps is close to the fulcrum, the biceps must create large amounts of force, but the arrangement enhances the speed with which the hand can move. It also translates a small movement of the biceps into a much larger movement of the hand. As shown here, when the biceps contracts and shortens 1 cm, the hand moves upward 5 cm. If the muscle shortens in 1 second, the hand then moves at a speed of 5 cm/sec.

at a speed of 5 cm/sec. Thus, the lever system of the arm amplifies both the distance and the speed of movement of the load.

In muscle physiology, the speed with which a muscle contracts depends on the type of muscle fiber (fast-twitch versus slow-twitch) and on the load that is being moved. Intuitively, you can see that you can flex your arm much faster with nothing in your hand than you can while holding a 7-kg weight in your hand. The relationship between load and velocity of contraction in a muscle fiber, determined experimentally, is graphed in Figure 12-21 ■. Contraction is fastest when the load on the muscle is zero. When the load on the muscle equals the ability of the muscle to create force, the muscle is unable to move the load and the velocity drops to zero. The muscle can still contract, but the contraction becomes isometric instead of isotonic. Because speed is a function of load and muscle fiber type, it cannot be regulated by the body except through recruitment of faster muscle fiber types. However, the arrangement of muscles, bones, and joints allows the body to amplify speed so that regulation at the cellular level becomes less important.

✓ One study found that many world-class athletes have muscle insertions that are farther from the joint than in the average person. Why would this trait translate into an advantage for a weight lifter?

Muscle Disorders Have Multiple Causes

Dysfunctions of skeletal muscles can arise from a problem with the signal from the nervous system, from miscommunication at the neuromuscular junction, or from defects in the muscle itself. Unfortunately, in many muscle conditions, even the simple ones, we do not fully understand the mechanism of the primary defect. As a result, we can treat the symptoms but may not be able to cure the problem.

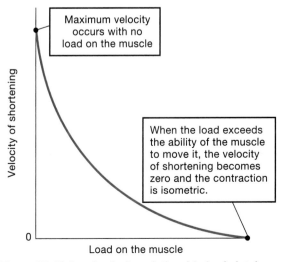

■ **Figure 12-21 Load-velocity relationship in skeletal muscle**

(a)

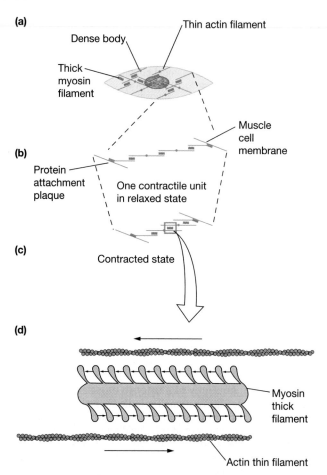

(b)

(c)

(d)

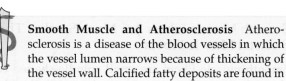

■ Figure 12-24 **Sliding filaments in smooth muscle** (a) The actin and myosin filaments of smooth muscle are longer than in skeletal muscle. (b) The long actin filaments attach to dense bodies in the cytoplasm and terminate at protein plaques in the cell membrane. (c) Their oblique arrangement allows myosin to slide along actin for long distances without encountering the end of a sarcomere. The overlap of the longer filaments also allows smooth muscle to be stretched more while retaining the ability to generate tension. (d) Smooth muscle myosin has hinged heads all along its length, in contrast to skeletal muscle myosin, which has no heads in the center of each filament.

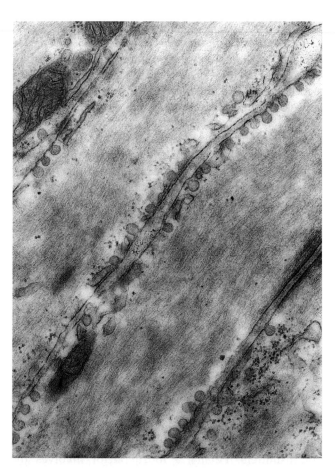

■ Figure 12-25 **Caveolae in smooth muscle** Caveolae are small vesicles close to the cell membrane. They concentrate Ca^{2+}, releasing it into the cytoplasm when Ca^{2+} channels in the vesicle membrane open.

amount varies from one type of smooth muscle to another. The sarcoplasmic reticulum in smooth muscle is chemically linked to the cell membrane rather than mechanically linked, with Ca^{2+} influx acting as the signal for sarcoplasmic Ca^{2+} release.

The calcium-storage function of the sarcoplasmic reticulum is supplemented by **caveolae,** small vesicles that cluster close to the cell membrane (Fig. 12-25 ■). The membranes of caveolae contain gated Ca^{2+} channels that open in response to either a change in membrane potential or the binding of a ligand, allowing Ca^{2+} concentrated inside the caveolae to enter the cell.

For years, it was believed that mature smooth muscle cells, like skeletal muscle, had lost the ability to undergo mitosis. Recent studies have shown, however, not only that smooth muscle fibers can reproduce themselves, but also

that their proliferation may play a significant role in *atherosclerosis*, commonly known as hardening of the arteries.

There are two types of smooth muscle in the human body. Most smooth muscle is **single-unit smooth mus-**

Smooth Muscle and Atherosclerosis Atherosclerosis is a disease of the blood vessels in which the vessel lumen narrows because of thickening of the vessel wall. Calcified fatty deposits are found in the extracellular matrix, leading to the hardened state that gives atherosclerosis its popular name, hardening of the arteries. As part of the disease process, smooth muscle cells in the blood vessel wall convert from a contractile state to a proliferative state. During proliferation, the muscle cells migrate toward the lumen, divide, and accumulate cholesterol. The change in smooth muscle function is regulated by growth factors and cytokines released by white blood cells in the region of the atherosclerotic lesion. If researchers can find a way to block the action of these growth factors and cytokines, they may have a powerful tool to help prevent atherosclerosis.

cle. This tissue is also called **visceral smooth muscle** because it forms the walls of the internal organs (viscera) such as the blood vessels, the intestinal tract, and the ureters. The muscle fibers of single-unit smooth muscle are electrically coupled by numerous gap junctions that allow depolarizations to pass rapidly from cell to cell (Fig. 12-26a ■). This arrangement permits an action potential in one cell to be spread rapidly throughout a sheet of tissue so that many cells contract as a single unit. Because the cells of single-unit smooth muscle are electrically coupled, it is not necessary to stimulate each muscle fiber to make the entire sheet contract.

(a) Single-unit smooth muscle

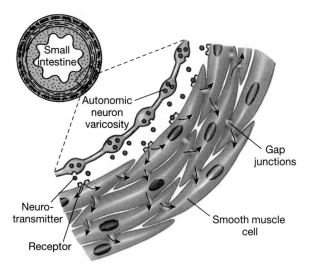

(b) Multi-unit smooth muscle

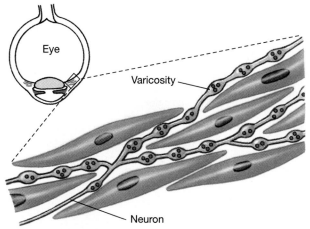

■ Figure 12-26 **Types of smooth muscle** (a) Single-unit smooth muscle is found in most organs of the body. When neurotransmitter from autonomic neurons depolarizes some of the cells, the wave of depolarization spreads throughout the sheet of tissue by way of the gap junctions that link the cells. In this fashion, the sheet of muscle tissue contracts as a single unit. (b) Multi-unit smooth muscle is found in the eye, uterus, and male reproductive tract. In multi-unit smooth muscle, the cells are not electrically connected. Each cell must be stimulated independently.

Multi-unit smooth muscle is found in the iris and ciliary body of the eye, in part of the male reproductive tract, and in the uterus, except just prior to labor and delivery. The cells of multi-unit smooth muscle are not linked electrically; therefore, each individual muscle fiber must be closely associated with an axon terminal or varicosity (Fig. 12-26b ■). This arrangement allows fine control and graded contractions to occur in these muscles by selective activation of individual muscle fibers. Interestingly, the multi-unit fibers of uterine muscle change and become single-unit just before labor and delivery. The addition of gap junctions to the cell membranes synchronizes the electrical signals and apparently allows the uterine muscle to contract more effectively while working to expel the baby.

Smooth Muscle Can Vary Its Force of Contraction

The unique organization of actin and myosin in smooth muscle is responsible for the ability of this tissue to maintain tension over a wide range of fiber lengths. Smooth muscles have longer actin and myosin filaments. The extra length allows the fibers to be stretched more, yet still maintain enough overlap to create optimum tension. In addition, once contraction begins, the actin filaments can slide along myosin for longer distances before the end is reached. This property is important for organs whose luminal volume may vary, such as the stomach, intestine, bladder, and uterus. The filling of the bladder with urine is a good example in which the smooth muscle must stretch as the lumen fills yet still maintain the ability to contract in order to expel the contents of the lumen.

Variable force of contraction in smooth muscle depends on the fiber type. Multi-unit smooth muscle resembles skeletal muscle because each cell responds to a stimulus independently. Increasing the force of contraction requires the recruitment of additional fibers. On the other hand, the fibers of single-unit smooth muscle are all electrically connected, so if one fiber fires, all fibers fire. No reserve units are left to be recruited. In single-unit smooth muscle, graded contractions vary with the amount of Ca^{2+} that enters the cell.

Smooth muscle uses less energy than skeletal muscle to generate a given amount of force. For example, it has been estimated that a smooth muscle cell can generate maximum tension with only 25%–30% of its crossbridges active. In addition, smooth muscle has low oxygen consumption rates yet does not fatigue because of anaerobic metabolism. We still do not fully understand the mechanisms through which this is accomplished, but some possibilities are discussed below.

✓ What is the difference between how contraction force is varied in multi-unit and single-unit smooth muscle?

Phosphorylation of Proteins Plays a Key Role in Smooth Muscle Contraction

The molecular events of contraction in smooth muscle are similar in many ways to those in skeletal muscle, but some important differences exist. Here is a summary of the key points as we currently understand them:

1. In both smooth muscle and skeletal muscle, the signal to initiate contraction is an increase in cytosolic Ca^{2+}.
2. In smooth muscle, Ca^{2+} enters from the extracellular fluid in addition to being released from the sarcoplasmic reticulum.
3. In smooth muscle, Ca^{2+} binds to **calmodulin,** a binding protein found in the cytosol. In skeletal muscle, Ca^{2+} binds to troponin. (Smooth muscle lacks troponin.)
4. In smooth muscle, Ca^{2+} binding to calmodulin is only the first step in a cascade that ends with contraction. In skeletal muscle, Ca^{2+} binding to troponin initiates contraction immediately.
5. In smooth muscle, the phosphorylation of proteins is an essential feature of the contraction process.
6. In smooth muscle, myosin regulation is the primary control of contraction. In skeletal muscle, actin is regulated.

Let us follow these steps in sequence (Fig. 12-27 ■). Smooth muscle contraction begins when a stimulus (electrical or chemical) opens Ca^{2+} channels in the sarcolemma and sarcoplasmic reticulum. Calcium ions enter the cytosol and bind to calmodulin. The Ca^{2+}-calmodulin then activates an enzyme called **myosin light chain kinase,** or **MLCK,** by complexing with it and removing a phosphate group. The resulting Ca^{2+}-calmodulin-MLCK complex in turn activates myosin.

The ATPase activity of smooth muscle myosin is modulated by the phosphorylation of light protein chains in the myosin head (Fig. 12-28a ■). When the myosin light chain is phosphorylated by myosin light chain kinase, ATPase activity is high, and crossbridge cycling can take place. If the myosin light chain is dephosphorylated, ATPase activity and contraction are inhibited. Thus, smooth muscle contraction is primarily controlled through myosin-linked regulatory processes, a feature not seen in skeletal muscle.

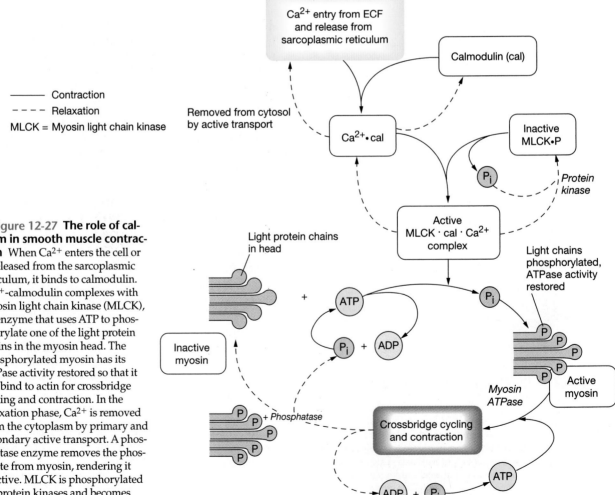

■ Figure 12-27 **The role of calcium in smooth muscle contraction** When Ca^{2+} enters the cell or is released from the sarcoplasmic reticulum, it binds to calmodulin. Ca^{2+}-calmodulin complexes with myosin light chain kinase (MLCK), an enzyme that uses ATP to phosphorylate one of the light protein chains in the myosin head. The phosphorylated myosin has its ATPase activity restored so that it can bind to actin for crossbridge cycling and contraction. In the relaxation phase, Ca^{2+} is removed from the cytoplasm by primary and secondary active transport. A phosphatase enzyme removes the phosphate from myosin, rendering it inactive. MLCK is phosphorylated by protein kinases and becomes inactive.

(a)

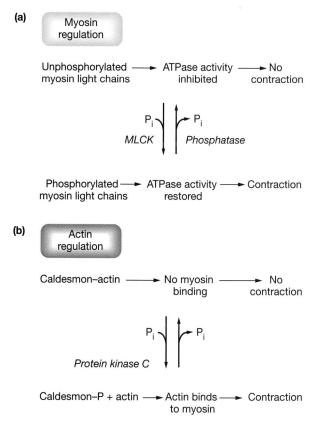

Myosin regulation

Unphosphorylated myosin light chains ⟶ ATPase activity inhibited ⟶ No contraction

P_i ⟍ MLCK   ⟋ P_i Phosphatase

Phosphorylated myosin light chains ⟶ ATPase activity restored ⟶ Contraction

(b)

Actin regulation

Caldesmon–actin ⟶ No myosin binding ⟶ No contraction

P_i ⟍   ⟋ P_i
Protein kinase C

Caldesmon–P + actin ⟶ Actin binds to myosin ⟶ Contraction

■ **Figure 12-28 Regulation of contraction in smooth muscle** Both actin and myosin can be regulated in smooth muscle by the phosphorylation of proteins associated with the contractile filaments.

Relaxation in a smooth muscle fiber is a multistep process: Ca^{2+} must be removed from the cytosol, and the myosin light chains must be dephosphorylated (Fig. 12-27 ■). Ca^{2+} is removed from the cytosol partially by a Ca^{2+}-Na^+ antiport exchange and partially by a Ca^{2+}-ATPase. Removal of the phosphate group is accomplished with the aid of **myosin light chain phosphatase.** The steps of smooth muscle contraction and relaxation are summarized in Table 12-4.

Curiously, dephosphorylation of myosin does not automatically result in relaxation. Under conditions that we do not fully understand, dephosphorylated myosin may remain attached to actin for a period of time in what is known as a **latch state.** This condition maintains tension in the muscle fiber without consuming ATP and it is a significant factor in the ability of smooth muscle to sustain contractions without fatiguing. The hinge muscles of certain bivalve mollusks such as oysters can enter a similar latch state that allows them to remain tightly closed under anaerobic conditions.

In some types of smooth muscle, myosin regulation is supplemented by regulation of actin. In these cells, actin binds to a regulatory protein called **caldesmon,** inhibiting actin-myosin binding (Fig. 12-28b ■). If caldesmon is phosphorylated, it can no longer bind to actin, allowing actin and myosin to interact. Phosphorylation of caldesmon takes place with the aid of enzymes known as **protein kinases.**

✓ What happens to contraction if a smooth muscle is placed in a saline bath from which all calcium has been removed?

✓ Tetrodotoxin (TTX) is a poison that blocks sodium channels. When TTX is applied to certain types of smooth muscle, it does not affect the spontaneous generation of action potentials. What conclusion can you draw about the action potential of this smooth muscle on the basis of this observation?

TABLE 12-4   Summary of Smooth Muscle Contraction and Relaxation

Contraction

1. Ca^{2+} enters smooth muscle fibers through:
   • voltage-gated channels opened when cell depolarizes
   • stretch-activated channels opened when cell membrane is deformed
   • chemically gated channels opened by neurotransmitters, hormones, and paracrines
2. Ca^{2+} triggers release of additional Ca^{2+} from the sarcoplasmic reticulum.
3. Ca^{2+} binds to calmodulin.
4. Ca^{2+}-calmodulin complexes with myosin light chain kinase (MLCK).
5. Activated MLCK phosphorylates light protein chains of the myosin heads, using energy and P_i from ATP.
6. Phosphorylation of myosin restores ATPase activity and allows crossbridge cycling and contraction.
7. In some smooth muscle, caldesmon must be phosphorylated before actin can bind to myosin.

Relaxation

1. Myosin phosphatase removes phosphate from myosin, decreasing its ATPase activity.
2. Ca^{2+} is removed from the cytoplasm using a Ca^{2+}-Na^+ antiport protein and a Ca^{2+}-ATPase.
3. Calmodulin releases Ca^{2+} and uncomplexes from MCLK.
4. MLCK is phosphorylated by protein kinases, inactivating it.
5. In some smooth muscle, phosphate removal from caldesmon allows caldesmon to inactivate actin.

Some Smooth Muscles Have Unstable Membrane Potentials

Many types of smooth muscle display unstable resting membrane potentials that vary between -40 and -80 mV. Cells that show cyclic depolarization and repolarization of their membrane potential are said to have **slow wave potentials.** Sometimes the cell simply cycles through a series of subthreshold slow waves. However, if the peak of the depolarization is above threshold, action potentials result, followed by contraction (Fig. 12-29a ■).

Other types of smooth muscle have a regular depolarization that always reaches threshold and fires an action potential (Fig. 12-29b ■). These depolarizations are called **pacemaker potentials** because they create regular rhythms of contraction. Pacemaker potentials are found in some cardiac muscles as well as in smooth muscle. Both slow wave and pacemaker potentials are due to ion channels in the cell membrane that spontaneously open and close.

Calcium Entry Is the Signal for Smooth Muscle Contraction

Action potentials in smooth muscle are different from action potentials of neurons and skeletal muscle because the depolarization phase is due to the entry of Ca^{2+} rather than Na^+. The Ca^{2+} influx also acts as the signal to initiate contraction. Interestingly, an action potential is not *required* to open voltage-gated Ca^{2+} channels in smooth muscle. Graded potentials open a few Ca^{2+} channels, allowing small amounts of Ca^{2+} into the cell. The entry of variable amounts of Ca^{2+} into the muscle fiber creates contractions whose force is graded according to the strength of the Ca^{2+} signal.

Calcium influx from the extracellular fluid initiates the release of Ca^{2+} from the sarcoplasmic reticulum, a process known as **Ca^{2+}-induced Ca^{2+} release.** However, since the Ca^{2+} stores in smooth muscle are limited, sustained contractions are dependent upon continued influx of Ca^{2+} from the extracellular fluid.

In addition to voltage-gated Ca^{2+} channels, the cell membranes contain stretch-activated Ca^{2+} channels and chemically gated Ca^{2+} channels. Stretch-activated channels open when pressure or other force distorts the cell membrane, allowing Ca^{2+} to enter and initiate contraction. Because this type of contraction originates from a property of the muscle fiber itself and not because of innervation or hormones, it is known as a **myogenic** contraction. Myogenic contractions are common in blood vessels that maintain a certain amount of tone at all times.

Chemically gated Ca^{2+} channels open when a ligand binds to them. In **pharmacomechanical coupling,** Ca^{2+} entry causes smooth muscle contraction without a significant change in membrane potential (Fig. 12-29c ■).

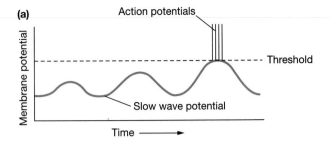

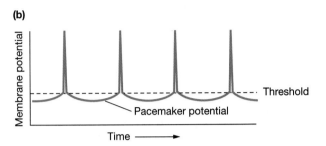

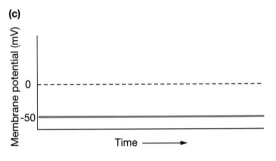

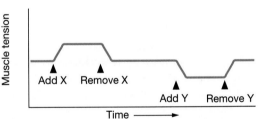

■ **Figure 12-29 Depolarizations in smooth muscle** (a) Some types of smooth muscle exhibit slow wave potentials in which the unstable membrane potential alternately depolarizes and hyperpolarizes. If a depolarization exceeds threshold, the cell fires action potentials. (b) Other types of smooth muscle exhibit pacemaker potentials, in which the cell steadily depolarizes until it reaches threshold and fires an action potential. (c) Some cells with chemically gated Ca^{2+} channels exhibit pharmacomechanical coupling, in which the muscle contracts or relaxes without a change in membrane potential. Chemical signal molecules open Ca^{2+} channels in the cell membrane, and Ca^{2+} entering the cell acts as a second messenger to influence muscle tension. The depolarization due to Ca^{2+} influx is offset by an increased rate at which Na^+ is pumped out of the cell.

Although the entry of Ca^{2+} would usually depolarize the cell, a simultaneous increase in the pumping of Na^+ out of the cell prevents a change in the membrane potential difference. As a result, contraction occurs without depolarization.

✓ How do pacemaker potentials differ from slow wave potentials?

Smooth Muscle Contraction Is Regulated by Chemical Signals

Smooth muscle contraction is controlled by a variety of chemical signals that may be either excitatory or inhibitory. In general, if the stimulus depolarizes the cell, it is more likely to contract. Hyperpolarization of the cell decreases the likelihood that it will contract.

In neurally controlled smooth muscle, neurotransmitter is released from varicosities of autonomic neurons onto the surface of the muscle fibers (∞ p. 309). Smooth muscle lacks specialized receptor regions such as the motor end plates found in skeletal muscle synapses, so the neurotransmitter simply diffuses across the cell surface.

Many smooth muscles have dual innervation: That is, they are controlled by both the sympathetic and parasympathetic divisions. However, other smooth muscles, such as those found in blood vessels, are innervated by only one of the two autonomic divisions. Single innervation is usually accompanied by tonic control so that the response can be graded by increasing or decreasing the amount of neurotransmitter (∞ p. 156).

Neurotransmitters can have different effects in different tissues, depending upon the receptors to which they bind. Thus, both the type of innervation and the receptor on the smooth muscle fiber determine the response of a smooth muscle to nervous stimulation.

An amazing variety of neurotransmitters such as vasoactive intestinal peptide and substance P are active in smooth muscle, in addition to norepinephrine and acetylcholine. In many instances, we know how these chemicals affect smooth muscle contraction, but we do not understand the reflex pathways that trigger their release.

Hormones and paracrines also control smooth muscle contraction, unlike skeletal muscle controlled only by the nervous system. Smooth muscles in the cardiovascular, gastrointestinal, urinary, respiratory, and reproductive systems all respond to blood-borne or locally released chemicals. For example, asthma is a condition in which the smooth muscle of the airways constricts in response to histamine release. This constriction can be reversed by the administration of a hormone, epinephrine, that relaxes smooth muscle and dilates the airway. Note that not all physiological responses are adaptive or favorable to the body: Constriction of the airways triggered during an asthma attack, if untreated, can be fatal.

Another important paracrine that affects smooth muscle contraction is nitric oxide (∞ p. 225). This gas is synthesized by the endothelial lining of blood vessels and relaxes adjacent smooth muscle that regulates the diameter of the blood vessels. For many years, the identity of

continued from page 344

Three weeks later, Paul has another attack of paralysis, this time at kindergarten after a game of tag. He is rushed to the hospital and is given glucose by mouth. Within minutes, he is able to move his legs and arms and asks for his mother.

Question 5: Recall that cells use glucose to produce energy in the form of ATP and that cells use ATP to run the Na^+/K^+-ATPase that actively exchanges K^+ and Na^+ across the cell membrane. Explain why oral glucose might help bring Paul out of his paralysis. (Hint: What happens to the extracellular K^+ level when Paul is given glucose?)

this *endothelium-derived relaxing factor*, or *EDRF*, eluded scientists even though its presence could be demonstrated experimentally. We know now that EDRF is nitric oxide, an important paracrine in many systems of the body.

Sometimes, no chemical signal is needed to initiate smooth muscle contraction. In these cases, the smooth muscle fibers can be induced to contract by stretching the muscle and opening stretch-activated Ca^{2+} channels. Some smooth muscles show adaptation to stretch if the muscle fiber is stretched for a period of time: As the stretch continues, the Ca^{2+} channels begin to close in a time-dependent fashion. As Ca^{2+} is pumped out of the cell, the muscle relaxes. This response explains why the bladder develops tension as it fills, then relaxes as it adjusts to the increased volume.

In skeletal muscle, contraction is controlled by the nervous system, and the muscle fiber always responds to an action potential with a twitch. In marked contrast, smooth muscle fibers can be directly inhibited by neurotransmitters, hormones, or paracrines. And, because multiple signals might reach the muscle fiber at one time, smooth muscle fibers also must act as integrating centers. For example, certain blood vessels may receive contradictory messages from two different sources: One message signals for contraction, the other for relaxation. The smooth muscle fibers must integrate the two signals and execute an appropriate response.

Although smooth muscles in the body do not have nearly the mass of skeletal muscles, they play a critical role in the function of most organ systems. Thus, you will have the opportunity to learn more about smooth muscle physiology in the chapters to come.

CARDIAC MUSCLE

Cardiac muscle, the specialized muscle of the heart, shares features with both smooth and skeletal muscle (Table 12-5). Like skeletal muscle fibers, cardiac muscle fibers are striated and have a sarcomere structure. However, cardiac muscle fibers are shorter than skeletal muscle fibers, may

TABLE 12-5 **Comparison of Three Muscle Types**

	Skeletal	*Smooth*	*Cardiac*
Appearance under light microscope	Striated	Smooth	Striated
Fiber arrangement	Sarcomeres	Longitudinal bundles	Sarcomeres
Fiber proteins	Actin, myosin; troponin and tropomyosin	Actin, myosin, tropomyosin	Actin, myosin; troponin and tropomyosin
Control	Voluntary Ca^{2+}-dependent via troponin	Involuntary Ca^{2+}-dependent via calmodulin	Involuntary Ca^{2+}-dependent via troponin
	Fibers independent	Fibers electrically linked via gap junctions	Fibers electrically linked via gap junctions
Nervous control	Somatic motor neuron	Autonomic neurons	Autonomic neurons
Hormonal influence	None	Multiple hormones	Epinephrine
Location	Attached to bones; a few sphincters close off hollow organs	Forms the walls of hollow organs and tubes; some sphincters	Heart muscle
Morphology	Multinucleate; long, thin cylindrical fibers	Uninucleate; small spindle-shaped fibers	Uninucleate; shorter branching fibers
Internal structure	T-tubule and sarco-plasmic reticulum	No t-tubules; sarcoplasmic reticulum reduced or absent	T-tubule and sarcoplasmic reticulum
Contraction speed	Fastest	Slowest	Intermediate
Contraction force of single fiber	All-or-none	Graded	Graded
Initiation of contraction	Requires input from motor neuron	Can be autorhythmic	Autorhythmic

be branched, and have a single nucleus (unlike multinucleate skeletal muscle fibers). Like single-unit smooth muscle, cardiac muscle fibers are electrically linked to each other. The gap junctions are contained in specialized cell junctions known as **intercalated disks.** Cardiac muscle, like smooth muscle, is under the control of both divisions of the autonomic nervous system (sympathetic and parasympathetic) as well as under hormonal control. In Chapter 14, you will learn more about cardiac muscle and how it functions within the heart.

CHAPTER REVIEW

Chapter Summary

1. Muscles generate motion, force, and heat. (p. 324)

2. The three types of muscle are **skeletal muscle, cardiac muscle,** and **smooth muscle.** Skeletal and cardiac muscles are **striated muscles.** (p. 324)

3. Skeletal muscles contract only in response to a signal from a somatic motor neuron. Cardiac and smooth muscle are controlled by autonomic innervation, paracrines, and hormones. Some types of smooth and cardiac muscle are autorhythmic and can contract spontaneously without outside influence. (p. 325)

Skeletal Muscle

4. Skeletal muscles are usually attached to bones by **tendons.** The **origin** is the end of the muscle attached closest to the trunk or to the more stationary bone. The **insertion** is the more distal or mobile attachment. (p. 326)

5. At a flexible **joint,** contraction of a muscle moves the skeleton. **Flexors** bring bones closer together; **extensors** move bones away from each other. Flexor-extensor pairs are called **antagonistic muscle groups.** (p. 326)

6. A skeletal **muscle** is a collection of **muscle fibers** arranged with their long axes in parallel. Skeletal muscle fibers are large cylindrical cells with several hundred nuclei. The cell membrane is called the **sarcolemma;** the cytoplasm is called the **sarcoplasm.** (p. 326)

7. The **sarcoplasmic reticulum** concentrates calcium ions. It is associated with the **t-tubules,** a continuation of the cell membrane that allows action potentials to move rapidly into the interior of the fiber. (p. 327)

8. The main intracellular structures are the **myofibrils,** bundles of contractile and elastic proteins. The **thick filaments** are made of **myosin,** and the **thin filaments** are made mostly of **actin. Titin** and **nebulin** serve as accessory proteins and contribute to the stability of the myofibril. (p. 327)

9. Myosin binds to actin much of the time, creating **crossbridges** between the parallel thick and thin filaments. (p. 327)

10. The basic contractile unit is the **sarcomere,** composed of two Z disks and the filaments found between them. Each sarcomere is divided into **I bands** (thin filaments only), an **A band** that runs the length of the thick filament, and a central **H zone** occupied by thick filaments only. The **M line** and **Z disks** represent the attachment sites for the thick filaments and thin filaments, respectively. (p. 327)

11. The force created by a contracting muscle is called muscle **tension.** The **load** is a weight or force that opposes contraction of a muscle. (p. 330)

12. The **sliding filament theory of contraction** states that during contraction, overlapping thick and thin filaments slide past each other in an energy-dependent manner owing to binding of actin and myosin and movement of the crossbridges. (p. 330)

13. Myosin is the motor protein that converts energy from ATP into motion. A contractile cycle starts with myosin tightly bound to actin but with no nucleotide bound to the second myosin site. When ATP binds, myosin releases from actin. The ATPase activity of myosin hydrolyses ATP to ADP and inorganic phosphate. Both products remain bound to myosin. The myosin head rotates and binds to a new actin. (p. 332)

14. Movement of the myosin head forms the **power stroke** that is the basis for muscle contraction. The power stroke begins when the myosin head releases the inorganic phosphate. As the myosin moves, it pushes actin toward the center of the sarcomere. Finally, myosin releases ADP and the cycle ends with myosin tightly bound to actin. (p. 332)

15. The regulatory protein **tropomyosin** is wound around actin so that tropomyosin blocks part of the myosin-binding site. With tropomyosin in this position, myosin cannot complete its power stroke. As contraction begins, **troponin** associated with tropomyosin binds to Ca^{2+}. The shape of troponin-tropomyosin changes, unblocking the myosin-binding sites, and allowing myosin to complete its power stroke. (p. 334)

16. During relaxation, Ca^{2+} unbinds from troponin, allowing troponin-tropomyosin to return to the position in which it covers the myosin-binding site. During relaxation, the sarcoplasmic reticulum pumps Ca^{2+} back into its lumen using a Ca^{2+}-ATPase. (p. 335)

17. An action potential in a somatic motor neuron releases ACh that initiates an action potential in the muscle fiber. The muscle action potential leads to contraction. This combination of events is called **excitation-contraction coupling.** (p. 335)

18. The muscle action potential moves into the t-tubules, where voltage-sensing **dihydropyridine receptors** are mechanically linked to Ca^{2+} channels in the sarcoplasmic reticulum. The action potential opens the Ca^{2+} channels and Ca^{2+} diffuses into the cytosol. (p. 335)

19. A single contraction-relaxation cycle in a skeletal muscle fiber is known as a **twitch.** The **latent period** between the end of the muscle action potential and the beginning of muscle tension development represents the time required for Ca^{2+} release and binding to troponin. (p. 335)

20. Muscle fibers store energy for contraction in **phosphocreatine.** Additional energy comes from metabolism. Anaerobic metabolism of glucose is a rapid source of ATP but is not efficient and makes the cell acidic through the production of lactic acid. Aerobic metabolism is very efficient but requires an adequate supply of oxygen to the muscles. (p. 338)

21. Muscle **fatigue** is a condition in which a muscle is no longer able to generate or sustain the expected power output. Fatigue is influenced by a variety of factors and has no single cause. Some factors known to play a role in fatigue include changes in the ionic composition of the muscle fiber, depletion of energy resources, and diminished neurotransmitter output from the somatic motor neuron. Fatigue probably results from failure in the excitation-contraction coupling mechanism. (p. 338)

22. Skeletal muscle fibers can be classified on the basis of their speed of contraction and resistance to fatigue into **fast-twitch glycolytic fibers, fast-twitch oxidative fibers,** and **slow-twitch oxidative fibers.** Oxidative fibers are the most fatigue-resistant. (p. 339)

23. **Myoglobin** is a red oxygen-binding pigment that transfers oxygen to the interior of the muscle fiber. (p. 339)

24. Within a single muscle fiber, the tension of a twitch is determined by the length of the sarcomeres before contraction begins. Fibers that are too long or too short will not generate maximum force. (p. 340)

25. The force generated by the contraction of a single muscle fiber can be increased by increasing the stimulus rate of the fiber. This causes the **summation** of twitches and an increase of tension up to a state of maximal contraction known as **tetanus.** (p. 341)

26. The basic unit of contraction in an intact skeletal muscle is a **motor unit,** composed of a group of muscle fibers and the somatic motor neuron that controls them. The number of muscle fibers in a motor unit varies, but all fibers in a single motor unit are of the same fiber type. (p. 342)

27. The force of contraction within a skeletal muscle can be increased by **recruitment** of additional motor units. (p. 343)

28. The nervous system avoids fatigue in sustained contractions through **asynchronous recruitment** of motor units. (p. 343)

29. A contraction that creates force and moves a load is known as an **isotonic contraction.** A contraction that creates force without movement is known as an **isometric contraction.** (p. 344)

30. Isometric contractions create force without movement owing to the stretch of **series elastic elements** that allow the fibers to maintain constant length even though the sarcomeres are shortening and creating tension. (p. 344)

31. A lengthening contraction is known as an **eccentric action.** Eccentric exercise is thought to contribute most to cellular damage after exercise and to lead to delayed muscle soreness. (p. 344)

32. The body uses its bones and joints as levers and fulcrums on which muscles exert force in order to move or resist a load. Most lever systems in the body maximize the distance and speed with which the lever can move the load but also require that the muscles do more work. (p. 345)

33. In muscle physiology, the speed with which a muscle contracts is a function of the type of muscle fiber and of the load that is being moved. Contraction is fastest when the load on the muscle is zero. (p. 347)

Smooth Muscle

34. Smooth muscle develops tension more slowly than skeletal muscle but can sustain contractions for extended periods of time without fatiguing. Some smooth muscles are **tonically contracted** and maintain tension at most times. A muscle that maintains a measurable level of tension is said to have **tone.** (p. 348)

35. Smooth muscle fibers are small spindle-shaped cells whose actin and myosin are arranged diagonally in a lattice around the central nucleus. When these fibers contract, they become globular. (p. 349)

36. Smooth muscle has relatively little sarcoplasmic reticulum, and calcium storage is supplemented by **caveolae.** (p. 349)

37. In contrast to skeletal muscles, smooth muscles have longer actin and myosin filaments that allow the fibers to be stretched more, yet maintain enough overlap to create optimum tension. (p. 349)

38. Most smooth muscle is **single-unit smooth muscle** that contracts as a single unit when depolarizations pass rapidly from cell to cell through numerous gap junctions. The cells of **multi-unit smooth muscle** are not linked electrically, and each individual muscle fiber is stimulated independently. (p. 350)

39. Smooth muscles exhibit summation and tetanus like skeletal muscles but only multi-unit smooth muscle has recruitment. Graded contractions in single-unit smooth muscle are a function of the amount of Ca^{2+} that enters the cell. (p. 351)

40. Smooth muscle uses less energy than skeletal muscle to generate a given amount of force. (p. 351)

41. The ATPase activity of myosin is modulated by phosphorylation of light protein chains in the myosin head. The phosphorylation state is controlled by **myosin light chain kinase (MLCK)** and **myosin light chain phosphatase.** (p. 352)

42. MLCK is activated by binding to a Ca^{2+}-**calmodulin** complex. This binding causes phosphorylation of myosin, activation of myosin ATPase, and contraction. (p. 352)

43. In smooth muscle, both actin and myosin regulate contraction. Actin has no troponin and instead is associated with a regulatory protein called **caldesmon.** When caldesmon is bound to actin, actin-myosin binding is inhibited. When caldesmon is phosphorylated, actin and myosin interact. (p. 353)

44. For relaxation to occur, Ca^{2+} must be removed from the cytosol, and the myosin light chains must be dephosphorylated. (p. 353)

45. Many types of smooth muscle display unstable membrane potentials. If these cycles of depolarization and repolarization do not reach threshold each time, they are known as **slow wave potentials.** If the depolarization always reaches threshold and fires an action potential, it called a **pacemaker potential.** (p. 354)

46. The rising phase of the smooth muscle action potential is due to the entry of Ca^{2+} instead of Na^+. The Ca^{2+} influx also initiates the release of Ca^{2+} from the sarcoplasmic reticulum in Ca^{2+}-induced Ca^{2+} release. (p. 354)

47. Smooth muscle fibers also contain stretch-activated Ca^{2+} channels and chemically gated Ca^{2+} channels. In **pharmacomechanical coupling,** Ca^{2+} entry causes smooth muscle contraction without a significant change in membrane potential. (p. 354)

48. Smooth muscle is controlled by a variety of chemical signals that can either stimulate or inhibit smooth muscle contraction. Many smooth muscles are controlled by both sympathetic and parasympathetic neurons, although some are innervated by only one autonomic division. Contraction in many smooth muscle fibers can also be induced simply by stretching the muscle and opening stretch-activated Ca^{2+} channels. (p. 355)

Cardiac Muscle

49. Cardiac muscle fibers are striated and have a single nucleus. They are also electrically linked through gap junctions. Cardiac muscle is under the control of both divisions of the autonomic nervous system as well as under hormonal control. (p. 355)

Questions

LEVEL ONE   Reviewing Facts and Terms

1. The three types of muscle tissue found in the human body are _____, _____, and _____. Which type is attached to the bones, enabling them to control body movement?

2. The two types of muscle tissue that are classified as striated are _____ and _____.

3. Which type of muscle tissue is controlled strictly by a signal from a somatic motor neuron?

4. Which statement is true about skeletal muscles?
   a. They constitute about 40% of a person's total body weight.
   b. They position and move the skeleton.
   c. The insertion of the muscle is more distal or mobile than the origin.
   d. They are often paired into antagonistic muscle groups called flexors and extensors.
   e. All of these statements are true.

5. Arrange these skeletal muscle components in order, from outermost to innermost: sarcolemma, connective tissue sheath, myofilaments, myofibrils.

6. The modified endoplasmic reticulum of skeletal muscle is called the _____. Its role is to sequester _____ ions.

7. T-tubules allow _____ to move to the interior of the muscle fiber.

8. Name the two components that fill the space between myofibrils. What does each one do?

9. List the proteins that make up the myofibrils. Which one has heads that move to form the power stroke?

10. List the letters used to label the elements of a sarcomere. Which band has a Z disk in the middle? Which is the darkest band? Why? Which element forms the boundaries of a sarcomere? Name the line that divides the A band in half. What is the function of this line?

11. Briefly explain the functions of titin and nebulin.

12. During contraction, the _____ band remains a constant length. This band is composed primarily of _____ molecules. Which lettered components approach each other during contraction?

13. The current theory of muscle contraction is called _____. Why?

14. The two regulatory proteins of muscle contraction are the elongated _____ and the complex _____, which reversibly binds _____ ions. Is it the presence or the absence of this ion that allows power strokes to occur?

15. Which neurotransmitter is released by somatic motor neurons?

16. What is the motor end plate, and what kind of receptors are found there? What effect does an influx of sodium ions have at the motor end plate?

17. A single contraction-relaxation cycle in a skeletal muscle fiber is known as a _____.

18. List the steps of the contraction cycle that require ATP.

19. For each characteristic listed below, decide which type of muscle it applies to: fast-twitch glycolytic fibers, fast-twitch oxidative fibers, slow-twitch oxidative fibers.
   a. largest diameter
   b. anaerobic metabolism, thus fatigues quickly
   c. most blood vessels
   d. have some myoglobin
   e. used for quick, fine movements
   f. also called red muscle
   g. use a combination of oxidative and glycolytic metabolism
   h. most mitochondria

20. The basic unit of contraction in an intact skeletal muscle is the _____. The force of contraction within a skeletal muscle is increased by _____ additional motor units.

21. The two types of smooth muscle are _____, also called visceral smooth muscle, and _____.

LEVEL TWO   Reviewing Concepts

22. **Concept map:** Make a map of muscle fiber structure using the following words or terms: actin, calcium ions, cell membrane, cell, crossbridges, cytoplasm, contractile protein, elastic protein, glycogen, mitochondria, muscle fiber, myosin, nucleus, regulatory protein, sarcolemma, sarcoplasm, sarcoplasmic reticulum, t-tubule, titin, tropomyosin, troponin.

23. List and describe the three types of muscle tissue that occur in the human body. Where is each type found? What are some alternative names for each type?

24. How does an action potential in the sarcolemma of the muscle fiber trigger calcium release inside the muscle cell?

25. Muscle cells depend on a supply of ATP constantly. How do the different types of muscle cells generate ATP? What is used for a backup energy source?

26. Define muscle fatigue. Summarize factors that play a role in its development. Compare it with central fatigue. How can muscle cells adapt in order to resist fatigue?

27. Explain how you vary the strength and effort made by your muscles in picking up a pencil versus a full gallon container of milk.

28. Compare and contrast the cellular anatomy and neural and chemical control of skeletal, smooth, and cardiac muscle.

29. What is the role of the sarcoplasmic reticulum in muscular contraction? How can smooth muscle contract when it has so little sarcoplasmic reticulum?

30. Compare and contrast
   a. fast-twitch oxidative, fast-twitch glycolytic, and slow-twitch muscle fibers

b. a twitch and tetanus
c. action potentials in motor neurons and action potentials in skeletal muscles
d. temporal summation in motor neurons and summation in skeletal muscles
e. isotonic contraction, isometric contraction, and eccentric action

31. What is a pacemaker potential? Which cells generate pacemaker potentials? How are they important?

32. Explain smooth muscle contraction and relaxation, particularly the roles of MLCK and calmodulin.

LEVEL THREE Problem Solving

33. One way that scientists study muscles is to put them into a state of rigor or stiffness by removing ATP. In this condition, the actin and myosin are strongly linked but unable to move. On the basis of what you know about muscle contraction, predict what would happen to these muscles in a state of rigor if you (a) added ATP but no free calcium ions; (b) added ATP with a substantial concentration of calcium ions.

34. When curare, a South American Indian arrow poison, is placed on a nerve-muscle preparation, the muscle will not contract when the nerve is stimulated, even though neu-

rotransmitter is still being released from the nerve. What is the neurotransmitter? What is a possible explanation for the action of curare?

35. On the basis of what you have learned about muscle fiber types and metabolism, predict what variations in structure you would find among these athletes.
a. a 7 foot, 2 inch tall, 325-pound basketball player
b. a 6 foot, 4 inch tall, 325-pound football lineman
c. a 5 foot, 10 inch tall, 180-pound steer wrestler
d. a 5 foot, 7 inch tall, 130-pound female figure skater
e. a 4 foot, 11 inch tall, 89-pound female gymnast

Problem Conclusion

In this running problem, you learned about hyperkalemic periodic paralysis, a condition caused by a defect in the Na^+ channels that are present on muscle cell membranes. You also learned how the attacks that characterize the disease are triggered by increased levels of K^+ in the blood.

Further check your understanding of this running problem by checking your answers against those in the summary table.

	Question	Facts	Integration and Analysis
1	What effect does the continuous influx of Na^+ have on the membrane potential of Paul's muscle fibers?	The resting membrane potential of cells is negative relative to the extracellular fluid.	The influx of positive charge will make the interior of the muscle fiber less negative (i.e., will depolarize the membrane potential).
2	In people with hyperkalemic periodic paralysis, attacks tend to occur during a period of rest after exercise. Why?	Each muscle twitch results from an action potential in the muscle fiber. In the repolarization phase of the action potential, K^+ leaves the cell.	During the repeated contractions of exercise, K^+ leaves the muscle fiber and accumulates in the t-tubules. The increase in K^+ concentration affects the Na^+ channels and triggers an attack.
3	Why does the Na^+ channel defect cause paralysis?	The gated channels that release Ca^{2+} from the sarcoplasmic reticulum are opened by the depolarization of an action potential. However, they close in a time-dependent manner, even if the cell remains depolarized.	During an attack, the Na^+ channels remain open and continuously admit Na^+ into the muscle cell, so the cell remains depolarized. The muscle fiber is unable to repolarize and fire additional action potentials. The first action potential causes a twitch, but the muscle then goes into a state of flaccid (uncontracted) paralysis.
4	Speculate on how diuretics prevent attacks in people with hyperkalemic periodic paralysis.	Diuretics cause the loss of water and ions from the body.	Because diuretics lower K^+ concentrations in the blood, they may prevent the higher than normal K^+ concentrations that trigger periodic attacks of paralysis.
5	Explain why oral glucose might help bring Paul out of his paralysis.	Cells use glucose to produce ATP, and they use ATP to run the Na^+/K^+-ATPase that actively exchanges K^+ and Na^+ across the cell membrane.	Providing glucose to cells may increase ATP available to run the Na^+/K^+-ATPase, which removes K^+ from the extracellular fluid. Insulin, a hormone that enhances glucose uptake into cells, also helps relieve the paralysis.

13
Integrative Physiology I: Control of Body Movement

BACKGROUND BASICS

In the preceding five chapters, you learned how neurons use electrical and chemical signals for communication, how the central nervous system integrates the information it receives from sensory neurons, and how efferent neurons control muscles and glands. This chapter puts these components together to show how the body controls movement and other muscle functions.

Think back to the baseball pitcher who was introduced in the last chapter. As he stands on the mound, watching the first batter, he is receiving sensory information from multiple sources: the sound of the crowd, the sight of the batter and the catcher, the feel of the ball within his hand, the alignment of his body as he prepares to throw. Many sensory receptors code this information and send it to the central nervous system, where it is integrated. Some information is acted on consciously: The pitcher decides that it is time to throw a fast ball. Other information is processed at the subconscious level and acted upon without conscious thought: As he thinks about starting to throw, he shifts his weight to offset the impending movement of his arm. The integration of sensory information into an involuntary response is the hallmark of a reflex (∞ p. 158). We look first at the characteristics of nervous reflexes, then discuss how the central nervous system uses reflexes to control muscle contraction and movement of the body.

Problem

Tetanus

"She has not been able to talk to us. We're afraid she may have had a stroke." That is how her neighbors described 77-year-old Cecile Evans when they brought her to the emergency room. But when a neurological examination revealed no problems other than Mrs. Evans's inability to open her mouth and stiffness in her neck, emergency room physician Dr. Doris Ling began to consider other diagnoses. She noticed some scratches healing on Mrs. Evans's arms and legs and asked the neighbors if they knew what had caused them. "Oh, yes. She was visiting us about 10 days ago, and our dog jumped up and knocked her against a rusty barbed wire fence." At that point, Dr. Ling realized she was probably dealing with her first case of tetanus.

continued on page 367

NERVOUS REFLEXES

All nervous reflexes begin with a stimulus that activates a sensory receptor. The receptor sends information in the form of action potentials through sensory neurons to the central nervous system (∞ p. 264). The central nervous system is the integrating center that evaluates all incoming information and selects an appropriate response. It then initiates action potentials in efferent neurons to direct the response of muscles and glands, the effectors.

A key feature of many reflex pathways is **negative feedback,** a concept introduced in Chapter 6 (∞ p. 162). Negative feedback signals from the muscles and joints of the body keep the central nervous system continuously informed of changing body position. Some reflexes have a **feedforward** component that allows the body to anticipate a stimulus and begin the response (∞ p. 163). Bracing yourself in anticipation of a collision would be an example of a feedforward response.

Nervous Reflex Pathways Can Be Classified Different Ways

Reflex pathways in the nervous system consist of chains or networks of neurons that link sensory receptors to muscles or glands. Neural reflexes can be classified in several ways (Table 13-1):

1. *By the efferent division of the nervous system that controls the response.* Reflexes that involve somatic motor neurons and skeletal muscles are known as **somatic reflexes.** Reflexes whose responses are controlled by autonomic neurons are called **autonomic, or visceral, reflexes.**

2. *By the central nervous system location where the reflex is integrated.* **Spinal reflexes** are integrated within the spinal cord. These reflexes may be modulated by higher input from the brain, but they can occur without that input. Reflexes that are integrated in the brain are called **cranial reflexes.**

3. *By whether the reflex is inborn or learned.* Many reflexes are innate, that is, we are born with them and they are genetically determined. For example, the knee jerk reflex, in which the lower leg kicks out when the lower edge of the kneecap is tapped, is an innate response. Other reflexes are acquired through experience. The example of Pavlov's dogs salivating upon hearing a bell is the classic example of a learned (conditioned) reflex.

4. *By the number of neurons in the reflex pathway.* The simplest reflex pathway is a **monosynaptic reflex**

TABLE 13-1   Classification of Neural Reflexes

Neural reflexes can be classified by:

1. Efferent division that controls the effector
   a. Somatic motor neurons control skeletal muscles.
   b. Autonomic neurons control smooth and cardiac muscle, glands, and adipose tissue.
2. Integrating region within the central nervous system
   a. Spinal reflexes do not require input from the brain.
   b. Cranial reflexes are integrated within the brain.
3. Time at which the reflex develops
   a. Innate (inborn) reflexes are genetically determined.
   b. Learned (conditioned) reflexes are acquired through experience.
4. The number of neurons in the reflex loop
   a. Monosynaptic reflexes have only two neurons: one afferent (sensory) and one efferent. Only somatic motor reflexes can be monosynaptic.
   b. Polysynaptic reflexes add one or more interneurons between the afferent and efferent neurons. All autonomic reflexes are polysynaptic because they have three neurons: one afferent and two efferent.

with only two neurons: an afferent sensory neuron and an efferent somatic motor neuron. These two neurons synapse within the spinal cord, allowing a signal from the receptor to go directly back to the skeletal muscle effector (Fig. 13-1a ■). The monosynaptic reflex gets its name from the single synapse between the neurons; in this instance, the synapse at the neuromuscular junction is ignored.

Most reflexes have three or more neurons in the pathway, leading to their designation as **polysynaptic reflexes** (Fig. 13-1b ■). Polysynaptic reflexes may be quite complex and often branch within the central nervous system to form networks with multiple interneurons.

Sometimes a single neuron branches, and its collaterals synapse on multiple target neurons. This pattern is known as **divergence** (Fig. 13-2a ■). If a larger number of presynaptic neurons provide input to a smaller number of postsynaptic neurons, the pattern is known as **convergence** (Fig. 13-2b ■). Convergence and divergence are key features in the complex control that the central nervous system exerts over its many targets.

Modulation of Neuronal Activity

One advantage of convergence in the central nervous system is that input from multiple sources can be used to influence the output of a single postsynaptic cell. When two or more presynaptic neurons converge on the dendrites or cell body of a single postsynaptic cell, the response of the postsynaptic cell will be determined by the summed input from the presynaptic neurons (∞ p. 211). Presynaptic neurons that release excitatory neurotransmitters create depolarizing graded potentials in the postsynaptic cell. Depolarization makes the postsynaptic neuron more likely to fire an action potential, so these graded potentials are called **excitatory postsynaptic potentials (EPSPs)** (∞ p. 210). In contrast, presynaptic neurons that release inhibitory neurotransmitters hyperpolarize the postsynaptic cell, making it less likely to fire an action potential. Hyperpolarizing graded potentials are therefore known as **inhibitory postsynaptic potentials (IPSPs).** When two or more graded potentials (EPSPs, IPSPs, or both) arrive at the trigger zone of an axon within a short period of time, their effects will be additive, and summation will occur.

(a) Monosynaptic reflex

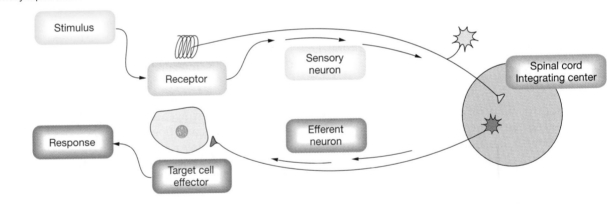

(b) Polysynaptic reflex

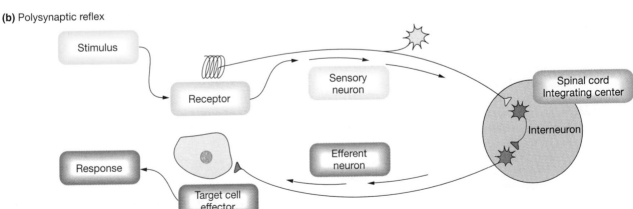

■ Figure 13-1 **Monosynaptic and polysynaptic reflexes** (a) A monosynaptic reflex has a single synapse between the afferent and efferent neurons. The synapse at the target organ is not counted. All monosynaptic reflexes are somatic motor reflexes. (b) Polysynaptic reflexes have two or more synapses. In somatic motor reflexes, the synapses occur in the central nervous system, as illustrated. In autonomic reflexes, a synapse always occurs between the pre- and the postganglionic efferent neurons.

(a) Divergence

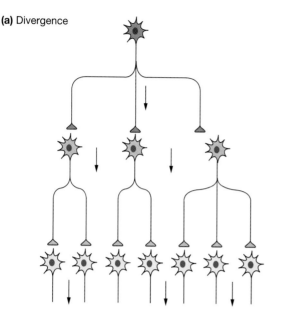

(b) Convergence

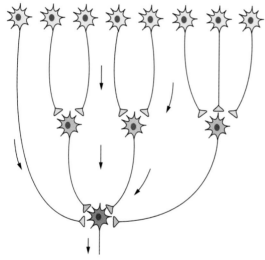

■ **Figure 13-2 Convergence and divergence** (a) In a divergent pathway, a presynaptic neuron branches to affect a larger number of postsynaptic neurons. (b) In convergent pathways, a large number of presynaptic neurons converge to influence a smaller number of postsynaptic neurons.

Modulation of neuronal pathways can occur either at the dendrites and cell body of the postsynaptic cell or at the axon terminal. Chemicals that act as neuromodulators change the responsiveness of the target cell to neurotransmitters. Most act through second messenger systems that alter existing proteins, and their effects last much longer than do those of neurotransmitters. Many neuromodulators are peptides, and some, such as somatostatin, are also known for their activity as hormones.

Postsynaptic modulation occurs when a neuron releases a neuromodulator, usually inhibitory, onto a postsynaptic cell (for example, a neuron or muscle) and alters its response. In Figure 13-3a ■, a modulatory neu-

> **Visualization Techniques in Sports**
> Presynaptic facilitation is now believed to be the physiological mechanism that underlies the success of visualization techniques in sports. Visualization enables athletes to maximize their performance by "psyching" themselves, picturing in their minds the perfect vault or the leaping catch for a touchdown. By pathways that we still do not understand, the visual image conjured up by the cerebral cortex is translated into signals that find their way to the muscles. This is only one example of many fascinating connections between the higher brain and the body.

ron synapses on the postganglionic autonomic neuron. In this example, the modulatory neuron alters the properties of the postsynaptic cell so that it no longer fires an action potential in response to neurotransmitter released by the presynaptic neuron.

A more selective form of modulation is **presynaptic modulation,** in which a modulatory neuron changes the amount of neurotransmitter released by another neuron. In presynaptic modulation, the axon terminal of the modulatory neuron terminates on or close to an axon terminal of the presynaptic cell (Fig. 13-3b ■). If activity in the modulatory neuron decreases neurotransmitter release by the presynaptic cell, the modulation is called *presynaptic inhibition. Presynaptic facilitation,* in which modulatory input increases neurotransmitter release by the presynaptic cell, can also occur.

Presynaptic modulation provides a more precise means of control than postsynaptic modulation. If the responsiveness of a neuron is changed at the dendrites and cell body (postsynaptic modulation), all target cells of the neuron will be affected equally (Fig. 13-3a ■). In contrast, presynaptic inhibition on a divergent neuron allows selective modulation of collaterals and their targets (Fig. 13-3b ■). One collateral can be inhibited while others remain unaffected.

> **Mechanisms of Presynaptic Inhibition**
> Several cellular mechanisms have been proposed for presynaptic inhibition. In the first mechanism, voltage-gated calcium channels in the axon terminal are inhibited, preventing the Ca^{2+} entry that triggers exocytosis of synaptic vesicles. In the second mechanism, chloride channels in the axon membrane open, allowing Cl^- into the cell. The hyperpolarization created by Cl^- entry offsets the depolarization of the action potential that arrives at the axon terminal. Hyperpolarization also occurs when K^+ channels are opened, allowing K^+ to leave the cell. In both instances, without depolarization, voltage-gated Ca^{2+} channels do not open and there is no Ca^{2+}-activated release of neurotransmitter. In the final mechanism, neurotransmitter release is directly inhibited by an unknown mechanism not associated with calcium entry.

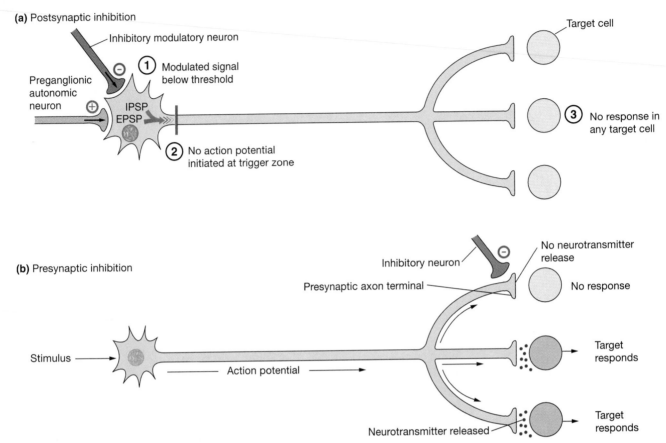

(a) Postsynaptic inhibition

Inhibitory modulatory neuron

Preganglionic autonomic neuron

① Modulated signal below threshold

IPSP
EPSP

② No action potential initiated at trigger zone

Target cell

③ No response in any target cell

(b) Presynaptic inhibition

Inhibitory neuron

Presynaptic axon terminal

No neurotransmitter release

No response

Stimulus

Action potential

Target responds

Neurotransmitter released

Target responds

■ Figure 13-3 **Postsynaptic and presynaptic inhibition** (a) In postsynaptic inhibition, the modulatory neuron decreases the likelihood that the postsynaptic cell will fire an action potential. It does so by creating a hyperpolarizing inhibitory postsynaptic potential (IPSP) that counteracts the depolarizing excitatory postsynaptic potential. In postsynaptic inhibition, all targets of the postsynaptic cell will be inhibited equally. (b) In presynaptic inhibition, the modulatory neuron synapses on one collateral of the presynaptic neuron. In this way, one target of the neuron can be selectively inhibited.

AUTONOMIC REFLEXES

Autonomic reflexes are also known as visceral reflexes because they often involve the internal organs of the body. Some visceral reflexes are spinal reflexes: Urination and defecation are examples. Often, these spinal reflexes are modulated by excitatory or inhibitory input from the brain, carried in descending tracts from higher brain centers. For example, urination may be voluntarily initiated by conscious thought, or it may be inhibited by emotion or a stressful situation, such as the presence of other people. Often, the higher control of a spinal reflex is a learned response. Toilet-training that we master as toddlers is an example of a learned reflex that our central nervous system uses to modulate the simple spinal reflex of urination.

Other autonomic reflexes are integrated in the brain, primarily in the hypothalamus, thalamus, and brain stem. These regions contain centers that coordinate body functions needed to maintain homeostasis, such as heart rate, blood pressure, breathing, eating, water balance, and maintenance of body temperature (see Fig. 11-9 ■, p. 316). The brain stem also contains the integrating centers for autonomic reflexes such as salivating, vomiting, sneezing, coughing, swallowing, and gagging.

An interesting type of autonomic reflex involves the conversion of emotional stimuli into visceral responses. The limbic system, the site of primitive drives such as sex, fear, rage, aggression, and hunger, has been called the "visceral brain" because of its role in these emotionally driven reflexes. We speak of "gut feelings" and "butterflies in the stomach"—all transformations of emotion into somatic sensation and visceral function. Other emotion-linked autonomic reflexes include urination, defecation, blushing, blanching, and *piloerection*, in which tiny muscles in the hair follicles pull the shaft of the hair erect ("I was so scared that my hair stood on end!").

Autonomic reflexes are all polysynaptic, with at least one synapse in the central nervous system between the sensory neuron and the preganglionic autonomic neuron, and an additional synapse found between the two autonomic neurons (Fig. 13-4 ■). Many autonomic reflexes are characterized by tonic activity, a continuous stream of action potentials that create an ongoing response in the effector. For example, the tonic control of blood vessels discussed in Chapter 6 is an example of

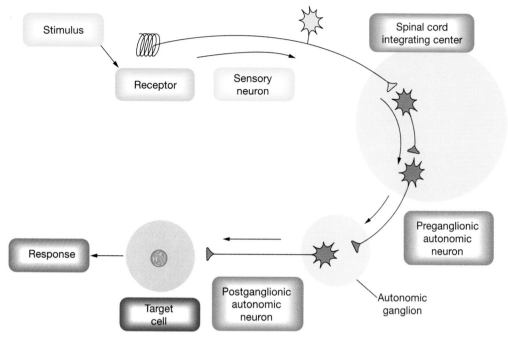

■ Figure 13-4 **Autonomic reflexes**

a continuously active autonomic reflex (∞ p. 315). You will encounter many autonomic reflexes as you continue your study of the systems of the body.

✓ Name the general steps of a reflex pathway and the anatomical structures in the nervous system that correspond to each step.

✓ If the membrane potential hyperpolarizes, does it become more positive or more negative? Move closer to threshold or move farther away from threshold?

SKELETAL MUSCLE REFLEXES

Although we are not always aware of them, skeletal muscle reflexes are involved in almost everything we do. Receptors that sense changes in muscle tension, stretch, and pressure feed information to the central nervous system, which responds in one of two ways. If muscle contraction is the appropriate response, action potentials pass along motor neurons to the muscle fibers. If muscles need to be relaxed to achieve the response, the central nervous system *inhibits* the motor neurons controlling the muscle. Recall that somatic motor neurons are always excitatory and always cause contraction in skeletal muscle. For relaxation to occur, the excitatory somatic motor neuron must be inhibited. Thus, relaxation is controlled at the level of the central nervous system, where output by the somatic motor neuron can be altered.

Receptors within the skeletal muscles themselves sense changes in muscle length and tension and activate muscle reflexes. Sensory neurons lead from these receptors

to the central nervous system. Once the input signal has been integrated, information is carried to skeletal muscles by two different types of motor neurons. **Alpha motor neurons** are efferent neurons that innervate the normal contractile fibers of the muscle, also known as **extrafusal muscle fibers.** Action potentials in extrafusal muscle fibers cause muscle contraction. **Gamma motor neurons** are smaller neurons associated with specialized muscle fibers within sensory receptors.

The three types of sensory receptors found in skeletal muscle are muscle spindles, Golgi tendon organs, and joint capsule mechanoreceptors (proprioreceptors). These receptors send information to the central nervous system about the relative positioning of bones linked by flexible joints. The next two sections detail the function of muscle spindles and tendon organs, two interesting and unique receptors. We will not discuss joint capsule mechanoreceptors.

---

continued from page 363

Tetanus, also known as lockjaw, is a devastating disease caused by the bacterium *Clostridium tetani.* These bacteria are commonly found in soil and enter the human body through a cut or wound. As the bacteria reproduce in the tissues, they release a protein neurotoxin. This toxin, called tetanospasmin, is taken up by motor neurons at the peripheral nerve endings. Tetanospasmin then travels along the axons until it reaches the motor neuron cell body in the spinal cord.

Question 1: Tetanospasmin is a protein. By what process is it taken up into neurons? (Hint: ∞ p. 126) By what process does it travel up the axon to the nerve cell body? (Hint: ∞ p. 205)

Muscle Spindles Respond to Muscle Stretch

The **muscle spindle** receptors send information about muscle length to the central nervous system. These small, elongated structures are scattered among and arranged parallel to the contractile extrafusal fibers of the muscle. Each spindle consists of a connective tissue sheath that encloses a group of **intrafusal fibers** [*intra-*, within + *fusus*, spindle]. Intrafusal fibers are modified muscle fibers that lack myofibrils in their central portions (Fig. 13-5b ■). Sensory neurons wrap around the middle of the intrafusal fibers and project to the spinal cord. These neurons fire when the center of the intrafusal fiber stretches.

Although the centers of the intrafusal fibers are not contractile, the ends do contain fibers that contract when stimulated by gamma motor neurons. Although this contraction does not affect overall muscle tension, it does stretch the central portion of the intrafusal fiber.

Muscle spindles act as stretch receptors. The sensory neurons within the spindle are tonically active, firing action potentials when the muscle is at its resting length (Fig. 13-6a ■). These signals travel to the spinal cord, where the afferent spindle neuron synapses with alpha

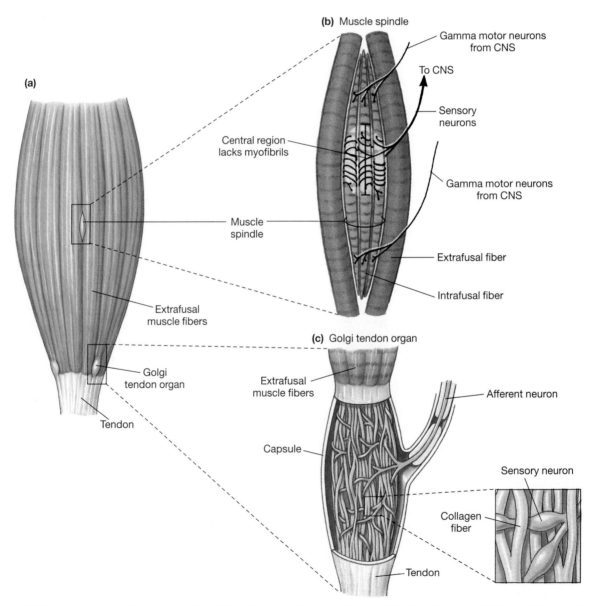

■ Figure 13-5 **Sensory receptors in muscle** (a) Buried among the normal contractile fibers (extrafusal fibers) of the muscle are stretch receptors known as muscle spindles. Golgi tendon organs are receptors that link the muscle and the tendon. Contraction in extrafusal fibers is controlled by alpha motor neurons, and contraction of the muscle spindles is controlled by gamma motor neurons. The Golgi tendon organ does not contract. (b) The central region of the muscle spindle lacks myofibrils and cannot contract. Sensory nerve endings wrap around the central region and fire when the central section of the muscle spindle stretches. The ends of the muscle spindle contain myofibrils that contract in response to commands carried by gamma motor neurons. (c) The Golgi tendon organ consists of sensory neurons interwoven among collagen fibers. If the collagen fibers are stretched, they pinch the sensory neurons and trigger action potentials.

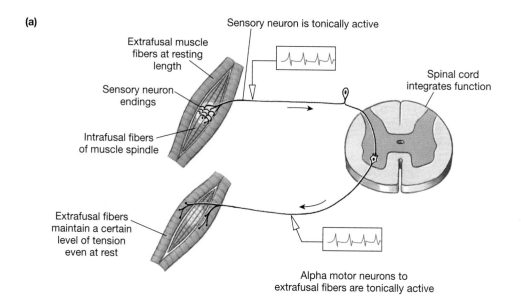

(a)

Extrafusal muscle fibers at resting length

Sensory neuron endings

Intrafusal fibers of muscle spindle

Sensory neuron is tonically active

Spinal cord integrates function

Extrafusal fibers maintain a certain level of tension even at rest

Alpha motor neurons to extrafusal fibers are tonically active

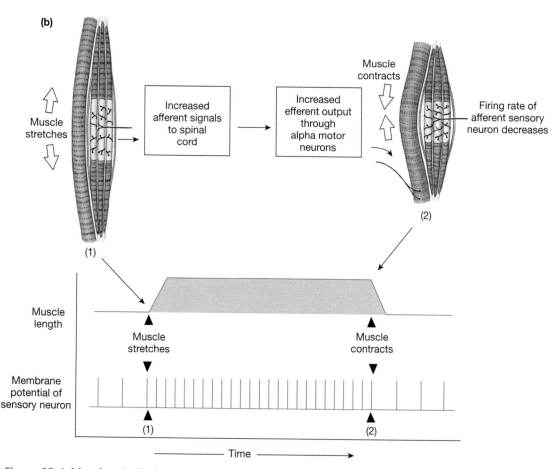

(b)

Muscle stretches

Increased afferent signals to spinal cord

Increased efferent output through alpha motor neurons

Muscle contracts

Firing rate of afferent sensory neuron decreases

(2)

(1)

Muscle length

Muscle stretches

Muscle contracts

Membrane potential of sensory neuron

(1)

(2)

Time

■ Figure 13-6 **Muscle spindle function** (a) When a muscle is at its resting length, the muscle spindle is slightly stretched and its associated sensory neuron shows tonic activity. As a result of tonic activity of the reflex, the associated muscle maintains a certain level of tension, or tone, even at rest. (b) If a muscle is stretched, its muscle spindles are also stretched. This stretching increases the firing rate of the spindle afferents, and the muscle contracts. Contraction relieves the stretch on the spindle and acts as negative feedback to diminish the reflex.

motor neurons. The alpha motor neurons in turn create tonic excitation and contraction in the extrafusal fibers associated with the muscle spindle. Because of this tonic activity, even a muscle at rest maintains a certain level of tension, known as muscle tone.

The sensory neurons in the muscle spindle change their activity depending on the length of the intrafusal fibers of the spindle. If the muscle stretches, the intrafusal fibers of the spindle also stretch, so the sensory neurons fire more rapidly (Fig. 13-6b ■). This more rapid firing initiates reflex contraction of the muscle, relieving the stretch on the muscle and preventing damage from overstretching. Because contraction also decreases stretch in the spindle fibers, their firing rate declines. This reflex pathway, in which muscle stretch initiates a contraction response, is known as a **stretch reflex.**

What happens to the firing rate of tonically active spindle afferents when a resting muscle contracts and shortens? You might predict that the afferent neurons slow their firing rate or stop firing altogether (Fig. 13-7a ■). But they neither slow nor stop firing, because of **alpha-gamma coactivation.** When alpha motor neurons to extrafusal fibers are activated for muscle contraction, the gamma motor neurons to the spindle are activated at the same time (Fig. 13-7b ■). The gamma motor neurons innervate the contractile ends of the intrafusal fibers, which contract and shorten. Contraction at both ends of the fibers lengthens the central section and maintains stretch on the sensory nerve endings, even though the bulk of the muscle has relaxed. Thus, when the extrafusal muscle fibers shorten during contraction, the intra-

fusal fibers remain stretched and continue to monitor tension in the muscle.

An example of how muscle spindles work during a stretch reflex is shown in Figure 13-8 ■. You can try this yourself with a friend. The subject stands with eyes closed, elbows at 90°, and one hand extended with the palm up. A small book or other flat weight is placed in the hand. The assistant suddenly drops a heavier load, such as another book, onto the hand. The added weight sends the hand downward, stretching the biceps muscle and activating its muscle spindles. Sensory input into the spinal cord then activates the alpha motor neurons of the biceps muscle. The biceps contracts, bringing the arm back to its original position.

Golgi Tendon Organs

A second type of receptor for stretch reflexes is the **Golgi tendon organ.** These receptors are found at the junction of the tendons and muscle fibers, placing them in series with the muscle fibers. Golgi tendon organs are com-

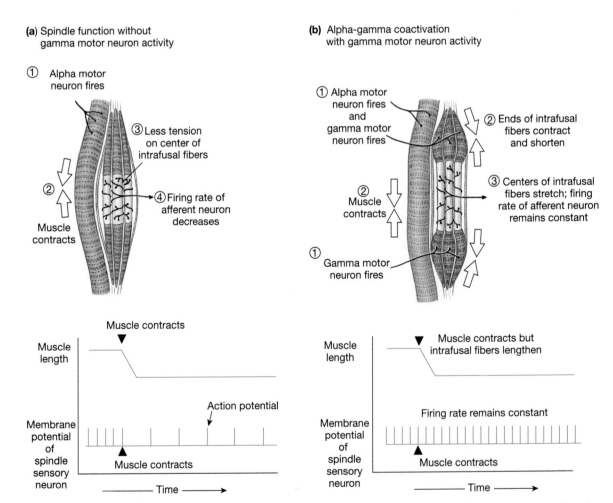

(a) Spindle function without gamma motor neuron activity

① Alpha motor neuron fires

② Muscle contracts

③ Less tension on center of intrafusal fibers

④ Firing rate of afferent neuron decreases

Muscle contracts

Muscle length

Action potential

Membrane potential of spindle sensory neuron

Muscle contracts

Time

(b) Alpha-gamma coactivation with gamma motor neuron activity

① Alpha motor neuron fires and gamma motor neuron fires

② Ends of intrafusal fibers contract and shorten

② Muscle contracts

③ Centers of intrafusal fibers stretch; firing rate of afferent neuron remains constant

① Gamma motor neuron fires

Muscle contracts but intrafusal fibers lengthen

Muscle length

Firing rate remains constant

Membrane potential of spindle sensory neuron

Muscle contracts

Time

■ Figure 13-7 **Gamma motor neurons** The gamma motor neurons allow the muscle spindle to maintain function when the muscle is contracting. (a) If the gamma motor neurons to the muscle spindle are cut, the spindle loses activity when the muscle contracts because the central region of the spindle loses its stretch. (b) Spindle function is maintained through alpha-gamma coactivation: When the alpha motor neurons are active, so are the gamma motor neurons to the spindle. Contraction of the ends of the spindle stretches the central region and maintains the tonic activity of the spindle.

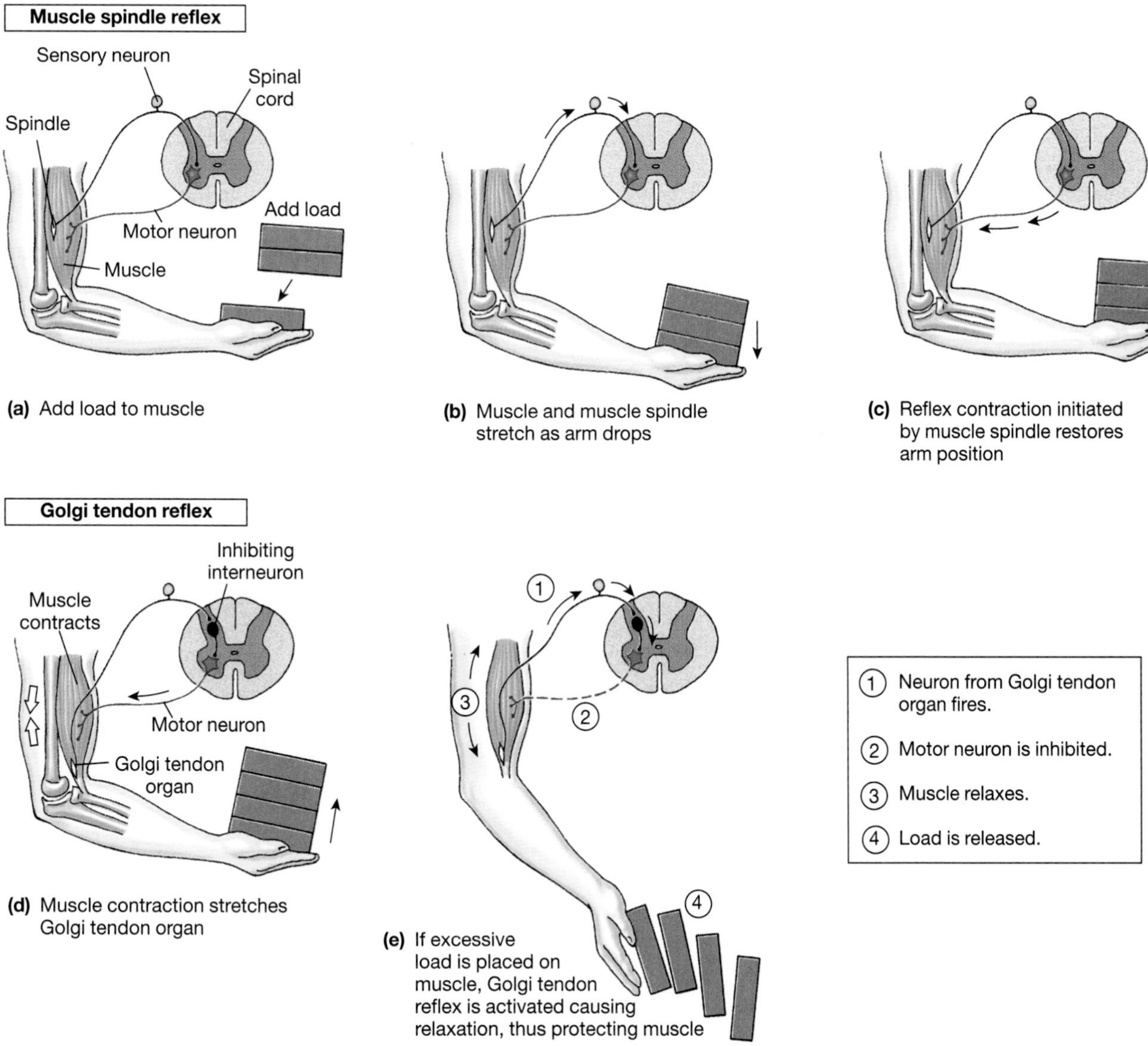

Muscle spindle reflex

(a) Add load to muscle

(b) Muscle and muscle spindle stretch as arm drops

(c) Reflex contraction initiated by muscle spindle restores arm position

Golgi tendon reflex

(d) Muscle contraction stretches Golgi tendon organ

(e) If excessive load is placed on muscle, Golgi tendon reflex is activated causing relaxation, thus protecting muscle

1. Neuron from Golgi tendon organ fires.
2. Motor neuron is inhibited.
3. Muscle relaxes.
4. Load is released.

■ **Figure 13-8 Muscle reflexes** (a–c) The muscle spindle reflex: The addition of a load to a muscle stretches the muscle and the spindles, creating a reflex contraction. (d, e) The Golgi tendon reflex protects the muscle from excessively heavy loads. If the muscle contraction initiated by the spindle reflex approaches maximum tension, the Golgi tendon organs fire, causing the muscle to relax and drop the load.

posed of free nerve endings that wind between collagen fibers inside a connective tissue capsule (Fig. 13-5c ■). Because Golgi tendon organs are in the elastic element of the muscle, they respond to both stretch and contraction of the muscle. If the muscle is stretched, the collagen fibers pull tight, pinching sensory endings of the afferent neurons and causing them to fire. If the muscle contracts, the tendons act as an elastic component and are pulled taut during the isometric phase of the contraction (∞ p. 344). Again, the sensory neurons in the Golgi tendon organ are stimulated and fire. Although both stretch and contraction stimulate the neurons, they respond must strongly to contraction. Activation of the Golgi tendon organ inhibits alpha motor neurons and decreases muscle contraction. Under most circumstances, this reflex slows muscle contraction as the force of contraction increases. In other instances, the Golgi

tendon organs prevent excessive contraction that might injure the muscle. In the example of the book placed on the outstretched hand, if the added weight requires more tension than the muscle can develop, the Golgi tendon organ will respond as muscle tension nears its maximum. It triggers reflex *inhibition* of the biceps muscle, which causes the arm to relax. Thus, the person will drop the added weight before the muscle fibers can be damaged (Fig. 13-8d, e ■).

Myotatic Reflexes and Reciprocal Inhibition

Movement around most flexible joints in the body is controlled by groups of synergistic and antagonistic muscles that act in a coordinated fashion. Afferent sensory neurons from a muscle and efferent motor neurons that control the muscle are linked together by diverging and converging pathways of interneurons within the spinal cord. The collection of pathways controlling a single joint is known as a **myotatic unit.**

The simplest reflex in a myotatic unit is the monosynaptic stretch reflex that involves only two neurons: the sensory neuron from the muscle spindle and the somatic motor neuron to the muscle. The knee jerk reflex is an example of a monosynaptic stretch reflex (Fig. 13-9 ■). To demonstrate the knee jerk reflex, the subject sits on the edge of a table so that the lower leg hangs relaxed. When the patellar tendon below the kneecap is tapped with a small rubber hammer, the pressure stretches the quadriceps muscle that runs up the front of the thigh. This stretching activates the muscle spindle stretch receptors within the muscle, sending an action potential through the sensory neuron to the spinal cord. The sensory neuron synapses directly onto the motor neuron that controls

contraction of the quadriceps muscle. The action potential in the motor neuron causes the muscle fibers of the quadriceps to contract, and the lower leg swings forward.

In addition to muscle contraction that extends the leg, the relaxation of antagonistic flexor muscles must also occur (**reciprocal inhibition**). In the leg, this requires relaxation of the hamstring muscles that run up the back of the thigh. The single stimulus of the blow to the tendon accomplishes both contraction of the quadriceps muscle and reciprocal inhibition of the hamstrings. The sensory neuron branches upon entering the spinal cord. One collateral activates the motor neuron innervating the quadriceps, while the other collateral synapses with an inhibitory interneuron. This interneuron inhibits the motor neuron controlling the hamstrings, preventing it from firing an action potential. The result is a relaxation of the hamstrings that allows contraction of the quadriceps to proceed unopposed.

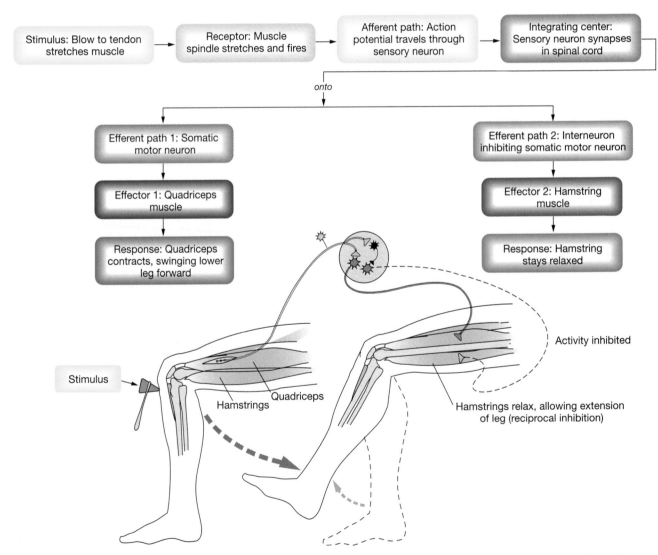

■ **Figure 13-9   The knee jerk reflex**   The knee jerk is a simple monosynaptic spinal reflex in which a blow to the patellar tendon causes contraction of the quadriceps muscle (pathway 1). An integral part of the reflex is the reciprocal inhibition of the hamstring muscles (pathway 2).

Flexion Reflexes and the Crossed Extensor Reflex

Flexion reflexes are polysynaptic reflex pathways that cause an arm or leg to be pulled away from a painful stimulus such as a pinprick or a hot stove. These reflexes, like the reciprocal inhibition reflex just described, rely on divergent pathways in the spinal cord. The afferent fiber from the nociceptor (pain receptor) activates several excitatory interneurons (Fig. 13-10 ■). Some of these interneurons excite alpha motor neurons and cause contraction in all the flexor muscles in the affected limb. The other interneurons simultaneously activate inhibitory interneurons that cause relaxation of the antagonistic muscle groups. With reciprocal inhibition, the limb is flexed, withdrawing it from the painful stimulus. This type of reflex takes longer to happen than a stretch reflex such as

the knee jerk, an indication that it is a polysynaptic rather than a monosynaptic reflex.

Flexion reflexes, particularly in the legs, are usually accompanied by the **crossed extensor reflex,** a postural reflex that helps maintain balance when one foot is lifted from the ground. In this reflex, also shown in Figure 13-10 ■, the quick withdrawal of the right foot from a painful stimulus (a tack) is matched by extension of the left leg so that it can support the sudden shift in weight. The extensors contract in the supporting left leg and relax in the withdrawing right leg while the opposite occurs in the flexor muscles. Note in the figure how the one sensory neuron synapses on multiple interneurons. Divergence of the sensory signal permits a single stimulus to control two sets of antagonistic muscle groups and also send sensory information to the brain. This type of complex reflex with multiple neuron interactions is

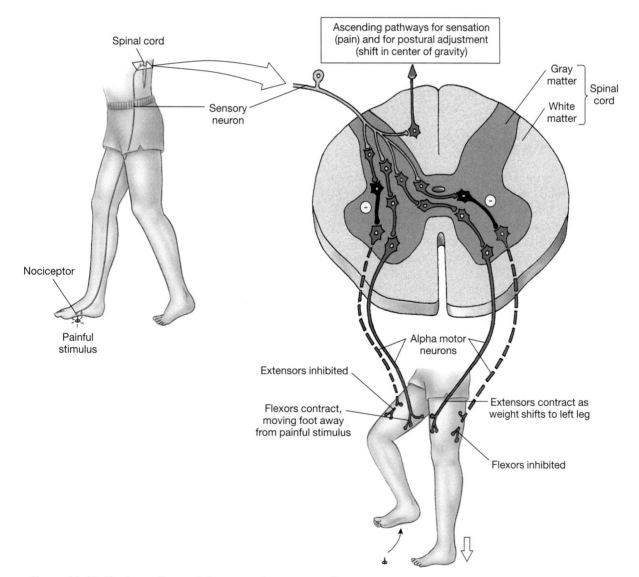

■ **Figure 13-10 Flexion reflex and the crossed extensor reflex** Coordination of reflexes with postural adjustments is essential for maintaining balance. When the right foot encounters a painful stimulus, a polysynaptic withdrawal reflex occurs when the right flexor muscles contract. Reciprocal inhibition keeps the right extensor muscles relaxed. The crossed extensor reflex, activated by the same sensory pathway, straightens the left leg so that it will support the sudden addition of weight.

more typical of our reflexes than the simple monosynaptic knee jerk stretch reflex.

The most complicated spinal reflex pathways are controlled by networks of neurons in the central nervous system called **central pattern generators.** Once activated, central pattern generators create spontaneous repetitive movement without further sensory input. In humans, rhythmic movements controlled by central pattern generators include locomotion and the unconscious rhythmicity of quiet breathing.

In the next section of the chapter, we look at how the central nervous system controls movements that range from involuntary reflexes to complex, voluntary movement patterns such as dancing, throwing a ball, or playing a musical instrument.

✓ As you pick up a heavy weight, which of the following are active in your biceps muscle: alpha motor neuron, gamma motor neuron, muscle spindle afferent neurons, Golgi tendon organ afferent neurons?

✓ What distinguishes a stretch reflex from an extensor reflex?

THE INTEGRATED CONTROL OF BODY MOVEMENT

Most of us never think about how our body translates thoughts into action. Even the simplest movement requires proper timing so that antagonistic and synergistic muscle groups contract in the appropriate sequence. In addition, the body must continuously adjust its position to compensate for differences between the intended movement and the actual one. For example, the baseball pitcher steps off the mound as he reaches for a ground ball, but in doing so, he slips on a wet patch of grass. His brain quickly compensates for the unexpected change in

 Reflexes and Muscle Tone　Clinicians use reflexes to investigate the condition of the nervous system and the muscles. For a reflex to be normal, there must be normal conduction through all neurons in the pathway, normal synaptic transmission at the neuromuscular junction, and normal muscle contraction. A reflex that is absent, abnormally slow, or greater than normal (hyperactive) suggests the presence of a disease. Interestingly, not all abnormal reflexes suggest neuromuscular disorders. For example, slowed relaxation of the ankle flexion reflex suggests hypothyroidism. Besides testing reflexes, clinicians assess muscle tone. Even when relaxed and at rest, muscles have a certain resistance to stretch that is the result of continuous (tonic) output by alpha motor neurons. If muscle tone is absent or if a muscle resists being passively stretched by the examiner (increased tone), there is probably a problem with the pathways that control muscle contraction.

position through reflex muscle activity, and he stays on his feet to intercept the ball.

Skeletal muscles cannot communicate with each other directly, so they send messages to the central nervous system, allowing that integrating center to take charge and direct movement. Most body movement is a highly integrated, coordinated response that requires input from multiple regions of the brain. This section examines the integrating centers within the central nervous system that are responsible for control of body movement.

Types of Movement

Movement can be loosely classified into three categories: reflex movement, voluntary movement, and rhythmic movement (Table 13-2). **Reflex movements** are the least complex and are integrated primarily in the spinal cord. However, like other spinal reflexes, reflex movements can be modulated by input from higher brain centers

---

TABLE 13-2　**Types of Movement**

	Reflex	Voluntary	Rhythmic
Stimulus that initiates movement	Primarily external via sensory receptors; minimally voluntary	External stimuli or at will	Initiation and termination voluntary
Example	Knee jerk, cough, postural reflexes	Playing piano	Walk, run
Complexity	Least complex; integrated at level of spinal cord with higher center modulation	Most complex; integrated in cerebral cortex	Intermediate complexity; integrated in spinal cord with higher center input required
Comments	Inherent, rapid	Learned movements that improve with practice; once learned, may become subconscious ("muscle memory")	Spinal circuits act as pattern generators; activation of these pathways requires input from brain stem

continued from page 367

Once in the spinal cord, tetanospasmin is released from the motor neuron. It then selectively blocks neurotransmitter release at inhibitory synapses. Patients with tetanus experience muscle spasms that begin in the jaw and may eventually affect the entire body. When the extremities become involved, the arms and legs may go into painful, rigid spasms.

Question 2: Using the reflex pathways diagrammed in Figures 13-9 and 13-10 ■, explain why inhibition of inhibitory interneurons might result in uncontrollable muscle spasms.

(Fig. 13-11 ■). In addition, the sensory input that initiates reflex movements, such as the input from muscle spindles and Golgi tendon organs, may go to the brain and participate in the coordination of voluntary movements or postural reflexes.

Postural reflexes that help us maintain body position as we stand or move through space are integrated in the brain stem. They require continuous sensory input from visual and vestibular (inner ear) sensory systems and from the muscles themselves (∞ p. 263). Muscle, tendon, and joint receptors provide information about the positions of various body parts relative to one another. You can tell if your arm is bent even when your eyes are closed because these receptors provide information about body position to the brain. Information from the vestibular apparatus of the ear and visual cues help us maintain our position in space. For example, we use the horizon to tell us our spatial orientation relative to the ground. In the absence of visual cues, we rely on tactile input. People trying to move in a dark room instinctively reach for a wall or piece of furniture to help orient themselves. Without visual and tactile cues, our orientation skills may fail. The lack of cues is what makes flying airplanes in clouds or fog impossible without instruments. The effect of gravity on the vestibular system is such a weak input when compared with visual or tactile cues that pilots may find themselves flying upside down relative to the ground.

Voluntary movements are the most complex type of movement. They require integration at the cerebral cortex, and they can be initiated at will without external stimuli. Learned voluntary movements improve with practice and some even become involuntary, like reflexes. Think about how difficult it was to learn to ride a bicycle. However, once you learned to pedal smoothly and to keep your balance, the movements became automatic. "Muscle memory" is the name dancers and athletes give the ability of the unconscious brain to reproduce voluntary, learned movements and positions.

Rhythmic movements such as walking or running are a combination of reflex movements and voluntary movements. Rhythmic movements must be initiated

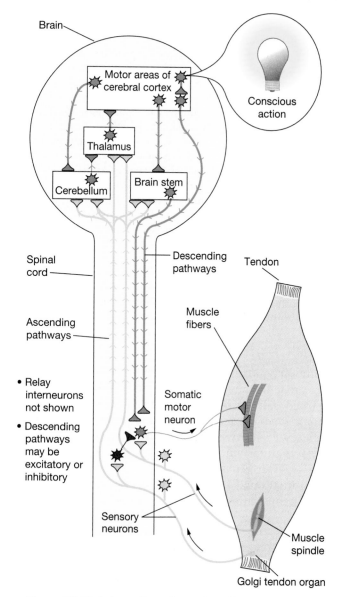

■ **Figure 13-11 Integration of muscle reflexes** Although many muscle reflexes are simple spinal reflexes, sensory information about them is transmitted to the brain through ascending pathways. In addition, the conscious and subconscious brain sends modulatory messages to the spinal integrating centers through descending pathways.

and terminated by input from the cerebral cortex, but once initiated, they can be sustained without further input from the brain. The rhythmic activity of skeletal muscles is maintained by groups of spinal interneurons. These act as central pattern generators, alternately contracting and relaxing muscles in a repetitive, rhythmic fashion until instructed to stop by signals from the brain. As an analogy, think of a battery-operated bunny. When the switch is thrown to "on," the bunny begins to hop. It continues its repetitive hopping until someone turns it off (or until the battery runs down).

The distinction between reflex, voluntary, and rhythmic movement is not always clear-cut. The precision of

Central Pattern Generators and Spinal Cord Injuries   The ability of CNS pattern generators, or central pattern generators, to sustain rhythmic movement without continued input from the central nervous system has proved important for research on spinal cord injuries. Scientists have stimulated rhythmic movements in experimental animals artificially when the spinal cord between the brain and the motor neurons has been severed. For example, an animal paralyzed by a spinal cord injury that blocks the "start walking" signal from the brain will walk if supported on a moving treadmill and given an electrical stimulus to activate the spinal pattern generator. As the treadmill moves the animal's legs, the spinal pattern generator, reinforced by sensory signals from muscle spindles, drives contraction of the leg muscles. Researchers are trying to take advantage of these rhythmic reflexes in people with spinal cord injuries. Some of the newest therapeutic techniques involve artificially stimulating portions of the spinal cord so that movement is restored to formerly paralyzed limbs.

continued from page 375

Dr. Ling admits Mrs. Evans to the intensive care unit. There Mrs. Evans is given tetanus antitoxin to deactivate any toxin that has not yet entered motor neurons; penicillin, an antibiotic that kills the bacteria; and drugs to help relax her muscles. Despite these treatments, by the third day Mrs. Evans is having difficulty breathing because of spasms in her chest muscles. Dr. Ling calls in the chief of anesthesiology to administer metocurine, a drug similar to curare. Curare and metocurine induce temporary paralysis of muscles by binding to ACh receptors on the motor end plate. Patients must be placed on respirators that breathe for them while receiving metocurine. For people with tetanus, however, metocurine can temporarily halt the muscle spasms and allow the body to recover.

Question 3: Why does the binding of metocurine to ACh receptors on the motor end plate induce muscle paralysis? (Hint: What is the function of ACh in synaptic transmission?) Is metocurine an agonist or an antagonist of ACh?

voluntary movements improves with practice, but so does that of some reflexes. Voluntary movements, once learned, can become reflexive. In addition, most voluntary movements require continuous input from postural reflexes. Feedforward reflexes allow the body to prepare for a voluntary movement, and feedback mechanisms are used to create a smooth, continuous motion. Coordination of movement thus requires cooperation from many parts of the brain.

Integration of Movement within the Central Nervous System

Three levels of the nervous system control movement: the spinal cord, the brain stem, and the motor areas of the cerebral cortex (Fig. 13-12 ■). The brain stem is influenced by input from the cerebellum, while the basal ganglia, clusters of neurons surrounding the thalamus,

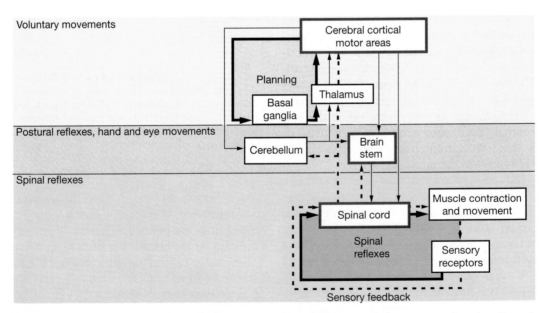

■ **Figure 13-12 Higher center control of movement** The simplest control of movement takes place through spinal reflexes. Descending pathways from the brain stem and cerebral cortex can initiate or modify movement. The cerebellum receives feedback information from sensory receptors and combines it with information from the cerebral cortex to modify the descending commands. The basal ganglia assist the cerebral cortex in planning movement.

and the cerebellum help cerebral cortical motor areas plan movement. The thalamus acts as a relay station for signals being sent to the cerebral cortex from the spinal cord, basal ganglia, and cerebellum (Table 13-3).

The simplest movements involve spinal reflexes integrated in the spinal cord. Sensory receptors such as muscle spindles, Golgi tendon organs, and joint proprioreceptors provide information to the spinal cord that can be acted upon without input from higher regions of the central nervous system. Some of this sensory information is sent through ascending pathways to the brain stem, along with sensory input from the eyes and vestibular apparatus. The brain stem is in charge of postural reflexes and hand and eye movements. It also gets commands from the cerebellum, the part of the brain responsible for "fine-tuning" movement.

Voluntary movements require the participation of the cerebral cortex, cerebellum, and the basal ganglia (Fig. 13-13 ■). The control of these behaviors can be divided into three steps: (1) decision making and planning, (2) initiating the movement, and (3) executing the movement. The cerebral cortex plays a key role in the first two steps. Behaviors such as movement require knowledge of the body's position in space (where am I?), a decision on what movement should be executed (what shall I do?), a plan for executing the movement (how shall I do it?), and the ability to hold the plan in memory long enough to carry it out. Let us return to our baseball pitcher and trace the process as he decides whether to throw a fast ball or a slow inside curve.

Standing out on the mound, the pitcher is acutely aware of his surroundings: the other players on the field, the batter in the box, the dirt beneath his feet. With the help of visual and somatosensory input, he is aware of his body position as he steadies himself for the pitch. Deciding which type of pitch to throw and anticipating the consequences occupy many pathways in his cortical association areas. Once the decision to throw a fast ball

Parkinson's Disease and the Basal Ganglia Our understanding of the role of the basal ganglia in the control of movement has been slow to develop because, for many years, animal experiments yielded little information. Randomly destroying portions of the basal ganglia did not affect research animals. However, research focusing on Parkinson's disease (Parkinsonism) in humans has been more rewarding. Parkinsonism is a condition characterized by abnormal movements and speech difficulties. These symptoms are associated with loss of the neurotransmitter dopamine in neurons of the basal ganglia. The cause of the disease is usually not known. However, a few years ago, a number of young drug users were diagnosed with Parkinsonism. Their disease was traced to the use of homemade heroin that contained a toxic contaminant that destroyed dopaminergic (dopamine-secreting) neurons. This contaminant has been isolated and now enables researchers to induce Parkinson's disease in experimental animals so that we have an animal model on which to test new treatments.

is made, this information is sent to the motor cortex, the primary region in charge of planning and organizing complex movements.

Now it is time to initiate the movement. Information from the association areas is sent to the cerebellum, which makes postural adjustments by combining feedback from peripheral sensory receptors with information from the cerebral cortex. This information also goes to the basal ganglia, which assist the cortical motor areas in planning the pitch. Once this network completes a plan of action, the motor cortex begins the execution of the movement.

The pitcher's decision to throw a fast ball now is translated into action potentials that travel down through the corticospinal tract, a group of interneurons that connect the cortex to the spinal cord. At the same time, *feedfor-*

TABLE 13-3 Neural Control of Movement

Location	Role	Receives Input from:	Sends Integrative Output to:
Spinal cord	Spinal reflexes	Sensory receptors	Brain stem, cerebellum, thalamus/cerebral cortex
Brain stem	Posture, hand and eye movements	Cerebellum, visual and vestibular sensory receptors	Spinal cord
Motor areas of cerebral cortex	Planning and coordinating complex movement	Thalamus	Brain stem, spinal cord, cerebellum, basal ganglia
Cerebellum	Monitors output signals from motor areas and adjusts movements	Spinal cord (sensory), cerebral cortex (commands)	Brain stem, cerebral cortex (Note: All output is inhibitory.)
Thalamus	Contains relay nuclei that pass messages to cerebral cortex	Basal ganglia, cerebellum, spinal cord	Cerebral cortex
Basal ganglia	Motor planning	Cerebral cortex	Cerebral cortex

■ **Figure 13-13 CNS control of voluntary movement** Voluntary movements can be divided into three phases: planning, initiation, and execution. Planning involves the exchange and coordination of information between the cortical association areas, the basal ganglia, and the cerebellum. Initiation is the responsibility of the motor cortex. Execution of voluntary movements is carried out through descending pathways to somatic motor neurons. Sensory information from the joints and muscles provides feedback to the brain that allows it to correct for any deviation between the planned movement and the actual movement.

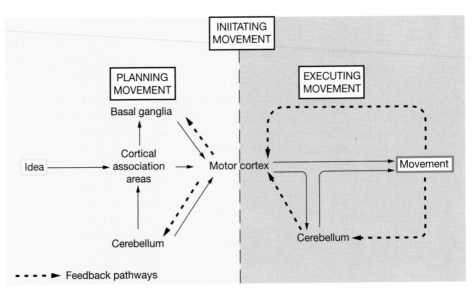

ward postural reflexes adjust the body position, shifting weight slightly in anticipation of the changes that are about to occur (Fig. 13-14 ■). Through divergent pathways, action potentials race down the interneurons to the somatic motor neurons that control the muscles used for the pitch: Some are excited, others are inhibited. Reciprocal inhibition allows precise control over antagonistic muscle groups as the pitcher flexes and retracts his right arm. His weight shifts onto his right foot as his arm moves back. Each of these movements activates sensory receptors that feed back information to the brain stem, activating postural reflexes. These reflexes adjust his body position so that the pitcher does not lose his balance and fall over backward. Finally, the pitcher releases the ball, catching his balance on the follow-through: another example of postural reflexes mediated through sensory feedback. His head stays erect and his eyes stay on the ball as it reaches the batter. WHACK . . . home run. As the pitcher's eyes follow the ball and he evaluates the result of his pitch, his brain is preparing for the next batter, hoping to use what it has learned from these pitches to improve those to come.

CONTROL OF MOVEMENT IN VISCERAL MUSCLES

Movement created by contracting smooth and cardiac muscles is very different from that created by skeletal muscles, in large part because smooth and cardiac muscle are not attached to bone. In the internal organs, or viscera, muscle contraction usually changes the shape of an organ, narrowing the lumen of a hollow organ or shortening the length of a tube. In many hollow internal organs, muscle contraction is used to push material through the lumen of the organ: The heart pumps blood, the digestive tract moves food, the uterus expels a baby.

Visceral muscle contraction is often reflexively controlled, but not always. Some types of smooth and cardiac muscle are capable of generating their own action

continued from page 376

Four weeks later, Mrs. Evans is ready to go home, completely recovered from her infection and showing no signs of lingering effects. Once she could talk, Mrs. Evans was able to tell Dr. Ling that she had never had immunization shots for tetanus or any other diseases. "Well, that made you one of only a handful of people in the United States who will develop tetanus this year," Dr. Ling told her. "You've been given your first two tetanus shots here in the hospital. Be sure to come back in six months for the last one so that this won't happen again." Because of national immunization programs begun in the 1950s, tetanus is now a rare disease in the United States. However, in developing countries without immunization programs, tetanus is still a common and serious condition.

Question 4: On the basis of what you know about who receives immunization shots in the United States, predict the age and background of people who are most likely to develop tetanus this year.

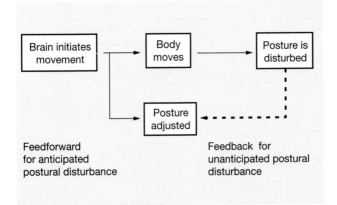

■ **Figure 13-14 Feedforward reflexes and feedback of information during movement**

potentials, independent of an external signal. Both the heart and digestive tract have spontaneously contracting muscle fibers that give rise to regular, rhythmic contractions.

Reflex control of visceral smooth muscle also varies from that of skeletal muscle. Skeletal muscles are controlled only by the nervous system, but in many types of visceral muscle, hormones are important in regulating contraction. In addition, some visceral muscle cells are connected to each other by gap junctions that allow electrical signals to pass directly from cell to cell.

Because smooth and cardiac muscle have such a variety of control mechanisms, we will discuss their control as we cover the appropriate organ system for each type of muscle. In the next chapter, we examine cardiac muscle and its function in the heart.

CHAPTER REVIEW

Chapter Summary

Nervous Reflexes

1. A nervous reflex consists of the following elements: stimulus → receptor → sensory neurons → central nervous system → efferent neurons → effectors (muscles and glands) → response. (p. 363)

2. Nervous reflexes can be classified in several different ways. **Somatic reflexes** involve somatic motor neurons and skeletal muscles. **Autonomic,** or **visceral, reflexes** are controlled by autonomic neurons. (p. 363)

3. **Spinal reflexes** are integrated within the spinal cord, and reflexes that are integrated in the brain are called **cranial reflexes.** (p. 363)

4. Many reflexes are inborn; others are acquired through experience. (p. 363)

5. The simplest reflex pathway is a **monosynaptic reflex** with only two neurons. **Polysynaptic reflexes** have three or more neurons in the pathway. (p. 363)

6. If a presynaptic neuron synapses on a larger number of postsynaptic neurons, the pattern is known as **divergence.** If several presynaptic neurons provide input to a smaller number of postsynaptic neurons, the pattern is known as **convergence.** (p. 364)

7. **Excitatory postsynaptic potentials (EPSPs)** are depolarizing graded potentials that make a neuron more likely to fire an action potential. **Inhibitory postsynaptic potentials (IPSPs)** are hyperpolarizing graded potentials that make a neuron less likely to fire an action potential. (p. 364)

8. Chemicals that alter the response of a neuron are called **neuromodulators.** (p. 365)

9. **Postsynaptic modulation** occurs when a modulatory neuron, usually inhibitory, synapses on a postsynaptic cell. **Presynaptic modulation** occurs when a modulatory neuron terminates on or close to an axon terminal of a presynaptic cell. Presynaptic modulation allows selective modulation of collaterals and their targets. (p. 365)

Autonomic Reflexes

10. Some visceral reflexes are spinal reflexes that are modulated by input from the brain. Other autonomic reflexes are integrated in the brain, primarily in the hypothalamus, thalamus, and brain stem. The brain controls reflexes needed to maintain homeostasis. (p. 366)

11. Autonomic reflexes are all polysynaptic, and many are characterized by tonic activity. (p. 366)

Skeletal Muscle Reflexes

12. Skeletal muscle relaxation must be controlled by the central nervous system because activation of a somatic motor neuron always causes contraction in skeletal muscle. (p. 367)

13. The normal contractile fibers of a muscle are known as **extrafusal muscle fibers.** Their contraction is controlled by **alpha motor neurons.** (p. 367)

14. **Muscle spindles** send information about muscle length to the central nervous system. These receptors consist of **intrafusal fibers,** modified muscle fibers that lack myofibrils in their central portions. Sensory neurons wrap around the noncontractile center, and **gamma motor neurons** innervate the contractile ends of the intrafusal fibers. (p. 368)

15. Muscle spindles are tonically active stretch receptors. Their output is translated into tonic excitation of extrafusal muscle fibers by the alpha motor neurons. Because of this tonic activity, a muscle at rest maintains a certain level of tension, known as muscle tone. (p. 368)

16. If an intact muscle stretches, the intrafusal fibers of the spindle stretch and initiate reflex contraction of the muscle. The contraction prevents damage from overstretching. This reflex pathway is known as a **stretch reflex.** (p. 370)

17. When a muscle contracts, **alpha-gamma coactivation** ensures that the muscle spindle remains active. Activation of gamma motor neurons causes contraction of the ends

of the intrafusal fibers. This contraction lengthens the central section of the intrafusal fiber and maintains stretch on the sensory nerve endings. (p. 370)

18. **Golgi tendon organs** are found at the junction of the tendons and muscle fibers. They consist of free nerve endings that wind between collagen fibers. Golgi tendon organs respond to both stretch and contraction of the muscle by causing a reflexive relaxation. (p. 370)

19. The synergistic and antagonistic muscles that control a single joint are known as a **myotatic unit.** When one set of muscles in a myotatic unit contracts, the antagonistic muscles must relax through a reflex known as **reciprocal inhibition.** (p. 371)

20. **Flexion reflexes** are polysynaptic reflexes that cause an arm or leg to be pulled away from a painful stimulus. Flexion reflexes are usually accompanied by the **crossed extensor reflex,** a postural reflex that helps maintain balance when one foot is lifted from the ground. (p. 373)

21. **Central pattern generators** are networks of neurons in the central nervous system that function spontaneously to control certain rhythmic muscle movements. (p. 374)

The Integrated Control of Body Movement

22. Movement can be loosely classified into three categories: reflex movement, voluntary movement, and rhythmic movement. **Reflex movements** are the least complex and are integrated primarily in the spinal cord. **Postural reflexes** help us maintain body position as we stand or move and are integrated in the brain stem. (p. 374)

23. **Voluntary movements** require integration at the cerebral cortex and can be initiated at will without external stimuli. Learned voluntary movements improve with practice and some even become involuntary, like reflexes. (p. 375)

24. **Rhythmic movements** such as walking are a combination of reflexes and voluntary movements. Rhythmic movements must be initiated and terminated by input from the cerebral cortex, but once initiated, they can be sustained by central pattern generators. (p. 375)

25. Feedforward reflexes allow the body to prepare for a voluntary movement; feedback mechanisms are used to create a smooth, continuous motion. (p. 376)

26. Three levels of the nervous system control movement: the spinal cord, the brain stem, and the motor areas of the cerebral cortex. The basal ganglia and the cerebellum help cortical motor areas plan movement. The thalamus acts as a relay station for signals being sent to the cerebral cortex. (p. 376)

27. Voluntary movements require the cerebral cortex, cerebellum, and basal ganglia. The control of voluntary behavior can be divided into decision making and planning, initiating the movement, and executing the movement. (p. 377)

28. Contraction in smooth and cardiac muscles may occur spontaneously or may be controlled by hormones or by the autonomic division of the nervous system. (p. 378)

Questions

LEVEL ONE   Reviewing Facts and Terms

1. All nervous reflexes begin with a _____ that activates a receptor.

2. Somatic reflexes involve _____ muscles; _____, or visceral, reflexes are controlled by autonomic neurons.

3. The pathway pattern that brings information from many neurons into a smaller number of neurons is known as _____.

4. Graded potentials in postsynaptic neurons are called _____ if they make that cell more likely to fire, or _____ if they make it less likely to fire.

5. When two or more graded potentials arrive at the trigger zone of a neuron's axon, their effects are additive and _____ occurs.

6. When the axon terminal of a modulatory neuron (cell M) terminates close to the axon terminal of a presynaptic cell (cell P) and decreases the amount of neurotransmitter released by cell P, the resulting type of modulation is called _____.

7. Autonomic reflexes are also called _____ reflexes. Why?

8. Some autonomic reflexes are spinal reflexes; others are integrated in the brain. List examples of each.

9. What part of the brain transforms emotions into somatic sensation and visceral function? List three autonomic reflexes that are linked to one's emotions.

10. How many synapses occur in the simplest autonomic reflexes? Where do the synapses occur?

11. List the three types of sensory receptors that convey information for muscle reflexes.

12. Match the organ to all correct statements about it.
    (a) muscle spindle
    (b) Golgi tendon organ
    (c) joint capsule mechanoreceptor

    1. strictly a sensory receptor
    2. has afferent neurons that bring information to the CNS
    3. associated with two types of efferent neurons
    4. convey information about the relative positioning of bones
    5. innervated by gamma motor neurons
    6. eventually synapses with alpha motor neurons that innervate extrafusal muscle fibers.

13. Because of neural tonic activity, even at rest a muscle maintains a low level of tension known as _____.

14. The stretching of an intact skeletal muscle causes sensory neurons to (increase/decrease) their rate of firing, causing the muscle to contract, thereby relieving the stretch. Why is this a useful reflex?

15. The Golgi tendon organ responds to both _____ and _____, although _____ elicits the stronger response.

Activation (increases/decreases) muscle contraction via the _____ neuron.

16. The simplest reflex requires a minimum of how many neurons? How many synapses? Give an example.

17. List and differentiate the three categories of movement. Give an example of each.

18. What does each of the listed abbreviations stand for: EPSP, IPSP, CNS? Define each term.

LEVEL TWO   Reviewing Concepts

19. The resting membrane potential of a given neuron is -67 mV. Threshold is at -50 mV. For each example listed below, sum the amplitude of the potentials arriving at the trigger zone and decide whether the neuron fires or not.
    a. an EPSP of $+20$ mV
    b. an IPSP of -2 mV
    c. a and b above arrive at the trigger zone simultaneously
    d. an EPSP of $+10$ mV and an EPSP of $+5$ mV
    e. an EPSP of $+10$ mV and an IPSP of -10 mV

20. For each lettered example in question 19, decide if the cell was depolarized, hyperpolarized, or not affected.

21. Explain the differences between presynaptic facilitation and postsynaptic excitation. Diagram two sets of neurons to illustrate these differences.

22. Name a neurotransmitter that could cause the EPSP described in question 19a to occur. What type of ion

would cross the cell membrane? Would the ions enter or leave the cell? Is the cell now depolarized or hyperpolarized? Now consider question 19b. Name a neurotransmitter that could cause the IPSP described to occur. What type of ion would cross the cell membrane? Would the ions enter or leave the cell? Is the cell now depolarized or hyperpolarized?

23. What is the purpose of alpha-gamma coactivation? Explain how it occurs.

24. Neuron M synapses on the axon terminal of neuron P, just before P synapses with the effector organ. If increased activity in neuron M changes the amount of neurotransmitter released by P, then M is a modulatory neuron. If M is an inhibitory neuron, what happens to neurotransmitter release by P? What effect does M's neurotransmitter have on the postsynaptic membrane potential of P?

LEVEL THREE   Problem Solving

25. Andy is working on improving his golf swing. He must watch the ball, swing the club back, then forward, twist his hips, straighten the left arm, then complete the follow-through, where the club arcs in front of him. What specific parts of the brain are involved in adjusting how hard he hits the ball, keeping all his body parts moving correctly, watching the ball, and then repeating these actions once he has verified that this swing is successful?

26. You are playing the clarinet, reading the music, and fingering the proper notes as they are called for in the music. You are even tapping your toe to keep time. Diagram five

different neural pathways involved in performing these actions, from the receptors to the muscles, and from the brain to the effectors.

27. Set up your own scenario involving a task familiar to you, such as driving a car and positioning it properly at the drive-through window, crocheting according to a printed pattern, playing tennis, or keyboarding and reproducing text from a page. Identify the types of movements involved (reflexive, voluntary, rhythmic, or a combination of these types) and sketch the neural pathways involved in completing the given task.

Problem Conclusion

In this running problem, you learned that the tetanus toxin is transported from peripheral wounds to the spinal cord, where it blocks the release of inhibitory neurotransmitters. You also learned that drugs such as curare can cause paralysis of skeletal muscles.

Further check your understanding of this running problem by checking your answers against those in the summary table.

Question	Facts	Integration and Analysis
1a By what process is tetanospasmin taken up into neurons?	Tetanospasmin is a protein.	Proteins are too large to cross cell membranes by mediated transport. Therefore, tetanospasmin must be taken up by endocytosis (∞ p. 126).
1b By what process does tetanospasmin travel up the axon to the nerve cell body?	Movement of substances from the axon terminal to the nerve cell body is called retrograde axonal transport (∞ p. 205).	Tetanospasmin is taken up by endocytosis, so it will be contained in endocytotic vesicles. These vesicles are "walked" up the axon along microtubules through retrograde axonal transport.
2 Using Figures 13-9 and 13-10, explain why inhibition of inhibitory interneurons might result in uncontrollable muscle spasms.	Muscles often occur in antagonistic pairs. When one muscle is contracting, its antagonist must be inhibited.	If the inhibitory interneurons are not functioning, both sets of antagonistic muscles can contract at the same time. This would lead to muscle spasms and rigidity because the bones attached to the muscles would be unable to move in any one direction.
3a Why does the binding of metocurine to ACh receptors on the motor end plate induce muscle paralysis?	Acetylcholine (ACh) is the somatic motor neuron neurotransmitter that initiates skeletal muscle contraction.	If metocurine binds to ACh receptors, it prevents ACh from binding. Without ACh binding, the muscle fiber will not depolarize and cannot contract, resulting in paralysis.
3b Is metocurine an agonist or an antagonist of ACh?	Agonists mimic the effects of a substance. Antagonists block the effects of a substance.	Metocurine blocks ACh action; therefore, it is an antagonist.
4 On the basis of what you know about who receives immunization shots in the United States, predict the age and background of people who are most likely to develop tetanus this year.	Immunizations are required for all children of school age. This practice has been in effect since about the 1950s. In addition, most people who suffer puncture wounds or dirty wounds are given tetanus booster shots when they are treated for those wounds.	Most cases of tetanus in the United States will occur in people over the age of 60 who have never been immunized, in immigrants (particularly migrant workers), and in newborn infants. Another source of the disease is contaminated heroin; injection of the drug under the skin may cause tetanus.

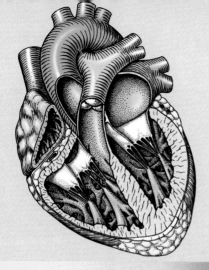

14

Cardiovascular Physiology

As life evolved, simple one-celled organisms began to band together, first into cooperative colonies and then into multicelled organisms. In most multicellular animals, only the surface layer of cells is in direct contact with the environment. This body structure presents a problem for the transfer of oxygen and nutrients from the environment to the interior cells. The rate of diffusion is severely limited by distance (∞ p. 118), and, in larger animals, oxygen consumption in interior cells exceeds the rate at which oxygen can diffuse from the body surface. One solution to this problem was the evolutionary development of circulatory systems that move fluid between the body surface and its deepest parts. In simple animals, fluid flow is created by muscular activity when the animal moves. More complex animals use muscular pumps called hearts to circulate internal fluid. In the most efficient circulatory systems, the heart pumps blood through a closed system of vessels. This one-way circuit steers blood flow along a specific route and ensures efficient distribution of gases, nutrients, signal molecules, and wastes. A circulatory system with a heart and blood vessels is known as a **cardiovascular system** [*kardia*, heart + *vasculum*, little vessel].

Although the idea of a closed cardiovascular system that cycles blood in an endless loop seems intuitive to us today, it has not always been so. **Capillaries,** the microscopic vessels in which blood exchanges material with the interstitial fluid, were not discovered until Marcello Malpighi, an Italian anatomist, observed them with a microscope in the middle of the seventeenth century. Then, European medicine was still heavily influenced by the ancient belief that the cardiovascular system distributed both blood and air. Blood was thought to be made in the liver and distributed throughout the body in the veins. Air went from the lungs to the heart, where it was "digested" and picked up "vital spirits." From the heart, air was distributed to the tissues through vessels called *arteries*. Anomalies such as the fact that a cut artery squirted blood rather than air were ingeniously explained by unseen links between arteries and veins that opened with injury.

According to this model of the circulatory system, the tissues consumed all blood delivered to them, so the liver continuously synthesized new blood. It took the calculations of William Harvey (1578–1657), court physician to King Charles I of England, to show that the weight of the blood pumped by the heart in a single hour is greater than the weight of the entire body! Once it became obvious that the liver could not make blood as rapidly as the heart pumped it, Harvey looked for an anatomical route that would allow the blood to recirculate rather than be consumed in the tissues. He showed that valves in the heart and veins created a one-way flow of blood and that veins carried blood back to the heart, not out to the limbs. He also showed that blood returning to the right side of the heart from the body had to go to the lungs before it could go to the left side of the heart. The results of these studies created a furor among Harvey's contemporaries, leading Harvey to say in a huff that no one under the age of 40 could understand his conclusions. Ultimately, Harvey's work became the foundation of modern cardiovascular physiology. Today, we understand the structure of the cardiovascular system at microscopic and molecular levels that Harvey never dreamed existed. Yet some things have not changed. Even now, with our sophisticated technology, we are searching for "spirits" in the blood, although we call them by names such as "hormone" and "cytokine."

OVERVIEW OF THE CARDIOVASCULAR SYSTEM

In the simplest terms, the cardiovascular system is a system of tubes (the blood vessels) filled with fluid (blood) and connected to a pump (the heart). Pressure generated in the heart propels blood through the system continuously. The blood picks up oxygen at the lungs and nutrients in the intestine, delivering them to cells, while simultaneously removing cellular wastes for excretion. In addition, the cardiovascular system plays an important role in cell-to-cell communication and in defending the body against foreign invaders. This chapter focuses on an overview of the cardiovascular system and on the heart as a pump. The properties of the blood vessels and

Problem

Myocardial Infarction

At 9:06 A.M., the blood clot that had been silently growing in Walter Parker's left coronary artery made its sinister presence known. The 53-year-old advertising executive had arrived at the Dallas Convention Center feeling fine, but suddenly a dull ache started in the center of his chest and he became nauseated. At first he brushed it off as aftereffects of the convention banquet the night before. However, when it persisted, he made his way to the Aid Station. "I'm not feeling very well," he told the medic. "I think it may be indigestion." The medic, on hearing Walter's symptoms and seeing his pale, sweaty face, immediately thought of a heart attack. "Let's get you over to the hospital and get this checked out."

continued on page 390

the homeostatic controls that regulate blood flow and blood pressure are described in Chapter 15.

The Cardiovascular System Transports Material throughout the Body

The primary function of the cardiovascular system is the transport of materials between all parts of the body. Substances transported by the cardiovascular system can be divided into nutrients, water, and gases that enter the body from the external environment, materials that are being moved from cell to cell within the body, and wastes that the cells are eliminating (Table 14-1).

Oxygen enters the blood in the lungs; nutrients and water are absorbed across the intestinal epithelium. Once in the bloodstream, these materials are distributed throughout the body. A steady supply of oxygen for the cells is particularly important because many cells deprived of oxygen become irreparably damaged within a short period of time. About 5–10 seconds after blood flow to the brain is stopped, a person loses consciousness. If oxygen delivery stops for 5–10 minutes, permanent brain damage results. Neurons of the brain have a very high rate of oxygen consumption and are unable to meet their metabolic need for ATP by using anaerobic pathways. Because of the sensitivity of the brain to *hypoxia* [*hypo-*, low + *-oxia*, oxygen], homeostatic controls do everything possible to maintain cerebral blood flow, even if it means that cells in other tissues will be deprived of oxygen.

Cell-to-cell communication is a key function of the cardiovascular system. For example, hormones secreted by endocrine glands are carried to their targets. Nutrients such as glucose from the liver or fatty acids released by adipose tissue are transported to metabolically active cells. Finally, the defense team of white blood cells and antibodies patrol the circulation to intercept foreign invaders.

TABLE 14-1 Transport in the Cardiovascular System

Substance Moved	From	To
Materials entering the body		
Oxygen	Lungs	All cells
Nutrients and water	Intestinal tract	All cells
Materials moved from cell to cell		
Wastes	Some cells	Liver for processing
Immune cells, antibodies, clotting proteins	Present in blood continuously	Available for any cell that needs them
Hormones	Endocrine cells	Target cells
Stored nutrients	Liver and adipose tissue	All cells
Materials leaving the body		
Metabolic wastes	All cells	Kidneys
Heat	All cells	Skin
Carbon dioxide	All cells	Lungs

The cardiovascular system also picks up material released by cells. Carbon dioxide and metabolic wastes are transported in the blood to the lungs and the kidneys for excretion. Some waste products are transported to the liver for processing before they are excreted in the urine and feces. Heat is also circulated through the blood, moving from the body core to the surface, where it dissipates.

The Cardiovascular System Consists of the Heart and Blood Vessels

The cardiovascular system is composed of the heart, the blood vessels (also known as the *vasculature*), and the cells and plasma of the blood. Blood vessels that carry blood away from the heart are called **arteries;** blood vessels that return blood to the heart are called **veins.** As blood moves through the cardiovascular system, a system of valves in the heart and veins ensures that blood flows in one direction. Like the turnstiles at an amusement park, the valves keep the blood from reversing its direction of flow. Figure 14-1 ■ is a schematic diagram that shows these components and the route that blood flow follows through the body.

The heart is divided by a central wall, or **septum,** into left and right halves. Each half consists of an **atrium** [*atrium*, central room; plural *atria*], which receives blood returning to the heart from the blood vessels, and a **ventricle** [*ventriculus*, belly], which pumps blood out into the blood vessels. The two halves of the heart are separate, and each side pumps blood independently. Consequently, the heart functions like two pumps working in series (one after the other). The right side of the heart receives blood from the tissues and sends it to the lungs for oxygenation. The left side of the heart receives the newly oxygenated blood from the lungs and pumps it to tissues throughout the body.

Starting in the right atrium, trace the path taken by blood as it flows through the cardiovascular system. Note that in Figure 14-1 ■, blood in the right side of the heart is colored blue. This is a convention used to show blood from which the tissues have extracted oxygen. Although this blood is often described as "deoxygenated," it is not completely devoid of oxygen; it simply has less oxygen than blood going from the lungs to the tissues. In living people, well-oxygenated blood is bright red; low-oxygen blood becomes darker red. Under some conditions, low-oxygen blood can impart a bluish color to certain areas of the skin, such as around the mouth and under the fingernails. This condition, known as *cyanosis* [*kyanos*, dark blue], is the reason blue is used in drawings to indicate blood with lower oxygen content.

From the right atrium, blood flows into the right ventricle of the heart. From there it is pumped through the **pulmonary arteries** to the lungs, where it is oxygenated. Note the color change from blue to red, indicating higher oxygen content. From the lungs, blood travels back to the left side of the heart through the **pulmonary veins.** The blood vessels that go from the right ventricle to the lungs and back to the left atrium are known collectively as the **pulmonary circulation.**

Blood from the lungs reenters the heart at the **left atrium** and passes into the **left ventricle.** Blood pumped out of the left ventricle enters the large artery known as the **aorta.** The aorta branches into a series of smaller and smaller arteries that finally lead into networks of capillaries. These are the tiny thin-walled vessels in which material is exchanged between the blood and the tissues. Notice the color change from red to blue in the capillaries, showing that oxygen has left the blood and diffused into the tissues. Blood then flows into the *venous* side of the circulation, moving from small veins into

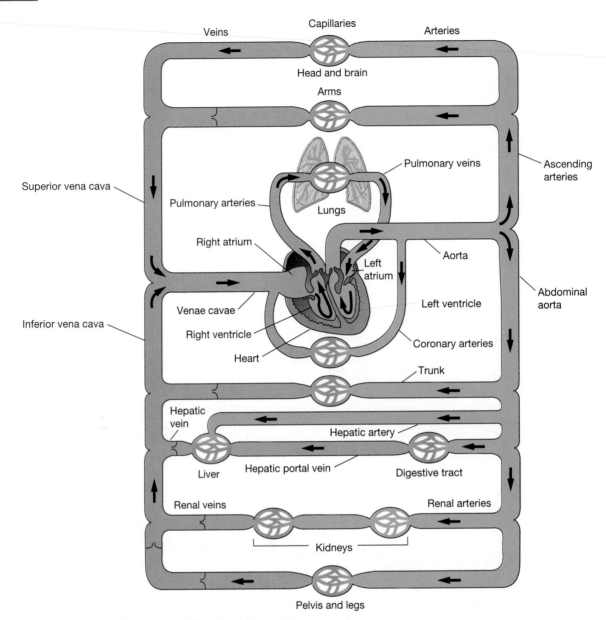

■ Figure 14-1 **General anatomy of the circulatory system**

larger and larger veins. The veins from the upper part of the body join together to form the **superior vena cava;** those from the lower part of the body form the **inferior vena cava.** The two **venae cavae** empty into the right atrium. The blood vessels that go from the left side of the heart to the tissues and back to the right side of the heart are collectively known as the **systemic circulation.**

Return to the aorta and follow its divisions. The first branch represents the *coronary arteries* that nourish the heart muscle itself. Ascending branches of the aorta go to the head, arms, and brain. The abdominal aorta supplies blood to the trunk, the legs, and the internal organs such as the kidneys (*renal arteries*), liver (*hepatic artery*), and digestive tract.

Notice two special regions of the circulation. One is the blood supply to the digestive tract and liver. Both regions receive well-oxygenated blood through their own arteries, but, in addition, blood leaving the digestive tract goes directly to the liver by means of the *hepatic portal vein*. The liver is an important site for nutrient processing and plays a major role in the detoxification of foreign substances. Most nutrients absorbed in the intestine are routed directly to the liver, allowing that organ to process the material before it is released into the general circulation. The two capillary beds of the digestive tract and liver, joined by the hepatic portal vein, are an example of a portal system. Another portal system, discussed earlier, is the hypothalamic-hypophyseal portal system that connects the hypothalamus and the anterior pituitary (∞ p. 185). The third major portal system is in the kidneys, where two capillary beds are connected in series. You will learn more about this special vascular arrangement in Chapter 18.

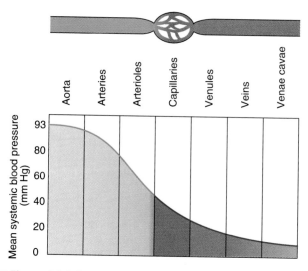

■ Figure 14-2 Pressure gradient in the blood vessels The mean blood pressure of the systemic circulation ranges from a high of 93 mm Hg (millimeters of mercury) in the arteries to a low of a few mm Hg in the venae cavae.

PRESSURE, VOLUME, FLOW, AND RESISTANCE

If you ask people why blood flows through the cardiovascular system, many of them respond, "So that oxygen and nutrients can get to all parts of the body." Although this is true, it is a teleological answer, one that describes the purpose of blood flow. In physiology, we are also concerned with how blood flows: the mechanisms or forces that create blood flow.

A simple mechanistic answer to "Why does blood flow?" is that liquids and gases flow down a **pressure gradient (ΔP)** from regions of higher pressure to regions of lower pressure. Therefore, blood can flow in the cardiovascular system only if one region develops higher pressure than other regions. In humans, high

pressure is created in the chambers of the heart when it contracts. Blood flows out of the heart (the region of highest pressure) into the closed loop of blood vessels (a region of lower pressure). As blood moves through the system, pressure is lost because of friction between the fluid and the blood vessel walls. Consequently, pressure falls continuously as blood moves farther from the heart (Fig. 14-2 ■). The highest pressure in the vessels of the circulatory system is found in the aorta and systemic arteries as they receive blood from the left ventricle. The lowest pressure is in the venae cavae, just before they empty into the right atrium.

This section reviews the laws of physics that explain the interaction of pressure, volume, flow, and resistance in the cardiovascular system. Many of these principles can be broadly applied to the flow of all types of liquids and gases, including that of air in the respiratory system. However, in this chapter we will focus on the flow of blood and its relevance to the function of the heart.

The Pressure of Fluid in Motion Decreases over Distance

Pressure in fluids is the force exerted by a fluid on the container holding the fluid. If the fluid is not moving, the pressure is called **hydrostatic pressure** (Fig. 14-3a ■), and force is exerted equally in all directions. For example, a column of fluid in a tube exerts hydrostatic pressure on the floor and sides of the tube. In the heart and blood vessels, pressure is commonly measured in millimeters of mercury (mm Hg, or torr).* This is equivalent to the hydrostatic pressure exerted by a 1-mm-high column of mercury on an area 1 cm^2.

In a system in which fluid is flowing, pressure falls over distance as energy is lost because of friction (Fig.

---

*1 mm Hg = 1 torr. Some physiological literature reports pressures in centimeters of water: 1 cm H_2O = 0.74 mm Hg.

(a)

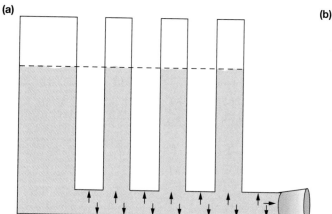

Hydrostatic pressure is the pressure exerted on the walls of the container by the fluid within the container. Hydrostatic pressure is proportional to the height of the water column.

(b)

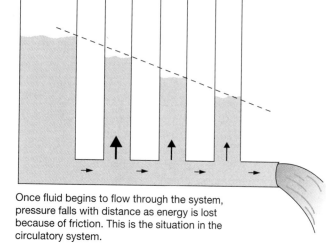

Once fluid begins to flow through the system, pressure falls with distance as energy is lost because of friction. This is the situation in the circulatory system.

■ Figure 14-3 Hydrostatic pressure

14-3b ■). In addition, the pressure exerted by the moving fluid has two components: a dynamic, flowing component that represents the kinetic energy of the system and a lateral component that represents the hydrostatic pressure (potential energy) exerted on the walls of the system. Pressure within our cardiovascular system is usually called hydrostatic pressure, although it is a system in which fluid is in motion. Some textbooks are beginning to replace the term *hydrostatic pressure* with the term *hydraulic pressure*. Hydraulics is the study of fluid in motion.

When Volume Decreases, Pressure Increases

If the volume of a fluid-filled container decreases, the pressure exerted by the fluid on the walls of the container increases. You can demonstrate this principle by filling a balloon with water and squeezing it in your hand. Water is not compressible, so the pressure applied to the balloon is transmitted throughout the fluid. As you squeeze, higher pressure in the fluid causes parts of the balloon to bulge. If the pressure becomes high enough, the stress on the balloon will cause it to pop.

In the human heart, contraction of the ventricles on the blood that fills them is like squeezing a water balloon: The pressure created by the contracting muscle is transferred to the blood. This high-pressure fluid within the ventricle then flows out of the ventricle and into the blood vessels, displacing fluid already in the vessels. The pressure created within the ventricles is called the **driving pressure** because it is the force that drives blood through the blood vessels.

If the volume of a fluid-filled container increases, the pressure of the fluid within the container decreases. Thus, when the heart relaxes and expands, pressure in the chambers falls.

Volume changes that affect pressure can also take place in the blood vessels. If blood vessels dilate, blood pressure inside them falls. If blood vessels constrict, blood pressure increases. Volume changes of the blood vessels and heart are major factors that influence blood pressure within the cardiovascular system.

Blood Flows from an Area of Higher Pressure to One of Lower Pressure

As stated earlier, in order for blood to flow through the circulatory system, a pressure gradient must exist. This pressure gradient, ΔP, is the difference in pressure between two ends of a tube through which fluid flows (Fig. 14-4a ■). Flow through the tube is directly proportional to ($\propto$) the pressure gradient (ΔP):

(1) Flow $\propto \Delta P$ where $\Delta P = P_1 - P_2$

This relationship says that the higher the pressure gradient, the greater the flow of fluid.

A pressure gradient is not the same thing as the absolute pressure in the system. For example, in Figure

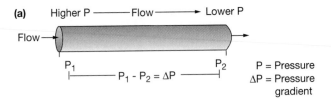

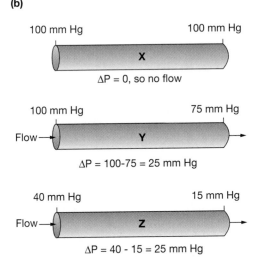

■ **Figure 14-4 Fluid flow through a tube** (a) Fluid flows as the result of a pressure gradient between two ends of the tube. (b) The flow rate depends on the pressure gradient, or driving pressure, rather than the absolute pressures in the tube. In the example shown, tube Y has much higher absolute pressures than tube Z, but the pressure gradient is the same, 25 mm Hg. Consequently, the flow through the two tubes is identical.

14-4b ■, tube X has an absolute pressure of 100 mm Hg at each end. However, because there is no pressure gradient between the two ends of the tube, there is no flow through the tube.

On the other hand, two tubes can have very different absolute pressures but the same flow. Tubes Y and Z are identical tubes. Tube Y has a hydraulic pressure of 100 mm Hg at one end and 75 mm Hg at the other end, with a pressure gradient between the ends of the tube equal to 25 mm Hg. An identical tube, Z, has hydraulic pressure of 40 mm Hg at one end and 15 mm Hg at the other end. Tube Z has lower absolute pressure all along its length, but it has the same pressure gradient as tube Y, 25 mm Hg. Because the pressure difference in the two tubes is identical, the fluid flow through the tubes will be the same.

Resistance Opposes Flow

In an ideal system, a substance in motion would remain in motion. However, no system is ideal because all movement creates friction. Just as a ball rolled across the ground loses energy to friction, blood flowing through blood vessels encounters friction from the walls of the blood vessels and from cells within the blood rubbing against each other as they flow. This tendency of the cardiovascular system to oppose blood flow is called its **resistance** to flow. Resistance is a term that most of us understand already from everyday life. We speak of people being resistant to change or of taking the path of least resistance. This concept translates well to the circulatory system because blood flow will also take the path of least resistance. An increase in the resistance of a blood vessel means a decrease in flow through that vessel. We can express that relationship as

(2) Flow $\propto 1/R$

Flow is inversely proportional to resistance. If resistance increases, flow decreases. If resistance decreases, flow increases.

What parameters determine resistance? For fluid flowing through a tube, resistance is influenced by three components: the length of the tube (L), the radius of the tube (r), and the viscosity (thickness) of the fluid flowing through the tube (η, eta). The following equation, derived by the French physician Jean Leonard Marie Poiseuille, shows the relationship between these factors:

(3) $R = 8L\eta/\pi r^4$

Because the value of $8/\pi$ is a constant, the relationship can be rewritten as

(4) $R \propto L\eta/r^4$

This expression says that resistance increases as the length of the tube and the viscosity of the fluid increase, but it decreases as the radius increases. To remember these relationships, think of drinking through a straw. You do not need to suck as hard on a short straw as on a long one (resistance increases with length). Drinking water through a straw is easier than drinking a thick milkshake (resistance increases with viscosity). And drinking the milkshake through a big fat straw is much easier than through a skinny cocktail straw (resistance increases as radius decreases).

How significant are length, viscosity, and radius to blood flow in a normal individual? The length of the systemic circulation is determined by the anatomy of the system and is essentially constant. The viscosity of blood is determined by the ratio of red blood cells to plasma and by how much protein is in the plasma. Normally, viscosity is constant, and small changes in either length or viscosity have little effect on resistance. This leaves changes in the radius of the blood vessels as the main contributor to variable resistance in the systemic circulation.

Let us return to the example of the straw and the milkshake to illustrate how changes in radius affect resistance. If we assume that the length of the straw and the viscosity of the milkshake do not change, this system is similar to the circulatory system—the radius of the tube has the greatest effect on resistance. If we consider only resistance and radius from equation 4, the relationship between resistance and radius can be expressed as follows:

(5) $R \propto 1/r^4$

If the skinny straw has a radius of 1, its resistance is proportional to $1/1^4$, or 1. If the larger straw has a radius of 2, its resistance is $1/2^4$, or 1/16th, that of the skinny straw. Since flow is inversely proportional to resistance, flow increases 16-fold when the radius doubles (Fig. 14-5 ■). As you can see from this example, a small change in the radius of a tube will have a large effect on the flow of a liquid through that tube. A small change in the radius of a blood vessel will have a large effect on the resistance to blood flow. A decrease in blood vessel diameter is known as **vasoconstriction** [*vas*, a vessel or duct]. An increase in blood vessel diameter is called **vasodilation.** Vasoconstriction will decrease blood flow

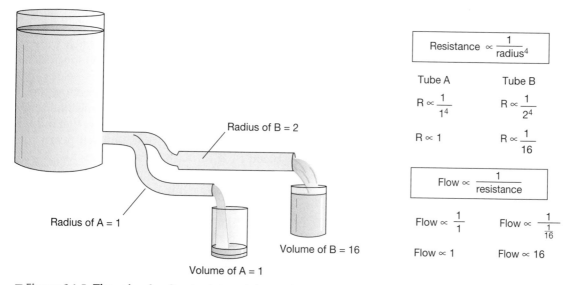

■ Figure 14-5 **The role of radius in determining resistance to flow** As the radius (*r*) of a tube increases, the resistance (*R*) of the tube to fluid flow decreases dramatically because $R \propto 1/r^4$.

through a vessel; vasodilation will increase blood flow through a vessel.

In summary, by combining equations 1 and 2 above, we get the following equation:

(6) Flow $\propto \Delta P/R$

The flow of blood within the circulatory system is directly proportional to the pressure gradient in the system and inversely proportional to the resistance of the system to flow. If the driving pressure remains constant, then flow will vary inversely with resistance.

Velocity of Flow Depends on the Flow Rate and the Cross-Sectional Area

The word *flow* is often used imprecisely in cardiovascular physiology, leading to confusion. Flow usually means **flow rate,** the volume of blood that passes one point in the system per unit time. In the circulation, flow is expressed in liters per minute (L/min) or milliliters per minute (mL/min). For instance, blood flow through the aorta in a 70-kg man at rest is about 5 L/min.

Flow rate should not be confused with **velocity of flow,** the distance a fixed volume of blood will travel in a given period of time. Velocity of flow is a measure of how fast blood is flowing; flow rate measures the volume of blood that flows past a point in a given period of time. In a tube of fixed size, velocity of flow is directly related to flow rate. In a tube of variable diameter, blood velocity will also vary inversely with the diameter. Thus, velocity will be faster through the narrow sections and slower in the wider sections. The relationship between flow rate (Q), cross-sectional area of the tube (A), and velocity of flow (v) is expressed by the following equation:

(7) Velocity = flow rate/cross-section area

Figure 14-6 ■ shows how the velocity of flow will vary according to the cross-sectional area of the tube. The vessel in the figure has two widths: narrow, with a small cross-sectional area of 1 cm², and wide, with a large cross-

---

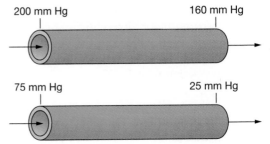

✓ The two identical tubes below have the pressures shown at each end. Which tube has the greater flow? Defend your choice.

200 mm Hg   160 mm Hg

75 mm Hg   25 mm Hg

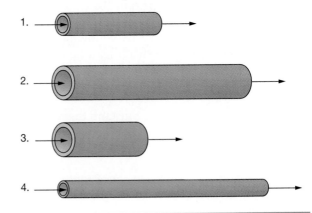

✓ All four tubes below have the same driving pressure. Which tube has the greatest flow? Which has the least flow? Defend your choices.

1.

2.

3.

4.

---

sectional area of 12 cm². The flow rate is identical in both parts of the vessel: 12 cm³ per minute.* This flow rate means that in one minute, 12 cm³ of fluid flows past point X in the narrow section, and 12 cm³ flows past point Y in the wide section. But *how fast* does the fluid need to flow to accomplish that rate? According to equation 7, the velocity of flow at point X is 12 cm/min, but at point Y it is only 1 cm/min. Thus, fluid flows more rapidly through narrow sections of a tube than through wide sections.

To see this principle in action, watch a leaf as it floats down a stream or gutter. Where the stream is narrow, the leaf moves rapidly, carried by the fast velocity of the water. In sections where the stream widens into a pool, the velocity of the water drops and the leaf meanders more slowly.

Table 14-2 summarizes the factors that influence blood flow. In the next chapter, you will learn more about blood flow through the cardiovascular system. In the remainder of this chapter, we explore the heart and how it creates the driving pressure for blood flow.

✓ Two canals in Amsterdam are identical in size, but the water flows faster through one than through the other. Which canal has the highest flow rate?

---

*1 cm³ = 1 mL

---

continued from page 384

The clot in Walter's coronary artery has restricted blood flow to his heart muscle, and heart muscle cells are beginning to die from lack of oxygen. In the next few hours, various interventions are critical to save Walter's life and prevent further damage. While waiting for the ambulance to arrive, the medic gives Walter oxygen, hooks him to a heart monitor, and starts an intravenous (IV) injection of normal (isotonic) saline. With an intravenous injection line in place, other drugs can be given rapidly if Walter's condition should suddenly get worse.

Question 1: Why is the medic giving Walter oxygen? What effect will the injection of isotonic saline have on Walter's extracellular fluid volume? On his intracellular fluid volume? On his total body osmolarity?

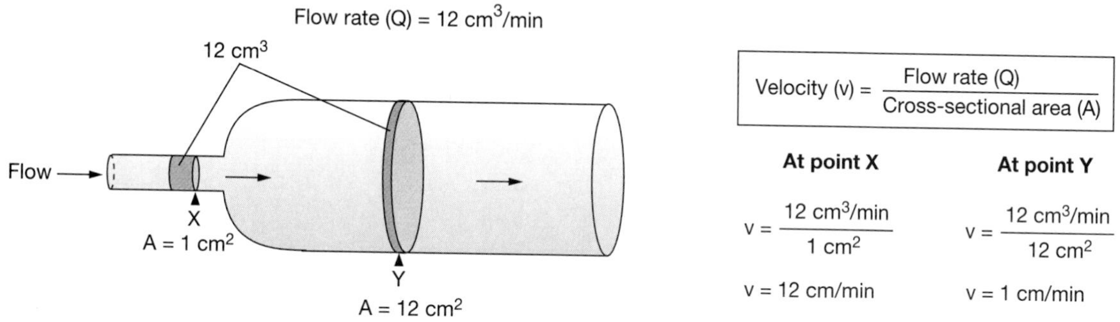

$$\text{Flow rate (Q)} = 12 \text{ cm}^3/\text{min}$$

$$12 \text{ cm}^3$$

Flow →

X
$$A = 1 \text{ cm}^2$$

Y
$$A = 12 \text{ cm}^2$$

$$\text{Velocity (v)} = \frac{\text{Flow rate (Q)}}{\text{Cross-sectional area (A)}}$$

At point X

$$v = \frac{12 \text{ cm}^3/\text{min}}{1 \text{ cm}^2}$$

$$v = 12 \text{ cm/min}$$

At point Y

$$v = \frac{12 \text{ cm}^3/\text{min}}{12 \text{ cm}^2}$$

$$v = 1 \text{ cm/min}$$

■ Figure 14-6 **Flow rate versus velocity of flow** This figure illustrates the fact that the more narrow the vessel, the faster the velocity of flow.

CARDIAC MUSCLE AND THE HEART

To ancient civilizations, the heart was more than a pump: It was "the seat of the mind." When ancient Egyptians were mummified after death, most of the viscera were removed, but the heart was left in place so that it could be weighed by the gods as an indicator of the owner's worthiness. Aristotle characterized the heart as the most important organ of the body, as well as "the seat of intelligence." Evidence of these ancient beliefs can still be found in modern expressions such as "heartfelt emotions." The link between the heart and mind is one that is still explored today.

The heart is the workhorse of the body, a muscle that contracts continually, resting only in the milliseconds' pause between beats. By one estimate, the work performed by the heart is equivalent to lifting a 5-pound weight up one foot, once a minute. The energy demands of contraction require a continuous supply of nutrients and oxygen to the heart muscle. In this part of the chapter, we look at the anatomy of the heart and at the electrical and mechanical events that take place as it pumps.

TABLE 14-2 Rules for the Cardiovascular System

1. Blood flows if a pressure gradient is present. Blood flows from areas of higher pressure to areas of lower pressure.
2. Blood flow is opposed by the resistance of the system. The three factors that affect resistance are the length of the system, the radius of the blood vessels, and the viscosity of the blood.
3. Flow rate (usually just called flow) is the volume of blood that passes one point in the system per unit time and is usually expressed in liters or milliliters per minute (L/min or mL/min). Flow rate is determined by the pressure gradient and by the resistance of the system.
4. Velocity of flow is the distance a volume of blood will travel in a given period of time. It is usually expressed as centimeters per minute (cm/min) or millimeters per second (mm/sec). The primary determinant of velocity of flow (when flow rate is constant) is the total cross-sectional area of the vessels at any given point in the system.

The Heart Has Four Chambers

The heart is a muscular organ, about the size of a fist, that lies in the center of the *thoracic cavity* (see Anatomy Summary, Fig. 14-7a, b, d ■). The pointed *apex* of the heart angles down to the left side of the body, while the broader *base* lies just behind the breastbone, or *sternum*. Because we usually associate the word *base* with the bottom, remember that the base of a cone is the broad end and the apex is the pointed end. The heart can be thought of as an inverted cone with apex down and base up. Within the thoracic cavity, the heart lies on the ventral side, sandwiched between the two lungs, with its apex resting on the diaphragm (Fig. 14-7b ■).

The heart is encased in a tough membranous sac, the **pericardium** [*peri*, around + *kardia*, heart] (Fig. 14-7d, e ■). A thin layer of clear pericardial fluid inside the pericardium lubricates the external surface of the heart as it beats within the sac. Inflammation of the pericardium (*pericarditis*) may reduce this lubrication to the point that the heart rubs against the pericardium, creating a sound known as a *friction rub*.

The heart itself is composed mostly of cardiac muscle, or **myocardium** [*myo-* muscle + *kardia*, heart], covered by thin outer and inner layers of epithelium and connective tissue. If seen from the outside, the bulk of the heart is the thick muscular walls of the ventricles, the two lower chambers (Fig. 14-7f ■). The thinner-walled atria lie above the ventricles. Emerging from the top of the heart are the major blood vessels: The aorta and pulmonary trunk (artery) take blood from the heart to the tissues and lungs, respectively; the venae cavae and pulmonary veins return blood to the heart (see Table 14-3). When the heart is viewed from the front (anterior view), as in Figure 14-7f ■, the pulmonary veins are hidden behind the other blood vessels. Running across the surface of the ventricles are three grooves containing the **coronary arteries** and **coronary veins** that supply blood to the heart muscle.

The relationship between the atria and ventricles can be seen in a cross-sectional view of the heart (Fig. 14-7g ■). Notice that in the figure the *right* side of the heart is on the *left* side of the page; that is, the heart is labeled as if you were viewing the heart of a person facing you. The

Anatomy Summary The Cardiovascular System

■ Figure 14-7

(a)

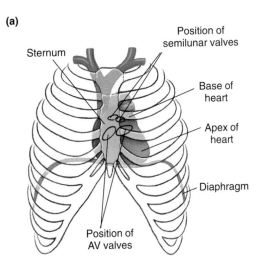

Sternum

Position of semilunar valves

Base of heart

Apex of heart

Diaphragm

Position of AV valves

(b)

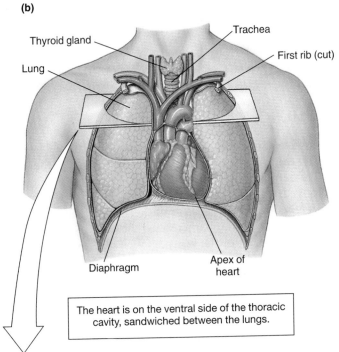

Thyroid gland

Trachea

First rib (cut)

Lung

Diaphragm

Apex of heart

The heart is on the ventral side of the thoracic cavity, sandwiched between the lungs.

(c)

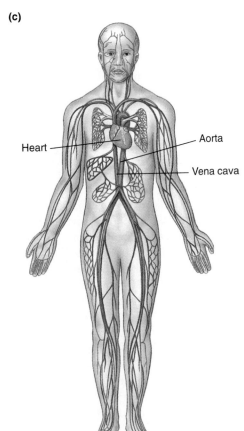

Heart

Aorta

Vena cava

Vessels that carry well-oxygenated blood are red; those with less well-oxygenated blood are blue.

(d)

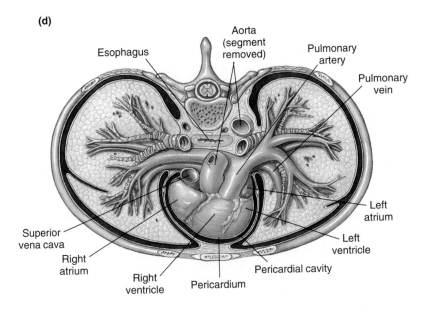

Esophagus

Aorta (segment removed)

Pulmonary artery

Pulmonary vein

Superior vena cava

Right atrium

Right ventricle

Pericardium

Left atrium

Left ventricle

Pericardial cavity

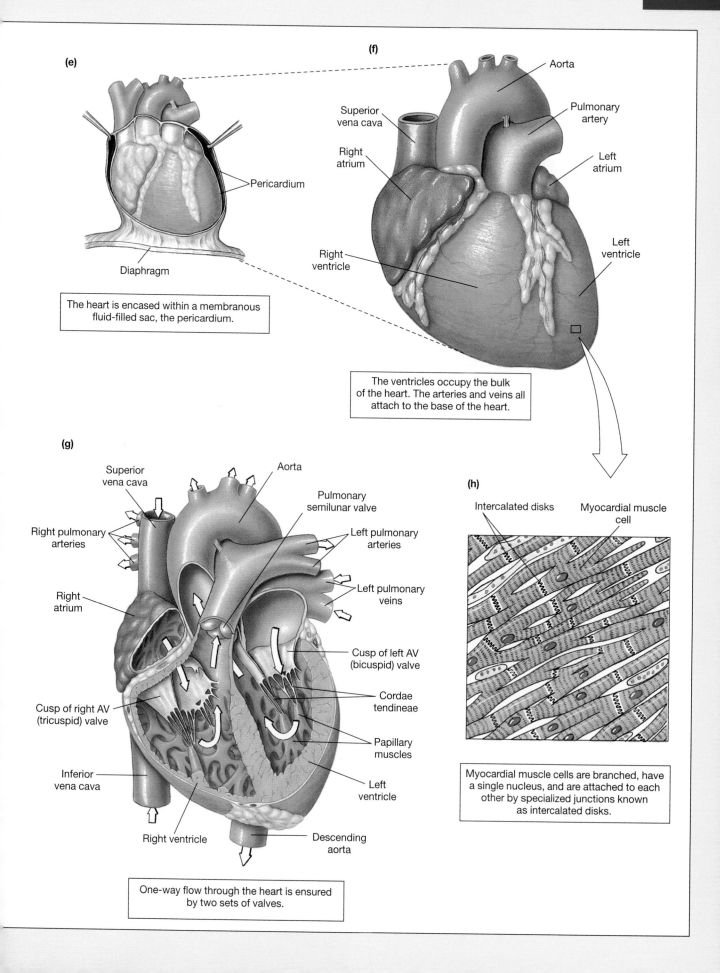

(e)

Pericardium

Diaphragm

The heart is encased within a membranous fluid-filled sac, the pericardium.

(f)

Aorta

Superior vena cava

Pulmonary artery

Right atrium

Left atrium

Left ventricle

Right ventricle

The ventricles occupy the bulk of the heart. The arteries and veins all attach to the base of the heart.

(g)

Superior vena cava

Aorta

Pulmonary semilunar valve

Right pulmonary arteries

Left pulmonary arteries

Left pulmonary veins

Right atrium

Cusp of left AV (bicuspid) valve

Cordae tendineae

Cusp of right AV (tricuspid) valve

Papillary muscles

Inferior vena cava

Left ventricle

Right ventricle

Descending aorta

One-way flow through the heart is ensured by two sets of valves.

(h)

Intercalated disks

Myocardial muscle cell

Myocardial muscle cells are branched, have a single nucleus, and are attached to each other by specialized junctions known as intercalated disks.

TABLE 14-3 The Heart and Major Blood Vessels.

Blue type indicates structures containing blood with lower oxygen content; red type indicates well-oxygenated blood.

	Receives Blood from	Sends Blood to
Heart		
Right atrium	Venae cavae	Right ventricle
Right ventricle	Right atrium	Lungs
Left atrium	Pulmonary veins	Left ventricle
Left ventricle	Left atrium	Body except for lungs
Vessels		
Venae cavae	Systemic veins	Right atrium
Pulmonary trunk (artery)	Right ventricle	Lungs
Pulmonary vein	Veins of the lungs	Left atrium
Aorta	Left ventricle	Systemic arteries

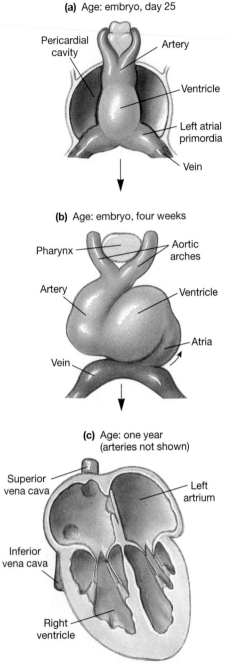

(a) Age: embryo, day 25

(b) Age: embryo, four weeks

(c) Age: one year (arteries not shown)

■ Figure 14-8 **Embryological development of the heart** (a) In the early embryo, the heart is a single tube. (b) By four weeks of development, the atria and ventricles can be distinguished. Eventually, the heart begins to twist so that the veins and arteries both emerge from the base of the heart. (c) By birth, the atria and ventricles are separated from each other.

left and right sides of the heart are separated from each other by the interventricular septum, so that blood on one side does not mix with blood on the other side. Although blood flow in the left and right heart is separated, the two sides of the heart contract in a coordinated fashion: The atria contract together and the ventricles contract together.

Blood entering the heart flows into the atria, the thin-walled upper chambers that serve as a reception area. From there, it passes through one-way valves into the ventricles, the pumping chambers. Blood leaves the heart via the pulmonary trunk from the right ventricle and via the aorta from the left ventricle. A second set of valves guards the exits of the ventricles so that blood cannot flow back into the heart once it has been ejected.

Notice that blood enters the ventricles at the top of the chamber but also leaves at the top. During development, the tubular embryonic heart twists back on itself so that arteries through which blood leaves the ventricles lie near the openings to the atria (Fig. 14-8 ■; see also Fig. 14-7g ■). Functionally, this means that the ventricles must contract from the bottom up so that blood is squeezed out of the top.

The heart contains four fibrous connective tissue rings that surround the openings to the major arteries and the openings between the chambers (Fig. 14-7a ■). These rings form both the origin and insertion for the cardiac muscle, an arrangement that pulls the apex and base of the heart together when the ventricles contract. In addition, the fibrous connective tissue acts as an insulator, blocking most transmission of electrical signals between the atria and the ventricles. This arrangement ensures that the electrical signal can be directed through special cells to the apex of the heart for the bottom-to-top contraction.

One-way flow in the heart is ensured by the heart valves   As the arrows in Figure 14-7g ■ indicate, blood flow through the heart proceeds in one direction. One-way flow is ensured by the location of the heart valves, one set between the atria and ventricles and the second set between the ventricles and the arteries. Although the two sets of valves are very different in structure, they serve the same function, preventing the backward flow of blood.

The passageways between the atria and ventricles are guarded by the **atrioventricular,** or **AV, valves** (Fig. 14-7g ■). These valves are thin flaps of tissue attached at their base to a ring of connective tissue. The flaps are slightly thickened at the edge and are attached on the ventricular side to collagenous cords, the **chordae tendineae.** The opposite ends of the cords are tethered to moundlike extensions of ventricular muscle known as the **papillary muscles** [*papilla,* nipple]. Most of the chordae attach to the edges of the valve flaps, but some attach to the middle and the base. When the ventricle contracts, blood pushes against the bottom side of the AV valve and forces it upward into a closed position (Fig. 14-9b ■). The tendons prevent the valve from being

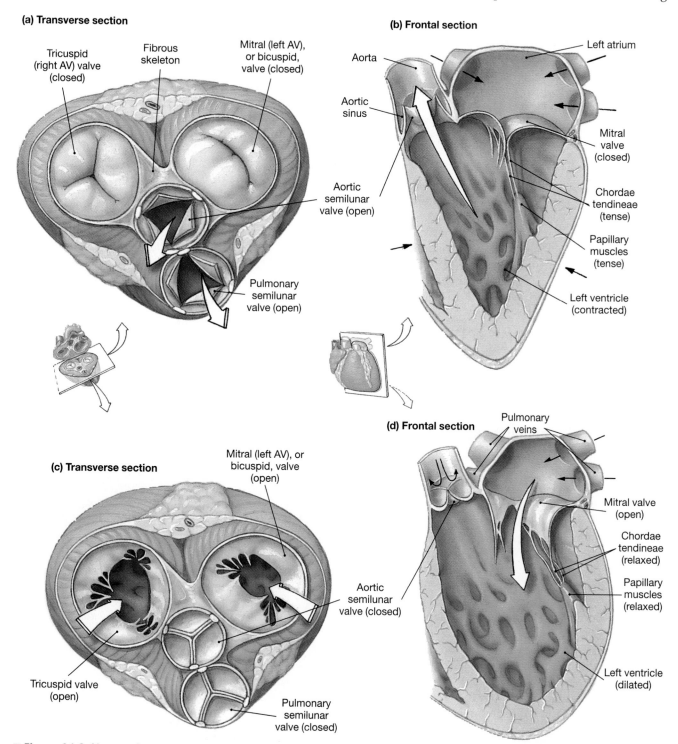

■ Figure 14-9 **Heart valves** The atrioventricular (AV), or cuspid, valves separate the atria from the ventricles. During ventricular contraction, these valves remain closed to prevent blood flow backward into the atria. The semilunar valves prevent blood that has entered the arteries from flowing back into the ventricles during ventricular relaxation. Views (a) and (c) show the valves as viewed from the atria and from inside the arteries.

pushed back into the atrium, just as the struts on an umbrella keep the umbrella from turning inside out in a high wind. Occasionally, the tendons fail and the valves are pushed back into the atrium during ventricular contraction, an abnormal condition known as *prolapse*.

The AV valves are not identical. The valve that separates the right atrium and right ventricle has three flaps and is called the **tricuspid valve** [*cuspis,* point] (Fig. 14-9a ■). The valve between the left atrium and left ventricle has only two flaps and is called the **bicuspid valve.** The bicuspid is also called the **mitral valve** because of its resemblance to the tall headdress, known as a miter, worn by popes and bishops.

The second set of valves lies at the openings from the ventricles into the arteries (aorta and pulmonary trunk). These two valves prevent blood pumped out of the heart from returning to it (Fig. 14-9d ■). They are called the **semilunar valves** because of their crescent-moon shape in cross section. They are referred to as the **aortic valve** or **pulmonary valve** according to their location. Both semilunar valves have three cuplike leaflets that fill with blood and snap closed when backward pressure is exerted on them (Fig. 14-9c ■). Because of their shape, the semilunar valves do not need connective tendons as do the AV valves.

✓ What prevents electrical signals from passing through the connective tissue within the heart?

✓ Trace a drop of blood from the superior vena cava to the aorta, naming all structures the drop of blood encounters.

Cardiac Muscle Cells Contract without Nervous Stimulation

The bulk of the heart is composed of cardiac muscle cells, or myocardium. Most cardiac muscle is contractile, but about 1% of the myocardial cells are specialized to generate action potentials spontaneously. These cells are responsible for a unique property of the heart: its ability to contract without any outside signal. For example, Spanish explorers in the New World wrote of witnessing human sacrifices in which hearts that had been torn from the chests of living victims and held aloft continued to beat for minutes. The heart can contract without a connection to other parts of the body because the signal for contraction is *myogenic,* originating within the heart muscle itself.

The signal for myocardial contraction comes not from the nervous system but from specialized myocardial cells known as **autorhythmic cells.** The autorhythmic cells are also called **pacemakers** because they set the rate of the heartbeat. Myocardial autorhythmic cells are anatomically distinct from contractile cells: They are smaller and contain few contractile fibers or organelles. Because they do not have organized sarcomeres, the autorhythmic cells do not contribute to the contractile force of the heart.

Most myocardial cells in the heart are typical striated muscle, with contractile fibers organized into sarcomeres

(∞ p. 327). Cardiac muscle fibers are much smaller than skeletal muscle fibers and have a single nucleus. About one-third of the cell volume of a cardiac contractile fiber is occupied by mitochondria, a reflection of the high energy demand of these cells. By one estimate, cardiac muscle extracts 70%–80% of the oxygen delivered to it by the blood, more than twice the amount removed by other cells in the body. This high oxygen demand does not leave much oxygen in the coronary blood to supply periods of increased activity. The only way to get more oxygen to the heart muscle is to increase its blood flow. Because the heart muscle depends on adequate blood flow, narrowing of a coronary vessel by a clot or fatty deposit that reduces myocardial blood flow can damage the myocardial cells.

Individual cardiac muscle cells branch and join to neighboring cells at their ends to form a complex network (Fig. 14-10b ■). The cell junctions are specialized

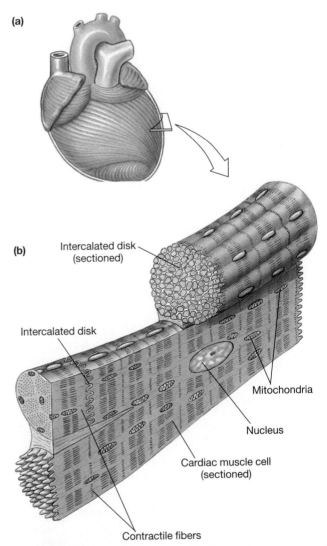

(a)

(b)

Intercalated disk (sectioned)

Intercalated disk

Mitochondria

Nucleus

Cardiac muscle cell (sectioned)

Contractile fibers

■ **Figure 14-10 Cardiac muscle** (a) The spiral arrangement of the ventricular muscle allows ventricular contraction to squeeze the blood upward from the apex of the heart. (b) Intercalated disks contain desmosomes that allow force to be transferred from cell to cell. Gap junctions in the intercalated disks allow electrical signals to pass rapidly from cell to cell.

regions known as **intercalated disks.** They consist of interdigitated membranes tightly linked by *desmosomes* that tie adjacent cells together (∞ p. 54). These strong connections allow force created in one cell to be transferred to the adjacent cell. In addition, the intercalated disks contain *gap junctions,* linked transmembrane proteins that allow direct movement of ions from the cytoplasm of one cell to the cytoplasm of the next cell (∞ p. 147). Gap junctions electrically couple the heart muscle so that waves of depolarization spread rapidly from cell to cell, allowing them to contract almost simultaneously.

Although cardiac muscle is a striated muscle, it differs in several ways from skeletal muscle. The t-tubules of myocardial cells are larger, and they branch inside the myocardial cell. The sarcoplasmic reticulum is smaller than that of skeletal muscle, reflecting the fact that cardiac muscle depends in part on extracellular Ca^{2+} to initiate contraction.

Excitation-Contraction Coupling in Cardiac Muscle Is Similar to Skeletal Muscle

Contraction in cardiac muscle takes place by the same type of sliding filament movement that occurs in skeletal muscle (∞ p. 330). As in skeletal muscle, an action potential in the muscle cell is the signal to initiate contraction. But there the similarity ends. In cardiac muscle, the action potential itself does not release Ca^{2+} from the sarcoplasmic reticulum. Instead, the depolarization opens voltage-gated Ca^{2+} channels in the cell membrane and allows Ca^{2+} to enter the cell from the extracellular fluid, moving into the cell down its concentration gradient (Fig. 14-11 ■). Through a process known as **calcium-induced calcium release,** the influx of Ca^{2+} triggers the release of stored Ca^{2+} from the sarcoplasmic reticulum.

The Ca^{2+} released from the sarcoplasmic reticulum provides about 90% of the Ca^{2+} needed for muscle con-

Gap Junctions Can Open and Close   The gap junctions between myocardial cells are normally open, but they can be closed by abnormally high concentrations of Ca^{2+} and H^+ levels in the cytosol. This situation occurs when blood supply to part of the heart is stopped, a condition known as **ischemia** [*ischien,* to suppress + *-emia,* blood]. The cells become oxygen-starved, produce lactic acid (H^+), and are unable to make ATP to pump Ca^{2+} out of the cytoplasm. The resultant increase in H^+ and Ca^{2+} closes gap junctions in the damaged cells. Closure electrically isolates the damaged cells, and they no longer contract. It also forces action potentials to find an alternate route from cell to cell. If the damaged area of myocardium is large, the pumping function of the heart may be severely impaired, even to the point of death. When people speak of "heart attacks," they are usually referring to large areas of ischemia caused by blocked blood vessels. In medical terms, a heart attack is called a **myocardial infarction,** referring to an area of tissue that is dying because of lack of blood supply.

continued from page 390

Upon Walter's arrival at the University of Texas Southwestern Medical Center emergency room, one of the first tasks is to determine whether he has actually had a heart attack. Walter's vital signs (pulse and breathing rates, blood pressure, temperature) are taken. A resident physician draws blood for an enzyme assay to determine the level of cardiac enzymes in Walter's blood. When heart muscle cells die, they release various enzymes that serve as markers of a heart attack. In addition, a second tube of blood is sent for an assay of its troponin I level. UT Southwestern is taking part in a clinical trial to determine if troponin I is a good indicator of heart damage following a heart attack.

Question 2: Some of the enzymes that are used as markers for heart attacks are also found in related forms in skeletal muscle. What are related forms of an enzyme called? (Hint: ∞ p. 80) What is troponin, and why would elevated blood levels of troponin indicate heart damage? (Hint: ∞ p. 334)

traction. In a process similar to that in skeletal muscle, Ca^{2+} diffuses through the cytosol to the contractile elements, where the ions bind to troponin and initiate the cycle of crossbridge formation and movement.

Relaxation in cardiac muscle occurs when Ca^{2+} unbinds from troponin. Calcium ions are transported back into the sarcoplasmic reticulum with the help of a *Ca^{2+}-ATPase.* They can also be removed from the cell in exchange for sodium ions. The latter process uses a Na^+-Ca^{2+} indirect active transporter. Calcium ions move out of the cell against their concentration gradient in exchange for Na^+ entering the cell down its concentration gradient. The Na^+ that enters the cell during this transfer is removed by the Na^+/K^+-ATPase.

✓ If a myocardial contractile cell is placed in fluid similar to the interstitial fluid and depolarized, the cell will contract. If Ca^{2+} is removed from the fluid surrounding the myocardial cell, and the cell is depolarized, it will not contract. If the experiment is repeated with a skeletal muscle fiber, it will contract when depolarized, whether Ca^{2+} is present in the extracellular fluid or not. What conclusion can you draw from the results of this experiment?

Cardiac Muscle Contraction Can Be Graded

A key property of cardiac muscle cells is the ability of a single muscle fiber to execute *graded contractions,* in which the fiber varies the amount of force it generates. (Recall that in skeletal muscle, contraction in a single fiber is all-or-none at any given fiber length.) The force generated by cardiac muscle is proportional to the number of crossbridges that are active. If cytosolic Ca^{2+} concentrations are low, some crossbridges will not be activated and force will be low. If

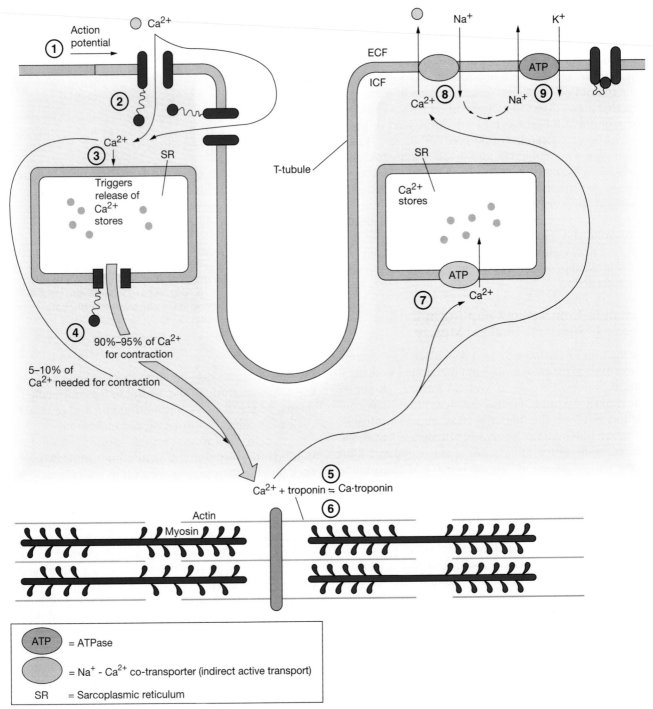

■ Figure 14-11 **Role of calcium in cardiac muscle contraction**   ① An action potential enters the contractile cell from an adjacent cell. ② The action potential opens voltage-gated Ca^{2+} channels in the cell membrane, allowing Ca^{2+} to enter the cell, moving down its electrochemical gradient. ③ Entry of Ca^{2+} triggers the release of additional Ca^{2+} from the sarcoplasmic reticulum. This process is known as calcium-induced calcium release. ④ The increase in cytoplasmic calcium initiates muscle contraction. Most of the Ca^{2+} for contraction comes from the sarcoplasmic reticulum. ⑤ Calcium ions bind to troponin in a process similar to that of skeletal muscle. Calcium binding allows strong crossbridge formation and movement of the sliding filaments. ⑥ Relaxation occurs when Ca^{2+} unbinds from troponin as free cytosolic Ca^{2+} concentrations decrease. ⑦ A Ca^{2+}-ATPase pumps Ca^{2+} back into the sarcoplasmic reticulum, where the Ca^{2+} is stored until the next contraction. ⑧ Some Ca^{2+} is extruded from the cell in exchange for Na^+. This co-transporter uses the potential energy of Na^+ moving into the cell down its concentration gradient to push Ca^{2+} out of the cell against its concentration gradient. ⑨ Sodium is restored to the extracellular fluid by the Na^+/K^+-ATPase.

additional Ca²⁺ enters the cell from the extracellular fluid, more Ca²⁺ is released from the sarcoplasmic reticulum. This additional Ca²⁺ release allows greater crossbridge formation and cycling, creating additional force.

The catecholamines epinephrine and norepinephrine are regulatory molecules that affect the amount of Ca²⁺ available for cardiac muscle contraction (∞ p. 181). As you may recall, epinephrine is the "fight-or-flight" hormone released by the adrenal medulla; norepinephrine is the neurotransmitter released by sympathetic neurons. Both molecules bind to and activate β₁ adrenergic receptors (∞ p. 314) on the membrane of the cardiac muscle cell (Fig. 14-12 ■). Once activated, β₁ receptors use a cyclic AMP second messenger system to phosphorylate specific intracellular proteins (∞ p. 154). Phosphorylation of the voltage-gated Ca²⁺ channels increases the probability that they will open; their opening allows more Ca²⁺ to enter the cell. In addition, phosphorylation of a regulatory protein known as **phospholamban** enhances Ca²⁺-ATPase activity in the sarcoplasmic reticulum. The ATPase concentrates Ca²⁺ in the sarcoplasmic reticulum so that more Ca²⁺ is available for calcium-induced calcium release. Because the force of contraction increases as the number of active crossbridges increases, the net result of catecholamine stimulation is a stronger contraction.

Besides increasing force, catecholamines shorten the duration of cardiac contraction. The enhanced activity of the Ca²⁺-ATPase transports Ca²⁺ back into the sarcoplasmic reticulum more rapidly. Both the time that Ca²⁺ is bound to troponin and the active time of the myosin crossbridges are thereby shortened.

When Cardiac Muscle Is Stretched, It Contracts More Forcefully

Another factor that affects the force of contraction in cardiac muscle is the sarcomere length at the beginning of contraction. For both cardiac and skeletal muscle, the tension generated is directly proportional to the length of the muscle fiber (∞ p. 340). As muscle fiber length and sarcomere length increase, tension increases, up to a maximum (Fig. 14-13 ■). For years, it was believed that the length-tension relationship in cardiac muscle strictly depended on the overlap between thick and thin filaments, as it does in skeletal muscle. But new evidence suggests that stretching the myocardial cell allows more Ca²⁺ to enter, contributing to a more forceful contraction, as discussed in the previous section. Thus, although it is traditional to separate changes in force due to fiber length from changes in force created by catecholamines, there is some overlap between the mechanisms underlying the two.

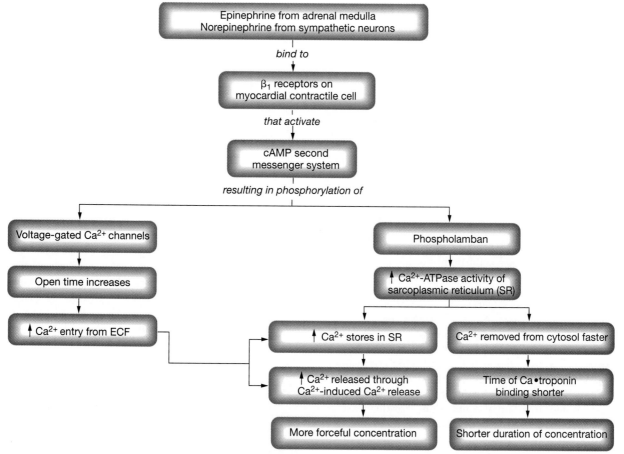

■ Figure 14-12 **Modulation of cardiac contraction**

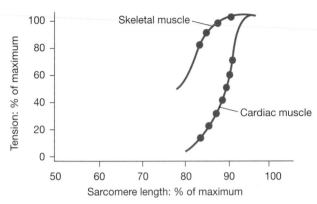

■ **Figure 14-13 Length-tension relationships in skeletal and cardiac muscle** The graph compares the amount of force developed by isolated skeletal and cardiac muscles that begin at different sarcomere lengths. The lines represent the range of sarcomere length within which the muscles can operate. The dots indicate the normal range of operation. Notice that skeletal muscles normally work close to their maximum, while cardiac muscles usually operate at less than their maximum.

In the intact heart, stretch on the individual fibers depends on how much blood is in the chambers of the heart. The relationship between force and ventricular volume is an important property of the intact heart that is discussed in detail later in this chapter.

✓ A drug that blocks all Ca^{2+} channels in the myocardial cell membrane is placed in the solution around the cell. What will happen to the force of contraction in that cell?

Foxglove for a Failing Heart Another mechanism for increasing the concentration of Ca^{2+} in the cytosol is preventing the removal of Ca^{2+} from the cell. This mechanism is not active under normal physiological conditions. However, it can be triggered by the administration of cardiac glycosides, a class of molecules that was first discovered in the plant *Digitalis purpurea* (purple foxglove). The cardiac glycosides include digitoxin and the related compound ouabain, a molecule used to inhibit sodium transport in research studies. Because all cardiac glycosides depress Na^+/K^+-ATPase activity, they are highly toxic; they poison the pumps on all cells, not just those of the heart. However, if administered in low doses, cardiac glycosides can partially block the removal of Na^+ from the myocardial cell. As the concentration of Na^+ builds up in the cytosol, the concentration gradient for Na^+ across the cell membrane diminishes. In turn, the potential energy available for indirect active transport diminishes. In the myocardial cell, cardiac glycosides decrease the cell's ability to remove Ca^{2+} by means of the Na^+-Ca^+ exchanger. The resultant increase in cytosolic Ca^{2+} causes stronger myocardial contractions. For this reason, cardiac glycosides have been used since the eighteenth century as a remedy for heart failure, a condition in which the heart is unable to contract forcefully.

Action Potentials in Myocardial Cells Vary According to Cell Type

Cardiac muscle, like skeletal muscle and neurons, is an excitable tissue with the ability to generate action potentials. Each of the two types of cardiac muscle cells has a distinctive action potential. In both types, Ca^{2+} plays an important role in the action potential, in contrast to the action potentials of skeletal muscle and neurons.

Myocardial contractile cells The action potentials of myocardial contractile cells are similar in several ways to those of neurons and skeletal muscle (∞ p. 213). The rapid depolarization phase of the action potential is due to Na^+ entry, and the steep repolarization phase is due to K^+ leaving the cell. The main difference between the action potential of the myocardial contraction cell and that of a skeletal muscle fiber or a neuron is a lengthening of the action potential that is due to Ca^{2+} entry.

Myocardial contractile cells have a stable resting potential of about -90 mV (phase 4, Fig. 14-14 ■). When a wave of depolarization moves into a contractile cell through the gap junctions, the membrane potential becomes more positive. Voltage-gated Na^+ channels open, allowing Na^+ to enter the cell and rapidly depolarize it (phase 0). The membrane potential reaches about $+20$ mV before the Na^+ channels close. At that point, the cell begins to repolarize as K^+ leaves through open K^+ channels (phase 1).

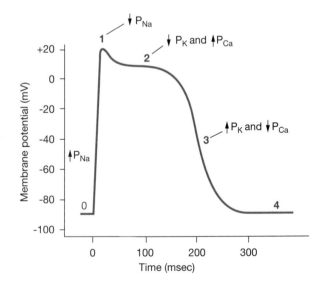

Phase	0	Na^+ channels open
	1	Na^+ channels close
	2	Ca^{2+} channels open; fast K^+ channels close
	3	Ca^{2+} channels close; slow K^+ channels open
	4	Resting potential

P_X = Permeability to ion X

■ **Figure 14-14 Action potential of a cardiac contractile cell**

Flattening of the action potential into the plateau of phase 2 occurs as the result of two events: a decrease in K⁺ permeability and an increase in Ca²⁺ permeability. Voltage-gated Ca^{2+} channels activated by depolarization have been slowly opening during phases 0 and 1. When they finally open, Ca^{2+} enters the cell. At the same time, some K⁺ channels close. The combination of Ca^{2+} influx and decreased K⁺ efflux causes the action potential to flatten out into a plateau (phase 2).

The plateau ends when the Ca^{2+} channels close and K⁺ permeability increases once more. The K⁺ channels responsible for this phase are similar to those in the neuron: They are activated by depolarization but are slow to open. When the delayed K⁺ channels open, K⁺ exits rapidly (phase 3), returning the cell to its resting potential (phase 4).

The influx of Ca^{2+} during phase 2 lengthens the duration of a myocardial action potential. A typical action potential in a neuron or skeletal muscle fiber lasts between 1 and 5 msec; in a contractile myocardial cell, it is typically 200 msec or more. The longer action potential in myocardial cells helps prevent the sustained contraction called tetanus. Prevention of tetanus is important in the heart because the muscles must relax between contractions to allow the ventricles to fill with blood.

To understand why a longer action potential prevents tetanus, compare the relationship between action potentials and contraction in skeletal and cardiac muscle cells. As you may recall from Chapter 12, the skeletal muscle action potential is ending as contraction begins (Fig. 14-15a ■). Thus, a second action potential fired immediately after the refractory period will cause summation of the contractions. If a series of action potentials occurs in rapid succession, the sustained contraction known as tetanus results (Fig. 14-15b ■). Tetanus cannot occur in cardiac muscle because the refractory period and the contraction end almost simultaneously due to the longer action potential (Fig. 14-15c ■). By the time a second action

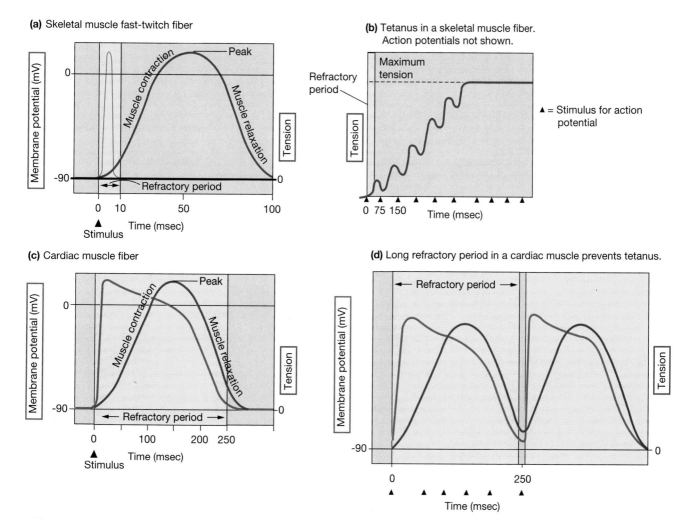

■ Figure 14-15 **Refractory periods in skeletal and cardiac muscle** (a) The refractory period of a skeletal muscle fiber is very short compared with the amount of time required for the development of tension in the muscle. Compare this with the cardiac muscle (c), in which the refractory period lasts almost as long as the entire muscle twitch. (b) Skeletal muscles that are stimulated repeatedly will exhibit summation and tetanus. (d) Because of the difference in the length of the refractory period, cardiac muscle will not show summation. If stimuli occur during the refractory period of the cardiac muscle fiber, nothing happens. There can be no subsequent contraction until the refractory period ends.

potential takes place, the myocardial cell has almost completely relaxed. Consequently, no summation will occur (Fig. 14-15d ■).

✓ Tetrodotoxin (TTX) is a molecule that blocks voltage-gated Na$^+$ channels. What will happen to the action potential of a myocardial contractile cell if TTX is applied to it?

Myocardial autorhythmic cells Myocardial autorhythmic cells have the unique ability to generate action potentials spontaneously in the absence of input from the nervous system. This property results from their unstable membrane potential that starts at −60 mV and slowly drifts upward toward threshold (Fig. 14-16 ■). Because the membrane potential never "rests" at a constant value, it is called a *pacemaker potential* rather than a resting membrane potential. Whenever the pacemaker potential depolarizes to threshold, the autorhythmic cell fires an action potential.

What causes the membrane potential of these cells to be unstable? Our current understanding is that the autorhythmic cells contain channels that are different from the channels of other excitable tissues. When the cell membrane potential is −60 mV, channels that are permeable to both K$^+$ and Na$^+$ are open. These channels are called **I$_f$ channels** because they allow current (I) to flow and because of their unusual properties. The researchers who first described these channels did not initially understand their behavior and named them "funny" channels, hence the subscript *f*. When I$_f$ channels are open at negative membrane potentials, Na$^+$ influx exceeds K$^+$ efflux. (This is similar to what happens at the neuromuscular junction when nonspecific cation channels open [∞ p. 319].) The net influx of positive charge slowly depolarizes the autorhythmic cell. As the membrane potential

becomes more positive, the I$_f$ channels gradually close and a few Ca^{2+} channels open. The subsequent influx of Ca^{2+} continues the depolarization, and the membrane potential moves steadily toward threshold.

When the membrane potential reaches threshold, many Ca^{2+} channels open. Calcium ions rush in, creating the steep depolarization phase of the action potential. Note that this process is different from that in other excitable cells, in which the depolarization phase is due to the opening of voltage-gated Na$^+$ channels. When the Ca^{2+} channels close at the peak of the action potential, slow K$^+$ channels have opened. The repolarization phase of the autorhythmic action potential is due to the resultant efflux of K$^+$. This phase is similar to repolarization in other types of excitable cells.

The timing of autorhythmic action potentials can be modified by altering the permeability of the autorhythmic cells to different ions. Norepinephrine from sympathetic neurons and epinephrine from the adrenal medulla increase the flow of ions through both the I$_f$ and Ca^{2+} channels. More rapid cation entry speeds up the rate of the pacemaker depolarization, causing the cell to reach threshold faster and increasing the rate of action potential firing. When the pacemaker fires action potentials more rapidly, heart rate increases (Fig. 14-17a ■). The catecholamines exert their effect by binding to and activating β$_1$ receptors on the autorhythmic cells. The β$_1$ receptors use a cAMP second messenger system to alter the transport properties of the ion channels.

The parasympathetic neurotransmitter acetylcholine (ACh) slows heart rate. Acetylcholine activates muscarinic receptors that influence K$^+$ and Ca^{2+} channels in the pacemaker membrane. Potassium permeability increases, hyperpolarizing the cell so that the pacemaker potential begins at a more negative value. At the same time, Ca^{2+} permeability of the pacemaker decreases. This decrease in Ca^{2+} permeability slows the depolarization

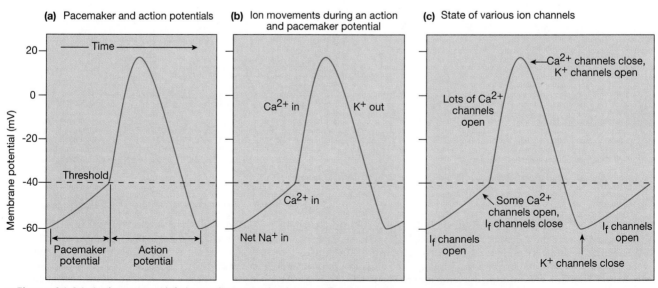

(a) Pacemaker and action potentials

(b) Ion movements during an action and pacemaker potential

(c) State of various ion channels

■ **Figure 14-16 Action potentials in cardiac autorhythmic cells**

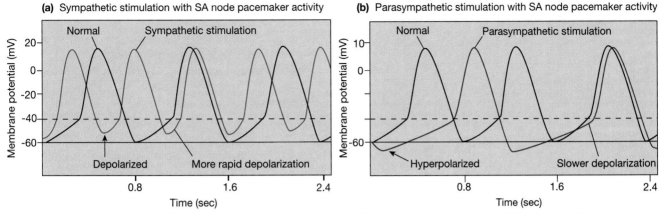

■ Figure 14-17 **Modulation of heart rate by the nervous system** (a) Sympathetic stimulation and circulating epinephrine depolarize the autorhythmic cell and speed up the depolarization rate. When the pacemaker reaches threshold more rapidly, the heart rate increases. (b) Parasympathetic stimulation hyperpolarizes the membrane potential of the autorhythmic cell and slows the depolarization rate. The pacemaker takes longer to reach threshold, slowing down the heart rate.

rate of the pacemaker potential. The combination of the two effects causes the cell to take longer to reach threshold, delaying the onset of the action potential in the pacemaker and slowing the heart rate (Fig. 14-17b ■).

As you have seen in this section, the electrical signals we call action potentials have many different properties. Table 14-4 compares the action potentials of skeletal muscle with those of the two types of myocardial muscle. In the next section, we will see how the action potentials of the autorhythmic cells spread throughout the heart to coordinate contraction of the contractile cells.

✓ Do you think that the Ca^{2+} channels in the autorhythmic cells are the same as the Ca^{2+} channels in the contractile cells? Defend your answer.

✓ What will happen to the action potential of a myocardial autorhythmic cell if tetrodotoxin (TTX), which blocks voltage-gated Na^+ channels, is applied to it?

✓ In an experiment, the vagus nerve to the heart was cut. The investigators noticed that heart rate increased. Why did this happen? (∞ p. 311)

TABLE 14-4 **Comparison of Action Potentials in Cardiac and Skeletal Muscle**

	Skeletal Muscle	Contractile Myocardium	Autorhythmic Myocardium
Membrane potential	Stable at −70 mV	Stable at −90 mV	Unstable pacemaker potential; usually starts at −60 mV
Events leading to threshold potential	Net Na^+ entry through nonspecific channels opened by ACh	Depolarization from adjacent cells enter via gap junctions	Net Na^+ entry through I_f channels that open at negative membrane potentials; reinforced with some Ca^{2+} entry
Rising phase of action potential	Na^+ entry	Na^+ entry	Ca^{2+} entry
Repolarization phase	Rapid; caused by K^+ efflux	Extended plateau caused by Ca^{2+} entry; rapid phase caused by K^+ efflux	Rapid; caused by K^+ efflux
Hyperpolarization	Due to excessive K^+ efflux at high K^+ permeability; when K^+ channels close, leak of K^+ and Na^+ restores potential to resting state	None; resting potential is −90 mV, the equilibrium potential for K^+, so no excess K^+ efflux	None; when repolarization hits −60 mV, the I_f channels open and the pacemaker potential begins
Duration of action potential	Short: 1–2 msec	Extended: 200+ msec	Variable; generally 150+ msec
Refractory period	Results from time required to reset Na^+ channel gates; generally brief	Resetting of Na^+ channel gates delayed until end of action potential, leading to long refractory period	None

continued from page 397

The results of the cardiac enzyme and troponin I assays will not come back for several hours. If Walter has had a blockage in one of his cardiac blood vessels, damage to the ischemic heart muscle could be severe by that time. In Walter's case, an electrocardiogram (ECG) shows an abnormal pattern of electrical activity in the myocardium. "He's definitely had an MI," says the ER physician, referring to a myocardial infarction, or heart attack. "Let's start him on t-PA." t-PA is short for *tissue plasminogen activator*. t-PA activates plasminogen, a substance produced in the body that dissolves blood clots. When t-PA is given within 1–3 hours of a heart attack, it can help dissolve clots that are blocking blood flow to the heart muscle and help limit the extent of ischemic damage.

Question 3: How do electrical signals move from cell to cell in the myocardium? What happens to contraction in a myocardial contractile cell if a wave of depolarization passing through the heart bypasses it?

THE HEART AS A PUMP

We now turn from electrical events and contraction in single myocardial cells to the electrical events and contraction of the intact heart during the **cardiac cycle,** the period of time from the start of one contraction, through relaxation. We look first at the electrical events of the cycle and at how an action potential in an autorhythmic cell initiates a wave of electrical activity that spreads throughout the heart. We then describe the mechanical events of the cycle: how sequential contraction of the atria and ventricles moves blood from the veins through the chambers of the heart and out into the arteries. A key feature of this process is the creation of high pressure in the ventricles that serves as the driving force for the circulation of blood throughout the cardiovascular system.

Electrical Conduction in the Heart Coordinates Contraction

If the heart is to create enough force to circulate the blood, individual myocardial cells must depolarize and contract in a coordinated fashion. An action potential that originates in an autorhythmic cell spreads rapidly from cell to cell through the gap junctions of the intercalated disks (Fig. 14-18 ■). The depolarization wave that results is followed by a wave of contraction that passes across the atria, then moves into the ventricles.

The **sinoatrial node (SA node),** a group of autorhythmic cells in the right atrium near the entry of the superior vena cava, is the main pacemaker of the heart (Fig. 14-19a ■). From the SA node, noncontractile autorhythmic fibers form a specialized conducting system that rapidly transmits action potentials through the heart. An **internodal pathway** connects the SA node to the **atrioventricular node (AV node),** a group of autorhythmic cells found near the floor of the right atrium. From the AV node, action potentials move into fibers known as the **bundle of His** ("hiss"), or **atrioventricular bundle.** The bundle passes from the AV node into the wall (*septum*) between the ventricles. A short way down the septum, the bundle divides into *left and right bundle branches*. These fibers continue downward to the apex where they divide into many small **Purkinje fibers** that spread outward among the contractile cells.

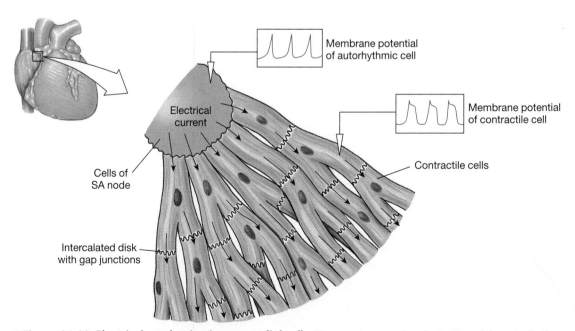

Membrane potential of autorhythmic cell

Membrane potential of contractile cell

Electrical current

Contractile cells

Cells of SA node

Intercalated disk with gap junctions

■ **Figure 14-18 Electrical conduction in myocardial cells** The spontaneous depolarization of the autorhythmic cell results in an action potential that rapidly spreads to adjacent contractile cells through gap junctions. The subsequent depolarization of the contractile cells causes them to contract.

The electrical signal that initiates contraction begins when the SA node fires an action potential (Fig. 14-19b ■). The depolarization of the pacemaker action potential spreads to adjacent cells through gap junctions. Conduction of the electrical signal is rapid through the internodal conducting pathways (Fig. 14-19c ■), but slower through the contractile cells of the atrium (Fig. 14-19d ■).

As action potentials spread outward across the atria, they encounter the fibrous skeleton of the heart at the junction of the atria and ventricles. This barricade prevents the transfer of electrical signals from the atria to the ventricles. Consequently, the AV node is the only pathway through which action potentials can reach the contractile fibers of the ventricles (Fig. 14-19d ■). The electri-

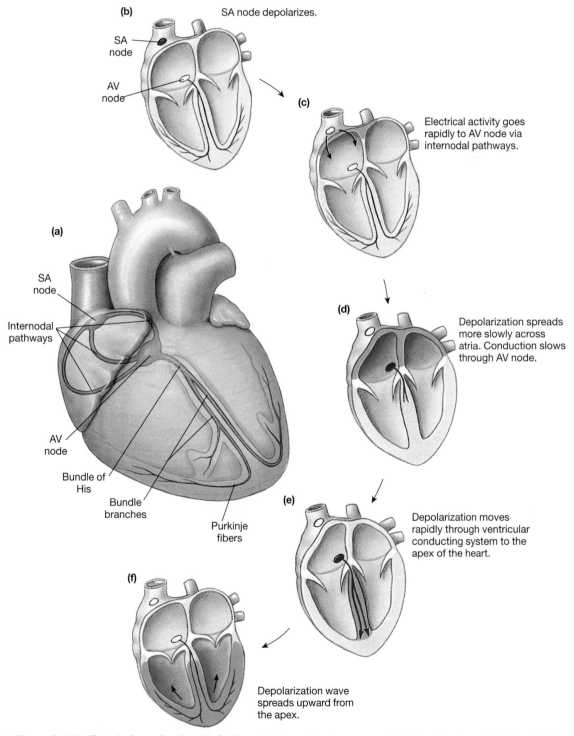

(b) SA node depolarizes.

SA node

AV node

(c) Electrical activity goes rapidly to AV node via internodal pathways.

(a)

SA node

Internodal pathways

AV node

Bundle of His

Bundle branches

Purkinje fibers

(d) Depolarization spreads more slowly across atria. Conduction slows through AV node.

(e) Depolarization moves rapidly through ventricular conducting system to the apex of the heart.

(f) Depolarization wave spreads upward from the apex.

■ Figure 14-19 **Electrical conduction in the heart** The conducting system of the heart begins at the SA node. The internodal pathway connects the SA and AV nodes, and the bundle of His and bundle branches carry signals to the Purkinje fibers in the apex of the heart. Depolarization in (b)–(f) is represented by purple shading.

cal signal passes from the AV nodes through the bundle of His and the bundle branches to the apex of the heart (Fig. 14-19c ■). At that point, the Purkinje fibers transmit impulses very rapidly, with speeds up to 4 m/sec, so that all contractile cells in the apex contract nearly simultaneously (Fig. 14-19f ■).

Why is it necessary to direct the electrical signals through the AV node? Why not allow them to spread downward from the atria? The answer lies in the fact that blood being pumped out of the ventricles leaves through openings at the top of the chambers. If the electrical signals from the atria were conducted directly into the ventricles, the ventricles would start contraction at the top. Then blood would be squeezed downward and would be trapped in the bottom of the ventricle (think of squeezing toothpaste from a tube starting at the top). The apex-to-base contraction squeezes blood toward the arterial openings at the base of the heart.

The ejection of blood from the ventricles is aided by the spiral arrangement of the muscles in the walls (see Fig. 14-10a ■). As these muscles contract, they pull the apex and base of the heart closer together, squeezing blood out openings at the top of the ventricles.

A second function of the AV node is to delay the transmission of action potentials slightly, allowing the atria to complete their contraction before ventricular contraction begins. The **AV node delay** is accomplished by slowing conduction through the AV node so that action potentials move at only 1/20 the rate of action potentials in the atrial internodal pathway.

Pacemakers Set the Heart Rate

The cells of the SA node set the pace of the heartbeat. Other cells in the conducting system, such as the AV node and the Purkinje fibers, have unstable resting potentials and can also act as pacemakers under some conditions. However, because their rhythm is slower than that of the SA node, they do not usually have a

chance to set the heartbeat. The Purkinje fibers, for example, can spontaneously fire action potentials, but their rate of firing is very slow, between 25 and 40 beats per minute.

Why is it that the fastest pacemaker determines the pace of the heartbeat? Consider the following analogy: A group of people are playing "Follow the Leader" as they walk. Initially, everyone is walking at a different pace: some fast, some slow. When the game starts, everyone must match his or her pace to the pace of the person who is walking the fastest. The fastest person in the group is the SA node, walking at 70 steps per minute. Everyone else in the group (autorhythmic and contractile cells) sees that the SA node is fastest, so they pick up their pace and follow the leader. In the heart, the cue to follow the leader is the electrical signal sent from the SA node to the other cells.

Now suppose that the SA node gets tired and drops out of the group. The role of leader defaults to the next fastest person, the AV node, who is walking at a rate of 50 steps per minute. The group slows to match the pace of the AV node, but they are still following the fastest leader.

What happens if the group divides? When they reach the corner, the AV node leader goes left, but a renegade Purkinje fiber decides to go right. Those people who follow the AV node continue to walk at 50 steps per minute, but the people who follow the Purkinje fiber slow down to match his pace of 35 steps per minute. Now there are two leaders, each walking at a different pace.

In the heart, the SA node is the fastest pacemaker and normally sets the heart rate. But if the SA node is damaged and cannot function, one of the slower pacemakers in the heart takes over. Heart rate then matches the rate of the new pacemaker. It is even possible for different parts of the heart to follow different pacemakers, just as the walking group split at the corner. In a condition known as *complete heart block,* the conduction of electrical signals from the atria to the ventricles through the AV node is disrupted. The SA node fires at its rate of 70 beats per minute, but those signals never reach the ventricles. So the ventricles coordinate with their fastest pacemaker. Because ventricular autorhythmic cells discharge only about 35 times a minute, the ventricles contract at a totally different rate than the atria. In conditions such as complete heart block, in which the rate of ventricular contraction is too slow to maintain adequate blood flow, it may be necessary for the heart's rhythm to be set artificially by a surgically implanted mechanical pacemaker. These battery-powered devices artificially stimulate the heart at a predetermined rate.

In the next section, we expand our look at electrical events in the heart to see how recordings taken from the surface of the skin can be used to explain activity within the heart itself.

Fibrillation Coordination of myocardial contraction is essential for normal cardiac function. In extreme cases in which the myocardial cells contract in an extremely disorganized manner, a condition known as **fibrillation** results. Ventricular fibrillation is a life-threatening emergency because without coordinated contraction of the muscle fibers, the ventricles cannot pump enough blood to supply adequate oxygen to the brain. One way to correct this problem is to administer an electrical shock to the heart. The shock creates a depolarization that triggers action potentials in all cells simultaneously, coordinating them again. You have probably seen this procedure on television hospital shows, when the doctors place flat paddles on the patient's chest and tell everyone to stand back ("Clear!") while they pass an electrical current through the body.

✓ Name two functions of the AV node.

✓ Occasionally an *ectopic pacemaker* [*ektopos,* out of place] will develop in part of the conducting system. What happens to heart rate if an ectopic pacemaker in the atria depolarizes at a rate of 120 times per minute?

The Electrocardiogram Reflects the Electrical Activity of the Heart

In the beginning of the twentieth century, physiologists discovered that they could use electrocardiograms to obtain information about the function of the heart. An **electrocardiogram,** or **ECG,** is a recording of the electrical activity of the heart made from electrodes placed on the surface of the skin. By simply placing electrodes on the arms and legs, we can record electrical activity that is taking place deep within the body. Salt solutions such as the NaCl-based extracellular fluid are good conductors of electricity and can transfer electrical activity to the skin's surface. Thus, external recordings allow us to monitor electrical activity not only in the heart but also in muscles (*electromyelogram,* or *EMG*) and in the brain (*electroencephalogram,* or *EEG*). The signals are very weak by the time they get to the skin. A ventricular action potential, for example, has a voltage change of 110 mV, but the ECG signal has an amplitude of only 1 mV by the time it reaches the surface of the body.

The first human electrocardiogram was recorded in 1887, but the procedure was not refined for clinical use until the first years of the twentieth century. The father of the modern ECG was a Dutch physiologist named Walter Einthoven. He named the parts of the ECG as we know them today and created "Einthoven's triangle," a hypothetical triangle created around the heart when electrodes are placed on both arms and the left leg (Fig. 14-20 ■). The sides of the triangle are numbered to correspond with the three leads, or pairs of electrodes, used for a recording.

An ECG tracing shows the sum of the electrical potentials generated by all the cells of the heart at any moment. Each component of the ECG reflects depolarization or repolarization of a portion of the heart. Because depolarization is the signal for contraction, portions of the ECG can be associated approximately to contraction or relaxation of the atria or ventricles (collectively referred to as the *mechanical events* of the cardiac cycle). We will follow an ECG through a single contraction-relaxation cycle.

There are three major components of an ECG: the P wave, the combined waves of the QRS complex, and the T wave (Fig. 14-21 ■). The first wave is the **P wave,** which corresponds to depolarization of the atria (Fig. 14-22 ■). The next trio of waves, the **QRS complex,** represents the progressive wave of ventricular depolariza-

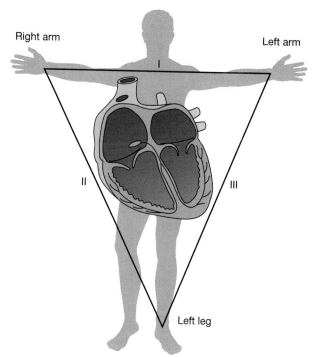

■ **Figure 14-20 Einthoven's triangle** Electrodes for recording electrical activity on the surface of the skin are attached to both arms and the left leg, forming a triangle. A lead consists of two electrodes. One electrode is assigned as the positive electrode and one as the negative electrode. Each lead gives a different electrical "view" of the heart, corresponding to the three sides of the triangle. In clinical practice, it is now common to make a 12-lead recording.

tion. The final wave, the **T wave,** represents the repolarization of the ventricles. Atrial repolarization is not represented by a special wave.

The mechanical events of the cardiac cycle lag slightly behind the electrical signals, just as the contraction of a single cardiac muscle cell follows its action potential. Atrial contraction begins during the latter part of the P wave and continues during the P-R segment. Ventricular contraction begins just after the Q wave and continues through the S-T segment.

One thing that many people find puzzling is that downward deflection may correspond to periods of depolarization in the heart. It is important to remember that the ECG is different from a single action potential (see Fig. 14-23 ■, p. 410). The single action potential is an electrical event in a single cell, recorded using an intracellular electrode. The ECG represents multiple action potentials taking place in the heart muscle at a given time and recorded from the surface of the body. When recording an ECG, one active surface electrode is the positive electrode, and the other active electrode is the negative electrode. (The third electrode is inactive). If net electrical current in the heart moves toward the positive electrode, the tracing of the ECG goes up from the baseline. If net current moves toward the negative electrode, the tracing moves downward.

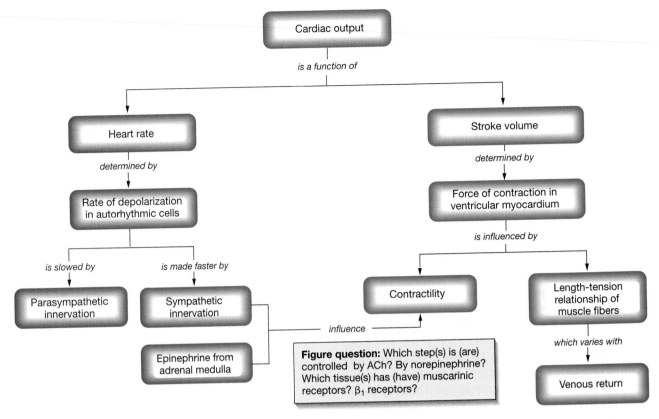

■ Figure 14-31 **Factors that affect cardiac output**

CHAPTER REVIEW

Chapter Summary

Overview of the Cardiovascular System

1. The human **cardiovascular system** consists of a heart that pumps blood through a closed system of blood vessels. (p. 383)

2. The primary function of the cardiovascular system is the transport of nutrients, water, gases, wastes, and chemical signals among all parts of the body. (p. 384)

3. Blood vessels that carry blood away from the heart are called **arteries;** blood vessels that return blood to the heart are called **veins.** A system of valves in the heart and veins ensures that blood flows in one direction. (p. 385)

4. The heart is divided into left and right halves. Each half consists of an **atrium** and a **ventricle,** separated with a set of closable valves. (p. 385)

5. The **pulmonary circulation** goes from the right side of the heart to the lungs and back to the heart. The **systemic circulation** goes from the left side of the heart to the tissues and back to the heart. (p. 385)

Pressure, Volume, Flow, and Resistance

6. To generate blood flow, the heart creates high pressure when it contracts. Blood moves down a **pressure gradient (∆P),** from the highest pressure in the arteries to the lowest pressure in the venae cavae and pulmonary veins. (p. 387)

7. **Pressure** in fluids is the force exerted by a fluid on its container. In a system in which fluid is flowing, pressure decreases over distance. (p. 387)

8. If the volume of a fluid-filled container decreases, the pressure exerted by the fluid on the walls of the container increases. The pressure created when the ventricles contract is called the **driving pressure** for blood flow. (p. 388)

9. The tendency of the cardiovascular system to oppose blood flow is called its **resistance** to flow. If resistance increases, flow decreases. If resistance decreases, flow increases. (p. 388)

10. Resistance of a fluid flowing through a tube increases as length of the tube and viscosity (thickness) of the fluid increase, but resistance increases as the radius of the tube

decreases. The radius of the tube (r) has the greatest effect on resistance (R): $R \propto 1/r^4$. (p. 389)

11. An increase in the radius of blood vessels is called **vasodilation;** a decrease in radius is **vasoconstriction.** Vasoconstriction increases blood pressure; vasodilation decreases it. (p. 389)

12. Fluid flow through a tube is proportional to the pressure gradient (ΔP): Flow $\propto \Delta P$, where $\Delta P = P_1 - P_2$. A pressure gradient is not the same thing as the absolute pressure in the system. (p. 390)

13. **Flow rate** (flow) is the volume of blood that passes one point in the system per unit time. (p. 390)

14. **Velocity of flow** is the distance a fixed volume of blood will travel in a given period of time. At a constant flow rate, blood will flow faster through a smaller tube than it will through a larger tube. (p. 390)

Cardiac Muscle and the Heart

15. The heart is composed mostly of cardiac muscle, or **myocardium.** Blood flows into the atria and leaves from the ventricles. (p. 391)

16. One-way flow is ensured by heart valves located between the atria and ventricles and between the ventricles and the arteries. (p. 394)

17. Most cardiac muscle is contractile and consists of typical striated muscle. (p. 396)

18. The signal for contraction originates in **autorhythmic cells** within the heart. The autorhythmic cells are smaller than contractile cells and contain few contractile fibers. (p. 396)

19. Myocardial cells are linked to each other by **intercalated disks** that contain gap junctions. Gap junctions allow waves of depolarization to spread rapidly from cell to cell. (p. 397)

20. In myocardial contractile cells, an action potential opens voltage-gated Ca^{2+} channels in the cell membrane. Calcium ions enter the cell and through **calcium-induced calcium release** trigger the release of Ca^{2+} from the sarcoplasmic reticulum. (p. 397)

21. Cardiac muscle contraction can be graded in force according to the amount of Ca^{2+} that enters the cell. Epinephrine and norepinephrine increase the force of myocardial contraction when they bind to β_1 adrenergic receptors on the myocardial cell. They also shorten the duration of cardiac contraction. (p. 397)

22. The amount of force generated by cardiac muscle is also proportional to the length of the muscle fiber: As muscle fiber length increases, force increases. In the intact heart, stretch on the individual fibers depends on the volume of blood in the chambers of the heart. (p. 399)

23. The action potentials of myocardial contractile cells have a rapid depolarization phase that is due to Na^+ influx and a steep repolarization phase that is due to K^+ efflux. The action potential also has a plateau phase that is due to Ca^{2+} influx. (p. 400)

24. Autorhythmic myocardial cells have an unstable membrane potential, called a pacemaker potential, that slowly drifts upward until it reaches threshold. The unstable membrane potential is due to I_f **channels** that allow net influx of positive charge. (p. 402)

25. The steep depolarization phase of the autorhythmic cell action potential is due to Ca^{2+} influx, whereas the repolarization phase is due to K^+ efflux. (p. 402)

26. The catecholamines norepinephrine and epinephrine act on β_1 receptors to speed up the rate of the pacemaker depolarization and increase heart rate. Acetylcholine activates muscarinic receptors and slows down heart rate. (p. 402)

The Heart as a Pump

27. The **cardiac cycle** is the period of time that includes one cycle of contraction and relaxation. (p. 404)

28. An action potential that originates in an autorhythmic cell spreads rapidly from cell to cell and is followed by a wave of contraction that passes across the atria, then moves into the ventricles. (p. 404)

29. The **sinoatrial node (SA node)** is the main pacemaker of the heart. The wave of cardiac depolarization passes from the SA node to the **atrioventricular node (AV node),** then into the **atrioventricular bundle** and bundle branches. At the apex of the heart, the fibers divide into **Purkinje fibers** that carry the depolarization to individual myocardial cells. (p. 404)

30. The cells of the SA node set the pace of the heartbeat. If the SA node malfunctions, other autorhythmic cells in the AV node or ventricles will take control of the heart rate. (p. 406)

31. An **electrocardiogram,** or **ECG,** is a surface recording of the electrical activity of the heart. The **P wave** corresponds to depolarization of the atria. The **QRS complex** represents ventricular depolarization. The **T wave** represents the repolarization of the ventricles. Atrial repolarization is not represented by a special wave. The contraction and relaxation of the atria and ventricles lag slightly behind the electrical signals. (p. 406)

32. An ECG provides information on heart rate and rhythm, conduction velocity, and the condition of tissues within the heart. (p. 408)

33. **Diastole** is the time during which cardiac muscle relaxes, and **systole** is the time when the muscle is contracting. (p. 409)

34. Most blood enters the ventricles while the atria are relaxed, but the last 20% of filling is accomplished when the atria contract. (p. 410)

35. Ventricular systole closes the AV valves so that blood cannot flow back into the atria. Vibrations following closure of the AV valves create the first heart sound. (p. 410)

36. During **isovolumic ventricular contraction,** the ventricles contract on a fixed volume of blood. When pressure in the ventricles opens the semilunar valves, blood is ejected into the arteries. (p. 411)

37. When the ventricles relax and ventricular pressure falls, the semilunar valves close, creating the second heart sound. (p. 411)

38. The amount of blood pumped by one ventricle during a contraction is known as the **stroke volume.** (p. 413)

39. **Cardiac output** is the amount of blood pumped per ventricle per unit time. It is equal to heart rate times stroke volume. The average cardiac output at rest is 5 L/min. (p. 415)

40. Homeostatic changes in cardiac output are accomplished by varying the heart rate or the stroke volume or both. (p. 415)

41. Parasympathetic activity slows heart rate; sympathetic activity speeds it up. (p. 415)

42. Stroke volume is affected by the length of the muscle fiber at the beginning of contraction and the contractility of the heart. As sarcomere length increases, force increases and, consequently, stroke volume increases. The Frank-Starling law of the heart says that the more blood there is in the ventricle at the beginning of contraction (increased end-diastolic volume), the greater the stroke volume is. (p. 416)

43. End-diastolic volume is determined by **venous return.** Factors that affect venous return include skeletal muscle contractions, the **respiratory pump,** and constriction of the veins by sympathetic activity. (p. 417)

44. Contractility of the heart is enhanced by catecholamines and certain drugs. (p. 417)

Questions

LEVEL ONE   **Reviewing Facts and Terms**

1. What contributions to understanding the cardiovascular system did each of these people make?

    a. William Harvey
    b. Frank and Starling
    c. Malpighi

2. List three functions of the cardiovascular system.

3. Put these structures in the order blood passes through them, starting and ending with the left ventricle:
    a. left ventricle of heart
    b. systemic veins
    c. pulmonary circulation
    d. systemic arteries
    e. aorta
    f. left ventricle of heart

4. The primary reason blood flows through the body is due to a _____ gradient. In humans, this value is highest at the _____ and in the _____; it is lowest in the _____. In a system in which fluid is flowing, pressure decreases over distance due to _____.

5. As a flexible container is squeezed, the volume decreases; thus the pressure must _____.

6. If vasodilation occurs in a blood vessel, volume (increases/decreases) whereas pressure (increases/decreases).

7. Match the anatomic terms with their description. Not all terms are used.

    (a) myocardium
    (b) endothelium
    (c) pericardium
    (d) apex
    (e) atrium
    (f) ventricle
    (g) artery
    (h) aorta
    (i) AV valves
    (j) bicuspid valve
    (k) tricuspid valve
    (l) semilunar valves
    (m) atria
    (n) base

    1. the narrow end of the heart; points downward and to the left
    2. the primary artery of the systemic circulation; other arteries branch from this one
    3. the lower chambers of the heart, which push blood out of the heart
    4. valves that have papillary muscles preventing their collapse
    5. the upper chamber of one side of the heart
    6. the valve between the left atrium and the left ventricle
    7. the muscular layer of the heart
    8. a vessel that carries blood away from the heart
    9. tough membranous sac that encases the heart
    10. valves between the ventricles and the main arteries

8. The heart contracts without a connection to any other part of the body because the signal is _____, meaning the signal originates within the heart muscle itself. This is due to _____ % of the myocardial cells that generate action potentials spontaneously.

9. Trace an action potential from the SA node through the conducting system of the heart.

10. Cell junctions between myocardial cells are specialized regions known as _____, which physically tie adjacent cells together. These areas also contain _____, which allow direct movement of ions from one cell to the next.

11. In cardiac muscle, an action potential opens ion channels in the cell membrane. Entry of _____ ion triggers release of _____ ions from the _____. Name two ways that these ions are removed from the cytosol.

12. What effect do the sympathetic and parasympathetic branches of the autonomic nervous system have on heart rate and force of contraction? Name the neurotransmitters for each division and their receptor types on the heart. In what part(s) of the heart are these receptors located? What effect does neurotransmitter binding to the receptor have on specific ion movements?

13. What is the name for the cluster of autorhythmic cells in the right atrium that functions as the main pacemaker? Name the electrical event it produces.

14. Match the terms with their definition:
    (a) stroke volume
    (b) end-diastolic volume
    (c) heart rate
    (d) blood volume
    (e) diastole
    (f) end systolic volume
    (g) cardiac output
    (h) systole

    1. the amount of blood left in the ventricle after the heart contracts
    2. the amount of blood in the entire body
    3. the number of contractions per minute
    4. the amount of blood in the ventricle before the heart contracts
    5. the contraction phase of the heart
    6. the amount of blood that enters the aorta with each contraction
    7. the amount of blood that leaves the heart in one minute
    8. the relaxation phase of the heart

15. What events cause the two principal heart sounds?

LEVEL TWO   Reviewing Concepts

16. **Concept map**: Create a map showing the factors that influence cardiac output. Include two ways to increase heart rate and two ways to increase stroke volume as well as factors that affect venous return. Where appropriate, include neurotransmitters or hormones and their target cell receptors.

17. List the events of the cardiac cycle in sequence, beginning with atrial and ventricular diastole. Note where valves open and close. Describe what happens to pressure and blood flow in each chamber at each step of the cycle.

18. a. Convert a pressure of 120 mm Hg to cm H_2O.
    b. Convert a pressure of 80 mm Hg to cm H_2O.
    c. Calculate the cardiac output if stroke volume is 65 mL and heart rate is 80 bpm.
    d. Calculate end-systolic volume if end-diastolic volume is 150 mL and stroke volume is 65 mL.

19. Compare and contrast the structure of a cardiac muscle cell with that of a striated muscle cell. What unique properties of cardiac muscle are essential to its function?

20. Explain why contractions in cardiac muscle cannot sum or exhibit tetanus.

21. Match the following ion movements with the appropriate phase of the action potential. More than one answer may apply to a single phase. Not all answers may be used.
    (a) K^+ from ECF to ICF
    (b) K^+ from ICF to ECF
    (c) Na^+ from ECF to ICF
    (d) Na^+ from ICF to ECF
    (e) Ca^{2+} from ECF to ICF
    (f) Ca^{2+} from ICF to ECF

    1. slow rising phase of autorhythmic cells
    2. plateau phase of contractile cells
    3. rapid rising phase of contractile cells
    4. rapid rising phase of autorhythmic cells
    5. rapid falling phase of contractile cells
    6. falling phase of autorhythmic cells

22. Correlate the waves of the ECG with what is happening in the atria and ventricles. Why are there only three electrical events even though there should be four cardiac components? What mechanical event follows each electrical event?

23. List and briefly explain four types of information about the heart that the ECG provides.

24. Define inotropic effect. Give examples of two drugs with a positive inotropic effect. How is calcium related?

LEVEL THREE   Problem Solving

25. Two drugs that are used to reduce cardiac output are calcium channel blockers and beta (receptor) blockers. What effect do these drugs have on the heart that explains how they decrease cardiac output?

26. Police Captain Jeffers has suffered a myocardial infarction. His ejection fraction (SV divided by EDV) is only 25%, with a stroke volume of 40 mL.
    a. What are his EDV, ESV, and CO? Show your calculations. Explain to his (nonmedically oriented) family what has happened to his heart.
    b. When you analyzed his ECG, you referred to several different leads, such as lead I. What are leads?
    c. Why is it possible to record an ECG on the body surface without direct access to the heart?

    d. What might cause a longer-than-normal P-R interval in an ECG?

27. The following paragraph is a summary of a newspaper article:

    A new treatment for atrial fibrillation due to an excessively rapid rate at the SA node involves a high-voltage electrical pulse administered to the AV node to destroy its autorhythmic cells. A ventricular pacemaker is then implanted in the patient.

    Briefly explain the physiological rationale for this treatment: Why is a rapid atrial depolarization rate dangerous, why is the AV node destroyed, and why must a pacemaker be implanted?

Problem Conclusion

In this running problem, you learned about the current treatments for heart attack. You also learned that many of these treatments depend on speed to work effectively.

Further check your understanding of this running problem by checking your answers against those in the summary table.

Question	Facts	Integration and Analysis
1a Why is the medic giving Walter oxygen?	The medic suspects that Walter has had a heart attack. A blood clot may be blocking blood flow and oxygen supply to the heart muscle.	If the heart is not pumping effectively, the brain may not receive adequate oxygen. Administration of oxygen will raise the amount of oxygen that reaches both the heart and the brain.
1b What effect will the injection of isotonic saline have on Walter's extracellular fluid volume? On his intracellular fluid volume? On his total body osmolarity?	An isotonic solution is one that does not change cell volume (∞ p. 134). Isotonic saline is isosmotic to the body.	The extracellular volume will increase because all of the saline administered will remain in that compartment. The intracellular volume and total body osmolarity will not change.
2a Some of the enzymes that are used as markers for heart attacks are also found in related forms in skeletal muscle. What are related forms of an enzyme called?	Related forms of an enzyme are called isozymes (∞ p. 80).	Although isozymes are variants of the same enzymes, their activity may vary with different conditions, and their structures are slightly different. Cardiac and skeletal muscle isozymes can be distinguished by their different structures.
2b What is troponin, and why would elevated blood levels of troponin indicate heart damage?	Troponin is the regulatory protein bound to tropomyosin (∞ p. 334). When Ca^{2+} binds to troponin, tropomyosin shifts and uncovers the myosin-binding site of actin.	Troponin is part of the contractile apparatus of the muscle cell. If troponin escapes from the cell and enters the blood, this is an indication that the cell has been damaged or is dead.
3a How do electrical signals move from cell to cell in the myocardium?	Electrical signals pass directly from cell to cell through gap junctions in the intercalated disks (∞ p. 147).	The cell-to-cell conduction of electrical signals in the heart allows rapid and coordinated contraction of the myocardium.
3b What happens to contraction in a myocardial contractile cell if a wave of depolarization passing through the heart bypasses it?	Depolarization in a muscle cell is the signal for contraction.	If a myocardial cell is not depolarized, it will not contract. Failure to contract creates a nonfunctioning region of heart muscle and impairs the pumping function of the heart.
4 If the ventricles of the heart are damaged, in which wave or waves of the electrocardiogram would you predict abnormal changes to occur?	The waves of the ECG are the P wave, QRS complex, and T wave. The P wave represents atrial depolarization; the QRS complex and T wave represent ventricular depolarization and repolarization, respectively.	The QRS complex and the T wave are mostly likely to show changes after a heart attack. Changes indicative of myocardial damage include enlargement of the Q wave, shifting of the S-T segment off of the baseline (elevated or depressed), and inversion of the T wave.
5 If Walter's heart attack has damaged the muscle of his left ventricle, what do you predict will happen to his left cardiac output?	Cardiac output equals stroke volume (the amount of blood pumped by one ventricle per contraction) times heart rate.	If the ventricular myocardium has been weakened, the stroke volume may decrease. Decrease in stroke volume in turn would decrease the cardiac output.

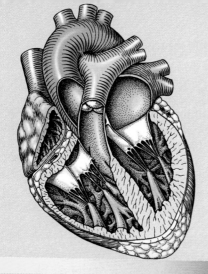

15

Blood Flow and the Control of Blood Pressure

In the previous chapter, you learned how the heart pumps blood and creates a region of high pressure that serves as the driving force for blood flow. Homeostatic regulation of the cardiovascular system is aimed at maintaining adequate blood flow (perfusion) to the heart and brain. Once that need is met, a secondary consideration is the appropriate distribution of blood flow among the remaining tissues according to their metabolic demands for oxygen and nutrients. In this chapter, we turn our attention to the vasculature, the blood vessels through which blood reaches the cells of the body.

A simplified model of the cardiovascular system (Fig. 15-1 ■) summarizes the key points that we will discuss in this chapter. The heart is shown as two separate pumps that work in series (one after the other), with the right heart pumping blood to the lungs, then to the left heart. The left heart then pumps blood through the rest of the body. Blood leaving the left heart enters the systemic arteries, shown here as an expandable, elastic region. The pressure produced by contraction of the left ventricle is stored in the elastic walls of the arteries and slowly released through elastic recoil. This mechanism maintains a continuous driving pressure for blood flow during the time when the ventricles are relaxing. The arterioles create a high-resistance outlet for arterial blood flow. In addition, the arterioles direct distribution of blood flow to individual tissues by selectively constricting and dilating. Once blood flows into the capillaries, a leaky epithelium allows exchange of material between the plasma, the interstitial fluid, and the cells of the body. At the distal end of the capillaries, blood flows into the venous side of the circulation and from there back to the right side of the heart.

Problem

Essential Hypertension

"Doc, I'm as healthy as a horse," says Kurt English, age 56, during his annual physical examination. "I don't want to waste your time. Let's get this over with." But to Dr. Arthur Cortez, Kurt does not appear to be the picture of health: He is about 20 pounds overweight. When Dr. Cortez asks about his diet, Kurt replies, "Well, I like to eat." Exercise? "Who has the time?" replies Kurt. Dr. Cortez slips a blood pressure cuff around Kurt's arm and takes a reading. "Your blood pressure is 164 over 100," says Dr. Cortez. "We'll take it again in 15 minutes. If it's still high, we'll need to discuss it further." Kurt stares at his doctor, flabbergasted. "But how can my blood pressure be too high? I feel fine!" he protests.

continued on page 430

Total blood flow through any level of the circulation is equal to the cardiac output. For example, if cardiac output is 5 L/min, blood flow through all the systemic capillaries is 5 L/min. In the same manner, blood flow through the pulmonary side of the circulation is equal to blood flow through the systemic circulation.

THE BLOOD VESSELS

The walls of the blood vessels are made from layers of smooth muscle, elastic connective tissue, and fibrous connective tissue. The inner lining of all blood vessels is a thin layer of **endothelium,** a type of epithelium. For years, this lining was thought to be simply a passive barrier, but now we know that endothelial cells play important roles in the regulation of blood pressure, blood vessel growth, and absorption of materials.

Surrounding the endothelium are layers of connective tissue and smooth muscle. The amount and type of connective tissue in the outer layers of the blood vessel wall vary with different vessels. The amount of smooth muscle also varies. The descriptions below apply to the vessels of the systemic circulation, although those of the pulmonary circulation are generally similar.

Blood Vessels Contain Vascular Smooth Muscle

The smooth muscle of blood vessels is known as **vascular smooth muscle.** Most blood vessels contain smooth muscle, arranged in either circular or spiral layers. Vasoconstriction narrows the diameter of the vessel lumen; vasodilation widens it.

In blood vessels, smooth muscle cells maintain a state of partial contraction at all times, creating the condition known as *muscle tone.* A variety of chemicals influence

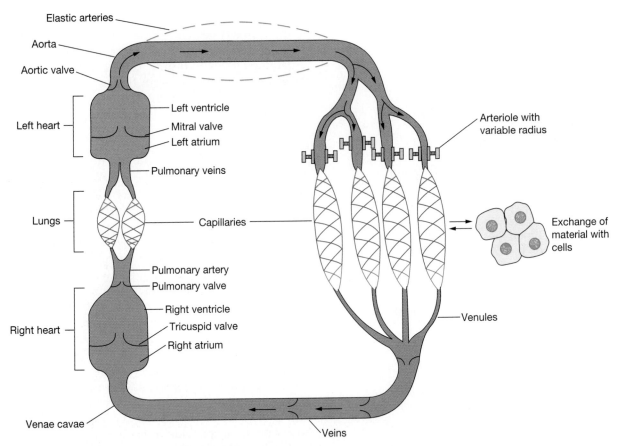

■ **Figure 15-1 Model of the cardiovascular system**

TABLE 15-1 Chemicals Mediating Vascular Smooth Muscle Contraction and Relaxation

Chemical	Physiological Role	Source	Type
Contraction			
Norepinephrine (alpha receptors)	Baroreceptor reflex	Sympathetic neurons	Neural
Endothelin	Paracrine mediator	Vascular endothelium	Local
Serotonin	Platelet aggregation, smooth muscle contraction	Neurons, digestive tract, platelets	Local, neural
Substance P	Pain, increases capillary permeability	Neurons, digestive tract	Local, neural
Vasopressin	Increased blood pressure in hemorrhage	Posterior pituitary	Hormonal
Angiotensin II	Increases blood pressure	Plasma hormone	Hormonal
Prostacyclin	Minimizes blood loss from damaged vessels before coagulation	Endothelium	Local
Relaxation			
Nitric oxide	Paracrine mediator	Endothelium	Local
Atrial natriuretic peptide	Reduces blood pressure	Atrial myocardium, brain	Hormonal
Vasoactive intestinal peptide	Digestive secretion, relaxes smooth muscle	Neurons	Neural, hormonal
Histamine	Increases blood flow	Mast cells	Local, systemic
Epinephrine (β_2 receptors)	Enhances local blood flow to skeletal muscle, heart, liver	Adrenal medulla	Hormonal
Acetylcholine (muscarinic receptors)	Erection of clitoris or penis	Parasympathetic neurons	Neural
Bradykinin	Increases blood flow via nitric oxide	Multiple tissue	Local
Adenosine	Enhances blood flow to match metabolism	Hypoxic cells	Local

vascular smooth muscle tone, including neurotransmitters, hormones, and paracrines (Table 15-1). Paracrines (∞ p. 147) are a particularly important component in the local control of blood flow. Many vasoactive paracrines are secreted by endothelial cells lining the blood vessels or by the tissues surrounding the vessels. Contraction of smooth muscle, like that of cardiac muscle, depends on the entry of Ca^{2+} from the extracellular fluid through Ca^{2+} channels (∞ p. 352).

Calcium Channel Blockers The calcium channels of both vascular smooth muscle and cardiac muscle can be blocked by a class of drugs known as calcium channel blockers. These drugs bind to the Ca^{2+} channel proteins, making it less likely that the channels will open in response to depolarization. With diminished Ca^{2+} entry, vascular smooth muscle dilates, while in the heart, the depolarization rate of the SA node decreases. Vascular smooth muscle is more sensitive than cardiac muscle to certain classes of calcium channel blockers, and it is possible to get vasodilation at low doses that have no effect on heart rate. Other tissues with Ca^{2+} channels, such as neurons, are only minimally affected by calcium channel blockers because their Ca^{2+} channels are a different subtype.

Arteries and Arterioles Carry Well-Oxygenated Blood to the Cells

The aorta and major arteries are characterized by walls that are both stiff and springy. Arteries have smooth muscle layers with large amounts of elastic and fibrous tissue (Fig. 15-2 ■). The stiffness of the fibrous tissue requires substantial amounts of energy to stretch the walls of an artery outward. This energy comes in the form of high-pressure blood ejected from the left ventricle. Once the artery is distended with blood, the energy stored by stretching elastic fibers is released through elastic recoil.

The arteries and arterioles are characterized by a divergent pattern of blood flow. As the major arteries divide into smaller and smaller arteries, the character of the wall changes, becoming less elastic and more muscular. Finally, the smallest arteries become **arterioles.** The walls of arterioles contain several layers of smooth muscle that contract and relax under the influence of various chemical signals.

Some arterioles branch into vessels known as **metarterioles** [*meta-*, beyond]. Metarterioles have only part of their wall surrounded by smooth muscle, in comparison with true arterioles, whose walls have a continuous muscle layer. Blood flowing through metarterioles

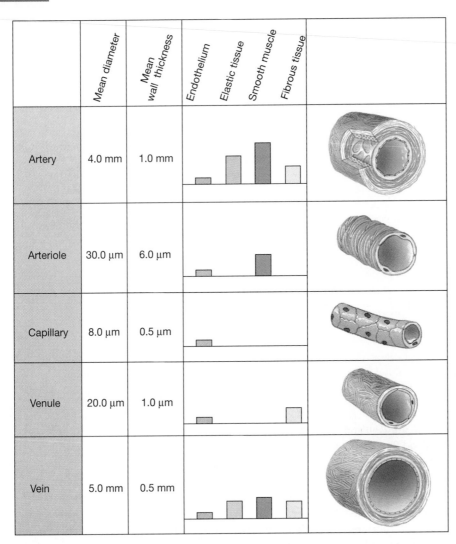

	Mean diameter	Mean wall thickness	Endothelium	Elastic tissue	Smooth muscle	Fibrous tissue	
Artery	4.0 mm	1.0 mm					
Arteriole	30.0 μm	6.0 μm					
Capillary	8.0 μm	0.5 μm					
Venule	20.0 μm	1.0 μm					
Vein	5.0 mm	0.5 mm					

Figure 15-2 Blood vessels The walls of the blood vessels vary in their diameter and composition. Arteries, which carry high-pressure blood, have the thickest walls. Capillaries have the thinnest walls, consisting of a single layer of exchange epithelium supported on an acellular basement membrane.

either can be directed into adjoining capillary beds (Fig. 15-3 ■) or can bypass the capillaries and go directly to the venous circulation if precapillary sphincters are contracted. In addition to regulating blood flow through the capillaries, the metarterioles allow white blood cells to go directly from the arterial to the venous circulation. Capillaries are barely large enough to let red blood cells through, much less white blood cells that are twice as large.

The arterioles, along with the capillaries and venules, are called the *microcirculation*. The regulation of blood flow through the microcirculation is an active area of physiological research.

Exchange between the Blood and Interstitial Fluid Takes Place in the Capillaries

Capillaries are the smallest vessels in the cardiovascular system and, along with postcapillary venules, are the site of exchange between the blood and the interstitial fluid. To facilitate exchange of material, capillary walls lack smooth muscle and elastic or fibrous tissue rein-

forcement (Fig. 15-2 ■). Instead, they consist of a flat layer of endothelium, one-cell thick, supported on an acellular matrix called the basement membrane (∞ p. 56). Because the capillary endothelium is an exchange epithelium, it usually has leaky junctions with spaces, or pores, between the cells. The notable exception to this pattern is the tight-junction endothelium of cerebral capillaries that creates the blood-brain barrier (∞ p. 241).

Blood Flow Converges in the Venules and Veins

Blood flows from the capillaries into small vessels called **venules.** The very smallest venules are similar to capillaries, with a thin exchange epithelium and little connective tissue (Fig. 15-2 ■). They are distinguished by a convergent pattern of flow. Smooth muscle begins to appear in the walls of larger venules. From venules, blood flows into veins that become larger in diameter as they travel toward the heart. Finally, the largest veins, the venae cavae, empty into the right atrium.

Veins are more numerous than arteries and have a larger diameter. As a result of their large volume, the

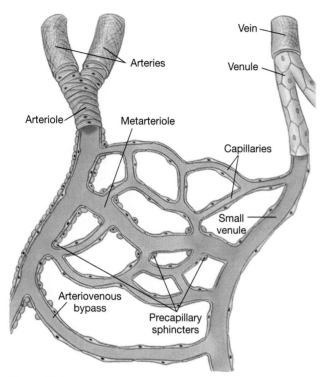

Figure 15-3 Metarterioles The metarterioles regulate blood flow into different sets of capillaries when the precapillary sphincters constrict and relax. They also act as a bypass channel that allows white blood cells to go directly from the arterial to the venous side of the circulation.

veins hold more than half of the blood in the circulatory system. Veins lie closer to the surface of the body than arteries, forming the bluish blood vessels that you see running just under the skin. Veins have thinner walls than arteries, with less elastic tissue. As a result, veins expand easily when they fill with blood. When you have blood drawn from your arm (*venipuncture*), the technician uses a tourniquet to exert pressure on the blood vessels. Blood flow coming into the arm through the high-

Pericytes Although capillaries are generally described as having walls composed of a single layer of endothelium, they are closely associated with a type of cell known as the **pericyte** [*peri-*, around]. These elongated, highly branched cells surround the capillaries, forming a meshlike outer layer between the endothelium and the interstitial fluid. Pericytes apparently contribute to the "tightness" of capillary permeability: the more pericytes, the less leaky the capillary endothelium. Pericytes are thought to contribute to many aspects of capillary function including growth, repair following injury, and local control of blood flow through secretion of vasoactive agents. Pericytes may also play a role in several diseases that affect the microcirculation, including diabetes, high blood pressure, and Alzheimer's disease. It is likely that in the coming years we will begin to see new therapies for these conditions that target the pericytes.

pressure arteries is not affected, but the pressure of the tourniquet stops flow through the low-pressure veins. As a result, blood collects in the surface veins, making them stand out against the underlying muscle tissue.

Angiogenesis

One topic of tremendous interest to researchers is **angiogenesis,** the process by which new blood vessels develop, especially after birth [*angeion*, vessel + *gignesthai*, to beget]. In children, capillary growth is necessary for normal development. In adults, angiogenesis takes place during wound healing and during growth of the uterine lining after menstruation. Angiogenesis also occurs with endurance exercise training, enhancing blood flow to the heart muscle and to skeletal muscles.

By learning the chemical signals that control angiogenesis, we may be able to develop new treatments for two major illnesses: cancer and coronary artery disease. As cancer cells invade tissues and multiply, they must grow new blood vessels in order to maintain a supply of nutrients and oxygen. Without these vessels, the interior cells of a cancerous mass would be unable to get adequate oxygen and nutrients and would die. We know that cancer cells are capable of triggering angiogenesis. A developing tumor causes nearby blood vessels to branch into it. If we can determine the signaling mechanisms used by the cancer cells, we may be able to find ways to block angiogenesis and literally starve tumors to death.

Coronary artery disease, in contrast, is a condition in which we would like to be able to selectively induce angiogenesis. In coronary artery disease, blood flow to the myocardium is decreased by fatty deposits that narrow the lumen of the coronary blood vessels. If we could induce the growth of new blood vessels to replace the vessels that are becoming blocked, we might be able to prevent the damage caused by lack of oxygen. This same technique might be applied to people who have suffered a *stroke*, in which the blood supply to regions of brain tissue is diminished because of blockage of a blood vessel in the brain.

VegF and Angiogenesis From studies of normal blood vessels and tumor cells, we have learned that a family of related growth factors appears to be the key regulator of angiogenesis. These growth factors, known collectively as **vascular endothelial growth factor (vegF),** are produced by normal smooth muscle cells and pericytes, as well as by tumor cells. VegF is a mitogen, meaning that it promotes mitosis, or cell division. In normal tissues, one form of vegF binds to receptors on endothelial cells with the aid of heparin (a naturally occurring anticoagulant), triggering the formation of new capillaries. If researchers can identify specific forms of vegF that are associated with angiogenesis in tumors, it may be possible to selectively block vegF action and prevent the tumors from creating a new blood supply.

BLOOD PRESSURE

The pressure created by ventricular contraction is the driving force for blood flow through the vessels of the system. As blood leaves the left ventricle, the aorta and arteries expand to accommodate the blood (Fig. 15-4a ■). When the ventricle relaxes and the semilunar valve closes, the elastic arterial walls recoil, propelling the blood forward into the smaller arteries and arterioles (Fig. 15-4b ■). By sustaining the driving pressure for blood flow during ventricular relaxation, the arteries produce continuous blood flow through the blood vessels. Flow in the arterial side of the circulation is pulsatile, reflecting the changes in arterial pressure throughout a cardiac cycle, but beyond the arterioles, blood flow is smooth.

Blood Pressure in the Systemic Circulation Is Highest in the Arteries and Lowest in the Veins

Blood pressure is highest in the arteries and falls continuously as blood flows through the circulatory system. The decrease in pressure occurs because energy is lost due to the resistance of the vessels to blood flow (Fig. 15-5 ■). In the systemic circulation, the highest pressure occurs in the aorta and reflects the pressure created by the left ventricle. Aortic pressure reaches an average high of 120 mm Hg during ventricular systole (**systolic pressure**), then falls steadily to a low of 80 mm Hg during ventricular diastole (**diastolic pressure**). Notice that although pressure in the ventricle falls to 0 mm Hg as the ventricle relaxes, the diastolic pressure in the large arteries remains relatively high. The high diastolic pressure in the arteries reflects the ability of those vessels to capture and store energy in their elastic walls.

The rapid pressure increase that occurs when the ventricles push blood into the aorta can be felt as a **pulse**, or pressure wave, that is transmitted through the fluid-filled arteries of the cardiovascular system. The pressure wave travels about 10 times as fast as the blood itself. Even so, a pulse felt in the arm is occurring slightly after the ventricular contraction that created the wave.

The amplitude of the pulse dies out over distance because of friction, finally disappearing at the capillaries (Fig. 15-5 ■). **Pulse pressure,** a measure of the amplitude, or strength, of the pulse pressure wave, is defined as the systolic pressure minus the diastolic pressure:

$$\text{Systolic pressure} - \text{diastolic pressure} = \text{pulse pressure}$$

For example, in the aorta:

$$120 \text{ mm Hg} - 80 \text{ mm Hg} = 40 \text{ mm Hg pulse pressure}$$

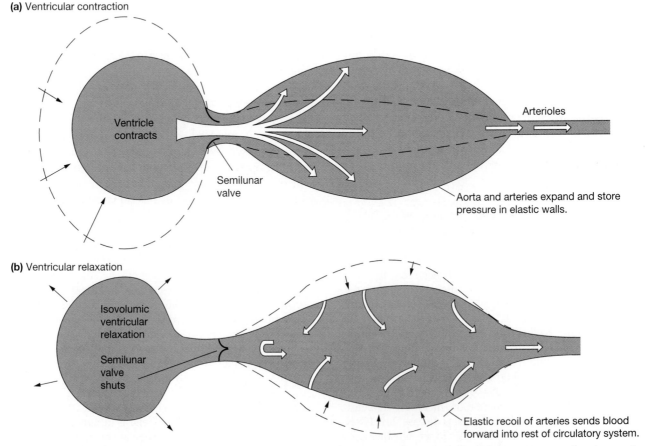

(a) Ventricular contraction

Ventricle contracts

Semilunar valve

Arterioles

Aorta and arteries expand and store pressure in elastic walls.

(b) Ventricular relaxation

Isovolumic ventricular relaxation

Semilunar valve shuts

Elastic recoil of arteries sends blood forward into rest of circulatory system.

Figure 15-4 Elastic recoil in the arteries (a) During ventricular contraction, the elastic walls of the arteries expand as the arteries fill with high-pressure blood. (b) During ventricular relaxation, once the semilunar valve shuts, elastic recoil of the arterial walls pushes the arterial blood forward into the circulatory system.

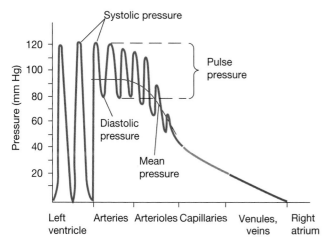

Figure 15-5 Pressure throughout the systemic circulation Pressure waves created by contraction and relaxation of the ventricle are reflected into the blood vessels. The pressure waves diminish in amplitude with distance until they disappear at the level of the capillaries.

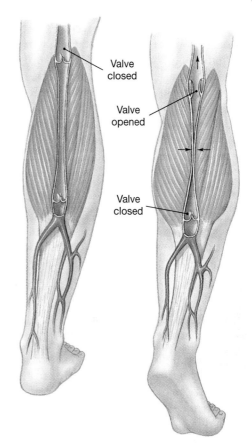

Figure 15-6 Valves create one-way flow in the veins When the skeletal muscles compress the veins, they force blood toward the heart (the skeletal muscle pump). A series of valves in the veins prevents the blood from flowing backward.

By the time blood flow reaches the veins, pressure within the vessels has fallen because of friction, and a pulse wave no longer exists. Low-pressure blood in veins below the heart must flow "uphill," or against gravity, to return to the heart. To assist venous flow, some veins have internal one-way valves (Fig. 15-6 ■). These valves, like those in the heart, ensure that blood passing the valve cannot flow backward. Venous flow is also aided by what is called the *skeletal muscle pump*. When muscles such as those in the calf of the leg contract, they compress the veins, forcing the blood upward past the valves.

Try holding your arm straight down without moving for several minutes and notice how the veins in the back of your hand begin to stand out as they fill with blood. (This effect may be more evident in older people, whose subcutaneous connective tissue has lost elasticity.) Then raise your hand so that gravity assists the venous flow and watch the bulging veins disappear. Once blood reaches the venae cavae, there are no valves. Blood flow is steady, pushed along by the continuous movement of blood into the venous circulation from the capillaries.

✓ Do you think that the veins from the brain to the heart have valves? Defend your answer.

✓ If you checked the pulse in a person's carotid artery and left wrist at the same time, would the pressure waves occur simultaneously? Explain.

Arterial Blood Pressure Reflects the Driving Pressure for Blood Flow

Arterial blood pressure, or simply "blood pressure," reflects the driving pressure created by the pumping action of the heart. Because ventricular pressure is difficult to measure, it is customary to assume that arterial blood pressure is indicative of the driving pressure for blood flow. Arterial pressure is pulsatile, so it is useful to have a single value for arterial pressure that is representative of the driving pressure. This number is the **mean arterial pressure** (MAP), defined as being approximately equal to the diastolic pressure plus one-third of the pulse pressure:

$$\text{MAP} = \text{diastolic P} + 1/3(\text{systolic P} - \text{diastolic P})$$

$$\text{MAP} = 80 \text{ mm Hg} + 1/3(120 - 80 \text{ mm Hg})$$

$$\text{MAP} = 93 \text{ mm Hg}$$

Mean arterial pressure is closer to diastolic pressure than to systolic pressure because diastole lasts twice as long as systole. This formula for MAP applies to a person whose heart rate is in the range of 60–80 beats per minute, a typical resting heart rate. If heart rate increases, the relative amount of time the heart spends in diastole decreases. In that case, the contribution of systolic pressure to mean arterial pressure increases.

Abnormally high or low arterial blood pressure can be indicative of a problem in the cardiovascular system. If blood pressure drops below a certain level, the driving force for blood flow will be unable to overcome opposition by gravity. In this instance, blood flow and oxygen supply to the brain are impaired, and the subject may become dizzy or faint. On the other hand, if blood pressure is chronically elevated (a condition known as *hypertension*, or high blood pressure), high pressure pushing on the walls of blood vessels may cause weakened areas to rupture, causing bleeding into the tissues. If a rupture occurs in the brain, it is called a *cerebral hemorrhage* and may cause the loss of neurological function commonly called a *stroke*. If a weakened area ruptures in a major artery such as the descending aorta, rapid blood loss into the abdominal cavity will cause blood pressure to fall below the critical minimum. Without prompt treatment, rupture of a major artery is fatal.

✓ Peter's systolic pressure is 112 mm Hg and his diastolic pressure is 68 mm Hg (written 112/68). What is his pulse pressure? His mean arterial pressure?

Blood Pressure Is Estimated by Sphygmomanometry

We estimate arterial blood pressure in the radial artery of the arm using a *sphygmomanometer,* an instrument consisting of an inflatable cuff and a pressure gauge [*sphygmus*, pulse + *manometer,* an instrument for measuring pressure of a fluid]. The cuff encircles the upper arm and is inflated until it exerts pressure higher than the systolic pressure driving the arterial blood. This pressure stops blood flow into the lower arm: The blood cannot flow past the region of higher pressure created by the cuff (Fig. 15-7a ■). The pressure on the cuff is gradually released until the cuff pressure drops below systolic blood pressure in the artery, and blood begins to flow again. As blood squeezes through the still-compressed artery, the turbulent flow makes a noise called a **Korotkoff sound** that can be heard through a stethoscope placed just below the cuff (Fig. 15-7b ■). A Korotkoff sound is heard with each heartbeat as long as the artery is compressed. Once the pressure exerted by the cuff no longer compresses the artery, blood flow is no longer turbulent and the sounds disappear (Fig. 15-7c ■).

![runner icon] *continued from page 424*

Kurt's second blood pressure reading is 158/98. Dr. Cortez asks him to take his blood pressure at home daily for two weeks, then return to the office. When Kurt comes back with his diary, the story is the same: His blood pressure continues to average 160/100. After running some tests, Dr. Cortez concludes that Kurt is one of approximately 50 million adult Americans with high blood pressure, also called hypertension. If not controlled, hypertension can lead to congestive heart failure, stroke, and kidney failure.

Question 1: Why are people with high blood pressure at higher risk for having a hemorrhagic, or bleeding, stroke?

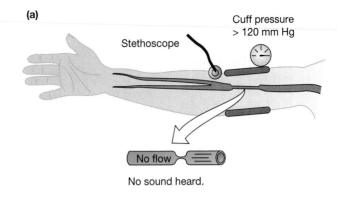

(a)

Stethoscope

Cuff pressure > 120 mm Hg

No flow

No sound heard.

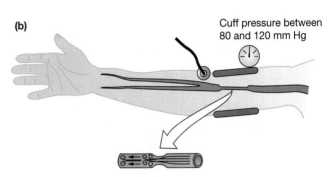

(b)

Cuff pressure between 80 and 120 mm Hg

Flow through compressed artery sets up vibrations that are heard as Korotkoff sounds.

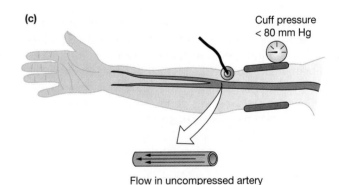

(c)

Cuff pressure < 80 mm Hg

Flow in uncompressed artery makes no sound.

Figure 15-7 Measurement of arterial blood pressure
Arterial blood pressure is measured with an instrument known as a sphygmomanometer, consisting of an inflatable cuff and a pressure gauge. (a) The cuff is placed around the upper arm and inflated so that it compresses the radial artery and stops blood flow. A stethoscope placed over the radial artery distal to the cuff will be unable to pick up any noise. (b) The pressure in the cuff is slowly lowered until a thumping noise known as a Korotkoff sound is heard through the stethoscope. The pressure at which the sound is first heard represents the systolic pressure, the highest pressure in the artery. The Korotkoff sound is created by blood flow through the compressed artery. (c) The pressure at which the sounds disappear is the diastolic pressure. This is the pressure at which the artery is no longer compressed and blood flow is silent.

blood-brain barrier, the selective capillaries in the brain that protect the neural tissue from toxins in the bloodstream.

Fenestrated capillaries [*fenestra,* window] have large pores that allow high volumes of fluid to pass rapidly between the plasma and interstitial fluid (Fig. 15-16b ■). These capillaries are found primarily in the kidney and the intestine, where they are associated with absorptive transporting epithelia. In the bone marrow, where newly synthesized blood cells enter the blood, and in the liver, where most plasma proteins are synthesized and enter the blood, the fenestrated capillaries have the ability to temporarily open gaps that are so large that both proteins and blood cells can squeeze through between adjacent cells.

The Velocity of Blood Flow Is Lowest in the Capillaries

The low velocity of blood flow through the capillaries is a useful characteristic that allows diffusion to go to equilibrium (∞ p. 117). In Chapter 14, you learned that at a constant flow rate, velocity of flow will be higher in a smaller vessel than it will be in a larger vessel. Because a capillary is the smallest type of blood vessel, you might conclude that blood moves very rapidly through a capillary. But the primary determinant for velocity of flow is not the diameter of an individual capillary but the *total* cross-sectional area of all the capillaries. If circles representing cross sections of all the capillaries were placed together, they would cover an area much larger than the total cross-sectional areas of all the arteries and veins combined. Since the total cross-sectional area of all the capillaries is so large, the velocity of their blood flow is low. Figure 15-17 ■ compares cross-sectional areas of different parts of the systemic circulation with the velocity of blood flow in each part. The fastest flow is in the relatively small-diameter arterial system, and the slowest flow is in the capillaries and venules, which collectively have the largest cross-sectional area.

Most Capillary Exchange Takes Place by Diffusion and Transcytosis

Most exchange between the plasma and interstitial fluid takes place by simple diffusion, either through pores in the capillary wall or through the cells of the endothelium. The diffusion rate for permeable solutes is determined primarily by the concentration gradient between the plasma and the interstitial fluid. Oxygen and carbon dioxide diffuse freely across the endothelium, reaching equilibrium with the interstitial fluid and cells by the time blood reaches the venous end of the capillary. Blood cells and most plasma proteins are unable to pass through the capillary wall. The few proteins that escape from the blood are returned to the plasma by the lymphatic system.

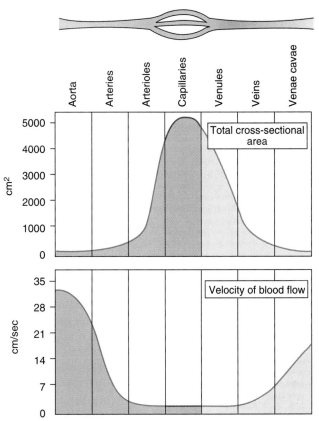

Figure 15-17 Vessel diameter, total cross-sectional area, and velocity of flow The velocity of flow through different types of vessels in the circulatory system depends on the total cross-sectional area of that vessel type, not on the diameter of a single vessel. Flow is slowest in the capillaries and most rapid in the aorta and major arteries.

If proteins cannot get through the pores, how do protein hormones and cytokines get from the plasma to the interstitial fluid? A few tissues such as the liver have capillary walls that can open gaps large enough for proteins to pass. In other tissues, transcytosis moves plasma proteins from one side of the endothelium cell to the other.

Capillary Filtration and Reabsorption Take Place by Bulk Flow

A third form of capillary exchange is the bulk flow of fluid into and out of the capillary. **Bulk flow** refers to the mass movement of water and dissolved solutes between the blood and the interstitial fluid as the result of hydraulic or osmotic pressure. If the direction of flow is out of the capillary, the fluid movement is known as **filtration.** If the direction of flow is into the capillary, it is called **absorption.** Most capillaries show a transition from net filtration at the arterial end to net absorption at the venous end. There are some exceptions to this rule. Capillaries in part of the kidney filter fluid along their entire length, whereas some capillaries in the intestine are only absorptive, picking up digested nutrients that have just been transported into the interstitial fluid from the lumen.

Two forces regulate bulk flow in the capillaries. One is hydraulic pressure, the lateral pressure component of blood flow that pushes fluid out through the capillary pores. The second factor is osmotic pressure. The main solute difference between the plasma and the interstitial fluid is proteins that are present in the plasma but mostly absent from the interstitial fluid. The osmotic pressure that is due to the presence of these proteins is known as the *colloid osmotic pressure* (π). Colloid osmotic pressure is *not* equivalent to the total osmotic pressure; it is simply a measure of the osmotic pressure created by the proteins. Because colloid osmotic pressure is higher in the plasma than in the interstitial fluid, osmotic movement of water is from the interstitial fluid into the capillary (Fig. 15-18a ■). The capillary endothelium is freely permeable to the other solutes in the plasma, and they therefore do not contribute to the osmotic gradient. For the purposes of our discussion, we will consider colloid osmotic pressure to be constant along the length of the capillary.

Capillary hydraulic pressure, on the other hand, decreases along the length of the capillary as energy is lost to friction. Average values for hydraulic pressure are 32 mm Hg at the arterial end of a capillary and 15 mm Hg at the venous end. The hydrostatic pressure of the interstitial fluid is very low, so we will consider it to be essentially zero. This means that water movement due to hydraulic pressure will be directed out of the capillary, with the magnitude decreasing from the arterial end to the venous end.

The difference between the opposing forces of colloid osmotic pressure and hydraulic pressure determines net fluid movement across the capillary, as shown in Figure 15-18a ■. At the arterial end, there is net filtration, and at the venous end, there is net absorption. If the point at which filtration equals absorption occurred in the middle of the capillary, there would be no net movement of fluid. All volume that was filtered at the arterial end would be absorbed at the venous end. However, filtration is usually greater than absorption, resulting in bulk flow of fluid out of the capillary into the interstitial space. By most estimates, that bulk flow amounts to about 3 liters per day, or the equivalent of the entire plasma volume! Unless this filtered fluid is returned to the plasma, the blood will turn into a sludge of blood cells and proteins. Restoring fluid lost from the capillaries to the circulatory system is one of the functions of the lymphatic system.

THE LYMPHATIC SYSTEM

The vessels of the lymphatic system interact with three physiological systems: the cardiovascular system, the digestive system, and the immune system. The functions of the lymphatic system include:

continued from page 433

After several weeks, Kurt returns to the office for a checkup. His blood pressure has not changed. "I swear, I'm trying to do better," says Kurt. "But it's difficult." Kurt confesses that although he now walks with his wife four mornings a week, he can't resist the salt shaker. "Food just tastes bland without it," he explains. Because lifestyle changes alone are not helping, Dr. Cortez prescribes an antihypertensive drug for Kurt. "This drug, called an ACE inhibitor, blocks production of a chemical called angiotensin II, a powerful vasoconstrictor. This medication should bring your blood pressure back to a normal value."

Question 3: Why would blocking the action of a vasoconstrictor lower blood pressure?

1. Returning fluid and proteins filtered out of the capillaries to the circulatory system
2. Picking up fat absorbed at the small intestine and transferring it to the circulatory system
3. Serving as a filter to help capture and destroy foreign pathogens

In this discussion, we focus on the role of the lymphatic system in fluid transport. The other two functions will be discussed in connection with digestion (Chapter 20) and immunity (Chapter 22).

The lymph system is designed for the one-way movement of interstitial fluid from the tissues into the circulation. Closed-end lymph vessels called *lymph capillaries* lie close to all blood capillaries, except those in the kidney and central nervous system (Fig. 15-18b ■). The walls of the lymph capillaries are composed of a single layer of flattened endothelium, even thinner than that of the blood capillaries. The walls of the lymph capillaries are anchored to the surrounding connective tissue by fibers that hold the thin-walled vessels open. Large gaps between cells allow fluid, interstitial proteins, and particulate matter such as bacteria to be swept into the lymph capillary by bulk flow. Once inside the lymphatic system, this clear fluid is simply called **lymph.**

Lymph capillaries in the tissues join to form larger collecting vessels that progressively increase in size (Fig. 15-19 ■). Finally, the largest lymph vessels empty into the venous circulation just under the collarbones, where the left and right subclavian veins join the internal jugular veins. At intervals along the way, vessels enter **lymph nodes,** bean-shaped nodules of tissue with a fibrous outer capsule and an internal collection of immunologically active cells, including lymphocytes and macrophages.

The lymphatic system has no pump like the heart. Lymph flow depends on contractile fibers in the endothelial cells, smooth muscle in the walls of larger lymph vessels, a system of one-way valves like those in

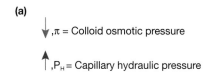

(a)

$\downarrow ,\pi$ = Colloid osmotic pressure

$\uparrow ,P_H$ = Capillary hydraulic pressure

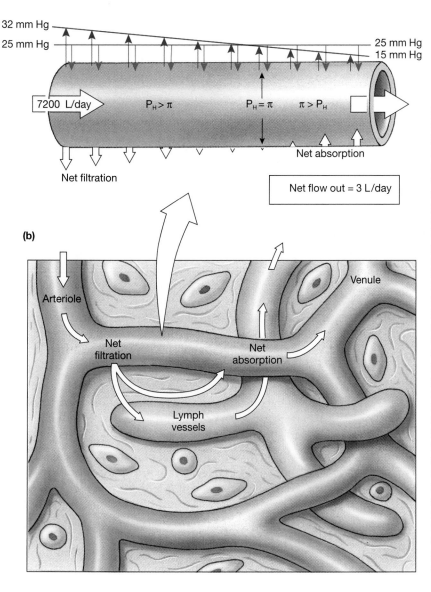

32 mm Hg

25 mm Hg

25 mm Hg
15 mm Hg

7200 L/day $P_H > \pi$ $P_H = \pi$ $\pi > P_H$

Net absorption

Net filtration

Net flow out = 3 L/day

(b)

Arteriole

Net filtration

Net absorption

Venule

Lymph vessels

Figure 15-18 Fluid exchange at the capillary (a) Two primary forces move fluid across the capillary epithelium. Hydrostatic pressure (blood pressure) of the fluid within the capillary forces fluid out of the capillary; the colloid osmotic pressure of proteins within the capillary pulls fluid into the capillary. (Fluid pressure created by the presence of fluid in the interstitial space is negligible and will be ignored.) The hydrostatic pressure is 32 mm Hg at the arterial end of the capillary and falls steadily to a low of 15 mm Hg at the venous end of the capillary. Colloid osmotic pressure is a constant 25 mm Hg. Net fluid flow depends on the difference between the two pressures. Thus, at the arterial end of the capillary, hydrostatic pressure exceeds osmotic pressure and there is net filtration of fluid out of the capillary. By the venous end, hydrostatic pressure is less than osmotic pressure and there is net absorption. Filtration exceeds reabsorption (7 "out" arrows vs. 3 "in" arrows), and a net flow of 3 L/day of fluid filters out of the capillaries. (b) The excess water and solutes that filter out of the capillary are picked up by the lymph vessels and returned to the circulation.

the venous circulation, and external compression created by skeletal muscles. The skeletal muscle pump plays a significant role in lymph flow, as you know if you have ever injured a wrist or ankle. An immobilized limb frequently swells from the accumulation of fluid in the interstitial space, a condition known as *edema* [*oidema*, swelling]. Edema is the reason patients are told to elevate an injured limb above the level of the heart: Gravity will assist lymph flow back to the blood.

An important component of returning filtered fluid to the circulation is the recycling of plasma proteins. It is critical that the body maintain a low concentration of protein in the interstitial fluid because colloid osmotic pressure is the only significant force that opposes capillary hydraulic pressure. Inflammation is a situation in which the balance of colloid osmotic and hydraulic pressures is disrupted. Histamine released in the inflammatory response makes capillary walls leakier and allows proteins to escape from the plasma into the interstitial fluid. Movement of proteins out of the plasma decreases the osmotic pressure gradient that opposes filtration, resulting in additional fluid movement into the interstitial space. Thus, the local swelling that accompanies a region of inflammation is an example of edema caused by redistribution of proteins from the plasma to the interstitial fluid.

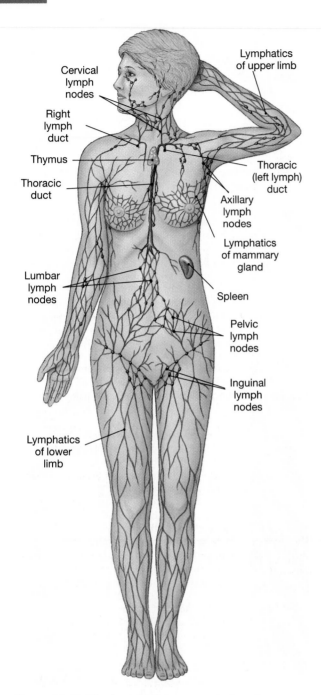

Figure 15-19 The lymphatic system The lymphatic system begins with blind-end lymph capillaries in the tissues. These join together to form larger and larger lymph vessels. At intervals, the fluid in the lymph vessels flows through lymph nodes. Finally, the lymph ducts empty into the venous side of the systemic circulation at the right and left lymphatic ducts.

Edema: Disruption of Capillary Exchange

Edema is a sign that normal capillary-lymph exchange has been disrupted. Edema can arise from several causes that result in inadequate drainage of lymph or capillary filtration that greatly exceeds capillary absorption.

Inadequate lymph drainage occurs with obstruction of the lymphatic system, particularly at the lymph nodes.

Parasites, cancer, or fibrotic tissue growth resulting from therapeutic radiation can block the movement of lymph through the system. For example, *elephantiasis* is a chronic condition marked by gross enlargement of the legs and lower appendages when parasites block the lymph vessels. Lymph drainage may also be impaired if lymph nodes are removed during surgery, a common procedure in the diagnosis and treatment of cancer.

Factors that disrupt the normal balance between capillary filtration and absorption include:

1. *An increase in capillary hydraulic pressure.* Increased capillary pressure is usually indicative of elevated venous pressure. An increase in arterial pressure is usually not noticeable at the capillaries because of the autoregulation of pressure in the arterioles. One common cause of increased venous pressure is heart failure, a condition in which the ventricle on one side of the heart is unable to pump all the blood sent to it by the other ventricle. As blood backs up, pressure rises first in the atrium, then in the veins and capillaries supplying blood to that side of the heart. When capillary hydraulic pressure increases, filtration greatly exceeds absorption, leading to edema. If the left ventricle fails to pump normally, edema may occur in the lungs. Oxygen exchange is impaired, and breathing may become difficult. This condition is known as *pulmonary edema.*

2. *A decrease in plasma protein concentration.* Plasma protein concentrations may decrease as a result of liver failure or severe malnutrition, because the liver is the main site for plasma protein synthesis.

3. *An increase in interstitial proteins.* As discussed earlier, excessive leakage of proteins out of the blood will cause the colloid osmotic pressure of the plasma to decrease and will increase net capillary filtration.

On occasion, changes in the balance between filtration and absorption help the body maintain homeostasis. For example, if arterial blood pressure falls, capillary hydraulic pressure also falls. This change increases fluid absorption. If pressure falls low enough, there will be net absorption in the capillaries rather than net filtration. This passive mechanism helps maintain blood volume in situations in which blood pressure is very low, such as hemorrhage or severe dehydration.

✓ Malnourished children who have inadequate protein in their diet often have grotesquely swollen bellies. This condition is known as ascites and can be described as edema of the abdomen. Use the information you have just learned about capillary filtration to explain why malnutrition causes ascites.

✓ Som
inter
whe
neur
refle:

CARD

Diseases
attacks a
deaths in
eases, **co**
heart dise
and won
the heart
cholester

REGULATION OF BLOOD PRESSURE

Because of the importance of maintaining adequate blood flow to the brain and heart, the central nervous system coordinates the reflex control of blood pressure. The integrating center for neural control of blood pressure resides in the medulla oblongata. However, because of the difficulty of studying neural networks in the brain, we still know relatively little about the nuclei, neurotransmitters, and interneurons of the **medullary cardiovascular control center.**

The Baroreceptor Reflex Is the Primary Homeostatic Control for Blood Pressure

Sensory input to the cardiovascular control center comes from a variety of peripheral sensory receptors. Stretch-sensitive mechanoreceptors known as **baroreceptors** are located in the walls of the carotid artery and aorta, where they monitor the pressure of blood flowing to the brain (carotid receptors) and to the body (aortic receptor). The carotid and aortic baroreceptors are tonically active stretch receptors that fire action potentials continuously at normal blood pressures.

The primary reflex pathway for homeostatic control of blood pressure is the **baroreceptor reflex.** When increased blood pressure in the arteries stretches the baroreceptor membrane, the firing rate of the receptor increases. If blood pressure drops, the firing rate of the receptor decreases. Action potentials from the baroreceptors travel to the cardiovascular control center via sensory neurons. The cardiovascular control center integrates the sensory input and initiates an appropriate response. The response of the baroreceptor reflex is quite rapid: Changes in cardiac output and peripheral resistance occur within two heartbeats of the stimulus.

Efferent output from the medullary cardiovascular control center is carried via autonomic neurons. Heart function is regulated by antagonistic control. Increased sympathetic activity increases heart rate at the SA node, shortens conduction time through the AV node, and enhances the force of myocardial contraction. Increased parasympathetic activity slows down heart rate but has no significant effect on ventricular contraction. Peripheral resistance is controlled by sympathetic neurons, with increased sympathetic discharge causing vasoconstriction.

The baroreceptor reflex is summarized in Figure 15-20 ■. Baroreceptors increase their firing rate as blood pressure increases, activating the cardiovascular control center. In response, the cardiovascular control center increases parasympathetic activity and decreases sympathetic activity to slow down the heart. When heart rate falls, cardiac output falls. In the circulation, decreased sympathetic activity causes dilation of the arterioles, allowing more blood to flow out of the arteries. The combination of reduced cardiac output and decreased peripheral resistance lowers the mean arterial blood pressure.

Cardiovascular function can be modulated by input from peripheral receptors other than the baroreceptors. For example, if arterial chemoreceptors are activated by low blood oxygen levels, they increase cardiac output. The cardiovascular control center also has reciprocal communication with centers in the medulla that control breathing. The integration of function between the respiratory and circulatory systems is adaptive: If tissues require more oxygen, it is supplied by the cardiovascular system working in tandem with the respiratory system. Consequently, increases in breathing rate are usually accompanied by increases in cardiac output.

Blood pressure is also subject to modulation by higher brain centers such as the hypothalamus and cerebral cortex. The hypothalamus is responsible for vascular responses involved in body temperature regulation (see Chapter 21) and for the fight-or-flight response. Learned and emotional responses may originate in the cerebral cortex and be expressed by cardiovascular responses such as blushing and fainting. One such reflex is the *vasovagal response,* which may be triggered in some people by the sight of blood or a hypodermic needle. In this pathway, increased parasympathetic activity and decreased sympathetic activity slow heart rate and cause widespread vasodilation. Cardiac output and peripheral resistance both fall, triggering a precipitous drop in blood pressure. With insufficient blood to the brain, the individual faints.

Regulation of blood pressure in the cardiovascular system is closely tied to the regulation of body fluid balance by the kidneys. Thus, certain hormones from the heart act on the kidneys, while hormones from the kidney act on the heart and blood vessels. Together, the heart and the kidney play a major role in maintaining the homeostasis of body fluids, the subject of Chapter 19. The overlap of these systems is an excellent example of the integration of organ system function that must take place in the body in order to maintain homeostasis.

✓ Baroreceptors have stretch-sensitive ion channels in their cell membrane. Increased pressure on the membrane opens the channels and allows ion flow that initiates action potentials. What kind of ion probably flows through these channels and in which direction?

Orthostatic Hypotension Triggers the Baroreceptor Reflex

The baroreceptor reflex functions every morning when you get out of bed. When you are lying flat, gravitational forces are distributed evenly up and down the length of the body, so blood also is distributed evenly throughout the circulation. When you stand up, gravity causes blood to pool in the lower extremities. This pooling creates an instantaneous decrease in venous return. As a result, less blood is in the ventricles at the beginning of the next con-

Figure
name th
type of r

Figure 1!

continued from page 440

Another few weeks go by, and Kurt again returns to Dr. Cortez for a checkup. Kurt's blood pressure is finally in the normal range and has been averaging 135/87. "But Doc, can you give me something for this dry, hacking cough I've been having? I don't feel bad, but it's driving me nuts." Dr. Cortez explains that a dry cough is an occasional side effect of taking ACE inhibitors. "It is more of a nuisance than anything else. But let's change your medicine. I'd like to try you on a calcium channel blocker this time."

Question 4: How do calcium channel blockers lower blood pressure?

As we increase our knowledge of cardiovascular function, we begin to understand how drugs that have been used for centuries work. A classic example is the drug digitalis (∞ p. 400), whose mechanism of action we finally understand. It is a humbling thought to realize that for many therapeutic drugs, we know *what* they do without fully understanding *how* they do it.

Risk Factors for Cardiovascular Disease Include Gender, Age, and Inheritable Factors

It is possible to predict the likelihood that a person will develop cardiovascular disease during his or her lifetime by examining the various risk factors that the person possesses. The list of risk factors is the result of many years following the medical histories of thousands of people, and, as more data become available, additional risk factors may be added. Risk factors are generally divided into those over which the person has no control and those that can be controlled. Medical intervention is aimed at reducing the risk from the controllable factors.

The risk factors that cannot be controlled include gender, age, family history of cardiovascular disease, and the presence of diabetes mellitus (to some extent). Men have a 3–4 times higher risk of developing coronary artery disease through middle age than do women. After age 50, when most women have entered menopause, the death rate from coronary artery disease is nearly equal in men and women who do not take estrogen replacement therapy. In general, the risk of coronary artery disease increases as people age. Heredity also plays an important role. If a person has one or more close relatives with coronary artery disease, his or her risk is elevated. Diabetes mellitus, an inherited metabolic disorder that contributes to the development of fatty deposits in the blood vessels, also puts a person at risk for developing coronary artery disease.

Risk factors that can be controlled include cigarette smoking, obesity, elevated serum cholesterol and triglycerides, untreated high blood pressure (hyperten-

sion), and diabetes mellitus (to some extent). The first three of these factors contribute directly to the development of *atherosclerosis*, also known as hardening of the arteries. In atherosclerosis, macrophages ingest LDL-cholesterol and form fatty streaks in the extracellular matrix just under the endothelial lining of the larger arteries (Fig. 15-22 ■). In addition, paracrines released by the macrophages attract smooth muscle cells that migrate from the underlying layer of vascular smooth muscle. If the condition progresses, the smooth muscle cells reproduce, and the streaks grow into bulging plaques that protrude into the lumen of the vessel, narrowing it. In the advanced stages of atherosclerosis, the plaques develop hard, calcified regions and fibrous scar tissue. As the plaques grow, their rough surface can trigger the formation of a blood clot (*thrombus*) that blocks the flow of blood through the vessel. If the thrombus forms in the vessels that supply the heart, a heart attack, or myocardial infarction, results. The region of tissue supplied by that artery is deprived of oxygen and may be irreversibly damaged. If the damaged region is large, the normal conduction of electrical signals through the ventricle will be disrupted. This disruption can lead to an irregular heartbeat (*arrhythmia*) and potentially result in cardiac arrest and death.

The role of elevated blood cholesterol levels in the development of coronary artery disease is well established. Cholesterol, like other lipids, is not very soluble in aqueous solutions such as the plasma. Therefore, when cholesterol in the diet is absorbed from the digestive tract, it combines with a lipoprotein molecule so that it dissolves in the plasma. There are multiple types of lipoprotein carriers for cholesterol, but clinicians generally are concerned with two: **high-density lipoprotein,** or **HDL,** and **low-density lipoprotein,** or **LDL.** HDL-cholesterol is the form that is more desirable, since the HDL carrier transports cholesterol to the liver, where it is either used for synthesis or excreted into the lumen of the digestive tract in the form of bile.

LDL-cholesterol is sometimes called the bad cholesterol, because elevated plasma LDL levels are associated with coronary artery disease. LDL-cholesterol in normal levels is not bad, however, because the LDL carrier is necessary for uptake of cholesterol into cells. In this process, the LDL portion of the complex combines with an LDL receptor on the cell membrane. The entire receptor-LDL-cholesterol complex is brought into the cell by endocytosis. Once freed from its carrier lipoprotein, cholesterol is used to make cell membranes or steroid hormones. Problems arise when LDL-cholesterol is not taken into cells and remains in the blood. Under those conditions, excess LDL-cholesterol moves into the walls of the arteries. There, it is apparently oxidized, becoming a toxic molecule that starts a series of events that lead to atherosclerosis. Because of the pivotal role that LDL-cholesterol plays in atherosclerosis, many forms of therapy, ranging from dietary modification to drugs, are aimed at lowering LDL-cholesterol.

The last controllable risk factor for coronary artery disease is hypertension. We will examine the link between hypertension and coronary artery disease in the next section.

Hypertension Represents a Failure of Homeostasis

Hypertension is defined as chronically elevated blood pressure, with either a systolic pressure greater than 140 mm Hg or a diastolic pressure greater than 90 mm Hg. Hypertension is a common disease in the United States

and is the single most common reason for visits to physicians and for the use of prescription drugs.

More than 90% of all patients with hypertension are considered to have essential or primary hypertension, with no clear-cut cause. Cardiac output is usually normal in these people, and the elevated blood pressure appears to be associated with increased peripheral resistance. Some investigators have speculated that the increased resistance may be due to a lack of nitric oxide, the locally produced vasodilator that is produced by endothelial cells in the arterioles. In the remaining 5%–10% of hypertensive cases, the cause is known, and the hypertension is considered to be secondary to an underlying pathology. For instance, the cause might be an endocrine disorder that causes fluid retention.

A key feature of hypertension from all causes is adaptation of the carotid and aortic baroreceptors to higher pressure, with subsequent down-regulation of their activity. Without input from the baroreceptors, the medullary cardiovascular control center interprets the high blood pressure as "normal," so no reflex reduction of pressure occurs.

Hypertension is a risk factor for atherosclerosis because high pressure in the arteries damages the endothelial lining of the vessels and promotes the formation of atherosclerotic plaques. In addition, high arterial blood pressure puts additional strain on the heart. When resistance in the arterioles is high, the myocardium must work harder to push the blood into the arteries. To understand why this occurs, think of a swinging door (the semilunar valves) between the kitchen (the ventricles) and the dining room (the arteries). A waiter pushing through the door uses a certain amount of force to push the door open each time he passes through. The increased resistance of essential hypertension is equivalent to piling fur-

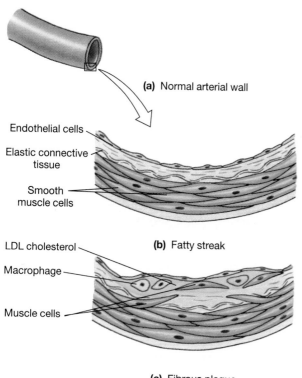

(a) Normal arterial wall

Endothelial cells
Elastic connective tissue
Smooth muscle cells

(b) Fatty streak

LDL cholesterol
Macrophage
Muscle cells

(c) Fibrous plaque

Platelets
Fibrous scar tissue
Thickening of smooth muscle

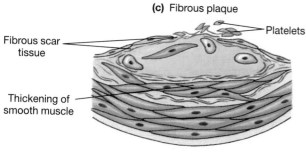

(d) Calcified plaque

Platelets
Calcifications

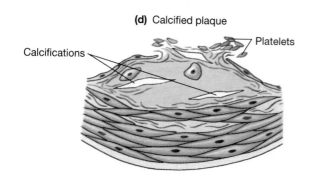

Figure 15-22 The development of atherosclerotic plaques (a) The normal arterial wall consists of smooth muscle and connective tissue with a layer of endothelial cells lining the lumen of the artery. (b) In early stages, excess LDL-cholesterol accumulates in the acellular layer between the endothelium and the connective tissue. There, it is oxidized and phagocytosed by macrophages, white blood cells that patrol the area. When the macrophages begin to fill with oxidized LDL-cholesterol, they produce paracrines that attract smooth muscle cells to the region. At this stage, the lesion is called a fatty streak. (c) As cholesterol continues to accumulate, fibrous scar tissue forms around it. In addition, the migrating smooth muscle cells divide, thickening the arterial wall and narrowing the lumen of the artery. This stage is known as a fibrous plaque. If the endothelium is damaged and collagen is exposed, platelets will stick to the damaged area and to each other. A blood clot (thrombus) may form on the wall of the vessel. If the clot completely blocks the lumen of the artery, blood flow to tissues beyond the clot will stop. If blood flow in the coronary blood vessel is stopped, a heart attack is the result. (d) In the advanced stages of atherosclerosis, cells in the fatty deposit may die, and calcified scar tissue will form. The deposits of calcium carbonate in the scar tissue create the "hardening of the arteries."

niture against the back of the door. When that happens, the waiter must use a lot more force to push through the door because he must push not only the door but also the furniture. The additional resistance created by the weight of the furniture is analogous to the increased resistance of the arterioles to blood flow.

Amazingly, stroke volume in hypertensive patients remains constant, up to a mean blood pressure of about 200 mm Hg, despite the increasing amount of work that the ventricle must perform as blood pressure increases. The cardiac muscle of the left ventricle responds to chronic high systemic resistance in the same way that skeletal muscle responds to a weight-lifting routine: The heart muscle hypertrophies, increasing the size and strength of the muscle fibers. However, if resistance remains high over time, the heart muscle is unable to meet the work load and begins to fail. At that point, cardiac output by the left ventricle decreases. If the cardiac output of the right heart remains normal while the output from the left side decreases, fluid collects in the lungs, creating pulmonary edema. At this point, a detrimental positive feedback loop begins. Oxygen exchange in the lungs diminishes because

of the pulmonary edema, leading to less oxygen in the blood. Lack of oxygen for aerobic metabolism further weakens the heart, so its pumping effectiveness diminishes even more. Unless treated, this condition, known as *congestive heart failure*, eventually leads to death.

Many of the treatments for hypertension have their basis in the cardiovascular physiology that you learned in these chapters. Calcium channel blockers decrease Ca^{2+} entry into vascular smooth muscle cells, causing arterioles to dilate and reducing peripheral resistance (see Focus on Calcium Channel Blockers, p. 425). Beta-blocking drugs that target β_1 receptors decrease catecholamine stimulation of cardiac output. A third major group of antihypertensive drugs, the ACE inhibitors, act by blocking an enzyme that promotes production of a powerful vasoconstrictor substance. You will learn more about this enzyme later in the book, when you study the integrated control of blood pressure by the cardiovascular and renal systems.

In the future, we may be seeing additional treatments for hypertension that are based on the molecular physiology of the heart and blood vessels.

CHAPTER REVIEW

Chapter Summary

1. Homeostatic regulation of the cardiovascular system is aimed at maintaining perfusion to the tissues, especially the heart and brain. (p. 423)

2. Total blood flow through any level of the circulation is equal to the cardiac output. (p. 424)

The Blood Vessels

3. Blood vessels are composed of layers of smooth muscle, elastic and fibrous connective tissue, and **endothelium.** Endothelial cells are important in regulation of blood pressure, growth of blood vessels, and absorption of materials. (p. 424)

4. **Vascular smooth muscle** maintains a state of muscle tone. A variety of chemicals influence vascular smooth muscle tone, including neurotransmitters, hormones, and paracrines. (p. 424)

5. The walls of the aorta and major arteries are both stiff and springy. This property allows them to absorb substantial amounts of energy from high-pressure blood ejected by the ventricles and release it through elastic recoil. (p. 425)

6. The walls of the **arterioles** contract and relax under the influence of various chemical signals. (p. 425)

7. **Metarterioles** regulate blood flow through the capillaries and allow white blood cells to go directly from the arterioles to the venous circulation. (p. 425)

8. Capillaries and postcapillary **venules** are the site of exchange between the blood and the interstitial fluid. Their walls consist of a leaky exchange epithelium supported on a basement membrane. (p. 426)

9. Veins are larger and more numerous than arteries, and they hold more than half of the blood in the circulatory system. Veins have thinner walls with less elastic tissue, so they expand easily when they fill with blood. (p. 426)

10. **Angiogenesis** is the process by which new blood vessels grow and develop, especially after birth. (p. 427)

Blood Pressure

11. The ventricles create high pressure that is the driving force for blood flow. The elastic walls of the aorta and arteries sustain this pressure during ventricular relaxation. (p. 428)

12. Blood pressure is highest in the arteries and falls continuously as blood flows through the circulatory system. In a resting individual, an average value for **systolic pressure** is 120 mm Hg, and an average **diastolic pressure** is 80

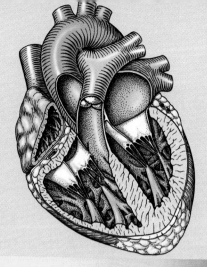

16
Blood

Blood, the fluid that circulates in the cardiovascular system, has occupied a prominent place throughout history as an almost mystical fluid. Humans undoubtedly made the association between blood and life by the time they began to fashion tools and hunt animals. A wounded animal that lost blood would weaken and die if the blood loss was severe enough. The logical conclusion was that blood was necessary for existence. This observation eventually led to the term *lifeblood*, meaning anything essential for existence.

Ancient Chinese physicians linked blood to energy flow in the body. They wrote about circulation of blood through the heart and blood vessels long before William Harvey described it in seventeenth-century Europe. In China, changes in blood flow were used as diagnostic clues to illness. Chinese physicians were expected to recognize some 50 variations in the pulse. Because blood was considered a vital fluid to be conserved and maintained, bleeding patients to cure disease was not a standard form of treatment.

In contrast, Western civilizations came to believe that disease-causing evil spirits circulated in the blood and that the way to remove these spirits was to remove the blood containing them. However, it was recognized that blood was an essential fluid, so bloodletting had to be done judiciously. Veins were opened with knives or sharp instruments (*venesection*), or blood-sucking leeches were applied to the skin. In ancient India, leeches were the treatment of choice because it was believed that they could distinguish between healthy and infected blood. There is no written evidence for venesection in ancient Egypt, but Galen of Pergamum in the second century A.D. influenced Western medicine for nearly 2000 years by advocating bleeding as treatment for many disorders. The location, timing, and frequency of the bleeding depended on the condition, and the physician was instructed to remove enough blood to bring the patient to the point of fainting. Over the years, this practice undoubtedly killed more people that it cured. What is even more remarkable is the fact that as late as 1923, there was an American medical textbook that advocated bleeding for treating certain infectious diseases such as pneumonia! Now that we better understand the importance of blood in the immune response, it is doubtful that modern medicine will ever again turn to blood removal as a nonspecific means of treating disease. It is still used, however, for selected *hematological disorders.*

What is this remarkable fluid that flows through the circulatory system? Blood makes up one-fourth of the extracellular fluid, the internal environment that bathes the cells and acts as a buffer between the cells and the external environment. Blood is the circulating portion of the extracellular fluid, responsible for carrying material from one part of the body to another. Total blood volume in the 70-kg man is equal to about 7% of his total body weight. Thus, if we assume that 1 kilogram of blood occupies a volume of 1 liter, then the 70-kg man has about 5 liters of blood (0.07 × 70 kg = 4.9 L). Of this volume, about 2 liters is composed of blood cells, while the remaining 3 liters is composed of plasma, the fluid portion of the blood.

This chapter presents an overview of the components of blood, which are summarized in Figure 16-1 ■. We will consider the functions of the plasma, red blood cells, and platelets in detail but will reserve the detailed discussion of white blood cells for Chapter 22.

PLASMA AND THE CELLULAR ELEMENTS OF BLOOD

Plasma Is Composed of Water, Ions, Organic Molecules, and Dissolved Gases

Plasma is the fluid portion of the blood, within which the cellular elements are suspended. Water is the main component of plasma and accounts for about 92% of its weight. Proteins account for another 7%. The remaining 1% is composed of other dissolved organic molecules (amino acids, glucose, lipids, nitrogenous wastes), ions (Na^+, K^+, Cl^-, H^+, Ca^{2+}, and HCO_3^-), trace elements and vitamins, and dissolved oxygen (O_2) and carbon dioxide (CO_2).

The composition of plasma is identical to that of the interstitial fluid except for the presence of **plasma proteins.** The liver makes most plasma proteins and secretes them into the blood. **Albumins** are the most prevalent type of protein in the plasma, but **globulins** and the clotting protein **fibrinogen** are also common. A few globulins, known as *immunoglobulins,* are synthesized and secreted by a particular type of white blood cell rather than by the liver.

Problem

Von Willebrand's Disease

Emily Richards, 20, had heard horror stories about getting wisdom teeth pulled, but none could compare with what she was currently experiencing. In addition to the pain and swelling, the extraction sites would not stop bleeding. After three days of lying in bed with gauze stuffed in her mouth, she finally called her doctor. "I think something weird is going on," she told the receptionist. "Could I have hemophilia?" Emily's doctor ran blood tests and called her the next day with the results: "You do have a bleeding disorder, Emily, but it is not hemophilia. You have von Willebrand's disease."

continued on page 456

The presence of proteins in the plasma makes the osmotic pressure of the blood higher than that of the interstitial fluid. This osmotic gradient pulls water from the interstitial fluid into the capillaries, as you learned in the last chapter (∞ p. 439).

Plasma proteins participate in many functions, including blood clotting and defense against foreign invaders. In addition, they act as carriers for steroid hormones, cholesterol, certain ions such as iron (Fe^{2+}), and drugs. Finally, some plasma proteins act as hormones or as extracellular enzymes. Table 16-1 is a summary of the functions of plasma proteins.

The Cellular Elements Include Red Blood Cells, White Blood Cells, and Platelets

Three main cellular elements are found in the blood: **red blood cells,** also called **erythrocytes** [*erythros,* red]; **white blood cells,** also called **leukocytes** [*leukos,* white]; and **platelets.** Only the white blood cells are fully functional cells in the circulation. Red blood cells have lost their nuclei by the time they enter the bloodstream. Platelets are cell fragments that have split off a parent cell known as a **megakaryocyte.**

Red blood cells play a key role in the transport of oxygen and carbon dioxide between the lungs and the tissues. Platelets are instrumental in *coagulation,* the process by which blood clots to prevent blood loss in damaged vessels. White blood cells play a key role in defending the body against foreign invaders such as parasites, bacteria, and viruses. Although most white blood cells circulate through the body in the blood, their work is usually carried out in the tissues rather than inside the circulatory system.

Blood contains five types of mature white blood cells: (1) **lymphocytes,** (2) **monocytes,** (3) **neutrophils,** (4) **eosinophils,** and (5) **basophils.** Monocytes that leave the circulation and enter the tissues develop into **macrophages.** The different white blood cells may be

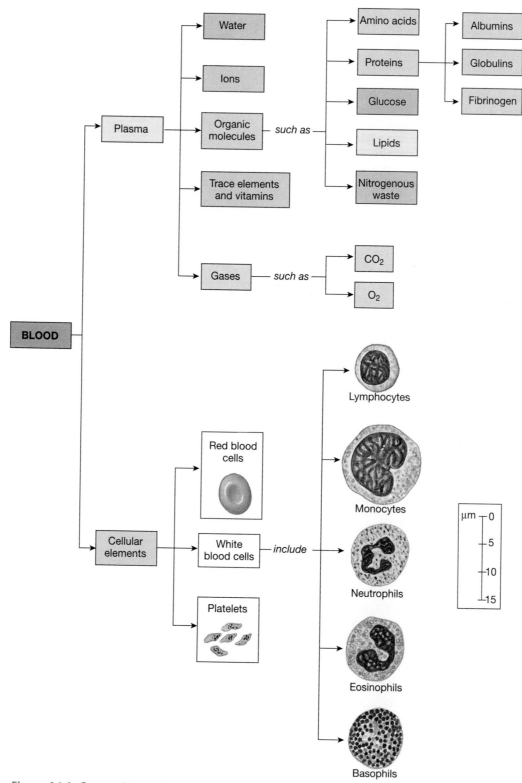

■ Figure 16-1 **Composition of blood**

grouped according to common morphological or functional characteristics. Neutrophils, monocytes, and macrophages are collectively known as **phagocytes** because of their ability to engulf and ingest foreign particles such as bacteria (phagocytosis; ∞ p. 126). Lymphocytes are sometimes called **immunocytes** because they are responsible for the specific immune responses that are directed against specific invaders. Basophils, eosinophils, and neutrophils are called **granulocytes** because they contain cytoplasmic inclusions that give them a granular appearance. The next section examines the process of blood cell synthesis.

The Differential White Cell Count   When the body's defense system is called on to fight off foreign invaders, both the absolute number of white blood cells and the relative proportions of the different types in the circulation will change. Clinicians often rely on a differential white cell count to help them arrive at a diagnosis (Fig. 16-2 ■). For example, a person with a bacterial infection will usually have a high total number of white blood cells in the blood and an increase in the percentage that are neutrophils. A person with a viral infection may have a high, normal, or low total white cell count but often shows an increase in the percentage of lymphocytes. Identifying the different types of white blood cells is not as simple as you might think from doing it in the student laboratory. In abnormal smears, accurate identification may be complicated by the presence of immature blood cells that are difficult to recognize. Medical technologists undergo months of special training in hematology, and those who work in laboratories certified by the College of American Pathologists are tested on cell identification three times each year.

✓ On the basis of what you have learned about the origin and role of plasma proteins, explain why patients with advanced degeneration of the liver frequently suffer from edema.

BLOOD CELL PRODUCTION

Blood Cells Are Produced in the Bone Marrow

All blood cells are descendants of a single precursor cell type known as the *multipotent progenitor* [*progignere*, to beget] that has the potential to become any type of blood cell. From this ancestral cell, found in the bone marrow, an assortment of uncommitted stem cells develop (Fig. 16-3 ■). One cell line differentiates into lymphocytes; the other cell line ultimately develops into red blood cells, white blood cells other than lymphocytes, and megakaryocytes, the parent cells of platelets. It is estimated that only about one out of every 10 thousand cells in the bone marrow is an uncommitted stem cell, making the isolation and study of these cells very difficult.

Hematopoiesis [*haima*, blood + *poiesis*, formation], the synthesis of blood cells, begins early in embryonic development and continues throughout a person's life. In about

continued from page 454

Von Willebrand's disease, named for the Finnish physician who discovered it in the late nineteenth century, is often diagnosed after an individual undergoes a dental or other surgical procedure. Besides prolonged bleeding, other symptoms include nosebleeds and bruising. The disease is caused by a deficiency in von Willebrand factor, a protein found in the blood that acts as a carrier for clotting factor VIII. Von Willebrand factor also promotes adherence of platelets to the endothelium and to each other during coagulation.

Question 1: Why does von Willebrand's disease cause prolonged bleeding after a dental or surgical procedure?

the third week of fetal development, special cells in the yolk sac of the embryo form clusters. Some of these cell clusters are destined to become the endothelial lining of blood vessels, while others become blood cells. The common embryological origin of the endothelium and blood cells perhaps explains why the cytokines that control hematopoiesis are released by the vascular endothelium.

As the embryo develops, blood cell production spreads from the yolk sac to the liver, spleen, and **bone marrow,** a soft tissue that fills the hollow centers of bones. By birth, the liver and spleen no longer produce blood cells. Hematopoiesis continues within all the bones of the skeleton until age five. However, as the child continues to age, the active regions of marrow decrease. In adults, only the spine, ribs, pelvis, and proximal ends of long bones produce blood cells. Bone marrow that is actively producing blood cells retains a red color because of the presence of hemoglobin, a compound found within red blood cells. Inactive marrow is yellow because of an abundance of adipocytes (fat cells). You can see the difference between red and yellow marrow the next time you look at bony cuts of meat in the grocery store. Although blood synthesis in adults is limited to small areas, the liver, spleen, and inactive (yellow) regions of marrow can resume blood cell production in times of need.

In the regions of marrow that are actively producing blood cells, about 25% of the cells are developing red blood cells, while 75% are destined to become white blood cells. The life span of white blood cells is considerably shorter than that of the red blood cells, and they must

TABLE 16-1	Functions of Plasma Proteins	
Name	*Source*	*Function*
Albumins (multiple types)	Liver	Major contributors to osmotic pressure of plasma; carriers for various substances
Globulins (multiple types)	Liver and lymphoid tissue	Clotting factors, enzymes, antibodies, carriers for various substances
Fibrinogen	Liver	Forms fibrin threads essential to blood clotting

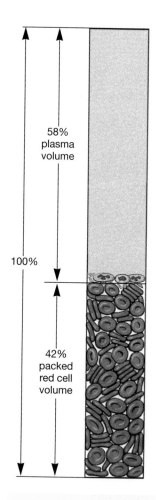

58% plasma volume

100%

42% packed red cell volume

1. The **hematocrit** is the percentage (%) of the total blood volume that is occupied by packed (centrifuged) red blood cells (see figure at left).

2. The **hemoglobin** (Hb) content of erythrocytes. This value is measured as total hemoglobin content of blood (g Hb/dL).

3. **Mean red cell volume** (MCV). In some disease states, the red blood cells may be either abnormally large or abnormally small. For example, they are abnormally small in iron-deficiency anemia.

4. **Red cell count** in millions of cells per microliter. A machine counts the cells as they stream through a beam of light.

5. The **morphology** of the red blood cells also gives clues to diseases. Sometimes the cells lose their flattened disk shape and become spherical (spherocytosis). In sickle cell anemia, the cells are shaped like a sickle or crescent moon.

6. **Total white cell count** tells the total number of all types of white blood cells but does not distinguish between the different types.

7. The **differential white cell** count estimates the relative numbers of the five types of white cells. The differential count is made by a medical technologist from a thin smear of blood that is stained with biological dyes.

8. **Platelet count** is suggestive of the ability of the blood to clot.

	Males	Females
Hematocrit (% of whole blood that is packed red cells)	40%–54%	37%–47%
Hemoglobin (g/dL blood)	14–17	12–16
Red cell count (cells/μL)	$4.5–6.5 \times 10^6$	$3.9–5.6 \times 10^6$
Total white cell count (cells/μL)	$4–11 \times 10^3$	$4–11 \times 10^3$
Differential white cell count		
Neutrophils	50%–70%	50%–70%+
Eosinophils	1%–4%	1%–4%
Basophils	<1%	<1%
Lymphocytes	20%–40%	20%–40%
Monocytes	2%–8%	2%–8%
Platelets (per μL)	$200–500 \times 10^3$	$200–500 \times 10^3$

■ **Figure 16-2 The blood count** The table at the bottom shows the normal range of values. A deciliter is equal to 100 mL.

be replaced more frequently. For example, neutrophils have a six-hour half-life, so the body must make 100 billion neutrophils each day in order to replace those that die. Red blood cells, on the other hand, live for nearly four months in the circulation.

Hematopoiesis Is Controlled by Cytokines, Growth Factors, and Interleukins

The reproduction and development of blood cells in the bone marrow are controlled by chemical factors known as cytokines. Cytokines are molecules released from one cell that affect the growth or activity of another cell (∞ p. 148). When cytokines are first discovered, they are often called *factors* and given a modifier that describes their actions: growth factor, differentiating factor, activating factor. Some of the best-known cytokines in hematopoiesis are the **colony-stimulating factors,** molecules made by endothelial cells and white blood cells. Once the amino acid sequence of a cytokine is determined, the cytokine may be given a numbered **interleukin** [*inter-*, between + *leuko*, white] designation. The name *interleukin* was first given to cytokines that are released by one white blood cell and act on another

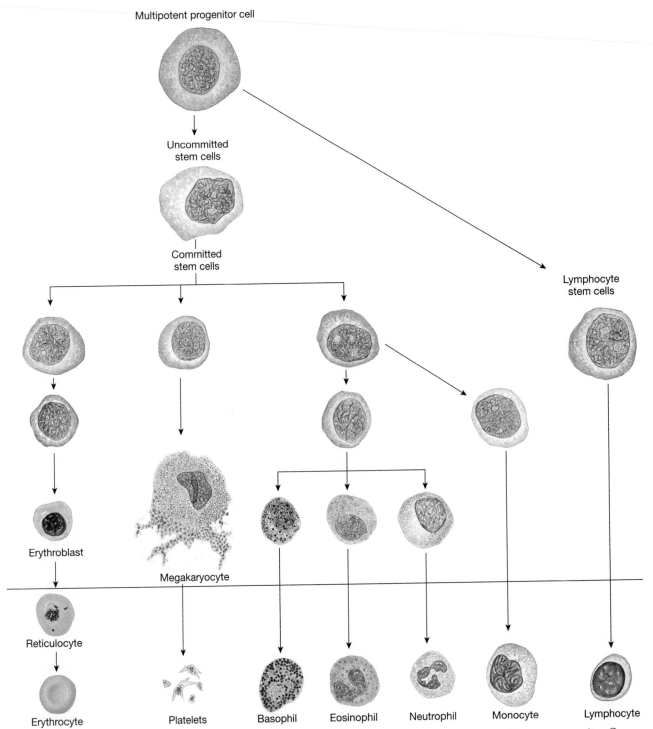

■ **Figure 16-3** **Hematopoiesis** All blood cells are derived from a single stem cell type known as the multipotent progenitor. Once stem cells become committed, they can develop into only one type of cell. Cells below the horizontal line are the predominant forms found circulating in the blood. Cells above the line are found mostly in the bone marrow.

white blood cell. The role of interleukins in the immune system will be discussed in Chapter 22.

Table 16-2 lists some known cytokines linked to hematopoiesis. Because many cytokines have multiple names and because research in this field is especially active, we will describe most cytokines in general terms rather than discussing specific molecules in detail.

Colony-Stimulating Factors Regulate Leukopoiesis

White blood cell reproduction and development are regulated by several colony-stimulating factors, cytokines made by endothelial cells, marrow fibroblasts, and white blood cells. These factors were identified and

TABLE 16-2 Cytokines Involved in Hematopoiesis

Name	Cell That Produces Cytokine	Influences Growth or Differentiation of
Erythropoietin (EPO)	Kidney cells*	Red blood cells
Granulocyte-macrophage colony-stimulating factor (GM-CSF)	Endothelial cells and fibroblasts of bone marrow; other white blood cells	Megakaryocytes, eosinophils, neutrophils, monocytes, red blood cells
Granulocyte colony-stimulating factor (G-CSF)	Endothelial cells and fibroblasts of bone marrow; monocytes	Granulocytes (neutrophils, eosinophils, basophils)
Macrophage colony-stimulating factor (M-CSF)	Endothelial cells and fibroblasts of bone marrow; monocytes	Monocytes
Interleukin-3 (IL-3; multi-CSF)	T lymphocytes	Primarily nonlymphocyte blood cells; growth factor for B lymphocytes

*There is still debate as to whether EPO is produced by kidney tubule cells or by the interstitial cells. The endothelial cells of renal blood vessels appear to be the oxygen-sensing cells.

named for their ability to stimulate the growth of white blood cell colonies in culture.

The normal development of white blood cells is not controlled by a single cytokine. Instead, a combination of protein factors interact to control the differentiation of immature cells into mature leukocytes. These cytokines are required to induce both cell division (mitosis) and cell maturation. Once a leukocyte matures, it loses its ability to undergo mitosis.

One fascinating aspect of hematopoiesis is that white blood cell production is regulated in part by the existing white blood cells. This form of control allows white blood cell development to be very specific and tailored to the body's needs. For example, if you have a bacterial infec-

Blood Cell Culture   The successful laboratory culture of isolated blood cells was first reported around 1961. By 1966, researchers had recognized that the isolated blood cells either must have growth factors added to the culture dish or must be raised close to other cells that secrete those factors. The fact that growth factors from other cells are necessary can be shown by the following experiment. In a petri dish, blood cells are grown in a semisolid layer of nutrient material, separated from a layer of "feeder cells" (in another semisolid layer) by a solid agar layer (Fig. 16-4 ■). The feeder cells are macrophages or other cytokine-secreting cells. The solid agar layer prevents direct cell contact between the blood cells and feeder cells. However, trophic growth factors secreted by the feeder cells diffuse through the solid agar layer. The growth factors were identified as a requirement for blood cell colony formation during those experiments, so they were named colony-stimulating factors. The colony-stimulating factors were secreted in such small amounts that it seemed impossible to get enough to use them clinically. But with technological advances, the factors were identified, and their genes were located and cloned. We now have recombinant factors available commercially and in widespread clinical use. The cost of developing colony-stimulating factors for therapeutic use is extraordinary. It has been estimated that the development of one hematopoietic factor from laboratory to bedside is greater than $500 million.

tion, cytokines released by active white blood cells fighting the infection stimulate the production of additional neutrophils and macrophages. The complex process by which leukocyte production is matched to need is still not well understood and is an active area of research.

Scientists hope to create a model for the control of hematopoiesis so that they can develop effective treatments for diseases characterized by either a lack or an excess of white blood cells. The *leukemias* are a group of diseases characterized by the abnormal growth and development of white blood cells. In *neutropenias* [*penia*, poverty], patients have too few white blood cells and are unable to fight off bacterial and viral infections as a result. Researchers hope to find better treatments for both leukemias and neutropenias by unlocking the secrets of how the body regulates cell growth and division.

Thrombopoietin Regulates Platelet Production

Thrombopoietin (TPO) is a glycoprotein that regulates the growth and maturation of megakaryocytes, the parent cells of platelets. It is produced primarily in the liver

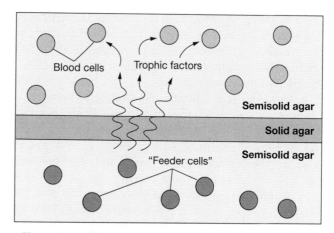

■ **Figure 16-4 Blood cell culture**   The blood cells are placed in the top compartment, separated from "feeder cells" by a layer of solid agar. The feeder cells are macrophages and other cells that secrete cytokines. These cytokines diffuse through the solid agar layer and act as trophic factors to control the growth and development of the blood cells in the top layer.

continued from page 456

"Is this disease serious?" Emily asks her doctor. "It can be," he replies. "Several different types of the disease exist, with varying degrees of severity. Since you have not had symptoms before this, you probably have one of the milder forms." Von Willebrand factor is present in megakaryocytes, endothelial cells, platelets, and plasma. In each location, the factor exists in three different "sizes," or multimers. The different types of von Willebrand disease vary according to the presence or absence of different multimers. The least serious type (type I) is characterized by decreased production of all von Willebrand multimers in all tissues. The most serious type (type III) is characterized by the complete absence of von Willebrand multimers in plasma.

Question 2: Why is type I the least serious form of von Willebrand's disease?

but is also present in the kidney. The gene for this cytokine, whose activity was first described in 1958, was cloned in 1994. Within a year, TPO was widely available to researchers through the use of recombinant DNA techniques, and there has been an explosion of papers describing its effects on megakaryocytes and platelet production. Although there is much that we still do not understand about the basic biology of the process, the synthesis of thrombopoietin is a huge step forward in the search for answers.

Erythropoietin Regulates Red Blood Cell Production

Red blood cell production (**erythropoiesis**) is regulated primarily by a hormone called **erythropoietin (EPO)** and is assisted by several cytokines. Erythropoietin is a glycoprotein hormone made primarily in the kidney of adults. The stimulus for EPO synthesis and release is *hypoxia*, low oxygen levels in the tissues. This pathway, like other endocrine pathways, is intended to help the body maintain homeostasis. Erythropoietin stimulates red blood cell synthesis and puts more hemoglobin into the circulation to carry oxygen.

The existence of a substance controlling red blood cell production was first suggested in the 1950s, but two decades passed before scientists succeeded in purifying the hormone. Scientists took another nine years to sequence the hormone and to isolate and clone the gene for it. However, an incredible leap in progress was made after cloning the gene for EPO: Only two years later, the hormone was produced by recombinant DNA technology and put into clinical use. Today, physicians can prescribe not only EPO (which is marketed under the names Epogen® and Procrit®) but also several colony-stimulating factors (for example, Neupogen®) that stimulate white blood cell synthesis. Cancer patients whose hematopoiesis has been suppressed by chemotherapy have benefited from injections of these hematopoietic hormones.

RED BLOOD CELLS

Red blood cells, or erythrocytes, are the most abundant cell type in the blood. A microliter of blood contains about 5 million red blood cells, compared with only 4000–11,000 white blood cells and 20,000–500,000 platelets per microliter. The primary function of red blood cells is to transport oxygen from the lungs to the cells and carbon dioxide from the cells to the lungs.

The ratio of red blood cells to plasma is indicated clinically by the **hematocrit** and is expressed as a percentage of the total blood volume (see Fig. 16-2 ■). Hematocrit is determined by drawing a blood sample into a narrow capillary tube and spinning it in a centrifuge so that the heavier red blood cells go to the bottom of the sealed tube, leaving the lighter white blood cells and platelets in the middle and the plasma on top. The column of *packed red cells* is measured and compared to the total sample volume to find the hematocrit. The normal range of hematocrit for a man is 40%–54%; for a woman, 37%–47%. This test provides a rapid and inexpensive way to estimate a person's red cell count because blood for a hematocrit can be collected by simply sticking a finger.

Mature Red Blood Cells Lack a Nucleus

In the bone marrow, committed stem cells differentiate through several stages into large, nucleated *erythroblasts*. As erythroblasts mature, the nucleus condenses and the cell shrinks in diameter from 20 μm to about 7 μm. In the last stage before maturation, the nucleus is pinched off and phagocytosed by marrow macrophages. At the same time, other membranous organelles such as mitochondria break down and disappear. The final immature cell form, called a *reticulocyte*, leaves the marrow and enters the circulation, where it matures into an erythrocyte in about 24 hours (see Fig. 16-5 ■).

Mature mammalian red blood cells are biconcave disks, shaped much like jelly doughnuts with the filling squeezed out of the middle (Fig. 16-6a ■). They lack a nucleus and membranous organelles and resemble a simple membranous "bag" filled with enzymes and hemoglobin, a red oxygen-carrying pigment.

Because red blood cells contain no mitochondria, they cannot carry out aerobic metabolism. Glycolysis is the primary source of ATP. Without a nucleus and endoplasmic reticulum to carry out protein synthesis, the cell is unable to make new enzymes or to renew its membrane components. This inability leads to an increasing loss of membrane flexibility, making older cells more fragile and likely to rupture. Once the red blood cell enters the circulation, it lives about 120 days.

The biconcave shape of the red blood cell is one of its most distinguishing features. The membrane is held in place by a complex cytoskeleton composed of filaments linked to transmembrane attachment proteins (Fig. 16-6c ■). Despite the cytoskeleton, red cells are remarkably flexible, like a partially filled water balloon that can

Focus on   Bone Marrow

The bone marrow, hidden within the bones of the skeleton, is easily overlooked as a tissue, although collectively, it is nearly the size and weight of the liver! Bone marrow consists of the blood cells in different stages of development and a supporting framework of tissue known as the **stroma** [mattress]. The stroma is composed of highly branched fibroblasts (⬭ p. 62), collagenous reticular fibers, and an extracellular matrix. (Fibroblasts are also known as reticular cells, or fibrocytes—an inactive form of the fibroblast that does not manufacture and secrete collagen fibers. Fibroblasts in bone marrow are some-times called stromal cells.) Scattered throughout the stroma are fixed macrophages, a mature form of the monocyte white blood cells. Some macrophages are responsible for removing aged red blood cells; others form centers around which stem cells cluster. Marrow is a highly vascular tissue, filled with branched blood vessels that open into **blood sinuses,** widened regions lined with endothelium. Mature blood cells squeeze out of the stroma through the endothelium to reach the circulation (Fig. 16-5 ■).

■ Figure 16-5 **Bone marrow**

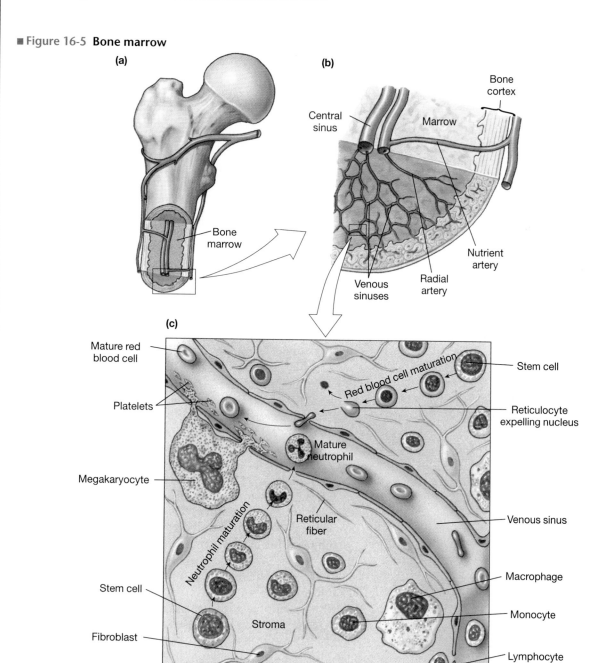

(a)

(b)

Central sinus

Bone cortex

Marrow

Bone marrow

Venous sinuses

Radial artery

Nutrient artery

(c)

Mature red blood cell

Platelets

Megakaryocyte

Neutrophil maturation

Stem cell

Fibroblast

Stroma

Reticular fiber

Mature neutrophil

Red blood cell maturation

Stem cell

Reticulocyte expelling nucleus

Venous sinus

Macrophage

Monocyte

Lymphocyte

(a)

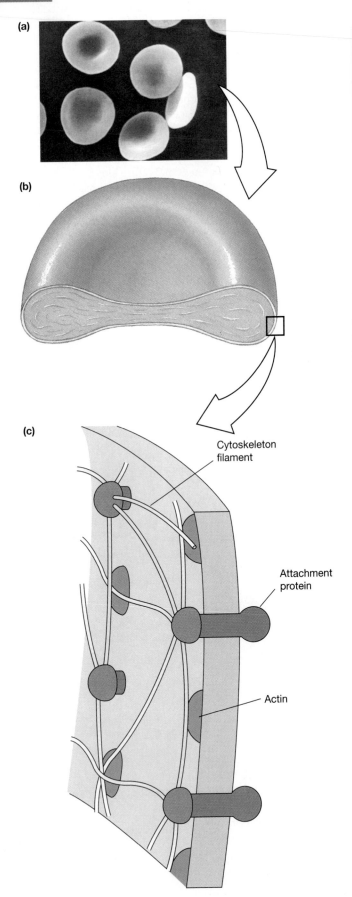

(b)

(c)

Cytoskeleton filament

Attachment protein

Actin

■ **Figure 16-6 Red blood cells** The extensive cytoskeleton of the red blood cell gives it its distinctive bioconcave disk shape.

Stem Cells from Umbilical Cord Blood
Scientists have been able to isolate and grow uncommitted stem cells to use as replacements for patients whose own stem cells have been killed by chemotherapy for cancer. Currently, these stem cells are obtained from the bone marrow or from peripheral blood. However, umbilical cord blood, collected at birth, has been found to be a rich source of stem cells. Besides being plentiful, umbilical cord stem cells are less likely to cause a foreign tissue rejection reaction than are stem cells obtained from a donor who is unrelated to the recipient. The potential for an easily collected source of stem cells for therapeutic use has led to the development of cord blood banking programs in the United States and Europe. In future years, umbilical cord blood may automatically be saved from each child at birth so that a source of stem cells will be available should the child need them in later life.

compress into various shapes. This flexibility allows the cells to change shape as they squeeze through the narrow capillaries of the circulation.

The disklike structure of red blood cells also allows them to modify their shape in response to osmotic changes in the blood. A red cell placed in slightly hypotonic medium will swell and form a sphere without disruption of its membrane integrity. In hypertonic media, red cells shrink up and develop a spiky surface where the membrane pulls tight against the cytoskeleton.

Hemoglobin Synthesis Requires Iron

Hemoglobin, the oxygen-carrying pigment of red blood cells, is a large complex molecule with a quaternary structure of four globular protein chains, each of which is wrapped around an iron-containing **heme** group (Fig. 16-7 ■). There are several forms of the **globin** protein chain. The most common forms are designated *alpha* (α), *beta* (β), *gamma* (γ), and *delta* (δ), depending on the structure of the chain. Most adult hemoglobin has two alpha and two beta chains (designated *HbA*). However, a small portion of adult hemoglobin (about 2.5%) has two alpha and two delta chains (*HbA$_2$*). The human fetus has a different form of hemoglobin adapted to "steal" oxygen from maternal blood in the placenta. Fetal hemoglobin (*HbF*) has two gamma chains in place of the two beta chains found in adult hemoglobin. Shortly after an infant's birth, fetal hemoglobin is replaced with the adult form as new red blood cells are made.

The four heme groups in a hemoglobin molecule are identical. Each consists of a carbon-hydrogen-nitrogen *porphyrin ring* with an iron atom (Fe) in the center (Fig. 16-7b ■). Some 70% of the iron in the body is found in the heme groups of red blood cells. Hemoglobin synthesis requires an adequate supply of iron in the diet. Absorbed iron is transported in the blood on a carrier protein called *transferrin* (Fig. 16-8 ■). In the bone marrow, iron is taken into the mitochondria of developing

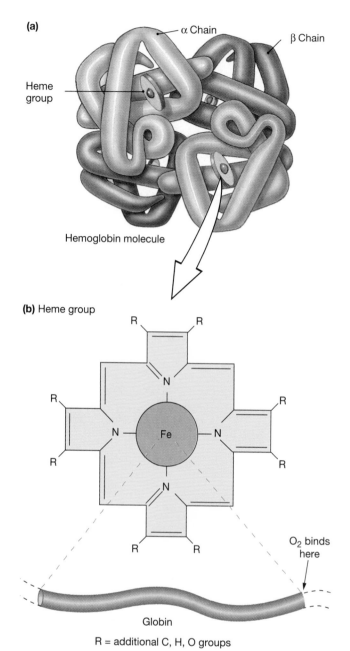

(a)

α Chain

β Chain

Heme group

Hemoglobin molecule

(b) Heme group

R = additional C, H, O groups

O_2 binds here

Globin

■ **Figure 16-7** **Hemoglobin** (a) A hemoglobin molecule is composed of four protein globin chains, each centered around a heme group. In most adult hemoglobin, there are two alpha chains and two beta chains as shown. (b) A heme group consists of a porphyrin ring with an iron atom in the center.

red blood cells and used to make heme. Excess iron in the body is stored, mostly in the liver, on the protein **ferritin** and its derivatives.

Red Blood Cells Live about Four Months

Red blood cells in the circulation live for 120 ± 20 days. The increasingly fragile older red blood cells may rupture as they try to squeeze through narrow capillaries, or they may be captured by scavenging macrophages as they pass through the spleen. The various components of hemoglobin are recycled. Amino acids from the pro-

tein chains become new proteins, while some iron from heme groups may be reused in new heme groups. The remainder of the old heme is converted by cells of the spleen and liver to a colored pigment called **bilirubin.** Bilirubin is carried by plasma albumin to the liver, where it is metabolized and incorporated into **bile** (Fig. 16-8 ■). Bile is secreted into the digestive tract, and the bilirubin metabolites leave the body in the feces. Small amounts of other bilirubin metabolites are filtered from the blood into the kidneys, where they contribute to the yellow color of the urine.

In some circumstances, bilirubin levels in the blood become elevated (hyperbilirubinemia). This condition, known as **jaundice,** causes the skin and whites of the eyes to take on a yellow cast. The accumulation of bilirubin can occur from multiple causes. Newborns whose fetal hemoglobin is being broken down and replaced with adult hemoglobin are particularly susceptible to jaundice, so doctors closely monitor bilirubin levels in the first weeks of life. Another common cause of jaundice is liver disease, in which the liver is unable to process or excrete bilirubin.

Red Blood Cell Disorders Decrease Oxygen Transport

Hemoglobin plays a key role in oxygen transport, so the hemoglobin content of the body is important. In some pathophysiological conditions, individual red blood cells contain a normal amount of hemoglobin but there are too few cells. In other disease states, the number of red blood cells is adequate but the hemoglobin content of the individual cells is too low. Pathologies may also result from abnormal forms of hemoglobin or from abnormal membrane structure that causes cells to rupture and be destroyed. Most of these pathologies result in **anemia,** a condition in which the blood's hemoglobin content is too low. Because oxygen binds to hemoglobin, individuals with anemia cannot transport enough oxygen to the tissues and they become tired and weak, especially after exercise. Some major causes of anemia are summarized in Table 16-3.

In the *hemolytic anemias* [*lysis*, rupture], the rate of red blood cell destruction exceeds the rate of cell production. The hemolytic anemias are usually hereditary defects in which the body makes fragile cells. For example, in *hereditary spherocytosis*, the cytoskeleton does not link properly because of defective or deficient cytoskeletal proteins. Consequently, the cells are shaped more like spheres than like biconcave disks. This disruption in the cytoskeleton results in cells that rupture easily and that are not able to withstand osmotic changes as well as normal cells.

Other anemias result from the failure of the bone marrow to make adequate amounts of hemoglobin. One of the most common examples of an anemia that results from insufficient hemoglobin synthesis is **iron-deficiency anemia.** If iron excretion exceeds iron intake, the marrow does not have adequate iron to make heme groups, and

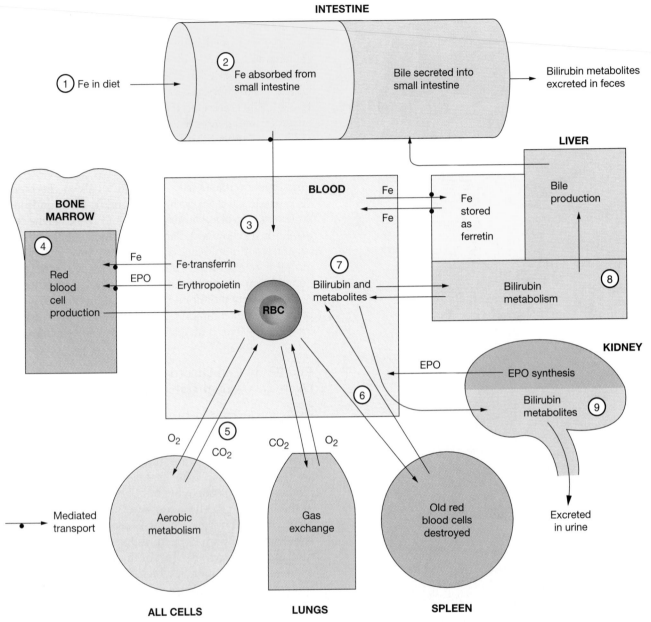

INTESTINE

① Fe in diet

② Fe absorbed from small intestine

Bile secreted into small intestine

Bilirubin metabolites excreted in feces

LIVER

BONE MARROW

④ Red blood cell production

Fe → Fe·transferrin

EPO → Erythropoietin

BLOOD

Fe
Fe

Fe stored as ferretin

Bile production

③

⑦ Bilirubin and metabolites

Bilirubin metabolism

⑧

RBC

KIDNEY

EPO → EPO synthesis

Bilirubin metabolites

⑨

⑥

O_2
CO_2 ⑤

CO_2 O_2

Excreted in urine

Mediated transport

Aerobic metabolism

Gas exchange

Old red blood cells destroyed

ALL CELLS   **LUNGS**   **SPLEEN**

■ Figure 16-8 **Iron metabolism and the life of a red blood cell**

hemoglobin synthesis slows. People with iron-deficiency anemia have a low red blood cell count or a low hemoglobin content in their blood. Their red blood cells are often smaller than usual (*microcytic* red blood cells), and the lower hemoglobin content may cause the cells to be paler than normal (*hypochromic*). Women of reproductive age who menstruate are likely to suffer from iron-deficiency anemia because of iron loss in menstrual blood.

Sickle cell disease is a genetic defect in which glutamine, the number 6 amino acid in the 146–amino acid beta chain of hemoglobin, is replaced by valine. The result is abnormal hemoglobin (HbS) that crystallizes and pulls the red blood cells into a sickle shape, like a crescent moon. The sickled cells get tangled with other sickled red cells as they pass through the smallest blood vessels, causing them to jam up and block blood flow to the tis-

Sickle Cell Disease and Hydroxyurea   One new treatment developed for sickle cell disease is the administration of hydroxyurea, a compound that inhibits DNA synthesis. In the bone marrow, hydroxyurea causes immature red blood cells to produce fetal hemoglobin (HbF), with alpha and gamma globin chains, instead of adult hemoglobin. Fetal hemoglobin interferes with the crystallization of sickle cell hemoglobin (HbS), so the red blood cells no longer develop a sickle shape. Hydroxyurea has been effective in alleviating the symptoms of sickle cell disease in many patients. However, its long-term effects, which are suspected to include inducing cancer or causing birth defects in the offspring of treated patients, have not been determined. The decision to use hydroxyurea is one in which the patient must weigh the benefits of the treatment against its possible risks.

TABLE 16-3	Causes of Anemia

Accelerated red blood cell loss

Blood loss: Cells are normal in size and hemoglobin content but low in number

Hemolytic anemias: Cells rupture at an abnormally high rate

Hereditary
    Membrane defects (example: hereditary spherocytosis)
    Enzyme defects
    Abnormal hemoglobin (example: sickle cell anemia)
Acquired
    Parasitic infections (example: malaria)
    Drugs
    Autoimmune reactions

Decreased red blood cell production

Defective red blood cell or hemoglobin synthesis in the bone marrow
  Aplastic anemia: Can be caused by certain drugs
  Inadequate dietary intake of essential nutrients
    Iron deficiency (iron is required for heme production)
    Folic acid deficiency (folic acid is required for DNA synthesis)
    Vitamin B_{12} deficiency (vitamin B_{12} is required for DNA synthesis)

Inadequate production of erythropoietin

sues. This blockage creates tissue damage from hypoxia and is also very painful.

Although the anemias are common, it is also possible to have too many red blood cells. *Polycythemia vera* [*vera*, true] is a stem cell dysfunction that produces too many blood cells, white as well as red. These patients may have hematocrits as high as 60%–70% (normal is 40%–54%). The increased number of cells causes the blood to become more viscous and thus more resistant to flow through the circulatory system (∞ p. 389).

✓ Erythropoietin (EPO) was first isolated from the urine of anemic patients who had high circulating levels of the hormone. However, despite the presence of the hormone that stimulates red blood cell production, these patients were unable to produce adequate amounts of hemoglobin or red cells. Give some possible reasons why the patients' own EPO was unable to correct their anemia.

✓ A person who goes from sea level to a city that is 5000 feet above sea level will show an increased hematocrit within two to three days. Draw the reflex pathway that links the hypoxia of high altitude to increased red blood cell production.

✓ An individual who becomes dehydrated will have an elevated hematocrit. Explain how dehydration leads to an elevated hematocrit although blood cell production is normal.

PLATELETS AND COAGULATION

Because of its fluid nature, blood flows freely throughout the circulatory system, but if there is a break in the "piping" of the system, blood will be lost unless steps are taken. One of the challenges of the body is to plug holes in damaged blood vessels while still maintaining blood flow through the vessel. It would be simple to block off a damaged blood vessel completely, like putting a barricade across a street full of potholes. But just as shopkeepers on that street lose business if traffic is blocked, cells downstream from the point of injury die from lack of oxygen and nutrients if the vessel is completely blocked. The body's task is to allow blood flow through the vessel while simultaneously repairing the damaged wall. This challenge is complicated by the fact that blood in the system is under pressure. If the repair "patch" is too weak, it will simply be blown out by the blood pressure. Thus, stopping blood loss involves several steps. First, the pressure in the vessel must be decreased long enough to create a secure mechanical seal in the form of a blood clot. Once the clot is in place and blood loss has been stopped, the body's repair mechanisms can take over. Then, as the wound heals, enzymes gradually dissolve the clot, while scavenger white blood cells ingest and destroy the debris.

The following sections examine the role of platelets and plasma proteins in this process.

Platelets Are Small Fragments of Cells

Platelets (*thrombocytes*) are cell fragments produced in the bone marrow from huge cells called megakaryocytes [*mega*, extremely large + *karyon*, kernel + *-cyte*, cell]. Megakaryocytes develop their formidable size by undergoing mitosis up to seven times without undergoing nuclear or cytoplasmic division. The result is a *polyploid* cell with a lobed nucleus (Fig. 16-9 ■). The outer edges of marrow megakaryocytes extend through the endothelium into the lumen of marrow blood sinuses, where the cytoplasmic extensions fragment into disklike platelets (see Fig. 16-5 ■). Platelets are smaller than red blood cells, are colorless, and have no nucleus. Their cytoplasm contains mitochondria, smooth endoplasmic reticulum, and many granules filled with clotting proteins and cytokines.

Platelets are present in the blood at all times, but they are not active unless damage has occurred to the walls of the circulatory system. The typical life span of a platelet is about 10 days. In the next section, we will examine how platelets are involved in the formation of blood clots.

Hemostasis Prevents Blood Loss from Damaged Blood Vessels

Hemostasis is the process of keeping blood within the blood vessels by repairing breaks without compromising the fluidity of the blood [*haima*, blood + *stasis*, stoppage]. (The opposite of hemostasis is *hemorrhage*.) The initial stages of hemostasis depend on mechanical blockage of

(a)

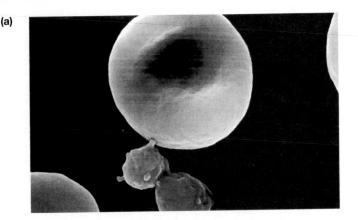

(b)

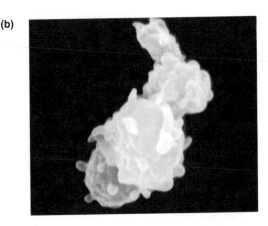

(c)

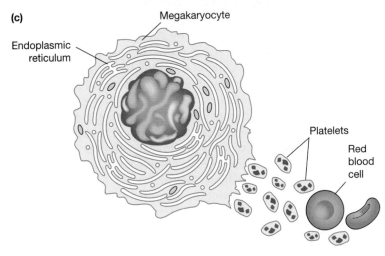

■ Figure 16-9 **Megakaryocytes and platelets** (a) Platelets are disklike cell fragments smaller than red blood cells. (b) When activated, platelets (shown enlarged) develop a spiky outer surface and become sticky, adhering to each other. (c) Megakaryocytes are giant cells with multiple copies of DNA in the nucleus. The edges of the megakaryocyte break off to form cell fragments called platelets.

the hole by a **platelet plug,** formed when platelets stick to the damaged blood vessel (**platelet adhesion**) and to each other (**platelet aggregation**). The platelet plug is converted into a clot when reinforced by protein fibers. Once the clot is in place, tissue repair begins, triggered by growth factors released by the platelets. As the tissue at the injury site heals, the clot is gradually removed by enzymatic action and phagocytic white blood cells.

The body must maintain the proper balance for hemostasis. Too little hemostasis results in excessive bleeding; too much will result in a **thrombus,** or blood clot that adheres to the unbroken wall of a blood vessel [*thrombos,* a clot or lump]. A large thrombus will block the lumen of the vessel and stop blood flow. Complicating the control of hemostasis is the fact that many steps in the process are positive feedback loops that stop only when the components are used up.

The steps of hemostasis can be summarized as follows:

- Damage to the wall of the blood vessel exposes the underlying collagen and releases chemical factors from the damaged endothelial cells.
- Platelets adhere to the damaged collagen (platelet adhesion) and become activated, releasing various factors into the area around the injury.
- The platelet factors cause local vasoconstriction and act in a positive feedback manner to activate more

platelets. The activated platelets stick to each other (platelet aggregation), forming a loose platelet plug. Vasoconstriction temporarily decreases flow and pressure within the vessel, helping the formation of the platelet plug.

- Simultaneously, exposed collagen and the factors released by the damaged endothelial cells initiate a series of reactions known as the **coagulation cascade.** In the first steps of the cascade, inactive plasma proteins are converted into active enzymes. The newly activated enzymes in turn activate other inactive

continued from page 460

Fortunately, Emily is diagnosed with type I disease. "From now on," explains her doctor, "you should take medication before any surgical procedure to prevent excessive bleeding." The medication consists of desmopressin and an antifibrinolytic agent. Desmopressin, a synthetic form of the hormone vasopressin (also called antidiuretic hormone), causes blood vessel constriction. The antifibrinolytic agent inhibits the action of plasmin in a blood clot.

Question 3: Explain the rationale for giving an antifibrinolytic agent to individuals with von Willebrand's disease.

plasma enzymes. In the last steps of the cascade, **thrombin** converts the plasma protein **fibrinogen** into **fibrin** fibers that intertwine with the platelet plug. Another chemical factor converts the fibrin into a cross-linked polymer that stabilizes the platelet plug. The reinforced platelet plug with fibrin and red blood cells trapped within it is called a clot.

- Some chemical factors involved in the coagulation cascade also promote platelet adhesion and aggregation to the damaged region. At the same time, the platelet plug and clot are limited to the area of tissue damage by inhibitory factors released by undamaged tissue. These factors prevent platelet adhesion and aggregation and also inhibit the coagulation cascade.

- As cell growth and division repairs the damaged vessel, the clot retracts and slowly dissolves, owing to the enzyme **plasmin,** which was trapped within the platelet plug when it was formed. The dissolution of fibrin by plasmin is known as **fibrinolysis.**

Figure 16-10 ■ summarizes the steps given above. Although hemostasis seems straightforward, there are still many unanswered questions at the cellular and molecular levels. Because inappropriate blood clotting plays an important role in strokes and heart attacks, this

Clot Busters The discovery of the factors controlling fibrinolysis was an important step in developing treatments for dissolving inappropriate clots. For example, the condition known as a heart attack is more properly called a myocardial infarction. An **infarct** is an area of tissue deprived of its blood supply because the vessel is blocked by an inappropriate clot (thrombus) or by an atherosclerotic plaque. Unless the blockage is removed promptly, the tissue normally supplied with oxygen by the vessel will die or be severely damaged. If the infarct occurs in a coronary artery, the heart may sustain so much damage that it can no longer function properly. Because thrombus formation is characteristic of myocardial infarctions, physicians can use fibrinolytic drugs to dissolve the clots and prevent some damage if the clot is removed soon after it occurs. The three drugs currently being used for this purpose are urokinase (extracted from kidneys), streptokinase (from bacteria), and genetically engineered **tissue plasminogen activator (t-PA),** a naturally occurring promoter of plasmin formation.

is a very active area of research. Research has led to the development and use of "clot busters," enzymes that can dissolve clots in the coronary arteries after heart attacks.

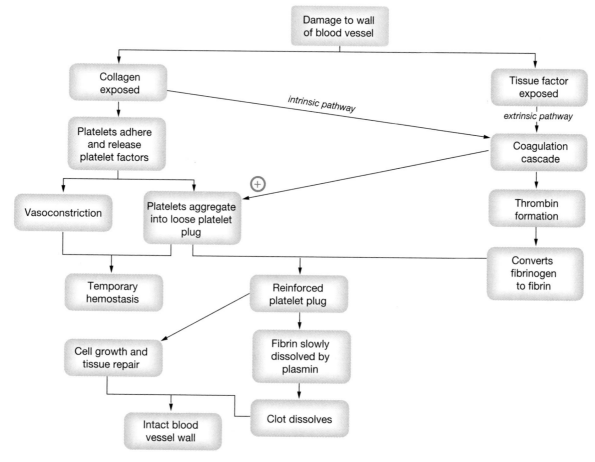

■ Figure 16-10 **Overview of hemostasis**

TABLE 16-4 Factors Involved in Hemostasis

Chemical factor	Source	Activated by or Released in Response to	Role in Platelet Plug Formation	Role in Coagulation or Fibrinolysis	Other Roles and Comments
von Willebrand factor (vWF)	Endothelium, megakaryocytes	Exposure to collagen	Links platelets to collagen	Regulates level of factor VIII	Deficiency or defect causes prolonged bleeding
Serotonin	Secretory vesicles of platelets	Platelet activation	Platelet aggregation	NA	Vasoconstrictor
Adenosine diphosphate (ADP)	Platelet mitochondria	Platelet activation, thrombin	Platelet aggregation	NA	NA
Thromboxane A_2	Phospholipids in platelet membranes	Platelet activating factor	Platelet aggregation	NA	Vasoconstrictor; eicosanoid
Platelet-derived growth factor (PDGF)	Platelets	Platelet activation	NA	NA	Promotes wound healing by attracting fibroblasts and smooth muscle cells
Platelet-activating factor (PAF)	Platelets, neutrophils, monocytes	Platelet activation	Platelet aggregation	NA	Plays role in inflammation; increases capillary permeability
Tissue factor, or tissue thromboplastin (factor III)	Most cells except platelets	Damage to tissue	NA	Starts extrinsic pathway	NA
Collagen	Subendothelial extracellular matrix	Injury exposes collagen to plasma clotting factors and platelets	Binds platelets to begin platelet plug	Starts intrinsic pathway	NA
Kininogen and kallikrein	Liver and plasma	Cofactors normally present in plasma	NA	Cofactors for contact activation of intrinsic pathway	Mediate inflammatory response; enhance fibrinolysis
Prothrombin and thrombin (factor II)	Liver and plasma	Platelet lipids, Ca^{2+}, and factor V	Enhances platelet plug formation	Fibrin production	NA
Fibrinogen and fibrin (factor I)	Liver and plasma	Thrombin	NA	Forms insoluble fibers that stabilize platelet plug	NA
Vitamin K	Diet	NA	NA	Needed for synthesis of factors II, VII, IX, X	NA
Fibrin-stabilizing factor (XIII)	Liver, megakaryocytes	Platelets	NA	Cross-links fibrin polymers to make stable mesh	NA
Plasminogen and plasmin	Liver and plasma	t-PA and thrombin	NA	Dissolves fibrin and fibrinogen	NA
Tissue plasminogen activator (t-PA)	Many tissues	Normally present; levels increase with stress, protein C	NA	Activates plasminogen	Recombinant t-PA used clinically to dissolve clots
Antithrombin III	Liver and plasma	NA	NA	Anticoagulant; blocks factors IX, X, XI, XII, thrombin, kallikrein	Facilitated by heparin; no effect on thrombin despite name
Prostacyclin (prostaglandin I, or PGI_2)	Endothelial cells	NA	Blocks platelet aggregation	NA	Vasodilator

Hemostasis involves many chemical factors, some of which play multiple roles and have multiple names (Table 16-4). Thus, learning about hemostasis may be especially challenging. For example, some factors participate in both platelet plug formation and coagulation. Another factor activates enzymes for both clot formation and clot dissolution. The fact that a single factor may have many different names can also create confusion. Scientists have assigned numbers to the coagulation factors to help reduce the confusion. However, the factors are not numbered according to the order in which they participate in the coagulation cascade; instead, they are numbered according to the order in which they were discovered. Because of the complexity of the coagulation cascade, we will discuss only a few aspects of hemostasis in additional detail.

Platelet Aggregation Begins the Clotting Process

Platelet activity is regulated by chemical factors released by damaged endothelial cells and by exposed collagen in the blood vessel wall (Fig. 16-11 ■). Normally, collagenous matrix fibers are separated from the circulating blood by the endothelial lining of the blood vessel. But if the vessel is damaged, collagen is exposed, and platelets rapidly begin to adhere to it.

If platelet aggregation is a positive feedback event, what prevents the platelet plug from continuing to form and, thus, spreading beyond the site of injury to other areas of the blood vessel wall? The answer lies in the fact that platelets will not adhere to normal endothelium. Intact vascular endothelial cells convert their membrane lipids into **prostacyclin,** an eicosanoid (∞ p. 29) that blocks platelet adhesion and aggregation. Nitric oxide

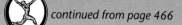

continued from page 466

"I can live a normal life, even with this disease?" asks Emily. "Yes," her doctor replies. "This coagulation disorder is different from hemophilia, in which a clotting factor is defective. Your clotting factor is normal." People with hemophilia experience spontaneous bleeding into joints or tissues. Death can occur if the bleeding is serious. Interestingly, people with hemophilia have normal amounts of von Willebrand factor.

Question 4: How can hemophilia be treated? (Hint: What is the defect in this disease?)

(NO) released by normal, intact endothelium also inhibits platelets from adhering. The combination of platelet attraction to the injury site and repulsion from the normal vessel wall creates a localized response that limits the platelet plug to the area of damage.

Coagulation Converts the Platelet Plug into a More Stable Clot

Coagulation is a complex process in which fluid blood forms a gelatinous clot. The coagulation cascade is a series of enzymatic reactions that end with formation of a fibrin mesh that stabilizes the platelet plug. Coagulation is divided into two pathways. An **intrinsic pathway** begins with collagen exposure and uses proteins already present in the plasma. An **extrinsic pathway** starts when damaged tissues expose **tissue factor,** a protein-phospholipid mixture, to the plasma proteins (Fig. 16-12 ■). The two pathways unite at the common path-

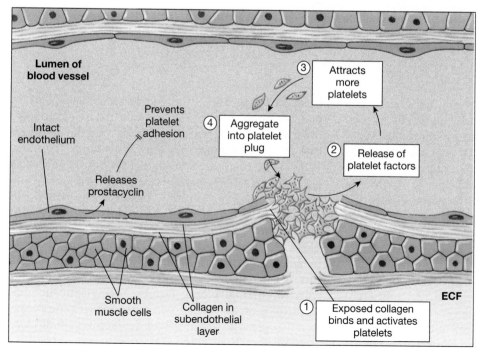

■ Figure 16-11 **Platelet adhesion and aggregation**

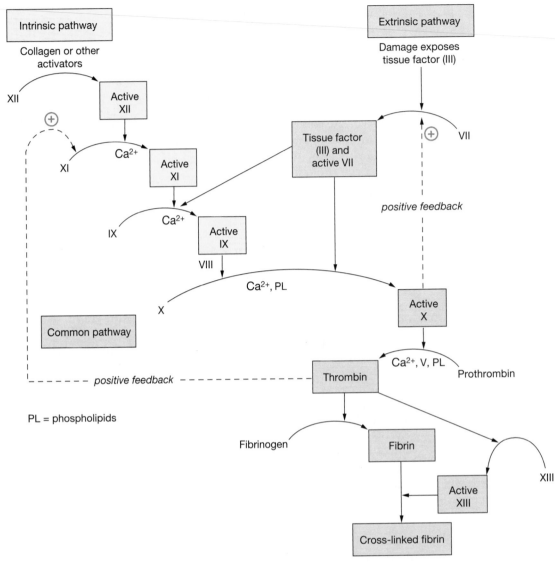

■ **Figure 16-12 The coagulation cascade** Inactive plasma proteins are converted into active enzymes in each step of the pathway. In the final step, the enzyme thrombin converts fibrinogen to fibrin, stabilizing the platelet plug. Two steps act as a positive feedback loop to sustain the cascade until one or more of the participating proteins is completely consumed. The activation of intrinsic pathway factor IX by extrinsic pathway factors is a recent finding.

way to initiate the formation of thrombin, the enzyme that converts fibrinogen into insoluble fibrin. Coagulation has traditionally been regarded as a cascade similar to the second messenger cascades diagrammed in Chapter 6 (p. 152), but recent studies have shown that the intrinsic and extrinsic pathways interact with each other, making the process more of a network than a simple cascade.

Anticoagulants Prevent Coagulation

Once coagulation begins, what keeps it from continuing until the entire circulation has clotted? Two different mechanisms limit the extent of blood clotting within a vessel: (1) inhibition of fibrin production and (2) inhibition of platelet adhesion. Protein factors present in the

blood vessel endothelium and plasma ensure that the platelet plug is restricted to the area of damage and that fibrin production does not spread beyond the injury site (see Table 16-4 for names of specific factors).

Chemicals that prevent coagulation from taking place are known as **anticoagulants.** Most act by blocking one or more of the steps in the cascade that forms fibrin. Anticoagulant drugs may be prescribed for people who are in danger of forming small blood clots that could block off critical vessels in the brain, heart, or lungs. Some of these drugs inhibit the synthesis of clotting factors; others enhance the anticoagulant activity of naturally occurring blood factors. Acetylsalicylic acid (aspirin) is an agent that prevents platelet plug formation. People who are at risk of developing small blood clots are sometimes told to take one aspirin every other day "to thin the blood." The aspirin does not actually dilute the blood, but it does prevent clots from forming by blocking platelet aggregation.

Several inherited diseases affect the coagulation process. The best-known is **hemophilia,** a name given to several diseases in which one of the factors in the coagulation cascade is defective or lacking. Hemophilia A, factor VIII deficiency, is the most common form, occurring in about 80% of all cases. Patients with coagulation disorders bruise easily and experience disabling bleeding into their joints and muscles. Spontaneous bleeding may occur throughout the body, and, if it occurs in the brain, it can be fatal.

In this chapter you have learned about the cellular elements of the blood, where the cellular elements are produced, and about hemostasis, the process that enables the body to conserve blood when a vessel is injured. In the next chapter, you will learn more about the critical role of the blood in transporting oxygen from the lungs to the cells.

CHAPTER REVIEW

Chapter Summary

1. Blood is the circulating portion of the extracellular fluid. (p. 453)

Plasma and the Cellular Elements of Blood

2. **Plasma** is the fluid portion of the blood. It is composed mostly of water and dissolved proteins. The remainder of the plasma is organic molecules, ions, and dissolved gases. (p. 454)

3. The **plasma proteins** consist of **albumins, globulins,** and the clotting protein **fibrinogen.** Plasma proteins function in blood clotting, defense, and as hormones, enzymes, or carriers for different substances. (p. 454)

4. The three cellular elements of blood are the red blood cells (**erythrocytes**), white blood cells (**leukocytes**), and **platelets.** Platelets are fragments of cells called **megakaryocytes.** (p. 454)

5. Blood contains five types of mature white blood cells: (1) **lymphocytes,** (2) **monocytes,** (3) **neutrophils,** (4) **eosinophils,** and (5) **basophils.** (p. 454)

Blood Cell Production

6. All blood cells are descendants of a single precursor cell type, the multipotent progenitor. Lymphocytes develop from one cell line, and the other blood cells develop from a different type of stem cell. (p. 456)

7. **Hematopoiesis** begins early in embryonic development and continues throughout a person's life. Most hematopoiesis takes place in the **bone marrow.** In adults, only the spine, ribs, pelvis, and proximal ends of long bones produce blood cells. (p. 456)

8. The reproduction and development of blood cells are controlled by cytokines. **Colony-stimulating factors** act with other cytokines to control white blood cell production. **Thrombopoietin** is the major factor that regulates the growth and maturation of megakaryocytes. Red blood cell production is regulated primarily by a hormone called **erythropoietin,** assisted by several cytokines. (p. 457)

Red Blood Cells

9. Red blood cells are the most abundant cell type in the blood. The primary function of red blood cells is to transport oxygen from the lungs to the cells and carbon dioxide from the cells to the lungs. (p. 460)

10. Mature mammalian red blood cells are biconcave disks that lack a nucleus and membranous organelles. Their cytoplasm is filled with enzymes and hemoglobin, a red oxygen-carrying pigment. (p. 460)

11. A hemoglobin molecule consists of four **globin** chains, each centered around a **heme** group. Most adult hemoglobin has two alpha and two beta chains. Fetal hemoglobin has two alpha and two gamma chains. (p. 462)

12. Hemoglobin synthesis requires iron in the diet. Iron is transported in the blood on transferrin and stored mostly in the liver, on **ferritin.** (p. 463)

13. When older red blood cells rupture, some heme from hemoglobin is converted into **bilirubin.** Bilirubin is incorporated into **bile** by the liver and excreted in the feces. Elevated bilirubin concentrations in the blood cause **jaundice.** (p. 463)

Platelets and Coagulation

14. **Platelets** are cell fragments that have no nucleus. Their cytoplasm contains mitochondria, smooth endoplasmic reticulum, and many granules filled with clotting proteins and cytokines. Platelets are present in the blood at all times, but they are not active unless damage has occurred to the walls of the circulatory system. (p. 465)

15. **Hemostasis** is the process of keeping blood within the blood vessels by repairing breaks without compromising the fluidity of the blood. Hemostasis begins with formation of a **platelet plug.** (p. 465)

16. When a blood vessel is damaged, exposed collagen triggers **platelet adhesion** and **platelet aggregation.** The platelet plug is converted into a clot when reinforced by protein fibrin. (p. 466)

17. **Fibrin** is made from **fibrinogen** through the action of **thrombin.** Fibrin formation is the last step of the **coagulation cascade.** (p. 467)

18. As cell growth and division repair the damaged vessel, **plasmin** trapped within the platelet plug dissolves fibrin and breaks down the clot. This process is called **fibrinolysis.** (p. 467)

19. Platelet plugs are restricted to the site of injury by **prostacyclin** found in the membrane of intact vascular endothelial cells. **Anticoagulants** limit the extent of blood clotting within a vessel. (p. 469)

Questions

LEVEL ONE Reviewing Facts and Terms

1. The fluid portion of the blood, called _____, is composed mainly of _____.

2. List the three types of plasma proteins in order of prevalence. Name at least one function of each type.

3. List the cellular elements found in blood and name at least one function of each one.

4. Blood cell production is called _____. When and where does it occur?

5. What role do colony-stimulating factors, cytokines, and interleukins play in blood cell production? How are these chemical signal molecules different? Give two examples of each.

6. List the technical terms for production of red blood cells, platelets, and white blood cells respectively.

7. The hormone that directs red blood cell synthesis is called _____. Where is it produced and what is the stimulus for its production?

8. How are the terms *hematocrit* and *packed red cells* related? What are normal hematocrit values for men and women?

9. Distinguish between an erythroblast and an erythrocyte. Give three distinct characteristics of erythrocytes.

10. What chemical element in the diet is important for hemoglobin synthesis?

11. Define the following terms and explain their significance in hematology.
    a. jaundice
    b. anemia
    c. transferrin
    d. hemophilia

12. Distinguish between these terms and tell how they are related to one another: fibrin, thrombin, plasmin, infarct, fibrinolysis, fibrinogen.

13. Chemicals that prevent blood clotting from occurring are called _____.

LEVEL TWO Reviewing Concepts

14. **Concept map:** Begin a map by placing each of the components of the blood across the top of a page. Complete the map by adding their functions and interactions.

15. Calculate the total blood volume in a 200-lb man and in a 130-lb woman (2.2 kg/lb).

16. Define hemostasis. Diagram the steps that result in its occurrence.

17. Diagram the clotting cascade, using the names and numbers of the factors.

18. Once platelets are activated to aggregate, what factors halt their activity?

19. What are the differences between the intrinsic pathway and the extrinsic pathway of coagulation?

LEVEL THREE Problem Solving

20. Rachel is undergoing chemotherapy for breast cancer. She has blood cell counts at regular intervals; below is the computer printout of her results:

	Normal Count	10 Days Post-chemotherapy	20 Days Post-chemotherapy
WBC ($\times 10^3$)	4.6–10.2	2.6	4.9
RBC ($\times 10^6$)	4.04–5.48	3.85	4.2
Platelets	142–424	133	151

At the time of the first test, Jen, the nurse, notes that Rachel, although pale and complaining of being tired, does not have any bruises on her skin. Jen tells Rachel to eat foods high in protein, take a multivitamin tablet with iron, and stay home and away from crowds as much as possible. How are Jan's observations and recommendations related to the results of the first and second blood tests?

Problem Conclusion

In this running problem, you learned about a coagulation disorder called von Willebrand's disease. You also learned how this disease is treated and how it differs from hemophilia.

Further check your understanding of this running problem by comparing your answers to those in the summary table.

	Question	Facts	Integration and Analysis
1	Why does von Willebrand's disease cause prolonged bleeding after a dental or surgical procedure?	Von Willebrand's disease is caused by a deficiency in von Willebrand factor, a protein found in the blood that promotes adherence of platelets to the endothelium and to each other during coagulation.	Because of a deficiency in von Willebrand factor, platelets do not stick together during coagulation, and blood clots do not easily form.
2	Why is type I the least serious form of von Willebrand's disease?	Type I is characterized by a deficiency in all multimers of von Willebrand factor at all sites.	Because at least some von Willebrand factor is present in all sites, some clotting does take place. Therefore, this form is the least serious.
3	Explain the rationale for giving an antifibrinolytic agent to individuals with von Willebrand's disease.	Antifibrinolytic agents inhibit the action of plasmin in a blood clot.	Plasmin dissolves blood clots in the latter stages of blood clotting. Inhibiting plasmin blocks this dissolution and allows clot formation to proceed.
4	How can hemophilia be treated? (Hint: What is the defect in this disease?)	Hemophilia is the name given to coagulation disorders characterized by deficient clotting factors. Classic hemophilia is caused by a defect in factor VIII.	Individuals with classic hemophilia are treated with infusions of factor VIII.

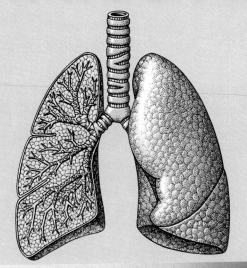

17

Respiratory Physiology

Imagine covering the playing surface of a racquetball court (about 75 m²) with thin plastic wrap, then crumpling up the wrap and stuffing it into a 3-liter soft drink bottle. Impossible? Maybe so, if you use plastic wrap and a drink bottle. But the lungs of a 70-kg man have a gas exchange surface the size of that plastic wrap, compressed into a volume that is less than the size of the bottle. This tremendous area for gas exchange is needed to supply the trillions of cells in the body with adequate amounts of oxygen.

Aerobic metabolism in animal cells depends on a steady supply of oxygen and nutrients from the environment, coupled with the removal of carbon dioxide and other wastes. In small, simple aquatic animals, these needs can be met by simple diffusion across the body surface. The rate of diffusion is limited by distance, however, and as a result, most multicelled animals use specialized respiratory organs associated with a circulatory system. Respiratory organs take a variety of forms, but all possess a large surface area compressed into a small space.

Besides needing a large exchange surface, humans and other terrestrial animals face an additional physiological challenge, that of dehydration. Their exchange surfaces must be thin and moist to allow gases to pass from air into solution, yet at the same time they must be protected to prevent drying out from exposure to air. Some terrestrial animals such as the slug, a shell-less snail, meet the challenge of dehydration with behavioral adaptations that restrict them to humid environments and nighttime activities. However, a more common solution is anatomical: an internalized respiratory epithelium. Human lungs are enclosed within the chest cavity to control their contact with the outside air. Internalization creates a humid environment for the exchange of gases with the blood and protects the exchange surface from damage.

Internalized lungs create another problem, however: how to exchange air between the atmosphere and the exchange surface deep within the body. Air flow requires a muscular pump to create pressure gradients. Thus, in more complex animals, the respiratory system consists of two separate components: a muscular pump and a thin, moist exchange surface. In humans, the pump is the musculoskeletal structure of the thorax, and the lungs themselves consist of the exchange epithelium and associated blood vessels.

The primary functions of the respiratory system are:

1. *Exchange of gases between the atmosphere and the blood.* Oxygen is brought into the body for distribution to the tissues, and carbon dioxide wastes produced by metabolism are eliminated.
2. *Homeostatic regulation of body pH.*
3. *Protection from inhaled pathogens and irritating substances.* Like all epithelia that contact the external environment, the epithelium found in the respiratory system is well supplied with mechanisms to trap and destroy potentially harmful substances before they can enter the body.
4. *Vocalization.* Air moving across the vocal cords creates vibrations used for speech, singing, and other forms of communication.

In addition to serving these functions, the respiratory system is also a source of significant losses of water and heat from the body. These losses must be balanced using homeostatic compensations.

This chapter examines how the respiratory system carries out these functions by exchanging air between the environment and the interior of the lungs. In addition, it examines how the circulatory and respiratory systems cooperate to move oxygen and carbon dioxide between the lungs and the cells of the body.

THE RESPIRATORY SYSTEM

The word *respiration* can have several meanings (Fig. 17-1 ■). **Cellular respiration** refers to the intracellular reaction of oxygen with organic molecules to produce carbon dioxide, water, and energy in the form of ATP (∞ p. 91). **External respiration,** the topic of this chapter, is the interchange of gases between the environment

Problem

Emphysema

"Diagnosis: COPD (blue bloater)," reads Edna Wilson's patient chart. COPD—chronic obstructive pulmonary disease—is a name given to diseases in which air exchange is impaired by narrowing of the airways. Most people with COPD have emphysema or chronic bronchitis or a combination of the two. Individuals in whom chronic bronchitis predominates are nicknamed "blue bloaters," owing to the bluish tinge of their skin and a tendency to be overweight. "Pink puffers" suffer more from emphysema. They tend to be thin, have normal (pink) skin coloration, and breathe shallow, rapid breaths. Because COPD is commonly caused by smoking, most people can avoid the disease simply by not smoking. Unfortunately, Edna has been a hard-core smoker for 35 of her 47 years.

continued on page 477

and the body's cells. External respiration can be subdivided into four integrated processes:

1. The exchange of air between the atmosphere and the lungs. This process is known as **ventilation,** or breathing. **Inspiration** is the movement of air into the lungs; **expiration** is the movement of air out of the lungs.
2. The exchange of oxygen and carbon dioxide between the lungs and the blood.
3. The transport of oxygen and carbon dioxide by the blood.
4. The exchange of the gases between blood and the cells.

External respiration requires the coordinated functioning of the respiratory and cardiovascular systems. The **respiratory system** is composed of the structures involved in ventilation and gas exchange (Fig. 17-2 ■). It consists of:

1. The **conducting system** of passages, or **airways,** that lead from the environment to the exchange surface of the **lungs.**
2. The **alveoli,** the exchange surface of the lungs, where oxygen and carbon dioxide transfer between air and the blood.
3. The bones and muscles of the **thorax** (chest) that assist in ventilation.

The respiratory system can be divided into two parts. The **upper respiratory tract** consists of the mouth, nasal cavity, pharynx, larynx, and trachea. The **lower respiratory tract** consists of the two bronchi, their branches, and the lungs. It is also known as the thoracic portion of the respiratory system because it is enclosed in the thorax.

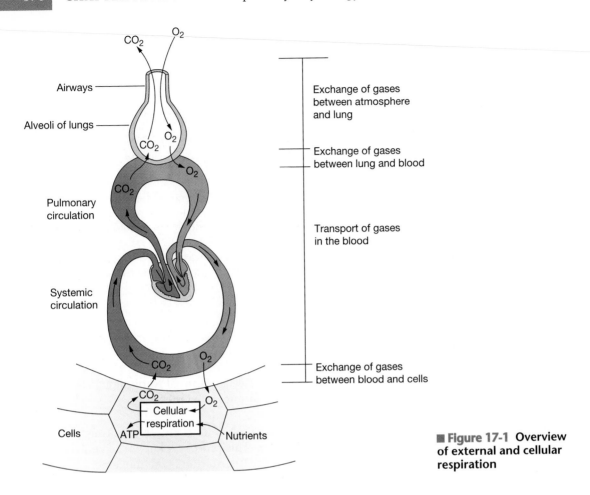

■ **Figure 17-1 Overview of external and cellular respiration**

The Bones and Muscles of the Thorax Surround the Lungs

The thorax, or chest cavity, is bounded by the bones of the spine and rib cage and their associated muscles. Together the bones and muscles are called the *thoracic cage*. The ribs and spine form the sides and top of the cage (the *chest wall*), and a dome-shaped sheet of skeletal muscle, the **diaphragm**, forms the floor (Fig. 17-2a ■). Two sets of **intercostal muscles,** internal and external, connect the 12 pairs of ribs. Additional muscles, the **sternocleidomastoids** and the **scalenes,** run from the head and neck to the sternum and first two ribs.

Functionally, the thorax is a sealed container filled with three membranous bags, or sacs. One, the pericardial sac, contains the heart. The other two bags, the *pleural sacs*, contain the lungs [*pleura*, rib, or side]. The esophagus and the thoracic blood vessels and nerves pass between the pleural sacs (Fig. 17-2d ■).

The Lungs Are Enclosed in the Pleural Sacs

The lungs (Fig. 17-2b ■) consist of light, spongy tissue whose volume is mostly occupied by air-filled spaces. These irregular cone-shaped organs nearly fill the thoracic cavity, with their bases resting on the curved

diaphragm. Rigid conducting airways, the bronchi, connect the lungs to the main airway, the trachea.

Each lung is contained within a double-walled pleural sac whose membranes line the inside of the thorax and cover the outer surface of the lungs (Fig. 17-3 ■, p. 480). The pleural membranes, or **pleura,** contain several layers of elastic connective tissue and numerous capillaries. The opposing layers of pleural membrane are held together by a thin film of **pleural fluid** whose total volume is only a few milliliters. The result is similar to an air-filled balloon (the lung) surrounded by a water-filled balloon (the pleural sac). Most illustrations exaggerate the volume of the pleural fluid, but you can appreciate its thinness if you imagine spreading 3 mL of water evenly over the surface of a 3-liter soft drink bottle.

Pleural fluid serves several purposes. First, it creates a moist, slippery surface so that the opposing membranes can slide across one another as the lungs move within the thorax. The second important function of pleural fluid is to hold the lungs tight against the thoracic wall. To visualize this arrangement, think of two panes of glass stuck together by a thin film of water. You can slide the panes back and forth across each other, but you cannot pull them apart because of the cohesiveness of the water (∞ p. 22). A similar fluid bond between the pleural membranes makes the lungs "stick" to the thoracic cage and holds them stretched in a partially inflated state, even at rest.

The Airways of the Conducting System Connect the Lungs with the Environment

Air enters the upper respiratory tract through the mouth and nose and passes into the **pharynx,** a common passageway for food, liquids, and air [*pharynx,* throat]. From the pharynx, air flows through the **larynx** into the **trachea,** or windpipe (Fig. 17-2b ■). The larynx contains the **vocal cords,** connective tissue bands that tighten to create sound when air moves past them.

The trachea is a semiflexible tube held open by 15 to 20 C-shaped cartilage rings (Fig. 17-2e ■). It extends down into the thorax, where it branches (division 1) into a pair of **primary bronchi,** one *bronchus* to each lung. Within the lungs, the bronchi branch repeatedly (divisions 2–11) into progressively smaller bronchi (Fig. 17-2b, e ■). Like the trachea, the bronchi are semirigid tubes supported by cartilage.

Within the lungs, the smallest bronchi branch to become **bronchioles,** small collapsible passageways with smooth muscle walls. The bronchioles continue branching (divisions 12–23) until the *terminal bronchioles* end at the exchange epithelium of the lung.

The diameter of the airways becomes progressively smaller from the trachea to the bronchioles, but as the individual airways get narrower, their numbers increase (Fig. 17-4 ■, p. 480). As a result, the total cross-sectional area increases with each division of the airways. Total cross-sectional area is lowest in the upper respiratory tract and greatest in the bronchioles, analogous to the increase in cross-sectional area that occurs from the aorta to the capillaries. Velocity of air flow is therefore highest in the trachea and lowest in the terminal bronchioles (∞ p. 390).

The Alveoli Are the Site of Gas Exchange

The bulk of lung tissue consists of exchange sacs known as alveoli [*alveus,* a concave vessel, singular *alveolus*]. The alveoli are grapelike clusters at the ends of the terminal bronchioles (Fig. 17-2f, g ■). Their primary function is the exchange of gases between air in the alveoli and the blood.

Each tiny alveolus is composed of a single layer of thin exchange epithelium (Fig. 17-2g ■). Two types of epithelial cells are found in the alveoli, and they occur in roughly equal numbers. The larger **type I alveolar cells** are very thin so that gases can diffuse rapidly through them. The smaller but thicker **type II alveolar cells** synthesize and secrete a chemical known as surfactant. Surfactant mixes with the thin fluid lining of the alveoli to ease the expansion of the lungs during breathing.

The thin walls of the alveoli do not contain muscle, because muscle fibers would block rapid gas exchange.

As a result, lung tissue itself cannot contract. However, connective tissue between the alveolar epithelial cells does contain many elastin fibers that contribute to elastic recoil when lung tissue is stretched.

The intimate link between the respiratory and cardiovascular systems is demonstrated by the close association of the alveoli with an extensive network of capillaries. These blood vessels cover 80%–90% of the alveolar surface, forming an almost continuous "sheet" of blood in close contact with the air-filled alveoli (Fig. 17-2f ■). Gas exchange in the lungs occurs by diffusion through the thin alveolar type I cells to the capillaries. In much of the exchange area, the basement membrane underlying the alveolar epithelium has fused with that of the capillary endothelium, and only a small amount of interstitial fluid is present. The proximity of capillary blood to air in the alveolus is essential for the rapid exchange of gases.

The Pulmonary Circulation Is a High-Flow, Low-Pressure System

The pulmonary circulation begins with the pulmonary trunk that receives low-oxygen blood from the right ventricle, then divides into two pulmonary arteries, one to each lung (∞ Fig. 14-1, p. 386). Oxygenated blood from the lungs returns to the left atrium via the pulmonary veins.

At any given moment, the pulmonary circulation contains about 0.5 liters of blood, or 10% of the total blood volume. About 75 mL of this amount is found in the capillaries, where gas exchange takes place, with the remainder in the pulmonary arteries and veins. The rate of blood flow through the lungs is quite high when compared with other tissues (∞ p. 436) because the lungs receive the entire cardiac output of the right ventricle, 5 L/min. This means that as much blood flows through the lungs in one minute as flows through the rest of the body in the same amount of time!

Despite the high flow rate, blood pressure in the pulmonary circulation is low. Pulmonary arterial blood pressure averages 25/8 mm Hg, compared with the average systemic arterial blood pressure of 120/80 mm Hg. The right ventricle does not have to pump as forcefully to create blood flow through the lungs because the resistance of the pulmonary circulation is low. This low

continued from page 475

Patients with chronic bronchitis have excessive mucus production and general inflammation of the entire respiratory tract. The mucus narrows the airways and makes breathing difficult.

Question 1: What does narrowing of the airways do to the resistance of the airways to air flow? (Hint: The relationship between radius and resistance is the same for air flow as it was for blood flow in the circulatory system; ∞ p. 387.)

Anatomy Summary **Respiratory System**

■ **Figure 17-2**

(a) Muscles used for ventilation
The muscles of inspiration include the diaphragm, external intercostals, sternocleidomastoids, and scalenes. The muscles of expiration include the internal intercostals and the abdominals.

(b) The respiratory system
The respiratory system consists of the upper respiratory system (mouth, nasal cavity, pharynx, larynx) and the lower respiratory system (trachea, bronchi, lungs). The lower respiratory system is enclosed in the thorax, bounded by the ribs, spine, and diaphragm.

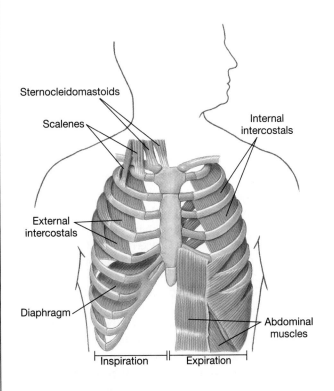

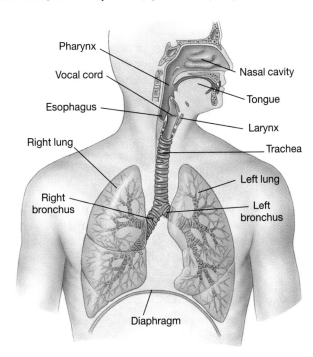

(c) External anatomy of lungs
Externally, the right lung is divided into three lobes and the left lung into two.

(d) Sectional view of chest
Each lung is enclosed in two pleural membranes. The pleural fluid and space is much smaller than illustrated. The esophagus and aorta pass through the thorax between the pleural sacs.

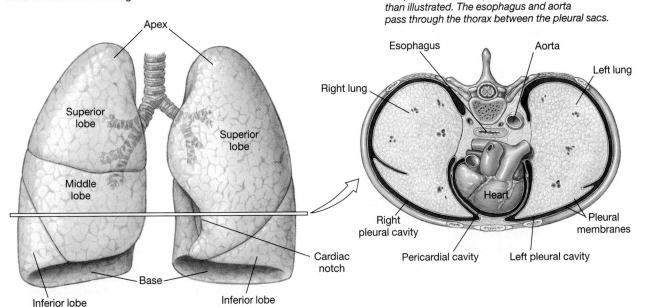

(e) Branching of airways
The trachea branches into two bronchi, one to each lung. Each bronchus branches 22 more times, finally terminating in a cluster of alveoli.

(f) Structure of lung lobule
Each cluster of alveoli is surrounded by elastic fibers and a network of capillaries.

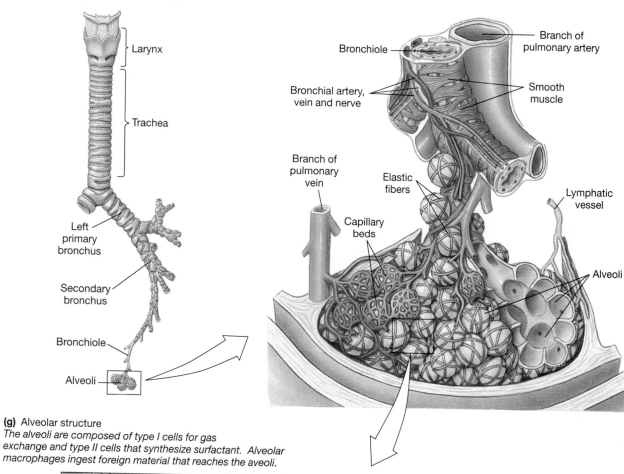

- Larynx
- Trachea
- Left primary bronchus
- Secondary bronchus
- Bronchiole
- Alveoli

- Bronchiole
- Bronchial artery, vein and nerve
- Branch of pulmonary vein
- Capillary beds
- Elastic fibers
- Branch of pulmonary artery
- Smooth muscle
- Lymphatic vessel
- Alveoli

(g) Alveolar structure
The alveoli are composed of type I cells for gas exchange and type II cells that synthesize surfactant. Alveolar macrophages ingest foreign material that reaches the aveoli.

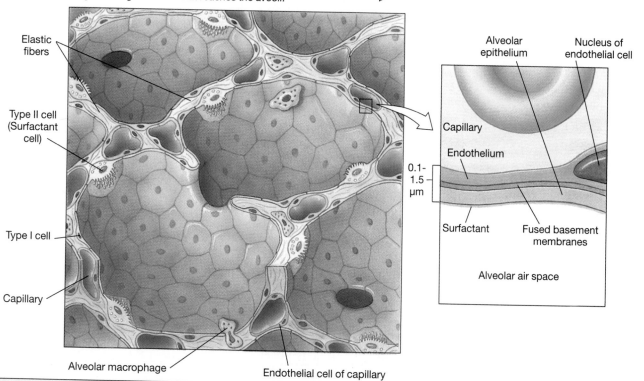

- Elastic fibers
- Type II cell (Surfactant cell)
- Type I cell
- Capillary
- Alveolar macrophage
- Endothelial cell of capillary

- Alveolar epithelium
- Nucleus of endothelial cell
- Capillary
- Endothelium
- 0.1-1.5 µm
- Surfactant
- Fused basement membranes
- Alveolar air space

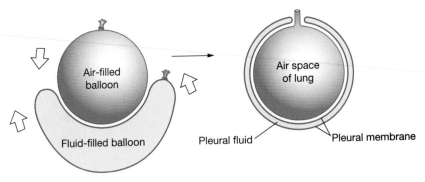

■ **Figure 17-3** **The relationship between the pleural sac and the lung** The pleural sac forms a double membrane surrounding the lung, similar to a fluid-filled balloon surrounding an air-filled balloon. The pleural fluid has a much smaller volume than is suggested by this illustration.

resistance can be attributed to the shorter total length of the pulmonary blood vessels and the distensibility and large cross-sectional area of pulmonary arterioles.

Because mean blood pressure is low in the pulmonary capillaries, the net hydraulic pressure forcing fluid out of the capillary into the interstitial space is also low (∞ p. 439). Filtered fluid from the pulmonary capillaries is efficiently removed by the lymphatic system, and only a small amount of fluid is found in the interstitial space of the lung. As a result, the distance between alveolus and capillary is small, and gases diffuse rapidly between them.

In the next section, we begin our discussion of respiratory physiology with a review of the laws of physics and chemistry that govern the behavior of gases.

✓ A person has left ventricular failure but normal right ventricular function. As a result, blood pools in the pulmonary circulation, doubling pulmonary capillary hydraulic pressure. What happens to fluid flow across the walls of the pulmonary capillaries?

GAS LAWS

Air flow in the respiratory system is very similar to blood flow in the cardiovascular system in many respects, although blood is a noncompressible liquid and air is a compressible mixture of gases. Blood pressure and environmental air pressure (**atmospheric pressure**) are both reported in millimeters of mercury (mm Hg).* At sea level, atmospheric pressure is 760 mm Hg.

One convention that we will follow in this book is the designation of atmospheric pressure as 0 mm Hg. Because atmospheric pressure varies with altitude and because very few people live exactly at sea level, this convention allows us to compare pressure differences that occur during ventilation without correcting for altitude. Subatmospheric pressures are designated by negative numbers, and higher-than-atmospheric pressures are written as positive numbers.

---

*Respiratory physiologists sometimes report gas pressures in units of centimeters of water: 1 cm H_2O = 1.3 mm Hg.

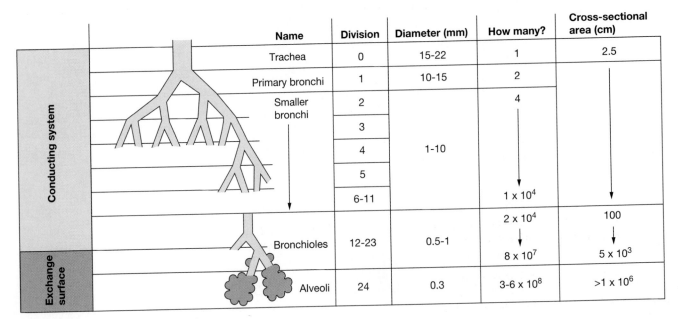

	Name	Division	Diameter (mm)	How many?	Cross-sectional area (cm)
	Trachea	0	15-22	1	2.5
	Primary bronchi	1	10-15	2	
	Smaller bronchi	2		4	
		3			
		4	1-10		
		5			
		6-11		1 x 10⁴	
	Bronchioles	12-23	0.5-1	2 x 10⁴ ↓ 8 x 10⁷	100 ↓ 5 x 10³
	Alveoli	24	0.3	3-6 x 10⁸	>1 x 10⁶

Conducting system

Exchange surface

■ **Figure 17-4** **Branching of the airways**

TABLE 17-1 Gas Laws

1. The total pressure of a mixture of gases is the sum of the pressures of the individual gases (Dalton's Law).
2. Gases, singly or in a mixture, move from areas of higher pressure to areas of lower pressure.
3. If the volume of a container of gas changes, the pressure of the gas will change in an inverse manner (Boyle's Law).
4. The amount of a gas that will dissolve in a liquid is determined by the partial pressure of the gas and the gas's solubility in the liquid.

The rules that govern the behavior of gases in air and in solution are summarized in Table 17-1. These rules provide the basis for the exchange of oxygen and carbon dioxide between the environment and cells. They are discussed briefly in the paragraphs below.

Air Is a Mixture of Gases

The atmosphere surrounding the earth is a mixture of gases and water vapor. **Dalton's Law** states that the total pressure of a mixture of gases is the sum of the pressures of the individual gases. Thus, in dry air at an atmospheric pressure of 760 mm Hg, 78% of the total pressure is due to nitrogen molecules, 21% to oxygen molecules, and so on (Table 17-2).

In respiratory physiology, we are concerned not only with total atmospheric pressure but also with the individual pressures of oxygen and carbon dioxide. The pressure of a single gas species is known as its **partial pressure,** abbreviated P_{gas}. To find the partial pressure of a gas, multiply the atmospheric pressure (P_{atm}) by the gas's relative contribution (%):

Partial pressure of an atmospheric gas = P_{atm} × % of gas in atmosphere

Partial pressure of oxygen = 760 mm Hg × 21%

P_{O_2} = 760 × 0.21 = 160 mm Hg

Thus, the P_{O_2}, or partial pressure of oxygen, in dry air at sea level is 160 mm Hg. The pressure of an individual gas is determined only by its relative abundance in the mixture and is independent of the molecular size or weight of the gas.

The actual partial pressure of gases varies slightly depending on how much water vapor is in the air. Water pressure "dilutes" the contribution of other gases to the total pressure. Table 17-2 compares the partial pressures of some atmospheric gases in dry air and at 100% humidity.

Gases Move from Areas of Higher Pressure to Areas of Lower Pressure

Air flow occurs whenever there is a pressure gradient. Air flow, like blood flow, moves from areas of higher pressure to areas of lower pressure. Meteorologists predict the weather by knowing that areas of high atmospheric pressure move in to replace areas of low pressure. In ventilation, flow of air down pressure gradients explains why air exchanges between the environment and the lungs. The movement of the thorax during breathing creates alternating conditions of high and low pressure within the lungs.

Movement down pressure gradients also applies to a single gas. Oxygen moves from areas of higher oxygen partial pressure (P_{O_2}) to areas of lower oxygen partial pressure. The simple diffusion of oxygen and carbon dioxide between lung and blood, or between blood and cells, depends on pressure gradients for these gases. Respiratory physiologists talk about the partial pressure of oxygen or carbon dioxide in the body rather than the concentration of the gas in solution because this measure allows direct comparison with the partial pressures of the gases in air.

Pressure-Volume Relationships of Gases Are Described by Boyle's Law

The pressure exerted by a gas or mixture of gases in a sealed container is created by the collisions of the moving gas molecules with the walls of the container and with each other. If the size of the container is reduced, the collisions will become more frequent and the pressure will rise. This relationship can be expressed by the following equation:

$P_1V_1 = P_2V_2$ where P represents pressure and V represents volume

For example, start with a 1-liter container (V_1) of a gas whose pressure is 100 mm Hg (P_1). One side of the container moves in to decrease the volume to 0.5 L (Fig. 17-5 ■).

TABLE 17-2 Partial Pressures of Some Atmospheric Gases at 25°C and 760 mm Hg

Gas	Partial Pressure in Atmospheric Air, Dry	Partial Pressure in Atmospheric Air, 100% Humidity
Nitrogen (N_2)	593 mm Hg	575 mm Hg
Oxygen (O_2)	160 mm Hg	152 mm Hg
Carbon dioxide (CO_2)	0.25 mm Hg	0.24 mm Hg
Water vapor	0 mm Hg	23.8 mm Hg

Boyle's Law: $P_1V_1 = P_2V_2$

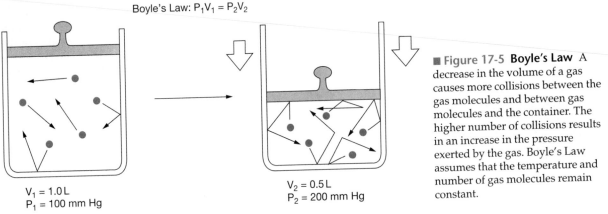

$V_1 = 1.0\,L$
$P_1 = 100\,mm\,Hg$

$V_2 = 0.5\,L$
$P_2 = 200\,mm\,Hg$

■ **Figure 17-5** **Boyle's Law** A decrease in the volume of a gas causes more collisions between the gas molecules and between gas molecules and the container. The higher number of collisions results in an increase in the pressure exerted by the gas. Boyle's Law assumes that the temperature and number of gas molecules remain constant.

What happens to the pressure of the gas? According to the equation,

$$P_1V_1 = P_2V_2$$

$$100\,mm\,Hg \times 1\,L = P_2 \times 0.5\,L$$

$$P_2 = 200\,mm\,Hg$$

If the volume is reduced by one-half, the pressure doubles. If the volume were to double, the pressure would be reduced by one-half. This relationship between pressure and volume was first noted by Robert Boyle in the 1600s and has been called **Boyle's Law** of gases.

In the respiratory system, changes in the volume of the chest cavity during ventilation cause pressure gradients that create air flow. When the chest increases in volume, the intrathoracic pressure drops and air flows into the respiratory system. When the chest decreases in volume, the pressure rises and air flows out into the atmosphere. This movement of air is called *bulk flow* because the entire gas mixture is moving rather than merely one or two gas species.

The Solubility of a Gas in a Liquid Depends on the Pressure and Solubility of the Gas and on the Temperature

When a gas is placed in contact with water, if there is a pressure gradient, the gas molecules move from one phase to the other. If the gas pressure is higher in the water than in the gaseous phase, gas molecules will leave the water. If the gas pressure is higher in the gaseous phase than in the water, the gas will dissolve into the water. The movement of gas molecules from air into solution is directly proportional to three factors: (1) the pressure gradient of the individual gas, (2) the solubility of the gas in the given liquid, and (3) temperature. Because temperature is relatively constant in mammals, we will ignore its contribution in this discussion.

The ease with which a gas dissolves in a solution is its **solubility.** If a gas is very soluble, large numbers of gas molecules will go into solution at low partial pressures. With less soluble gases, high partial pressures may cause only a few molecules of the gas to dissolve.

For example, imagine a closed container half-filled with water and half-filled with oxygen (Fig. 17-6 ■). Initially, the gas has a P_{O_2} of 100 mm Hg and the water has no oxygen dissolved in it ($P_{O_2} = 0$ mm Hg). As the gas phase stays in contact with the water, the moving oxygen molecules in the gas diffuse into the water and dissolve. This process will continue until equilibrium is reached. At equilibrium, the movement of oxygen into the water is equal to the movement of oxygen back into the air. We refer to the amount of oxygen that dissolves in the water at any given P_{O_2} as the partial pressure of the gas in solution. Thus, if the gaseous phase has a P_{O_2} of 100 mm Hg, at equilibrium the water will also have a P_{O_2} of 100 mm Hg.

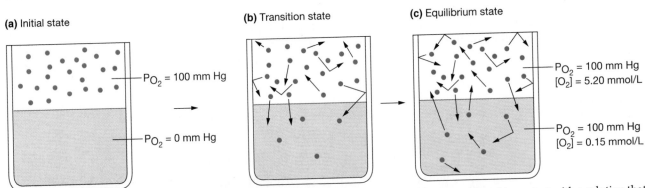

(a) Initial state   **(b)** Transition state   **(c)** Equilibrium state

$P_{O_2} = 100\,mm\,Hg$

$P_{O_2} = 0\,mm\,Hg$

$P_{O_2} = 100\,mm\,Hg$
$[O_2] = 5.20\,mmol/L$

$P_{O_2} = 100\,mm\,Hg$
$[O_2] = 0.15\,mmol/L$

■ **Figure 17-6** **Gases in solution** (a) Gas containing oxygen molecules ($P_{O_2} = 100$ mm Hg) is placed in contact with a solution that contains no oxygen molecules ($P_{O_2} = 0$ mm Hg). (b) Oxygen molecules move down their pressure gradient and dissolve in the solution. (c) At equilibrium, the P_{O_2} of the gas and solution are equal, and net movement of oxygen molecules between the gas and liquid phases stops. But the concentration of oxygen in the two phases is not equal. The amount of oxygen that dissolves depends on the solubility of oxygen in the solution as well as on the partial pressure of the gas.

Note that this does *not* mean that the concentration of oxygen is the same in the air and in the water. The concentration of dissolved oxygen also depends on the solubility of oxygen. For example, when the P_{O_2} of both air and water is 100 mm Hg, the concentration of oxygen in the air is 5.2 mmoles/L air, whereas the concentration of oxygen in the water is only 0.15 mmoles/L water. As you can see, oxygen is not very soluble in aqueous solutions; its insolubility is one reason for the evolution of oxygen-carrying molecules in blood.

Carbon dioxide is about 20 times as soluble in water as oxygen is. At a P_{CO_2} of 100 mm Hg, the CO_2 concentration in air is 5.2 mmoles/L of air and its concentration in solution is 3.0 mmoles/L of water.

✓ A saline solution is exposed to a gas mixture with equal partial pressures of O_2 and CO_2. What information do you need to know to predict if equal amounts of O_2 and CO_2 will dissolve in the saline?

✓ If nitrogen is 78% of atmospheric air, what is the partial pressure of nitrogen (P_{N_2}) when the dry atmospheric pressure is 720 mm Hg?

VENTILATION

The first exchange in respiratory physiology is ventilation, or breathing, the movement of air between the environment and the alveoli.

The Airways Warm, Humidify, and Filter Inspired Air

The upper airways and the bronchi do more than simply serve as passageways for air. They play an important role in conditioning air before it reaches the alveoli. Conditioning has three components:

1. Warming air to 37° C so that core body temperature will not change and alveoli will not be damaged by cold air,
2. Adding water vapor until the air reaches 100% humidity so that the moist exchange epithelium will not dry out,
3. Filtering out foreign material so that viruses, bacteria, and inorganic particles will not reach the alveoli.

Inhaled air is warmed and moistened by heat and water from the mucosal lining of the airways. Under normal circumstances, by the time air reaches the trachea, it has been conditioned to 37° C and 100% humidity. Breathing through the mouth is not nearly as effective at warming and moistening air as breathing through the nose. If you exercise outdoors in very cold weather, you may be familiar with the ache in your chest that results from breathing cold air through your mouth.

Filtration of air takes place in the trachea and bronchi as well. These airways are lined with a ciliated epithelium that secretes both mucus and a dilute saline solution

(Fig. 17-7 ▪). The cilia themselves are bathed in a watery saline layer covered by a sticky layer of mucus. The mucus is secreted by *goblet cells* in the epithelium (∞ p. 60). Mucus traps most inhaled particles larger than 2 μm, and its immunoglobulins disable many inhaled microorganisms. The mucus layer is continuously moved toward the pharynx by the upward beating of the cilia, a process called the *mucus escalator*. Once mucus reaches the pharynx, it is swallowed, so acid and enzymes in the stomach can destroy any remaining microorganisms.

Secretion of the watery layer beneath the mucus is a critical step in the mucus escalator. Without the watery layer, the cilia would become trapped in the thick, sticky mucus and cease to function. This is what happens in the inherited disease *cystic fibrosis*. A genetic error in one amino acid creates a defective Cl^- channel protein that is unable to secrete chloride ions, an essential step in saline secretion (∞ p. 113).

✓ Cigarette smoking paralyzes the cilia in the epithelial lining of the airways. Why would paralysis of the cilia cause smokers to develop a cough?

During Ventilation, Air Flows because of Pressure Gradients

Air flows into and out of the lungs because of pressure gradients created by a pump, just as blood flows in the cardiovascular system because of the pumping action of the heart. In the respiratory system, most lung tissue is thin exchange epithelium, so the muscles of the thoracic cage and diaphragm function as the pump. When these muscles contract, moving the rib cage and diaphragm, the lungs move also, held to the inside of the thoracic cage by the pleural fluid.

Breathing is an active process that uses muscle contraction to create pressure gradients. The primary muscles involved in quiet breathing are the diaphragm, the intercostals, and the scalenes. During forced breathing, other muscles of the chest and abdomen may be recruited to assist. Examples of situations in which breathing is forced include exercise, playing a wind instrument, and blowing up a balloon.

Air flow in the respiratory tract obeys the same rule as blood flow in the cardiovascular system:

Flow $\propto \Delta P/R$

continued from page 477

Smokers usually develop chronic bronchitis before they develop emphysema. Cigarette smoke paralyzes the cilia that sweep debris and mucus out of the airways. Without the action of cilia, mucus and debris pool in the airways, leading to a chronic cough. Eventually, breathing becomes difficult.

Question 2: Why do people with chronic bronchitis have a higher-than-normal rate of respiratory infections?

(a)

Cilia move mucus to pharynx

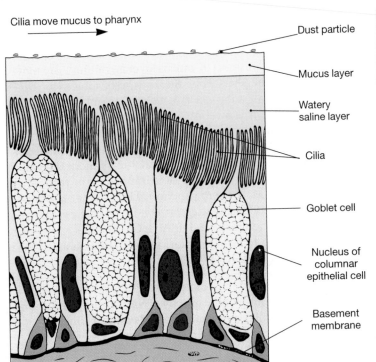

Dust particle

Mucus layer

Watery saline layer

Cilia

Goblet cell

Nucleus of columnar epithelial cell

Basement membrane

(b)

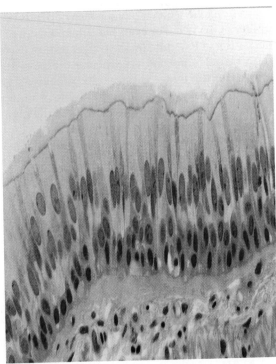

■ **Figure 17-7** **Ciliated respiratory epithelium** (a) Goblet cells in respiratory epithelium secrete a thick mucus layer that traps inhaled particles. The mucus floats on top of a saline solution that allows the cilia to push the mucus upward toward the pharynx, where it is swallowed. (b) This scanning electron micrograph shows the cilia of the trachea and some mucus drops.

This equation means that (1) air flows in response to a pressure gradient (ΔP) and that (2) flow decreases as the resistance (R) of the system to flow increases. Before we discuss resistance, let us consider how the respiratory system creates a pressure gradient.

Pressures in the respiratory system can be measured either in the air spaces of the lungs (**intrapulmonary pressure**) or within the pleural fluid (**intrapleural pressure**). Because atmospheric pressure is relatively constant, pressure in the lungs must be higher or lower than atmospheric pressure in order for air to flow between the environment and the alveoli. The respiratory system ends in a dead end, so the direction of air flow reverses. Air flows into the lungs when you inhale (inspiration) and out of the lungs when you exhale (expiration). A single **respiratory cycle** consists of an inspiration followed by an expiration. The pressure-volume relationships of Boyle's Law provide the basis for pulmonary ventilation.

Inspiration Occurs When Intrapulmonary Pressure Decreases

During inspiration, somatic motor neurons trigger contraction of the diaphragm and the inspiratory muscles. When the diaphragm contracts, it loses its dome shape and drops down toward the abdomen. In quiet breathing, the diaphragm drops about 1.5 cm. This movement increases the volume of the thoracic cavity by flattening its floor (Fig.

17-8 ■). Between 60% and 75% of the inspiratory volume change during normal quiet breathing is caused by contraction of the diaphragm. The remaining 25% to 40% of the volume change is due to movement of the rib cage. Contraction of the external intercostal and scalene muscles swings the ribs upward and out. The movement of the ribs during inspiration has been likened to a pump handle lifting up and away from the pump (the ribs moving up and away from the spine) and to the movement of a pair of bucket handles as they lift away from the sides of a bucket (ribs moving outward in a lateral direction). The combination of these two movements broadens the rib cage in all directions (Fig. 17-9 ■). As the volume of the thoracic cavity increases, pressure decreases and air flows into the lungs.

For many years, quiet breathing was attributed solely to the action of the diaphragm and the external intercostal muscles. It was thought that the scalenes and sternocleidomastoid muscles were active only during deep breathing. But in recent years, studies of patients with neuromuscular disorders have changed our understanding of how these accessory muscles contribute to quiet breathing. It now appears that without lifting of the sternum and upper ribs by the scalenes, contraction of the diaphragm during inspiration pulls the lower ribs inward. This action decreases the volume of the thoracic cage and works against inspiration. Because we know that the lower ribs move up and out in normal quiet breathing, the scalenes must assist quiet inspiration. New

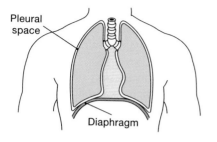

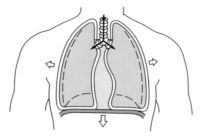

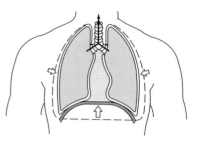

Pleural space

Diaphragm

(a) At rest, diaphragm is relaxed.

(b) Diaphragm contracts, thoracic volume increases.

(c) Diaphragm relaxes, thoracic volume decreases.

■ Figure 17-8 **Movement of the diaphragm**

evidence also downplays the role of the external inter- costal muscles during quiet breathing. However, the exter- nal intercostals play an increasingly important role as res- piratory activity increases. Because the exact contribution of the external intercostals and scalenes varies depending upon the type of breathing, we will group these muscles together and call them the *inspiratory muscles*.

Now let us follow intrapulmonary pressure as it changes during a single inspiration. At the start of an inspiration, the brief pause between breaths, intrapul-

monary pressure is equal to atmospheric pressure and there is no air flow (Fig. 17-10 ■, point A_1). As inspira- tion begins, the muscles of the thoracic cage contract and the volume of the thorax increases. With the increase in volume, intrapulmonary pressure falls about 1 mm Hg below atmospheric pressure (-1 mm Hg) and air begins to flow into the alveoli. The volume changes faster than air can flow, so intrapulmonary pressure reaches its low- est value about halfway through inspiration (point A_2). As air flows into the alveoli, the pressure gradually rises

(a) "Pump handle" motion increases anterior-posterior dimension of rib cage

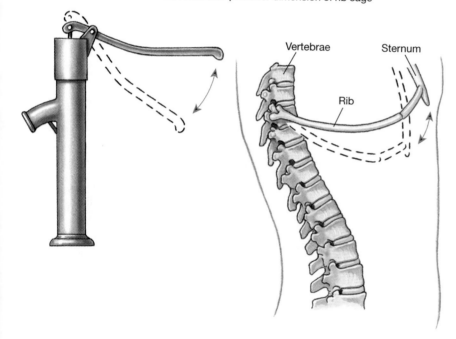

Vertebrae   Sternum

Rib

(b) "Bucket handle" motion increases lateral dimension of rib cage

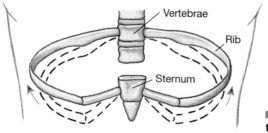

Vertebrae

Rib

Sternum

■ Figure 17-9 **Movement of the rib cage during inspiration**

until the thoracic cage stops expanding, just before the end of inspiration. Air movement continues for a fraction of a second longer, until the pressure inside the lungs equalizes with atmospheric pressure (point A_3). At the end of inspiration, the volume of air in the lungs is at its maximum for the respiratory cycle (point C_2) and intrapulmonary pressure is equal to atmospheric pressure.

You can demonstrate this phenomenon by taking a deep breath and stopping the movement of your chest at the end of inspiration. (Do not "hold your breath," because doing so causes the opening of the pharynx to close and prevents air flow). If you do this correctly, you will notice that air flow stops after you freeze the inspiratory movement. This exercise shows that, at the end of inspiration, air pressure in the alveoli is equal to atmospheric pressure.

Expiration Occurs When Intrapulmonary Pressure Exceeds Atmospheric Pressure

At the end of inspiration, impulses from somatic motor neurons to the inspiratory muscles cease and the muscles relax. The elastic recoil of the stretched muscle fibers returns the diaphragm and rib cage to their original relaxed positions, just as a stretched elastic waistband recoils when released. Muscle recoil is reinforced by the elastic recoil of the pleura and the lung tissue itself. Because expiration during quiet breathing involves passive elastic recoil rather than active muscle contraction, it is called **passive expiration.**

As the volume of the thorax decreases during expiration, air pressure in the lungs increases, reaching about 1 mm Hg above atmospheric pressure at its maximum (Fig. 17-10 ■, point A_4). Intrapulmonary pressure is now higher than atmospheric pressure, so air flow reverses and air moves out of the lungs. At the end of expiration, air movement ceases when the intrapulmonary pressure is again equal to atmospheric pressure (point A_5). The volume of air in the lungs reaches its minimum for the respiratory cycle (point C_3). At this point, the respiratory cycle has ended and is ready to begin again with the next breath.

The pressure differences shown in Figure 17-10 ■ apply to quiet breathing. During exercise or forced heavy breathing, these values will become proportionately larger. **Active expiration** occurs during voluntary exhalations and when ventilation exceeds 30–40 breaths per minute. (Normal resting ventilation rate is 12–20 breaths per minute for an adult.) Active expiration uses a different set of muscles from those used during inspiration, namely the internal intercostal muscles and the

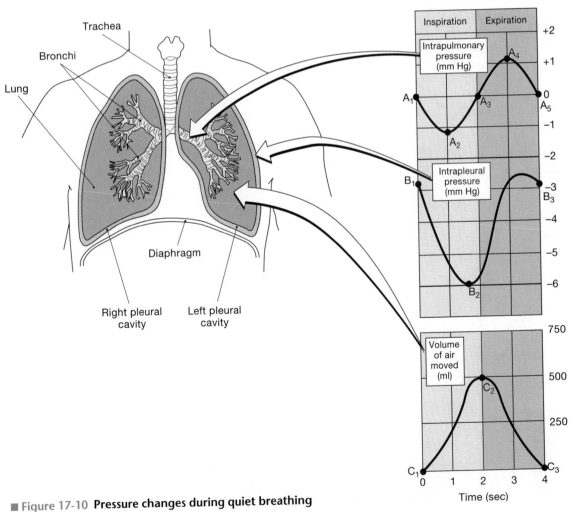

■ Figure 17-10 **Pressure changes during quiet breathing**

abdominal muscles. These muscles are collectively called the *expiratory muscles.*

The internal intercostal muscles line the inside of the rib cage. When they contract, they pull the ribs inward, reducing the volume of the thoracic cavity. To feel this action, place your hands on your rib cage. Forcefully blow as much air out of your lungs as you can, noting the movement of your hands as you do so.

The internal and external intercostals act as antagonistic muscle groups (∞ p. 326) to alter the position and volume of the rib cage during ventilation, but the diaphragm has no antagonistic muscles. Therefore, the abdominal muscles become active during active expiration to supplement the activity of the intercostals. Contraction of abdominal muscles during active expiration pulls the lower rib cage inward and decreases abdominal volume. The displaced intestines and liver push the diaphragm up into the thoracic cavity, passively decreasing the chest volume even more. This is why when you are doing abdominal exercises in an aerobics class, the instructor tells you to blow air out as you lift your head and shoulders. The active process of blowing air out helps you contract your abdominals, the very muscles you are trying to strengthen. The diaphragm stays relaxed during active expiration because contraction would move it downward and would work against the abdominal muscles that are trying to push the diaphragm upward.

Any neuromuscular disease that weakens skeletal muscles or damages their motor neurons can adversely affect ventilation. With decreased ventilation, less fresh air enters the lungs. In addition, loss of the ability to cough increases the risk of pneumonia and other infections. Examples of diseases that affect the motor control of ventilation include **myasthenia gravis,** an illness in which the acetylcholine receptors of the motor end plate of skeletal muscles are destroyed, and **polio** (poliomyelitis), a viral illness that sometimes paralyzes the respiratory muscles.

✓ Scarlett O'Hara was trying to squeeze herself into a corset with an 18-inch waist. Was she more successful when she took a deep breath and held it or when she blew all the air out of her lungs? Why?

✓ Why would loss of the ability to cough increase the risk of respiratory infections? (Hint: What does coughing do to mucus in the airways?)

Intrapleural Pressure Changes during Ventilation

Ventilation requires that the lungs, which are unable to expand and contract on their own, move in association with the contraction and relaxation of the thorax. As discussed earlier in this chapter, the lungs are "stuck" to the thoracic cage by the pleural fluid, the fluid between the two pleural membranes.

The intrapleural pressure, pressure within the fluid between the pleural membranes, is normally subatmo-

spheric. This subatmospheric pressure arises during development, when the thoracic cage with its associated pleural membrane grows more rapidly than the lung with its associated pleural membrane. The two pleural membranes are held together by the pleural fluid bond, so the elastic lungs are forced to stretch to conform to the larger volume of the thoracic cavity. At the same time, however, elastic recoil of the lungs creates an inwardly directed force that attempts to pull the lungs away from the chest wall (Fig. 17-11a ■). The combination of the outward pull of the thoracic cage and inward recoil of the elastic lungs creates an intrapleural pressure of about −3 mm Hg.

You can create a similar situation with a syringe half-filled with fluid and capped with a plugged-up needle. Begin with the fluid inside the syringe at atmospheric pressure. Now pick up the syringe and hold the barrel (the chest wall) in one hand while you try to withdraw the plunger (the elastic lung pulling away from the chest wall). As you pull on the plunger, the volume inside the barrel increases slightly, but the cohesive forces between the water molecules cause the fluid inside the syringe to resist expansion. The pressure within the barrel, which was initially equal to atmospheric pressure, decreases slightly as you pull on the plunger. If you release the plunger, it snaps back to its resting position, restoring atmospheric pressure inside the syringe.

But what happens to subatmospheric intrapleural pressure if an opening is made between the sealed pleural cavity and the atmosphere? A knife thrust between the ribs, a broken rib that punctures the pleural membrane, or any event that opens the pleural cavity to the atmosphere will allow air to flow in down its pressure gradient, just as air enters when you break the seal on a vacuum-packed can. Air in the pleural cavity breaks the fluid bond holding the lung to the chest wall. The elastic lung collapses to an unstretched state, like a deflated balloon, while the chest wall expands outward (Fig. 17-11b ■). This condition, called **pneumothorax** [*pneuma,* air + *thorax,* chest], results in a collapsed lung that is unable to function normally. Pneumothorax can occur spontaneously if a congenital *bleb,* or weakened section of lung tissue, ruptures, allowing air from inside the lung to enter the pleural cavity. Correction of a pneumothorax has two components: removing as much air from the pleural cavity as possible with a suction pump, and sealing the hole to prevent more air from entering. Any air remaining in the cavity will gradually be absorbed into the blood, restoring the pleural fluid bond and reinflating the lung.

Pressures in the pleural fluid vary during a respiratory cycle. At the beginning of inspiration, intrapleural pressure is about −3 mm Hg (Fig. 17-10 ■, point B$_1$). As inspiration proceeds, the pleural membranes and lungs follow the thoracic cage because of the pleural fluid bond. But the elastic lung tissue resists being stretched. The lungs attempt to pull farther away from the chest wall, causing the intrapleural pressure to become even more negative (Fig. 17-10 ■, point B$_2$). Because this

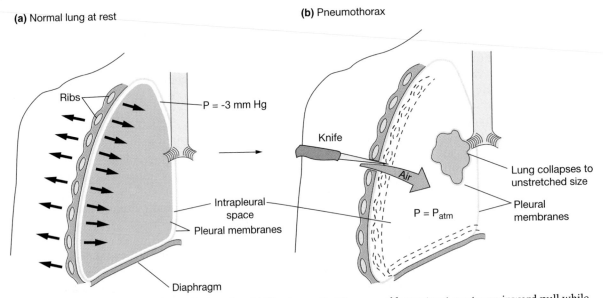

(a) Normal lung at rest

Ribs

P = -3 mm Hg

Intrapleural
space

Pleural membranes

Diaphragm

(b) Pneumothorax

Knife

Air

P = P$_{atm}$

Lung collapses to
unstretched size

Pleural
membranes

■ Figure 17-11 **Pressure in the pleural cavity** (a) Elastic recoil of the normal lung at rest creates an inward pull while the elastic recoil of the chest wall tries to pull the chest wall outward. Pressure in the fluid between the pleural membranes is subatmospheric. (b) If the sealed pleural cavity is opened to the atmosphere, air flows into the cavity. The lung collapses to its unstretched size, and the rib cage expands slightly. This condition is called pneumothorax.

process is difficult to visualize, return to the analogy of the fluid-filled syringe with the blocked needle. You can pull the plunger out a small distance without much effort, but if you try to pull it out farther, it is harder to do because of the cohesiveness of the fluid. The increased amount of work you do trying to pull out the plunger is paralleled by the work your muscles must do when they contract for inspiration. The bigger the breath, the more work is required to "stretch" the pleural fluid and the elastic lung.

By the end of quiet inspiration, when lungs are fully expanded, intrapleural pressure drops to around −6 mm Hg (point B$_2$). During exercise or other powerful inspirations, intrapleural pressure may reach −18 mm Hg.

With expiration, the thoracic cage returns to its resting position. The lungs are released from their extra-stretched position, and the intrapleural pressure returns to its normal value of −3 mm Hg (point B$_3$). Notice that intrapleural pressure never equilibrates with the atmosphere because the pleural cavity is a closed compartment.

The pressure gradients required for air flow are created by the work of skeletal muscle contraction. Normally, about 3%–5% of the body's energy expenditure is used for

✓ A person has periodic spastic contractions of the diaphragm, otherwise known as hiccups. What happens to intrapleural and intrapulmonary pressure when a person hiccups?

✓ A stabbing victim is brought into the emergency room with a knife wound between the ribs on the left side of his chest. What has probably happened to his left lung? To his right lung? Why does the left side of his rib cage seem larger than the right side?

quiet breathing. During exercise, the energy required for breathing increases substantially. The two factors that have the greatest influence on the amount of work needed for breathing are (1) the stretchability, or compliance, of the lungs and (2) the resistance of the airways to air flow.

Lung Compliance and Elastance May Change in Disease States

Adequate ventilation depends on the ability of the lungs to expand normally. Most of the work of breathing goes into overcoming the resistance of the elastic lungs and the thoracic cage to stretch. Clinically, the ability of the lung to stretch is called **compliance.** A high-compliance lung stretches easily, just as a compliant person is easy to persuade. A low-compliance lung requires more force from the inspiratory muscles to stretch it. Compliance is different from **elastance** (elasticity). The fact that a lung stretches easily (high compliance) does not necessarily mean that it will return to its resting volume when the stretching force is released (elastance). For example, you may have had a pair of gym shorts in which the elastic waistband was easy to stretch but lacking in elasticity, making the shorts impossible to keep around your waist. Analogous problems may occur in the respiratory system. For example, *emphysema* is a disease in which the elastin fibers normally found in lung tissue are destroyed. Destruction of the elastin fibers results in lungs that exhibit high compliance and stretch easily during inspiration. However, these lungs also have decreased elastance, and therefore they do not recoil to their resting position during expiration. To understand the importance of elastic recoil to expiration, think of an inflated balloon and an inflated plastic bag. The balloon is similar to the normal lung. Its elastic walls squeeze on the air inside the balloon, thereby increasing the internal air pressure. When the neck of the balloon is opened to the atmosphere, elastic recoil causes air to flow out of the balloon. The inflated plastic bag, on the other hand, is like the lung of an individual with emphysema. It has high compliance and is easily inflated, but it has little elastic recoil. If the inflated plastic bag is opened to the atmosphere, most of the air remains inside the bag.

A decrease in lung compliance will also affect ventilation because more work must be expended to stretch a stiff lung. Pathological conditions in which compliance is reduced are called **restrictive lung diseases.** In these conditions, the energy expenditure required to stretch stiff, less-compliant lungs can far exceed the normal range. Two common causes of decreased compliance are inelastic scar tissue formed in fibrotic lung diseases and inadequate production of surfactant in the alveoli.

Surfactant Decreases the Work of Breathing

For years, physiologists assumed that elastin and other elastic fibers were the primary source of resistance to stretch in the lung. But studies comparing the work required to expand air-filled and saline-filled lungs showed that air-filled lungs are much harder to inflate, so the lung tissue itself must contribute less to resistance than we thought. Another property of the normal air-filled lung, not present in a saline-filled lung, creates most of the resistance to stretch.

This property is **surface tension,** created by the thin fluid layer between the alveolar cells and the air. At any air-fluid interface, the surface of a liquid behaves as if it is under tension, like a thin membrane being stretched. This surface tension arises because of the hydrogen bonds between water molecules. The surface molecules are attracted to other water molecules beside and beneath them, but not to the air. If water is isolated in a drop, it will take on a rounded shape (∞ Fig. 2-10, p. 24). If water is in the shape of a bubble, as it is when it lines the alveoli, surface tension exerts a force directed toward the center of the bubble.

The **Law of LaPlace** is an expression of the force, or pressure, created by a fluid sphere or bubble. In physiology, the sphere is analogous to a fluid-lined alveolus. The Law of LaPlace states that the pressure within a fluid-lined alveolus depends on two factors: the surface tension of the fluid and the radius of the alveolus. This relationship is expressed by the equation

$P = 2 \cdot T/r$ where P is the pressure inside the alveolus

T is the surface tension of the fluid lining the alveolus

r is the radius of the alveolus

If two alveoli have different diameters but are lined by fluids with the same surface tension, the pressure inside the smaller alveolus will be greater (Fig. 17-12a ■). If the two alveoli are connected to each other, air will flow from the higher-pressure small alveolus to the lower-pressure large alveolus. This movement of air causes the smaller alveolus to collapse, while the larger alveolus increases in volume.

Thus, the presence of fluid lining the alveoli creates surface tension that increases the resistance of the lung to stretch. This surface tension also makes the alveoli behave like elastic bubbles or inflated balloons that, once expanded, tend to collapse. Consequently, surface tension has the potential to increase the work needed to expand the alveoli with each breath.

Normally, however, our lungs secrete a chemical called a surfactant that reduces surface tension. **Surfactants** are molecules that disrupt cohesive forces between water molecules. For example, in dishwasher rinses, surfactants keep rinse water from beading up on the dishes. In the lungs, surfactant decreases the surface tension of the fluid lining the alveoli and prevents small alveoli from collapsing. Surfactant is more concentrated in smaller alveoli, making the surface tension in the smaller alveoli less than that in larger alveoli (Fig. 17-12b ■). Lower surface tension equalizes the pressure among different sizes of alveoli and keeps the smaller alveoli from collapsing when their air flows into larger, lower-pressure alveoli. With lower surface tension, the work needed to expand the alveoli with each breath is also greatly reduced.

Human surfactant is a mixture containing lipoproteins such as *dipalmitoylphosphatidylcholine.* Surfactant is manufactured and secreted into the alveolar air space by the type II alveolar cells (see Fig. 17-2g ■). Normally, surfactant synthesis begins about the twenty-fifth week of fetal development under the influence of various hormones. Production usually reaches adequate levels by the thirty-second week (about eight weeks before normal delivery). Babies who are born prematurely without adequate concentrations of surfactant in their alveoli

Fibrotic Lung Disease   Fibrotic lung diseases often result from the chronic inhalation of fine particulate matter that escapes the mucus lining of the airways and reaches the exchange epithelium of the alveoli. The only protective mechanism in that region of the respiratory system is removal by wandering alveolar macrophages (Fig. 17-2g ■). These phagocytic cells patrol the alveoli, engulfing any airborne particles that have managed to reach the exchange surface. If the particles are organic, the macrophages digest them with lysosomal enzymes. But if the particles are inorganic and cannot be digested, an inflammatory process ensues. Intracellular accumulation of the particles causes the macrophage to secrete growth factors that stimulate fibroblasts in the connective tissue of the lung. The fibroblasts produce collagen that forms inelastic, fibrous scar tissue. Large amounts of scar tissue reduce the compliance of the lung. The proliferation of inelastic scar tissue in the lung is called **fibrotic lung disease,** or **fibrosis.** Inorganic particles that can trigger fibrosis include asbestos, coal dust, silicon, and even dust and pollutants from urban areas.

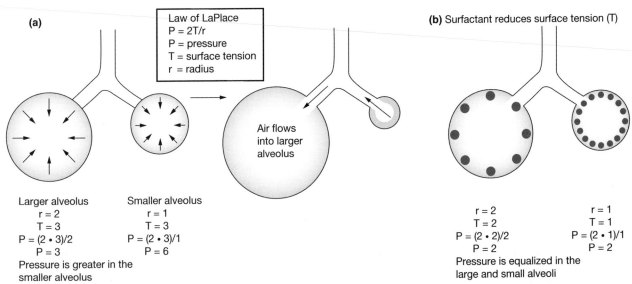

(a)

Law of LaPlace
$P = 2T/r$
P = pressure
T = surface tension
r = radius

(b) Surfactant reduces surface tension (T)

Air flows into larger alveolus

Larger alveolus
r = 2
T = 3
P = (2 • 3)/2
P = 3
Pressure is greater in the smaller alveolus

Smaller alveolus
r = 1
T = 3
P = (2 • 3)/1
P = 6

r = 2
T = 2
P = (2 • 2)/2
P = 2
Pressure is equalized in the large and small alveoli

r = 1
T = 1
P = (2 • 1)/1
P = 2

■ **Figure 17-12** **Surfactant prevents the collapse of alveoli** (a) Air pressure within the alveoli is determined by the Law of LaPlace. If two alveoli have the same surface tension, the small alveolus will have higher pressure and will collapse as its air flows into the larger alveolus. (b) Surfactant decreases surface tension in the fluid lining the alveoli. Smaller alveoli have a higher concentration of surfactant so their surface tension is lower. The higher concentration of surfactant in smaller alveoli equalizes the pressure between smaller and larger alveoli.

develop *newborn respiratory distress syndrome (NRDS).* In addition to "stiff" (low-compliance) lungs that require tremendous effort to expand with each breath, babies with NRDS also have alveoli that collapse.

Resistance of the Airways to Air Flow Is Determined Primarily by Airway Diameter

The other factor besides compliance that influences the work of breathing is the resistance of the respiratory system to air flow. According to Poiseuille's Law (∞ p. 389), parameters that contribute to resistance are:

✓ Coal miners who spend years inhaling fine coal dust have much of their alveolar surface area covered with scarlike tissue. What happens to their lung compliance as a result?

✓ When a baby takes its first breath, what happens to the alveoli if surfactant is not present?

1. The length of the system (L)
2. The viscosity of the substance flowing through the system (η)
3. The radius of the tubes in the system (r)

The equation that expresses the relation of those parameters is

$$R \propto L\eta/r^4$$

Because the length of the respiratory system is constant, we can ignore that variable. The viscosity of air is also

Respiratory Distress Syndrome and Surfactant
The clue to what creates additional resistance in an air-filled alveolus came from studies of premature babies with newborn respiratory distress syndrome (NRDS). From their first breath, NRDS babies have difficulty keeping their lungs inflated. With each breath, their lungs collapse and they must use a tremendous amount of energy to expand them again for the next breath. Unless treatment is initiated rapidly, about 50% of these infants die. When President and Mrs. John Fitzgerald Kennedy's son was born with NRDS in the 1960s, the physicians could do little but administer oxygen and hope that the baby's lungs would mature before hypoxia and exertion overwhelmed him. Now the prognosis for NRDS babies is much better. Amniotic fluid can be sampled to assess if the baby's lungs are producing adequate amounts of surfactant. If they are not, and if delivery cannot be delayed, the NRDS baby can be treated by artificial ventilation that forces air into the lungs and keeps the alveoli open. The latest treatment is the aerosol administration of artificial surfactant until the baby's lungs mature enough to produce their own.

continued from page 483

Emphysema is characterized by a loss of elastin, the elastic fibers that help the alveoli recoil during expiration. Researchers believe that elastin is destroyed by proteases released by immune system cells, which work overtime in smokers to rid the lungs of irritants. People with emphysema have no difficulty inhaling air; however, because their alveoli have lost elastic recoil, expiration—normally a passive process—requires conscious effort. They literally have to work to push air out of their lungs.

Question 3: Name the muscles that patients with emphysema use to exhale forcefully.

almost constant, although you may have noticed that it feels harder to breathe in a sauna filled with steam than in a room with normal humidity. Water droplets in the steam increase the viscosity of the steamy air, thereby increasing its resistance to flow. Viscosity also changes slightly with atmospheric pressure, increasing as pressure increases. Divers breathing high-pressure compressed air may feel a little more resistance to air flow, whereas someone at high altitude may feel less resistance. Despite these exceptions, viscosity plays very little role in resistance to air flow.

Because length and viscosity are essentially constant for the respiratory system, the radius (or diameter) of the airways is the primary determinant of airway resistance. Normally, however, the work needed to overcome the resistance of the airways to air flow is small compared with the work needed to overcome the resistance of the lungs and thoracic cage to stretch.

Nearly 90% of airway resistance normally can be attributed to the trachea and bronchi, rigid structures with the smallest total cross-sectional area of the airways. These structures are supported by cartilage and bone, so their diameters normally do not change, and their resistance to air flow is constant. However, mucus accumulation from allergies or infections can dramatically increase resistance. If you have ever tried breathing through your nose when you have a cold, you can appreciate how the narrowing of an upper airway limits air flow!

The bronchioles normally do not contribute significantly to airway resistance because their total cross-sectional area is about 2000 times that of the trachea. But, because the bronchioles are collapsible tubes, a decrease in their diameter can suddenly turn them into a significant source of airway resistance. **Bronchoconstriction** increases resistance to air flow and decreases the amount of fresh air that reaches the alveoli. Bronchioles, like arterioles, are subject to reflex control by the nervous system and by hormones. However, most minute-to-minute changes in bronchiolar diameter occur in response to paracrines.

Carbon dioxide in the airways is the primary paracrine that affects bronchiolar diameter. Increased concentrations of carbon dioxide in expired air relax bronchiolar smooth muscle and cause **bronchodilation.** On the other hand, *histamine* is a paracrine that acts as a powerful bronchoconstrictor. This chemical is released by mast cells (∞ p. 147) in response to tissue damage or allergic reactions. In severe allergic reactions, large amounts of histamine may lead to widespread bronchoconstriction and difficulty breathing. Immediate medical treatment in these patients is imperative.

The primary neural control of bronchioles comes from parasympathetic neurons that cause bronchoconstriction, a reflex designed to protect the lower respiratory tract from inhaled irritants. There is no significant sympathetic innervation of the bronchioles in humans. However, smooth muscle in the bronchioles is well supplied with β_2 receptors that respond to epinephrine. Stimulation of β_2 receptors relaxes airway smooth muscle and results in bronchodilation. This reflex is used therapeutically in the treatment of asthma and various allergic reactions characterized by histamine release and bronchoconstriction. Factors that alter airway resistance are summarized in Table 17-3.

✓ A cancerous lung tumor has grown into the walls of a group of bronchioles, narrowing their lumens. What has happened to the resistance to air flow in these bronchioles?

✓ Why does it feel more difficult to breathe in a steam room than outside in drier air?

Pulmonary Function Tests Measure Lung Volume during Ventilation

One method that physiologists use to assess a person's pulmonary function is to measure how much air the person moves during quiet breathing and with maximum effort. Many **pulmonary function tests** use a **spirometer,** an instrument that measures the volume of air moving with each breath (Fig. 17-13 ■). Although pulmonary function tests are relatively simple to perform, they have considerable diagnostic value. For example, in asthma, the bronchioles are constricted.

TABLE 17-3 Factors That Affect Airway Resistance

Factor	Affected by	Mediated by
Length of the system	Constant; not a factor	
Viscosity of air	Usually constant. Humidity and altitude may alter slightly.	
Diameter of airways		
Upper airways	Physical obstruction	Mucus and other factors
Bronchioles	Bronchoconstriction	Parasympathetic neurons (muscarinic receptors) Histamine
	Bronchodilation	Carbon dioxide Epinephrine (β_2 receptors)

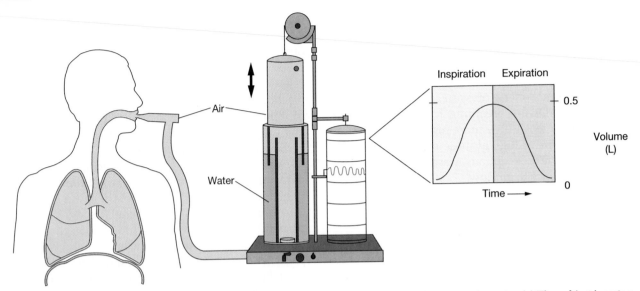

■ **Figure 17-13 The recording spirometer** Lung volumes and air flow are recorded using a spirometer. (a) The subject inserts a mouthpiece that is attached to an inverted bell filled with air or oxygen. The volume of the bell and the volume of the subject's respiratory tract create a closed system. When the subject inhales, air moves from the bell into the lungs. The volume of the bell decreases, and the pen rises on the tracing. When the subject exhales, air moves from the lungs back into the bell, and its volume increases. This increase causes the pen to drop on the tracing. In addition to measuring lung volumes, spirometers can be used to determine the *rate* of air movement during forced expiration. Most clinical spirometers today have been computerized, but the spirometer illustrated is still widely used in student laboratories.

They tend to collapse and close off before a forced expiration is completed, reducing both the amount and rate of air flow. Diseases in which air flow during expiration is diminished owing to narrowing of the bronchioles are known as **obstructive lung diseases.** Emphysema and chronic bronchitis are sometimes called *chronic obstructive pulmonary disease,* or *COPD,* because of their ongoing, or chronic, nature.

Lung volumes The air moved during breathing can be divided into four lung volumes: tidal volume, inspiratory reserve volume, expiratory reserve volume, and residual

Asthma Asthma is an obstructive lung disease characterized by bronchoconstriction and airway edema, which increase airway resistance and decrease air flow. Patients complain of shortness of breath (*dyspnea*), and when they are asked to exhale forcefully, a wheezing sound is heard as air whistles through the narrowed lower airways. The severity of asthma attacks ranges from mild to life-threatening. Asthma is an inflammatory condition that is often associated with allergies. It can also be triggered by exercise (exercise-induced asthma) and by rapid changes in the temperature or humidity of inspired air. At the cellular level, a variety of chemical signals may be responsible for inducing bronchoconstriction, including acetylcholine, histamine, substance P (a neuropeptide), and leukotrienes secreted by mast cells, macrophages, and eosinophils. Asthma is treated with inhaled and oral medications that include β_2-adrenergic agonists, anti-inflammatory drugs, and a new leukotriene antagonist.

volume. These volumes are diagrammed in Figure 17-14 ■ and described below. The numerical values given represent average volumes for the 70-kg man; the volumes for women are typically about 20% less. Each paragraph below begins with the instructions that you would be given if you were being tested for these volumes.

"Breathe quietly." The volume of air that moves in a single normal inspiration or expiration is known as the **tidal volume** (V_T). Average tidal volume during quiet breathing is 500 mL. Tidal volume will vary with the age, sex, and height of the individual.

"Now, at the end of a quiet inspiration, take in as much additional air as you possibly can." The additional volume you inspire above the tidal volume represents your **inspiratory reserve volume** (IRV). In a 70-kg man, this volume is about 2500 mL, a fivefold increase over the normal tidal volume.

"Now stop at the end of a normal exhalation, then exhale as much air as you possibly can." The amount of air exhaled after the end of a normal expiration is the **expiratory reserve volume** (ERV), which averages about 1000 mL.

The fourth volume is one that cannot be measured directly. Even if you blow out as much air as you can, air still remains in the lungs and the airways. The volume of air in the respiratory system after maximal exhalation, about 1200 mL, is called the **residual volume** (RV). Most of this residual volume exists because the lungs are held stretched against the thoracic wall by the pleural fluid.

Lung capacities Sums of two or more lung volumes are called **capacities.** Adding the inspiratory reserve vol-

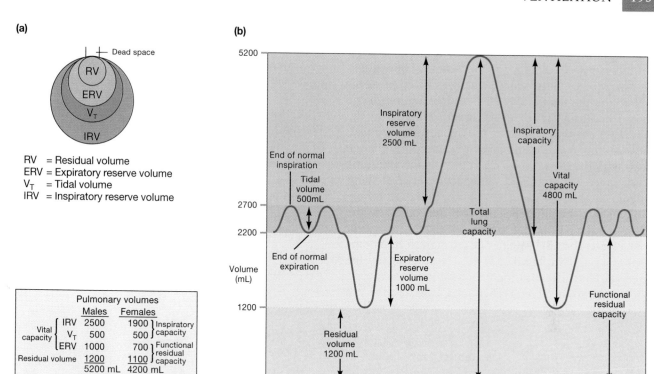

■ Figure 17-14 **Lung volumes and capacities** (a) The four lung volumes. (b) A spirometer tracing showing the lung volumes and capacities.

ume, the expiratory reserve volume, and the tidal volume gives a measure of the **vital capacity** (VC). Vital capacity represents the maximum amount of air that can be voluntarily moved into or out of the respiratory system with one breath. To measure vital capacity, you would instruct the person to take in as much air as possible, then blow it all out. Vital capacity decreases with age.

The vital capacity plus the residual volume yields the **total lung capacity** (TLC). Other capacities of importance in pulmonary medicine include the inspiratory capacity (tidal volume plus inspiratory reserve volume) and the functional residual capacity (expiratory reserve volume plus residual volume).

Auscultation of breath sounds Auscultation of breath sounds is an important diagnostic technique in pulmonary medicine, just as auscultation of heart sounds is used in cardiovascular diagnosis (∞ p. 413). But breath sounds are more complicated to interpret owing to a wide range of normal variation. In general, breath sounds are distributed evenly over the lungs and resemble a quiet "whoosh" made by flowing air. In conditions in which air flow is reduced, such as pneumothorax, breath sounds may be diminished or absent. Abnormal sounds include a variety of squeaks, pops, wheezes, or bubbling sounds caused by fluid and secretions in the airways or alveoli. Inflammation of the pleural membrane results in a crackling or grating sound known as a *friction rub.* It is caused by swollen, inflamed pleural membranes rubbing against each other, and it disappears when fluid again separates them.

✓ Restrictive lung disease decreases the compliance of the lung. How will the inspiratory reserve volume change in patients with a restrictive lung disease?

✓ Chronic obstructive lung disease causes patients to lose the ability to exhale fully. How does residual volume change in these patients?

✓ If vital capacity decreases with age but total lung capacity does not change, what volume must be changing? In which direction?

The Effectiveness of Ventilation Is Determined by the Rate and Depth of Breathing

You may recall that the efficiency of the heart is measured by the cardiac output, which is calculated by multiplying heart rate by stroke volume. Likewise, we can estimate the effectiveness of ventilation by calculating the **total pulmonary ventilation,** the volume of air moved into and out of the lungs each minute. Total pulmonary ventilation, also known as the *minute volume,* is calculated as follows:

Ventilation rate × tidal volume (V_T) = total pulmonary ventilation

The normal ventilation rate for an adult is 12–20 breaths per minute. Using the average tidal volume of 500 mL and the slowest ventilation rate, we get

12 breaths/min × 500 mL/breath = 6000 mL/min, or 6 L/min

Total pulmonary ventilation represents the physical movement of air into and out of the respiratory tract. But it is not necessarily a good indicator of how much fresh air reaches the alveolar exchange surface. Some air that enters the respiratory system does not reach the alveoli because part of every breath remains in the conducting airways, such as the trachea and bronchi. Because the conducting airways do not exchange gases with the blood, they are known as the **anatomic dead space.** Anatomic dead space averages about 150 mL.

To illustrate the difference between air that enters the airways and fresh air that reaches the alveoli, let us consider a typical breath bringing 500 mL of fresh air into the respiratory system (Fig. 17-15b ■). The entering air displaces the 150 mL of stale air in the anatomic dead space, sending it into the alveoli. The first 350 mL of the fresh

air also enters the alveoli. The last 150 mL of inspired air remains in the dead space. Consequently, only 350 mL of fresh air enters the alveoli with each breath. This 350 mL mixes with 2200 mL of air already in the lungs (the functional residual capacity; see Fig. 17-14 ■).

Because a significant portion of inspired air never reaches an exchange surface, a more accurate indicator of the efficiency of ventilation is **alveolar ventilation,** the amount of air that reaches the alveoli each minute. Alveolar ventilation is calculated by multiplying the ventilation rate by the volume of air that reaches the alveoli:

$$\text{Ventilation rate} \times (\text{tidal volume} - \text{dead space volume})$$
$$= \text{alveolar ventilation}$$

Using the same ventilation rate and tidal volume as above and a dead space volume of 150 mL yields an alveolar ventilation of

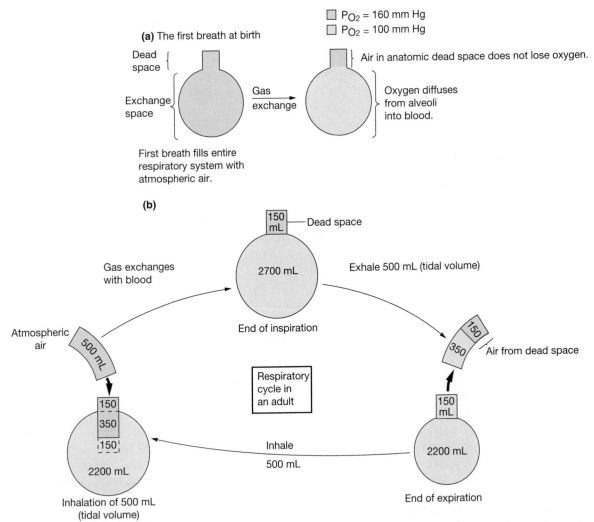

(a) The first breath at birth

☐ P_{O_2} = 160 mm Hg
☐ P_{O_2} ≈ 100 mm Hg

Air in anatomic dead space does not lose oxygen.

Oxygen diffuses from alveoli into blood.

First breath fills entire respiratory system with atmospheric air.

(b)

Dead space — 150 mL

Gas exchanges with blood

2700 mL

Exhale 500 mL (tidal volume)

End of inspiration

Atmospheric air — 500 mL

Gas exchange

Respiratory cycle in an adult

150 / 350 — Air from dead space

150 / 350 / 150 — 2200 mL

Inhalation of 500 mL (tidal volume)

Inhale 500 mL

150 mL — 2200 mL

End of expiration

■ **Figure 17-15 Total pulmonary and alveolar ventilation** (a) Air that remains in the conducting airways does not exchange gases with the blood. The conducting airways are also called the dead space for this reason. (b) In adults, the dead space is about 150 mL. When a breath of 500 mL is exhaled, the first 150 mL of air comes from the dead space and the remaining 350 mL is alveolar air. At the end of expiration, the dead space is filled with alveolar air. The next inspiration brings 500 mL of fresh air into the respiratory system. However, the first 150 mL of air to move into the lungs is the 150 mL of "stale" alveolar air that was in the dead space. Only 350 mL of fresh air enters the alveoli. The remaining 150 mL of fresh air remains in the dead space.

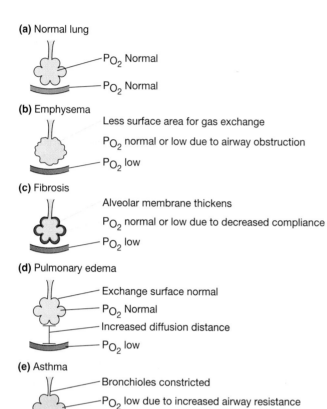

(a) Normal lung
- P_{O_2} Normal
- P_{O_2} Normal

(b) Emphysema
- Less surface area for gas exchange
- P_{O_2} normal or low due to airway obstruction
- P_{O_2} low

(c) Fibrosis
- Alveolar membrane thickens
- P_{O_2} normal or low due to decreased compliance
- P_{O_2} low

(d) Pulmonary edema
- Exchange surface normal
- P_{O_2} Normal
- Increased diffusion distance
- P_{O_2} low

(e) Asthma
- Bronchioles constricted
- P_{O_2} low due to increased airway resistance
- P_{O_2} low

■ Figure 17-19 **Pulmonary pathologies that affect alveolar ventilation and gas exchange** (a) Normal lung. (b) In emphysema, destruction of alveoli results in less surface area for gas exchange. (c) In fibrotic lung diseases, thickening of the alveolar membrane from scar tissue will slow gas exchange. In addition, the loss of lung compliance may decrease alveolar ventilation, causing lower alveolar P_{O_2}. (d) In pulmonary edema, excess fluid in the interstitial space will increase the diffusion distance between the alveoli and the blood. P_{CO_2} in the arterial blood may be normal because of the higher solubility of carbon dioxide, but P_{O_2} is likely to be decreased. In addition, edema may cause a loss of compliance. (e) In asthma, narrowing of the bronchioles will cause a decrease in alveolar ventilation.

the point that the lymphatics become unable to remove all the fluid (Fig. 17-19d ■). Excess fluid accumulates in the pulmonary interstitial space and may even leak across the alveolar membrane, collecting inside the alveoli. The increased diffusion distance does not usually affect carbon dioxide exchange because carbon dioxide is relatively soluble in body fluids. Oxygen, however, is much less soluble and is unable to cross the increased diffusion distance as easily. Subsequently, oxygen exchange diminishes. In these cases, it is not unusual to find normal arterial P_{CO_2} accompanying decreased arterial P_{O_2}.

If the diffusion of oxygen from alveolus to blood is significantly impaired, **hypoxia,** or too little oxygen in the cells, results. Hypoxia frequently goes hand in hand with **hypercapnia,** elevated concentrations of carbon dioxide. These two conditions are clinical signs, not diseases, and the clinician must gather additional information to pinpoint their cause.

The Pulse Oximeter   One important clinical indicator of the effectiveness of gas exchange in the lungs is the amount of oxygen in arterial blood. Previously, obtaining an arterial blood sample was difficult and painful because it meant finding an accessible artery (most blood is drawn from superficial veins rather than from arteries, which lie deeper within the body). Recently, scientists developed instruments that quickly and painlessly measure blood oxygen levels through the surface of the skin on a finger. The pulse oximeter clips onto the tip of a finger and, within seconds, gives a digital reading of arterial hemoglobin saturation by measuring light absorbance of the tissue at two wavelengths. Transcutaneous oxygen sensors measure dissolved oxygen (P_{O_2}) using a variant of traditional gas-measuring electrodes. Both methods have their limitations but are popular because they give a rapid, noninvasive means of estimating arterial oxygen concentration.

✓ Why does left ventricular failure or mitral valve dysfunction cause elevated pulmonary blood pressure? (Hint: ∞ p. 415.)

✓ If alveolar ventilation increases, what do you predict will happen to arterial P_{O_2}? To arterial P_{CO_2}? To venous P_{O_2} and P_{CO_2}? Explain.

GAS EXCHANGE IN THE TISSUES

Diffusion of gases between the blood and the cells also depends on the pressure gradient between the two compartments. Arterial blood reaching the systemic capillaries has a P_{O_2} of 100 mm Hg and a P_{CO_2} of 40 mm Hg (Fig. 17-18 ■). The cells are continuously using oxygen and producing carbon dioxide through cellular respiration, so their P_{O_2} is ≤40 mm Hg and their P_{CO_2} is ≥ 46 mm Hg. The lower P_{O_2} in the cells sets up a gradient favoring diffusion of oxygen from the plasma into the cells. Conversely, higher P_{CO_2} in the cells than in the capillary blood allows carbon dioxide to diffuse out of the cells into the capillary. Just as at the alveoli, exchange is rapid and goes to equilibrium. Blood in the systemic venous circulation has an average P_{O_2} of 40 mm Hg and a P_{CO_2} of 46 mm Hg. Gas exchange at the tissue level may be affected slightly by tissue edema, but it usually proceeds without interference.

continued from page 490

Edna has been admitted to the hospital for tests related to her COPD. One of the tests is a hematocrit, an indicator of the number of red blood cells in her blood. The results of this test show that Edna has higher-than-normal numbers of red blood cells.

Question 4: Why does Edna have an increased hematocrit? (Hint: Because of Edna's COPD, her arterial P_{O_2} is low.)

GAS TRANSPORT IN THE BLOOD

Now that we have examined the role of the respiratory system in gas exchange, let us turn our attention to the transport of oxygen and carbon dioxide in the blood. The gases are transported either dissolved in the plasma or within the red blood cells.

Hemoglobin Transports Most Oxygen to the Tissues

The presence of adequate amounts of hemoglobin in the blood is essential for normal gas transport. Oxygen in the blood is transported two ways: dissolved in the plasma and bound to hemoglobin (Hb). This fact can be summarized in the following statement:

> Total blood oxygen content = amount dissolved in plasma + amount bound to hemoglobin

Because of the low solubility of oxygen in aqueous solutions, only 3 mL of O_2 will dissolve in the plasma fraction of 1 liter of arterial blood (Fig. 17-20 ■). Thus, with a typical cardiac output of 5 L blood/min, about 15 mL of dissolved oxygen reaches the systemic tissues each minute. This amount cannot begin to meet the needs of the tissues, however. Oxygen consumption at rest is about 250 mL/min, and that figure increases dramati-

cally with exercise. Thus, the body is heavily dependent on the oxygen carried by hemoglobin.

More than 98% of the oxygen in a given volume of blood is transported inside the red blood cells, where it is bound to hemoglobin. At normal hemoglobin levels, oxygen in the red blood cells amounts to about 197 mL/L blood. When the oxygen bound to hemoglobin is added to that dissolved in the plasma, the total oxygen content of whole blood jumps from 3 mL/L to 200 mL/L. If the cardiac output remains 5 L/min, oxygen delivery jumps to almost 1000 mL/min, nearly four times the oxygen consumption of the tissues at rest.

The amount of oxygen that binds to hemoglobin depends on two factors: (1) the P_{O_2} of the plasma surrounding the red blood cells and (2) the number of potential oxygen-binding sites available within the red blood cells. The P_{O_2} of the plasma is the primary factor that determines how many hemoglobin binding sites are occupied by oxygen. Arterial plasma P_{O_2} is established by the composition of the inspired air, the alveolar ventilation rate, and the efficiency of gas exchange between lung and blood, as you learned in previous sections.

The number of potential binding sites depends on the total number of hemoglobin molecules in the blood. Clinically, this number can be estimated either by count-

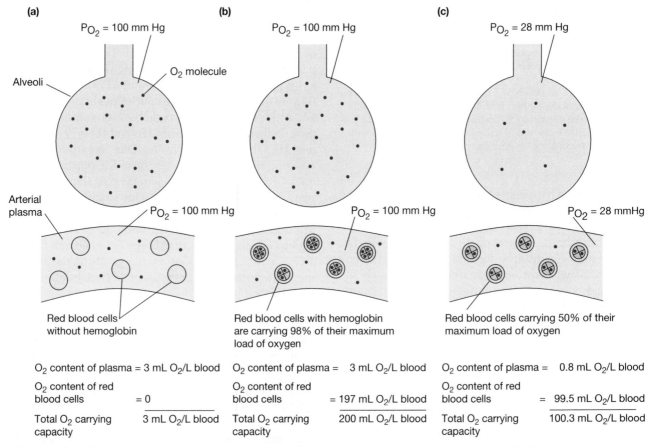

(a) P_{O_2} = 100 mm Hg — Alveoli — O_2 molecule — Arterial plasma — P_{O_2} = 100 mm Hg — Red blood cells without hemoglobin

| O_2 content of plasma = 3 mL O_2/L blood |
| O_2 content of red blood cells = 0 |
| Total O_2 carrying capacity = 3 mL O_2/L blood |

(b) P_{O_2} = 100 mm Hg — P_{O_2} = 100 mm Hg — Red blood cells with hemoglobin are carrying 98% of their maximum load of oxygen

| O_2 content of plasma = 3 mL O_2/L blood |
| O_2 content of red blood cells = 197 mL O_2/L blood |
| Total O_2 carrying capacity = 200 mL O_2/L blood |

(c) P_{O_2} = 28 mm Hg — P_{O_2} = 28 mmHg — Red blood cells carrying 50% of their maximum load of oxygen

| O_2 content of plasma = 0.8 mL O_2/L blood |
| O_2 content of red blood cells = 99.5 mL O_2/L blood |
| Total O_2 carrying capacity = 100.3 mL O_2/L blood |

■ **Figure 17-20 Hemoglobin and oxygen transport** (a) Oxygen transport in blood without hemoglobin. (b) Oxygen transport at normal P_{O_2} in blood with hemoglobin. (c) Oxygen transport at reduced P_{O_2} in blood with hemoglobin.

ing the red blood cells and quantifying the amount of hemoglobin per red blood cell (*mean corpuscular hemoglobin*) or by simply determining the blood hemoglobin content (g Hb/dL whole blood). Any pathology that decreases the amount of hemoglobin in the cells or the number of red blood cells will adversely affect the blood's oxygen-transporting capacity.

People who have lost large amounts of blood need to replace hemoglobin for oxygen transport. The ideal replacement for blood loss is a blood transfusion, but in emergencies this is not always possible. Saline infusions can replace lost blood volume, but saline, like plasma, cannot transport sufficient quantities of oxygen and carbon dioxide. Researchers are currently trying to find artificial oxygen carriers to replace hemoglobin. In times of large-scale disasters, these hemoglobin substitutes will eliminate the need to establish blood type before giving transfusions for blood loss.

Each Hemoglobin Molecule Binds Up to Four Oxygen Molecules

Less than 2% of the oxygen in arterial blood is carried dissolved in the plasma. The remaining oxygen is bound to hemoglobin molecules inside circulating red blood cells. A hemoglobin molecule is composed of four globin subunits, each centered around a heme group whose central iron atom binds reversibly with an oxygen molecule (∞ p. 462). The iron-oxygen interaction is a weak bond, and the two molecules can be easily separated

Blood Substitutes Physiologists have been attempting to find a substitute for blood ever since 1878, when an intrepid scientist named T. Gaillard Thomas transfused whole milk in place of blood. Although milk seems an unlikely replacement for blood, it has two important properties: proteins to provide colloid osmotic pressure and molecules (emulsified lipids) capable of binding to oxygen. In the development of hemoglobin substitutes, oxygen transport is the most difficult property to mimic. A hemoglobin solution would seem to be the obvious answer, but hemoglobin that is not compartmentalized in red blood cells behaves differently than hemoglobin that is. First, the quaternary structure changes so that the four-subunit (tetramer) form is in equilibrium with a smaller two-subunit (dimer) form. The smaller version of hemoglobin is easily excreted by the kidneys, so almost half a dose of hemoglobin solution disappears from the blood in 2–4 hours. Second, hemoglobin found outside the red blood cells does not release oxygen as easily in peripheral tissues. Investigators are making progress by polymerizing hemoglobin into larger, more stable molecules and loading these hemoglobin polymers into phospholipid liposomes (∞ p. 109). By encapsulating hemoglobin into these artificial red blood cells, researchers hope to extend the life span of the blood substitute and make its properties approach those of real blood.

without altering either the hemoglobin or the oxygen. Because there are four iron atoms in a hemoglobin molecule, each hemoglobin molecule has the potential to bind four oxygen molecules. Hemoglobin (Hb) bound to oxygen is known as *oxyhemoglobin*, abbreviated HbO_2. (It would be more accurate to show the actual number of oxygen molecules carried on each heme molecule: $Hb(O_2)_{1-4}$, but, because the number bound varies from molecule to molecule, we will use the simpler abbreviation.) The reversible reaction of oxygen binding to hemoglobin at the lungs can be summarized as

$$Hb + O_2 \rightleftharpoons HbO_2$$

The amount of oxygen bound to hemoglobin depends primarily on the P_{O_2} of the surrounding plasma (Fig. 17-20 ■). Dissolved O_2 in the plasma diffuses into the red blood cell, where it binds to hemoglobin. Binding effectively removes dissolved O_2 from the plasma so that more oxygen can diffuse in from the alveoli. The transfer of oxygen from air to plasma into red blood cells and onto hemoglobin occurs so rapidly that blood leaving the alveoli normally picks up as much oxygen as the P_{O_2} of the plasma and number of red blood cells permit.

It is possible to calculate what percentage of potential oxygen-binding sites are actually carrying O_2 by knowing how much oxygen is actually bound to hemoglobin and the maximum amount of oxygen that it can bind. The amount of oxygen bound to hemoglobin at any given P_{O_2} is expressed as a percentage:

(Actual amount of O_2 bound/maximum that could be bound) $\times 100$ = percent saturation of hemoglobin

The **percent saturation of hemoglobin** refers to the percent of available binding sites that are bound to oxygen. If all binding sites of all hemoglobin molecules are occupied by oxygen molecules, the blood is 100% oxygenated, or *saturated* with oxygen. If half the available binding sites are carrying oxygen, the hemoglobin is 50% saturated, and so on.

The relationship between the plasma P_{O_2} and oxygen-hemoglobin binding can be explained with the following analogy. The hemoglobin molecules carrying oxygen are like students moving books from an old library to a new one. Each student (a hemoglobin molecule) can carry a maximum of four books (100% saturation). The librarian in charge controls how many books (O_2 molecules) each student will carry, just as P_{O_2} determines the amount of oxygen that binds to hemoglobin. At the same time, the total number of books being carried depends on the number of available students, just as the amount of oxygen delivered to the tissues depends on the number of available hemoglobin molecules. For example, if there are 100 students and the librarian gives each of them four books, then 400 books will be carried to the new library. If the librarian has only three books for each (decreased P_{O_2}), then only 300 books will go to the new library, even though each student can carry four (decreased percent saturation of hemoglobin). If the librarian is handing out four books per student but only 50 students show up (fewer hemoglobin molecules), then only 200 books will get to the new library, even though the students will be carrying the maximum number of books they can carry.

Once arterial blood reaches the tissues, the exchange process that took place in the lungs reverses. As dissolved oxygen diffuses out of the plasma into the cells, the resultant drop in plasma P_{O_2} disturbs the equilibrium of the oxygen-hemoglobin binding reaction by removing oxygen from the left side of the equation. The reaction shifts to the left according to the Law of Mass Action (p. 85), and the hemoglobin molecules release their oxygen stores. Like oxygen loading at the lungs, this process takes place very rapidly and goes to equilibrium. The amount of oxygen unloaded from hemoglobin at a cell is determined primarily by the P_{O_2} of the cell, which in turn reflects its metabolic activity.

The Oxygen-Hemoglobin Dissociation Curve Shows the Relationship between P_{O_2} and Hemoglobin Binding of Oxygen

The physical relationship between P_{O_2} and oxygen binding to hemoglobin can be studied *in vitro* in the laboratory. Samples of hemoglobin are exposed to varying P_{O_2} levels, and the amount of oxygen that binds is determined quantitatively. The results of these *in vitro* binding studies are graphically represented by **oxyhemoglobin dissociation curves,** such as the one shown in Figure 17-21 ■. The shape of the oxyhemoglobin dissociation curve reflects the properties of the hemoglobin molecule and its affinity for oxygen. If you look at the curve, you will see that at the normal alveolar and arterial P_{O_2} of 100 mm Hg, 98% of the hemoglobin is bound to oxygen. Because the curve is nearly flat at higher P_{O_2} levels (i.e., slope approaches zero), large changes in P_{O_2} will cause only minor changes in the percent saturation. Although hemoglobin is about 98% saturated at a P_{O_2} of 100 mm Hg, 100% saturation does not occur until the P_{O_2} reaches nearly 250 mm Hg, a partial pressure far higher than anything we encounter in everyday life.

The flattened portion of the dissociation curve at higher dissolved oxygen levels also means that P_{O_2} at the alveoli can drop significantly without having a major effect on hemoglobin saturation. As long as P_{O_2} in the pulmonary capillaries stays above 60 mm Hg, hemoglobin will be more than 90% saturated and near-normal levels of oxygen transport will be maintained.

Once the P_{O_2} drops below 60 mm Hg, the dissociation curve develops a steeper slope. The steep slope means that a small decrease in P_{O_2} causes a relatively large amount of oxygen to be released by hemoglobin. For example, if P_{O_2} falls from 100 mm Hg to 60 mm Hg, the percent saturation of hemoglobin goes from 98% saturation to about 88% saturation, a decrease of 10%. This is equivalent to a saturation change of 2.5% for each 10

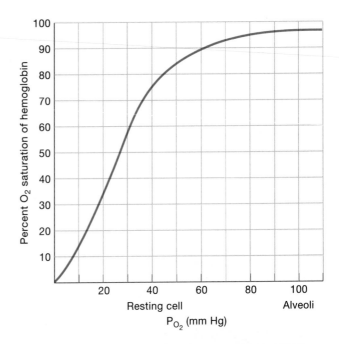

Graph question:
(a) When the P_{O_2} is 20 mm Hg, what is the percent O_2 saturation of hemoglobin?
(b) At what P_{O_2} is hemogoblin 50% saturated with O_2?

■ Figure 17-21 **Oxygen-hemoglobin dissociation curve**

mm Hg change. If P_{O_2} falls further, from 60 to 40 mm Hg, the percent saturation goes from 88% to 75%, a decrease of 6.5% for each 10 mm Hg. Notice that the slope of the curve is greater in the 40–60 mm Hg range than in the 60–100 mm Hg range. If P_{O_2} decreases from 40 mm Hg to 20 mm Hg, the slope of the curve becomes even steeper. Hemoglobin saturation declines from 75% to 35%, a change of 20% for each 10 mm Hg.

Note that at a P_{O_2} of 40 mm Hg, the partial pressure of resting cells, hemoglobin is still 75% saturated and has released only one-fourth of the oxygen it is capable of carrying. This is another example of the built-in reserve capacity of the body. The 75% of the oxygen that remains bound serves as a reservoir that cells can draw upon to meet their needs as metabolism increases. If metabolically active tissues use additional oxygen and their P_{O_2} decreases, additional oxygen is released by hemoglobin. At a P_{O_2} of 20 mm Hg, an average value for exercising muscle, hemoglobin saturation drops to about 35%. With a 20 mm Hg decrease in P_{O_2} (40 mm Hg to 20 mm Hg), hemoglobin releases an additional 40% of the oxygen it is capable of carrying.

Temperature, pH, and Metabolites Affect Oxygen-Hemoglobin Binding

Although P_{O_2} is the primary factor influencing oxygen transport by hemoglobin, any factor that changes the configuration of the hemoglobin protein may affect its

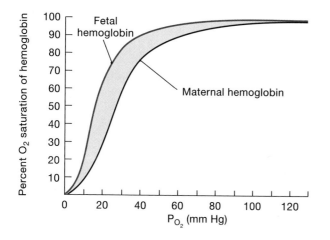

ability to bind oxygen. In humans, physiological changes in pH, P_{CO_2}, and temperature of the blood all affect the oxygen-binding capability of hemoglobin.

Changes in binding affinity are reflected by changes in the shape of the oxygen-hemoglobin dissociation curve. For example, fetal hemoglobin has a unique protein chain for two of its subunits, giving it a different binding affinity for oxygen that is more appropriate for binding oxygen in the low-oxygen environment of the placenta. The different binding affinity is reflected by the different shape of the fetal oxygen-hemoglobin dissociation curve (Fig. 17-22 ■).

Physiological changes in temperature, P_{CO_2}, or H^+ concentration also affect the oxygen-binding capacity of hemoglobin. Increases in temperature, P_{CO_2}, or H^+ concentration decrease the affinity of hemoglobin molecules for oxygen and shift the oxygen-hemoglobin dissociation curve to the right (Fig. 17-23 ■). Decreases in those parameters increase binding affinity and shift the curve to the left. Notice, however, that the changes are much more pronounced in the steep part of the curve. This means that oxygen binding at the lungs will not be greatly affected by temperature, pH, and P_{CO_2}, but oxygen delivery at the tissues will be significantly altered.

Let us examine one specific example, the shift that takes place when pH goes from 7.4 (normal) to 7.2 (more

Graph question:
• Because of incomplete gas exchange across the thick membranes of the placenta, hemoglobin in fetal blood leaving the placenta is 80% saturated with oxygen. What is the P_{O_2} of that placental blood?
• Blood in the vena cava of the fetus has a P_{O_2} around 10 mm Hg. What is the percent O_2 saturation of maternal hemoglobin at the same P_{O_2}?

■ **Figure 17-22 Differences in the oxygen-binding properties of maternal and fetal hemoglobin**

(a) Effect of pH

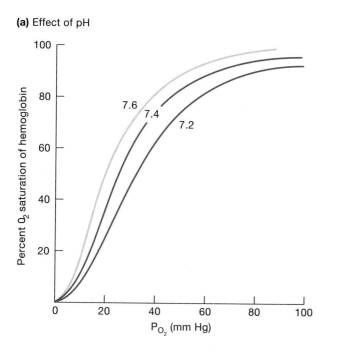

(b) Effect of temperature

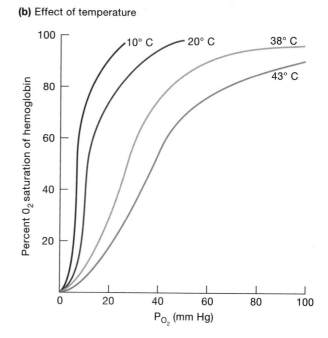

Graph question: At a P_{O_2} of 20 mm Hg, how much more oxygen is released at an exercising muscle cell whose pH is 7.2 than by a cell with a pH of 7.4? What happens to oxygen release when the exercising muscle cell warms up?

■ Figure 17-23 **Physical factors such as pH and temperature affect oxygen binding to hemoglobin.**

acidic). A pH of 7.2 is lower than normal for the body (normal range 7.38–7.42) but is compatible with life. At a P_{O_2} of 40 mm Hg and pH of 7.2, hemoglobin molecules release 15% more oxygen than they do at a pH of 7.4. Lesser changes in pH also shift the curve to the right but not to the same extent. Anaerobic exercise is one example of a situation in which body pH decreases. Anaerobic metabolism in exercising muscle fibers produces lactic acid, which in turn releases H^+ into the cytoplasm and extracellular fluid. As pH falls owing to increasing H^+ concentrations, the affinity of hemoglobin for oxygen decreases and the oxygen-hemoglobin dissociation curve shifts to the right. This shift shows that more oxygen is being released at the tissues as the blood becomes more acidic (pH decreases). A shift in the hemoglobin saturation curve that results from a change in pH is called the **Bohr effect.**

One other factor that affects oxygen-hemoglobin binding is **2,3-diphosphoglycerate (2,3-DPG),** a compound made from an intermediate of the glycolysis pathway. By mechanisms not well understood, **chronic hypoxia** (extended periods of low oxygen) triggers an increase in 2,3-DPG production in red blood cells. Like H^+ and CO_2, 2,3-DPG lowers the binding affinity of hemoglobin and causes the dissociation curve to shift to the right. Ascent to high altitude and anemia are two situations that trigger increased 2,3-DPG production.

Figure 17-24 ■ summarizes the factors that influence oxygen transport in the blood.

✓ A muscle that is actively contracting may have an intracellular P_{O_2} of 25 mm Hg. What happens to oxygen binding to hemoglobin at this P_{O_2}? What is the P_{O_2} of the venous blood leaving the active muscle?

High Altitude   In 1981 a group of 20 physiologists, physicians, and climbers, supported by 42 Sherpa assistants, formed the American Medical Research Expedition to Mt. Everest. The purpose of the expedition was to study human physiology at extreme altitudes, starting with the base camp at 5400 M and continuing on to the summit at 8848 M. Starting with the work of these scientists and adding studies of people who reside at high altitudes, we now have a good picture of the physiology of high-altitude acclimatization. The body's immediate homeostatic response to the hypoxia of high altitude is hyperventilation. Hyperventilation enhances alveolar ventilation, raising the alveolar and arterial P_{O_2}. However, it also causes lower plasma P_{CO_2} and increases pH, causing a state of alkalosis. The pH change in turn increases the affinity of hemoglobin for oxygen, seen on the oxygen-hemoglobin saturation curve as a left shift (Fig. 17-23■). Increased oxygen-hemoglobin binding might seem counterproductive, but in reality it allows the hemoglobin to pick up more oxygen at the lower P_{O_2} of the lungs. After an acclimatization period of hours to days, the production of 2,3-DPG by red blood cells increases, shifting the oxygen-hemoglobin saturation curve back to the right and offsetting the effects of the respiratory alkalosis. Thus, at altitudes up to 6300 M, the hemoglobin saturation curve is essentially normal.

The hypoxia of high altitude also triggers the release of the hormone erythropoietin from the kidney and liver. This hormone stimulates red blood cell production. Even though the P_{O_2} of the blood remains low, the total oxygen-carrying capacity is thus increased. In some individuals, the increased hematocrit (polycythemia) that results from living at high altitudes may be maladaptive. Hematocrits greater than 45% increase the viscosity of the blood enough to impede blood flow to the brain and peripheral tissues. The decrease in blood flow counteracts the increased oxygen-carrying capacity of the blood. Altitude-induced polycythemia is considered to be the cause of the condition known as chronic mountain sickness, with symptoms of headache, fatigue, mental status changes, and poor exercise tolerance. People who exhibit symptoms of elevated hematocrits may be helped by phlebotomy (blood withdrawal). In Leadville, Colorado (3100 M), the 60 people who suffer from chronic mountain sickness are the main donors to the local blood bank!

Carbon Dioxide Is Transported Dissolved in Plasma, Bound to Hemoglobin, and as Bicarbonate Ions

Gas transport in the blood is a two-way process. Carbon dioxide produced as a by-product of cellular respiration is picked up in the tissues and conveyed to the lungs. Removal of carbon dioxide from the body is an essential part of external respiration. Elevated P_{CO_2} (hypercapnia) leads to increased plasma H^+ concentrations and the pH disturbance known as *acidosis* (see Chapter 19). Abnormally high P_{CO_2} levels can also depress the function of the central nervous system, causing confusion, coma, or even death.

Although carbon dioxide is more soluble in body fluids than is oxygen, the cells produce more carbon dioxide than can be carried dissolved in the plasma. Venous plasma with a P_{CO_2} of 46 mm Hg contains about 0.3 mL CO_2/100 mL blood. This represents only about 7% of the carbon dioxide carried by venous blood. The remaining 93% of carbon dioxide that enters the systemic capillaries diffuses into the red blood cells. There, it either binds directly to hemoglobin ($Hb \cdot CO_2$) or is converted to bicarbonate ion, as explained below. Figure 17-25 ■ summarizes carbon dioxide transport in the blood.

About 70% of the carbon dioxide molecules that enter the circulation are transported to the lungs as bicarbonate ions (HCO_3^-) dissolved in the plasma. The conversion of carbon dioxide to bicarbonate serves two purposes: (1) It provides an additional means by which carbon dioxide can be transported from cells to lungs, and (2) the bicarbonate is available to act as a buffer for metabolic acids (∞ p. 26), thereby helping stabilize the body's pH. Rapid production of bicarbonate from car-

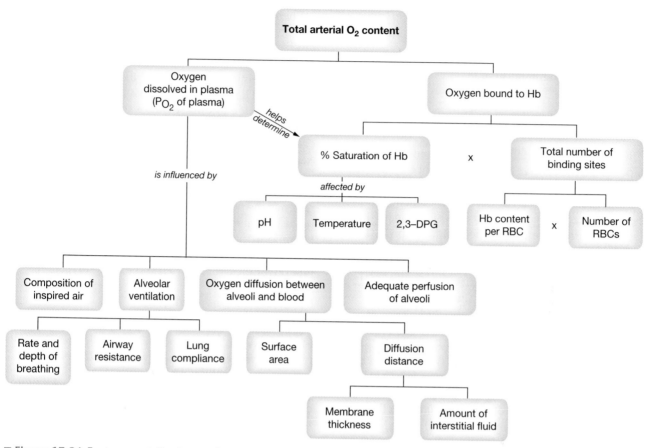

■ Figure 17-24 **Factors contributing to the total oxygen content of arterial blood**

bon dioxide depends on the presence of **carbonic anhydrase,** an enzyme found concentrated in red blood cells.

Dissolved carbon dioxide in plasma diffuses into the red blood cells, where it reacts with water in the presence of carbonic anhydrase to form carbonic acid. Carbonic acid then dissociates into a hydrogen ion and a bicarbonate ion. The equation for the reaction is written as follows:

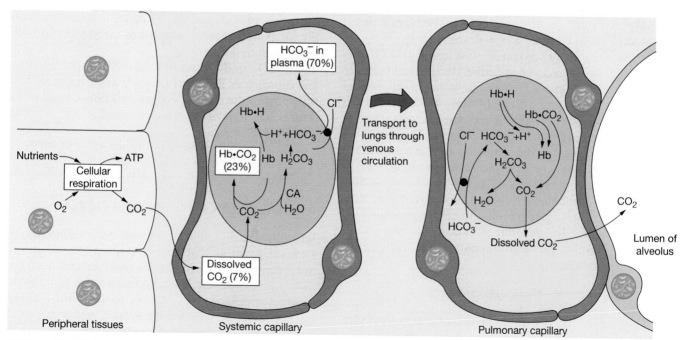

■ Figure 17-25 **Carbon dioxide transport**

$$CO_2 + H_2O \underset{\text{carbonic acid}}{\overset{\text{carbonic anhydrase}}{\rightleftharpoons}} H_2CO_3 \rightleftharpoons H^+ + HCO_3^-$$

Because of the dissociation of carbonic acid, we may ignore the intermediate step and summarize the reaction as:

$$CO_2 + H_2O \overset{\text{carbonic anhydrase}}{\rightleftharpoons} H^+ + HCO_3^-$$

This reaction is reversible. The rate in either direction depends on the relative concentrations of the substrates and obeys the Law of Mass Action. When carbon dioxide moves into the red blood cells, the reaction goes to the right, producing more acid and bicarbonate until a new equilibrium state is reached (Fig. 17-25 ■). (Water is always in excess in the body, so water concentration plays no role in the dynamic equilibrium of this reaction.) In order to keep the reaction going, the products of the reaction, H^+ and HCO_3^-, must be removed from the cytoplasm of the red blood cell.

Removal of free H^+ and HCO_3^- from the red blood cells is accomplished in two separate steps. Hemoglobin is a buffer and binds hydrogen ions (Hb · H). This binding keeps the intracellular concentration of free H^+ low. At the same time, bicarbonate ions produced from carbon dioxide move out of the red blood cell via an antiport transport protein (∞ p. 122). This transport process, known as the **chloride shift,** exchanges one HCO_3^- for one Cl^-. The one-for-one exchange maintains electrical neutrality so that the cell's membrane potential is not affected. As H^+ and HCO_3^- levels in the red blood cell fall, carbon dioxide continues to form carbonic acid. The resultant drop in plasma P_{CO_2} allows more carbon dioxide to diffuse out of cells and into the plasma.

The buffering of H^+ by hemoglobin is an important step that prevents large changes in the body's pH. If blood P_{CO_2} is elevated much above normal, the hemoglobin buffer is not adequate to soak up all of the H^+ produced from the reaction of carbon dioxide and water. In those cases, the excess H^+ remains in the plasma, causing the condition known as **respiratory acidosis.** Further information on the role of the respiratory system in maintaining pH homeostasis is found in Chapter 19.

Although most carbon dioxide that enters the red blood cells is converted to bicarbonate ions, about 23% of the carbon dioxide molecules in venous blood bind directly to hemoglobin. When oxygen leaves its binding sites on the hemoglobin molecule, carbon dioxide binds with the free hemoglobin at exposed amino groups (NH_2), forming **carbaminohemoglobin.** This reaction can be summarized as

$$CO_2 + Hb \rightleftharpoons Hb \cdot CO_2 \text{ (carbaminohemoglobin)}$$

The formation of carbaminohemoglobin is facilitated by the presence of H^+ produced from carbon dioxide because decreased pH in red blood cells decreases the binding affinity of hemoglobin for oxygen.

When venous blood reaches the lungs, the processes that took place in the systemic capillaries proceed in reverse (Fig. 17-25 ■). The P_{CO_2} of the alveoli is lower than that in the venous blood. Carbon dioxide diffuses out of the plasma into the alveoli, and the plasma P_{CO_2} begins to decline. The decrease in plasma P_{CO_2} allows dissolved carbon dioxide to diffuse out of the red blood cells. As carbon dioxide levels in the red blood cells decrease, the equilibrium of the CO_2-bicarbonate reaction is disturbed. Removal of carbon dioxide causes the reaction to proceed from right to left. Hydrogen ions leave the hemoglobin molecules. The chloride shift reverses: Chloride ions return to the plasma in exchange for bicarbonate ions that move back into the red blood cells. The bicarbonate and H^+ re-form into carbonic acid that is converted into water and carbon dioxide. The carbon dioxide is then free to diffuse out of the red blood cell and into the alveoli.

Summary of Gas Transport

Figure 17-26 ■ summarizes the transport of carbon dioxide and oxygen in the blood. Oxygen diffuses down its pressure gradient from the alveoli into the plasma and, from there, into the red blood cells. Hemoglobin binds oxygen molecules, increasing the amount of oxygen that can be transported to the cells. At the cells, the process reverses. P_{O_2} in the cells is lower than that in the arterial blood, so O_2 diffuses from the plasma to the cells. The drop in plasma P_{O_2} causes hemoglobin to release oxygen, giving up additional oxygen to enter the cells. Carbon dioxide from aerobic metabolism simultaneously enters the blood, dissolving in the plasma. From there, it enters the red blood cells, where most is converted to bicarbonate ion and H^+. The bicarbonate is returned to the plasma in exchange for a chloride ion, while the H^+ binds to hemoglobin. A small fraction of the carbon dioxide also binds directly to hemoglobin. At the lungs, the process reverses as carbon dioxide diffuses out into the alveoli.

In order to understand fully how the respiratory system coordinates delivery of oxygen to the lungs with its transport in the circulation, we will now consider the central nervous system control of ventilation.

✓ How would an obstruction of the airways affect the body's pH?

REGULATION OF VENTILATION

Breathing is a rhythmic process that occurs without conscious thought. In that respect, it resembles the rhythmic beating of the heart. However, skeletal muscles, unlike autorhythmic cardiac muscles, are not able to contract spontaneously. Instead, skeletal muscle contraction must be initiated by somatic motor neurons, which in turn are controlled by the central nervous system. In the respiratory system, contraction of the diaphragm and intercostals is initiated by control neu-

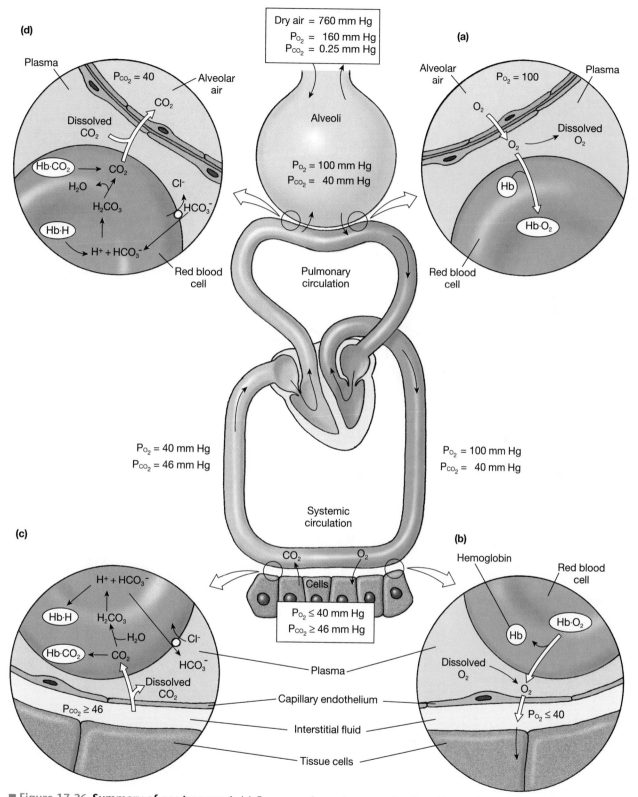

■ Figure 17-26 Summary of gas transport (a) Oxygen exchange between alveoli and blood. (b) Oxygen exchange from blood to cells. (c) Carbon dioxide exchange from cells to blood. (d) Carbon dioxide exchange from blood to alveoli.

rons in the medulla oblongata (Fig. 17-27 ■). These control neurons take the form of a network, or **central pattern generator,** that has intrinsic rhythmic activity (∞ p. 374). It has been postulated that the rhythmic activity arises from a pacemaker neuron, a single neuron with an unstable membrane potential that periodically reaches threshold. However, to date, no pacemaker has been definitively identified, and scientists now believe that the rhythmicity arises within a network of neurons with unstable membrane potentials.

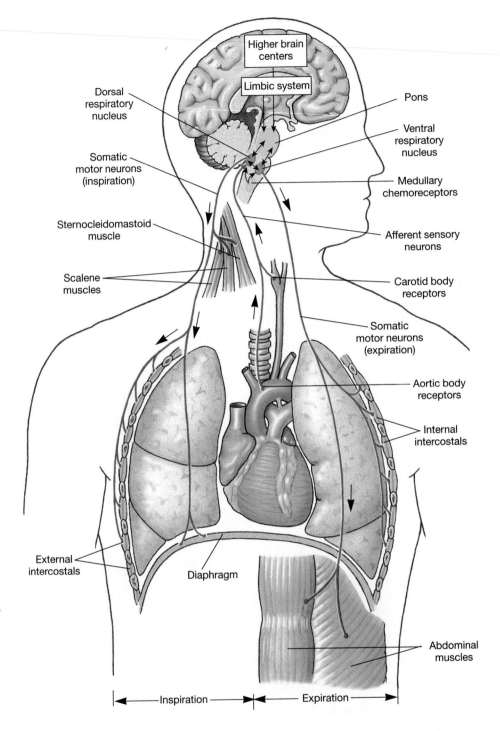

■ Figure 17-27 Reflex control of ventilation Control centers in the brain stem regulate activity in somatic motor neurons leading to the muscles that control ventilation. Chemoreceptors in the brain stem and arteries monitor blood gases and H$^+$ concentrations and influence ventilation.

Direct study of the brain centers controlling ventilation is difficult because of the complexity of the neuronal network and its anatomical location. Consequently, some of our understanding of the control of ventilation has come from observing patients with brain damage. Other information has come from experiments in which the neural connections between major parts of the brain stem were severed. In order to better understand the complex relationships between the parts of the brain that control ventilation, let us review some of these clinical and experimental findings and see what deductions researchers made about the control of rhythmic breathing.

1. *Observation.* If the brain stem is severed below the medulla, all respiratory movement ceases. If the brain stem is cut above the level of the pons, ventilation is normal.

   Hypothesis 1. The control centers for ventilation lie in the medulla or the pons or both.

2. *Observation.* If the medulla is completely separated from the pons and higher brain centers, ventilation becomes gasping and irregular in depth, but the respiratory rhythm remains.

Hypothesis 2. The primary control center for ventilation lies in the medulla.

Hypothesis 3. The medulla contains neurons that set the rhythm of ventilation.

Hypothesis 4. The normal smooth pattern of ventilation depends on communication between neurons in the pons and the medulla.

From these observations and hypotheses, the following model for the control of ventilation has been proposed. Some parts of the model are well supported with experimental evidence; other aspects are still under investigation. The model states that:

1. Respiratory neurons in the medulla control inspiration and expiration.
2. Neurons in the pons influence the rate and depth of ventilation.
3. The rhythmic pattern of breathing arises from a network of spontaneously discharging neurons.
4. Ventilation is subject to modulation by various chemical factors and by higher brain centers. (The experimental evidence for this point will be discussed later.)

Respiratory Neurons in the Medulla Control Inspiration and Expiration

The classic descriptions of the brain's respiratory control of ventilation divided groups of neurons into various control centers. The most recent descriptions have become less specific about assigning function to particular centers and now simply refer to the network of neurons in the brain stem as the central pattern generator. The central pattern generator is unique in that it functions automatically throughout a person's life, and yet it can be controlled voluntarily, up to a point. Complicated synaptic interactions between neurons in the network create the rhythmic cycles of inspiration and expiration, influenced continuously by sensory input from receptors for CO_2, O_2, and H^+. Ventilation pattern depends in large part on the levels of those three substances in the blood.

Although there is still much to be learned about the central pattern generator, we do know that respiratory neurons are found concentrated in two nuclei in the medulla oblongata. The **dorsal respiratory group** (DRG) contains mostly **inspiratory neurons** (I neurons) that control the external intercostal muscles and the diaphragm (Fig. 17-27 ■). The **ventral respiratory group** (VRG) contains neurons that control the muscles used for active expiration (E neurons) and for greater-than-normal inspiration (I+ neurons), such as occurs during vigorous exercise.

During quiet respiration, the inspiratory neurons of the dorsal respiratory group gradually increase stimulation of the inspiratory muscles for 2 seconds. This increase is sometimes called ramping because of the shape of the graph of inspiratory neuron activity (Fig. 17-28 ■). A few inspiratory neurons fire to begin the ramp. The firing of these neurons recruits other inspiratory neurons to fire in an apparent positive feedback loop. As more neurons fire, more skeletal muscle fibers are recruited. The rib cage expands smoothly as the diaphragm contracts. At the end of 2 seconds, the inspiratory neurons abruptly stop firing and the respiratory muscles relax. Over the next 3 seconds, passive expiration occurs because of elastic recoil of the inspiratory muscles and the elastic lung tissue. However, there is some motor neuron activity during passive expiration, suggesting that perhaps muscles in the upper airways contract to slow the flow of air out of the respiratory system.

The expiratory neurons and I+ neurons of the ventral respiratory group remain mostly inactive during quiet respiration. They function primarily during forced breathing, when inspiratory movements are exaggerated, or during active expiration. In forced breathing, the increased activity of dorsal respiratory group inspiratory neurons in some way activates the I+ neurons of the ventral respiratory group. These I+ neurons in turn stimulate accessory inspiratory muscles such as the sternocleidomastoid. Contraction of the accessory inspiratory muscles enhances expansion of the thorax by raising the sternum and upper ribs.

In active expiration, expiratory (E) neurons from the ventral respiratory group activate the internal intercostal and abdominal muscles. There seems to be reciprocal inhibition between the inspiratory and expiratory neurons. The I neurons inhibit the E neurons during inspiration, whereas the E neurons inhibit I neurons during active expiration.

The cycling of respiratory activity is controlled through the neurons of the brain stem. However, activity in these neurons is subject to continuous modulation as gas levels in the blood change.

Carbon Dioxide, Oxygen, and pH Influence Ventilation

Sensory input from two sets of chemoreceptors modifies the rhythmicity of the central pattern generator, triggering reflex changes in ventilation. Carbon dioxide is the primary stimulus for changes in ventilation, with oxygen and plasma pH playing lesser roles. The chemoreceptors for oxygen and carbon dioxide are strategically associated with the arterial circulation. If too little oxygen is present in the arterial blood destined for the tissues, the rate and depth of breathing increase. Or, if the rate of carbon dioxide production by the cells exceeds the rate of carbon dioxide removal by the lungs, ventilation is intensified to match carbon dioxide removal to production. These homeostatic reflexes operate constantly, keeping arterial P_{O_2} and P_{CO_2} within a narrow range.

Peripheral chemoreceptors located in the carotid and aortic bodies sense changes in the oxygen concentration and pH of the plasma (Fig. 17-27 ■). The carotid and

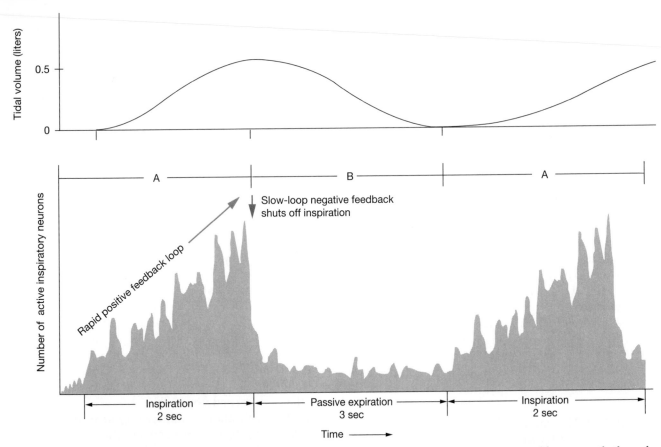

■ **Figure 17-28 Rhythmic breathing** During inspiration, the activity of inspiratory neurons increases steadily, apparently through a positive feedback mechanism. At the end of inspiration, the activity shuts off abruptly and expiration takes place through elastic recoil of the muscles and elastic lung tissue.

aortic bodies also contain the baroreceptors involved in reflex control of blood pressure (∞ p. 443). The **central chemoreceptors** monitor cerebrospinal fluid (CSF) composition and respond to changes in the concentration of CO_2 in the cerebrospinal fluid. These receptors lie on the ventral surface of the medulla, close to neurons involved in respiratory control.

The carotid and aortic bodies The carotid and aortic chemoreceptors are sensitive to the P_{O_2} and pH of arterial blood. When these receptors are activated by decreases in P_{O_2} or pH, they send action potentials through sensory neurons to the brain stem. Sensory information is integrated within the medullary control centers, and the centers respond by sending signals through somatic motor neurons to the skeletal muscles that control ventilation. The response to decreased P_{O_2} and decreased pH is the same: an increase in ventilation.

Under most circumstances, oxygen is not an important factor in modulating ventilation. Arterial P_{O_2} must drop below 60 mm Hg before ventilation is stimulated. This decrease in P_{O_2} is equivalent to ascending to an altitude of 3000 m. (For reference, Denver is located at an altitude of 1609 m.) Because the peripheral chemoreceptors respond only to dramatic changes in arterial P_{O_2}, they

The Carotid Oxygen Sensors The carotid and aortic bodies contain chemoreceptors that adjust ventilation in response to changes in blood levels of oxygen. But what is the mechanism that translates dissolved oxygen concentrations into electrical signals? In the last decade, researchers have come closer to answering that question. Cells in the carotid bodies called **glomus cells** [*glomus*, a ball-shaped mass] have been identified as the chemoreceptors (Fig. 17-29 ■). The glomus cells contain gated potassium channels, called K_{O_2} channels, that are oxygen-sensitive. An oxygen sensor on the extracellular fluid side of the cell membrane is associated with the channel. When the sensor is combined with oxygen, the channel stays open and K^+ leaves the cell, hyperpolarizing it. If oxygen levels in the blood decrease, fewer of the sensors are combined with O_2 and more of the K_{O_2} channels close. The resultant decrease in potassium permeability depolarizes the cells, causing them to fire repetitive action potentials. The action potentials open voltage-gated calcium channels in the glomus cell membrane, allowing Ca^{2+} to enter the cells. Just as in the axon terminal, calcium entry triggers exocytosis of secretory vesicles containing a neurotransmitter, in this case dopamine. Dopamine initiates action potentials in sensory neurons leading to the central nervous system, signaling the respiratory control centers to increase ventilation.

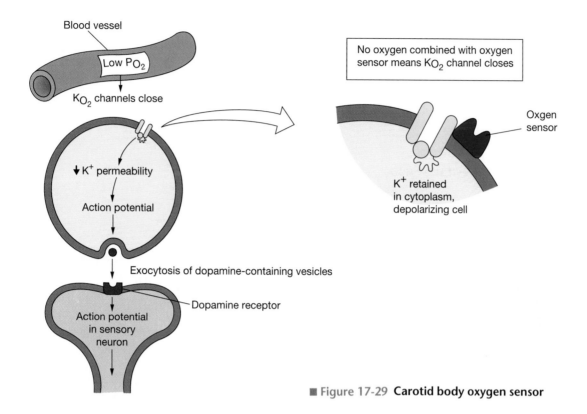

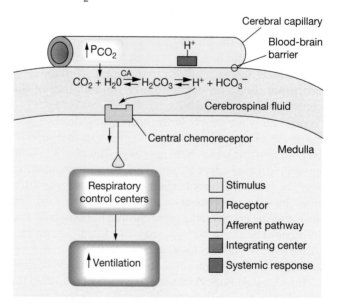

■ Figure 17-29 **Carotid body oxygen sensor**

do not play a role in the everyday regulation of ventilation. However, unusual physiological conditions (such as ascending to high altitude) and pathological conditions such as chronic obstructive pulmonary disease (COPD) can reduce arterial P_{O_2} to low levels that activate the peripheral chemoreceptors.

The peripheral chemoreceptors in the carotid and aortic bodies are more responsive to increases in plasma H^+ concentrations than to changes in P_{O_2}. Any condition that reduces plasma pH, including an increase in plasma P_{CO_2}, stimulates ventilation by way of peripheral chemoreceptors.

Central chemoreceptors The most important chemical controller of ventilation is carbon dioxide, mediated through central H^+ chemoreceptors located in the medulla (Fig. 17-30 ■). These receptors set the respiratory pace, providing continuous input into the central pattern generator.

When arterial P_{CO_2} increases, carbon dioxide crosses the blood-brain barrier quite rapidly, resulting in activation of the central chemoreceptors. The receptors signal the central pattern generator to increase the rate and depth of ventilation, thereby increasing alveolar ventilation and removing carbon dioxide from the blood (Fig. 17-31 ■).

Although we say that the central chemoreceptors respond to carbon dioxide, they actually sense changes in the H^+ concentration of the cerebrospinal fluid. These H^+ ions are formed from carbon dioxide that has diffused across the blood-brain barrier into the cerebrospinal fluid ($CO_2 + H_2O \rightleftharpoons H^+ + HCO_3^-$). Free H^+ in the plasma is unable to cross the blood-brain barrier, however, so H^+ from lactic acid and other metabolic acids has no direct effect on the central chemoreceptors.

Decreases in arterial P_{CO_2} also influence ventilation. If alveolar P_{CO_2} falls, as it might owing to hyperventilation, plasma P_{CO_2} declines, as does P_{CO_2} in the cerebrospinal fluid. As a result, central chemoreceptor activity declines, and the central pattern generator slows the ventilation rate. With decreased ventilation, carbon dioxide begins to accumulate in alveoli and the plasma. Eventually, the arterial P_{CO_2} rises above the threshold level for the

■ Figure 17-30 **Central chemoreceptors** The central chemoreceptors are located in the medulla, facing the cerebrospinal fluid. Carbon dioxide can diffuse across the blood-brain barrier, but free H^+ in the plasma cannot. Once in the cerebrospinal fluid, the carbon dioxide is converted to bicarbonate and H^+. The hydrogen ions thus produced stimulate the central chemoreceptors, causing an increase in ventilation.

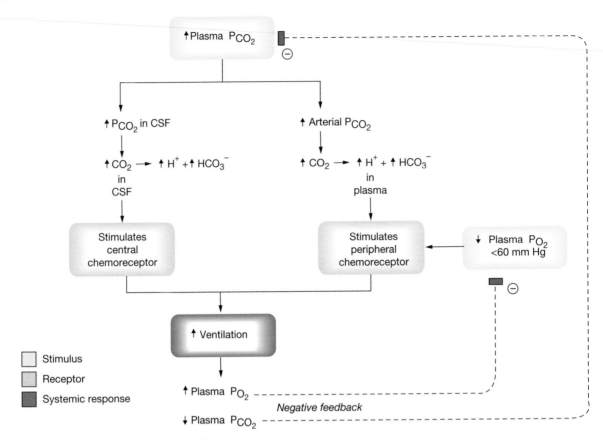

■ Figure 17-31 **Chemoreceptor reflex**

chemoreceptors. At that point, the receptors fire, and the central pattern generator again increases ventilation.

The central chemoreceptors initially respond strongly to an increase in plasma P_{CO_2} by increasing ventilation. However, if P_{CO_2} remains elevated for several days, ventilation decreases. This adaptation to elevated P_{CO_2} occurs because the blood-brain barrier begins to transport bicarbonate ions into the cerebrospinal fluid. The bicarbonate acts as a buffer, removing H^+ from the cerebrospinal fluid and decreasing H^+ stimulation of the central pattern generator.

Fortunately for people with chronic lung diseases, the response of peripheral chemoreceptors to low arterial oxygen remains intact over time, even though the central chemoreceptors adapt to high P_{CO_2}. In some situations, low P_{O_2} becomes the only chemical stimulus for ventilation. For example, patients with severe chronic lung disease such as emphysema have chronic hypercapnia and hypoxia. Their arterial P_{CO_2} may rise to 50–55 mm Hg, while their P_{O_2} falls to 45–50 mm Hg. Because these levels are chronic, the central chemoreceptors gradually adapt to the elevated P_{CO_2}. Most of the chemical stimulus for ventilation then comes from low P_{O_2}, sensed by the carotid and aortic chemoreceptors. If these patients are given too much oxygen, they may stop breathing because their chemical stimulus for ventilation is eliminated.

On occasion, homeostatic compensation for decreased P_{O_2} will adversely affect plasma P_{CO_2} levels. For example, at altitudes above 3000 m, the P_{O_2} of inspired air drops

below 60 mm Hg. This decrease in P_{O_2} activates the carotid and aortic bodies and increases ventilation in an attempt to bring arterial P_{O_2} back into a normal range. But, because the inspired air is low in oxygen, hyperventilation cannot significantly increase arterial P_{O_2}. However, hyperventilation will cause arterial P_{CO_2} to decrease. The decrease in arterial P_{CO_2}, monitored by the central chemoreceptors, then *depresses* ventilation. Within a few days, acclimatization to the lower arterial P_{CO_2} takes place, so ventilation again increases.

Mechanoreceptor Reflexes Protect the Lungs from Inhaled Irritants

In addition to the chemoreceptor reflexes that help regulate ventilation, there are protective reflexes that respond to physical injury or irritation of the respiratory tract and to overinflation of the lungs. The major protective reflex is bronchoconstriction, mediated through parasympathetic neurons that innervate bronchiolar smooth muscle. Inhaled particles or noxious gases stimulate **irritant receptors** in the airway mucosa. The irritant receptors send signals through sensory neurons to control centers in the central nervous system that trigger bronchoconstriction. Protective reflex responses also include coughing and sneezing.

The **Hering-Breuer inflation reflex** is designed to prevent overexpansion of the lungs during strenuous exercise. If tidal volume exceeds 1 liter, stretch receptors in the lung

continued from page 499

Edna is discharged from the hospital after three days with prescriptions for bronchodilator drugs (to keep her airways open) and the nicotine patch (to help her stop smoking). Four days later, however, she becomes dizzy and confused. She falls in her kitchen, but manages to dial 9-1-1. The paramedics who arrive ask her if she has any medical conditions. She replies that she has COPD. The paramedics start Edna on high-flow oxygen and prepare to transport her to the hospital. A paramedic-in-training, along for the ride, asks one of the paramedics about the danger of giving oxygen to Edna. "Thinking has changed on that," the paramedic replies. "Experts think it's more dangerous to withhold oxygen from these patients, especially if they have severe hypercapnia, like this patient."

Question 5: Without performing blood tests, why does the paramedic suspect that Edna has severe hypoxia and hypercapnia (abnormally high levels of P_{CO_2})?

signal the brain stem to terminate inspiration. However, because normal tidal volume is only 500 mL, this reflex does not operate during quiet breathing and mild exertion.

Higher Brain Centers Affect Patterns of Ventilation

Conscious and unconscious thought processes also affect respiratory activities. Higher centers in the hypothalamus and cerebrum can alter the activity of the central pattern generator and change ventilation rate and depth. Voluntary control of ventilation falls into this cat-

egory. But higher brain center control is not a *requirement* for ventilation. Even if the brain stem above the pons is severely damaged, essentially normal respiratory cycles continue.

Respiration can also be affected by stimulation of portions of the limbic system. As a result, emotional and autonomic activities such as fear and excitement may affect the pace and depth of respiration. In some of these situations, the neural pathway goes directly to the somatic motor neurons, bypassing the central pattern generator in the brain stem.

Although we can temporarily alter our respiratory performance, we cannot override the chemoreceptor reflexes. Holding one's breath is a good example. We can hold our breath voluntarily only until elevated P_{CO_2} in the blood and cerebrospinal fluid activates the chemoreceptor reflex, forcing us to inhale. Small children having temper tantrums sometimes attempt to manipulate parents by threatening to hold their breath until they die. However, the chemoreceptor reflexes make it impossible for the children to carry out that threat. Extremely strong-willed children can continue holding their breath until they turn blue and pass out from hypoxia, but once they are unconscious, normal breathing will automatically resume.

Breathing is intimately linked to cardiovascular function. The integrating centers for both functions are located in the brain stem, and interneurons project between the two centers, allowing signaling back and forth. In Chapter 19, we will examine the integration of cardiovascular, respiratory, and renal function as the three systems work together to maintain fluid and acid-base homeostasis. But first, in Chapter 18, we examine the physiology of the kidneys.

CHAPTER REVIEW

Chapter Summary

1. Aerobic metabolism in living cells consumes oxygen and produces carbon dioxide. (p. 474)

2. Gas exchange requires a thin, moist, internalized exchange surface with a large surface area, a pump to move air between the atmosphere and the exchange sur-

face, and a circulatory system to transport gases between the exchange surface and the cells. (p. 474)

3. The functions of the respiratory system in addition to gas exchange include pH regulation, vocalization, and protection from foreign substances. (p. 475)

The Respiratory System

4. **Cellular respiration** refers to the metabolic processes of the cell that consume oxygen and nutrients and produce energy. **External respiration** is the exchange of gases between the atmosphere and the cells. It includes ventilation, gas exchange at the lung surface and at the cells, and transport of gases in the blood. **Ventilation** is the movement of air into and out of the lungs. (p. 475)

5. The **respiratory system** is composed of the anatomical structures involved in ventilation and gas exchange in the lungs. (p. 475)

6. The **upper respiratory tract** includes the mouth, nasal cavity, **pharynx**, **larynx**, and **trachea**. The **lower respira-**

tory tract includes the bronchi, bronchioles, and exchange surfaces of the alveoli. (p. 475)

7. The thoracic cage is bounded by the ribs, spine, and **diaphragm.** Two sets of **intercostal muscles** connect the ribs. The respiratory muscles are skeletal muscles innervated by somatic motor neurons. (p. 476)

8. The lungs are paired air-filled organs, each contained within a double-walled pleural sac. A small quantity of **pleural fluid** lies between the pleural membranes. (p. 476)

9. Air passes through the upper respiratory system to the **trachea.** In the **thorax,** the trachea divides to form the two **primary bronchi** that enter the lungs. Each primary

bronchus divides into progressively smaller bronchi and finally into **bronchioles,** small collapsible passageways that deliver air to the respiratory exchange surfaces. (p. 477)

10. Bronchioles terminate in **alveoli,** air sacs composed mostly of thin-walled **type I alveolar cells** for gas exchange. **Type II alveolar cells** produce surfactant. A network of capillaries surrounds each alveolus. (p. 477)

11. Blood flow through the lungs is equal to the cardiac output. Resistance to blood flow in the pulmonary circulation is low, and pulmonary arterial pressure averages 25/8 mm Hg. (p. 477)

Gas Laws

12. The total pressure of a mixture is the sum of the pressures of the individual gases in the mixture (**Dalton's Law**). The pressure contributed by a single gas such as oxygen or carbon dioxide is known as its **partial pressure** (P_{O_2} and P_{CO_2}, respectively). (p. 480)

13. Bulk flow of air occurs down pressure gradients, as does the movement of individual gas species. Gases move from areas of higher pressure of the gas to areas of lower pressure. (p. 481)

14. The body creates pressure gradients and air flow by changing the volume of the thorax. The inverse relationship between pressure and volume is called **Boyle's Law.** (p. 482)

15. The amount of a gas that will dissolve in a liquid is proportional to the partial pressure of the gas and to the **solubility** of the gas in the liquid. Carbon dioxide is 20 times more soluble in body fluids than oxygen. (p. 482)

Ventilation

16. The upper respiratory system is lined with a ciliated epithelium with mucus-secreting goblet cells. The sticky mucus traps dust and microorganisms, filtering out harmful particles. The cilia move the mucus toward the pharynx, where it is swallowed. Evaporation of water from the mucus humidifies the incoming air. (p. 483)

17. Air flow in the respiratory system is directly proportional to the pressure gradient and inversely related to the resistance of the airways to flow. (p. 483)

18. A single **respiratory cycle** consists of an inspiration and an expiration. (p. 484)

19. During **inspiration,** air flows into the lungs as the thoracic cavity enlarges and the **intrapulmonary pressure** drops. Inspiration requires contraction of the inspiratory muscles and the diaphragm. (p. 484)

20. Air flows out during **expiration,** when the volume of the thoracic cavity decreases and the intrapulmonary pressure increases. Expiration is usually passive, resulting from elastic recoil of the skeletal muscles and lungs. (p. 486)

21. **Active expiration** requires contraction of the internal intercostal and abdominal muscles. (p. 486)

22. Intrapleural pressures during ventilation normally remain subatmospheric because the pleural cavity is a sealed compartment. (p. 487)

23. **Compliance** is the ease with which the chest wall and lungs expand. Loss of compliance increases the work of breathing. **Elastance** is the ability of a stretched lung to resume its normal volume. (p. 488)

24. **Surfactant** is a chemical secreted by type II alveolar cells. It decreases surface tension in the fluid that lines the alveoli and prevents the smaller alveoli from collapsing into the larger ones. (p. 489)

25. The diameter of the bronchioles is an important factor affecting air flow. Smaller diameter causes higher resistance to flow. (p. 490)

26. Carbon dioxide in expired air is the main controller of bronchiolar diameter. Increased CO_2 dilates bronchioles. Parasympathetic neurons cause **bronchoconstriction** in response to irritant stimuli in the airways. There is no significant sympathetic innervation of bronchioles, but epinephrine will cause **bronchodilation.** (p. 491)

27. The **tidal volume** is the amount of air taken in during a single normal inspiration. **Vital capacity** is the tidal volume plus the **expiratory** and **inspiratory reserve volumes.** The air left in the lungs at the end of maximum expiration is the **residual volume.** (p. 492)

28. **Total pulmonary ventilation** is the tidal volume times the ventilation rate. **Alveolar ventilation** is a more accurate indicator of air supply to the alveoli. Alveolar ventilation is ventilation rate times (tidal volume minus dead space volume). (p. 493)

29. Gas composition in the alveoli changes very little during a normal respiratory cycle. **Hyperventilation** increases alveolar P_{O_2} and decreases alveolar P_{CO_2}. **Hypoventilation** has the opposite effect. (p. 495)

30. Air flow is matched to blood flow around the alveoli by local mechanisms. Carbon dioxide dilates bronchioles, and decreased oxygen levels constrict pulmonary arterioles. The response of pulmonary arterioles to low oxygen is opposite that of the systemic arterioles. (p. 495)

Gas Exchange in the Lungs

31. Oxygen and carbon dioxide move between the alveoli and blood by simple diffusion. Factors that affect partial pressure gradients at the alveoli include the composition of the inspired air and the effectiveness of alveolar ventilation. (p. 496)

32. Gas exchange between the alveoli and blood is affected by changes in the surface area available for diffusion, the thickness of the alveolar membrane, and the diffusion distance across the interstitial space. (p. 498)

33. Normal alveolar and arterial P_{O_2} is 100 mm Hg, and normal P_{CO_2} is 40 mm Hg. (p. 499)

Gas Exchange in the Tissues

34. Gas exchange in the tissues depends on the pressure gradient between the blood and the cells. Venous blood leaving the tissues has an average P_{O_2} of ≤ 40 mm Hg and an average P_{CO_2} of ≥ 46 mm Hg. (p. 499)

Gas Transport in the Blood

35. Oxygen and carbon dioxide that enter the blood first dissolve in the plasma. Their solubility, however, is limited, and red blood cells assist with gas transport. (p. 500)

36. More than 98% of the oxygen carried in the blood is found inside the red blood cells bound to hemoglobin. (p. 501)

37. The P_{O_2} of the plasma determines how much oxygen will bind to hemoglobin. At alveolar P_{O_2}, hemoglobin is almost fully saturated with oxygen. At the peripheral tissues, hemoglobin releases variable amounts of oxygen depending on the P_{O_2} of the tissue. (p. 501)

38. Oxygen-hemoglobin binding is affected by pH, temperature, and **2,3-diphosphoglycerate (2,3-DPG),** a red blood cell metabolite. More oxygen will be released if the pH drops (**Bohr effect**), the temperature rises, or the red blood cells generate 2,3-DPG owing to hypoxia. (p. 502)

39. About 7% of the carbon dioxide transported in the blood is dissolved in the plasma. Another 23% is bound as **carbaminohemoglobin** within the red blood cells. (p. 504)

40. The remaining 70% of blood carbon dioxide is converted to carbonic acid by the enzyme **carbonic anhydrase** inside the red blood cells. Carbonic acid then dissociates into H^+ and HCO_3^-. The H^+ ions are buffered by hemoglobin. The HCO_3^- leaves the red blood cell in exchange for a chloride ion (the **chloride shift**). Once in the plasma, HCO_3^- acts as a buffer. (p. 505)

Regulation of Ventilation

41. Respiratory control resides in a network of neurons in the pons and medulla oblongata. This network is known as the **central pattern generator.** (p. 506)

42. Inspiratory neurons are concentrated in the **dorsal respiratory group** of the medulla. They control somatic motor neurons to the inspiratory muscles and the diaphragm. The **ventral respiratory group** of neurons assists in greater-than-normal inspiration and active expiration. Normal passive expiration occurs when the inspiratory neurons cease firing. (p. 509)

43. Carbon dioxide is the primary stimulus for changes in ventilation. **Central chemoreceptors** in the medulla facing the cerebrospinal fluid respond to changes in P_{CO_2} via changes in CSF H^+ concentrations. An increase in arterial P_{CO_2} increases ventilation. (p. 509)

44. **Peripheral chemoreceptors** in the carotid and aortic bodies monitor P_{O_2} and pH in the blood. Ventilation increases with a decrease in pH or if P_{O_2} falls below 60 mm Hg. (p. 510)

45. Protective reflexes monitored by peripheral mechanoreceptors prevent injury to the lungs from overinflation or irritants. (p. 512)

46. Conscious and unconscious thought processes can affect respiratory activity. (p. 513)

Questions

LEVEL ONE **Reviewing Facts and Terms**

1. List four functions of the respiratory system.

2. Give two different definitions for the word *respiration.*

3. Which sets of muscles are used for normal quiet inspiration? For normal quiet expiration? For active expiration?

4. What is the function of pleural fluid?

5. Name the anatomical structures that an oxygen molecule passes on its way from the atmosphere to the blood.

6. Diagram the structure of an alveolus and give the function of each part. How are capillaries associated with an alveolus?

7. Trace the path of the pulmonary circulation. About how much blood is found here at any given moment? What is a typical arterial blood pressure for the pulmonary circuit and how does this compare to that of the systemic circulation?

8. List three factors that influence the movement of gas molecules from air into solution. Which of these factors is usually not significant in humans?

9. What happens to inspired air as it is conditioned during its passage through the airways?

10. During inspiration, most of the volume change of the chest is due to movement of the _____.

11. Describe the changes in intra-alveolar and intrapleural pressure during one respiratory cycle.

12. What is the function of surfactant?

13. Of the three factors that contribute to the resistance of air flow through a tube, which plays the largest role in changing resistance in the human respiratory system?

14. Match the following items with their correct effect on the bronchioles:
    (a) histamine
    (b) epinephrine
    (c) acetylcholine
    (d) increased P_{CO_2}

    1. bronchoconstriction
    2. bronchodilation
    3. no effect

15. Define these four terms and explain how they relate to one another: tidal volume, inspiratory reserve volume,

residual volume, expiratory reserve volume. Give an average value for each volume in a young male. List some factors that explain differing values among different individuals.

16. More than _____% of the oxygen in arterial blood is transported bound to hemoglobin. How is the remaining oxygen transported to the cells?

17. Name four factors that affect the amount of oxygen that binds to hemoglobin. Which of these four factors is the most important?

18. Describe the structure of a hemoglobin molecule. What element is essential for hemoglobin synthesis?

19. The centers for control of ventilation are found in the _____ and _____ of the brain. The dorsal and ventral respiratory groups of neurons control what processes, respectively? What is a central pattern generator?

20. Name the chemoreceptors that influence ventilation and explain how they do so. What chemical is the most important controller of ventilation?

21. Describe the protective reflexes of the respiratory system. What does the Hering-Breuer reflex prevent? How is it initiated?

LEVEL TWO   Reviewing Concepts

22. **Concept map:** Draw an alveolus and adjacent capillary similar to that shown in Figure 17-20. Write in the partial pressures of oxygen and carbon dioxide for the arterial and venous blood and the air. Draw arrows showing diffusion of the gases. Relate the four rules governing diffusion to the figure. Indicate how at least three different pulmonary pathologies can interfere with gas exchange.

23. A container of gas with a moveable piston has a volume of 500 mL and a pressure of 60 mm Hg. The piston is moved and the new pressure is measured as 150 mm Hg. What is the new volume of the container?

24. You have a mixture of gases in dry air, with an atmospheric pressure of 760 mm Hg. Calculate the partial pressure of each gas for each of the examples below:
    a. 21% oxygen, 78% nitrogen, 0.3% carbon dioxide
    b. 40% oxygen, 13% nitrogen, 45% carbon dioxide, 2% hydrogen
    c. 10% oxygen, 15% nitrogen, 1% argon, 25% carbon dioxide

25. Compare and contrast the following sets of concepts:
    a. compliance and elastance
    b. inspiration and expiration
    c. intrapleural and intra-alveolar pressures

    d. total pulmonary ventilation and alveolar ventilation
    e. transport of oxygen and carbon dioxide in arterial blood

26. Neelesh is helping his mother clean a dusty attic. As he watches the dust particles dance in a sunbeam, it occurs to him that he is inhaling many particles like these. Where do these particles end up in his body? Explain.

27. Define the following terms: pneumothorax, spirometer, hypoxia, COPD, auscultation, hypoventilation, hypercapnia, bronchoconstriction, minute volume.

28. The cartoon coyote is blowing up a balloon as part of an attempt to once more catch a roadrunner. He first breathes in as much air as he can, then blows out all that he can into the balloon. The volume of air in the balloon is equal to the _____ _____ of the coyote's lungs. This volume can be measured directly, as above, or by adding what respiratory volumes together? In 10 years, when the coyote is still chasing the roadrunner, will he still be able to put as much air into the balloon in one breath? Explain.

29. Define these terms and explain how they differ: oxyhemoglobin, carbaminohemoglobin, hemoglobin saturation, carbonic anhydrase, polycythemia, erythropoietin.

LEVEL THREE   Problem Solving

30. Li is a tiny woman, with a tidal volume of 400 mL and a respiratory rate of 12 breaths per minute at rest. What is her total pulmonary ventilation? Just before a physiology exam, her ventilation increases to 18 breaths per minute from nervousness. Now what is her total pulmonary ventilation? Assuming her dead space is 120 mL, what is her alveolar ventilation in each case?

31. Marco tries to hide at the bottom of a swimming hole by breathing through a garden hose that greatly increases his dead space. What happens to the following parameters in his arterial blood, and why?
    a. P_{CO_2} b. P_{O_2} c. bicarbonate ion   d. pH

32. A hospitalized patient with severe chronic obstructive lung disease has a P_{CO_2} of 55 mm Hg and a P_{O_2} of 50 mm Hg. To elevate his blood oxygen, he is given pure oxygen through a nasal tube. The patient immediately stops breathing. Explain why this might occur.

33. You are a physiologist on the manned space flight to a distant planet. You find intelligent humanoid creatures inhabiting the planet, and they willingly submit to your tests. Some of the data you have collected are described below.
    a. The graph below shows the oxygen dissociation curve for the oxygen-carrying pigment in the blood of the humanoid named Bzork. Bzork's normal alveolar P_{O_2} is 85 mm Hg. His normal cell P_{O_2} is 20 mm Hg but it drops to 10 mm Hg with exercise.

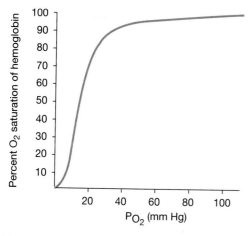

(1) What is the percent saturation for Bzork's oxygen-carrying pigment in blood at the alveoli? At an exercising cell?

(2) What conclusions can you draw about Bzork's oxygen requirement during normal activity and during exercise, based on the graph?

b. The next experiment on Bzork involves his ventilatory response to different conditions. The data from that experiment are graphed below. Line A represents ventilation with an alveolar P_{O_2} of 50 mm Hg. Line B represents ventilation with an alveolar P_{O_2} of 85 mm Hg. Line C represents ventilation with an alveolar P_{O_2} of 85 mm Hg after Bzork has been given beer to drink. Interpret the results of experiments A and C.

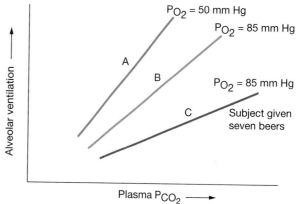

Problem Conclusion

In this running problem, you learned about chronic obstructive pulmonary disease.

Further check your understanding of this running problem by checking your answers against those in the summary table.

Question	Facts	Integration and Analysis
1 What does narrowing of the airways do to the resistance of the airways to air flow?	The relationship between radius and resistance is the same for air flow as it was for blood flow in the circulatory system: As radius decreases, resistance increases (∞ p. 389).	When resistance increases, the body must use more energy to create air flow or blood flow.
2 Why do people with chronic bronchitis have a higher-than-normal rate of respiratory infections?	Cigarette smoke paralyzes the cilia that sweep debris and mucus out of the airways. Without the action of cilia, mucus and trapped particles pool in the airways.	Because of the lack of the ciliary "housekeeping," bacteria can become established in the lower respiratory tract more easily.
3 Name the muscles that patients with emphysema use to exhale forcefully.	Normal expiration depends on elastic recoil of muscles and elastic tissue in the lungs.	Forceful expiration uses the internal intercostal muscles and the abdominal muscles.
4 Why does Edna have an increased hematocrit?	The body is heavily dependent on the oxygen carried by hemoglobin in red blood cells. Because of Edna's COPD, her arterial P_{O_2} is low. The major stimulus for red blood cell synthesis is hypoxia.	Low arterial oxygen levels will trigger the synthesis of additional red blood cells. This provides more binding sites for oxygen transport.
5 Without performing blood tests, why does the paramedic suspect that Edna has severe hypoxia and hypercapnia (abnormally high levels of P_{CO_2})?	Abnormally high P_{CO_2} or low P_{O_2} can depress the function of the central nervous system.	Edna's altered mental state suggests lack of oxygen to the brain or excessively high CO_2 levels.

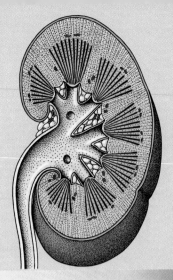

18
The Kidneys

BACKGROUND BASICS

This chapter introduces the **urinary system** [*ouron*, urine], composed of the kidneys and accessory structures. The study of kidney function is called **renal physiology,** from the Latin word *renes*, meaning "kidneys."

The kidneys are best known for the production of urine, a fluid waste product whose composition and volume reflect the functions of many organs and systems of the body. Physicians since ancient times have used urine as a diagnostic tool. *Uroscopy*, inspection of the urine, has been a standard procedure for centuries, and physicians used to carry special flasks for that purpose. The first characteristic they examined was the color of the urine. Was it dark yellow (concentrated), pale straw (dilute), red (indicating the presence of blood), or black (indicating the presence of hemoglobin metabolites)? One form of malaria was known as blackwater fever because the abnormal breakdown of red blood cells caused hemoglobin to turn victims' urine black or dark red. Urine was also examined for clarity, smell, taste, and froth (abnormal presence of proteins). Physicians who did not want to taste the urine themselves would allow their students the "privilege" of tasting it for them. The physician without students might expose insects to the urine and study their reaction. Probably the most famous example of using urine for diagnosis is the taste test for diabetes mellitus, historically known as the honey-urine disease. Diabetes is an endocrine disorder characterized by the presence of glucose in the urine. The urine of diabetics tasted sweet and attracted insects, making the diagnosis clear. Although today we have much more sophisticated tests for glucose in the urine, the first step of a urinalysis is still to examine the color, clarity, and odor of the urine. In this chapter, you will learn why we can tell so much about the body's function by what is present in the urine.

Problem

Gout

Michael Moustakakis, 43, had spent the last two days on the sofa in teeth-gritting pain. But it was not a migraine or stomach pain that laid him low: It was a throbbing, relentless pain in his left big toe. When the pain began, Michael thought he had a mild sprain or perhaps the beginnings of arthritis. But then the pain intensified, and the toe joint became hot and red. Finally Michael hobbled into his doctor's office, feeling a little silly about the nature of his problem. But upon hearing his symptoms, the doctor seemed to know instantly what was wrong. "Sounds to me like you have gout," said Dr. Garcia.

continued on page 525

FUNCTIONS OF THE KIDNEYS

If you asked people on the street, "What is the most important function of the kidney?" they would probably say, "The removal of wastes." Actually, the most important function of the kidney is the homeostatic regulation of the water and ion content of the blood, also called "salt and water" or "fluid and electrolyte" balance. Waste removal is also important, but disturbances in blood volume or ion levels will cause serious medical problems before the accumulation of metabolic wastes reaches toxic levels. The kidneys maintain normal concentrations of ions and water by balancing the intake of those substances with their excretion in the urine, obeying the principle of mass balance (∞ p. 6).

The functions of the kidneys can be classified into six general areas:

1. *Regulation of extracellular fluid volume.* If extracellular fluid volume and plasma volume fall below a certain level, blood pressure falls to levels that cannot sustain adequate blood flow to the brain and other essential organs. The kidneys work in an integrated fashion with the cardiovascular system to ensure that blood pressure and tissue perfusion remain within an acceptable range.

2. *Regulation of osmolarity.* A major function of the kidneys is the maintenance of blood osmolarity at a value close to 290 mOsM. The reflex pathways involved in the regulation of volume and osmolarity are the subject of Chapter 19.

3. *Maintenance of ion balance.* The kidneys keep the concentrations of key ions within a normal range by balancing dietary intake with urinary loss. Sodium (Na^+) is the major ion involved in the regulation of extracellular fluid volume and osmolarity. Potassium (K^+) and calcium (Ca^{2+}) concentrations are also closely regulated. The regulation of Na^+ and K^+ is discussed in Chapter 19. The discussion of Ca^{2+} is deferred until Chapter 21, when we look at the overall picture of calcium homeostasis in the body.

4. *Homeostatic regulation of pH.* The pH of plasma is determined by the concentration of hydrogen ions (H^+). If extracellular fluid becomes too acidic, the kidneys remove H^+ and conserve bicarbonate ions (HCO_3^-), which act as a buffer (∞ p. 26). Conversely, when the extracellular fluid becomes too alkaline, the kidneys remove HCO_3^- and conserve H^+. Although the kidneys do not correct pH disturbances as rapidly as the lungs, the kidneys play a significant role in the homeostasis of blood pH.

5. *Excretion of wastes and foreign substances.* The kidneys remove two types of wastes: those that are by-products of metabolism, and foreign substances such as drugs and environmental toxins. Metabolic wastes include creatinine from muscle metabolism (∞ p. 338) and the nitrogenous wastes *urea* and *uric acid*. Examples of foreign substances that are actively removed by the kidneys include the artificial sweetener *saccharin* and the anion *benzoate*, part of the preservative *potassium benzoate* that you ingest each time you drink a diet soft drink.

6. *Production of hormones.* Although the kidneys are generally not thought of as endocrine glands, two major hormones are synthesized and released by renal cells: *erythropoietin*, the regulator of red blood cell synthesis (∞ p. 460), and *renin*, a hormone linked to salt and water balance. The kidneys also help convert vitamin D_3 into a hormone that regulates Ca^{2+} balance.

The kidneys, like many organs, are supplied with a tremendous reserve capacity. It is estimated that you must lose nearly three-fourths of your kidney function before homeostasis begins to be affected. Many people function perfectly normally with only one kidney, including the 1 in 1000 people born with only one kidney (the mate fails to develop during gestation) or those people who donate a kidney for transplantation.

The next section of the chapter examines the anatomical organization of the urinary system.

✓ Ion regulation is a key feature of kidney function. What happens to the resting membrane potential of a neuron if extracellular K^+ levels decrease? (∞ p. 223)

✓ What happens to the force of cardiac contraction if plasma Ca^{2+} levels decrease substantially? (∞ p. 397)

Anatomy Summary The Urinary System

■ Figure 18-1

(a) The urinary system

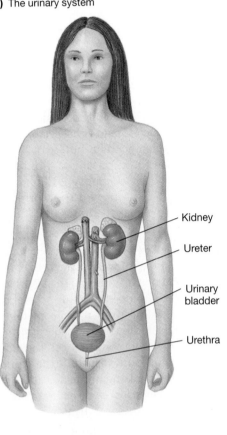

Kidney

Ureter

Urinary bladder

Urethra

(c) In cross section, the kidney is divided into an outer cortex and an inner medulla. Urine leaving the nephrons flows into the renal pelvis prior to passing through the ureter into the bladder

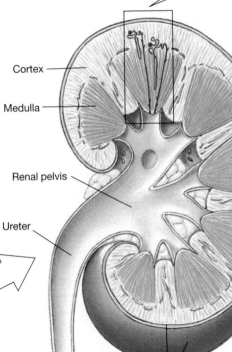

Cortex

Medulla

Renal pelvis

Ureter

Capsule

(b) The kidneys are located retroperitoneally at the level of the lower ribs.

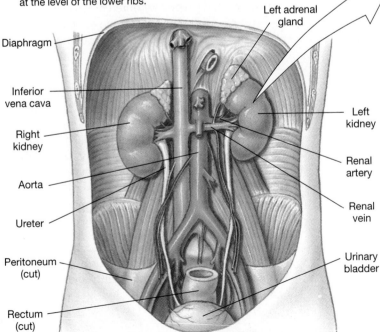

Diaphragm

Inferior vena cava

Right kidney

Aorta

Ureter

Peritoneum (cut)

Rectum (cut)

Left adrenal gland

Left kidney

Renal artery

Renal vein

Urinary bladder

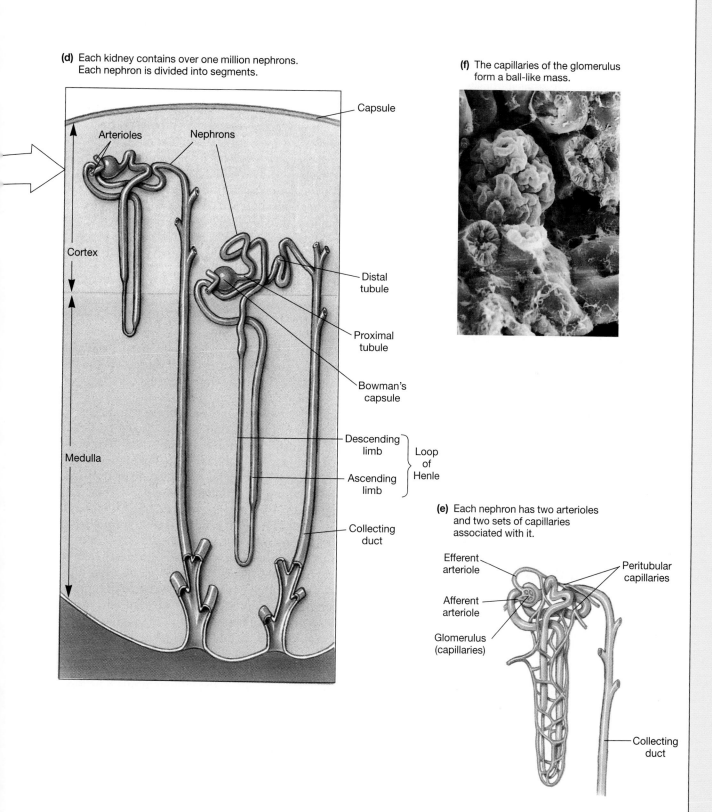

(d) Each kidney contains over one million nephrons. Each nephron is divided into segments.

Capsule

Arterioles

Nephrons

Cortex

Medulla

Distal tubule

Proximal tubule

Bowman's capsule

Descending limb

Ascending limb

Loop of Henle

Collecting duct

(f) The capillaries of the glomerulus form a ball-like mass.

(e) Each nephron has two arterioles and two sets of capillaries associated with it.

Efferent arteriole

Afferent arteriole

Glomerulus (capillaries)

Peritubular capillaries

Collecting duct

ANATOMY OF THE URINARY SYSTEM

The Urinary System Consists of Kidneys, Ureters, Bladder, and Urethra

The urinary system is composed of two **kidneys** linked to the urinary bladder by a pair of tubes called **ureters** (Fig. 18-1a ■). The bladder is connected to the external environment through a single tube, the urethra. The kidneys lie on either side of the spine at the level of the eleventh and twelfth ribs, just above the waist (Fig. 18-1b ■). Although they are below the diaphragm, they are technically outside the abdominal cavity (**retroperitoneal**), sandwiched between the membranous **peritoneum,** which lines the abdomen, and the bones and muscles of the back.

In humans, the concave surface of each kidney faces the spine. The renal blood vessels, nerves, lymphatics, and the ureters all emerge from this surface. The blood supply to the kidneys comes from **renal arteries,** which branch off the abdominal aorta. Blood leaving the kidneys is carried back to the inferior vena cava through **renal veins.** At any given time, 20%–25% of the cardiac output goes to the kidneys, although they are only 0.4% of the total body weight (4.5–6 ounces each). This high rate of blood flow through the kidneys is critical to renal function.

One ureter leads from each kidney to the **urinary bladder,** a hollow, expandable organ that collects and stores urine. The walls of both the ureters and bladder have well-developed layers of smooth muscle. Urine produced in the kidneys is moved down the ureters by periodic rhythmic contractions. The bladder fills with urine until, by reflex action, the urine is released to the environment through a single passageway, the **urethra.** The urethra in males exits the body through the shaft of the penis. In females, the urethral opening is found anterior to the openings of the vagina and the anus.

The Nephron Is the Functional Unit of the Kidney

A cross section through a kidney shows that the interior is arranged in two layers: an outer **cortex** and inner **medulla** (Fig. 18-1c ■). The layers are formed by the organized arrangement of microscopic tubules called **nephrons** (Fig. 18-1d ■). The nephron is the functional unit* of the kidney, with about 1 million nephrons in each kidney. Each nephron has both **vascular elements,** the blood vessels associated with the nephron, and **tubular elements** (Fig. 18-1d, e ■).

Vascular elements of the nephron Blood enters the kidney through the renal artery, moving into smaller arteries and then into arterioles. The organization of arterioles and capillaries within the kidney is unique. Blood enters an arteriole and a set of capillaries, then flows into a *second* arteriole and a second set of capillaries.

Arterial blood arriving at a nephron flows into the **afferent arteriole,** then into a ball-like network of capillaries known as the **glomerulus** [*glomus,* a ball-shaped mass; plural *glomeruli*] (Fig. 18-1f ■). The glomerulus is the site where plasmalike fluid filters out of the capillaries and into the lumen of the nephron tubule. This process is similar to the filtration of water and molecules out of systemic capillaries in other tissues, as you will see.

Blood leaving the glomerulus flows into an **efferent arteriole.** It then enters a series of **peritubular capillaries** [*peri-,* around] that surround the tubule (Fig. 18-1e ■). Additional exchange of water and solutes with the nephron takes place in these capillaries. Finally, the peritubular capillaries join to form venules and small veins, exiting the kidney through the renal vein.

Tubular elements of the nephron The nephron begins in the cortex of the kidney with a hollow ball-like structure called **Bowman's capsule** that surrounds the glomerulus. The capillary endothelium of the glomerulus is fused to the epithelium of Bowman's capsule so that fluid filtering out of the capillaries passes directly into the lumen of the tubule. The combination of glomerulus and Bowman's capsule is called the **renal corpuscle.**

From Bowman's capsule, filtered fluid flows into the **proximal tubule** [*proximal,* close or near], then into the **loop of Henle,** a hairpin-shaped segment that dips down into the medulla and then goes back up to reenter the cortex. The loop of Henle is divided into two *limbs,* a **descending limb,** in which fluid flows into the medulla, and an **ascending limb,** in which fluid flows back to the cortex. Fluid leaving the loop of Henle goes into the **distal tubule** [*distal,* distant or far]. Notice in Figure 18-1d ■ the manner in which the kidney tubule twists and folds back on itself so that the distal tubule passes between the afferent and efferent arterioles, forming a region known as the **juxtaglomerular apparatus** [*juxta-,* beside]. The association of the distal tubule with the arterioles allows paracrine communication between the two structures, a key feature of kidney autoregulation.

The distal tubules of up to eight nephrons drain into a single larger tube called the **collecting duct.** (The distal

Urinary Tract Infections Because of the shorter length of the female urethra and its proximity to bacteria from the large intestine, women are more prone than men are to develop bacterial infections of the bladder and kidneys, or **urinary tract infections** (UTIs). The most common cause of UTIs is the bacterium *Escherichia coli,* a normal inhabitant of the human large intestine. *Escherichia coli* is not harmful while restricted to the lumen of the large intestine, but it is pathogenic [*patho-,* disease + *-genic,* causing] if it gets into the urethra. The most common symptoms of a UTI are pain or burning with urination and frequency of urination. A urine sample from a patient with a UTI will often contain many red and white blood cells, neither of which is commonly found in normal urine. UTIs are treated with antibiotics.

---

*A *functional unit* is the smallest structure that can carry out all the functions of the organ.

tubules and collecting duct together form the **distal nephron.**) Collecting ducts pass from the cortex through the medulla and drain into the **renal pelvis.** From the renal pelvis, the filtered and modified fluid, now called **urine,** goes through the ureter to the urinary bladder.

That is the basic pattern of blood and fluid flow through the kidney. We will reserve detailed descriptions of these structures for later, when you will examine the function and structure of each segment of the tubule. Because the twisted configuration of the kidney tubule makes it difficult to follow fluid flow, we will unfold the nephron in the remaining figures in this chapter so that fluid flows from left to right across the figure (see Figure 18-2 ■).

PROCESSES OF THE KIDNEYS

Imagine drinking a 12-ounce soft drink every three minutes around the clock; by the end of 24 hours, you would have consumed the equivalent of 90 2-liter bottles. The thought of putting that much fluid through our intestinal tract is staggering, but that is how much plasma moves into the nephrons in a day! Because the average volume of urine leaving the kidneys is only 1.5 L/day, more than 99% of the fluid that enters the nephrons finds its way back into the blood rather than into the bladder for excretion.

The Four Processes of the Kidney Are Filtration, Reabsorption, Secretion, and Excretion

There are four basic processes of the urinary system: filtration, reabsorption, secretion, and excretion (Fig. 18-2 ■). **Filtration** is the movement of fluid from the blood into the lumen of the nephron. Filtration takes place only in the renal corpuscle, where the epithelium of Bowman's capsule is modified to allow bulk flow of fluid across it. After filtered fluid has passed into the lumen of the nephron, it is considered to be outside the body, just as substances in the lumen of the intestinal tract are not part of the internal environment (∞ Fig. 1-2 ■ p. 3). Thus, anything that filters into the nephron is destined for removal in the urine unless reabsorbed into the body.

As filtered fluid flows from Bowman's capsule through the tubule, it is modified by the addition of solute and the removal of solutes and water. The process of returning filtered material from the lumen of the nephron to the blood is known as **reabsorption.** Additional material can be added to the lumen by **secretion,** the movement of selected molecules from the blood into the nephron. Both reabsorption and secretion take place across transporting epithelium that forms the walls of the kidney tubule (∞ p. 57). Although secretion and filtration both move substances from the blood into the lumen, secretion is a more selective process that usually uses mem-

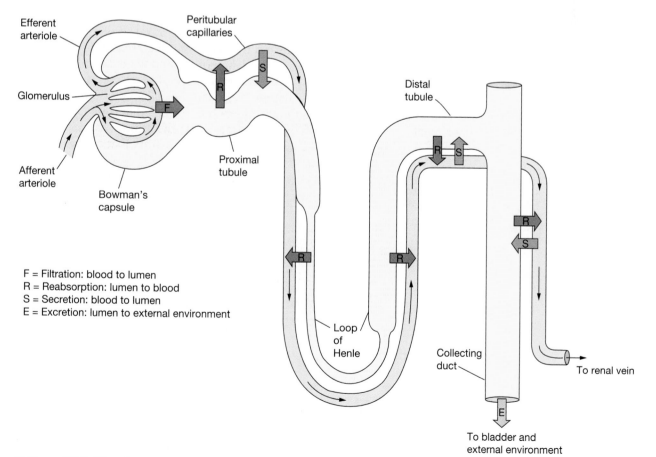

F = Filtration: blood to lumen
R = Reabsorption: lumen to blood
S = Secretion: blood to lumen
E = Excretion: lumen to external environment

■ Figure 18-2 **Filtration, reabsorption, secretion, and excretion**

brane proteins to move molecules across the tubule epithelium.

Once filtrate reaches the end of the collecting duct, it can no longer be modified. The fluid, now called urine, passes through the renal pelvis and down the ureters into the bladder, where it is stored. In humans, no further changes in the composition of the urine take place in the bladder. **Micturition,** or urination, is the process by which urine is **excreted,** or removed from the body. A word of caution here: It is very easy to confuse *secretion* with *excretion.* Try to remember the origins of the two prefixes: *Se-* means *apart,* as in to separate something from its source; *ex-* means *out,* or *away,* as in out of or away from the body. Excretion refers to the removal of a substance from the body. Other organs that carry out excretory processes include the lungs (CO_2) and intestines (undigested food, bilirubin).

Figure 18-2 ■ summarizes the four processes. Filtration takes place in the renal corpuscle as fluid moves from the capillaries of the glomerulus into Bowman's capsule. Reabsorption and secretion occur along the remainder of the tubule. The quantity and composition of the substances being reabsorbed or secreted vary in different segments of the nephron. Fluid that remains in the lumen at the end of the nephron is excreted in the urine.

The amount of a substance excreted in the urine is a reflection of how it was handled during its passage through the nephron (Fig. 18-3 ■). The amount excreted is equal to the amount filtered into the tubule, minus the amount reabsorbed, plus any additional amount of the substance secreted into the lumen:

> Amount excreted =
> amount filtered − amount reabsorbed + amount secreted

This equation summarizes the processes of the kidney and is a useful way to think about renal handling of solutes.

Volume and Osmolarity Change as Fluid Flows through the Nephron

Now that you understand the processes by which fluid enters and leaves the nephron, follow some fluid from the plasma to the urine to learn what happens to the fluid in various segments of the tubule. The fluid that filters into Bowman's capsule is almost identical in composition to plasma and is nearly isosmotic to plasma, about 300 mOsM (∞ p. 454). The filtrate formed in a day averages 180 liters (Table 18-1). As the filtrate moves through the proximal tubule, both water and solutes are reabsorbed. By the end of the proximal tubule, only about 54 liters of the original 180 liters remain. Seventy percent of the fluid that filtered into Bowman's capsule was reabsorbed into the peritubular capillaries while passing through the proximal tubule. The osmolarity of the filtrate remains isosmotic at 300 mOsM. Thus, the primary function of the proximal tubule is the bulk reabsorption of isosmotic fluid.

Filtered fluid passes from the proximal tubule into the loop of Henle. When it exits the loop of Henle, the volume has decreased from 54 L/day to 18 L/day. At this point, 90% of the fluid that entered the nephron has been reabsorbed. A significant change in osmolarity occurs in the loop of Henle because more solute is reabsorbed than water. This difference results in filtrate that is more dilute than the plasma, averaging 100 mOsM as it reenters the cortex. The loop of Henle is the primary site in the nephron that makes dilute fluid.

From the loop of Henle, filtrate passes into the distal tubule and the collecting duct. In these two segments, the fine regulation of salt and water balance takes place under the control of several hormones, to be discussed in Chapter 19. It is in the collecting duct that the final concentration of the urine is determined. By the end of the collecting duct, the filtrate volume averages 1.5 L/day and has an osmolarity that can range from 50

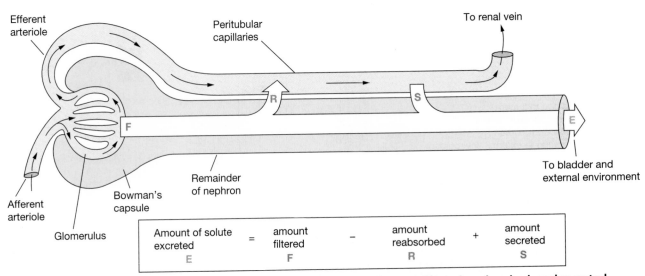

Amount of solute excreted	=	amount filtered	−	amount reabsorbed	+	amount secreted
E		F		R		S

■ Figure 18-3 **The excretion of a substance depends on the amount that was filtered, reabsorbed, and secreted**

TABLE 18-1	Changes in Volume and Osmolarity along the Nephron	
Location in Nephron	*Volume of Fluid*	*Osmolarity of Fluid*
Bowman's capsule	180 L/day	300 mOsM
End of proximal tubule	54 L/day	300 mOsM
End of loop of Henle	18 L/day	100 mOsM
End of collecting duct (final urine)	1.5 L/day (average)	50–1200 mOsM

The Renal Corpuscle Consists of the Glomerulus and Bowman's Capsule

Substances leaving the plasma must pass through three layers before reaching the lumen of Bowman's capsule: the glomerular capillary endothelium, the basal lamina (basement membrane), and the epithelium of Bowman's capsule.

The glomerular capillaries are fenestrated capillaries with large pores (∞ p. 438) that allow most components of plasma to filter out of the blood (Fig. 18-4d ■). However, the pores are small enough to prevent blood cells from leaving the capillary, and their negatively charged surface proteins help repel negatively charged plasma proteins.

An acellular layer called the **basal lamina** separates the capillary endothelium from the epithelial lining of Bowman's capsule. The basal lamina is composed of glycoproteins and a collagen-like material that act like a coarse sieve, excluding plasma proteins.

The third filtration barrier is the epithelium of Bowman's capsule. The portion of the capsule epithelium that surrounds each capillary consists of specialized cells called **podocytes** [*podos,* foot]. The podocytes have long, finger-like cytoplasmic extensions that extend from the main cell body (Fig. 18-4a, b ■). Adjacent "fingers," or **foot processes,** wrap around the capillaries and interlace, leaving narrow slits closed by a membrane. The size of the **filtration slits** can be varied by the contraction of **mesangial cells** in the basal lamina. When the mesangial cells contract, the slits close. Their closing decreases the surface area for filtration, and glomerular filtration rate declines.

When you visualize plasma filtering out of the glomerular capillaries, it is easy to imagine that all the plasma in the capillary moves into Bowman's capsule. However, total filtration of the plasma would leave behind a sludge of blood cells and proteins that could not flow out of the glomerulus. Normally, only about one-fifth of the plasma that flows through the kidneys at any one time filters into the nephrons. The remaining four-fifths of the plasma, along with the plasma proteins

to 1200 mOsM. Both volume and osmolarity of the urine depend on the body's need to conserve or excrete water and solute.

This completes our overview of kidney function. In the remainder of the chapter, we examine the processes of the kidneys in more detail.

✓ Name one way in which filtration and secretion are alike. Name one way in which they differ.

✓ A water molecule enters the renal corpuscle from the blood and ends up in the urine. Name all the anatomical structures that the molecule passes through on its trip to the outside world.

✓ What would happen to the body if filtration continued at a normal rate but reabsorption dropped to one-half of normal?

FILTRATION

The first step in urine formation is the transfer of fluid from the blood into the kidney tubule by filtration. This relatively nonspecific process creates a filtrate much like interstitial fluid or like plasma without the plasma proteins. Under normal conditions, blood cells are not filtered. Filtered fluid is composed only of water and dissolved solutes.

continued from page 519

continued from page 519

Gout is a metabolic disease characterized by high blood concentrations of uric acid (*hyperuricemia*). If uric acid concentrations reach a critical level (7.5 mg/dL), crystals of monosodium urate form in peripheral joints, particularly in the feet, ankles, and knees. These crystals trigger an inflammatory reaction and cause periodic attacks of excruciating pain. Uric acid crystals may also form kidney stones in the renal pelvis.

Question 1: Trace the route followed by kidney stones when they are excreted.

Mesangial Cells and Glomerular Disease The mesangial cells of the kidney are contractile cells with bundles of actinlike filaments in their cytoplasm. The role of the mesangial cells in the regulation of glomerular filtration is still being investigated, but it now appears that several hormones known to alter glomerular filtration do so by their action on mesangial cells. In addition to being contractile, mesangial cells are phagocytic and apparently ingest debris that accumulates as a result of filtration. Evidence links abnormal activity in mesangial cells to several types of glomerular diseases. In some diseases, abnormal proliferation of mesangial cells compresses the glomerular capillaries and impedes blood flow. In other conditions, the mesangial cells manufacture excessive amounts of extracellular matrix that retard filtration.

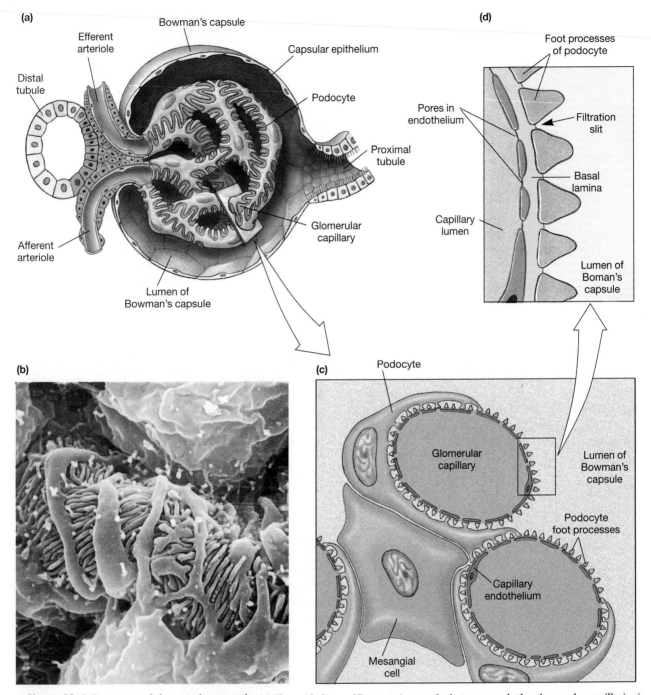

(a)
Bowman's capsule
Efferent arteriole
Capsular epithelium
Distal tubule
Podocyte
Proximal tubule
Afferent arteriole
Glomerular capillary
Lumen of Bowman's capsule

(d)
Foot processes of podocyte
Pores in endothelium
Filtration slit
Basal lamina
Capillary lumen
Lumen of Boman's capsule

(b)

(c)
Podocyte
Glomerular capillary
Lumen of Bowman's capsule
Podocyte foot processes
Capillary endothelium
Mesangial cell

■ Figure 18-4 **Structure of the renal corpuscle** (a) The epithelium of Bowman's capsule that surrounds the glomerular capillaries is modified into special cells known as podocytes. (b, c) The foot processes of the podocytes surround each capillary, leaving slits through which filtration takes place. Contraction of mesangial cells decreases the size of the filtration slits. (d) Filtered substances must pass through pores in the capillary endothelium, the acellular basal lamina, and the filtration slits before reaching the lumen of Bowman's capsule.

and blood cells, flow into the peritubular capillaries (Fig. 18-5 ■). The percentage of total plasma volume that filters is called the **filtration fraction.**

Filtration Occurs because of Hydraulic Pressure in the Capillaries

Filtration across the walls of the glomerular capillaries is similar in many ways to filtration of fluid out of systemic capillaries (∞ p. 439). Glomerular filtration takes place because of the following events:

1. The hydraulic pressure of blood flowing through the capillaries forces blood out of the leaky epithelium of the capillaries. Hydraulic pressure averages 55 mm Hg and favors filtration. Although hydraulic pressure declines along the length of the glomerular capillaries, it remains higher than the opposing pressures. Consequently, filtration takes place along nearly the entire length of the glomerular capillaries.

2. The osmotic pressure inside the capillaries is higher than the osmotic pressure of fluid within Bowman's

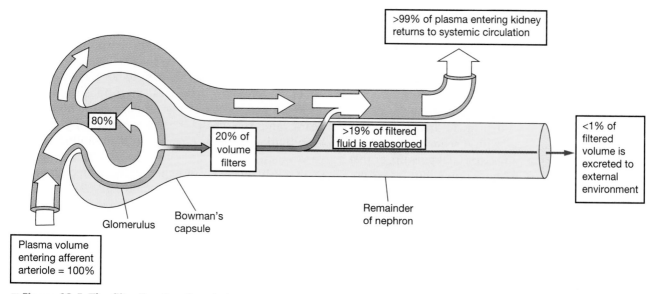

Figure 18-5 The filtration fraction Only 20% of the plasma that passes through the glomerulus is filtered. Less than 1% of the filtered fluid is eventually excreted.

capsule owing to the presence of proteins in the plasma. The osmotic pressure gradient averages 30 mm Hg and favors fluid movement back into the capillaries.

3. Bowman's capsule is an enclosed space (unlike the interstitial fluid), so the presence of fluid within the capsule creates **fluid pressure.** Fluid filtering out of the capillaries must displace the fluid already in the lumen. Fluid pressure averages 15 mm Hg, and, like osmotic pressure, it opposes fluid movement into Bowman's capsule.

The three pressures—hydraulic, osmotic, and fluid—that influence glomerular filtration are summarized in Figure 18-6 ■. The net driving force is 10 mm Hg, favor-

ing filtration. Although this pressure may not seem very high, when combined with the very leaky nature of the fenestrated capillaries, it results in the filtration of 180 L of fluid per day into the nephrons.

Glomerular Filtration Rate Averages 180 Liters per Day

Filtration efficiency is described by the **glomerular filtration rate (GFR),** the amount of fluid that filters into Bowman's capsule per unit time. The GFR in turn is influenced by two factors: the net filtration pressure just described and the surface area available for filtration. In this respect, GFR is similar to gas exchange at the alveoli, where the rate of gas exchange depends on the partial pressure difference and the surface area of the alveoli.

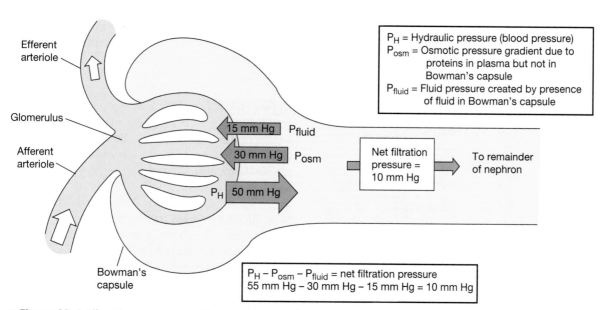

Figure 18-6 Filtration pressure in the renal corpuscle

Average GFR is 125 mL/min, or 180 L/day, an incredible rate considering that the total plasma volume is only about 3 liters. This rate means that the kidneys filter the entire plasma volume 60 times a day, or 2.5 times every hour. If most of the filtrate were not reabsorbed during its passage through the nephron, we would run out of plasma in only 24 minutes of filtration!

Blood pressure provides the primary driving force for glomerular filtration. So, it would seem reasonable to assume that if blood pressure rises, GFR goes up, or that if blood pressure falls, GFR goes down. But that is not always the case. GFR is remarkably constant over a wide range of blood pressures. As long as mean arterial blood pressure remains between 80 and 180 mm Hg, GFR averages 180 L/day (Fig. 18-7 ■). The minute-to-minute control of GFR is accomplished by autoregulation of blood flow in the afferent arteriole. In some situations, autoregulation is assisted or overridden by neural and hormonal reflexes.

GFR Is Regulated by Paracrines, Hormones, and the Nervous System

Regulation of glomerular filtration can be divided into autoregulatory mechanisms that are at work constantly and reflex control mechanisms that are called upon when homeostasis is threatened. Autoregulation of GFR is a complex process that is not completely understood. Two different mechanisms appear to be at work. The **myogenic response** is the intrinsic ability of vascular smooth muscle to respond to pressure changes. **Tubuloglomerular feedback** is a poorly understood process in which changes in fluid flow through the distal tubule influence GFR.

Myogenic response   The myogenic response is similar to autoregulation in other systemic arterioles. When smooth muscle in the arteriole wall stretches because of increased blood pressure, stretch-sensitive ion channels

open, and the muscle cells depolarize, then contract (∞ p. 355). Vasoconstriction increases resistance to flow, so blood flow through the arteriole diminishes. The decrease in blood flow lowers hydraulic pressure within the glomerulus, and GFR returns to normal.

If blood pressure decreases, the opposite response occurs. The tonic level of arteriolar contraction disappears, and the vessel becomes maximally dilated. Vasodilation is not as effective at maintaining GFR as vasoconstriction because the afferent arteriole is normally fairly relaxed. Consequently, if mean blood pressure drops below 80 mm Hg, GFR will decrease. This decrease is adaptive in the sense that if less plasma is filtered, the potential for fluid loss in the urine is decreased. In other words, a decrease in GFR helps the body conserve blood volume.

Tubuloglomerular feedback   Fluid flow in the distal tubule of the nephron also influences GFR. The configuration of the nephron causes the distal tubule to pass between the afferent and efferent arterioles (Fig. 18-8 ■). The walls of both the tubule and the arterioles are modified in the regions where they contact each other. The distal tubule wall is modified into a region known as the **macula densa.** The adjacent walls of the arterioles have specialized cells called **juxtaglomerular cells** (JG cells; also known as granular cells). The JG cells are modified for the synthesis and release of renin, a hormone involved in salt and water balance. The complex of macula densa and JG cells is called the *juxtaglomerular apparatus.*

When fluid flow through the distal tubule increases as a result of increased GFR, the macula densa cells send a chemical message to the neighboring afferent arteriole (Fig. 18-9 ■). The afferent arteriole constricts, increasing resistance and decreasing GFR. This type of autoregulation involving both the kidney tubule and the arteriole is known as tubuloglomerular feedback. The exact stimulus that signals increased tubular fluid flow is unclear. Experimental evidence suggests that the absorption of NaCl at the macula densa is important, because inhibitors of NaCl transport decrease the response. The paracrine signal that passes between the macula densa and the arteriolar smooth muscle is also uncertain.

Reflex control of GFR   Reflex control of GFR is mediated through systemic signals such as hormones and the autonomic nervous system. Both afferent and efferent arterioles are innervated by sympathetic neurons that terminate on alpha receptors. Activation of these receptors causes vasoconstriction. If the amount of sympathetic activity is moderate, there is little effect on GFR. If systemic blood pressure drops sharply, as occurs with hemorrhage or severe dehydration, sympathetically induced vasoconstriction of the arterioles decreases GFR.

A variety of hormones influence arteriolar resistance and GFR. Among the most important are angiotensin II, a potent vasoconstrictor, and prostaglandins, which act

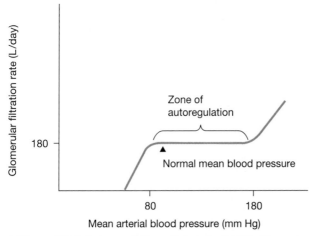

■ **Figure 18-7 Autoregulation of glomerular filtration rate**
Autoregulation of glomerular filtration by myogenic control and tubuloglomerular feedback maintains a nearly constant GFR as long as mean arterial blood pressure is between 80 and 180 mm Hg.

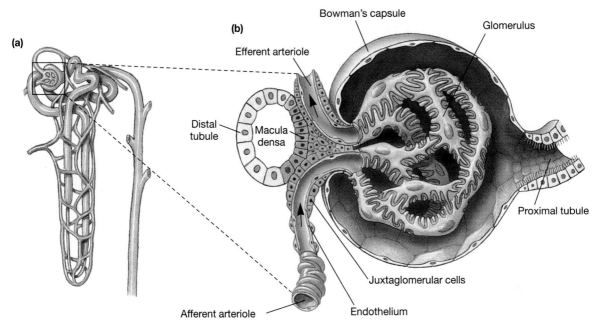

■ Figure 18-8 **The juxtaglomerular apparatus** The close association between the distal tubule and the arterioles of the nephron is an essential factor in tubuloglomerular feedback. Part of the distal tubule is modified into cells known as the macula densa that are able to sense fluid flow through the tubule. The macula densa cells release paracrines that affect both arteriolar resistance and the secretion of the hormone renin by juxtaglomerular cells in the wall of the arterioles.

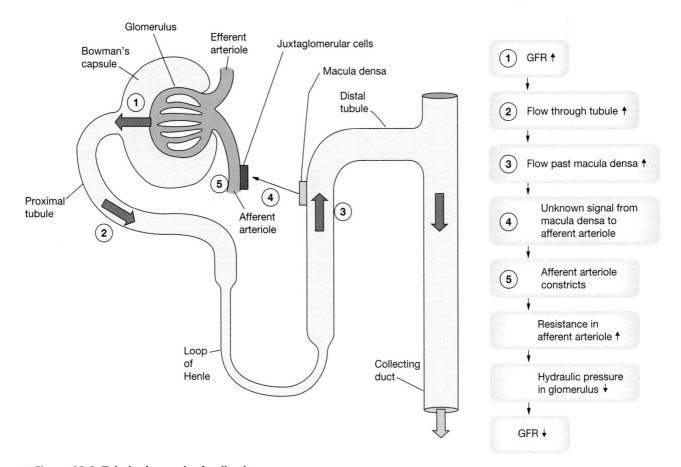

■ Figure 18-9 **Tubuloglomerular feedback**

as vasodilators. In addition, some hormones appear to alter GFR by changing the size of the filtration slits. If the slits widen, more surface area is available for filtration, and GFR increases. Reflex control of GFR is discussed further in the next chapter.

✓ Systemic blood pressure remains constant but the afferent arteriole of a nephron constricts. What happens to glomerular blood flow and GFR in that nephron?

✓ A person with cirrhosis of the liver has lower-than-normal levels of plasma proteins and a higher-than-normal GFR. Explain why a decrease in plasma protein concentration would cause an increase in GFR.

REABSORPTION

Each day, 180 liters of filtered fluid pass from the glomerular capillaries into the nephrons, yet only about 1.5 liters is excreted in the urine. More than 99% of what filters therefore must be reabsorbed as the fluid moves through the nephron. There are two patterns of reabsorption. Bulk reabsorption is aimed at recovering most of the filtered fluid. Bulk reabsorption takes place in the proximal tubule. Regulated reabsorption takes place in the later segments of the nephron and allows the kidneys to return ions and water to the plasma selectively as needed to maintain homeostasis.

One question you might be asking is, "Why bother to filter 180 L/day and then reabsorb 99% of it? Why not simply filter and excrete the 1% that needs to be eliminated?" There are two reasons. First, many foreign substances filter into the nephron but are not reabsorbed. The high daily filtration rate helps clear such substances from the plasma very rapidly. Second, the regulation of ions and water is simplified by filtering them into the tubule. If the portion of filtrate that reaches the distal nephron is not needed to maintain homeostasis, it passes into the urine. With a high GFR, this excretion can occur quite rapidly. However, if the ions and water are needed, they are reabsorbed.

The peritubular capillaries that surround the nephron reabsorb fluid along their entire length. In contrast to the capillaries of the glomerulus, where hydraulic pressure averages 55 mm Hg, the peritubular capillaries have an average hydraulic pressure of 10 mm Hg. Osmotic pressure, which favors movement of fluid into the capillaries, is 30 mm Hg. As a result, the net pressure in the peritubular capillaries is 20 mm Hg, favoring the reabsorption of fluid into the capillaries. Fluid that is reabsorbed then passes into the venous circulation and returns to the heart.

The next section looks at the mechanisms by which reabsorption takes place, with emphasis on bulk reabsorption in the proximal tubule. Regulated reabsorption is discussed in the next chapter.

Reabsorption May Be Active or Passive

Reabsorption of water and solutes in the kidney tubule is dependent upon the active transport of substances from the lumen to the extracellular fluid. Filtrate in Bowman's capsule has the same solute concentrations as extracellular fluid. Therefore, active transport must be used to create concentration gradients. Water osmotically follows solutes as they are reabsorbed.

Most reabsorption involves transepithelial transport, in which substances cross both the apical and basolateral membranes of the tubule epithelial cell. (A few substances can pass between the cells because this is a leaky epithelium, but we will ignore them in this discussion.) The mechanism used to transport a molecule across a membrane depends on the concentration gradient of the molecule (∞ p. 121). Molecules moving down their concentration (or electrochemical) gradient use open leak channels or facilitated diffusion carriers. Molecules that need to be pushed against their concentration gradient are moved by either primary or secondary active transport. The following paragraphs look at some specific examples of how molecules are reabsorbed in the nephron.

Active transport of sodium   Filtrate entering the proximal tubule is similar in composition to interstitial fluid, with a Na^+ concentration that is much higher than that within the cells. Subsequently, Na^+ in the filtrate moves into the proximal tubule cell at the apical membrane through open leak channels, moving down its electrochemical gradient (Fig. 18-10 ■). Once inside the cell, Na^+ is actively transported into the extracellular fluid by the Na^+/K^+-ATPase on the basolateral membrane. The result is Na^+ reabsorption across the epithelium.

Secondary active transport: symport with sodium   Sodium-linked secondary active transport in the nephron is responsible for the reabsorption of multiple substances, including glucose, amino acids, ions, and various organic metabolites. Figure 18-11 ■ shows Na^+-dependent glucose movement across the epithelium (∞ Fig. 5-26 ■, p. 129). The apical membrane contains a Na^+-glucose co-transporter that brings glucose into the cytoplasm against its concentration gradient by harnessing the energy of Na^+ moving down its electrochemical gradient. On the basolateral side of the cell, Na^+ is pumped out by the Na^+/K^+-ATPase, while glucose diffuses out with the aid of a facilitated diffusion carrier. The same basic pattern holds for the other molecules absorbed by Na^+-dependent transport: an apical symport protein and a basolateral facilitated diffusion carrier.

Passive reabsorption: urea reabsorption   Urea has no membrane transporters to move it in the proximal tubule, but it can diffuse freely through open leak channels if there is a urea concentration gradient. Initially, urea concentrations in the filtrate and extracellular fluid

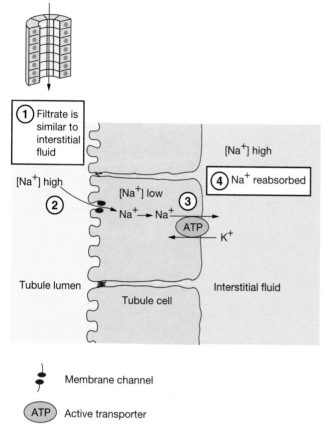

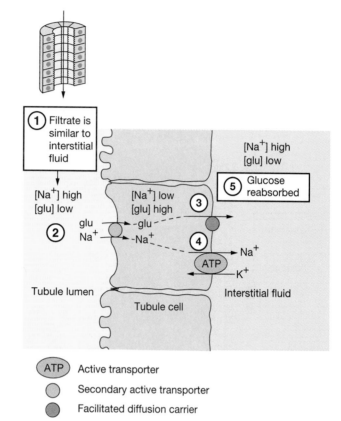

■ **Figure 18-10 Sodium reabsorption** Sodium ion concentration in the filtrate is about the same as that in the extracellular fluid. Sodium ions enter the tubule cell at the apical membrane by diffusion through open leak channels. On the basolateral side of the tubule cell, Na^+ is pumped into the extracellular fluid by the Na^+/K^+-ATPase.

■ **Figure 18-11 Sodium-linked glucose reabsorption** Glucose enters the apical membrane on a co-transporter with Na^+, using the energy of Na^+ moving down its electrochemical gradient. Glucose diffuses out the basolateral side on a facilitated diffusion carrier, while Na^+ is pumped out on the Na^+/K^+-ATPase.

are equal. However, the active transport of Na^+ and other solutes creates a urea concentration gradient by the following process.

When Na^+ and other solutes are reabsorbed from the lumen of the nephron, the transfer of osmotically active particles makes the extracellular fluid more concentrated than the fluid remaining in the lumen (Fig. 18-12 ■). In response to the osmotic gradient, water moves by osmosis across the epithelium. Up to this point, no urea molecules have moved out of the lumen because there has been no urea concentration gradient. Now, however, when water leaves the lumen, the *concentration* of urea goes up because the same number of urea molecules is contained within a smaller volume. Once a concentration gradient for urea exists, urea diffuses out of the lumen into the extracellular fluid.

Transcytosis: plasma proteins Filtration of plasma at the glomerulus normally excludes most plasma proteins, but some smaller protein hormones and enzymes can pass through the filtration barrier. However, most filtered proteins are reabsorbed in the proximal tubule, and only trace amounts of protein appear in normal urine. Proteins are too large to cross cell membranes on

carriers or through channels, so they are brought into the epithelial cells by endocytosis at the apical membrane. Once in the cell, they are either digested and released as amino acids or delivered intact to the extracellular fluid by transcytosis (∞ p. 130).

continued from page 525

Uric acid, the molecule that causes gout, is a normal product of purine metabolism. Increased uric acid production may be associated with cell and tissue breakdown, or it may occur as a result of inherited enzyme defects. Uric acid in the blood filters freely into Bowman's capsule but is almost totally reabsorbed in the initial section of the proximal tubule. Normally, the uric acid that appears in the urine is secreted into the tubule lumen in the last half of the proximal tubule.

Question 2: Why would uric acid levels in the blood go up when cell breakdown increases? (Hint: Purines are part of what biomolecules?)

Question 3: Based on what you have learned about uric acid, predict two different ways a person may develop hyperuricemia.

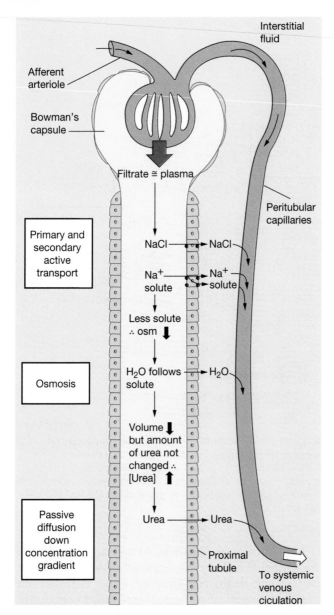

■ **Figure 18-12 Passive reabsorption of urea in the proximal tubule**

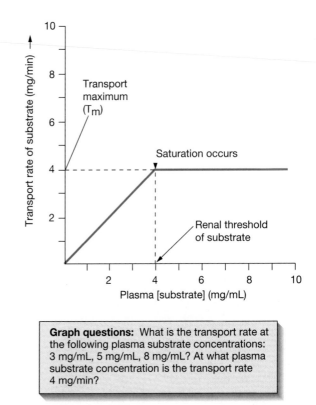

Graph questions: What is the transport rate at the following plasma substrate concentrations: 3 mg/mL, 5 mg/mL, 8 mg/mL? At what plasma substrate concentration is the transport rate 4 mg/min?

■ **Figure 18-13 Saturation of mediated transport** The transport rate of a substance is proportional to the plasma concentration of the substance, up to the point at which all transporters are working at their maximum capacity (i.e., have become saturated with substrate). Once saturation occurs, the transport rate reaches a maximum, known as the transport maximum. The plasma concentration of substrate at which the transport maximum occurs is called the renal threshold.

Saturation of Renal Transport Plays an Important Role in Kidney Function

Most transport in the nephron is mediated by membrane proteins and exhibits the three characteristics of mediated transport discussed in Chapter 5: saturation, specificity, and competition (∞ p. 120).

Saturation refers to the maximum rate of transport that occurs when all available carriers are occupied with substrate (i.e., have become saturated with substrate). At substrate concentrations below the saturation point, the rate of transport is directly related to the concentration of the substrate (Fig. 18-13 ■). At substrate concentrations equal to or above the saturation point, transport occurs at a maximum rate. The transport rate at saturation is called the **transport maximum,** or T_m.

Glucose reabsorption in the nephron provides us with an excellent example of the consequences of saturation.

At normal plasma glucose concentrations, all glucose that enters the nephron is reabsorbed before it reaches the end of the proximal tubule. The tubule epithelium is well supplied with carriers to capture the glucose as it flows past. But if blood glucose concentrations become excessive, as they do in the disease diabetes mellitus, glucose is filtered faster than the carriers can reabsorb it. The carriers become saturated and are unable to reabsorb all the glucose that flows through the tubule. As a result, some glucose escapes reabsorption and is excreted in the urine.

Consider the following analogy. Assume that the carriers are like seats on a train at Disney World. Instead of boarding the stationary train from a stationary platform, passengers step onto a moving sidewalk that rolls them past the train. As the passengers see an open seat, they grab it. But if more people are allowed onto the moving sidewalk than there are seats in the train, some people will not find seats. And because the sidewalk is moving these people past the train toward an exit, the extra people cannot wait for the next train. Instead, they are transported out the exit.

In the same fashion, glucose molecules entering the kidney tubule as part of the filtrate are like passengers stepping onto the moving sidewalk. In order to be reabsorbed back into the blood, each glucose must bind to a transporter as the filtrate flows through the proximal tubule. If only a few glucose filter at a time, each one will find a free transporter and be reabsorbed, just as a small number of people on the moving sidewalk all find seats on the train. However, if glucose filters into the tubule faster than the glucose carriers can transport them, some glucose molecules remain in the lumen and are excreted in the urine.

Figure 18-14 ■ is a graphic representation of glucose handling by the kidney. Graph (a) shows that the filtration rate of glucose from plasma into Bowman's capsule is proportional to the plasma concentration of glucose. Because filtration does not exhibit saturation, the graph continues infinitely in a straight line. The concentrations of glucose in plasma and in filtrate are always equal.

Figure 18-14b ■ plots the reabsorption rate of glucose in the proximal tubule against the plasma concentration of glucose. Reabsorption exhibits a maximum transport rate (T_m) when the carriers reach saturation. The plasma concentration at which saturation of the transporters occurs is called the **renal threshold.**

Figure 18-14c ■ plots the excretion rate of glucose in relation to the plasma concentration of glucose. When plasma glucose concentrations are low enough that 100% of the filtered glucose is reabsorbed, no glucose is excreted. Once the carriers reach saturation, glucose excretion begins.

Figure 18-14d ■ is a composite graph that compares filtration, reabsorption, and excretion of glucose. Recall from our earlier discussion that

Amount excreted =
amount filtered − amount reabsorbed + amount secreted

For glucose, which is not secreted, the equation can be rewritten as

Amount glucose excreted =
amount glucose filtered − amount glucose reabsorbed

Under normal conditions, all filtered glucose is reabsorbed; that is, filtration is equal to reabsorption. Notice in Figure 18-14d ■ that the lines representing filtration and reabsorption are identical up to the plasma glucose concentration that equals the renal threshold. If filtration equals reabsorption, the algebraic difference between the two is zero, and there is no excretion.

Once the renal threshold is reached, filtration begins to exceed reabsorption. Notice on the graph that the filtration and reabsorption lines diverge once the transport maximum for reabsorption is reached. The difference between the filtration line and the reabsorption line represents the excretion rate:

Excretion = filtration − reabsorption
(increasing) (constant)

Excretion of glucose in the urine is called **glucosuria,** or **glycosuria** [-*uria,* in the urine], and is usually indicative of an elevated blood glucose concentration. Sometimes, there is a genetic disorder in which the tubule does not make enough carriers, causing glucose to appear in the urine although the blood glucose concentrations are normal.

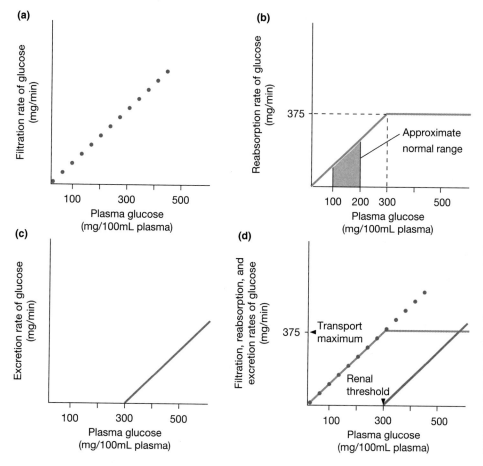

■ Figure 18-14 **Glucose handling by the nephron** (a) Filtration of glucose is directly proportional to the plasma concentration of glucose. (b) The reabsorption of glucose is proportional to the plasma concentration until the transport maximum is reached. (c) Glucose excretion is zero until the renal threshold is reached. At that point, glucose that has been filtered but not reabsorbed begins to appear in the urine. (d) A composite graph shows the relationship between filtration, reabsorption, and excretion of glucose.

SECRETION

Secretion is the transfer of molecules from the extracellular fluid into the lumen of the nephron. Like reabsorption, secretion depends mostly on membrane transport systems. The secretion of K^+ and H^+ by the nephron is an important component in the homeostatic regulation of those ions (see Chapter 19). In addition, many organic compounds are secreted in the kidney. These compounds include both metabolites produced in the body and substances brought into the body. In both instances, the secreted substances are those that the kidney is trying to excrete rapidly.

Secretion provides a mechanism through which the nephron can speed excretion by removing a substrate as it moves through the peritubular capillaries. Secretion is an active process because it requires moving substrates against their concentration gradients. Most organic compounds are transported across the tubule epithelium by secondary active transport.

Competition and Penicillin Secretion

An interesting and important example of an organic molecule secreted by the kidney tubule is the antibiotic **penicillin.** Many people today take antibiotics for granted, but until the early decades of the twentieth century, infections were a leading cause of death. In 1928, Alexander Fleming discovered a substance in the *Penicillium* bread mold that retarded the growth of bacteria. But the antibiotic was difficult to isolate, so it did not become available for clinical use until the late 1930s. During World War II, penicillin made a major difference in the number of deaths and amputations caused by infected wounds. The only means of producing penicillin, however, was to isolate it from bread mold, so supplies were limited.

Demand for the drug was heightened by the fact that kidneys secrete penicillin. Renal secretion is so efficient at clearing foreign molecules from the blood that within three to four hours of administering a dose of penicillin, about 80% has been excreted in the urine. During the war, the drug was in such short supply that it was common procedure to collect the urine from patients being treated with penicillin so that the antibiotic could be isolated and reused. This was not a satisfactory solution, however, so researchers looked for a way to slow penicillin secretion.

They hoped to find a molecule that could compete with penicillin for the organic acid transporter responsible for secretion (∞ p. 120). That way, when presented with both drugs, the carrier would bind preferentially to the competitor and secrete it, leaving penicillin behind in the blood. A synthetic compound named *probenecid* was the answer. When administered concurrently with penicillin, probenecid is secreted on the transporter, and the activity of penicillin is prolonged. Once mass-produced synthetic penicillin became avail-

continued from page 531

Michael finds it amazing that a metabolic problem could lead to pain in his big toe. "There's a way to treat gout, right?" he asks. Dr. Garcia explains that the treatment includes anti-inflammatory agents, lots of water, and avoidance of alcohol, which can trigger gout attacks. "In addition, I would like to put you on a *uricosuric agent*, like probenecid, that will enhance renal excretion of uric acid," replies Dr. Garcia. "By enhancing excretion, we can reduce uric acid levels in your blood and thus provide relief." Michael agrees to try these measures, and on his way home, stops off at a convenience store for a case of his favorite bottled water.

Question 4: Uric acid is reabsorbed into the proximal tubule on a general organic acid transporter. Uricosuric agents are also organic acids. Knowing those two facts, explain how uricosuric agents enhance excretion of uric acid.

able and supply was no longer a problem, the medical use of probenecid declined.

EXCRETION

Urine output is the result of all the processes that take place in the kidney. By the time fluid reaches the end of the nephron, it bears little resemblance to the filtrate that started in Bowman's capsule. Glucose, amino acids, and useful metabolites are gone, having been reabsorbed into the blood; organic wastes are more concentrated. The concentrations of ions and water in the urine are highly variable depending on the state of the body.

Although excretion tells us what the body is eliminating, excretion by itself cannot tell us the details of renal function. Recall that for any substance,

Amount excreted =
amount filtered − amount reabsorbed + amount secreted

Simply looking at the excretion rate of a substance tells us nothing about how that substance is being handled by the kidney. The excretion rate of a substance depends on (1) its filtration rate and (2) whether the substance is reabsorbed or secreted as it passes through the nephron.

These three processes in the kidney are often of interest, however. For example, clinicians use information about a person's glomerular filtration rate as an indicator of overall kidney function. Pharmaceutical companies developing drugs must provide the Food and Drug Administration with complete information on how the human kidney handles each new compound. But investigators dealing with living humans have no direct way to assess filtration, reabsorption, and secretion at the level of the individual nephron. Scientists therefore had to develop a technique that would allow them to assess renal function using only analysis of the urine and blood.

The result of their work was the concept known as clearance. **Clearance** is an abstract concept that describes how many milliliters of plasma passing through the kidneys have been totally cleared of a substance in a given period of time. The equation for clearance is

Clearance of a substance =
$$\frac{\text{urine excretion rate of the substance (mg/min)}}{\text{plasma concentration of the substance (mg/mL)}}$$

where the urine excretion rate is calculated by

Urine excretion rate of a substance =
urine flow rate × urine concentration of the substance

As an example, consider the clearance of **inulin,** a polysaccharide isolated from the tuberous roots of dahlia plants. Once injected into the plasma, inulin is freely filtered into the nephron. From experiments using isolated nephrons, scientists discovered that inulin is neither reabsorbed nor secreted as it passes through the lumen of the kidney tubule.

Assume that inulin is injected so that the plasma concentration is 100 mg/dL plasma. Its excretion rate in the urine is 125 mg inulin/min. Using the equation for clearance,

Clearance of inulin =
$$\frac{\text{inulin excretion (mg/min)}}{\text{plasma inulin (mg/mL)}}$$

Clearance of inulin =
$$\frac{125 \text{ mg inulin/min}}{100 \text{ mg inulin/100 mL plasma}}$$

Clearance of inulin = 125 mL plasma/min

The answer says that in one minute, 125 mL of plasma was totally cleared of inulin.

What good is this information? For one thing, it can be used to calculate the glomerular filtration rate. If a substance such as inulin is neither reabsorbed nor secreted, its excretion rate must be equal to the rate at which the substance was filtered into the nephron. In other words, the only way that inulin appears in the urine is by filtration, and 100% of what is filtered is excreted.

To prove that the clearance rate of inulin is equal to the GFR, we calculate filtration rate by another method:

Filtration rate of a substance =
plasma concentration of the substance × GFR

Filtration rate of inulin =
plasma concentration of inulin × GFR

For inulin, filtration rate is equal to excretion rate; therefore,

Excretion rate of inulin =
plasma concentration of inulin × GFR

This equation can be rearranged to read

$$\frac{\text{inulin excretion}}{\text{plasma inulin}} = \text{GFR}$$

The left side of the equation is identical to the clearance equation for inulin:

$$\frac{\text{inulin excretion (mg/min)}}{\text{plasma inulin (mg/mL)}} = \text{clearance of inulin}$$

By using algebra (if A = B and A = C, then B = C),

GFR = clearance of inulin

In other words, for any substance such as inulin that is freely filtered but neither reabsorbed nor secreted, the clearance rate is equal to the GFR.

Estimation of GFR using inulin requires that the polysaccharide be administered by continuous intravenous infusion. This is not practical for routine clinical applications, and, as a result, inulin is restricted to research use. Unfortunately, no substance that occurs naturally in the human body is handled by the kidney exactly like inulin.

In clinical settings, creatinine is used to measure GFR. **Creatinine** is the breakdown product of phosphocreatine, an energy-storage compound found primarily in muscles (p. 338). Normally, the production and breakdown rates of phosphocreatine are relatively constant, and plasma concentrations of creatinine do not vary much. Although it is always present and is easy to measure, creatinine is not the perfect molecule for estimating GFR because a small amount is secreted into the urine. However, the error is small enough that, in most people, creatinine clearance is routinely used to assess GFR.

Once a person's GFR is known, it is possible to calculate how much of a substance is filtered into the nephron (assuming that the substance is not excluded at the glomerulus):

Filtration rate of a substance =
plasma concentration of the substance × GFR

To determine how the nephron handles the substance once it is filtered, compare the filtration rate of the substance with its excretion rate (Table 18-2). If less of the substance appears in the urine than was filtered, then some was reabsorbed (excreted = filtered − reabsorbed + secreted). If *more* of the substance appears in the urine

TABLE 18-2 Renal Handling of Solutes

For any molecule X that is freely filtered at the glomerulus:

If filtration rate is greater than excretion rate,	there is net reabsorption of X.
If excretion rate is greater than filtration rate,	there is net secretion of X.
If filtration and excretion rate are the same,	X passes through the nephron without net reabsorption or secretion.
If the clearance of X is less than the inulin clearance,	there is net reabsorption of X.
If the clearance of X is equal to the inulin clearance,	X is neither reabsorbed nor secreted.
If the clearance of X is greater than the inulin clearance,	there is net secretion of X.

than was filtered, there must have been net secretion of the substance into the lumen. If the same amount of the substance is filtered and excreted, then the substance is handled like inulin: neither reabsorbed nor secreted.

For example, suppose glucose is present in the plasma at 100 mg glucose/dL plasma, and the GFR is calculated from creatinine clearance to be 125 mL plasma/min:

Filtration rate of a substance =
plasma concentration of the substance × GFR

Filtration rate of glucose =
(100 mg glucose/100 mL plasma) × 125 mL plasma/min

Filtration rate of glucose = 125 mg glucose/min

But there is no glucose present in the urine; that is, the excretion rate is zero. Because glucose was filtered at a rate of 125 mg/min but excreted at a rate of 0 mg/min, it must have been reabsorbed.

Clearance Can Be Used to Determine Renal Handling of a Substance

Clearance values are also used to determine how the nephron handles a substance filtered into it. In this method, researchers calculate creatinine or inulin clearance, then compare the clearance rate of the substance being investigated with the creatinine or inulin clearance (Table 18-2). If the clearance of the substance is less than the inulin clearance, the substance has been reabsorbed. On the other hand, if the clearance rate of the substance is higher than the inulin clearance, some substance has been secreted into the nephron. More plasma was cleared of the substance than was filtered, so the additional material must have been removed from the plasma by secretion.

Figure 18-15 ■ shows how clearance would be calculated for four different molecules: inulin, glucose, urea, and penicillin. In this example, all four substances have the same concentration, 4 molecules/100 mL plasma (point 1). As 100 mL of plasma filter into the nephron in one minute, all of the molecules in that plasma also filter (point 2). Now assume that all 100 mL of the water volume is reabsorbed into the plasma (point 3). The number of solute molecules reabsorbed varies according to how the kidney tubule handles each compound. For example, four molecules of inulin filter with each 100 mL of plasma, and none is reabsorbed (Fig. 18-15a ■, point 4). This means that 100 mL of plasma are cleared of inulin. Therefore, the clearance rate for inulin is 100 mL of plasma cleared per minute.

At the other extreme is glucose (Fig. 18-15b ■). Normally, 100% of the glucose that is filtered is reabsorbed. Consequently, no glucose is cleared from the plasma (point 5), and the clearance rate for glucose is 0 mL of plasma cleared per minute.

Urea is an example of a molecule that is partially reabsorbed (Fig. 18-15c ■, point 6). Four molecules filter, but only two are reabsorbed. Consequently, only 50 mL of plasma are cleared of urea in one minute.

continued from page 534

Three weeks later, Michael is back in Dr. Garcia's office. The anti-inflammatory drugs and probenecid had eliminated the pain in his toe, but last night, he had gone to the hospital with a very painful kidney stone. "We'll have to wait until the analysis comes back," says Dr. Garcia, "But I will guess that it is a uric acid stone. Did you drink as much water as I told you to?" Sheepishly, Michael admits that he had good intentions, but never could seem to find the time at work to get out his bottled water. "You have to drink enough water while on this drug to produce 3000 milliliters or more of urine a day. That's over three quarts. Otherwise, you may end up with another uric acid kidney stone." Michael agrees that this time, he will follow instructions to the letter.

Question 5: Explain why not drinking enough water while taking uricosuric agents may cause uric acid stone formation in the urinary tract.

Penicillin (Fig. 18-15d ■) is filtered and not reabsorbed, but additional penicillin molecules are secreted from plasma in the peritubular capillaries (point 7). In this example, an extra 50 mL of plasma have been cleared of penicillin in addition to the original 100 mL that filtered. The clearance rate for penicillin therefore is 150 mL plasma cleared per minute.

Note that a comparison of clearance values tells you only the *net* handling of the substance. It does not tell you if a molecule is both reabsorbed and secreted. For example, nearly all K^+ filtered is reabsorbed in the proximal tubule and loop of Henle, then a small amount is secreted back into the lumen by the distal nephron. On the basis of the clearance rate, it appears that there was only net reabsorption.

Clearance calculations are useful because all you need to know are the urine excretion rates of various substances and their plasma concentrations; both parameters are easily obtained. If you can calculate the clearance of either inulin or creatinine, then you can determine the renal handling of any other compound.

MICTURITION

Once fluid from the nephrons leaves the collecting ducts, it flows into the renal pelvis of the kidney and down the ureter to the bladder. In the bladder, urine is stored until released in the process known as urination, voiding, or more formally, micturition [*micturire*, to desire to urinate].

The bladder is a hollow sac that can expand to hold a volume of about 500 mL. The neck of the bladder is continuous with the urethra, a single tube through which urine passes to the external environment. The opening

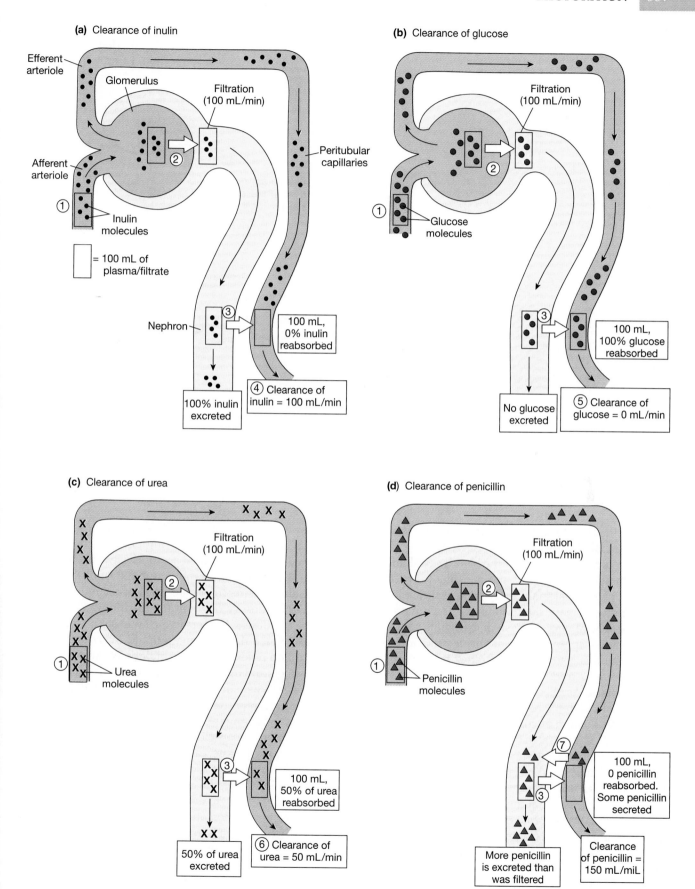

(a) Clearance of inulin

Efferent arteriole

Glomerulus

Filtration (100 mL/min)

②

Peritubular capillaries

Afferent arteriole

① Inulin molecules

= 100 mL of plasma/filtrate

Nephron

③

100 mL, 0% inulin reabsorbed

④ Clearance of inulin = 100 mL/min

100% inulin excreted

(b) Clearance of glucose

Filtration (100 mL/min)

②

① Glucose molecules

③

100 mL, 100% glucose reabsorbed

No glucose excreted

⑤ Clearance of glucose = 0 mL/min

(c) Clearance of urea

Filtration (100 mL/min)

②

① Urea molecules

③

100 mL, 50% of urea reabsorbed

⑥ Clearance of urea = 50 mL/min

50% of urea excreted

(d) Clearance of penicillin

Filtration (100 mL/min)

②

① Penicillin molecules

⑦

③

100 mL, 0 penicillin reabsorbed. Some penicillin secreted

More penicillin is excreted than was filtered

Clearance of penicillin = 150 mL/miL

■ **Figure 18-15 The relationship between clearance and excretion** The figure represents the events taking place in one minute. For simplicity, 100% of the filtered volume was reabsorbed.

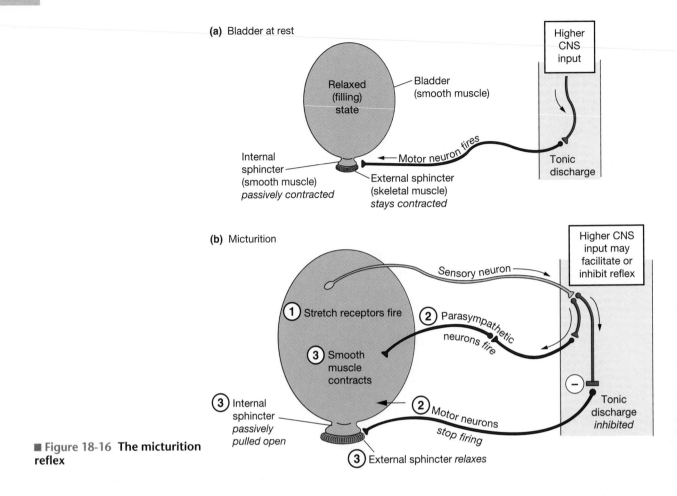

(a) Bladder at rest

Higher CNS input

Bladder (smooth muscle)

Relaxed (filling) state

Internal sphincter (smooth muscle) *passively contracted*

Motor neuron fires

External sphincter (skeletal muscle) *stays contracted*

Tonic discharge

(b) Micturition

Higher CNS input may facilitate or inhibit reflex

Sensory neuron

① Stretch receptors fire

② Parasympathetic neurons fire

③ Smooth muscle contracts

③ Internal sphincter *passively pulled open*

② Motor neurons stop firing

③ External sphincter *relaxes*

Tonic discharge *inhibited*

■ **Figure 18-16 The micturition reflex**

between the bladder and urethra is closed by two rings of muscle called **sphincters** (Fig. 18-16a ■). The internal sphincter is a continuation of the bladder wall and consists of smooth muscle. Its normal tone keeps it contracted. The external sphincter is a ring of skeletal muscle controlled by somatic motor neurons. Tonic stimulation from the central nervous system maintains contraction of the external sphincter except during urination.

Micturition is a simple spinal reflex that is subject to both conscious and unconscious control from higher brain centers. As the bladder fills with urine and its walls expand, stretch receptors send signals via sensor neurons to the spinal cord (Fig. 18-16b ■). There, the information is integrated and transferred to two different sets of neurons. The stimulus of a full bladder *excites* parasympathetic neurons leading to the smooth muscle in the bladder wall. The smooth muscle contracts, increasing the pressure on the bladder contents. Simultaneously, somatic motor neurons leading to the external sphincter are *inhibited*. Contraction of the bladder occurs in a wave that pushes urine downward toward the urethra. Pressure exerted by urine forces the internal sphincter open while the external sphincter relaxes. Urine passes into the urethra and out of the body, aided by gravity.

This simple micturition reflex occurs primarily in infants who have not yet been toilet trained. A person who has been toilet trained acquires a learned reflex that keeps the micturition reflex inhibited until the person consciously desires to urinate. The learned reflex involves additional sensory fibers in the bladder that signal the degree of fullness. Centers in the brain stem and cerebral cortex receive that information and override the basic micturition reflex by direct inhibition of the parasympathetic fibers and by reinforcing contraction of the external sphincter. When an appropriate time to urinate arrives, those same centers remove the inhibition and facilitate the reflex by inhibiting the external sphincter.

In addition to conscious control of urination, various subconscious factors can affect the micturition reflex. "Bashful bladder" is a condition in which a person is unable to urinate in the presence of other people despite the conscious intent to do so. The sound of running water facilitates micturition and is often used to help patients urinate if the urethra is irritated from insertion of a *catheter*, a tube inserted into the bladder to drain it passively.

This concludes our discussion of the basic processes of the kidneys. The next chapter looks in detail at the role of the kidneys in the homeostatic regulation of blood pressure, osmolarity, and acid-base balance.

CHAPTER REVIEW

Chapter Summary

1. The **urinary system** is composed of the kidneys, bladder, and accessory structures. (p. 518)

2. The kidneys produce urine, a fluid waste product whose composition and volume vary. (p. 519)

Functions of the Kidneys

3. The six functions of the kidneys are regulation of extracellular fluid volume, regulation of osmolarity, maintenance of ion balance, homeostatic regulation of pH, excretion of wastes and foreign substances, and production of hormones. The most important function of the kidneys is the homeostatic regulation of the water and ion content of the blood. (p. 519)

Anatomy of the Urinary System

4. Each **kidney** has about 1 million **nephrons.** Each nephron has **vascular elements** and **tubular elements.** (p. 522)

5. Arterial blood arriving at a nephron flows into an **afferent arteriole,** then into capillaries known as the **glomerulus.** Blood leaving the glomerulus flows into an **efferent arteriole** and a series of **peritubular capillaries.** (p. 522)

6. Fluid filters from the glomerulus into **Bowman's capsule.** From there, it flows through the **proximal tubule, loop of Henle, distal tubule,** and **collecting duct.** Fluid from the collecting ducts drains into the **renal pelvis.** From there, **urine** goes through the **ureter** to the **urinary bladder.** (p. 522)

Processes of the Kidneys

7. There are four basic processes of the urinary system: filtration, reabsorption, secretion, and excretion. **Filtration** is the movement of fluid from the glomerulus into Bowman's capsule. **Reabsorption** is the movement of filtered material from the nephron to the blood. **Secretion** is the movement of selected molecules from the blood into the nephron. **Excretion** is the removal of a substance from the body. (p. 523)

8. The amount of a substance that is excreted equals the amount filtered into the tubule, minus the amount reabsorbed, plus the amount that is secreted into the lumen. (p. 524)

9. Average urine volume is 1.5 L/day, with an osmolarity between 50 and 1200 mOsM. (p. 524)

Filtration

10. Substances filtering from the plasma pass through the glomerular capillary endothelium, the **basal lamina,** and the epithelium of Bowman's capsule before reaching the lumen of Bowman's capsule. (p. 525)

11. Filtration allows most components of plasma to enter the tubule but excludes blood cells and most plasma proteins. (p. 525)

12. The capsule epithelium has specialized cells called **podocytes** that wrap around the capillaries and create **filtration slits** whose size can be varied by the contraction of **mesangial cells.** Alterations in the size of filtration slits change the glomerular filtration rate. (p. 525)

13. One-fifth of the plasma that flows through the kidneys at any one time filters into the nephrons. The percentage of total plasma volume that filters is called the **filtration fraction.** (p. 526)

14. Hydraulic pressure in the glomerular capillaries is 55 mm Hg, favoring filtration. Opposing filtration is osmotic pressure of 30 mm Hg and **fluid pressure** averaging 15 mm Hg. The net driving force is 10 mm Hg, favoring filtration. (p. 526)

15. The **glomerular filtration rate (GFR)** is the amount of fluid that filters into Bowman's capsule per unit time. Average GFR is 125 mL/min, or 180 L/day. (p. 527)

16. Autoregulation of glomerular filtration is accomplished by a **myogenic response** (the intrinsic ability of vascular smooth muscle to respond to pressure changes) and by **tubuloglomerular feedback.** If fluid flow through the distal tubule increases as a result of increased GFR, the **macula densa** cells send a chemical message to the afferent arteriole, which constricts. (p. 528)

17. Reflex control of GFR is mediated through systemic signals such as hormones and the autonomic nervous system. (p. 528)

Reabsorption

18. Bulk reabsorption of filtered fluid takes place in the proximal tubule. Regulated reabsorption takes place in the later segments of the nephron. (p. 530)

19. The peritubular capillaries reabsorb fluid along their entire length because there is a net pressure of 20 mm Hg that favors the reabsorption of fluid into the capillaries. (p. 530)

20. Reabsorption of water and solutes in the kidney tubule is dependent upon the active transport of substances from the lumen to the extracellular fluid. Most reabsorption involves transepithelial transport. (p. 530)

21. Sodium ions in the filtrate diffuse into the tubule cell through open leak channels on the apical membrane, then are actively transported into the extracellular fluid by the Na^+/K^+-ATPase on the basolateral membrane. (p. 530)

22. Sodium-linked secondary active transport is responsible for the reabsorption of glucose, amino acids, ions, and various organic metabolites. (p. 530)

23. The active transport of Na^+ and other solutes creates concentration gradients for urea and other substances that allow them to be passively reabsorbed. (p. 530)

24. Most transport in the nephron is mediated by membrane proteins and exhibits saturation, specificity, and competition. The transport rate at saturation is called the **transport maximum,** or T_m. (p. 532)

25. The plasma concentration at which saturation of the transporters occurs is called the **renal threshold.** (p. 533)

Secretion

26. Secretion provides a mechanism through which the nephron can speed excretion by removing a substrate as it passes through the peritubular capillaries. Secreted substances include K^+, H^+, and a variety of organic compounds. (p. 534)

27. Molecules that compete for renal carriers can be used to slow the secretion of a molecule. (p. 534)

Excretion

28. The excretion rate of a substance depends on (1) its filtration rate and (2) whether it is reabsorbed or secreted as it passes through the nephron. (p. 534)

29. **Clearance** is an abstract concept that describes how many milliliters of plasma passing through the kidneys have been totally cleared of a substance in a given period of time. The clearance of **inulin** can be used to calculate the glomerular filtration rate. In clinical settings, **creatinine** is used to measure GFR. (p. 535)

30. If a person's GFR is known, it is possible to calculate the filtration rate of a substance. (p. 535)

31. If less of a substance appears in the urine than was filtered, then some was reabsorbed. If more of the substance appears in the urine than was filtered, there is net secretion of the substance. If the same amount of the substance is filtered and excreted, then the substance is neither reabsorbed nor secreted. (p. 535)

32. Clearance values are also used to determine how the nephron handles a substance filtered into it. If the clearance of the substance is less than the inulin clearance, the substance has been reabsorbed. If the clearance rate of the substance is higher than the inulin clearance, some substance has been secreted into the nephron. (p. 536)

Micturition

33. Urine is stored in the bladder until released by urination, also known as **micturition.** (p. 536)

34. The opening between the bladder and urethra is closed by two rings of muscle called **sphincters.** The external sphincter is skeletal muscle that is tonically contracted except during urination. (p. 538)

35. Micturition is a simple spinal reflex that is subject to both conscious and unconscious control from higher brain centers. Stretch receptors in the distended bladder initiate the reflex. Parasympathetic neurons cause contraction of the smooth muscle in the bladder wall, while simultaneously somatic motor neurons leading to the external sphincter are inhibited. (p. 538)

Questions

LEVEL ONE   Reviewing Facts and Terms

1. List and explain the significance of the five characteristics of urine that can be found by physical examination.

2. List and explain the six major kidney functions.

3. At any given time, what percentage of the cardiac output goes to the kidneys?

4. List and give a brief function for each of the major structures of the urinary system. Place these organs in sequence from the kidneys to the urine leaving the body.

5. Arrange these structures in the order that a drop of water entering the nephron would encounter them:

a. afferent arteriole
b. Bowman's capsule
c. collecting duct
d. distal convoluted tubule
e. glomerulus
f. loop of Henle
g. proximal convoluted tubule
h. renal pelvis

6. Which components from the list in Question 5 make up the renal corpuscle? In which component does a drop of filtrate become a drop of urine?

7. Name the three layers that substances pass through as they move from the plasma to the lumen of Bowman's capsule. What components of blood are usually excluded by these layers?

8. What force(s) promotes glomerular filtration? What force(s) opposes it? What is meant by the term *net driving force*?

9. What does the abbreviation GFR stand for? What is a typical numerical value for GFR in milliliters per minute? In liters per day?

10. Identify the following structures, then explain their significance in renal physiology:
    a. juxtaglomerular apparatus
    b. macula densa cells
    c. mesangial cell
    d. podocyte
    e. sphincters in the bladder
    f. renal cortex

11. The GFR is 180 liters of fluid per day, but average urine volume is 1.5 liters per day. What happens to most filtered fluid and solutes?

12. In which segment of the nephron does most reabsorption take place?

13. When a molecule or ion is reabsorbed from the lumen of the nephron, where does it go? If a substance is filtered and not reabsorbed from the tubules, where does it go?

14. Match the substance with its primary mode(s) of transport across the kidney epithelium.
    1. sodium
    2. glucose
    3. urea
    4. plasma proteins
    5. water
    (a) transcytosis
    (b) primary active transport
    (c) secondary active transport
    (d) facilitated diffusion
    (e) diffusion through open channels
    (f) simple diffusion through the phospholipid bilayer

15. List three substances secreted by the kidney tubule.

16. What substance, normally present in the body, is used to estimate GFR in humans?

17. What is micturition?

LEVEL TWO Reviewing Concepts

18. Map the factors that influence or control the glomerular filtration rate.

19. Define, compare, and contrast the following sets of terms and/or events:
    a. filtration, secretion, and excretion
    b. saturation, transport maximum, and renal threshold
    c. probenecid, creatinine, inulin, and penicillin
    d. clearance, excretion, and glomerular filtration rate

20. Diagram tubuloglomerular feedback.

21. What are the advantages of a kidney that filters a large volume of fluid and then reabsorbs 99% of it?

22. If the afferent arteriole of a nephron constricts, what happens to GFR in that nephron? If the efferent arteriole of a nephron constricts, what happens to GFR in that nephron? Assume that no autoregulation takes place.

23. Diagram the micturition reflex. How is this reflex altered by toilet training? How do higher brain centers influence micturition?

LEVEL THREE Problem Solving

24. Darlene weighs 50 kg. Assume that her total blood volume is 8% of her body weight, her heart pumps her total blood volume once a minute, and her renal blood flow is 25% of her cardiac output. Calculate the volume of blood that flows through Darlene's kidneys each minute.

25. Dwight was competing for a spot on the Olympic equestrian team. As his horse, Nitro, cleared a jump, the footing gave way, causing the horse to somersault, landing on Dwight and crushing him. The doctors feared kidney damage and ran the following test. Dwight's serum creatinine level was 2 mg/100 mL. His 24-hour urine specimen contained 1 L and had a creatinine concentration of

20 mg/mL. How many milligrams of creatinine are in the specimen? What is Dwight's creatinine clearance? What is his GFR? Evaluate these results and comment on them.

26. You are a physiologist taking part in an archeological expedition to search for Atlantis. One of the deep-sea submersibles has come back with a mermaid on which you are running a series of experiments. You have determined that the mermaid's GFR is 250 mL/min. The mermaid's kidney reabsorbs glucose with a transport maximum (T_m) of 50 mg/min. What is the mermaid's renal threshold for glucose? When the mermaid's plasma concentration of glucose is 15 mg/mL, what is its glucose clearance?

Problem Conclusion

In this running problem, you learned that gout, which often presents as a debilitating pain in the big toe, is a metabolic problem whose cause and treatment may be linked to kidney function.

Further check your understanding of this running problem by comparing your answers against those in the summary table.

	Question	Facts	Integration and Analysis
1	Trace the route followed by kidney stones when they are excreted.	Kidney stones often form in the renal pelvis.	From the renal pelvis, a stone passes down the connecting ureter and into the urinary bladder. From there, it leaves the body through the urethra.
2	Why would uric acid levels in the blood go up when cell breakdown increases?	Purines include adenine and guanine, which are components of DNA, RNA, and high-energy molecules such as ATP.	When a cell dies, the nucleus and other components are broken down into biomolecules. Degradation of DNA, RNA, and ATP would increase purine production, which in turn would increase uric acid.
3	Based on what you have learned about uric acid, predict two different ways a person may develop hyperuricemia.	Hyperuricemia is a condition of elevated uric acid levels in the blood. Uric acid is made from purines. It is filtered, totally reabsorbed, and then secreted by the kidney tubule.	Hyperuricemia results from over-production of uric acid in the body or from failure of the kidneys to eliminate uric acid. Because uric acid normally gets into the urine by secretion, a defect in the renal secretory mechanism would lead to hyperuricemia.
4	Uric acid is reabsorbed into the proximal tubule on a general organic acid transporter. Uricosuric agents are also organic acids. Knowing those two facts, explain how uricosuric agents enhance excretion of uric acid.	Competition is one property of mediated transport systems. In competition, related molecules compete for the same transporter. Usually, one molecule will bind preferentially and therefore inhibit transport of the second molecule.	Uricosuric agents compete with uric acid for the organic acid transporter in the proximal tubule. Uricosuric agents such as probenecid are transported preferentially, leaving uric acid in the tubule lumen and increasing its excretion.
5	Explain why not drinking enough water while taking uricosuric agents may cause uric acid stone formation in the urinary tract.	Uric acid crystals form when uric acid concentrations exceed a critical level.	If a person drinks large volumes of water, the excess water will be excreted by the kidneys. Large amounts of water will dilute the urine and thus prevent high concentrations of uric acid.

19

Integrative Physiology II: Fluid and Electrolyte Balance

The American businesswoman in Tokyo finished her workout and stopped at the snack bar of the fitness club for a drink. When she asked for Gatorade®, the attendant handed her a bottle labeled "Pocari Sweat®." Although the thought of drinking sweat is not very appealing, the physiological basis for the name is sound. In order to maintain homeostasis, the body must replace any substance that is lost to the external environment. During exercise, the body sweats, losing water and ions, particularly Na^+, K^+, and Cl^-. Therefore, the replacement fluid should resemble sweat. In this chapter, we explore how the body maintains salt and water balance, also known as fluid and electrolyte balance. The homeostatic control mechanisms for fluid and electrolyte balance are aimed at maintaining four parameters: volume, osmolarity, concentrations of individual ions, and pH of body fluids.

The process of fluid and electrolyte balance is truly integrative, involving not only renal and behavioral responses but also the respiratory and cardiovascular systems. Small changes in blood pressure due to fluid gain or loss are quickly compensated under the direction of the cardiovascular control center of the brain, as you learned in Chapter

15 (∞ p. 443). A decrease in plasma pH (increased H^+ concentration) is offset by increased elimination of carbon dioxide (CO_2) at the lungs (∞ p. 504). If the pH or volume changes are persistent or of large magnitude, the kidneys step in to help maintain homeostasis. Adjustments made by the lungs and cardiovascular system are primarily under nervous control and can be made quite rapidly. Homeostatic compensation by the kidneys occurs more slowly because the kidneys are primarily under endocrine and neuroendocrine control. Renal function is integrated with the cardiovascular and respiratory systems to keep blood pressure, osmolarity, and pH within a normal range.

Because of the overlap in their function, a change made by one of these three systems is likely to have consequences that affect one or both of the other systems. Endocrine pathways initiated by the kidneys have direct effects on the cardiovascular system, and hormones released by myocardial cells act on the kidneys. Sympathetic pathways from the cardiovascular control center affect not only cardiac output and vasoconstriction but also glomerular filtration and hormone secretion by the kidneys. The maintenance of blood pressure, blood volume, and extracellular fluid osmolarity forms a network of interwoven control pathways. This integration of function in multiple systems is one of the more difficult concepts in physiology but, at the same time, one of the most exciting.

Problem

Diabetes Insipidus

"Urine without flavor." That is the literal meaning of the term *diabetes insipidus,* a disorder in which the mechanism that concentrates urine is inadequate or absent. The main symptoms of this rare disease are intense thirst, particularly for ice water, and copious excretion of dilute urine. Christopher Godell, age 2, has both these symptoms. His mother has brought him to the pediatrician because of problems she noted during her attempts to toilet train her son. "Christopher is always thirsty, and always urinating," she explains. "I know this isn't normal."

continued on page 546

HOMEOSTASIS OF VOLUME AND OSMOLARITY

The kidneys influence extracellular fluid volume and osmolarity in two ways: (1) They can alter the amount of water excreted, and (2) they can alter the amount of sodium excreted. The amount of water excreted is usually reflected by the urine concentration. When the kidneys are conserving water, the urine becomes quite concentrated, up to four times as concentrated as the blood (1200 mOsM vs. 300 mOsM). When excess water needs to be eliminated, the kidneys put out copious amounts of dilute urine with osmolarities as low as 50 mOsM. Removal of water in the urine is known as **diuresis** [*diourein,* to pass in urine]; drugs that cause diuresis are known as diuretics.

The excretion of water by the kidneys is controlled primarily by the hormone **vasopressin,** also known as **antidiuretic hormone,** or **ADH** (∞ p. 185). The presence of ADH increases the permeability of the collecting duct to water in a graded fashion. When ADH concentrations are maximal, water reabsorption is maximal. Without ADH, water reabsorption in the final sections of the nephron is minimal.

The electrolytes whose concentrations in the body are regulated by the kidneys include sodium, potassium, hydrogen, calcium, phosphate, and bicarbonate ions. Excretion of these ions is controlled by several different hormones, to be discussed in detail later.

Fluid balance is closely tied to the renal regulation of sodium because the amount of sodium in the body is the key determinant of extracellular fluid volume. Strangely enough, however, the body appears to have few special receptors for Na^+. Instead, the body's Na^+ content is monitored indirectly by receptors that are sensitive to the volume and osmolarity of the extracellular fluid.

Although physiological mechanisms that maintain fluid and electrolyte balance are important, behavioral mechanisms also play an essential role. Thirst is a critical part of fluid and electrolyte balance because drinking is the normal means of replacing water loss. Salt appetite is a behavior that leads people and animals to seek and ingest salt (sodium chloride, NaCl). The first half of this chapter examines the homeostasis of fluid and electrolyte balance, concentrating on renal handling of water and Na^+.

Bed-Wetting and ADH Bed-wetting, or **nocturnal enuresis,** is a problem that affects a significant number of school-age children. For many years, enuresis was thought to be of psychological origin, but now it is felt that most enuretic children have a physiological cause for their bed-wetting. In normal children, ADH secretion shows a circadian rhythm, with levels increasing at night. The result is nocturnal production of a smaller volume of more concentrated urine. Many children with enuresis fail to increase ADH secretion at night, with the result that their urine output stays elevated to the point that the bladder fills to its maximum capacity and empties spontaneously during sleep. These children can be successfully treated with a nasal spray of desmopressin, an ADH derivative, at bedtime.

TABLE 19-2 Factors Affecting Aldosterone Release

Direct, at the adrenal cortex

Increased extracellular K^+ concentration	Stimulates
Increased osmolarity	Inhibits

Indirect, through the RAAS pathway

Decreased blood pressure	Stimulates
Decreased flow past the macula densa	Stimulates

K^+). The other stimulus that acts directly on the adrenal gland is osmolarity: An increase in extracellular fluid osmolarity inhibits aldosterone secretion. Less aldosterone means increased Na^+ excretion, which helps decrease osmolarity. The direct inhibitory effect of high osmolarity on the adrenal cortex is an important part of the body's response to dehydration.

The renin-angiotensin-aldosterone pathway   Aldosterone release is mediated primarily by a trophic hormone known as angiotensin II (AGII). AGII is one component of the **renin-angiotensin-aldosterone system (RAAS),** a complex, multistep pathway for maintaining blood pressure.

The RAAS pathway begins when **juxtaglomerular granular cells** (JG cells) in the afferent and efferent arterioles of the nephron secrete a peptide called **renin** (Fig. 19-12 ■). Renin converts an inactive plasma protein, **an-**

Rapid Action of Aldosterone   For years, physiologists divided hormones into two groups: those with intracellular receptors that act by initiating new protein synthesis (known as genomic effects because the hormone acts on DNA) and those whose activity depends on plasma membrane receptors and second messenger systems that modify existing proteins (nongenomic effects). One of the most interesting discoveries in recent years has been the blurring of the distinction between these two groups. There is now good evidence that aldosterone, a steroid hormone, has both types of receptors: a cytoplasmic receptor that initiates a slow response and a cell membrane receptor that allows aldosterone to affect cell function within 1–2 minutes. (The aldosterone membrane receptor has not yet been described in Na^+-reabsorbing epithelia such as the kidney tubule.) In addition, the cytoplasmic receptor also seems to control a two-phase response. In the early phase, the apical Na^+ channels increase their open time under the influence of an as-yet-unidentified signal molecule. As intracellular Na^+ levels rise, the Na^+/K^+-ATPase speeds up, transporting cytoplasmic Na^+ out into the extracellular fluid. The net result is a rapid increase in Na^+ reabsorption that does not require the synthesis of new channel or ATPase proteins. In the later phase of aldosterone action, new channels and pumps are inserted into the epithelial cell membranes.

continued from page 546

Nephrogenic diabetes insipidus is associated with low water permeability at the apical membrane of collecting duct cells despite adequate levels of ADH. Normally, ADH binds to ADH receptors and stimulates a cAMP second messenger pathway, which in turn prompts insertion of water pores in the apical membranes of the tubule cells. Without ADH stimulation, the tubule cells store the water pores on vesicles in the cytoplasm.

Question 2: Scientists have recently discovered that nephrogenic diabetes insipidus can be caused by two different defects. One defect affects the ADH receptor, leading to the inability of ADH to activate the receptor and second messenger pathway. In the other defect, the ADH receptors are normal. What could be the problem in the other defect?

giotensinogen, into **angiotensin I** (AGI). (The suffix *-ogen* indicates an inactive precursor.) When AGI in the blood encounters an enzyme called **angiotensin converting enzyme (ACE),** AGI is converted into angiotensin II. This conversion was originally thought to take place only in the lungs, but ACE is now known to occur on the endothelium of blood vessels throughout the body. When AGII in the blood reaches the adrenal gland, it causes synthesis and release of aldosterone. Finally, at the distal nephron, aldosterone initiates a series of intracellular reactions that cause the tubule to reabsorb Na^+.

The stimuli that begin the RAAS pathway are all related directly or indirectly to low blood pressure (Fig. 19-13 ■):

1. The JG cells in the walls of the arterioles are directly sensitive to pressure. They respond to low pressure in the renal arterioles by secreting renin.
2. Sympathetic neurons, activated by the cardiovascular control center when blood pressure decreases, also stimulate renin secretion.
3. A third stimulus for renin release is a signal from the macula densa in the wall of the distal tubule (∞ p. 528). If fluid flow in the distal tubule decreases because of a decrease in blood pressure, GFR, or both, the macula densa cells release a paracrine signal that tells the JG cells to increase renin secretion. The mechanism by which the macula densa cells sense fluid flow and the nature of the paracrine signal are still being debated by researchers.

Sodium reabsorption does not directly raise low blood pressure, but retention of Na^+ does help stimulate fluid intake and volume expansion. When blood volume increases, blood pressure also increases.

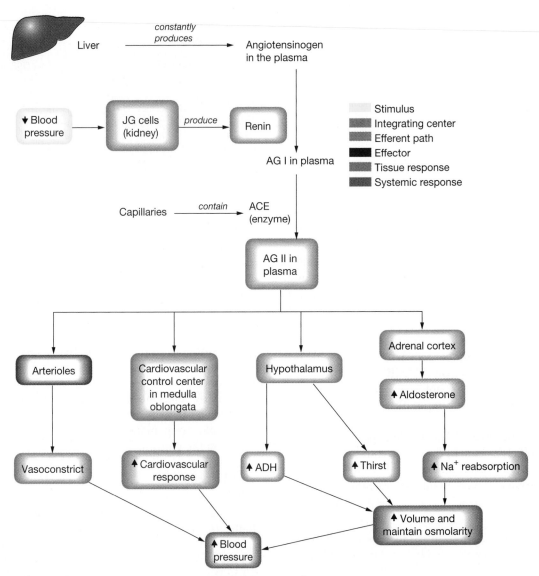

■ Figure 19-12 **The renin-angiotensin-aldosterone pathway**

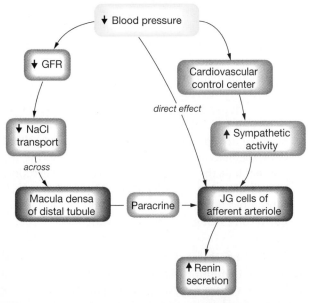

■ Figure 19-13 **Stimuli that cause secretion of renin are primarily related to decreased blood pressure**

The RAAS pathway does not end here, however. Angiotensin II is a remarkable hormone with additional effects that are directed at raising blood pressure. These actions of AGII make it an important hormone in its own right, not merely an intermediate step in the aldosterone control pathway.

Angiotensin II Helps Regulate Blood Pressure

Angiotensin II has significant effects on fluid balance and blood pressure beyond stimulating aldosterone secretion (Fig. 19-12 ■):

1. Activation of AGII receptors in the brain increases ADH secretion. Fluid retention in the kidney under the influence of ADH helps conserve blood volume.
2. AGII stimulates thirst. Fluid ingestion is a behavioral response that expands blood volume and raises blood pressure.
3. AGII is one of the most potent vasoconstrictors known in humans. Vasoconstriction decreases the

volume of the circulatory system, causing blood pressure to increase without a change in blood volume (∞ p. 431).

4. AGII receptors are found in the cardiovascular control center; their activation increases sympathetic output to the heart and blood vessels. This increased sympathetic output elevates blood pressure by increasing cardiac output and vasoconstriction.

Through these four pathways, AGII increases blood pressure both directly and indirectly. It is not surprising then that pharmaceutical companies started looking for drugs to block AGII once its blood pressure–raising effects became known. From their research came a new class of antihypertensive drugs called *ACE (angiotensin converting enzyme) inhibitors*. These drugs block the ACE-mediated conversion of AGI to AGII, thereby helping to relax blood vessels and lower blood pressure. Less AGII means less aldosterone release, resulting in a decrease in Na^+ reabsorption and, ultimately, a decrease in extracellular fluid volume. All these responses contribute to lowering blood pressure.

As you have learned in this section, aldosterone is the primary hormonal control for Na^+ reabsorption. Different hormones are involved if the body needs to eliminate Na^+. The next section looks at another hormone involved in renal regulation of salt and water balance.

✓ A man comes to the doctor with high blood pressure. Tests show that he also has elevated levels of renin in his blood and atherosclerotic plaques that have nearly blocked blood flow through his renal arteries. How does decreased blood flow in his renal arteries cause renin secretion to increase?

✓ Describe the pathways through which elevated renin causes high blood pressure in this man.

Atrial Natriuretic Peptide

Once the effects of aldosterone and ADH on Na^+ and water reabsorption were known, scientists speculated that other hormones might cause Na^+ loss (**natriuresis**) [*natrium,* sodium + *ourein,* to urinate] and water loss (diuresis) in the urine. If found, these hormones might be used clinically to lower blood volume and blood pressure in patients with essential hypertension (∞ p. 447). Despite years of searching, however, evidence for the other hormones was not forthcoming. Then, in 1981, a group of Canadian researchers found that injections of homogenized rat atria caused rapid but short-lived excretion of Na^+ and water in the urine. They hoped that they had found the missing hormone, one whose activity would complement that of aldosterone and ADH. As it turned out, they had discovered the first in an entire family of hormones, some produced in the myocardial cells of the heart and others found in the brain.

Atrial natriuretic peptide (ANP; also known as atrial natriuretic factor and atriopeptin) is a peptide hormone produced in specialized myocardial cells in the atria of the heart. It is synthesized as part of a large pro-hormone molecule that is cleaved into several active hormone fragments (∞ p. 179).

Atrial natriuretic peptide is secreted when atrial cells stretch more than is normal, as would occur with increased blood volume. At the systemic level, ANP enhances Na^+ excretion and urinary water loss, but the exact mechanisms by which it does so still are not clear. Atrial natriuretic peptide increases the glomerular filtration rate (GFR), apparently by relaxing the contractile cells surrounding the filtration slits (∞ p. 525). Relaxation of the contractile cells opens the pores in the glomerular membrane and makes more surface area available for filtration. In addition, ANP has been reported to decrease NaCl and water reabsorption in the collecting duct (Table 19-3). The mechanism by which ANF affects tubular reabsorption is not known.

Atrial natriuretic peptide also has several indirect effects on kidney function that alter Na^+ and water excretion. It inhibits the release of renin, aldosterone, and ADH. These actions all reinforce its natriuretic-diuretic effect. At this time, however, it is not clear that ANP is the primary hormone responsible for regulation of Na^+ excretion, so the search for other natriuretic factors continues.

Atrial natriuretic peptide and related peptides are also secreted by neurons in the brain. These neurons originate in the hypothalamus and terminate in the cardiovascular control center of the medulla. When secreted as a neurotransmitter or neuromodulator in this location, ANP apparently helps lower blood pressure, the very effect that makes it of interest to drug companies. Unfortunately, ANP has a short half-life in the body, and patients would have to take ANP continuously to maintain its effects. Researchers are hoping to learn more about the cellular mechanism of ANP action so that they can produce a longer-acting agonist that can be used to treat hypertension.

POTASSIUM BALANCE

The regulation of body potassium levels is essential to maintain a state of well-being. Most K^+ is found within the cells; only about 2% is in the extracellular fluid.

TABLE 19-3	Mechanisms by Which Atrial Natriuretic Peptide Increases Urinary Na^+ and Water Loss

1. Increases glomerular filtration rate by relaxing filtration slits at glomerulus
2. Inhibits renin release and aldosterone secretion
3. Inhibits ADH release
4. Possibly, inhibits tubular reabsorption of Na^+ and water

Guanylin and Uroguanylin Scientists have observed that, if large amounts of salt are ingested, the body responds by excreting more Na^+ in the urine. On the other hand, if the same salt load is given intravenously, Na^+ excretion barely increases. From this observation, scientists proposed a feedforward mechanism, in which sensors in the intestine or hepatic portal system monitor the amount of Na^+ ingested and, when appropriate, initiate a reflex that increases urinary Na^+ output. There are two lines of evidence to support this theory. One group of researchers described a set of Na^+ receptors in the hepatic portal system that, when activated, cause the central nervous system to inhibit renin secretion and decrease intestinal Na^+ absorption. In related studies, other investigators have found two candidate hormones, guanylin and uroguanylin, that decrease intestinal Na^+ absorption. Both of these lines of investigation promise to expand our understanding of Na^+ homeostasis in the body.

Although only a small fraction of the body's K^+ is in the extracellular fluid, potassium homeostasis is designed to keep plasma K^+ concentrations within a narrow range. As you learned in Chapter 8, changes in the concentration of extracellular K^+ affect the resting membrane potential of cells by altering the K^+ concentration gradient between the cytoplasm and the extracellular fluid (∞ p. 223 and Fig. 8-23 ■). If plasma (and extracellular fluid) K^+ increases (**hyperkalemia**), the concentration gradient decreases and more K^+ remains in the cell, depolarizing it. If plasma K^+ concentrations decrease (**hypokalemia**), the concentration gradient becomes larger, more K^+ leaves the cell, and the resting membrane potential becomes more negative. Because of the effect of plasma K^+ on excitable tissues such as the heart, clinicians are always concerned about keeping plasma K^+ within the normal range (2.5–6.0 meq/L).

If K^+ drops below 2.5 meq/L or goes above 6.0 meq/L, the excitable tissues of muscle and nerve begin to show altered function. Hypokalemia causes muscle weakness because it is more difficult for hyperpolarized neurons and muscles to fire action potentials. The danger in this condition lies in the failure of respiratory muscles and the heart. Fortunately, skeletal muscle weakness usually occurs before cardiac problems and is significant enough to lead the patient to seek treatment. Correction of the hypokalemia generally involves increasing oral intake of potassium with K^+ supplements and foods that are rich in K^+, such as orange juice and bananas.

Hyperkalemia is a more dangerous potassium disturbance because depolarization of excitable tissues makes them more excitable initially. Subsequently, the cells are unable to repolarize fully and actually become *less* excitable. In this state, they have action potentials that are either smaller than normal or nonexistent. Cardiac muscle excitability affected by changes in plasma K^+

and hyperkalemia can lead to life-threatening cardiac arrhythmias.

Under normal conditions, the body very carefully matches K^+ excretion to K^+ ingestion. If plasma K^+ levels go up, aldosterone is secreted because of the effect of hyperkalemia on the adrenal cortex. Aldosterone prompts the kidneys to excrete more K^+ while simultaneously retaining Na^+. (Remember that when plasma K^+ concentrations change, anions such as Cl^- are also added or subtracted to the extracellular fluid in a 1:1 ratio, maintaining overall electrical neutrality.)

Disturbances in potassium balance may be caused by kidney disease, loss of K^+ in diarrhea, or the use of certain types of diuretics that prevent the kidneys from fully reabsorbing K^+. Inappropriate correction of dehydration can also cause problems with K^+ concentrations. For example, a golfer went out to play a round of golf when the temperature was above 100° F. He was aware of the risk of dehydration, so he took water along to replace his fluid loss. The replacement of sweat loss with pure water kept the golfer's extracellular fluid volume normal but dropped his total blood osmolarity as well as his K^+ and Na^+ concentrations. He was unable to finish the round of golf because of muscle weakness, and he required medical attention that included potassium replacement therapy.

Potassium balance is also closely tied to acid-base balance, as you will learn in the last section of the chapter. Correction of a pH disturbance requires close attention to plasma potassium levels. Similarly, correction of potassium imbalance may alter body pH.

BEHAVIORAL MECHANISMS IN SALT AND WATER BALANCE

Although neural, neuroendocrine, and endocrine reflexes play key roles in salt and water homeostasis, behavioral responses are critical for the restoration of the normal state, especially when extracellular fluid volume decreases or osmolarity increases. Drinking water is normally the only way to restore lost water, and eating salt is the only way to raise the body's Na^+ content. Both behaviors are essential for normal salt and water balance. The clinician

continued from page 555

"Christopher is always thirsty," Mrs. Godell tells the doctor. "And he constantly urinates. With all the water he drinks, sometimes I think he'll drown!" The doctor explains that people with diabetes insipidus will not "drown." On the contrary, they run the risk of severe dehydration, despite the copious amounts of water they drink.

Question 3: Why do people with diabetes insipidus become dehydrated even though they are drinking a lot?

Question 4: Explain why intense thirst is a main symptom of diabetes insipidus.

chemoreceptors of the medulla oblongata cannot respond directly to changes in plasma pH because H^+ does not cross the blood-brain barrier (∞ p. 241). But increases or decreases in pH change CO_2, the trigger for ventilation mediated by the central chemoreceptors. Dual control through the central and peripheral chemoreceptors helps the body respond rapidly to changes in either pH or plasma CO_2.

✓ In Figure 19-17 ■, name the muscles of ventilation that might be involved in this reflex.

✓ In equation 6 above, HCO_3^- is shown as increasing in amount. Why doesn't the HCO_3^- buffer the increased H^+ and prevent acidosis from occurring?

Renal Compensation in Acid-Base Disturbances Takes Place through Excretion or Reabsorption of H^+ and HCO_3^-

The kidneys help compensate for pH changes in two ways. They affect pH (1) directly by retaining or excreting H^+ and (2) indirectly by changing the reabsorption or excretion of HCO_3^- buffer. In acidosis, the kidney secretes H^+ into the lumen of both the proximal and distal tubules using direct and indirect active transport (Fig. 19-18a ■). Ammonia and phosphate ions in the urine act as urinary buffers, trapping the H^+ and allowing more H^+ to be secreted. Even with these buffers, urine can become quite acidic, down to a pH of about 4.5. While H^+ is being secreted, the kid-

neys make new HCO_3^- from CO_2 and H_2O. The HCO_3^- is reabsorbed into the blood to act as a buffer and increase pH.

In times of alkalosis, the kidney reverses the process described above, secreting HCO_3^- into the lumen and reabsorbing H^+ in an effort to bring pH back into the normal range (Fig. 19-18b ■). Renal compensations are slower than respiratory compensations, and their effect on pH may not be noticed for 24–48 hours. However, once activated, renal compensations effectively handle all but severe acid-base disturbances.

The cellular mechanisms for renal handling of H^+ and HCO_3^- are described below. These mechanisms involve some membrane transporters that you have not encountered before:

1. The **apical Na^+-H^+ antiporter** is an indirect active transporter that brings Na^+ into the epithelial cell in exchange for moving H^+ against its concentration gradient into the lumen of the tubule.

2. The **basolateral Na^+-HCO_3^- symporter** moves Na^+ and HCO_3^- out of the epithelial cell and into the interstitial fluid. This indirect active transporter couples the energy of HCO_3^- diffusing down its concentration gradient to the uphill movement of Na^+ from the cell to the extracellular fluid.

3. The **H^+-ATPase** uses energy from ATP to acidify the urine, pushing H^+ against its concentration gradient into the lumen of the nephron.

4. The **H^+/K^+-ATPase** puts an H^+ into the urine in exchange for a reabsorbed K^+. This exchange explains

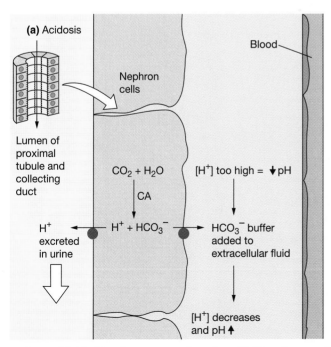

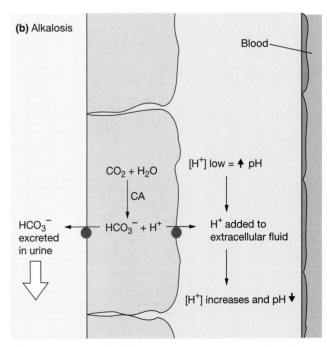

■ **Figure 19-18 Renal compensations for pH disturbances** The transporters shown in this figure are generic transporters. The actual specific transporters are shown in Figures 19-19 and 19-20.

why acid-base disturbances are often accompanied by problems with potassium balance and vice versa.

In addition to these transporters, the renal tubule also uses the ubiquitous Na^+/K^+-ATPase and the same HCO_3^--Cl^- antiport protein that is responsible for the chloride shift in red blood cells.

The proximal tubule: hydrogen ion secretion and bicarbonate reabsorption   The proximal tubule is responsible for the reabsorption of most of the HCO_3^- that enters the nephron at the glomerulus. There appears to be no membrane transporter that can bring HCO_3^- into a cell, so the process of HCO_3^- reabsorption across the epithelium takes place by an indirect route (Fig. 19-19 ■). In the proximal tubule: ① H^+ is secreted from the cells into the lumen in exchange for a filtered Na^+, using the Na^+-H^+ antiport protein; ② the secreted H^+ then combines with filtered HCO_3^- to form CO_2 in the lumen; ③ the CO_2 diffuses into the proximal tubule cell and combines with water to form H_2CO_3, which dissociates to form H^+ and HCO_3^- in the cytoplasm; ④ the H^+ can be secreted again, replacing the H^+ that combined with the filtered HCO_3^-. The HCO_3^- is transported out of the cell on the basolateral side by the HCO_3^--Na^+ symport protein ⑤. The net result of this process is that filtered Na^+ and HCO_3^- are reabsorbed and H^+ is secreted.

The distal nephron: H^+ and HCO_3^- handling depend on the acid-base state of the body   The distal nephron (the distal tubule and collecting duct) plays a significant role in the fine regulation of acid-base balance. Special cells known as **intercalated cells,** or **I cells,** are interspersed among the principal cells in this section of the nephron. In times of acidosis, type A intercalated cells secrete H^+ and reabsorb bicarbonate. During periods of alkalosis, type B intercalated cells secrete HCO_3^- and reabsorb H^+.

Intercalated cells are characterized by high concentrations of carbonic anhydrase in their cytoplasm. This enzyme allows them to convert large amounts of CO_2 into H^+ and HCO_3^-. The H^+ ions are pumped out of the intercalated cell by the H^+-ATPase or by an ATPase that exchanges one H^+ for one K^+. The HCO_3^- leaves the cell by means of the HCO_3^--Cl^- antiport exchanger.

Figure 19-20a ■ shows how the type A intercalated cell works in times of acidosis, secreting H^+ and reabsorbing HCO_3^-. The process is similar to H^+ secretion in the proximal tubule except for the specific H^+ transporters. The distal nephron uses a H^+-ATPase and a H^+/K^+-ATPase rather than a Na^+-H^+ antiport protein.

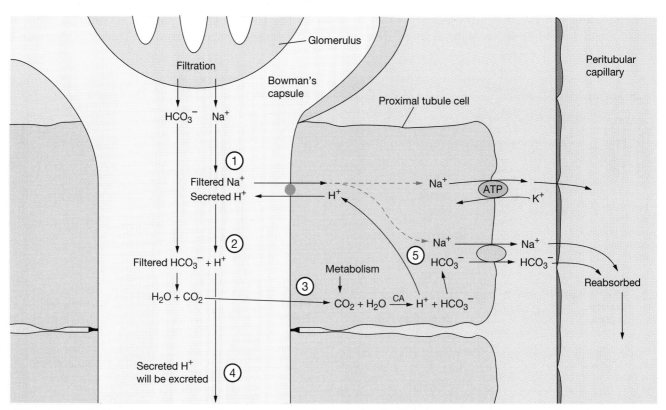

■ **Figure 19-19   Reabsorption of filtered bicarbonate and secretion of hydrogen ions in the proximal tubule**   See text for explanations of the numbers.

In times of alkalosis, when the H$^+$ concentration of the body is too low, H$^+$ is reabsorbed and HCO$_3^-$ buffer is excreted in the urine (Fig. 19-20b ■). Once again, the ions are produced by the dissociation of H$_2$CO$_3$ formed from H$_2$O and CO$_2$. Hydrogen ions are reabsorbed into the extracellular fluid on the basolateral side, while HCO$_3^-$ is secreted into the lumen. The same transporters are used simply by reversing the polarity of the cell and inserting the transport proteins on opposite membranes.

The H$^+$/K$^+$-ATPase of the distal nephron provides the link between H$^+$ and K$^+$ that creates parallel disturbances of acid-base and potassium balance. In acidosis, when plasma H$^+$ is high, the kidney secretes H$^+$ into the urine and therefore reabsorbs K$^+$. Thus, acidosis is often accompanied by hyperkalemia. The reverse is true for

---

✓ Why does the K$^+$-H$^+$ transporter of the distal nephron require ATP in order to secrete H$^+$ but the Na$^+$-H$^+$ exchanger of the proximal tubule does not?

✓ In hypokalemia, the intercalated cells of the distal nephron reabsorb K$^+$ from the tubule lumen. What happens to pH in hypokalemia as a result?

✓ Draw a map similar to the one in Figure 19-17 ■ that shows the renal compensation for acidosis. Draw another map for alkalosis.

---

alkalosis, when blood H$^+$ levels are low. The mechanism that allows the distal nephron to retain H$^+$ simultaneously causes it to secrete K$^+$, so alkalosis goes hand in hand with hypokalemia.

Disturbances of Acid-Base Balance

The three compensatory mechanisms (buffers, ventilation, and renal excretion) take care of most variations in plasma pH. But under some circumstances, the production or loss of H$^+$ or HCO$_3^-$ is so extreme that compensatory mechanisms fail to maintain pH homeostasis. In these states, the pH of the blood moves out of the normal range of 7.38–7.42. If the body fails to keep pH within a range of pH 7.00–7.70, acidosis or alkalosis will be fatal. This section looks at the four primary acid-base disturbances and their causes.

Acid-base problems are classified both by the direction of the pH change (acidosis or alkalosis) and by the underlying cause (metabolic or respiratory). As you learned earlier, changes in P$_{CO_2}$ due to hyper- or hypoventilation cause pH to shift. These disturbances are said to be of respiratory origin. If the pH problem arises from acids or bases of non-CO$_2$ origin, the problem is said to be a metabolic problem.

One point to note is that by the time an acid-base disturbance shows up as a change in plasma pH, the buffer capacity of the body has been exceeded. The loss of

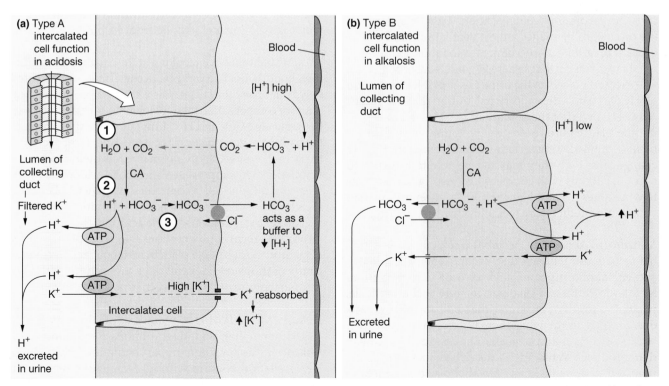

■ Figure 19-20 **Role of the intercalated cell in acidosis and alkalosis** The intercalated cells secrete or reabsorb H$^+$ and bicarbonate according to the needs of the body.

buffer capacity leaves the body with only two options: respiratory compensation and renal compensation. If the problem is of respiratory origin, only one homeostatic compensation is available—the kidneys. On the other hand, metabolic acid-base disturbances can be compensated using both respiratory and renal mechanisms. The combination of an initial pH disturbance and compensatory changes is one of the factors that make analysis of acid-base disorders in the clinical setting so difficult. In this book, we will concentrate on the simple disturbances.

Respiratory acidosis

A state of respiratory acidosis occurs when ventilation is inadequate to remove CO_2 produced by cells, resulting in retention of CO_2 and elevated plasma P_{CO_2}. This is also known as alveolar hypoventilation. Some of the situations in which this occurs include respiratory depression due to drugs or alcohol, increased airway resistance in asthma, impaired gas exchange in fibrosis or severe pneumonia, or muscle weakness in muscular dystrophy and other muscle diseases. The most common cause of respiratory acidosis is chronic obstructive pulmonary disease such as emphysema, in which inadequate gas exchange is compounded by loss of alveolar exchange area. No matter what the cause of respiratory acidosis, plasma CO_2 levels increase (shown in blue), leading to elevated H^+ and HCO_3^-:

$$\Uparrow CO_2 + H_2O \Rightarrow \Rightarrow \Uparrow H^+ + \Uparrow HCO_3^-$$

The hallmark of simple respiratory acidosis is decreased pH and elevated bicarbonate levels (Table 19-5). Because the problem is of respiratory origin, the body cannot carry out respiratory compensation. (However, depending on the problem, mechanical ventilation can be used to assist breathing.) Any compensation must be through renal mechanisms that secrete H^+ and reabsorb HCO_3^-. The excretion of H^+ in the urine raises pH. Reabsorption of HCO_3^- provides additional buffer that combines with H^+, lowering the amount of free H^+ and also raising the pH. In chronic obstructive pulmonary disease, renal compensation mechanisms can moderate the pH change, although they may not be able to return it to normal. Patients with compensated respiratory acidosis have a higher pH and higher HCO_3^- concentration than you would predict from their elevated P_{CO_2} values.

Metabolic acidosis

Metabolic acidosis is a disturbance of mass balance that occurs when the dietary and metabolic input of acid exceeds acid excretion. Metabolic causes of acidosis include lactic acidosis, which is a result of anaerobic metabolism, and ketoacidosis, which occurs when there is excessive breakdown of fats or certain amino acids. The metabolic pathway that produces ketoacids is described in Chapter 21. Ingested toxins that cause metabolic acidosis include methanol, aspirin, and ethylene glycol (antifreeze).

Metabolic acidosis can also occur if the body loses buffer capacity. The most common example is loss of

TABLE 19-5	Plasma P_{CO_2}, Ions, and pH in Acid-Base Disturbances			
Disturbance	P_{CO_2}	H^+	pH	HCO_3^-
Acidosis				
Respiratory	$\Uparrow$	$\Uparrow$	$\Downarrow$	$\Uparrow$
Metabolic	Normal or $\Downarrow$*	$\Uparrow$	$\Downarrow$	$\Downarrow$
Alkalosis				
Respiratory	$\Downarrow$	$\Downarrow$	$\Uparrow$	$\Downarrow$
Metabolic	Normal or $\Uparrow$*	$\Downarrow$	$\Uparrow$	$\Uparrow$

*These values are different from what you would expect from the Law of Mass Action because almost instantaneous respiratory compensation occurs to keep P_{CO_2} from changing significantly.

HCO_3^- from the intestines with diarrhea. The exocrine pancreas produces HCO_3^- for secretion into the small intestine by a mechanism that is similar to the renal mechanism illustrated in Figure 19-18b ■. The H^+ that is made at the same time is released into the blood. Normally, the HCO_3^- secreted into the small intestine is reabsorbed, buffering the H^+ that was released. However, if the HCO_3^- is not reabsorbed from the small intestine owing to diarrhea, a state of acidosis results.

Metabolic acidosis is expressed in the equation below. Hydrogen ion concentration increases because of H^+ contributed by the metabolic acids. This increase shifts the equation to the left, increasing CO_2 and using up HCO_3^- buffer:

$$\Uparrow CO_2 + H_2O \Leftarrow \Leftarrow \Uparrow H^+ + \Downarrow HCO_3^-$$

The decrease in HCO_3^- concentration is an important way to distinguish metabolic from respiratory acidosis (Table 19-5).

You would think from looking at the equation above that metabolic acidosis would be accompanied by elevated P_{CO_2}. But unless the subject has a lung disease as well, respiratory compensation takes place almost instantaneously. Both elevated CO_2 and H^+ stimulate ventilation through the pathways described earlier. As a result, P_{CO_2} decreases to normal or even below-normal levels owing to hyperventilation. Uncompensated metabolic acidosis is rarely seen clinically. Indeed, a common sign of metabolic acidosis is hyperventilation, evidence of respiratory compensation occurring in response to the acidosis.

The same renal compensations that were discussed for respiratory acidosis take place in metabolic acidosis. Renal compensations take several days to reach full effectiveness, so they are not usually seen in acute disturbances.

Respiratory alkalosis

States of alkalosis are much less common than acidotic conditions. Respiratory alkalosis occurs as a result of hyperventilation, when alveolar ventilation goes up without a matching increase in metabolic CO_2 production. Consequently, plasma P_{CO_2} drops and alkalosis results:

$$\Downarrow CO_2 + H_2O \Leftarrow \Leftarrow \Downarrow H^+ + \Downarrow HCO_3^-$$

The decrease in CO_2 pulls the equation to the left, so both H^+ and HCO_3^- concentrations drop. Low HCO_3^- levels in alkalosis indicate a respiratory disorder.

The primary clinical cause of respiratory alkalosis is excessive artificial ventilation. Fortunately, this condition is easily corrected by adjusting the ventilator. The most common physiological cause of respiratory alkalosis is hysterical hyperventilation due to anxiety. In these situations, the neurological symptoms caused by alkalosis can be partially reversed by having the patient breathe into a paper bag. In doing so, the patient rebreathes exhaled CO_2, which raises arterial P_{CO_2} and corrects the problem.

Metabolic alkalosis Metabolic alkalosis has two common causes: excessive vomiting of acidic stomach contents and excessive ingestion of bicarbonate-containing antacids. In both cases, the resulting alkalosis reduces H^+ concentration:

$$\Downarrow CO_2 + H_2O \Rightarrow \Rightarrow \Rightarrow \Downarrow H^+ + \Uparrow HCO_3^-$$

The decrease in H^+ shifts the reaction to the right. Carbon dioxide levels decrease, and HCO_3^- levels go up. Just as in metabolic acidosis, respiratory compensation is very rapid. The increase in pH and drop in P_{CO_2} depress ventilation. Less CO_2 is blown off, raising the P_{CO_2} and creating more H^+ and HCO_3^-. The ventilatory compensation helps correct the pH problem but elevates HCO_3^- levels even more. However, the ventilatory compensation is limited because hypoventilation causes hypoxia. Once the arterial P_{O_2} drops below 60 mm Hg, hypoventilation ceases.

The renal response to metabolic alkalosis is the same as that for respiratory alkalosis: HCO_3^- is excreted and H^+ is reabsorbed.

This chapter has used fluid and acid-base balance as examples of the integration of function between various systems of the body. Changes in body fluid volume, reflected by changes in blood pressure, trigger both cardiovascular and renal homeostatic responses. Disturbances of acid-base balance are met with compensatory responses from both the pulmonary and renal systems. Because of the interwoven responsibilities of these three systems, a pathology in one system is likely to cause disturbances in the other two. Recognition of this fact is an important aspect of treatment for many clinical conditions.

In the next unit of the book, we turn to the metabolic processes of the body and to the endocrine control of body function.

CHAPTER REVIEW

Chapter Summary

1. The homeostatic control mechanisms for fluid and electrolyte balance involve the renal, respiratory, and cardiovascular systems. Behaviors such as drinking play an important role in fluid balance. (p. 543)

2. Adjustments made by the lungs and cardiovascular system are primarily under nervous control and are more rapid than compensation by the kidneys. (p. 544)

Homeostasis of Volume and Osmolarity

3. The kidneys influence extracellular fluid volume and osmolarity by altering the amount of water and sodium excreted. (p. 544)

Water Balance and the Regulation of Urine Concentration

4. Homeostatic control of body water maintains osmolarity around 300 mOsM and blood volume and blood pressure at a level that provides an adequate supply of oxygen to the tissues. (p. 544)

5. Most daily water intake comes from food and drink. The major route of water loss in the body is the urine, with a daily volume of 1.5 liters. Smaller amounts are lost in the feces, by evaporation from the skin, and in the humidified air exhaled from the respiratory tract. (p. 545)

6. An increase in extracellular fluid osmolarity or a decrease in blood pressure is the stimulus for increased water reabsorption in the kidney. Osmolarity is sensed by hypothalamic **osmoreceptors.** Blood pressure and blood volume are sensed by receptors in the atria and carotid and aortic bodies. (p. 546)

7. To produce dilute urine, the kidney must reabsorb solute without allowing water to follow osmotically. This reabsorption occurs primarily in the ascending limb of the loop of Henle. To produce concentrated urine, the nephron must reabsorb water while leaving salt behind in the lumen. This reabsorption occurs primarily in the collecting duct and is regulated by **antidiuretic hormone (ADH).** (p. 547)

8. The loop of Henle is a **countercurrent system** that makes the interstitial fluid of the kidney medulla very hyperosmotic by actively transporting Na^+ and Cl^- out of the tubule without allowing water to follow. This high medullary osmolarity is necessary for the formation of concentrated urine as fluid flows through the collecting duct. (p. 548)

9. The excretion of water by the kidneys is controlled by antidiuretic hormone, or ADH, which controls the perme-ability of the collecting duct to water in a graded fashion. When ADH is absent, water permeability is nearly zero. (p. 550)

10. The secretion of ADH is stimulated by increased osmolarity and decreased blood pressure. (p. 551)

11. Water reabsorption in the kidneys conserves water and decreases osmolarity, but it cannot restore lost volume. (p. 551)

Sodium Balance and the Regulation of Extracellular Fluid Volume

12. Sodium chloride is the main extracellular solute, and the total amount of Na^+ in the body is a primary determinant of the extracellular fluid volume. (p. 552)

13. The reabsorption of Na^+ in the distal tubule and collecting duct is regulated by the steroid hormone **aldosterone:** the more aldosterone, the more Na^+ reabsorption. Aldosterone also increases K^+ secretion. (p. 553)

14. The target cell of aldosterone is the **principal cell, or P cell.** Principal cells have Na^+/K^+-ATPase pumps on the basolateral membrane and leak channels for Na^+ and K^+ on the apical membrane. Aldosterone stimulates the synthesis of channels and pumps. (p. 554)

15. Aldosterone secretion is controlled directly at the adrenal cortex and indirectly by renin. Increased K^+ stimulates aldosterone production and results in K^+ secretion. Increased extracellular fluid osmolarity inhibits aldosterone secretion. (p. 554)

16. Aldosterone release is mediated primarily by **angiotensin II (AGII).** The pathway begins when JG cells in the nephron secrete **renin.** Renin converts **angiotensinogen** to **angiotensin I.** A membrane-bound enzyme called **angiotensin converting enzyme (ACE)** converts AGI to AGII. Angiotensin II causes synthesis and release of aldosterone. (p. 555)

17. The stimuli that release renin are related directly or indirectly to low blood pressure. (p. 555)

18. Angiotensin II has additional effects that raise blood pressure. These include increased ADH secretion, stimulation of thirst, direct vasoconstriction, and activation of the cardiovascular control center. (p. 556)

19. **Atrial natriuretic peptide** (ANP) is a peptide hormone secreted when atrial cells stretch because of increased blood volume. Atrial natriuretic peptide enhances Na^+ excretion and urinary water loss by increasing GFR, inhibiting tubular reabsorption, and inhibiting the release of renin, aldosterone, and ADH. (p. 557)

Potassium Balance

20. Potassium homeostasis keeps plasma K^+ concentrations in a narrow range. **Hyperkalemia** and **hypokalemia** cause problems with excitable tissues, especially the heart. (p. 557)

Integrated Control of Volume and Osmolarity

21. Homeostatic compensations for changes in salt and water balance attempt to follow the law of mass balance: Fluid and solute added to the body must be removed, or fluid and solute lost must be replaced. However, perfect compensation is not always possible. (p. 559)

Acid-Base Balance

22. The concentration of H^+ in the body is closely regulated because pH affects intracellular proteins such as enzymes and membrane channels. pH also affects the activity of the nervous system. (p. 563)

23. Normally, the body is challenged by intake and production of acids rather than bases. Acids include foods and metabolic intermediates that are organic acids. However, the biggest source of acid is CO_2 from respiration that combines with water to form carbonic acid (H_2CO_3). (p. 564)

24. The body copes with changes in pH by using buffers, ventilation, and renal excretion of H^+ and HCO_3^-. (p. 565)

25. Bicarbonate produced from CO_2 is the most important extracellular buffer system of the body. Bicarbonate is useful in buffering acids produced by metabolism. (p. 565)

26. Ventilation can correct disturbances in acid-base balance because changes in plasma P_{CO_2} affect both the H^+ and the HCO_3^- content of the blood. Ventilation is affected by pH through the carotid and aortic chemoreceptors and the central chemoreceptors. An increase in plasma H^+ stimulates breathing, allowing the lungs to excrete CO_2 and convert H^+ to water. (p. 566)

27. The kidneys affect pH directly by retaining or excreting H^+ and indirectly by changing the reabsorption or excretion of HCO_3^- buffer. In **acidosis,** the kidney secretes H^+ and reabsorbs HCO_3^-. In **alkalosis,** the kidney secretes HCO_3^- and reabsorbs H^+. (p. 567)

28. **Intercalated cells** in the collecting duct of the nephron are responsible for the fine regulation of acid-base balance. (p. 568)

Questions

Reviewing Facts and Terms

1. What is an electrolyte? Name five electrolytes whose concentrations must be regulated by the body.

2. List five organs and four hormones important in maintaining fluid and electrolyte balance.

3. Compare the routes by which water enters the body to the ways the body loses water.

4. List the receptors that regulate osmolarity, blood volume, blood pressure, ventilation, and pH. Where are they located, what stimulates them, and what compensatory mechanisms are triggered by these receptors?

5. How do the two limbs of the loop of Henle differ in their permeability? How is this possible?

6. Which ion is a primary determinant of extracellular fluid volume? Which ion is the determinant of extracellular pH?

7. What happens to the resting membrane potential of excitable cells when plasma concentrations of K^+ decrease? Which organ is most likely to be affected by changes in K^+ concentration?

8. Appetite for what two substances is important in regulating fluid volume and osmolarity?

9. Write out the words for the following abbreviations: ADH, ANP, ACE, AGII, JG cell, P cell, I cell.

10. Make a list of all the different membrane transporters in the kidney. For each transporter, tell (a) which section(s) of the nephron contain the transporter; (b) whether the transporter is on the apical or basolateral membrane, or both; (c) whether it participates in reabsorption, secretion, or both.

11. List and briefly explain three reasons why monitoring and regulating the pH of extracellular fluid are important. What three mechanisms does the body use to cope with changing pH?

12. Which is more likely to accumulate in the body, acids or bases? List some sources of each one.

13. What is a buffer? List three intracellular buffers. Name the primary extracellular buffer.

14. Name two ways that the kidneys alter the plasma pH. Which compounds serve as urinary buffers?

15. Write the equation that shows how CO_2 is related to pH. What enzyme increases the rate of this reaction? Name two different cell types that possess high concentrations of this enzyme. What might cause arterial P_{CO_2} to increase?

16. When ventilation increases, what happens to arterial P_{CO_2}? To pH? To plasma H^+ concentration?

Reviewing Concepts

17. **Concept map:** Map the homeostatic reflexes that occur in response to each of the following situations:
    a. decreased volume, normal osmolarity
    b. increased volume, increased osmolarity
    c. normal volume, increased osmolarity

18. Create a map that shows how the kidneys and lungs interact to regulate plasma pH.

19. Explain how the loop of Henle and vasa recta work together to create dilute fluid that enters the distal tubule.

20. Diagram the mechanism by which ADH alters the composition of the urine.

21. Make a table and specify for each substance listed (a–f): hormone or enzyme? steroid or peptide? cell or tissue that produces it? target cell or tissue? response of the target to the substance?

a. ANP	b. aldosterone
c. renin	d. AGII
e. ADH	f. angiotensin-converting enzyme

22. Name the four main compensatory mechanisms for restoring low blood pressure to normal. Why do you think there are so many homeostatic pathways for raising low blood pressure?

23. Compare and contrast the following sets of terms: (a) principal and intercalated cells; (b) renin, angiotensin II, aldosterone, ACE; (c) respiratory acidosis with metabolic acidosis, including their causes and compensations; (d) water reabsorption in the proximal tubule, distal tubule, and ascending limb of the loop of Henle; (e) respiratory alkalosis with metabolic alkalosis, including their causes and compensations.

Problem Solving

24. Karen has bulimia, in which she induces vomiting to avoid weight gain. When the doctor sees her, her weight is 89 lb and respiration is 6 breaths/min (normal 12). Her blood HCO_3^- is 62 meq/L (normal 24–29). Her arterial blood pH is 7.61 and her P_{CO_2} is 61 mm Hg. What is her acid-base condition called? Explain why her plasma bicarbonate level is so high. Why is she hypoventilating? What effect will this have on the pH and total oxygen content of her blood? Explain your answers.

25. Draw a graph showing the general relationship between thirst and plasma osmolarity. Label the axes.

26. Hannah, a 31-year-old woman, decided to have colonic irrigation, a procedure during which large volumes of distilled water were infused into her rectum. Over the course of the treatment she absorbed 3000 mL of water. About 12 hours later, her roommate found her in convulsions and took her to the emergency room. Her blood pressure was 140/90, her plasma Na^+ concentration was 106 meq/L (normal is 135 meq/L), and her plasma osmolarity was 270 mOsM. In a concept map or flow chart, diagram all the homeostatic responses that her body was using to compensate for the changes in blood pressure and osmolarity.

Problem Conclusion

In this running problem, you learned about a rare disease characterized by the inability to concentrate urine. You also learned that the inability to concentrate urine can be due to defects either in the brain or in the kidney.

Further check your understanding of this running problem by checking your answers against those in the summary table.

	Question	Facts	Integration and Analysis
1	Does Christopher have central or nephrogenic diabetes insipidus?	Central diabetes insipidus is caused by a defect in the pituitary or hypothalamus that leads to decreased levels of ADH. Nephrogenic diabetes insipidus is caused by a defect in the kidneys. Christopher has normal blood levels of ADH.	Because nephrogenic diabetes insipidus is caused by a defect in the kidney, ADH levels are normal. Christopher has nephrogenic diabetes insipidus.
2	Nephrogenic diabetes insipidus can be caused by two different defects. One defect affects the ADH receptor. What could the other defect be?	ADH binding to ADH receptors in the collecting duct initiates a cAMP second messenger pathway and prompts the insertion of water pores into the apical membranes of the tubule cells.	The second defect could be any part of the process that occurs after the ADH receptor is activated. For example, problems with the cytoskeleton might prevent membrane recycling, or the water pores themselves could be defective.
3	Why do people with diabetes insipidus become dehydrated even though they are drinking a lot?	In diabetes insipidus, the collecting duct is unable to reabsorb water.	Inability to reabsorb water in the collecting duct leads to the excretion of large volumes of dilute urine. If patients do not continuously drink water, they can lose enough fluid to go into a state of severe dehydration.
4	Explain why intense thirst is a main symptom of diabetes insipidus.	Thirst is triggered when osmolarity rises above 280 mOsM or by a decrease in blood pressure.	Excretion of large volumes of dilute urine decreases blood volume and blood pressure and increases osmolarity. In addition, decreased blood pressure stimulates aldosterone secretion and renal Na^+ reabsorption. The increase in blood osmolarity triggers the osmoreceptors.

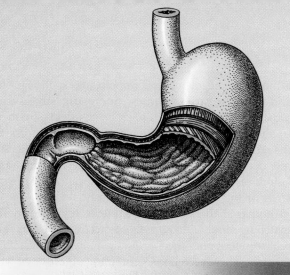

20

Digestion

BACKGROUND BASICS

Composition of biomolecules (p. 26)
Smooth muscle structure and contraction (p. 348)
Anatomy of the lymphatics (p. 440)
Secondary active transport (p. 122)
Cell-to-cell junctions (p. 53)
Renal transport of sodium (p. 530)
Positive feedback loops (p. 163)
Microvilli (p. 47)
Endocrine and exocrine gland formation (p. 61)
Transporting and secretory epithelia (p. 57)
Enzymes (p. 78)
Micelles (p. 109)
Protein synthesis and storage (p. 98)
Acidification of the urine (p. 567)
Exocytosis and transcytosis (p. 127)
Portal systems (p. 185)

A shotgun wound to the stomach seems an unlikely beginning to the scientific study of digestive processes. But in 1822 at Fort Mackinac, young Canadian trapper Alexis St. Martin narrowly escaped death when a gun discharged only three feet from him, tearing open his chest and abdomen and leaving a hole in his stomach wall. He was attended by American army surgeon William Beaumont, who nursed St. Martin back to health over the next two years. The gaping wound over the stomach failed to heal properly, leaving a *fistula*, or opening, into the lumen. St. Martin was destitute and unable to care for himself, so Beaumont "retained St. Martin in his family for the special purpose of making physiological experiments." In a legal document, St. Martin even agreed to "obey, suffer, and comply with all reasonable and proper experiments of the said William [Beaumont] in relation to . . . the exhibiting . . . of his said stomach and the power and properties . . . and states of the contents thereof."

Beaumont's observations on digestion and the state of St. Martin's stomach under various conditions created a sensation. In 1832, just before Beaumont's observations were published, the nature of gastric juice [*gaster*, stomach] and digestion in the stomach was a subject of much debate. Beaumont's careful observations went far toward solving the mystery. Like physicians of old who tasted urine when making a diagnosis, Beaumont tasted the mucous lining of the stomach and the gastric juices. He described them both as "saltish," but mucus was not at all acid and gastric fluid was very acid. Beaumont collected copious amounts of gastric fluid through the fistula, and in controlled experiments he confirmed that gastric fluid digested meat, aided by a combination of hydrochloric acid and another active factor that we now know is the enzyme pepsin. These observations and others about motility and digestion within the stomach became the foundation of what we know about digestive physiology.

Although research today is being conducted more at the cell and molecular level, researchers still create surgical fistulas in experimental animals in order to observe and sample the contents of the lumen of the digestive tract. Why is the digestive system of such great interest? Gastrointestinal diseases today account for nearly one-tenth of the money spent on health care. Many of these conditions, such as heartburn, indigestion, gas, and constipation, are troublesome rather than major health risks, but their significance should not be underestimated. Go into any drugstore and look at the number of over-the-counter medications for digestive disorders to get a feel for the impact that gastrointestinal diseases have on our society. This chapter examines the digestive system and the remarkable way that it transforms the food we eat into nutrients for the body's use.

Problem

Peptic Ulcers

Your stomach is one of the most hostile environments on earth. Bathed in acid strong enough to digest nails, its walls protected with a sticky mucus, and constantly churning, the stomach was once thought to be uninhabitable by any life form. Recently, however, scientists have discovered a remarkable organism that actually thrives in the stomach's formidable environment. This hardy bacterium, called *Helicobacter pylori,* has evolved a set of defenses so unique that it escaped detection until 1982. But this bacterium is not a silent passenger in the stomach: It is believed to cause most cases of peptic ulcers worldwide. Proper drug treatment kills *H. pylori* and allows ulcers in the stomach and duodenum to heal. That is good news to Tonya Berry, who has just been informed by her physician that she has a peptic ulcer. "I thought it was stress," she said.

continued on page 581

FUNCTION AND PROCESSES OF THE DIGESTIVE SYSTEM

The gastrointestinal tract, or GI tract, is a long tube whose function is to move nutrients, water, and electrolytes from the external to the internal environment. The food we eat

is mostly in the form of macromolecules such as proteins and complex carbohydrates, so our digestive systems must secrete powerful enzymes to digest them into smaller molecules, or nutrients, that can be absorbed. At the same time, however, the enzymes must not digest the cells of the GI tract itself. If protective mechanisms against autodigestion fail, we develop raw patches known as *ulcers* on the walls of the gastrointestinal tract.

Another challenge the digestive system faces daily is matching input with output. Various exocrine glands and cells secrete about 7 liters of fluid per day into the lumen of the tract. These secretions, containing digestive enzymes, mucus, and water, are critical to proper digestive function, but they must be reabsorbed or the body will rapidly dehydrate. *Diarrhea,* or watery stools, can become an emergency if fluid loss from the GI tract depletes the extracellular fluid volume to the point that the circulatory system is unable to maintain adequate blood pressure.

These physiological challenges are met by coordinating the four basic processes of the digestive system: digestion, absorption, motility, and secretion (Fig. 20-1 ■). **Digestion** is the chemical and mechanical breakdown of foods into smaller units that can be taken across the intestinal epithelium into the body. **Absorption** is the active or passive transfer of substances from the lumen of the gastrointestinal tract to the extracellular fluid. **Motility** involves the movement of material through the GI tract. **Secretion** refers to the release of hormones from endocrine cells or of enzymes, mucus, and paracrines from exocrine glands or cells. For most nutrients, absorption is not regulated, so "what you eat is what you get." In contrast, motility and secretion are continuously regulated to maximize the availability of absorbable material. Motility is regulated because if food moves through the system too rapidly, there is not enough time for everything in the lumen to be digested and absorbed. Secretion is regulated because if digestive enzymes are not secreted in adequate amounts, food in the GI tract cannot be broken down into an absorbable form.

When digested nutrients have been absorbed and reach the cells, cellular metabolism, closely regulated by a complex system of hormones, directs their use or storage (see Chapter 21). Some hormones, such as somatostatin, that alter digestive motility and secretion also participate in the control of metabolism, providing an integrating link between the two steps.

Although we tend to think of the digestive system in terms of its digestive function, repelling foreign invaders is another challenge the gastrointestinal tract faces. The GI tract passes through the interior of the body, but the lumen and its contents are part of the external environment (∞ p. 2). Indeed, the digestive system provides the largest amount of contact between the internal environment and the outside world, with an area estimated to be the size of a tennis court. The GI tract faces daily conflict between the need to absorb water and nutrients and the need to keep bacteria,

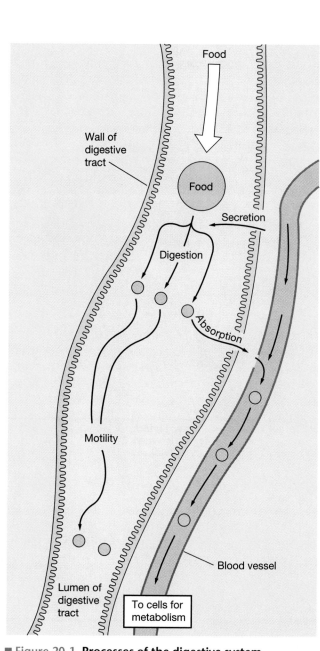

■ Figure 20-1 Processes of the digestive system

viruses, and other pathogens from entering the body. As a result, the transporting epithelium of the GI tract is combined with an array of physiological defense mechanisms including mucus, digestive enzymes, acid, and the largest collection of lymphoid tissue in the body, the **gut-associated lymphoid tissue** (GALT). By one estimate, 80% of all immunoglobulin-producing cells in the body are found in the small intestine. Our discussion of digestive physiology begins with a review of the anatomy of the digestive system.

✓ What is the difference between digestion and metabolism (∞ p. 87)?

✓ What is the difference between absorption and secretion?

ANATOMY OF THE DIGESTIVE SYSTEM

The digestive system begins with the oral cavity (mouth and pharynx), which serves as a receptacle for food and is where the first stages of digestion begin with chewing and secretion of saliva by the **salivary glands.** Once swallowed, food moves into the gastrointestinal tract, a long tube composed of layers of smooth muscle and lined with a transporting epithelium. The tube is closed off by skeletal muscle sphincters at the ends. Food is moved through the GI tract by waves of muscle contraction. Along the way, secretions are added to the food by the accessory glandular organs, creating a soupy substance known as **chyme.** Digestion takes place primarily in the lumen of the tube. The products of digestion are absorbed across the epithelium of the wall and pass into the extracellular space. From there, they move into the blood or lymph for distribution throughout the body. Any waste remaining at the end of the GI tract leaves the body through the opening known as the **anus.**

The Digestive System Consists of the Gastrointestinal Tract and Accessory Glandular Organs

A piece of food that enters the mouth passes from the pharynx through an esophageal sphincter into the **esophagus,** a narrow tube that travels through the thorax to the abdomen (Fig. 20-2a ■). Just below the diaphragm, the esophagus ends with a second sphincter at its junction with the **stomach,** a baglike organ that can hold 2–3 liters of food and fluid when fully (but uncomfortably) expanded. The stomach is roughly divided into three sections: the upper **fundus,** central **body,** and lower **antrum** (Fig. 20-2b ■). The stomach begins digestion and stores food, mixing it with acid and enzymes to create chyme. The opening between the stomach and **small intestine** is guarded by the **pyloric sphincter,** sometimes called the pylorus, or pyloric valve. This tonically contracted ring of smooth muscle relaxes in response to certain signals, allowing limited amounts of chyme into the small intestine at any one time. By regulating the rate at which chyme reaches the intestine, the stomach can ensure that the intestine will not be overwhelmed with more than it can digest and absorb. The stomach therefore acts as an intermediary between the behavioral act of eating and the physiological acts of digestion and absorption in the intestine.

Most digestion takes place in the small intestine, which is divided into three sections: the **duodenum** (the first 25 cm), **jejunum,** and **ileum** (the last two together are about 260 cm long). Digestion is carried out by intestinal enzymes, aided by exocrine secretions from the **pancreas** and **liver** that enter the initial section of the duodenum through a duct. A tonically contracted sphincter keeps pancreatic fluid and bile from entering the small intestine except during a meal.

Anatomy Summary   The Digestive System

Figure 20-2

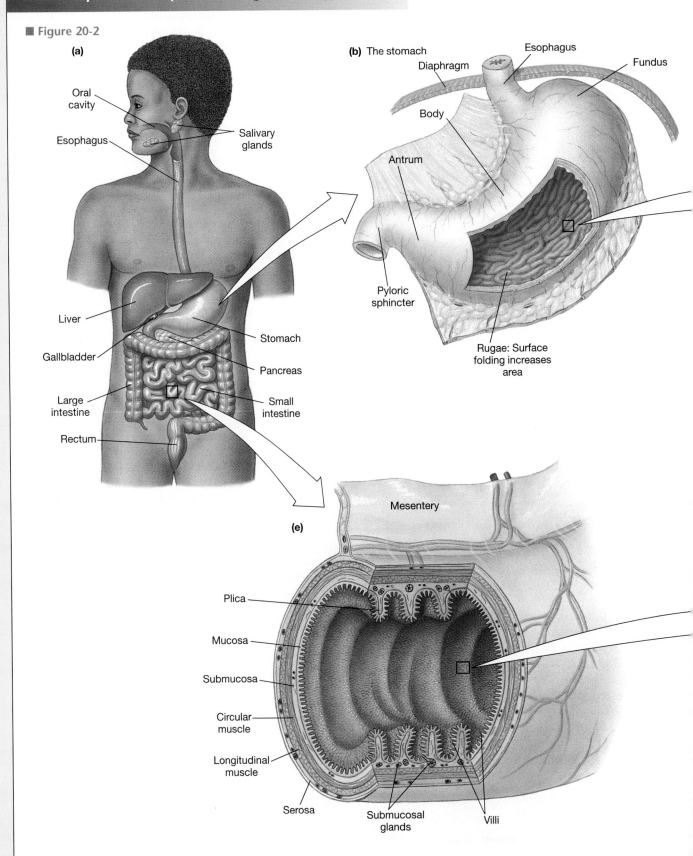

(a)

Oral cavity

Esophagus

Salivary glands

Liver

Gallbladder

Large intestine

Rectum

Stomach

Pancreas

Small intestine

(b) The stomach

Diaphragm

Body

Antrum

Pyloric sphincter

Esophagus

Fundus

Rugae: Surface folding increases area

Mesentery

(e)

Plica

Mucosa

Submucosa

Circular muscle

Longitudinal muscle

Serosa

Submucosal glands

Villi

material short distances in both directions (Fig. 20-4b ■). In the contracting segment, the circular muscles contract while the longitudinal muscles relax. In the receiving segment, the opposite pattern of contraction and relaxation occurs. Segmental contractions churn the intestinal contents back and forth, mixing them and keeping them in contact with the absorptive epithelium.

Movement of Food through the Gastrointestinal Tract

This section follows a bolus of food through the digestive tract, from the time it is ingested to the point at which it is excreted. Mechanical digestion of food begins in the oral cavity with chewing. The lips, tongue, and teeth all contribute to the **mastication** of food, creating a softened, moistened mass that can be easily swallowed. Swallowing, or **deglutition,** is a reflex action that pushes a bolus of food or liquid into the esophagus (Fig. 20-5 ■). The stimulus for swallowing is pressure created when the bolus is pushed against the soft palate and back of the mouth with the tongue. Sensory input to the swallowing center in the medulla oblongata begins a reflex reaction. First, the **glottis,** the opening between the pharynx and the larynx, is closed when the **epiglottis** folds down over it to prevent food and liquid from entering the airways. At the same time, respiration is inhibited and the upper esophageal sphincter relaxes. Then, peristaltic waves of muscle contraction push the food downward through the esophagus, aided by gravity. The musculature of the esophagus changes from skeletal muscle to smooth muscle about one-third of the way along the length of the esophagus.

The lower end of the esophagus lies just below the diaphragm and is separated from the stomach by the lower esophageal sphincter. This area relaxes when food is swallowed, allowing it to pass into the stomach. Normally, this sphincter acts as a barrier between the esophagus and the stomach, preventing acid in the stomach from moving back up into the esophagus. If the lower esophageal sphincter is not competent, stomach acid will be sucked into the esophagus during inspiration, when the intrapleural pressure drops. The walls of the esophagus expand with inspiration, creating subatmospheric pressure in the esophageal lumen. The churning action of the stomach when filled with food can also allow acid to back up into the esophagus if the sphincter is not contracted fully. In either case, gastric acid irritates the lining of the esophagus, leading to a condition known as **reflux esophagitis,** or heartburn.

Food in the stomach is churned by peristaltic waves that sweep along the walls from the lower esophageal sphincter to the pyloric sphincter at the distal end. As digestion proceeds, the pyloric sphincter relaxes slightly with each wave of contraction, allowing small spurts of chyme to pass into the duodenum of the small intestine. The peristaltic waves of the stomach are coordinated by autorhythmic slow waves that persist even if the stom-

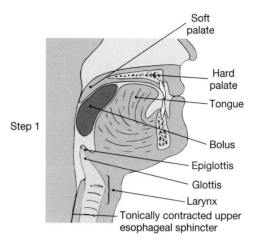

Step 1

Soft palate
Hard palate
Tongue
Bolus
Epiglottis
Glottis
Larynx
Tonically contracted upper esophageal sphincter

1. Tongue pushes bolus against soft palate and back of mouth, triggering swallowing reflex.

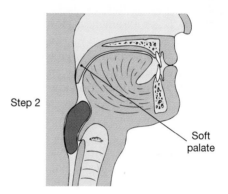

Step 2

Soft palate

2. Upper esophageal sphincter relaxes while glottis and epiglottis close to keep swallowed material out of the airways.

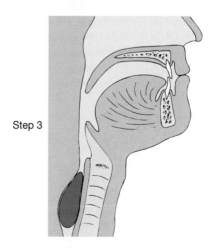

Step 3

3. Food moves downward into the esophagus, propelled by peristaltic waves and aided by gravity.

■ Figure 20-5 **The swallowing reflex**

continued from page 581

The digestion of nutrients that occurs in the stomach results in the release of large amounts of urea. Urea is a nitrogenous breakdown product of amino acids. *Helicobacter pylori* takes advantage of this sea of urea and uses it for protection against the acidic environment of the stomach. The outer membrane of *H. pylori* is studded with enzymes called ureases. Urease converts urea into carbon dioxide and ammonia, which is a base.

Question 2: How does the conversion of urea into carbon dioxide and a base protect *H. pylori* from the hostile environment of the stomach?

ach is empty. The growling and rumbling of your stomach when you are hungry are the result of peristaltic contractions in the absence of food.

Once food enters the small intestine, it is slowly propelled forward by a combination of peristaltic and segmental contractions. This movement is slow enough to allow digestion and absorption to go to completion, but eventually the unabsorbed portion of chyme reaches the large intestine. In the colon, forward movement continues, aided by a third type of contraction known as **mass movement.** This contraction sends a substantial bolus of material forward and is responsible for the sudden distension of the rectum that triggers defecation.

The **defecation reflex** removes undigested feces from the body. Defecation resembles urination in that it is a spinal reflex triggered by distension of the organ wall. The anus, like the urinary bladder, has two sphincters: an internal smooth muscle sphincter under reflex control and an external skeletal muscle sphincter under voluntary control. Sensory input from the rectum to the conscious brain allows children to develop control of their bowel movements through voluntary contraction of the external anal sphincter. Defecation, like urination, is subject to emotional influence. Stress may cause increased intestinal motility and psychosomatic diarrhea in some individuals but decreased motility and constipation in others.

SECRETION

In a typical day, 9 liters of fluid pass through the lumen of an adult's gastrointestinal tract—the equivalent of three 3-liter soft drink bottles! Only about 2 liters of that volume enters the system through the mouth. The remaining 7 liters of fluid come from body water secreted along with enzymes and mucus (Fig. 20-6 ■). About half of the secreted fluid comes from accessory organs and glands such as the salivary glands, pancreas, and liver. The remaining 3.5 liters is secreted by the epithelial cells of the digestive tract itself.

The significance of the secreted volume is more apparent if you realize that the secreted fluid equals one-sixth of the entire body water volume, or more than twice the plasma volume. In other words, if the fluid secreted into the lumen is not reabsorbed as it passes along the GI tract, the body will rapidly dehydrate. In this section, we examine the general patterns of secretion in the gastrointestinal tract.

Digestive Enzymes Are Secreted in the Mouth, Stomach, and Intestine

Digestive enzymes are selected either by exocrine glands (salivary glands and the pancreas) or by epithelial cells in the mucosa of the stomach and small intestine (Table 20-1). Enzymes are proteins, so they are packaged by the Golgi apparatus into secretory vesicles and stored within the cell until needed. On demand, they are released into the extracellular space by exocytosis (∞ p. 127). A few secreted enzymes bind to apical membranes of the epithelial cells, anchored to transmembrane protein "stalks." These associated proteins are then considered part of the **brush border,** the name given to the brushlike microvilli on the epithelial cells. Brush border enzymes are stationary and will not be swept out of the small intestine along with the chyme as it is propelled forward.

Some digestive enzymes are secreted in an inactive **proenzyme** form and are known collectively as **zymogens.** These enzymes must be activated in the GI lumen before they can carry out digestion. Late activation prevents the enzymes from digesting the cells in which they are synthesized; it also allows them to be stockpiled within the cells until needed. Among the more important zymogens are **pepsinogen** in the stomach and a variety of pancreatic enzymes (Fig. 20-7 ■). Pancreatic **trypsinogen** is activated to trypsin by **enteropeptidase,** a brush border enzyme. **Trypsin** then converts other inactive pancreatic enzymes to their active forms.

The control pathways for enzyme release vary. Salivary secretion is under neural control and can be triggered by multiple stimuli such as the sight, smell, touch, and even thought of food. In the stomach and intestine, enzyme secretion is due to a combination of neural and hormonal signals. Usually, stimulation of parasympathetic neurons in the vagus nerve enhances enzyme secretion.

Mucus Is Secreted by Specialized Cells

Mucus is a viscous secretion composed primarily of glycoproteins that are collectively called **mucins.** The primary functions of mucins in the digestive system are to form a protective coating over the GI mucosa and to lubricate the contents of the gut. Mucus is made in specialized **mucus cells** in the stomach and **goblet cells** in the intestine. Goblet cells comprise between 10% and 24% of the intestinal cell population. In the mouth, the minor salivary glands secrete about 70% of the salivary mucus.

The release signals for mucus include parasympathetic innervation, a variety of neuropeptides found in the enteric nervous system, and cytokines from immunocytes. Parasitic infections and inflammatory processes in the gut both cause substantial increases in mucus output as the body attempts to increase its protective barrier.

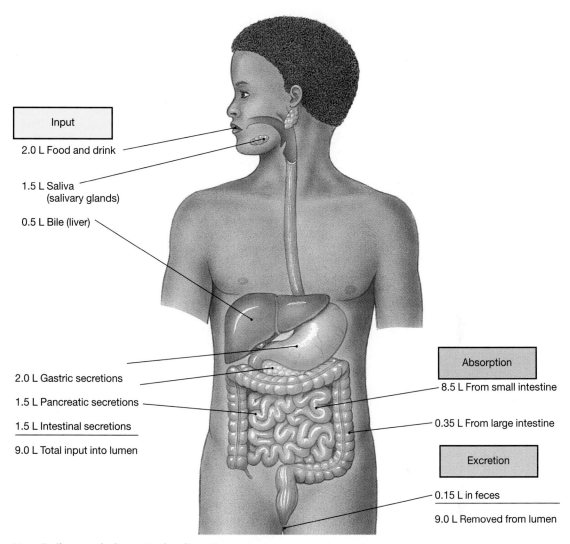

Input

2.0 L Food and drink

1.5 L Saliva
(salivary glands)

0.5 L Bile (liver)

2.0 L Gastric secretions

1.5 L Pancreatic secretions

1.5 L Intestinal secretions
_____
9.0 L Total input into lumen

Absorption

8.5 L From small intestine

0.35 L From large intestine

Excretion

0.15 L in feces
_____
9.0 L Removed from lumen

■ Figure 20-6 **Daily mass balance in the digestive system**

TABLE 20-1	**Digestive Enzymes**	
Location	*Enzyme*	*Digests*
Salivary glands	Amylase	Starch
	Lingual lipase	Triglycerides
Stomach	Pepsin (pepsinogen)	Proteins
	Gastric lipase	Triglycerides
Pancreas	Amylase	Starch
	Lipase and colipase	Triglycerides
	Phospholipase	Phospholipids
	Trypsin (trypsinogen)	Peptides
	Chymotrypsin (chymotrypsinogen)	Peptides
Intestinal epithelium	Enterokinase	Activates trypsin
	Disaccharidases	
	Sucrase	Sucrose
	Maltase	Maltose
	Lactase	Lactose
	Peptidases	Peptides
	Endopeptidases	Interior peptide bonds
	Exopeptidases	Terminal peptide bonds
	Aminopeptidase	Works at NH_2-terminal end
	Carboxypeptidase	Works at COOH-terminal end

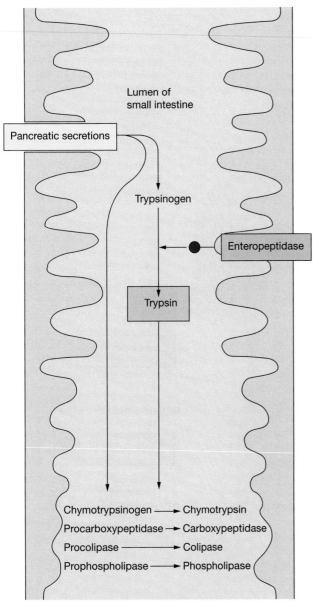

■ **Figure 20-7 Activation of pancreatic proenzymes**
Several pancreatic enzymes are secreted in inactive proenzyme
form. Trypsinogen is activated to trypsin by the brush border
enzyme enteropeptidase. Trypsin in turn activates the other
proenzymes.

The Digestive System Secretes Substantial Amounts of Fluid and Ions

A large portion of the 7 liters of fluid secreted by the
digestive system each day is composed of water and
ions, particularly Na^+, K^+, Cl^-, HCO_3^-, and H^+. The
cellular mechanisms by which these substances are first
secreted into the lumen of the tract and then reabsorbed
are variations on the processes you learned for renal
transport of fluid and electrolytes. Intestinal epithelial
cells, like those in the kidney, have distinct apical and
basolateral membranes, each with appropriate transporters for active transport, facilitated diffusion, and diffusion through open channels. Water follows osmotic

gradients created by the transfer of solutes from one side
of the epithelium to another.

Sodium, potassium, and chloride movement The
transport of Na^+, K^+, and Cl^- from one side of the
secretory epithelium to the other uses many membrane
transporters that are also present in the renal tubule (∞
p. 530). The basolateral membrane contains the ubiquitous Na^+/K^+-ATPase, with various cotransporters
(Na^+-K^+-$2Cl^-$ symport, Cl^--HCO_3^- antiport), pumps
(H^+/K^+-ATPase), and diffusion channels arranged on
the cell surface according to the function of the cell.

One type of transepithelial movement we have not discussed before is the movement of Na^+ *between* the cells.
Because most of the GI epithelia (except the stomach)
are leaky epithelia, Na^+ secretion is accomplished by
simple diffusion of Na^+ through the cell junctions (Fig.
20-8a ■). Consequently, the secreted fluid usually has a
Na^+ concentration equal to that of the plasma. Sodium
is then actively reabsorbed into the body by processes
similar to sodium reabsorption in the kidney (∞ Fig.
18-10 ■, p. 531).

To create electrochemical gradients to drive the passive
movement of Na^+, the epithelial cells transport Cl^-
from the extracellular fluid into the lumen (Fig. 20-8a ■).
The basolateral membrane contains a Na^+-K^+-$2Cl^-$
symport protein that brings Cl^- into the cell. On the apical membrane, a gated channel known as the **cystic
fibrosis transmembrane regulator,** or **CFTR chloride
channel,** allows Cl^- to enter the lumen. The movement
of negative charge from the extracellular fluid to the
lumen attracts positive Na^+, which diffuses between the
cells. As solute moves to the lumen, creating an osmotic
gradient, water follows. The net result is secretion of an
isosmotic NaCl solution.

In cystic fibrosis, an inherited defect causes the CFTR
channel to be defective or absent from the membrane.
As a result, secretion of chloride and fluid ceases, but
goblet cells continue to secrete mucus. The result is
thickened mucus that clogs the small pancreatic ducts
and accumulates in the airways of the respiratory system, where the CFTR channel is also found (∞ p. 113).

In the intestine, a related mechanism exists for the
secretion of K^+ and Cl^- (Fig. 20-8b ■). In these cells,
however, K^+ leaves the apical membrane through K^+
leak channels. Some bacterial toxins such as cholera
toxin and *Escherichia coli* enterotoxin cause diarrhea by
enhancing KCl and water secretion. When coupled with
increased motility, this fluid secretion creates diarrhea,
and the fluid is lost from the body. The secreted fluid is
high in K^+, so replacement fluids for diarrhea should
include electrolytes. Sports drinks such as Gatorade™
are useful for do-it-yourself rehydration therapy.
Diarrhea in response to intestinal infection can be
viewed as adaptive because it helps flush pathogens out
of the lumen, but it also has the potential to cause dehydration if the loss of secreted fluid is excessive.

environment of the stomach lumen, which denatures proteins, unfolding them and destroying their tertiary structure. When the peptide chain unfolds, more peptide bonds in the interior of the molecule are exposed to pepsin's action, facilitating digestion of the molecule.

As chyme created in the stomach moves forward into the duodenum of the small intestine, gastric acid is neutralized by bicarbonate secretion from the pancreas, and the luminal pH increases. Protein digestion by pepsin ceases when the enzyme is inactivated at higher pH. At this point, pancreatic proteases and at least 17 different proteases and peptidases in the brush border create free amino acids plus di- and tripeptides, all of which can be absorbed into the epithelial cell.

There are multiple amino acid transport systems owing to the structural variations among amino acids (∞ Figure 2-15, p. 30). Free amino acids are carried across both apical and basolateral membranes by Na^+-dependent cotransport proteins similar to those in the proximal tubule of the kidney (Fig. 20-14 ∎). A few basolateral transporters are not Na^+-dependent.

Small peptides can be absorbed intact into the mucosal cell using a cotransport protein that is H^+-dependent rather than Na^+-dependent. Once inside the epithelial cell, these small peptides have two possible fates. Most of them are digested into amino acids by cytoplasmic peptidases. The resultant amino acids are then transported across the basolateral membrane and into the circulation.

Not all small peptides are digested inside the epithelial cells, however. Some are carried intact across the basolateral membranes on protein transporters. Other peptides are transported by transcytosis (∞ p. 130) after binding to a receptor on the luminal surface.

The discovery that significant amounts of ingested protein are absorbed as peptides has implications in medicine. Peptides may act as *antigens*, substances that stimulate antibody formation and result in allergic reactions. Consequently, the intestinal absorption of peptides may be a significant factor in the development of food allergies and intolerances. It is known that peptide absorption takes place primarily in crypt cells in the intestine. At birth, intestinal villi are very small, and the crypts are well exposed to the luminal contents. As the villi grow

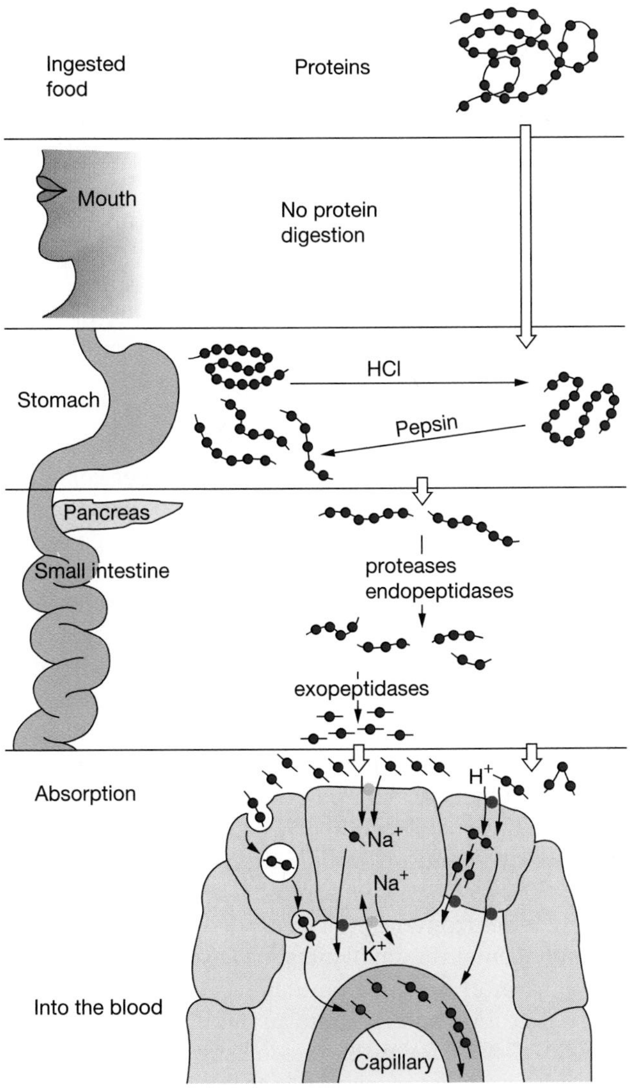

■ Figure 20-13 **Protein digestion**

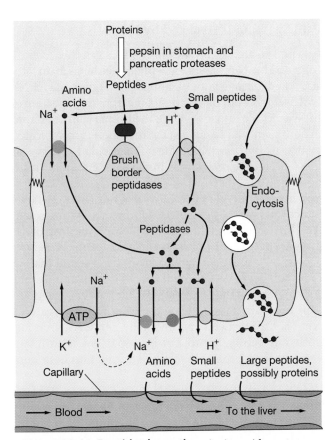

■ Figure 20-14 **Peptide absorption** Amino acids are transported by secondary active transport with Na^+ or by facilitated diffusion. Peptides are transported by secondary active transport with H^+ or are carried intact across the cell by transcytosis.

and the crypts have less access to chyme, the high peptide absorption rates present at birth decline steadily. If parents delay the ingestion of allergy-inducing peptides, the gut has a chance to mature, lessening the likelihood of antibody formation.

One of the most common antigens responsible for food allergies is gluten, a component of wheat. Interestingly, the incidence of childhood gluten allergies has decreased since the 1970s, when the recommendation was made that infants not be fed gluten-based cereals until they are several months of age. Currently, antigen ingestion by infants is being investigated as a precipitating factor for schizophrenia and autism.

In another medical application, intestinal absorption of intact peptides is being used by pharmaceutical companies to develop indigestible peptide drugs that can be given orally instead of by injection. Probably the best-known example is DDAVP (1-deamino-8-D-arginine vasopressin), the synthetic analog of vasopressin (antidiuretic hormone, or ADH) used clinically. If the natural hormone ADH is ingested, it is digested rather than absorbed intact. By changing the structure of the hormone slightly, scientists were able to create a synthetic hormone with the same activity that can be absorbed without being digested.

Fat Digestion Is Assisted by Bile

Fats and related molecules in the Western diet include triglycerides, cholesterol, phospholipids, long-chain fatty acids, and the fat-soluble vitamins (Figure 2-14, p. 29). Nearly 90% of our fat calories come from the ingestion of triglycerides because they are the primary form of lipid in both plants and animals. Fat digestion is complicated by the fact that most lipids are not particularly water-soluble and form large clumps in the aqueous chyme solution of the intestinal tract. As a result, enzymatic fat digestion must be aided by two non-enzyme secretions: bile, which coats the fat droplets and breaks them up into smaller particles, and colipase, which allows the pancreatic enzyme lipase to act on the bile-coated droplets.

In recent decades, most textbooks have stated that fat digestion begins when food reaches the small intestine. As long ago as the 1880s, researchers observed significant lipolysis by gastric juices, but these observations were not confirmed experimentally and were soon forgotten. Recently, the role of lipases in the mouth and stomach has been reinvestigated and found to contribute perhaps as much as 30% of all fat digestion.

Lipases secreted by glands in the tongue (**lingual lipase**) and by the stomach (**gastric lipase**) are closely related molecules. Lingual lipase comes from serous glands at the base of the tongue rather than from the salivary glands. Gastric lipase is co-secreted from the chief cells along with pepsinogen. Because these two lipases work best at low pH, they are collectively called

continued from page 584

Tonya undergoes endoscopy, a procedure in which a tube with a camera at one end is inserted into her esophagus and maneuvered into her stomach. Small pincers attached at the end of the tube pluck a piece of her stomach lining, which is sent to the pathology laboratory for study. Tonya is awake but groggy from a tranquilizer during the procedure. That afternoon, she feels fine, but her stomach hurts. She chews a tablet of antacid for relief.

Question 3: Some antacids contain sodium bicarbonate ($NaHCO_3$), while many others contain aluminum hydroxide ($Al(OH)_3$). Antacids neutralize hydrochloric acid (HCl) secreted by the stomach by acting as buffers. For both types of antacids, write the chemical equation showing how the antacids buffer hydrochloric acid.

the acid lipases. The other significant source of lipase in the digestive system is the pancreas. All three lipases—lingual, gastric, and pancreatic—have the same activity. They remove two fatty acids from triglycerides, resulting in a monoglyceride and two free fatty acids (Fig. 20-15). Phospholipids are digested by pancreatic phospholipase, but cholesterol does not have to be digested before being absorbed.

Absorption of lipids occurs primarily by simple diffusion across the apical membrane. The rate-limiting step for fat digestion is making the fats available for absorption, a process hindered by the low solubility of lipids in water. Enhancing this availability is the primary role of bile secreted by the liver.

The roles of bile and colipase   **Bile** is a non-enzyme solution secreted from **hepatocytes,** or liver cells (see Focus on the Liver, p. 595). The key components of bile are salts for fat digestion; bile pigments, such as bilirubin, that are the waste products of hemoglobin degradation; and cholesterol. **Bile acids** are steroid detergents with polar side chains that allow them to interact with both lipids and water, creating water-soluble droplets (Fig. 20-16). In the liver, bile acids combine with amino acids to form **bile salts.**

Bile is secreted into hepatic ducts that lead to the **gallbladder,** which stores and concentrates the solution. During a meal, contraction of the gallbladder sends bile into the duodenum through the **common bile duct,** along with a watery solution of bicarbonate and digestive enzymes from the pancreas.

Fats that enter the small intestine from the stomach are in the form of a coarse emulsion created by the acid lipase digestion and mechanical mixing in the stomach (Fig. 20-17). An **emulsion** consists of small droplets suspended in a liquid, similar to the emulsion you create when you shake up a bottle of vinegar and oil for salad

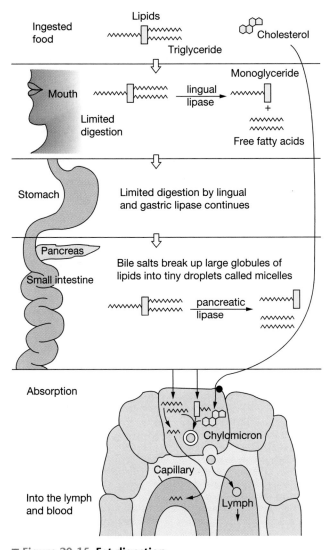

■ Figure 20-15 **Fat digestion**

dressing. In the duodenum, bile salts coat the droplets of the coarse emulsion, allowing them to form smaller droplets with greater surface area as they are churned about in the intestine. As enzymatic and mechanical digestion proceed, the droplets get smaller and smaller until they form **micelles** (∞ p. 109).

The digestion of triglycerides in the lipid droplets is carried out by pancreatic lipase. Lipase alone is unable to penetrate the coating of bile salts and requires the presence of **colipase,** a protein cofactor secreted from the pancreas. Colipase allows lipase to break through the bile salt coating of the emulsion. Lipase digests the triglycerides trapped in the interior of the droplet into monoglycerides and free fatty acids (∞ p. 29). At the brush border surface, the fatty acids and monoglycerides diffuse into the epithelial cell. Cholesterol is transported on a specific, energy-dependent membrane transporter.

Once in the cytoplasm of the epithelial cell, monoglycerides and fatty acids move to the smooth endoplasmic reticulum, where they are resynthesized into triglycerides. They then combine with cholesterol and proteins into large droplets called **chylomicrons.** Because of their size, chylomicrons have to be packaged into secretory vesicles in order to leave the epithelial cells by exocytosis. Once they are in the extracellular space, their size also prevents them from crossing the basement membrane that surrounds the capillaries. Instead, chylomicrons are absorbed into the lymph vessels of the villi, the **lacteals.** They pass through the lymphatic system and finally enter the venous blood just before it flows into the heart (∞ p. 440). Some shorter fatty acids (10 or fewer carbons) are not assembled into chylomicrons; they can cross the capillary basement membrane and go directly into the blood.

Bile salts are not altered during fat digestion. When they reach the terminal section of the small intestine, the

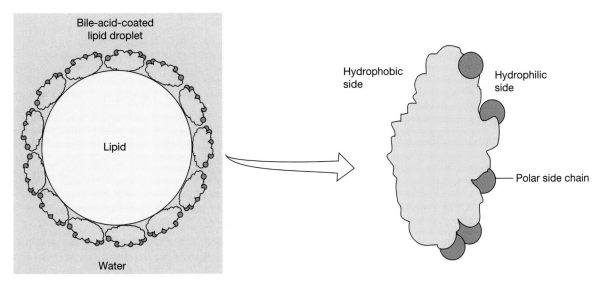

■ Figure 20-16 **Bile acids** Bile acids have a hydrophobic side that associates with lipids and a hydrophilic side that associates with water.

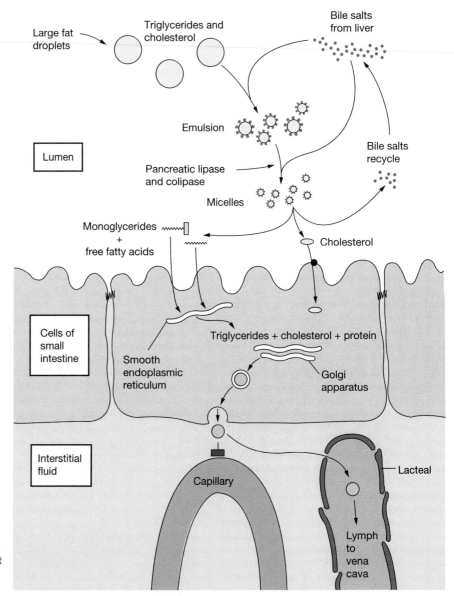

■ Figure 20-17 The role of bile in fat digestion Bile salts coat fat droplets to create emulsions and micelles.

ileum, they encounter cells that transport them back into the hepatic portal circulation. From there, they are returned to the liver, taken back up into the hepatocytes, and resecreted. This recirculation of bile salts is essential to maintain fat digestion because the body's pool of bile salts will cycle from two to five times for each meal.

Nucleic Acids Are Digested into Bases and Monosaccharides

The nucleic acid polymers DNA and RNA do not make up a significant part of most diets. They are digested by pancreatic and intestinal enzymes, first into their component nucleotides and then into bases and monosaccharides (∞ Figure 2-17, p. 32). The bases are absorbed by active transport, whereas the monosaccharides are absorbed by facilitated diffusion Dand secondary active transport as described previously.

Vitamins and Minerals Are Absorbed Along with Nutrients

In general, the fat-soluble vitamins (A, D, E, and K) are absorbed along with fats, whereas the water-soluble vitamins are absorbed by mediated transport. The major exception is vitamin B_{12}, also known as cobalamin. This vitamin is made by bacteria; we obtain most of our dietary supply from seafood, meat, and milk products. Intestinal absorption of vitamin B_{12} depends on the presence of a protein known as **intrinsic factor** that is secreted by the stomach. The intestinal transporter for B_{12} will recognize the vitamin only when it is complexed with intrinsic factor. In the absence of intrinsic factor, a state of vitamin B_{12} deficiency causes the condition known as *pernicious anemia*. In this state, red blood cell synthesis (erythropoiesis), which depends on vitamin B_{12}, is severely diminished. Lack of intrinsic factor can-

Focus on   **The Liver**

■ Figure 20-18

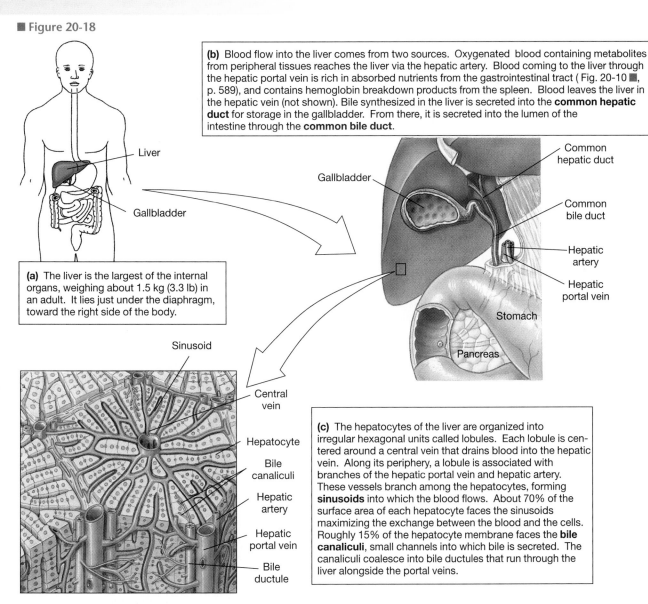

(b) Blood flow into the liver comes from two sources. Oxygenated blood containing metabolites from peripheral tissues reaches the liver via the hepatic artery. Blood coming to the liver through the hepatic portal vein is rich in absorbed nutrients from the gastrointestinal tract (Fig. 20-10 ■, p. 589), and contains hemoglobin breakdown products from the spleen. Blood leaves the liver in the hepatic vein (not shown). Bile synthesized in the liver is secreted into the **common hepatic duct** for storage in the gallbladder. From there, it is secreted into the lumen of the intestine through the **common bile duct**.

(a) The liver is the largest of the internal organs, weighing about 1.5 kg (3.3 lb) in an adult. It lies just under the diaphragm, toward the right side of the body.

(c) The hepatocytes of the liver are organized into irregular hexagonal units called lobules. Each lobule is centered around a central vein that drains blood into the hepatic vein. Along its periphery, a lobule is associated with branches of the hepatic portal vein and hepatic artery. These vessels branch among the hepatocytes, forming **sinusoids** into which the blood flows. About 70% of the surface area of each hepatocyte faces the sinusoids maximizing the exchange between the blood and the cells. Roughly 15% of the hepatocyte membrane faces the **bile canaliculi**, small channels into which bile is secreted. The canaliculi coalesce into bile ductules that run through the liver alongside the portal veins.

(d) Blood entering the liver brings nutrients and foreign substances from the digestive tract, bilirubin from hemoglobin breakdown, and metabolites from peripheral tissues of the body. In turn, the liver excretes some of these in the bile and stores or metabolizes others. Some of the liver's products are wastes to be excreted by the kidney; others are essential nutrients, such as glucose. In addition, the liver synthesizes an assortment of plasma proteins.

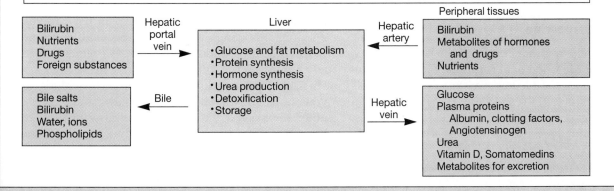

Olestra, the No-Calorie Fat Substitute Although nutritionists have determined that the typical Western diet contains too much fat, many consumers are reluctant to give up fat in their diet. As a result, food manufacturers such as Procter and Gamble have been developing low-calorie fat substitutes. In 1996, a fat substitute named Olestra® was approved by the Food and Drug Administration. Olestra is sucrose polyester: a sucrose molecule with 6–8 fatty acids attached to it. This unusual arrangement of naturally occurring substances tastes and feels like dietary fats but cannot be digested by intestinal enzymes or absorbed across the intestinal epithelium. Foods containing Olestra were introduced with considerable fanfare, but, within weeks, unhappy consumers were reporting unpleasant side effects. Intestinal cramping, gas, and diarrhea were significant enough in some people to cause them to discontinue use of Olestra. In addition, nutritionists are concerned that excessive consumption of Olestra will lead to deficiencies of the fat-soluble vitamins (A, D, E, and K) that are normally absorbed along with dietary fats. Procter and Gamble is continuing studies on the long-term effects of eating Olestra, and the Food and Drug Administration will hold another set of hearings on the food additive in 1999.

not be remedied directly, but patients with pernicious anemia can be given vitamin B_{12} shots.

Mineral absorption usually occurs by active transport. Iron and calcium are two of the few substances whose intestinal absorption is linked to their concentration in the body. For both minerals, a decrease in body concentrations of the mineral leads to enhanced uptake at the intestine. (Calcium balance is described in detail in the next chapter.)

Water and Electrolyte Absorption Is Similar in the Kidney and Intestine

Water and electrolyte absorption in the intestines is very similar to the process in the nephron. Ions are moved by active transport, with water osmotically following solute relocation. Net transport in the intestines is usually absorptive, but water and electrolytes may be secreted, as discussed previously.

Very Little Digestion Occurs in the Large Intestine

According to the traditional view of the large intestine, no significant digestion or absorption of organic molecules takes place there. However, in recent years this view has been revised. We now know that the colon can absorb significant amounts of amino acids and peptides, although normally these molecules are completely absorbed in the small intestine. We have also learned that the digestion of some complex carbohydrates in foods

such as rice and beans is not completed in the small intestine. Instead, significant amounts of carbohydrate from these foods pass into the large intestine, where they are metabolized by bacteria into short-chain fatty acids that can be absorbed by the body. The bacteria that inhabit the large intestine also produce significant amounts of absorbable vitamins, as well as gases that escape from the gastrointestinal tract (**flatus**). Some foods, such as beans, are notorious for producing intestinal gas.

REGULATION OF GI FUNCTION

Although the digestive system is not closely involved with maintaining homeostasis, it has remarkably complex regulatory mechanisms that we are nowhere close to understanding. The enteric nerve plexus acts as a "little brain," allowing some reflexes to begin and end in the gastrointestinal tract without communicating with the rest of the nervous system. At the same time, the influence that emotions have on the GI tract illustrates the link between the enteric nervous system and the "big brain," or cephalic brain. Emotional responses manifested by the digestive system range from travelers' constipation to "butterflies in the stomach" to psychologically induced diarrhea.

Between the two extremes of gastrointestinal control are reflexes that start with stimuli in the gastrointestinal tract, are integrated in the central nervous system, and then are carried out by autonomic neurons synapsing with neurons of the enteric nerve plexus. Complicating these neural pathways are a host of unusual neurotransmitters, hormones, and other regulatory neuropeptides.

One particularly difficult part of mastering the complexities of GI regulation is learning the terminology. Some of the hormones and neurotransmitters involved in digestive regulation were first identified in other systems of the body. As a result, their names have nothing to do with their function in the gastrointestinal system.

The Enteric Nervous System Is Known as the Little Brain

The enteric nervous system was first recognized over a century ago, when scientists noted that an increase in intraluminal pressure caused a reflex wave of peristaltic contraction to sweep along sections of isolated intestine removed from the body. In the 1920s, British physiologist Johannis Langley speculated that the nerve networks (plexuses) and ganglia of the GI tract were actually a third division of the autonomic nervous system. He named the intestinal nerve network the enteric nervous system (ENS). The peristaltic reflex continued to be noted in physiology texts over the years, but Langley's hypothesis of a third autonomic division was largely ignored until the early 1970s. Since then, physiologists have come to recognize that the enteric nervous system can receive stimuli, integrate sensory information, and act upon it by changing motility or secretion, despite the absence of an identifiable integrating center like the brain or spinal cord.

In this respect, the enteric nervous system is much like the nervous systems of jellyfish and sea anemones (the phylum Cnidaria). You might have seen sea anemones being fed at an aquarium. As the piece of shrimp or fish drifts down close to the tentacles, they begin to wave, picking up chemical "odors" through the water. Once the food contacts the tentacles, it is directed toward the mouth, passed from one tentacle to another until it disappears into the digestive cavity. This purposeful reflex is accomplished without a brain, eyes, or a nose. The anemone's nervous system consists of a network of neurons composed of sensory receptors and neurons, interneurons, and efferent neurons that control the muscles and secretory cells of the anemone's body. Somehow, the neurons of the network are linked in a way that allows them to integrate information and act upon it.

In the same way that the anemone captures its food, the enteric nervous system receives stimuli and acts upon them. Reflexes that originate within the enteric nervous system and are integrated there without outside input are called **short reflexes** (Fig. 20-19 ■). The primary behaviors controlled by the enteric nervous system are related to motility, secretion, and growth.

Although the enteric nervous system can work in isolation, it is also in contact with the cephalic brain. Signals originating in the GI tract can be sent to the central nervous system via sensory neurons. Following CNS integration of the information, several hundred autonomic efferent neurons carry signals back to the millions of neurons in the enteric nervous system.

Gastrointestinal reflexes can also originate completely outside the enteric nervous system. These reflexes, sometimes called the **cephalic phase** of digestion, begin with stimuli such as the sight, smell, sound, or thought of food. An example of a cephalic reflex is the way your mouth waters and your stomach growls when you smell dinner cooking. Cephalic reflexes act in a feedforward manner, preparing the digestive system to receive the food that the sensory system is anticipating. Cephalic reflexes and other digestive reflexes integrated in the CNS are called **long reflexes** (Fig. 20-19 ■).

The autonomic divisions of the nervous system are active in long reflexes. In general, we say that parasympathetic cholinergic neurons to the GI tract are excitatory and enhance GI functions, whereas sympathetic adrenergic neurons inhibit GI functions.

Although many physiologists follow Langley's classification of the enteric nervous system as a third subdivision of the autonomic nervous system, that classification may change in coming years. Both anatomically and functionally, the ENS is much more like the brain, so much so that it has been nicknamed the little brain. Some similarities between the ENS and the brain are:

■ The neurons of the enteric nervous system release more than 20 neurotransmitters and neuromodulators, many of which are identical to molecules found

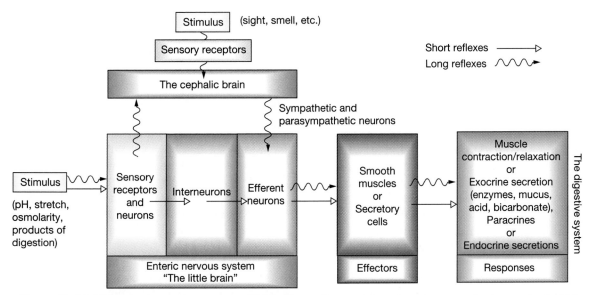

■ **Figure 20-19 The enteric nervous system (ENS)** Short reflexes originate in and are completely contained within the digestive system. An example of a short reflex is the peristaltic motility reflex, in which distension of the intestine wall causes a reflex wave of contraction to pass down the length of the intestine. Long reflexes either originate outside the digestive system or are partially integrated outside the ENS. An example is the Pavlovian conditioned reflex of salivation at the sight of food.

in the brain. These neurotransmitters are sometimes called "non-adrenergic, non-cholinergic" to distinguish them from the traditional autonomic neurotransmitters, norepinephrine and acetylcholine. Among the best-known neurotransmitters and neuromodulators in the enteric nervous system are serotonin, vasoactive intestinal peptide (VIP), and nitric oxide (NO).

■ The support cells of neurons within the enteric nervous system are more similar to the astroglia of the brain than to the Schwann cells of the peripheral nervous system.

■ The capillaries that surround ganglia within the enteric nervous system are not very permeable and create a diffusion barrier that is similar to the blood-brain barrier of cerebral blood vessels.

■ Reflexes that originate with sensory receptors in the GI tract can be integrated and acted upon without neural signals leaving the enteric nervous system. Thus, the neuron network of the enteric nervous system is its own integrating center, much like the brain and spinal cord.

The enteric nervous system has become a subject of much interest in recent years. It was thought that if we could explain how it integrates simple behaviors, we could use the gut as a simple model system for central nervous system function. But study of enteric nervous system function is difficult because there is no discrete command center for enteric reflexes. Instead, in an interesting twist, GI physiologists are applying information gleaned from studies of the brain and spinal cord to the function of the enteric nervous system. The complex interactions between the enteric and central nervous systems, the endocrine system, and the immune system promise to provide scientists with questions to investigate for many years to come.

Digestive Hormones Control GI Function, Metabolism, and Eating Behavior

The hormones of the gastrointestinal tract occupy an interesting place in the history of endocrinology. In 1902, two Canadian physiologists, W. M. Bayliss and E. H. Starling, discovered that acidic chyme entering the small intestine from the stomach caused the release of pancreatic juices even if all nerves to the pancreas were cut. Because the only communication remaining between the intestine and the pancreas was the blood supply that ran between them, they postulated the existence of some blood-borne (*humoral*) factor that was released by the intestine. When extracts of the duodenum applied to the pancreas also stimulated secretion, they knew that they were dealing with a chemical produced in and released from the duodenum. They named the substance *secretin*. Starling further proposed that the general name *hormone*, from the Greek word meaning "I excite," be given

to all humoral agents that act at a site distant from their release. A few years later, in 1905, J. S. Edkins postulated the existence of a gastric hormone that stimulated gastric acid secretion. It took more than 30 years for a relatively pure extract of the gastric hormone to successfully be made, and it was 1964 before the hormone named *gastrin* was finally purified.

Why was research on the digestive hormones so slow to develop? A major reason is the fact that the hormones of the gastrointestinal tract are not secreted by discrete glands such as the adrenal or pituitary glands. Instead, they are secreted by isolated endocrine cells scattered among the other cells of the mucosal epithelium. The only way to obtain these hormones in the early years was to make a crude extract of the entire epithelium, a procedure that also liberated digestive enzymes and paracrines made in adjacent cells. Thus, it was very difficult to tell if the physiological effect elicited by the extract came from one hormone, from more than one hormone, or from a paracrine such as histamine.

Today, although we have sequenced a variety of peptides from the GI mucosa, we still do not have the entire picture of how they help regulate digestive function. In addition to six substances that are now generally accepted to have endocrine effects, candidate hormones include pancreatic polypeptide and peptide YY. Certain reflexes in the digestive system are known to be humorally mediated even though the hormone involved has not been identified. Because of the uncertainty associated with the field, we will restrict our focus to the six accepted hormones.

Classification of GI hormones The gastrointestinal hormones are generally divided into three groups:

1. The gastrin family includes the hormones **gastrin** and **cholecystokinin** (CCK), with several variants of each.
2. The secretin family includes **secretin, vasoactive intestinal peptide** (VIP), and GIP, a hormone known originally as *gastric inhibitory peptide* because in experiments, it inhibited gastric acid secretion. In recent studies, however, GIP was not found to block acid secretin when administered in lower physiological doses. An alternative name has been proposed, **glucose-dependent insulinotropic peptide.** This name retains the original initials of the hormone (*GIP*) and more accurately describes its actions (see below).
3. The third group includes any hormones that do not fit into the two families above. The primary member of this group is the hormone **motilin.**

The sources, targets, and effects of these and other digestive hormones are summarized in Table 20-2. In the following sections we focus on common themes that govern the release and function of digestive hormones.

Synthesis sites of GI hormones   Gastrin is the primary hormone secreted by the stomach. Secretin, CCK, GIP, and motilin are secreted by cells in the intestinal mucosa. Vasoactive intestinal peptide is secreted by neurons in the enteric nervous system. However, with the introduction of immunohistochemical techniques for identifying specific chemicals within cells, we have discovered that many GI peptides are not found solely in the gastrointestinal tract. For example, CCK and VIP are found in the brain as well as in the gut. Another family of candidate GI hormones, the *enteroglucagons,* is closely related to the pancreatic hormone *glucagon.* And the hormone candidate *pancreatic polypeptide* was first isolated from the endocrine pancreas, as its name suggests. The physiological significance of multiple release sites for these hormones has yet to be determined in most cases.

Physiological effects of GI hormones   As a group, the gastrointestinal hormones excite or inhibit motility and secretion (Fig. 20-20 ■). Motility effects include changes in peristaltic activity, contraction of the gallbladder for bile release, and delay of gastric emptying. Gastric emptying is a useful method for regulating the amount of food that enters the small intestine, allowing digestion and absorption to be maximally effective.

Secretory processes under endocrine control include the release of enzymes from the stomach and the release of pancreatic juices. This is a good place to note that secretion of pancreatic enzymes and secretion of bicar-

---

TABLE 20-2   The Digestive Hormones

	Gastrin	Cholecystokinin (CCK)	Secretin	Vasoactive Intestinal Peptide (VIP)
Secreted by	G cells in stomach antrum	Endocrine cells of small intestine; neurons of brain and gut	Endocrine cells in small intestine	Neurons of brain, gut, and elsewhere
Target(s)	Enterochromaffin cells (histamine); parietal cells	Gallbladder; pancreas; gastric smooth muscle	Pancreas; stomach	Vascular smooth muscle; pancreas; intestine
Effects				
Endocrine secretion	None	None	None	Stimulates somatostatin release
Exocrine secretion	Stimulates acid secretion in the stomach	Stimulates pancreatic enzyme secretion; inhibits gastric acid secretion; potentiates bicarbonate secretion	Stimulates bicarbonate secretion and pepsin release; inhibits gastric acid	None
Motility	None	Stimulates gallbladder contraction for bile release; inhibits gastric emptying; enhances intestinal peristalsis	Inhibits gastric emptying	Decreases motility
Other	Enhanced mucosal cell growth	Stimulates satiety	None	Stimulates electrolyte secretion in intestine; stimulates bicarbonate secretion from pancreas
Stimulus for release	Peptides and amino acids in lumen; gastrin-releasing peptide and ACh in nervous reflexes	Fatty acids and some amino acids	Acid in small intestine	Uncertain
Release inhibited by	pH < 1.5; somatostatin	NA	Somatostatin	NA
Other information	NA	Some effects may be due to CCK as a neuropeptide rather than as a hormone	NA	NA

(continued on next page)

TABLE 20-2	**The Digestive Hormones (continued)**			
	Glucose-Dependent Insulinotropic Peptide (GIP)	*Motilin*	*Enteroglucagons*	*Somatostanin*
Secreted by	Endocrine cells in small intestine	Endocrine cells of upper small intestine	Endocrine cells of small intestine	D cells of stomach; intestine and endocrine pancreas; neurons of CNS
Target(s)	Beta cells of endocrine pancreas	Smooth muscle of antrum and duodenum	Endocrine pancreas; possibly stomach	Pancreas; stomach; GI smooth muscle
Effects				
Exocrine secretion	Stimulates insulin release (feedforward mechanism)	None	Stimulates insulin release	Inhibits insulin, glucagon, and gastrin release
Exocrine secretion	Possibly inhibits acid secretion	None	Possibly inhibits acid secretion	Inhibits gastric acid, pepsin, pancreatic enzymes, and bicarbonate secretion
Motility	None	Stimulates gastric contractions that travel down the intestine	Possibly inhibits gastric emptying	None
Other	None	None	None	None
Stimulus for release	Glucose and amino acids in small intestine	Fasting: periodic release every 1.5–2 hours; stimulus unknown	Mixed meal that includes carbohydrates or fats in the lumen	Acid in stomach
Release inhibited by	NA	NA	NA	NA
Other information	Acid inhibition questionable at physiological concentrations	Changes associated with both constipation and diarrhea, but relationship is unclear	Related to but not identical to pancreatic glucagon; may act together with GIP	Hormone release and effects not well understood

Stimuli

Cephalic phase of digestion (feedforward)
Distension
Acid
Presence of food

⇩

Endocrine cells of the stomach and small intestine

⇩

Receptor/
integrating
center

Efferent
pathway

GI hormones
(secreted into bloodstream)

⇩

Effectors

GI smooth muscle
Exocrine cells of stomach, pancreas, intestine
Other endocrine cells
Nervous system

⇩

Responses

Changes in GI motility
Release of bile and pancreatic secretions
Enzyme, acid, and bicarbonate
  synthesis/release
Hunger/satiety

■ **Figure 20-20 Endocrine reflexes in the digestive system**

bonate-rich pancreatic juice are controlled separately. In this way, it is possible to get pancreatic juice that has lots of bicarbonate and few enzymes, or vice versa.

Some GI hormones have trophic effects on other hormones. For example, GIP release is stimulated by glucose entering the duodenum. Glucose-dependent insulinotropic peptide then acts in a feedforward fashion to enhance insulin release so that the body is prepared to handle the glucose load as soon as it is absorbed.

The most interesting and difficult endocrine effects to study are those in which hormones act on the brain to change behavior. In the digestive system, the main candidate for behavioral effects is the hormone CCK. In several studies, CCK was found to enhance **satiety,** the state of feeling that hunger has been satisfied. If these data are confirmed, CCK has the potential to become a "natural" appetite suppressant that could be used for the control of obesity.

Signals for hormone release The stimuli for GI hormone secretion arise primarily from the ingestion of food. The biomolecules that stimulate release are those that you might predict by looking at the effects of the hormone. For example, CCK, which stimulates gall-bladder contraction for bile release to aid fat digestion, is released by fatty meals. Cholecystokinin also slows gastric emptying into the small intestine, allowing the slower digestion and absorption of fats to go to completion by not dumping too much chyme into the intestine at one time. As mentioned above, GIP, which causes insulin release, is stimulated by glucose in the lumen. Secretin triggers pancreatic bicarbonate secretion in response to acid in the small intestine. In each of these examples, the stimulus for hormone release and the action of the hormone are directly related.

In some cases, input from the nervous system can influence hormone release. One example is the release of gastrin by increased parasympathetic activity following ingestion of a meal.

Paracrines in the GI Tract Have Diverse Effects on Digestion

New information about paracrines in the gastrointestinal tract is being published almost daily. Paracrines may be (1) molecules found in the lumen that combine with cell membrane receptors to elicit a response or (2) molecules secreted into the extracellular fluid by epithelial cells in the wall of the gut. An interesting example of a paracrine found in the lumen is a molecule formed from inactive procolipase. In the lumen, the proenzyme splits into colipase and a peptide that has tentatively been named **enterostatin**. Enterostatin, like CCK, is under a great deal of scrutiny because it seems to act as a satiety signal for fat ingestion. At this time, its site of action and mechanism of action are unknown.

Paracrines are also secreted into the extracellular fluid on the basolateral side of the GI epithelium. Histamine is a well-studied paracrine that has multiple effects on the stomach, including stimulation of acid secretion. Another paracrine is 5-hydroxytryptamine (5-HT), also known as serotonin. This molecule, better known for its role as a CNS neurotransmitter, is released from epithelial cells in response to increased pressure in the lumen. The 5-HT then diffuses over to the nerve plexus and stimulates sensory neurons, triggering the peristaltic reflex that is characteristic of the enteric nervous system.

Figure 20-21 ■ is a summary that relates the four processes of the digestive system to the main sections of the gastrointestinal tract.

✓ The recipes for oral rehydration therapy usually include sugar (sucrose) and various ions. The sugar not only acts as an energy source but also enhances the intestinal absorption of an ion. Which ion? Explain how sugar enhances this ion's transport.

✓ Does bile digest triglycerides into monoglycerides and free fatty acids?

---

continued from page 592

Tonya's stomach still hurts later that night. Her roommate suggests that she try one of the new "acid blockers" that she has seen advertised on TV. These drugs are known as H_2 or histamine receptor blockers because they bind competitively to H_2-type histamine receptors on parietal cells. Fortunately, Tonya's roommate has just bought a package of these drugs. Two hours after taking the drug, Tonya feels much better.

Question 4: How does blocking H_2 receptors stop acid production in the stomach?

INTEGRATION OF GI FUNCTION: THE STOMACH

We conclude this chapter with a focus on the stomach to show how the four processes of the digestive system work together in an integrated fashion. The stomach is the gatekeeper of the gastrointestinal tract. Because eating, like drinking, is often a social behavior, at times we ingest more than we need from a nutritional standpoint. In these situations, it falls to the stomach to regulate the rate at which food enters the small intestine. Without such regulation, the small intestine would not be able to digest and absorb the load with which it is presented, and significant amounts of unabsorbed chyme would pass into the large intestine. Because the epithelium of the colon is not as well adapted for absorption as the small intestine, much of this chyme would pass out in the feces, resulting in diarrhea. This "dumping syndrome" is one of the less pleasant side effects of surgery in which portions of either the stomach or small intestine must be removed.

Secretion Is a Major Function of the Stomach

The stomach secretes a variety of substances from an assortment of cells (Fig. 20-22 ■):

■ **Mucus cells** secrete both mucus and bicarbonate. These substances create a barrier that protects the stomach from digesting itself (autodigestion). The mucus forms a physical barrier, while the bicarbonate creates a chemical buffer barrier underlying the mucus (Fig. 20-23 ■). Researchers using micro pH electrodes have shown that the bicarbonate layer just above the cell surface in the stomach has a pH that is close to 7, even when the pH in the lumen is highly acidic at pH 2. Mucus secretion is increased when the stomach is irritated, such as by the ingestion of aspirin (acetylsalicylic acid) or alcohol.

■ **Parietal cells** deep in the gastric glands secrete gastric acid. Acid secretion in the stomach averages 1–3 liters per day of isosmotic hydrochloric acid, which

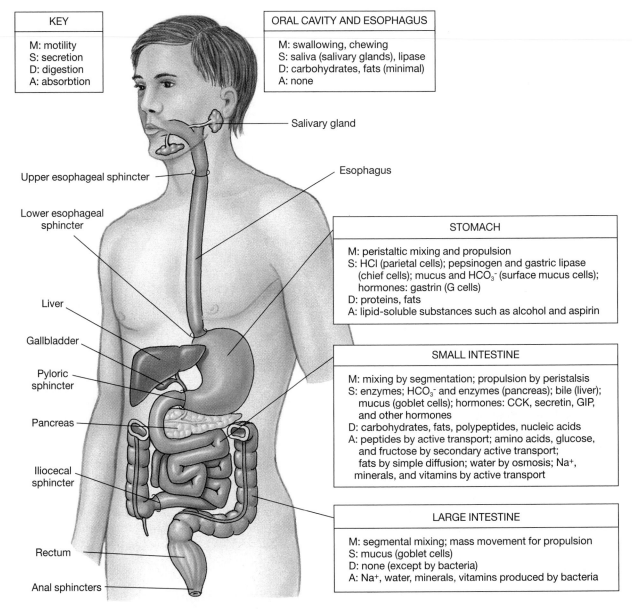

KEY

M: motility
S: secretion
D: digestion
A: absorbtion

ORAL CAVITY AND ESOPHAGUS

M: swallowing, chewing
S: saliva (salivary glands), lipase
D: carbohydrates, fats (minimal)
A: none

Salivary gland

Upper esophageal sphincter

Esophagus

Lower esophageal sphincter

STOMACH

M: peristaltic mixing and propulsion
S: HCl (parietal cells); pepsinogen and gastric lipase (chief cells); mucus and HCO_3^- (surface mucus cells); hormones: gastrin (G cells)
D: proteins, fats
A: lipid-soluble substances such as alcohol and aspirin

Liver

Gallbladder

Pyloric sphincter

Pancreas

SMALL INTESTINE

M: mixing by segmentation; propulsion by peristalsis
S: enzymes; HCO_3^- and enzymes (pancreas); bile (liver); mucus (goblet cells); hormones: CCK, secretin, GIP, and other hormones
D: carbohydrates, fats, polypeptides, nucleic acids
A: peptides by active transport; amino acids, glucose, and fructose by secondary active transport; fats by simple diffusion; water by osmosis; Na^+, minerals, and vitamins by active transport

Iliocecal sphincter

LARGE INTESTINE

M: segmental mixing; mass movement for propulsion
S: mucus (goblet cells)
D: none (except by bacteria)
A: Na^+, water, minerals, vitamins produced by bacteria

Rectum

Anal sphincters

■ Figure 20-21 **Summary of the processes in the digestive system**

can cause the pH in the lumen of the stomach to fall as low as 1. The cytoplasmic pH of the parietal cell averages about 7.2, so the cells must pump out H^+ on the H^+/K^+-ATPase against a gradient that is 2.5 *million* times as concentrated as the H^+ concentration in the cell.

■ **Chief cells** in the gastric glands secrete pepsinogen. Inactive pepsinogen is cleaved to active pepsin in the lumen of the stomach by the action of H^+. Pepsin activation is the first step of a positive feedback loop in which pepsin activates additional pepsinogen molecules (Fig. 6-17 ■, p. 164). Pepsin is a protease that carries out the initial digestion of proteins. It is particularly effective on collagen and therefore plays an important role in meat diges-

tion. The most significant stimulus for pepsin secretion is cholinergic innervation. Gastric lipase is cosecreted with pepsin and begins fat digestion.

■ **D cells,** which are closely associated with the parietal cells, secrete somatostatin. Somatostatin inhibits gastric acid, gastrin, pepsin, and pancreatic exocrine secretion.

■ **Enterochromaffin cells** secrete histamine. Histamine is a paracrine that stimulates acid secretion by combining with histamine receptors (H_2 receptors) on the parietal cells.

■ **G cells,** found deep within the gastric glands, secrete gastrin. The release of gastrin is stimulated by the presence of amino acids and peptides in the stomach, by distension of the stomach, and by ner-

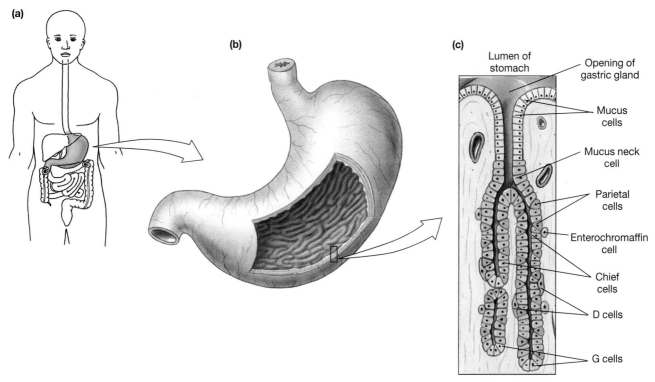

■ **Figure 20-22 Cell types of the gastric mucosa** Gastric acid secretion by parietal cells is stimulated by histamine from enterochromaffin cells and gastrin from G cells. Somatostatin from D cells inhibits gastric acid secretion. Chief cells secrete pepsinogen in response to proteins and peptides in the lumen. The mucus cells secrete mucus and bicarbonate to protect the stomach from digestive enzymes and acid.

vous reflexes mediated by **gastrin-releasing peptide.** Coffee, even decaffeinated, is also a good stimulant of gastrin release, one reason that people with excess acid secretion syndromes are advised to avoid coffee. Gastrin release is inhibited by somatostatin and by a luminal pH below 1.5.

In the condition known as *Zollinger-Ellison syndrome,* patients have excessive levels of gastrin, usually from gastrin-secreting tumors in the pancreas. As a result, hyperacidity in the stomach overwhelms the normal protective mechanisms and irritates the mucosa, causing a gastric ulcer. However, not all ulcers are caused by excess acid. Irritants such as non-steroidal anti-inflammatory drugs (NSAIDs) and a bacterium are also associated with the development of ulcers.

The various secretions of the stomach and their stimuli are summarized in Table 20-3. The next section follows the functioning of the stomach in response to ingestion of a meal.

Gastric Events Following Ingestion of a Meal

In this section, we will assume that a subject consumes a mixed meal with protein, carbohydrates, and fat. Digestion is traditionally divided into three phases: a cephalic phase, a gastric phase, and an intestinal phase.

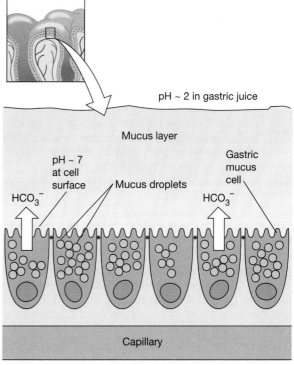

■ **Figure 20-23 The mucus and bicarbonate barrier of the gastric mucosa** Mucus provides a physical barrier, while an underlayer of bicarbonate provides a chemical barrier to neutralize acid that gets close to the mucosal cells.

TABLE 20-3 Secretions in the Stomach

Substance Secreted	Source	Stimulus for Release	Function
Mucus	Surface mucus cells	Tonic secretion; increased with irritation of the mucosa	Physical barrier between lumen and epithelium
Bicarbonate	Surface mucus cells	Secreted with mucus	Buffers gastric acid to prevent damage to epithelium
Gastric acid (HCl)	Parietal cells	Acetylcholine, gastrin, histamine	Activates pepsin; kills bacteria
Intrinsic factor	Parietal cells	Acetylcholine, gastrin, histamine	Complexes with vitamin B_{12} to permit absorption
Pepsin(ogen)	Chief cells	Acetylcholine; acid?	Digests proteins
Gastric lipase	Chief cells	Acetylcholine; acid?	Digests fats
Histamine	Enterochromaffin cells	Tonic secretion	Stimulates gastric acid secretion
Somatostatin	D cells	Acid in the stomach	Inhibits gastric acid secretion
Gastrin	G cells	Acetylcholine, peptides, and amino acids	Stimulates gastric acid secretion

The cephalic phase of gastric function Digestive processes in the stomach begin with the cephalic phase, when a person sees and smells food. The long reflexes of the cephalic phase create a feedforward, anticipatory response (Fig. 20-24 ■). Sensory stimuli activate parasympathetic neurons in the medulla oblongata, which in turn project very long axons through the vagus nerve to the enteric nervous system. (Recall that ganglia for the parasympathetic division lie in the peripheral organs; ∞ p. 309). The **cephalic vagal reflex** then causes the stomach to begin acid secretion. The parasympathetic neurons innervate (1) parietal cells, which release gastric acid; (2) enterochromaffin cells, which release histamine that triggers gastric acid

release; and (3) G cells, which release the hormone gastrin. Gastrin in turn promotes acid release, both directly and indirectly through histamine release. By the time food reaches the stomach, the lumen has become acidified. When acid is being secreted, mucosal bicarbonate secretion increases to protect the mucosa from autodigestion.

The gastric phase of gastric function When food reaches the stomach, stimuli within the gastric lumen initiate a series of short reflexes. These reflexes are often called the **gastric phase,** although some responses of the gastric phase begin during the preceding cephalic phase. In gastric phase reflexes, luminal receptors activate sensory neurons and endocrine cells within the enteric nervous system. The two main stimulants for the gastric phase are distension of the stomach and the presence of peptides and amino acids in the lumen. These stimuli trigger pepsinogen release from the chief cells and reinforce acid secretion. Gastric acid converts pepsinogen into pepsin, and protein digestion begins.

Distension also changes motility in the stomach. When food arrives, the stomach relaxes and expands to hold the increased volume. The upper portion of the stomach remains relatively quiet, holding food until it is needed. In the distal half of the stomach, a series of waves pushes the food down toward the pyloric valve, mixing it with acid and pepsin. As large food particles are digested to the more uniform texture of chyme, each contractile wave squirts a small amount of chyme through the pyloric valve into the duodenum. Enhanced gastric motility during a meal is primarily under neural control and is stimulated by distension of the stomach.

Helicobacter pylori and GI Disease Two common diseases of the stomach and duodenum are **gastritis** [-*itis*, suffix used for inflammation or disease of a part of the body] and **peptic ulcer disease.** In both conditions, the protective barriers that insulate the gastrointestinal epithelium from acid and enzymes fail to do their job. The result is damage to the epithelium, inflammation, and sometimes bleeding. Until 1982, most clinicians believed that peptic ulcer disease was caused by excessive gastric acid secretion. But in that year, a new bacterium, *Helicobacter pylori*, was isolated from patients suffering from gastritis. *Helicobacter pylori* is found in more than 90% of patients with duodenal ulcers, but only 70% of people with gastric ulcers are infected, so *H. pylori* is not the only causative agent in these conditions. Current treatments for peptic ulcer disease include antimicrobial therapy to eradicate *H. pylori*, histamine H_2 receptor antagonists (cimetadine, ranitidine) to block the histamine stimulation of acid secretion, and proton pump inhibitors (omeprazole) to decrease the activity of the H^+/K^+-ATPase.

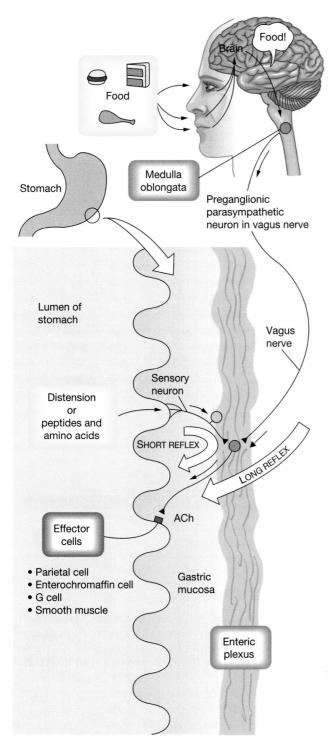

■ Figure 20-24 Long and short reflexes in the stomach
The sight, smell, and taste of food initiate the long cephalic reflexes that prepare the stomach for food that is about to arrive. Distension of the stomach and peptides or amino acids in the stomach initiate short reflexes. Both types of reflexes result in acid and enzyme secretion and increased motility.

The intestinal phase of gastric function and feedback signals As chyme enters the small intestine, the **intestinal phase** of digestion begins. Some reflexes from

this phase feed back to regulate gastric function, serving as a means by which the intestine influences the delivery of chyme from the stomach.

In general, the presence of acidic chyme in the duodenum inhibits gastric motility and secretion. These reflexes are mediated through the enteric nervous system and through various hormones, one of which is still unidentified (Fig. 20-25 ■). Acid in the intestine releases the hormone secretin; secretin then inhibits acid production and gastric motility, and slows gastric emptying. In addition, secretin stimulates the production of pancreatic bicarbonate to neutralize the acidity of chyme that has entered the intestine.

If a meal contains fats, then CCK is also secreted into the bloodstream. Cholecystokinin slows gastric motility and acid secretion. Because fat digestion is slower than either protein or carbohydrate digestion, it is crucial that the intestine communicate with the stomach so that only small amounts of fat enter the intestine at one time. It is because fats are digested and absorbed so slowly that fatty meals seem to be more filling than meals with little fat content.

If the meal contains carbohydrates, then GIP is released. In experiments, large doses of GIP inhibit gastric acid secretion, but there is some question whether GIP release reaches these levels with a normal meal.

The mixture of acid, enzymes, and digested food in chyme usually forms a hyperosmotic solution. Osmoreceptors in the wall of the intestine are sensitive to the osmolarity of the entering chyme. When stimulated by high osmolarity, they inhibit gastric emptying in a reflex mediated by some unknown blood-borne substance.

The net result of all three phases of gastric function is the digestion of proteins in the stomach by pepsin, the formation of chyme by the action of acid and churning, and the controlled entry of chyme into the small intestine so that further digestion and absorption can take place. The next chapter examines the fate of the nutrients absorbed from the intestinal tract and follows them as they enter the metabolic pathways of the body.

continued from page 601

Tonya's lab results are in. Tests performed on her stomach tissue confirm that she has an *H. pylori* infection that is probably the cause of her duodenal ulcer. Tonya's physician gives her an antibiotic and changes her H_2 blocker to a drug that inhibits H^+/K^+-ATPase.

Question 5: Explain why Tonya was given these two drugs to treat her ulcer.

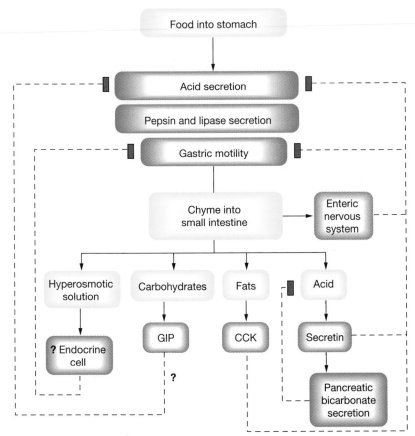

■ **Figure 20-25 The intestinal phase of gastric function** Once food passes into the small intestine, feedback loops slow gastric motility and acid secretion so that intestinal digestion will not be overwhelmed with too much food at once.

CHAPTER REVIEW

Chapter Summary

Function and Processes of the Digestive System

1. The gastrointestinal (GI) tract is a long tube whose function is to move nutrients, water, and electrolytes from the external to the internal environment. (p. 576)

2. The four processes of the digestive system are digestion, absorption, motility, and secretion. **Digestion** is the chemical and mechanical breakdown of foods into units that can be absorbed into the body. **Absorption** is the transfer of substances from the lumen of the gastrointestinal tract to the extracellular fluid. **Motility** involves the movement

of material through the gastrointestinal tract. **Secretion** refers to the release of hormones, enzymes, mucus, and paracrines from glands to cells. (p. 576)

3. The GI tract provides the largest amount of contact between the internal environment and the outside world and therefore contains the largest collection of lymphoid tissue in the body, the **gut-associated lymphoid tissue.** (p. 577)

Anatomy of the Digestive System

4. **Chyme** is a soupy substance created as ingested food is broken down by mechanical and chemical digestion. (p. 577)

5. Food entering the digestive system proceeds along the following path: mouth, pharynx, **esophagus, stomach (fundus, body, antrum), small intestine (duodenum, jejunum, ileum), large intestine, rectum, anus.** (p. 577)

6. Exocrine secretions with enzymes and mucus are added to the lumen by the salivary glands, **pancreas,** and **liver.** (p. 577)

7. The wall of the GI tract consists of four layers: an inner mucosa, a middle submucosa, a smooth muscle, and an outer layer of connective tissue. (p. 580)

8. The **mucosa** consists of epithelium, collections of lymphoid tissue known as **Peyer's patches,** the connective tissue **lamina propria,** and the smooth muscle **muscularis mucosa.** The gut wall is folded and invaginated to increase surface area. The intestinal mucosa has small **villi,** and the apical membranes of individual cells have **microvilli.** (p. 580)

9. The submucosa contains blood and lymph vessels and the **submucosal plexus,** a nerve network of the **enteric nervous system.** (p. 580)

10. The outer wall of the intestinal tract consists of an inner layer of circular muscle and an outer layer of longitudinal muscle. Between the two muscle layers is the **myenteric plexus.** The **serosa** covers the outside surface of the GI tract. (p. 580)

Motility

11. Motility in the gastrointestinal tract moves food from the mouth to the anus and mechanically mixes food to maximize its exposure to digestive enzymes and the absorptive epithelium. (p. 581)

12. Muscle in the GI tract is single-unit smooth muscle whose cells are electrically connected by gap junctions. Some intestinal muscle cells exhibit spontaneous depolarizations called **slow wave potentials.** When the slow wave reaches threshold, it fires action potentials that spread to adjacent muscle cells, creating a wave of contraction. (p. 581)

13. **Pharmacomechanical coupling** occurs when hormones or drugs initiate contraction without a significant change in membrane potential. (p. 581)

14. The sphincters that separate different sections of the digestive system are tonically contracted. The upper esophageal and external anal sphincters are composed of skeletal muscle; the other five sphincters are smooth muscle. (p. 582)

15. **Peristaltic contractions** are progressive waves of contraction that propel material from the esophagus to the rectum. Peristalsis may be mediated through the enteric nervous system (the peristaltic reflex), but it is also subject to control by hormones, paracrines, and the autonomic nervous system. (p. 582)

16. **Segmental contractions** are mixing contractions that knead material back and forth without propelling it forward at a very fast rate. (p. 582)

17. Mechanical digestion of food begins with chewing, or **mastication.** Swallowing, or **deglutition,** is a reflex integrated by a medullary center. (p. 583)

18. Digested material entering the large intestine moves forward by **mass movement.** The **defecation reflex,** a spinal reflex subject to higher control, is triggered by sudden distension of the rectum. (p. 584)

Secretion

19. In a typical day, 2 liters of fluid enter the GI tract through the mouth. Another 7 liters is secreted by the body into the lumen. Nearly all of this volume is reabsorbed. (p. 584)

20. Digestive enzymes are secreted either by exocrine glands or by epithelial cells of the stomach and small intestine. A few secreted enzymes bind to the epithelial cells and become part of the **brush border.** Some enzymes are secreted in an inactive proenzyme form (zymogens). (p. 584)

21. Mucus forms a protective coating over the GI mucosa and lubricates the contents of the gut. Release signals for mucus include parasympathetic innervation, the enteric nervous system, and cytokines from immunocytes. (p. 584)

22. The digestive system secretes water and ions, using processes similar to those of the kidney. Sodium diffuses through leaky cell junctions, followed by water. The epithelial cells transport Cl^- from the extracellular fluid into the lumen with the aid of the apical **cystic fibrosis transmembrane regulator,** or **CFTR chloride channel.** (p. 586)

23. The GI tract secretes H^+ and HCO_3^- by processes similar to those used in the kidney for acid-base balance. (p. 587)

Digestion and Absorption

24. Most nutrient absorption takes place in the small intestine. Water and ions are absorbed in the large intestine. Only a few lipid-soluble substances are absorbed in the stomach. (p. 587)

25. Most digested nutrients are absorbed into the hepatic portal system and delivered directly to the liver. Most digested fats go into the lymphatic system. (p. 588)

26. Carbohydrate digestion takes place in the mouth (salivary amylase) and the small intestine (**amylase** and **disaccharidases**). Glucose absorption uses an apical Na^+-glucose symport protein and a basolateral facilitated diffusion carrier for glucose. Fructose moves by facilitated diffusion across both membranes of the epithelial cell. (p. 588)

27. Protein digestion takes place in the stomach and small intestine. **Proteases** break proteins up into smaller peptides. **Peptidases** break peptides into smaller pieces. (p. 590)

28. Amino acids are absorbed using Na^+-dependent cotransporters. Small peptides are absorbed intact using H^+-dependent cotransporters. Some small peptides are absorbed intact by transcytosis. (p. 591)

29. A small amount of fat digestion takes place in the mouth and stomach with the aid of the acid **lipases.** Most fat digestion is accomplished by pancreatic lipase, aided by **colipase.** Absorption of fats occurs primarily by simple diffusion. (p. 592)

30. Fat digestion is aided by bile from the liver, which emulsifies the fats in order to create greater surface area. As enzymatic and mechanical digestion proceed, the droplets form **micelles.** (p. 593)

31. Absorbed monoglycerides and fatty acids are reassembled into triglycerides in the intestinal cell, then combined with cholesterol and proteins into **chylomicrons.** Chylomicrons are absorbed into the lymph. (p. 593)

32. Fat-soluble vitamins are absorbed along with fats; water-soluble vitamins are absorbed by mediated transport. Vitamin B_{12} absorption requires **intrinsic factor** secreted by the stomach. Mineral absorption usually occurs by active transport. (p. 594)

Regulation of GI Function

33. Three patterns of neural reflexes occur in the GI tract. **Short reflexes** originate within the enteric nervous system and are integrated there without outside input. **Long reflexes** are integrated in the central nervous system. Some long reflexes, such as the feedforward reflexes of the cephalic phase, originate outside the enteric nervous system. Other long reflexes originate in the GI tract. (p. 597)

34. Usually, parasympathetic cholinergic neurons are excitatory and enhance GI functions, whereas sympathetic adrenergic neurons inhibit GI functions. (p. 597)

35. The digestive hormones are generally divided into the gastrin family (**gastrin, cholecystokinin**), the secretin family (**secretin, vasoactive intestinal peptide** [VIP], and gastric inhibitory peptide, also called **glucose-dependent insulinotropic peptide**), and hormones that do not fit into either of those two families (**motilin**). (p. 598)

36. Digestive hormones either excite or inhibit motility and secretion. Some hormones have trophic effects on other hormones. Other hormones act on the brain to change eating behavior. (p. 599)

37. The stimuli for GI hormone secretion arise primarily from the ingestion of food. (p. 600)

38. Paracrines in the gastrointestinal tract, such as histamine, interact with hormones and neural reflexes. (p. 601)

Integration of GI Function: The Stomach

39. The stomach secretes a variety of substances including mucus and bicarbonate from **mucus cells,** gastric acid from **parietal cells,** pepsinogen from **chief cells,** somatostatin from **D cells,** histamine from **enterochromaffin cells,** and gastrin from **G cells.** (p. 601)

40. Digestion can be divided into **cephalic, gastric,** and **intestinal phases.** The phases overlap and influence secretion and motility throughout the digestive tract. (p. 604)

Questions

LEVEL ONE   Reviewing Concepts

1. Briefly define the four basic processes of the digestive system.

2. For most nutrients, _____ is not regulated, whereas _____ and _____ are continuously regulated. Why do you think these differences exist? Defend your answer.

3. A number of functions are closely linked to the digestive system. (a) Once nutrients reach the cells of the body, what name is given to the chemical reactions that govern the body's use or storage of nutrients? (b) What group of regulatory compounds helps determine whether nutrients are used or stored? (c) What system is assigned to keep pathogens from entering the body via the gastrointestinal tract?

4. Match each of the following terms with an appropriate description:

    1. Appendix
    2. Chyme
    3. Colon
    4. Duodenum
    5. Ileum
    6. Jejunum
    7. Pancreas
    8. Pylorus
    9. Rectum
    10. Small intestine
    11. Stomach

    (a) Chyme is produced and waits to be released from here.
    (b) Organ where most digestion occurs.
    (c) Initial section of small intestine.
    (d) An organ that adds exocrine secretions to the duodenum via a duct.
    (e) Sphincter that prevents premature emptying of the stomach.
    (f) The soupy substance that results when fluids and enzymes are added to food as it travels through the digestive tract.
    (g) Section where chyme is processed to remove water and electrolytes, leaving waste products of digestion.
    (h) Distension of the walls of this organ triggers the defecation reflex.

5. How long is the gastrointestinal tract in a living person? Does it lengthen or shorten on autopsy? Why?

6. List the four layers found in the GI tract walls. What type of tissue predominates in each?

7. Describe the functional types of epithelium found lining the stomach and intestines.

8. What are Peyer's patches?

9. Describe how the location and function of the submucosal plexus differ from those of the myenteric plexus.

10. How are the serosa, peritoneal membrane, and mesentery related, both structurally and functionally?

11. What purposes does motility serve in the gastrointestinal tract? What types of tissue contribute to gut motility? What types of contraction do the tissues undergo?

12. Seven sphincters separate regions of the digestive system. Name them in order (starting at the mouth, ending at the rectum) and describe the location of each. Specify what type of muscle tissue each is composed of. Each of these sphincters is _____ contracted.

13. Describe how peristaltic contractions differ from segmental contractions.

14. What is a zymogen? What is a proenzyme? List two specific examples.

15. Match each product with the cell that secretes it.
   1. parietal cells   (a) enzymes
   2. goblet cells   (b) HCl
   3. brush border   (c) HCO_3^-
   4. pancreatic cells   (d) mucus

16. Which of these factors does not significantly affect digestion? For each factor involved in digestion, briefly explain how and where it exerts its effects.

   emulsification   neural activity
   pH   hormonal activity
   surface area of food   enzymatic activity
   temperature   muscular activity

17. Most digested nutrients are absorbed into the _____ of the _____ system, delivering nutrients to the _____ (organ). However, digested fats go into the _____ system because intestinal capillaries have a _____ around them that most lipids are unable to cross.

18. Give at least three examples of each chemical class of nutrient and list enzymes that digest nutrients in each class: carbohydrates, proteins, and lipids.

19. What is the enteric nervous system and what is its function?

20. What are short reflexes? What types of responses do they regulate? What is meant by the term *long reflex*?

21. What role do paracrines play in digestion? Give specific examples.

22. List the types of epithelial cells found associated with gastric glands. Briefly describe the role of each one.

LEVEL TWO Reviewing Concepts

23. **Concept map:** Diagram the activation and interactions of the cephalic, gastric, and intestinal phases of digestion.

24. Define, compare, and contrast the following terms:
   a. Mastication, deglutition
   b. Microvilli, villi
   c. Peristalsis, reflux esophagitis, mass movements, defecation
   d. Chyme, feces
   e. Short reflexes, long reflexes

25. How are sodium, potassium, and chloride ions transported out of the gut? What is the CFTR channel and its significance? How are H^+ ions secreted into the stomach?

26. Compare the ENS with the cephalic brain. Give some specific examples concerning neurotransmitters, neuromodulators, and supporting cells.

27. List and briefly describe the actions of the members of each of the three groups of GI hormones.

LEVEL THREE Problem Solving

28. Erica's baby, Justin, has had a severe bout of diarrhea and is now dehydrated. Is his blood more likely to be acidotic or alkalotic? Why?

29. Mary Littlefeather arrives in her physician's office complaining of severe, steady pain in the upper right quadrant of her abdomen. The pain began shortly after she ate a meal of fried chicken, french fries, and peas. Lab tests and an ultrasound reveal the presence of gallstones in the common bile duct that drains the liver, gallbladder, and pancreas into the small intestine.
   a. Why was Mary's pain precipitated by the meal she ate?
   b. Which of the following will be affected by the gallstones: micelle formation in the intestine, carbohydrate digestion in the intestine, protein absorption in the intestine. Explain your reasoning.

Problem Conclusion

In this running problem, you learned about peptic ulcers and an unusual bacterium that is believed to be the cause of most duodenal ulcers. You also learned that antacids and histamine-receptor blockers reduce the level of acid in the stomach.

Further check your understanding of this running problem by checking your answers against those in the summary table.

	Question	Facts	Integration and Analysis
1	Why don't high levels of acid and enzyme in the stomach cause irritation of the stomach mucosa in everyone?	The epithelium of the intestinal tract contains goblet cells that secrete mucus.	Mucus secreted by the walls of the intestinal tract forms a protective barrier that keeps the cells from being damaged by acid and enzymes.
2	How does the conversion of urea into carbon dioxide and a base protect *H. pylori* from the hostile environment of the stomach?	A base is an anion that combines with H^+. Ammonia combines with H^+ to create ammonium ions, NH_4^+. This reduces acidity (increases pH).	Because *H. pylori* is studded with enzymes that produce ammonia, a base, the acidity of the environment immediately surrounding the bacteria is less. In this way, the bacteria protect themselves from being killed by the acidic environment of the stomach.
3	Write the chemical equation showing how the antacids sodium bicarbonate ($NaHCO_3$) and aluminum hydroxide ($Al(OH)_3$) buffer hydrochloric acid (HCl).	Buffers are molecules that combine with H^+.	$NaHCO_3 + HCl \rightarrow NaCl + H_2CO_3$ $Al(OH)_3 + 3\,HCl \rightarrow AlCl_3 + 3\,H_2O$
4	How does blocking H_2 receptors stop acid production in the stomach?	Histamine is a paracrine that stimulates acid secretion by the stomach.	If histamine receptors are blocked by H_2 blockers, histamine cannot combine with the receptors and stimulate acid secretion.
5	Explain why Tonya was given an antibiotic and a H^+/K^+-ATPase inhibitor to treat her ulcer.	Tonya's ulcer is associated with presence of the bacterium *H. pylori* and is aggravated by acid secretion in the stomach.	Antibiotics kill bacteria, so they will kill the *H. pylori*. The apical membranes of acid-secreting parietal cells contain a H^+/K^+-ATPase that pumps acid into the stomach lumen. Blocking this transporter will reduce acid secretion.

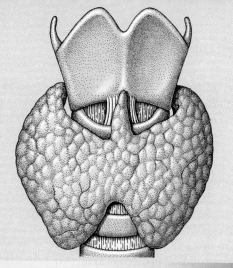

21
Endocrine Control of Metabolism and Growth

The magazine covers at the grocery store checkout stands tell visitors to this country a lot about Americans. Headlines screaming "Lose 10 pounds in a week without dieting" or "CCK: The hormone that makes you thin" vie for attention with glossy full-color photos of luscious desserts dripping with chocolate, whipped cream, fat, and calories. As one magazine article put it, we are a nation obsessed with staying trim and with eating, mutually exclusive occupations. You learned in the last chapter that "what you eat is what you get" in terms of digestion and absorption. Most of the food that we put into our mouths ends up in our bloodstream. It is the body's chemical reactions, known as metabolism, that determine what happens to those nutrients: Are they destined to burn off as heat, become muscle, or turn into extra pounds that make it difficult to zip up blue jeans? This chapter follows the nutrients absorbed from the intestine as they travel through the pathways of metabolism to their fate.

Problem

Hyperparathyroidism

Broken bones, kidney stones, abdominal groans, and psychic moans. Medical students memorize this saying when they learn about a condition called hyperparathyroidism, a disease in which the parathyroid glands (∞ p. 176) work overtime and produce excess parathyroid hormone (PTH). Dr. Adiaha Spinks suddenly recalls this saying as she examines Professor Bob Magruder, who has arrived at the office in pain from a kidney stone lodged in his ureter. When questioned about his symptoms, Dr. Bob also mentions pain in his shin bones, muscle weakness, stomach upset, and a vague feeling of depression. "I thought it was all just the stress of getting my textbook published," he says. But to Dr. Spinks, his combination of symptoms sounds suspiciously like Dr. Bob might be suffering from hyperparathyroidism.

continued on page 618

ENERGY BALANCE AND METABOLISM

Metabolism and energy were introduced in Chapter 4, Cell Processes. Recall that **metabolism** is the sum of all the chemical reactions in the body. These reactions extract energy from nutrients, use energy for work, and store excess energy so that it can be used later. Some reactions of metabolism release energy in small amounts, whereas other reactions require the input of energy (exergonic and endergonic reactions; ∞ p. 76).

Metabolic pathways that synthesize small molecules into larger ones are called **anabolic** pathways [*ana-*, completion + *metabole*, change]. Metabolic pathways that break large molecules down into smaller components are considered **catabolic** [*cata-*, down or back]. Notice that the classification of a pathway is its end result, not what happens in one or two steps.

Energy Input Equals Energy Output

The first law of thermodynamics states that energy in the universe is constant. By extension, this statement means that all energy that goes into a biological system such as the human body can be accounted for:

(1)   Energy intake = energy output

Energy intake for humans consists of the nutrients we eat, digest, and absorb. However, because the digestive system itself does not regulate energy intake, we must depend on behavioral mechanisms such as hunger and satiety to tell us when and how much to eat. Psychological and social aspects of eating, such as parents who say "Clean your plate," complicate the physiological control of food intake.

Energy output by the body either does work (which includes storing energy) or is returned to the environment as heat. Therefore we can rewrite equation 1 to read

(2)   Energy intake = heat + work

In the human body, at least half the energy released in the chemical reactions of metabolism is lost as heat. Much of this heat production is unregulated "waste" heat. But because humans are homeothermic animals, we also regulate heat production and heat loss in order to maintain a relatively constant body temperature. Thermoregulation is discussed later in this chapter.

Biological work takes one of three forms (∞ p. 74):

1. Transport work, which is used to move molecules from one side of a membrane to another
2. Mechanical work, which uses intracellular fibers and filaments to create movement
3. Chemical work, which is used for growth, maintenance, and storage of information and energy

Transport processes are used to bring material into and out of the body and to transfer them between compartments within the body. Movement includes both external work, such as that done by skeletal muscle contraction, and internal work, such as the beating of cilia or the cytoplasmic movement of vesicles. Chemical work in the body can be subdivided into synthesis and storage (Fig. 21-1 ■). Storage includes both short-term energy storage in high-energy phosphate compounds such as ATP and long-term energy storage in the chemical bonds of glycogen and fat. Cells have a limited capacity for glycogen storage, so most excess energy ends up as fat droplets in adipose tissue.

The pattern of fat deposition in our bodies is greatly affected by age, gender, and genetics. Despite the uncontrollable nature of these factors, many Americans try to achieve the "ideal" figure as portrayed in the media. For many, this means losing weight and fat. Shrinking the size of adipose tissue and moving energy out of fat stores requires that energy output exceed energy intake. Because the synthesis, transport, and heat production components of energy output are not under conscious control, the only way that people can voluntarily increase energy output is with movement such as exercise. The other factor that people can control is their energy intake, by watching what they eat. Although energy balance is a very simple concept, it is a difficult one for people to accept. Behavior modifications such as eating less and exercising more are among the most frequently given and most difficult to follow instructions from health care professionals to patients.

Body Temperature Is Regulated by Balancing Heat Production, Gain, and Loss

The development of obesity may be linked to the efficiency with which the body converts food energy into cell and tissue components. According to one theory, people

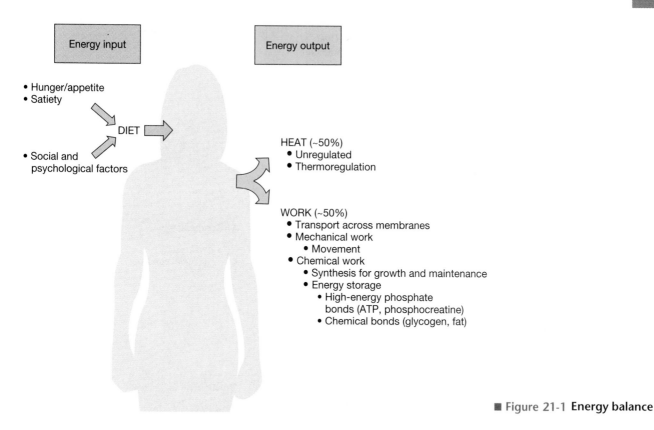

Energy input

- Hunger/appetite
- Satiety

DIET

- Social and psychological factors

Energy output

HEAT (~50%)
- Unregulated
- Thermoregulation

WORK (~50%)
- Transport across membranes
- Mechanical work
  - Movement
- Chemical work
  - Synthesis for growth and maintenance
  - Energy storage
    - High-energy phosphate bonds (ATP, phosphocreatine)
    - Chemical bonds (glycogen, fat)

■ Figure 21-1 **Energy balance**

who are more efficient in transferring chemical bond energy from food to fat are the ones who put on weight. In contrast, people who are less metabolically efficient can eat the same number of calories and not gain weight because more of the food energy is released as heat. Much of what we know about the regulation of energy balance comes from studies on temperature adaptation.

Body temperature in humans   Temperature regulation in humans is linked to metabolic heat production (thermogenesis). Humans are **homeothermic** animals who regulate their internal body temperatures within a relatively narrow range despite fluctuating environmental temperatures. Average body temperature is 37° C (98.6° F) with a normal range of 35.5°–37.7° C (96°–99.9° F). However, these values are subject to considerable variation, both from one individual to the next and throughout the day in a single individual. The site at which temperature is measured also makes a difference, because the temperature at the core of the body may be much higher than that at the surface of the skin. Oral temperatures are about 0.5° C less than rectal temperatures. The newest digital thermometers record temperature at the tympanic membrane of the ear, an advantageous method to use if working with a squirmy child.

Several factors affect body temperature in a given individual. Body temperature increases with exercise or after a meal because of diet-induced thermogenesis. Temperature also varies throughout the day in a cyclical fashion: The lowest (basal) temperature occurs in the early morning and the highest in the early evening. In women of reproductive age, a monthly cycle of temperature is also observable: Basal body temperatures are about 0.5° higher in the second half of the menstrual cycle, after ovulation, than before ovulation.

Heat gain and loss   Temperature balance in the body, like energy balance and mass balance, depends on a dynamic equilibrium between heat input, from both internal and external sources, and heat output (Fig. 21-2 ■). Internal heat production comes from the normal metabolic pathways for energy assimilation and from heat released as a result of muscle contraction. Mechanical work is not very efficient, so as much as 80% of the energy released when ATP is hydrolyzed may be given off as heat. In **shivering thermogenesis,** the body uses **shivering** (rhythmic tremors caused by skeletal muscle contraction) to generate heat for temperature regulation. **Nonshivering thermogenesis** is the metabolic production of heat specifically for temperature regulation.

External heat input comes from the environment through radiation or conduction. All objects with a temperature above absolute zero give off radiant energy (radiation) with infrared or visible wavelengths; this energy can be absorbed by other objects. You absorb radiant energy each time you go out and sit in the sun or sit in front of a fire on a chilly night. **Conductive heat gain** is the transfer of heat between objects that are in contact with each other, such as the skin and a heating pad or the skin and water.

Heat loss from the body takes place by radiation, conduction, convection, and evaporation. **Radiant heat loss**

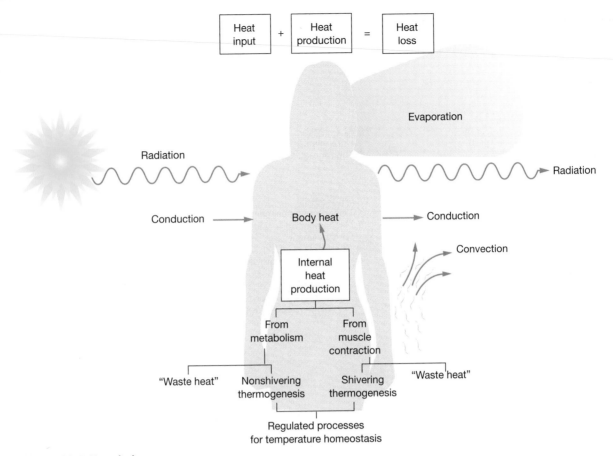

■ Figure 21-2 **Heat balance**

from the human body is estimated to account for nearly half the heat lost from a person at rest in a normal room. **Thermography** is a technique in diagnostic medicine that measures radiant heat loss. Cancerous tumors and regions with higher metabolic activity give off more heat than surrounding tissues and can be visually identified. "Night vision" glasses that allow people to see in the dark use infrared radiation given off by living organisms to create a visible image.

Radiant and conductive heat loss are enhanced by **convective heat loss,** heat that is carried away by warm air rising from the surface of the body. Convective air currents are created wherever there is a temperature difference in the air: Hot air rises and is replaced by cooler air. Convection helps move warmed air away from the skin's surface. Clothing, which traps air and prevents convective air currents, therefore helps retain heat close to the body.

The fourth type of heat loss in the body is **evaporative heat loss,** which takes place as water evaporates at the skin's surface or in the respiratory tract. The conversion of water from a liquid to a gaseous (vapor) state requires the input of substantial amounts of energy, so when water on the body evaporates, it acts as a "sink" for heat. You can demonstrate the effect of evaporative cooling by wetting one arm but not the other. The arm that is wet will feel much cooler as it dries because of heat drawn out of the

skin to vaporize the water. Similarly, the 500 mL of water vapor that leave the body through the lungs and skin each day take with it a significant amount of body heat. Evaporative heat loss across the skin's surface is affected by the humidity of the surrounding air, with less evaporation taking place at higher humidities.

Conductive, convective, and evaporative heat loss from the body are aided by the bulk flow of air across the body, such as the air moved across the skin by a fan or breeze. The effect of wind on body temperature regulation is described in the winter as the "wind chill factor," a combination of absolute environmental temperature and the effect of convective heat loss.

Regulation of body temperature   Although the human body is usually warmer than its environment, the energy generated by normal metabolism is sufficient to maintain normal body temperature when the environmental temperature stays between 27.8° and 30° C (82° and 86° F). This range is known as the **thermoneutral zone.** Above that temperature range, the body gains heat because heat production exceeds heat loss; below that range, heat loss exceeds heat production. Outside the thermoneutral zone, the body must use homeostatic compensation to lower or raise internal temperature.

The human body without benefit of clothing can thermoregulate at environmental temperatures between 10°

muscle mass, people raise their basal metabolic rate and burn off more calories at rest.

An increase in metabolic rate over the basal rate arises from one of two conditions: Energy is being used to fuel activity (mechanical work) or to digest and assimilate food. We now know that resting (nonactive) metabolic rate increases after a meal. In other words, there is an energetic cost to the digestion and assimilation of food. This phenomenon has been termed **diet-induced thermogenesis** and is measured by an increase in the body's heat production after eating. Diet-induced thermogenesis is related to the type and amount of food ingested. Fats cause relatively little diet-induced thermogenesis; proteins increase heat production the most. This phenomenon may support the claim of some nutritionists that eating one calorie of fat is different from eating one calorie of protein, although they contain the same amount of energy when measured by direct calorimetry. Most of the energy in fat calories is stored by the body; a smaller fraction of the energy in protein calories will be stored because more is released as heat.

Scientists are now looking for links between cold-induced nonshivering thermogenesis in laboratory animals and diet-induced thermogenesis in humans. The fact that both are mediated by the sympathetic nervous system and the hormone thyroxine (T_4) suggests a possible common mechanism. If we can understand the mechanism by which energy is diverted from fat storage into heat production, we may be able to develop a new class of thermogenic drugs for the treatment of obesity.

Another factor that influences metabolic rate is physical activity. You are probably familiar with tables that show the caloric expenditure per hour of various activities. Sitting and lying down consume relatively little energy; rowing in a race and bicycling are among the energetically most expensive activities.

In the end, the two factors that a person can control to balance energy expenditure and avoid weight gain are energy intake (eating) and activity. If a person's activity includes strength training that increases lean muscle mass, resting metabolic rate goes up. The addition of lean muscle mass to the body creates additional energy use, which in turn decreases the number of calories that go into storage.

Daily energy requirements   The daily energy intake needed to do different types of work varies with the needs and activity of the body. A person's daily energy requirement is usually expressed in terms of caloric intake. For example, an average man may have a daily caloric requirement of 2000 kcal. If we assume that his total energy requirement could be met by ingesting glucose, a carbohydrate with an energy content of 4 kcal/g, then he would have to consume 500 g, or 1.1 pounds, of glucose in a day. However, our bodies cannot absorb crystalline glucose, so those 500 g of glucose must be dissolved in water. If the glucose is ingested as an isosmotic 5% solution, his daily caloric requirement would have to be dissolved in 10 liters of water: a substantial volume to drink over the course of a day!

Fortunately, we do not usually ingest glucose as our primary fuel. Proteins, complex carbohydrates, and fats also contain energy in their chemical bonds. Glycogen, a glucose polymer, is a more compact form of energy than an equal number of individual glucose molecules; it also requires less water for hydration. Our cells therefore store glucose in the form of glycogen granules (∞ p. 27). The liver contains about 100 g of glycogen, and the skeletal muscles contain about 200 g. But even 300 g of glycogen can provide only enough energy for 10 to 15 hours of the body's energy needs. The brain alone requires 150 g of glucose per day. Consequently, the body keeps most of its energy reserves in compact, high-energy fat molecules.

One gram of fat has 9 kcal, more than twice the energy content of an equal amount of carbohydrates or proteins. Furthermore, fat storage is not associated with water because it is hydrophobic. This feature makes adipose tissue a very efficient way for the body to keep considerable amounts of energy stored in a minimal amount of space. Metabolically, however, the energy in fat is harder to access, and the metabolism of fats is slower than that of carbohydrates.

Energy for Work May Come from Absorbed Nutrients or from Storage

In the human body, we generally divide metabolism into two states based on whether or not a meal has been recently consumed. The period of time following a meal, when the products of digestion are being absorbed, used, and stored, is called the **fed,** or **absorptive, state.** This is an anabolic state in which the energy of nutrient biomolecules is transferred to high-energy compounds or is stored in the chemical bonds of other molecules. Once nutrients from a recent meal are no longer in the bloodstream and available for use by the tissues, the body enters a **fasted,** or **postabsorptive, state.** As the pool of available nutrients in the blood decreases, the body taps

Fat That Releases Heat   Diet-induced thermogenesis has several components. One component is the heat released by the processes of digestion, absorption, and storage of ingested nutrients. A second component of diet-induced thermogenesis is the activation of **brown adipose tissue,** also known as brown fat (∞ p. 63). The mitochondria of brown fat cells contain a protein that dissociates ATP production and the flow of electrons and H^+ through the electron transport system. In this uncoupled state, the energy that would normally be trapped in the high-energy phosphate bond of ATP is released as heat. Factors known to induce brown fat thermogenesis in animal models include sympathetic innervation, epinephrine, and exposure to cold. However, adult humans have only small amounts of brown fat, and its role in diet-induced thermogenesis is still being investigated.

into its stored reserves. The postabsorptive state is a catabolic state, in which the cells degrade large molecules into smaller molecules. The energy released by breaking chemical bonds is used to do work.

The specific fate of absorbed nutrients depends on which class of biomolecule the nutrient belongs to: carbohydrates, proteins, or fats. (This discussion of energy metabolism will not consider the role of vitamins, minerals, and trace elements, although they play a significant role in other aspects of human nutrition.) In general, biomolecules meet one of three fates in the body:

1. They are metabolized immediately so that the energy in their chemical bonds is trapped in the high-energy phosphate bonds of ATP, phosphocreatine, and other high-energy compounds. This energy can then be used to do mechanical or transport work.
2. They are used for synthesis of basic components needed for growth and maintenance of cells and tissues, or,
3. If the food ingested exceeds the body's requirements for energy and synthesis, the excess nutrients go into storage as glycogen and fat. Storage makes energy available for times of fasting.

One of the most important aspects of metabolism is the interaction between carbohydrates, proteins, and fats. Figure 21-5 ■ is a schematic diagram that follows these

biomolecules from the diet into the three nutrient pools of the body. (The nutrient pools are nutrients that are available for immediate use.) Fats that enter pathways for energy metabolism are absorbed primarily as fatty acids and glycerol. Fatty acids form the primary pool of lipids in the blood. They can be used as an energy source by many tissues but are also easily stored as fat in the adipose tissues of the body. Fatty acids, when metabolized, contribute two-carbon units to the citric acid cycle. The carbons are excreted as CO_2, and the energy from their chemical bonds is trapped as ATP. Another fat, cholesterol, is important for synthetic pathways but is not generally metabolized for energy.

Carbohydrates are absorbed mostly in the form of glucose. The plasma glucose concentration is the most closely regulated of the three nutrient pools. One reason is that glucose is the only fuel that the brain can metabolize except in times of starvation. Notice in Figure 21-5 ■ that if the glucose pool drops below a certain level, the only tissue that has access to glucose is the brain. This is a conservation measure to ensure that the body's control center has adequate amounts of energy. Just as the circulatory system gives top priority to blood supply to the brain, metabolism also gives top priority to the brain. If the glucose pool is above the minimum, then most tissues can use glucose for their energy source. If the glucose pool increases even further, then glucose goes into storage as glycogen. The synthesis of glycogen

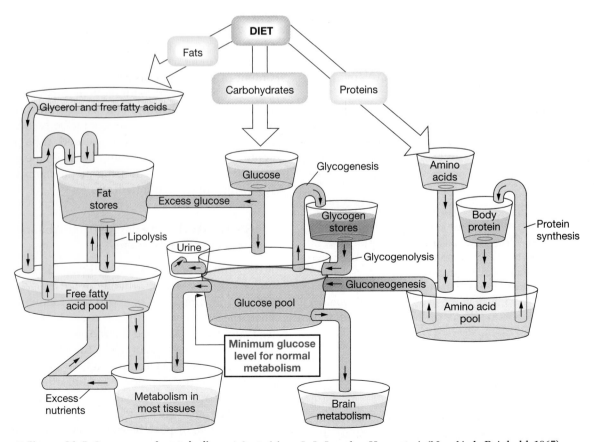

■ **Figure 21-5 Summary of metabolism** Adapted from L. L. Langley, *Homeostasis* (New York: Reinhold, 1965).

from glucose is known as **glycogenesis.** If plasma glucose concentrations begin to drop in the fasted state, the body can break down glycogen through the process called **glycogenolysis.**

Under normal conditions, excess ingested glucose is synthesized into fatty acids and stored as fat. If for some reason plasma glucose concentrations rise faster than fat can be made, excess glucose is excreted in the urine when the renal threshold for glucose reabsorption is exceeded (∞ p. 532). The balance between oxidative metabolism, glycogen synthesis and breakdown, and fat synthesis allows the body to maintain plasma glucose concentrations within a relatively narrow range.

The amino acid pool of the body is used primarily for protein synthesis. However, if glucose intake is low, amino acids can be converted into glucose through the pathways known as **gluconeogenesis.** This word literally means "the

birth of new glucose" and refers to the synthesis of glucose from a noncarbohydrate precursor. Amino acids are the main source for glucose through the gluconeogenesis pathways. Both gluconeogenesis and glycogenolysis are important backup sources for glucose during periods of fasting.

Metabolic Pathways Are Regulated by Changes in Enzyme Activity

The most important biochemical pathways for energy production are illustrated in Figure 21-6 ■. This figure does not include all of the metabolic intermediates in each pathway (∞ p. 91). Instead, it emphasizes the points at which different pathways intersect, because these intersections are often key points at which metabolism is controlled.

One significant feature of metabolic regulation is the use of different enzymes to catalyze forward and reverse reac-

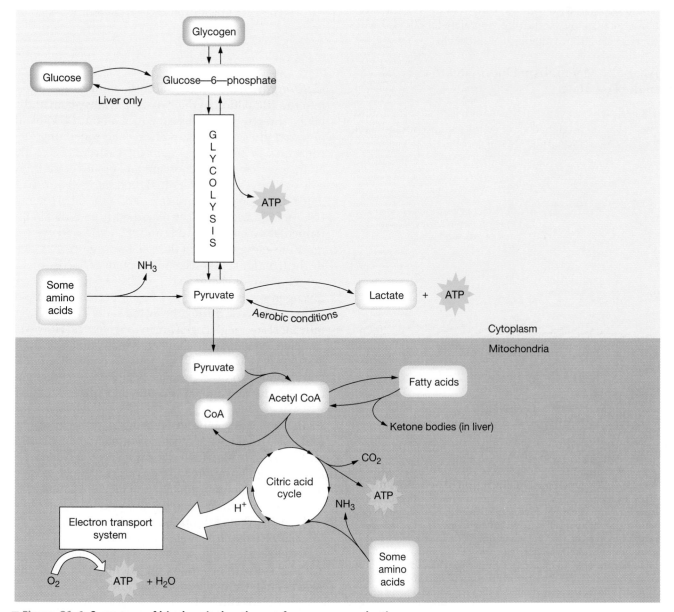

■ Figure 21-6 **Summary of biochemical pathways for energy production**

tions. This dual control, sometimes called *push-pull control*, allows close regulation of a reaction. In the example shown in Figure 21-7 ■, enzyme 1 catalyzes the reaction of A to B; enzyme 2 catalyzes the reverse reaction. If the activity of the two enzymes is roughly equal, as soon as A is converted into B, B is converted back into A. Turnover of the two substrates is rapid, but there is no net production of product B. Figure 21-7b ■ shows an example in which the activity of one set of enzymes is enhanced while the enzymes catalyzing the reverse reaction are inhibited. This type of dual control is exerted by the pancreatic hormone insulin, which stimulates enzymes for glycogen synthesis and inhibits enzymes for glycogen breakdown. The net result is glycogen synthesis from glucose. The reverse pattern is shown in Figure 21-7c ■, where glucagon, another pancreatic hormone, stimulates the enzymes of glycogenolysis while inhibiting the enzymes for glycogenesis. The net effect of glucagon on the metabolic enzymes is glucose synthesis from glycogen. In the next two sections, we turn to how the direction of metabolic pathways changes with the availability of nutrients.

Anabolic Metabolism Dominates in the Fed State

The fed state is an anabolic state in which absorbed nutrients are being used for energy, synthesis, and storage. Table 21-1 summarizes the fates of nutrients in the fed state.

(a)   Enzyme 1
A ⇌ B   Without regulation of enzymatic
Enzyme 2   activity, the pathway will simply cycle back and forth.

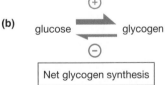

(b)   glucose ⊕ glycogen ⊖

Net glycogen synthesis

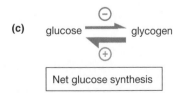

(c)   glucose ⊖ glycogen ⊕

Net glucose synthesis

■ **Figure 21-7 Dual (push-pull) control of metabolism**
(a) Many metabolic reactions use different enzymes to catalyze forward and reverse reactions. If the two enzymes are equally active, the pathway will cycle rapidly without net gain or loss of products and reactants. (b) Insulin stimulates the enzymes that favor glycogen synthesis from glucose while simultaneously inhibiting the enzymes that regulate glycogen breakdown to glucose. The result is net glycogen synthesis. (c) Glucagon has the opposite effect of insulin. It inhibits the enzymes that favor glycogen synthesis while stimulating the enzymes that favor glycogen breakdown. The result is net glucose synthesis.

TABLE 21-1	Fate of Nutrients in the Absorptive, or Fed, State

Carbohydrates, absorbed primarily as glucose

1. Used immediately for energy through aerobic pathways*
2. Used for lipoprotein synthesis in the liver and secreted into blood
3. Stored as glycogen in liver and muscle
4. Excess converted to fat and stored in adipose tissue

Proteins, absorbed primarily as amino acids

1. Most amino acids go to tissues for protein synthesis*
2. If needed for energy, amino acids convert in liver to intermediates that feed into aerobic metabolism
3. Excess converted to fat and stored as adipose tissue

Fats, absorbed primarily as triglycerides

1. Stored as fats primarily in liver and adipose tissue*

---

*Primary fate.

Carbohydrates   Glucose absorbed after a meal goes into the circulation of the hepatic portal system and is taken directly to the liver, where about 30% of all ingested glucose is metabolized. The remaining 70% continues in the bloodstream for distribution to the brain, muscles, and other organs and tissues of the body. Glucose moves from interstitial fluid into cells via facilitated diffusion glucose transporters of the GLUT family (∞ p. 589). Most glucose absorbed from a meal goes immediately into glycolysis and the citric acid cycle to meet the energy needs of the cell. Some glucose is used by the liver for synthesis of lipoproteins. Glucose that is not required for energy and synthesis is stored either as glycogen or as fat. The liver cells become the major source of glucose during the fasted state. The other significant stores of glycogen are in the muscles. However, muscle glycogen cannot be shared with other cells of the body and is used only for muscle contraction. The body's ability to store glycogen is limited, so excess glucose is converted into fat and stored in adipose tissue.

Proteins   Most amino acids absorbed from a meal go to the tissues for protein synthesis. Like glucose, amino acids are taken first to the liver by the hepatic portal system. The liver is a significant site for the synthesis of plasma proteins such as albumin and clotting factors. Amino acids not taken up by the liver enter the systemic venous blood and are distributed throughout the body. They are used by cells to create structural or functional proteins such as cytoskeletal elements, enzymes, and hormones. Amino acids are also incorporated into nonprotein molecules such as amine hormones and neurotransmitters.

Amino acids can be used for energy if glucose intake is low. On the other hand, if more protein is ingested than is needed for synthesis and energy needs, the excess amino acids are converted into fat. Some body-builders spend large amounts of money on special amino acid supplements that are advertised to build bigger muscles. What the athletes may not realize is that these amino acids do not automatically go into protein synthesis. If amino acid intake exceeds the body's need for protein synthesis, excess amino acids are burned to create energy or are stored as fat.

Fats   Most ingested fats are assembled into lipoprotein complexes called chylomicrons in the intestinal epithelium (∞ p. 593). Chylomicrons reach the venous circulation via the lymphatics, bypassing the liver (Fig. 21-8 ■). As they circulate through the blood, **lipoprotein lipase** bound to the capillary endothelium of muscles and adipose tissue converts triglycerides in the chylomicron to free fatty acids and monoglycerides. These molecules then diffuse into cells, particularly adipose tissue, where they are reassembled into triglycerides for storage.

Chylomicron remnants that remain in the circulation are composed primarily of lipoproteins and cholesterol. The combination of lipid with proteins makes cholesterol more soluble in plasma, but the complexes are unable to diffuse through cell membranes as the free fatty acids can. Instead, lipoprotein complexes must be brought into cells using receptor-mediated endocytosis (∞

p. 126). The chylomicron remnants, for example, are taken up and metabolized by the liver.

Once in the liver cells, cholesterol from the remnants joins the liver's pool of lipids. If cholesterol is in excess, some may be excreted in the bile. The remainder is added to newly made cholesterol and fatty acids, then packaged again into lipoprotein complexes for secretion back into the blood. There are several forms of lipoprotein used to form these complexes, ranging from a *very low-density lipoprotein (VLDL)* to a *high-density lipoprotein (HDL)*. The complexes contain varying ratios of protein, triglyceride, phospholipids, and cholesterol. The more protein the complex contains, the heavier its weight.

Most lipoprotein in the blood is **low-density lipoprotein, or LDL.** LDL-cholesterol is the form in which most cells get cholesterol through receptor-mediated endocytosis. LDL-cholesterol is sometimes known as the "bad cholesterol" because elevated concentrations of plasma LDL are associated with the development of atherosclerosis. The second most common lipoprotein is **high-density lipoprotein, or HDL.** HDL-cholesterol is sometimes called the "good cholesterol" because HDL takes cholesterol into the liver cells, where it is metabolized or excreted. A simpler way to remember the difference between the two types is to let *L* stand for *lethal* and *H* stand for *healthy*. The ratio of HDL-cholesterol to LDL-cholesterol in the plasma is as important as the absolute

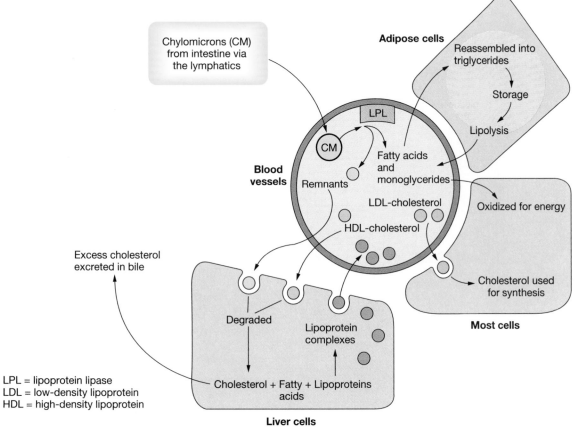

LPL = lipoprotein lipase
LDL = low-density lipoprotein
HDL = high-density lipoprotein

■ Figure 21-8 **Transport and fate of dietary fats**

concentration. The ratio is used by clinicians to predict a person's risk of developing atherosclerosis.

Metabolism in the Fasted State

Once all the nutrients from a meal have been digested, absorbed, and distributed to various cells, plasma concentrations of glucose begin to fall. This is the signal for the body to shift from a fed (absorptive) state into a fasted (postabsorptive) state. The goal of the fasted state is to maintain plasma glucose concentrations within an acceptable range so that the brain and neurons have adequate fuel.

The liver is the primary source of glucose production during the fasted state. Liver glycogen can provide enough glucose through glycogenolysis to meet four to five hours of the body's energy needs (Fig. 21-9 ■). The glycogen stores of skeletal muscle can contribute additional plasma glucose in the fasted state but not through direct conversion of glycogen to glucose. Muscle cells, like most other cells, lack the enzyme that makes glucose from glucose-6-phosphate, so they metabolize glucose-6-phosphate to pyruvate (aerobic conditions) or lactate (anaerobic conditions). Pyruvate and lactate are then transported to the liver, which uses them to make glucose.

Apolipoproteins, Cholesterol Metabolism, and Heart Disease The protein components of the lipoproteins are called **apoproteins**, or **apolipoproteins**. LDL complexes contain apoprotein B (apo B), a molecule that combines with specific receptors to allow cholesterol and triglycerides to be endocytosed into most cells of the body. HDL complexes contain apo E, which activates receptors in the liver. Several inherited forms of hypercholesterolemia (elevated plasma cholesterol levels) have been linked to defective forms of either apo E or apo B. These abnormal apoproteins may help explain the accelerated development of atherosclerosis in people with hypercholesterolemia.

Additional glucose can be made from amino acids and glycerol. A major source of amino acids is muscle protein. Muscle proteins are degraded into amino acids, which are then deaminated, producing intermediates that are processed into pyruvate. The amino groups removed from amino acids during deamination are converted into urea and excreted. Pyruvate goes to the liver and is converted to glucose, as previously described. Although not all amino acids can be converted to glucose, most can be used directly by cells for ATP produc-

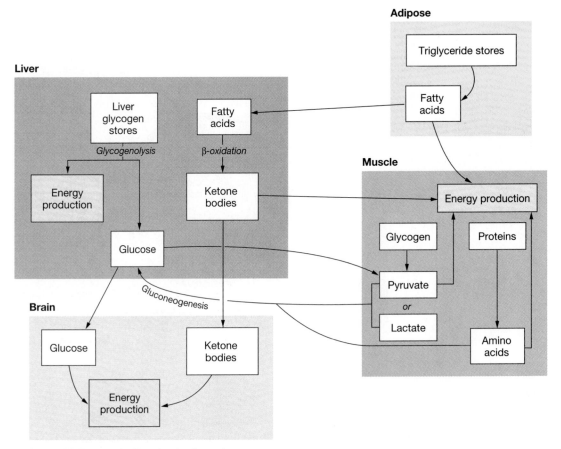

■ Figure 21-9 **Metabolism in the fasted state**

continued from page 618

High concentrations of calcium in the blood lead to high concentrations of calcium in the kidney filtrate. Calcium-based kidney stones occur when calcium phosphate or calcium oxalate crystals form and aggregate with organic material in the lumen of the kidney tubule. Once Dr. Bob's kidney stone passes into the urine, Dr. Spinks sends it for a chemical analysis.

Question 3: Only the free calcium ions in the blood can filter into Bowman's capsule at the nephron. A significant portion of plasma calcium cannot be filtered. Use what you have learned about filtration at the glomerulus to speculate on why some calcium cannot filter (∞ p. 525).

tion. This alternative fuel source thus spares plasma glucose for use by the brain.

In the fasted state, adipose tissue breaks down its stores of triglycerides into fatty acids and glycerol. Glycerol goes to the liver and can be converted to glucose. Fatty acids are released into the blood and are used as a source of energy by many tissues. The long carbon chains of fatty acids are chopped into two-carbon units through the process of β-oxidation (∞ p. 96). These two-carbon units then feed into the citric acid cycle through acetyl CoA. If there is a large demand for fatty acid breakdown, β-oxidation in the liver results in the formation of molecules known as **ketone bodies.** Ketone bodies are transported from the liver to other cells that convert them to acetyl CoA. Acetyl CoA then enters the citric acid cycle.

Ketone bodies are the only fuel besides glucose that the brain can use, so in cases of prolonged starvation, ketones become a significant source of energy. Unfortunately, the ketones *acetoacetic acid* and *β-hydroxybutyric acid* are moderately strong acids, so excessive ketone production leads to a state of acidosis known as **ketoacidosis.** People in ketoacidosis have a fruity odor to their breath due to acetone, a volatile ketone whose odor you might recognize from nail polish remover.

With this summary of metabolic pathways as background, we now turn to the regulation of the biochemical pathways of metabolism.

ENDOCRINE CONTROL OF METABOLISM: PANCREATIC HORMONES

The regulation of the metabolic pathways described in the previous sections is predominantly under control of the endocrine system, although the nervous system does have some influence. Hour-to-hour regulation depends on the ratio of insulin to glucagon, two hormones secreted by the endocrine cells of the pancreas. In this section, we examine the structure of the endocrine pancreas and then consider how its hormones supervise the movement of biomolecules through the complex pathways of metabolism.

The Endocrine Pancreas Secretes Several Hormones

The endocrine cells of the pancreas are less than 2% of the organ's total mass, with most of the pancreatic tissue devoted to the production and exocrine secretion of digestive enzymes and bicarbonate (∞ p. 584). In 1869, the German anatomist Paul Langerhans described small clusters of cells scattered throughout the body of the pancreas (Fig. 21-10 ■). These clusters, now known as the **islets of Langerhans,** contain four distinct cell types, each associated with secretion of a different peptide hormone. Nearly three-quarters of the islet cells are **β (beta) cells,** which produce insulin. Another 20% are **α (alpha) cells,** which secrete glucagon. Most of the remaining cells are somatostatin-secreting **D cells,** but a few cells produce a mystery peptide known as pancreatic polypeptide. These rare cells are called either **PP cells** or **F cells.**

Like all endocrine glands, the islets are closely associated with capillaries into which the hormones pass after they are released from the cell. Both sympathetic and parasympathetic neurons terminate on the islets, providing a means by which the nervous system can influence metabolism.

The Ratio of Insulin to Glucagon Is a Key to Metabolic Regulation

Insulin and glucagon act in antagonistic fashion to keep plasma glucose concentrations within an acceptable range. Both hormones are present in the blood most of the time, and it is the ratio of the two hormones that determines which hormone dominates.

In the fed state, when the body is absorbing nutrients, insulin dominates and the body undergoes net anabolism (Fig. 21-11a ■). Ingested glucose is used for energy production, and excess glucose is stored as glycogen through glycogenesis or as fat. Proteins go primarily to protein synthesis. In the fasted state, the goal of metabolic regulation is to prevent low plasma glucose concentrations (**hypoglycemia**). When glucagon predominates (Fig. 21-11b ■), the liver uses glycogen and nonglucose intermediates to synthesize glucose for release into the blood.

Figure 21-12 ■ is a graph showing plasma glucose, glucagon, and insulin concentrations over the course of a day. In a normal person, fasting plasma glucose is maintained around 90 mg/dL plasma. After absorption of nutrients from a meal, plasma glucose rises. The increase in glucose level is a stimulus for insulin secretion. The main action of insulin is to enhance the movement of glucose from the blood into the cells. In response to insulin stimulation, plasma glucose concentrations fall back toward the fasting level shortly after each meal. During an overnight fast, plasma glucose concentrations fall to a low but steady value, and insulin secretion also decreases. Glucagon secretion remains rel-

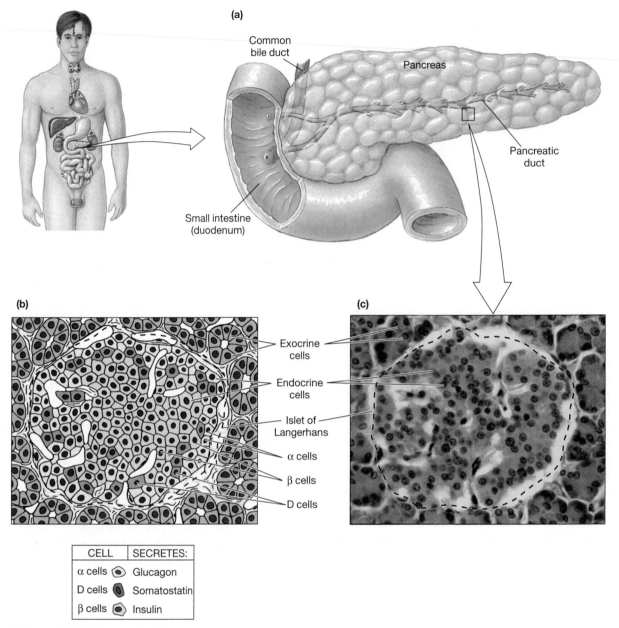

(a)

Common
bile duct

Pancreas

Pancreatic
duct

Small intestine
(duodenum)

(b)

(c)

Exocrine
cells

Endocrine
cells

Islet of
Langerhans

α cells

β cells

D cells

CELL		SECRETES:
α cells	◉	Glucagon
D cells	◕	Somatostatin
β cells	◉	Insulin

■ Figure 21-10 **The endocrine pancreas**

atively steady during the 24-hour period, supporting the theory that it is the ratio of insulin:glucagon that determines the direction of metabolism.

Insulin Is the Hormone of the Fed State

Insulin is a typical peptide hormone, produced as an inactive prohormone and activated prior to exocytosis from the β cells (∞ p. 179). Like other peptide hormones, it combines with a membrane receptor on its target cells. The insulin receptor of target tissues has tyrosine kinase activity that initiates intracellular cascades (∞ p. 153). The actual intracellular signals that link insulin receptors to the cellular responses are still not well understood, despite the tremendous amount of money being spent on insulin research.

Insulin release may be triggered by multiple stimuli: increased plasma glucose or amino acid concentrations, the intestinal hormone known as glucose-dependent insulinotropic peptide, or GIP (∞ p. 598), and increased parasympathetic input. Insulin secretion is inhibited by the sympathetic division of the nervous system.

When a meal containing protein or carbohydrate is eaten, the first signal that reaches the β cell to initiate insulin release is GIP. This hormone is released from cells in the small intestine in response to the presence of glucose or amino acids in the lumen. It is carried in the hepatic portal circulation to the β cells, where it releases

TABLE 21-3 Glucagon

Cell of origin	α cells of pancreas
Chemical nature	29–amino acid peptide
Biosynthesis	Typical peptide
Transport in the circulation	Dissolved in plasma
Half-life	3–4 minutes
Factors affecting release	Plasma [glucose] < 200 mg/dL; maximum secretion below 50 mg/dL; ↑ blood amino acids
Target cells or tissues	Liver primarily
Target receptor	Membrane receptor
Whole body or tissue action	↑ Plasma [glucose] by glycogenolysis and gluconeogenesis; ↑ lipolysis leads to ketogenesis in liver
Action at molecular level (including second messenger)	Acts via cAMP; alters existing enzymes and stimulates synthesis of new enzymes
Feedback regulation	↑ Plasma [glucose] shuts off glucagon secretion
Other information	Glucagon is structurally similar to the digestive hormones secretin, VIP, GIP, and enteroglucagon

✓ Why must insulin be administered as a shot and not as an oral pill?

✓ Patients who are admitted to the hospital with acute diabetic ketoacidosis and dehydration are given insulin and fluids that contain K⁺ and other ions. The patient's acidosis is usually accompanied by hyperkalemia, so why is K⁺ included in the rehydration fluids? (Hint: Dehydrated patients may have a high *concentration* of K⁺, but a low total amount. If fluid is given without K⁺, what happens to the concentration of K⁺ as the volume returns to normal?)

acid production and mobilization to the liver. At the liver, glucagon orchestrates disposition of the fatty acids, converting them into ketones that peripheral tissues and the brain can burn for fuel.

NEURALLY MEDIATED ASPECTS OF METABOLISM

The nervous system is closely integrated with metabolism in multiple ways. Signals from the brain are translated into autonomic output to various organs such as the endocrine pancreas. In addition, the brain regulates

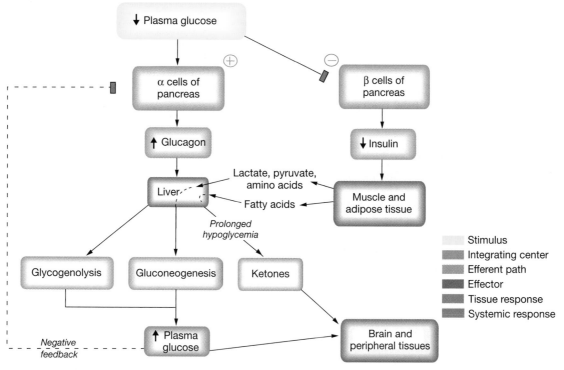

■ Figure 21-17 **Endocrine response to hypoglycemia**

the secretion of hypothalamic neurohormones that have widespread effects. The nervous system provides the link between external stimuli such as environmental temperature and internal stimuli such as emotions. It also adds another level of complexity to metabolic regulation.

Autonomic neurons are closely involved with both digestive function and metabolism. During the absorptive period following a meal, parasympathetic activity is increased, stimulating insulin secretion. In times of stress, sympathetic input to the endocrine pancreas is emphasized and reinforced by catecholamine release from the adrenal medulla (Table 21-4). Epinephrine and norepinephrine inhibit insulin secretion and switch metabolism to gluconeogenesis to provide extra fuel for the nervous system and skeletal muscles. Adrenergic mechanisms are also involved in the metabolic aspects of temperature regulation.

Regulation of Food Intake

One of the most interesting areas in which the central nervous system interacts with metabolism is the control of food intake. Digestion and metabolism are not selective processes: If food is eaten, it will be digested, absorbed, and assimilated, even if it is in excess and simply gets stored as fat. So the act of eating is the main point at which the body exerts control of its energy input.

The hypothalamus contains two centers that regulate food intake: a **feeding center** that is tonically active and a **satiety center** that stops food intake by inhibiting the feeding center. If the feeding center is destroyed, animals will cease to eat. If the satiety center is destroyed, animals will overeat and become obese.

Efferent messages from the feeding and satiety centers cause changes in eating behavior and create sensations of hunger and fullness. But hunger and satiety are also affected by psychological, environmental, and social factors. As a result, we still do not fully understand what controls food intake. A number of theories have been proposed, and it is likely that many, if not all, of these factors will be proved to play a role.

The two theories for which we have the most supportive evidence are the glucostatic theory and the involvement of the intestinal hormone cholecystokinin (CCK). The **glucostatic theory** states that glucose utilization by the hypothalamic centers regulates food intake. When blood glucose concentrations decrease, the satiety center is suppressed and the feeding center is dominant. When glucose utilization is up, the satiety center inhibits the feeding center. This theory would explain the polyphagia (excessive eating) associated with untreated type I diabetes mellitus. Although most neurons in the brain are not affected by insulin, glucose utilization in the satiety center is insulin-sensitive. Therefore, in the absence of insulin, the satiety center, like most other cells of the body, is unable to use the high concentrations of plasma glucose. It perceives the absence of glucose as starvation and allows the feeding center to increase food intake.

The **CCK theory** is now well established. Cholecystokinin is a hormone released from the small intestine in response to fatty acids and amino acids in the small intestine (∞ p. 598). It is also found as a neuropeptide in neurons of both the brain and the gut. When CCK is released during a meal, it apparently acts as a satiety signal that decreases appetite through both central and peripheral pathways. Most CCK is released by endocrine cells or enteric neurons in the intestine; it travels through the blood to combine with specific sensory receptors that activate the satiety center (Fig. 21-18 ■). Gastric distension and motility activate additional sensory receptors that stimulate release of CCK directly from brain neurons. This CCK also acts on the satiety center. No matter what its origin, CCK inhibits food intake.

A number of other peripheral peptides have been implicated in the control of eating, including glucagon and a molecule called **enterostatin** that is co-secreted with colipase (∞ p. 593). One of the newest peptides

TABLE 21-4 Catecholamines (Epinephrine and Norepinephrine)

Origin	Adrenal medulla (90% epinephrine and 10% norepinephrine)
Chemical nature	Amines made from tyrosine
Biosynthesis	Typical peptide
Transport in the circulation	Some bound to sulfate
Half-life	2 minutes
Stimulus for release	Primarily fight-or-flight reaction through CNS and autonomic nervous system; hypoglycemia
Target cells or tissues	Mainly neurons, pancreatic endocrine cells, heart and blood vessels
Target receptor	α and β adrenergic membrane receptors
Whole body or tissue action	↑ Plasma [glucose]; activate fight-or-flight and stress reactions; ↑ glucagon and ↓ insulin secretion
Action at molecular level (including second messenger)	α receptors via ↑ intracellular Ca^{2+} levels; β receptors via cAMP
Onset and duration of action	Rapid and brief

linked to appetite is **leptin** [*leptos*, thin], a protein synthesized in adipocytes under control of the *obese (ob) gene*. Mice who lack the *ob* gene and leptin protein become obese, leading investigators to hope that a drug based on leptin could be used to treat human obesity. Studies show, however, that obese humans have *elevated* leptin concentrations in their blood. This finding suggests that human target tissues are not responsive to leptin, so a leptin-like drug would have no beneficial effect.

Appetite and eating are influenced by sensory input through the nervous system in addition to chemical signals. The simple acts of swallowing and chewing food help create a sensation of fullness; the sight, smell, and taste of food can either stimulate or suppress appetite. An interesting study was done recently to assess if chocolate craving can be attributed to psychological factors or physiological stimuli. Subjects were given either dark chocolate, white chocolate with none of the pharmacological agents of cocoa, cocoa capsules, or placebo capsules. The results showed that white chocolate was the best substitute for the real thing and that aroma plays a significant role in satisfying chocolate craving.

Psychological factors, such as eating when stressed, are known to play a significant role in regulating food intake. The eating disorder **anorexia nervosa** probably has both a psychological component and a physiological one. The concept of **appetite** is closely linked to the psychology of eating, and it may explain why dieters who crave smooth, cold ice cream cannot be satisfied by a crunchy carrot stick. A considerable amount of money is directed to research on eating behaviors.

Another theory on control of food intake is based on a setpoint for body weight, the **lipostat theory.** According to this theory, adipocytes secrete a protein called **adipsin.** Adipsin is known to affect triglyceride synthesis and to be deficient in genetically obese rodents, but it has not yet been directly linked to feeding behavior. This theory, if supported, would attribute the pathology for obesity to the adipocyte rather than to the central regulatory mechanisms.

Regardless of which theories are supported by experimental evidence, one of the strongest factors in determining body weight is heredity. According to some studies, genes are responsible for as much as 70% of the way your body handles the food you eat.

✓ Give two explanations for why a target cell might not be responsive to leptin.

LONG-TERM ENDOCRINE CONTROL OF METABOLISM

In a normal person, the hormones that influence metabolism are difficult to study because their effects are subtle and their interactions with one another are complex. As a result, much of what we know about these hormones comes from studying pathological conditions in which a hormone is either over- or undersecreted. From the symptoms of patients with endocrine pathologies, researchers formulate hypotheses about the functions that the hormones control. They then try to test their hypotheses in the laboratory. Yet, even today, there is still much we do not understand about how many hormones work on the cellular and molecular levels.

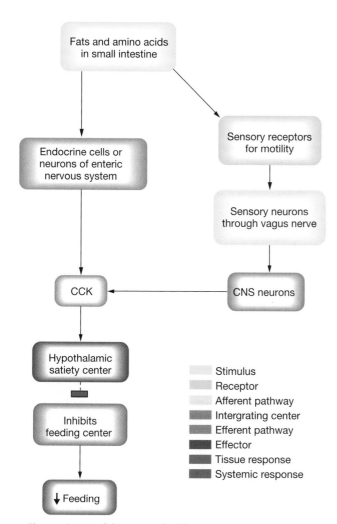

■ Figure 21-18 CCK control of food intake

continued from page 625

Dr. Bob's blood work reveals that his blood calcium level is 12.3 mg/dL plasma. A normal value is between 8.5 and 10.5 mg/dL. These results support the suspected diagnosis of hyperparathyroidism. "Do you take vitamin D or use a lot of antacids?" Dr. Spinks asks. "Those could raise your blood calcium." Dr. Bob denies the use of either substance. "Well, we need to do one more test before we can say conclusively that you have hyperparathyroidism," Dr. Spinks says.

Question 4: Speculate on what one test could definitively prove that Dr. Bob has hyperparathyroidism.

Insulin and glucagon, with their short half-lives, are responsible for the minute-to-minute control of plasma glucose concentration. Reinforcing them and playing a role in long-term regulation of metabolism are four other major groups of hormones: cortisol from the adrenal cortex, catecholamines from the adrenal medulla, thyroid hormones from the thyroid gland, and growth hormone from the anterior pituitary. All these hormones except the catecholamines are controlled by hypothalamic releasing hormones that act on the anterior pituitary. Growth hormone, secreted from the anterior pituitary, acts both directly on tissues and as a trophic hormone. Other trophic hormones from the anterior pituitary regulate secretion of cortisol from the adrenal cortex and thyroid hormones from the thyroid glands. The full control pathways for those hormones were introduced in Chapter 7 (∞ p. 185).

Additional hormones and cytokines influence metabolism in ways that we have yet to understand. For example, the gastrointestinal tract hormones cholecystokinin (CCK) and glucose-dependent insulinotropic peptide (GIP) are involved with aspects of metabolism independent of their role in digestion. But because these hormones are secreted from isolated cells and are present in the blood at the same time as insulin and glucagon, their effects are difficult to study. Some aspects of their activity are easier to study than others, such as their interaction with the hunger and satiety centers of the brain discussed in the previous section.

The Adrenal Glands Secrete Both Peptide and Steroid Hormones

The adrenal glands were introduced in connection with the autonomic division of the nervous system (Chapter 11, ∞ p. 308) and the hormone aldosterone (Chapter 19, ∞ p. 553). The adrenal glands are two embryologically distinct glands that merged during development. The central portion, the adrenal medulla, occupies a little over a quarter of the mass and is composed of modified sympathetic ganglia (∞ p. 311). The adrenal medulla secretes catecholamines, mostly epinephrine, into the blood in response to signals from the brain. Adrenal catecholamines act in concert with norepinephrine from sympathetic neurons to mediate rapid metabolic responses to stress. The adrenal corticosteroid hormones modulate stress in a slower fashion.

The adrenal cortex forms the outer three-quarters of the adrenal gland. This region secretes three major types of steroid hormones: aldosterone (sometimes called a mineralocorticoid because of its effect on the minerals sodium and potassium), glucocorticoids, and sex hormones. Histologically, the adrenal cortex is divided into three layers, or zones (Fig. 21-19 ■). The outer zone, or *zona glomerulosa*, secretes only aldosterone. The **glucocorticoids** are secreted primarily by the middle zone (*zona fasciculata*), and the sex hormones (**androgens** and **estrogens**) are produced primarily by the inner zone (*zona reticularis*).

The name *glucocorticoid* comes from the ability of these hormones to raise plasma glucose concentrations. **Cortisol** is the main glucocorticoid secreted by the adrenal cortex (Table 21-5). Older students may be familiar with its metabolite **cortisone,** a compound that was widely used as an anti-inflammatory drug before the development of nonsteroidal anti-inflammatory drugs such as ibuprofen. Cortisol is a typical steroid hormone synthesized in the adrenal gland from cholesterol. Its release is regulated by adrenocorticotropic hormone (ACTH) from the anterior pituitary (∞ Fig. 7-17 ■, p. 191). ACTH secretion in turn is controlled by the brain, which acts through hypothalamic corticotropin-releasing hormone (CRH). Cortisol is continuously secreted with a strong diurnal rhythm (Fig. 21-20 ■). Secretion normally peaks in the morning and diminishes during the night. Cortisol acts as a negative feedback signal, shutting down ACTH secretion from the anterior pituitary.

Cortisol is considered essential for life. Animals whose adrenal glands have been removed are unable to withstand any significant environmental stress. Overall, the net effects of cortisol are catabolic:

1. Cortisol promotes gluconeogenesis in the liver. The glucose produced is both released into the blood and stored as glycogen.
2. Cortisol causes the breakdown of skeletal muscle proteins to provide substrate for gluconeogenesis.
3. Cortisol enhances fat breakdown (lipolysis) so that fatty acids are available to peripheral tissues for energy use. Paradoxically, cortisol causes fat deposition in certain tissues, particularly in the trunk and cheeks. This deposition leads to the classic appearance of patients with excess cortisol concentrations: thin arms and legs, obesity in the trunk, and a "moon face," with plump checks.
4. Cortisol suppresses the immune system through multiple pathways.
5. Cortisol is catabolic on bone tissue, causing breakdown of the calcified bone matrix. People who take cortisol for extended periods therefore have a higher than normal incidence of broken bones.

The most important metabolic effect of cortisol is its protective effect against hypoglycemia. If plasma glucose falls below a certain concentration, the cells of the adrenal cortex begin to secrete cortisol even without stimulation from ACTH. In the absence of cortisol, glucagon alone is unable to respond adequately to a hypoglycemic challenge, one reason that cortisol is essential for life. The presence of cortisol is critical to ensure that glucagon and catecholamines are effective. Because cortisol is required for full activity of those hormones, it is said to have a permissive effect on them.

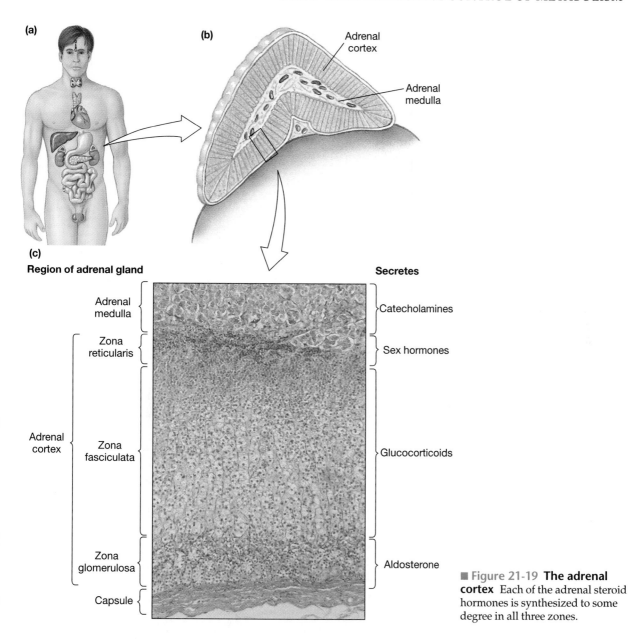

(a)

(b)

Adrenal
cortex

Adrenal
medulla

(c)

Region of adrenal gland		Secretes
Adrenal medulla		Catecholamines
	Zona reticularis	Sex hormones
Adrenal cortex	Zona fasciculata	Glucocorticoids
	Zona glomerulosa	Aldosterone
	Capsule	

■ Figure 21-19 **The adrenal cortex** Each of the adrenal steroid hormones is synthesized to some degree in all three zones.

TABLE 21-5	**Cortisol**
Origin	Adrenal cortex
Chemical nature	Steroid
Biosynthesis	From cholesterol; made on demand, not stored
Transport in the circulation	On corticosteroid-binding globulin (made in liver)
Half-life	60–90 minutes
Factors affecting release	Circadian rhythm of tonic secretion; stress enhances
Control axis	Corticotropin-releasing hormone (CRH) (hypothalamus) → adrenocorticotropic hormone (ACTH) (anterior pituitary) → cortisol (adrenal cortex)
Target cells or tissues	Most tissues except brain and heart
Target receptor	Intracellular
Whole body or tissue action	↑ Plasma [glucose]; ↓ immunocyte activity; permissive for glucagon and catecholamines
Action at cellular level	↑ Gluconeogenesis and glycogenolysis; ↑ protein catabolism; blocks cytokine production by immunocytes
Action at molecular level (including second messenger)	Initiates transcription, translation, and new protein synthesis
Feedback regulation	Negative feedback to anterior pituitary and hypothalamus

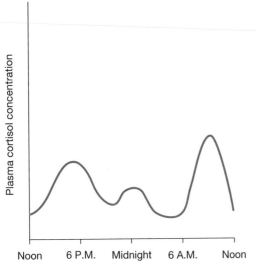

■ Figure 21-20 **Circadian rhythm of cortisol secretion**

✓ The illegal use of anabolic steroids by body-builders and athletes periodically receives much attention. Do these illegal steroids include cortisol? Explain.

Thyroid Hormones Are Important for Quality of Life

The thyroid gland is a butterfly-shaped gland that lies across the trachea at the base of the throat, just below the larynx (Fig. 21-21 ■). It is one of the larger endocrine glands, weighing between 15 and 20 g. The thyroid gland has two distinct endocrine cell types: the C ("clear") cells, which secrete a calcium-regulating hormone called calcitonin, and the cells of the thyroid follicle. The **follicles** (also called *acini*) are spherical structures in which a single layer of epithelial cells forms the wall. The hollow center of each follicle is filled with a glycoprotein material called **colloid.** Colloid is manufactured in the follicular cells and secreted into the lumen. Colloid consists of a large protein named **thyroglobulin** and various enzymes necessary for the synthesis of thyroid hormones.

The thyroid gland secretes amine hormones derived from the amino acid tyrosine (Fig. 21-22 ■). Thyroid hormones are unusual in that they contain iodine (Table 21-6). Currently, thyroid hormones are the only known use for iodine in the body, although a few other tissues are known to concentrate the mineral.

The synthesis of thyroid hormones is summarized in Figure 21-23 ■. Iodine from the diet is absorbed from the digestive tract in its ionic form, **iodide,** I⁻. Iodide is actively concentrated in the thyroid follicles and transported into the colloid. There, enzymes add the iodide to tyrosine residues of thyroglobulin. Addition of one iodide to each tyrosine creates **monoiodotyrosine**

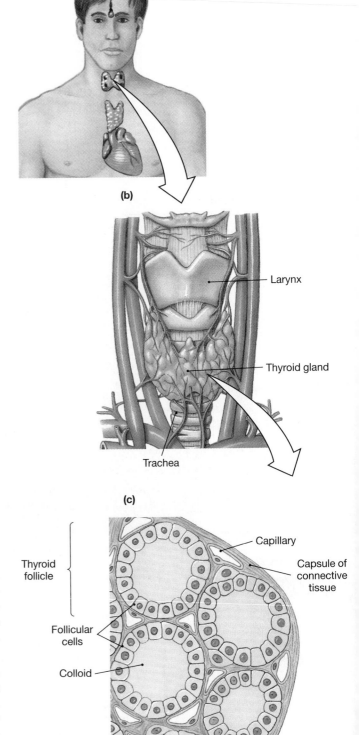

■ Figure 21-21 **The thyroid gland** (a) The thyroid gland is located in the neck. (b) It is a butterfly-shaped gland, located just below the pharynx. (c) The gland consists of C cells, which secrete calcitonin, and follicles. Each follicle is filled with glycoprotein colloid.

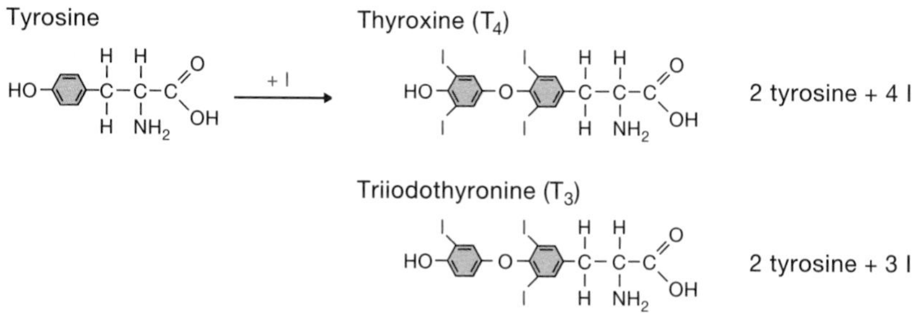

Tyrosine
Thyroxine (T$_4$)

$+$ I $\longrightarrow$

2 tyrosine + 4 I

Triiodothyronine (T$_3$)

2 tyrosine + 3 I   ■ **Figure 21-22 Thyroid hormones**

(MIT); addition of a second iodide forms **diiodotyrosine** (DIT). MIT and DIT then undergo coupling reactions to create either **triiodothyronine,** called **T$_3$** (MIT + DIT), or **tetraiodothyronine,** also known as **thyroxine,** or **T$_4$** (DIT + DIT). Note the name change from *tyrosine* to *thyronine.* At this point, the hormones are still attached to thyroglobulin, along with uncoupled MIT and DIT.

The secretion of thyroid hormones requires that thyroglobulin in the colloid be taken back into the follicular cells by endocytosis. Enzymes then free T$_3$ and T$_4$ from the protein. Both hormones are lipophilic, so they diffuse out of the follicular cells and into the plasma. The colloid of the follicles holds a 2–3 month supply of thyroid hormones at any one time.

Because T$_3$ and T$_4$ are lipophilic molecules, they have limited solubility in plasma. As a result, most thyroid hormone in the blood is found bound to plasma proteins such as **thyroxine-binding globulin (TBG).** Most plasma hormone is in the form of T$_4$. For years, it was thought that T$_4$ (thyroxine) was the active hormone, but we now know that T$_3$ is three to five times more active biologically and is the primary mediator of thyroid activity in target cells. About 85% of active T$_3$ is made in the target tissues by enzymatic removal of an iodine from T$_4$. This tissue activation of the hormone adds

another layer of control because regulation of the tissue enzyme **deiodinase** allows individual target tissues to control their exposure to active thyroid hormone.

Once in the blood, thyroid hormones behave like steroid hormones. They are transported in the plasma bound to carrier proteins, diffuse across cell membranes, and combine with nuclear receptors in their target cells. Receptor binding initiates transcription, translation, and synthesis of new proteins.

The thyroid hormones are not essential for life as cortisol is, but they do allow the body to maintain an optimum metabolic rate. In adults, oxygen consumption in most tissues is controlled by thyroid hormones, although the exact mechanism is still unclear.

In addition, thyroid hormones interact with other hormones to control protein, carbohydrate, and fat metabolism. In children, thyroid hormones are necessary for full expression of growth hormone and for normal growth and development. Deficient thyroid hormone secretion in infancy causes **cretinism,** a condition marked by short stature and decreased mental capacity. In adults, hyposecretion of thyroid hormone causes brittle nails, thinning hair, and thin skin due to decreased protein synthesis.

One interesting effect of thyroid hormones is their influence on the nervous system. Like many actions of hor-

TABLE 21-6 Thyroid Hormones

Cell of origin	Thyroid follicle cells
Chemical nature	Iodinated amine
Biosynthesis	From iodine and tyrosine; formed and stored on parent protein thyroglobulin in colloid of follicle
Transport in the circulation	Bound to thyroxine-binding globulin and albumins
Half-life and degradation	6–7 days for thyroxine (T$_4$); about 1 day for triiodothyronine (T$_3$)
Stimulus for release	Tonic release
Control axis	Thyrotropin-releasing hormone (TRH) (hypothalamus) → thyroid-stimulating hormone (TSH) (anterior pituitary) → T$_3$ and T$_4$ (thyroid) → T$_4$ deiodinates in tissues to form more T$_3$
Target cells or tissues	Most cells of the body
Target receptor	Nuclear receptor
Whole body or tissue action	↑ Oxygen consumption (thermogenesis); protein catabolism in adults but anabolism in children; normal development of nervous system
Action at cellular level	Increases activity of metabolic enzymes and Na$^+$/K$^+$-ATPase
Action at molecular level (including second messenger)	Production of new enzymes
Feedback regulation	T$_3$ has negative feedback effect on anterior pituitary and hypothalamus

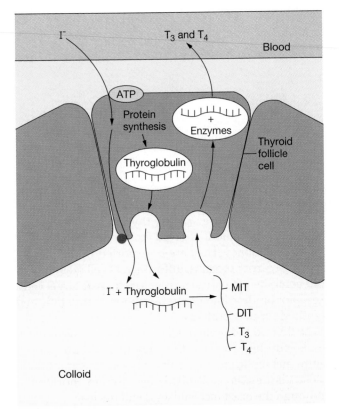

■ **Figure 21-23 Thyroid hormone synthesis** Thyroid hormone synthesis takes place in the central colloid of the thyroid follicle. Follicle cells synthesize thyroglobulin and actively take up iodide from the blood. Iodide is added to tyrosines on thyroglobulin, producing MIT and DIT. MIT and DIT couple to produce the thyroid hormones T_3 and T_4. The follicle cells then take up thyroglobulin by endocytosis. Intracellular enzymes separate the thyroid hormones from the protein so that they can then diffuse into the blood.

mones, these effects are subtle in normal people, but they become exaggerated in patients with either hyper- or hypothyroidism. For example, hyperthyroid patients may have psychological disturbances ranging from irritability and insomnia to psychosis. Patients with hypothyroidism often complain of fatigue, and they may exhibit slow speech and thought processes. The mechanism for the psychological disturbances is unclear, but morphological changes in the hippocampus and effects on beta-adrenergic receptors have been suggested.

Another effect of thyroid hormones on beta-adrenergic receptors has been well documented. A significant symptom of hyperthyroidism is rapid heartbeat due to up-regulation of the beta-adrenergic receptors on the myocardium. As a result of additional β_1 receptors, the heart is more sensitive to the effects of epinephrine or norepinephrine, resulting in increased heart rate and force of contraction (∞ p. 415).

Thyroid gland function is controlled by the anterior pituitary hormone **thyrotropin,** also known as **thyroid-stimulating hormone,** or **TSH** (Fig. 21-24 ■). TSH secretion in turn is regulated by **thyrotropin-releasing hor-**

mone, or **TRH,** from the hypothalamus. The thyroid hormones normally act as a negative feedback signal to prevent oversecretion.

The trophic action of TSH on the thyroid gland causes enlargement, or **hypertrophy,** of the follicular cells. In certain pathological conditions in which TSH levels are elevated, the thyroid gland can enlarge to weights of hundreds of grams, forming a huge mass that almost encircles the neck and extends downward behind the sternum. An enlarged thyroid gland is known as a **goiter** and is indicative of thyroid disease. But goiter can arise from a number of different causes, so an enlarged thyroid gland alone does not suggest what kind of thyroid disease the patient has.

For example, primary hypothyroidism (∞ p. 193) may be caused by a lack of iodine in the diet. Without iodide, the thyroid gland cannot make thyroid hormones. When the concentrations of T_3 and T_4 in the blood fall, negative feedback on the hypothalamus and anterior pituitary is removed. TSH secretion rises dramatically, and

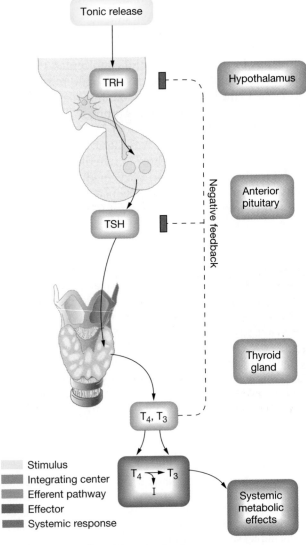

■ **Figure 21-24 Thyroid control axis**

TSH stimulation of the thyroid causes the gland to enlarge. But despite the hypertrophy, the gland cannot obtain iodine to make hormone, so the patient remains hypothyroid (Fig. 21-25a ■). These patients will exhibit the signs of hypothyroidism previously described.

An enlarged thyroid gland may also be indicative of hypersecretion. In Graves' disease, the body produces antibodies called **thyroid-stimulating immunoglobulins,** or TSI. These antibodies combine with the TSH receptors on the thyroid gland and activate them, stimulating thyroid hormone production. In Graves' disease, the usual negative feedback pathways fail to slow thyroid hormone secretion. As thyroid hormone concentrations increase, negative feedback shuts down TRH and TSH secretion (Fig. 21-25b ■). However, the TSH-like activity of the TSI antibodies continues to stimulate the thyroid gland unchecked. The result is an enlarged thyroid gland, hypersecretion of thyroid hormone, and symptoms of hormone excess.

✓ A patient who has had her thyroid gland removed owing to cancer is given pills containing only T_4 and not the more-active form of the hormone, T_3. Why does she not show signs of being hypothyroid?

✓ Why would excessive production of thyroid hormone, which uncouples (dissociates) mitochondrial ATP production and proton transport, cause a person to become intolerant of heat?

ENDOCRINE CONTROL OF GROWTH

Growth in human beings is a continuous process that begins before birth. Growth rates in children are not steady. The first two years of life and the adolescent years are marked by spurts of rapid growth and devel-

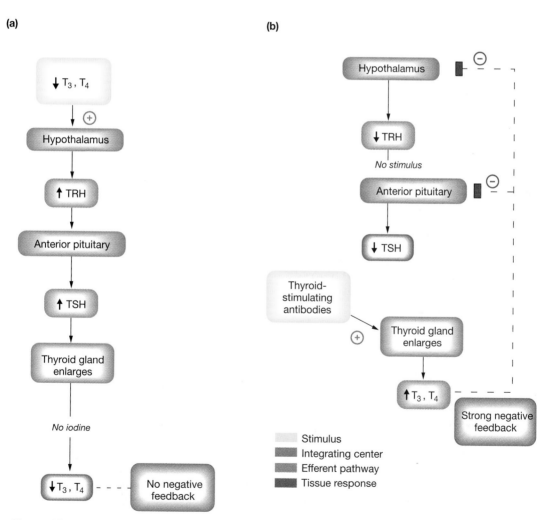

■ **Figure 21-25 Thyroid disease** (a) If thyroid hormone concentrations are low because of inadequate iodine in the diet, there is no negative feedback to the hypothalamus and anterior pituitary. TRH and TSH levels are therefore elevated. High levels of TSH cause the thyroid gland to enlarge. (b) In Graves' disease, thyroid-stimulating antibodies cause the thyroid gland to oversecrete thyroid hormone. Strong negative feedback shuts down TRH and TSH secretion, but the thyroid gland enlarges under the influence of the thyroid-stimulating antibodies.

opment. Normal growth is a complex process that depends on a number of factors, including:

1. *Growth hormone and other hormones.* Without adequate amounts of growth hormone, children simply will not grow. Thyroid hormones, insulin, and the sex hormones at puberty also play both direct and permissive roles. A deficiency in any one of these hormones will lead to abnormal growth and development.

2. *An adequate diet that includes protein, sufficient energy (caloric intake), vitamins, and minerals.* Many amino acids can be manufactured in the body from other precursors, but the essential amino acids (∞ p. 29) needed for protein synthesis must come from dietary sources. Among the minerals, calcium in particular is needed for proper bone formation.

3. *An absence of stress.* Cortisol from the adrenal cortex is released in times of stress and has significant catabolic, antigrowth effects. Children who are subjected to stressful environments may exhibit a condition known as *failure to thrive* that is marked by abnormally slow growth.

4. *Genetics.* Each human's potential adult size is genetically determined at conception.

Growth Hormone Is Essential for Normal Growth in Children

Growth hormone, also known as **somatotropin,** is secreted by the anterior pituitary. It is unique among the anterior pituitary hormones because it acts directly on some target tissues but is also a trophic hormone (Table 21-7). The trophic action of growth hormone stimulates secretion of **somatomedins** from the liver (Fig. 21-26 ■). Somatomedins, also called **insulin-like growth factors (IGFs)** because they are structurally related to insulin, act with growth hormone to stimulate both bone and soft tissue growth.

The stimuli for growth hormone release are complex and not well understood, but they include circulating nutrients and other hormones. These stimuli are integrated in the hypothalamus, which releases two trophic factors: growth hormone–releasing hormone (GHRH) and **growth hormone–inhibiting hormone,** better known as **somatostatin.** Hypothalamic somatostatin is identical to somatostatin secreted by the endocrine pancreas and intestinal mucosa (∞ p. 600). The hypothalamic trophic factors together regulate growth hormone secretion.

Growth hormone (GH) is released throughout life, although its biggest role is in children. Peak GH secretion occurs during the teenage years. On a daily basis, GH is released in response to pulses of GHRH from the hypothalamus. In adults, the largest pulse of GH release occurs in the first two hours of sleep. It is speculated that GHRH has sleep-inducing properties, but the role of GH itself in sleep cycles is unclear.

Growth hormone is a peptide hormone, and GH-secreting cells account for more than a third of the endocrine cells of the anterior pituitary. For years, scientists believed that GH was transported as free hormone in the plasma, but in the mid-1980s, researchers discovered that nearly half the GH in the blood is bound to a plasma **growth hormone binding protein.** Interestingly, this plasma protein has the same structure as growth hormone receptors on target cells. The binding protein protects plasma growth hormone from being filtered into the urine, extending its half-life from 7 to 18 minutes. In addition, the bound portion acts as a growth hormone reservoir in the blood. Researchers have hypothesized that genetic determination of the concentration of binding protein in the blood may play a key role in determining adult height.

Oversecretion of growth hormone in children leads to **giantism.** Once bone growth stops in late adolescence, growth hormone cannot further increase height, but it will act on cartilage and soft tissues. Adults with excessive concentrations of growth hormone develop a condi-

TABLE 21-7	**Growth Hormone**
Origin	Anterior pituitary
Chemical nature	191–amino acid peptide; several closely related forms
Biosynthesis	Typical peptide
Transport in the circulation	Half is dissolved in plasma, half is bound to a binding protein whose structure is identical to that of the GH receptor
Half-life	18 minutes
Factors affecting release	Circadian rhythm of tonic secretion; influenced by circulating nutrients and other hormones in a complex fashion
Target cells or tissues	Trophic on liver for somatomedin (insulin-like growth factors) production; also acts directly on many cells
Target receptor	Membrane receptor with tyrosine kinase activity
Whole body or tissue action	Bone and cartilage growth; soft tissue growth; insulin-like growth factors increase plasma [glucose]
Action at cellular level	Protein synthesis
Action at molecular level (including second messenger)	Receptor linked to unknown second messenger system

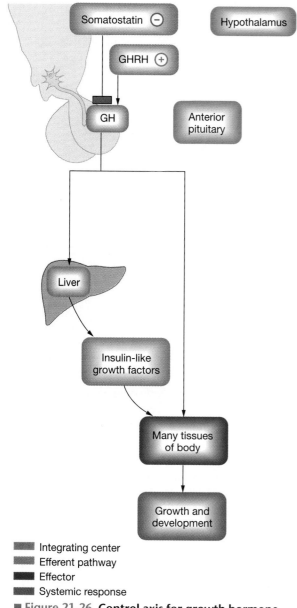

Integrating center
Efferent pathway
Effector
Systemic response

■ Figure 21-26 **Control axis for growth hormone**

children with normal growth hormone secretion be treated with growth hormone? What about children whose parents want them to be given the hormone so that they will be better athletes? Will giving growth hormone to the elderly help prevent some of the physical deterioration that comes with age? And, most important, what are the negative side effects of giving growth hormone to people with normal physiological concentrations already in their blood? These questions remain largely unanswered.

Metabolically, growth hormone is anabolic for proteins, directing energy and amino acids into protein synthesis, an essential part of tissue growth. It increases fat breakdown and hepatic glucose output, thereby raising plasma fatty acid and glucose concentrations. Growth hormone also acts with the insulin-like growth factors to stimulate bone and cartilage growth.

The endocrine control of growth can be divided into two general areas: soft tissue growth and bone growth. In children, bone growth is usually assessed by measuring height, and tissue growth by measuring weight. Multiple hormones have direct or permissive effects on growth. In addition, we are just beginning to understand how paracrine growth factors interact with classic hormones to influence the development and differentiation of tissues.

✓ Growth hormone elevates plasma glucose concentrations. Why, then, is plasma glucose normal in people with growth hormone deficiency?

Tissue Growth Is under the Control of Several Hormones

Soft tissue growth requires adequate amounts of growth hormone, thyroid hormone, and insulin. Growth hormone and insulin-like growth factors are required for tissue protein synthesis and cell division. Under the influence of these hormones, cells undergo both hypertrophy (increased cell size) and **hyperplasia** (increased cell number).

Thyroid hormones play a permissive role in growth and contribute directly to the development of the nervous system. Thyroid-stimulating hormone (TSH) is needed in addition to GHRH for adequate growth hormone secretion. At the target tissue level, thyroid hormone interacts synergistically with growth hormone for protein synthesis and development of the nervous system. Children with untreated hypothyroidism (cretinism) will not grow to a normal height. In addition, they usually exhibit some degree of mental retardation, which can be attributed directly to the effects of thyroid hormones on neuronal growth and development. Especially in the first few years after birth, T_3 and T_4 are needed for proper myelination of neurons and for formation of synapses. Cytological studies suggest that

tion known as **acromegaly** that is characterized by coarsening of facial features and growth of hands and feet.

Growth hormone deficiency in childhood leads to **dwarfism.** Dwarfism can result from a problem with either growth hormone synthesis or with the tissue hormone receptors. However, unlike insulin therapy for diabetes, bovine or porcine hormone cannot be used for replacement therapy: Only primate growth hormone is active in humans. Until genetically engineered human growth hormone became available, donated human pituitaries harvested at autopsy were the only source for growth hormone. Fortunately, true growth hormone deficiency is a relatively rare condition.

Once synthetic human growth hormone became available in the mid-1980s, the medical profession was faced with a number of dilemmas. Should genetically short

thyroid hormones regulate microtubule assembly, which is an essential part of neuronal growth.

Insulin supports growth by stimulating protein synthesis and by providing energy to cells in the form of glucose. Because insulin is permissive for growth hormone, children who are insulin-deficient will fail to grow normally even though they may have normal growth hormone concentrations.

Bone Growth Requires Adequate Amounts of Calcium in the Diet

Bone growth, like soft tissue development, requires the proper hormones and adequate amounts of protein and calcium. Bone has large amounts of calcified extracellular matrix formed when calcium phosphate crystals precipitate and attach to a collagenous lattice support. The most common form of calcium phosphate is **hydroxyapatite,** $Ca_{10}(PO_4)_6(OH)_2$.

Although the large amount of inorganic matrix makes some people think of bone as nonliving, in reality it is a dynamic tissue, constantly being formed and broken down, or **resorbed.** Spaces in the collagen and calcium matrix are occupied by living cells, well supplied with oxygen and nutrients by blood vessels that run through adjacent channels. Bone is generally formed into one of two patterns: dense, or **compact,** bone and spongy, or **trabecular,** bone, which has many open spaces between struts of calcified lattice (Fig. 21-27 ■).

Bone growth occurs when matrix is deposited faster than bone is resorbed. Special bone-forming cells called **osteoblasts** produce collagen, enzymes, and other proteins that make up the organic portion of bone matrix. Bone diameter increases when matrix deposition occurs on the outer surface of the bone. Linear growth of long bones occurs at special regions called the **epiphyseal plates,** found between the ends (epiphyses) and shaft (diaphysis) of the bone (Fig. 21-28 ■). On the epiphyseal side of the plate are continuously dividing columns of **chondrocytes,** cells that produce a layer of cartilage. As the layer thickens, the collagen calcifies and the older chondrocytes degenerate, leaving spaces that osteoblasts invade. The osteoblasts then lay down bone matrix on top of the original cartilage base. As new bone is added at the ends, the shaft continuously lengthens as long as the epiphyseal plate is active. Once osteoblasts have completed their work, they revert to a less active form known as **osteocytes.**

Growth of long bone is under the influence of both growth hormone and the insulin-like growth factors. In the absence of either growth hormone or insulin-like growth factors, normal bone growth will not occur. Growth of long bone is also influenced by sex steroid hormones. The growth spurt of adolescence is attributed to the increase in androgen production in boys. For girls, the picture is less clear-cut: In experiments, estrogens

■ Figure 21-27 **Compact and spongy bone**

both stimulate and inhibit linear growth. In all adolescents, the sex steroid hormones eventually cause the epiphyseal plate to become inactive and close so that long bone growth stops. Because the epiphyseal plates of various bones close in a constant, ordered sequence, X-rays that show which plates are open and which have closed can be used to calculate a child's "bone age."

Linear bone growth ceases in adults, but bones are dynamic tissues that are constantly being remodeled throughout life under the control of hormones that regulate calcium metabolism in the body.

Calcium Concentrations in the Blood Are Closely Regulated

Although 99% of all calcium in the body, nearly 2.5 pounds, is found in the bones, the small fraction of nonbone calcium is the most critical to physiological functioning of the body (Table 21-8). A tremendous concentration gradient exists between the extracellular Ca^{2+} (2.5 mmol/L) and the free cytoplasmic Ca^{2+} (0.001 mmol/L). If cell membrane channels for Ca^{2+} open, Ca^{2+} flows into the cell. The movement of Ca^{2+} out of cells requires active transport to push the ion against its concentration gradient.

Intracellular Ca^{2+} (0.9% of the body's Ca^{2+}) is important as an intracellular signal in second messenger pathways.

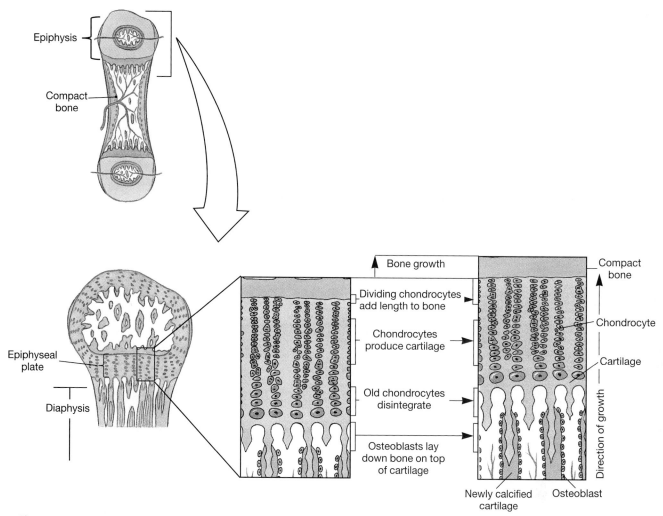

■ Figure 21-28 **Bone growth at the epiphyseal plate**

Changes in cytoplasmic Ca^{2+} concentrations initiate exocytosis of synaptic and secretory vesicles, contraction in muscle fibers, and altered activity of enzymes and transporters. Calcium is also part of the intercellular cement that holds cells together at tight junctions and a cofactor in several steps of the coagulation cascade (∞ p. 470). The most significant effect of altered plasma Ca^{2+} concentrations is related to the excitability of the nervous system. If plasma Ca^{2+} falls too low (**hypocalcemia**), neuronal permeability to Na^+ increases, neurons depolarize, and the nervous system becomes hyperexcitable. In its most extreme form, hypocalcemia causes sustained contraction (tetany) of the respiratory muscles, resulting in asphyxiation. **Hypercalcemia** has the opposite effect, causing depressed neuromuscular activity.

Because calcium is critical to so many physiological functions, the body's plasma Ca^{2+} concentration is very closely regulated. Calcium is absorbed by active trans-

TABLE 21-8 Functions of Calcium in the Body

Compartment	Function	Percentage of All Calcium in Body
Extracellular	Calcified matrix of bone	99%
Extracellular fluid	"Cement" for tight junctions	0.1%
	Myocardial and smooth muscle contraction	
	Release of neurotransmitters at synapses	
	Excitability of neurons due to effect on Na^+ permeability	
	Cofactor for coagulation cascade	
Intracellular	Signal in second messenger pathways	0.9%
	Muscle contraction	

port in the small intestine. In the plasma, about half is carried bound to proteins. The protein-bound Ca^{2+} is too large to be filtered at the kidneys, so it is not excreted. The remaining plasma Ca^{2+} is in the form of free ions. Free Ca^{2+} in the plasma is filtered at the glomerulus and reabsorbed along the length of the nephron. The hormonal regulation of calcium reabsorption takes place only in the distal nephron.

Bone is the largest reservoir of Ca^{2+} in the body, so it forms a pool that can be tapped if plasma Ca^{2+} concentrations fall. A small fraction of bone Ca^{2+} is readily exchangeable and in equilibrium with the Ca^{2+} of the interstitial fluid. **Osteoclasts,** large, mobile, multinucleate cells derived from monocytes, are responsible for bone resorption. Osteoclasts attach around their periphery to a section of calcified matrix, much like a suction cup (Fig. 21-29 ■). The central region of the cell next to the bone secretes acid (with the aid of a H^+-ATPase) and protease enzymes that work at low pH. The combination of acid and enzymes dissolves both the calcified matrix and its collagen support, freeing Ca^{2+}, which can enter the blood. The bulk of bone is more slowly exchanged with the extracellular fluid.

Endocrine control of calcium balance   Plasma concentrations of calcium are regulated by three hormones influencing the movement of Ca^{2+} between bone, kidney, and intestine: parathyroid hormone, vitamin D_3, and calcitonin (Fig. 21-30 ■). Changes in plasma Ca^{2+} concentrations are monitored by the four **parathyroid glands,** which are found on the dorsal surface of the thyroid gland (Fig. 21-31 ■). Their significance was first recognized in the 1890s by physiologists studying the role of the thyroid gland. The scientists noted that dogs and cats died a few days after total removal of the thyroid, but rabbits died only if the little parathyroid "glandules" alongside the thyroid were removed. If the parathyroid glands were left behind when the thyroid was surgically removed from dogs and cats, the animals lived. The scientists concluded that the parathyroids contained a substance that was essential for life, whereas the thyroid gland did not.

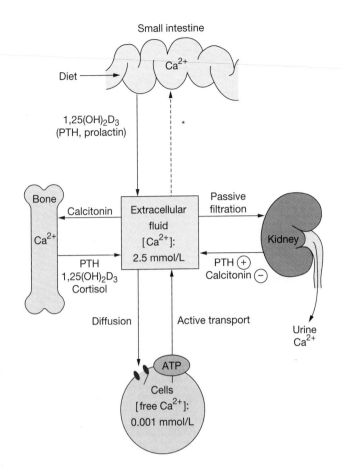

*Some calcium is secreted into the small intestine

■ **Figure 21-30 Calcium balance in the body**

We now know that the parathyroid glands secrete **parathyroid hormone** (parathormone, or PTH), a peptide whose main effect is to increase plasma Ca^{2+} concentrations (Table 21-9). The stimulus for PTH release is a decrease in plasma Ca^{2+}. Parathyroid hormone raises calcium concentrations in three ways:

1. *PTH mobilizes calcium from bone by enhancing the movement of calcium from matrix to the blood.* Increased bone resorption by osteoclasts takes about 12 hours to become measurable. Curiously, although osteoclasts are responsible for dissolving the calcified matrix and would be logical targets for PTH, they do not have PTH receptors. Researchers speculate that osteoclasts are affected by some PTH-mediated paracrine, perhaps secreted by osteocytes or osteoblasts.

2. *PTH enhances renal reabsorption of calcium.* This takes place in the distal nephron and apparently is accomplished by enhancing movement through calcium channels and transporters, although the exact mechanism is still unclear. Simultaneously, PTH enhances

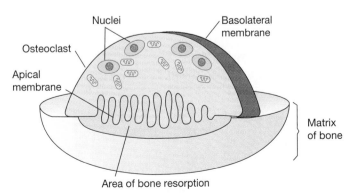

■ **Figure 21-29 Osteoclasts and bone resorption**
Osteoclasts secrete acid and enzymes at their apical membrane, dissolving the adjacent bone.

dependent diabetes mellitus, the body attacks its own β cells with antibodies and destroys them. In **non-insulin-dependent diabetes mellitus,** insulin levels in the blood are normal or even elevated, but the target tissues fail to respond normally to the hormone. (p. 628)

31. Insulin-dependent diabetes is marked by **hyperglycemia,** catabolism of muscle and adipose tissue, glucosuria, polyuria, and metabolic acidosis due to high ketone concentrations. In non-insulin-dependent diabetes, the symp-

toms are not as severe, but abnormal glucose and fat metabolism leads to complications that include atherosclerosis, neurological changes, and problems with the eyes and kidneys. (p. 630)

32. Glucagon stimulates glycogenolysis and gluconeogenesis. Glucagon secretion is also stimulated by an increase in plasma amino acid concentration. It protects the body from hypoglycemia after ingestion of a pure protein meal. (p. 632)

Neurally Mediated Aspects of Metabolism

33. Eating is the main point at which the body exerts control of its energy input. The hypothalamus contains a tonically active **feeding center** and a **satiety center** that inhibits the feeding center. (p. 634)

34. The **glucostatic theory** states that glucose utilization by

the hypothalamic centers regulates food intake. The **CCK theory** states that when CCK is released during a meal, it acts as a satiety signal. A third theory on control of food intake is the **lipostat theory,** based on a setpoint for body weight. (p. 634)

Long-Term Endocrine Control of Metabolism

35. Four other hormones play a role in long-term regulation of metabolism: **cortisol,** catecholamines, thyroid hormones, and growth hormone. (p. 636)

36. The adrenal cortex secretes three types of steroid hormone: aldosterone, **glucocorticoids** such as cortisol, and sex hormones. (p. 636)

37. Cortisol release is regulated by adrenocorticotrophic hormone (ACTH) and corticotropin-releasing hormone (CRH). Cortisol is secreted with a strong diurnal rhythm. (p. 636)

38. Cortisol is considered to be catabolic. It promotes gluconeogenesis in the liver, causes breakdown of skeletal muscle proteins and adipose tissue, suppresses the immune system, and causes breakdown of calcified bone matrix. (p. 636)

39. The thyroid gland secretes calcitonin and thyroid hormones. The thyroid **follicle** consists of a single layer of

epithelial cells with a hollow center filled with **colloid.** Colloid consists of **thyroglobulin** and enzymes. (p. 638)

40. Thyroid hormones are derived from the amino acid tyrosine and contain iodine. **Triiodothyronine** (T_3) and **tetraiodothyronine** (**thyroxine,** or T_4) are the hormones found in the blood. T_3 is 3–5 times more active than T_4. Most T_3 is made in the target tissues by enzymatic removal of an iodine from T_4. (p. 639)

41. Thyroid hormones are not essential for life, but they do allow the body to maintain an optimum metabolic rate. Thyroid hormones also interact with other hormones to control protein, carbohydrate, and fat metabolism. (p. 639)

42. Thyroid hormone secretion is controlled by **thyrotropin** (**thyroid-stimulating hormone,** or TSH) and **thyrotropin-releasing hormone,** or **TRH.** (p. 640)

Endocrine Control of Growth

43. Normal growth depends on adequate amounts of growth hormone, thyroid hormones, insulin, and the sex hormones at puberty. Growth also requires an adequate diet and absence of stress. (p. 642)

44. Growth hormone (**somatotropin**) is secreted by the anterior pituitary. Growth hormone stimulates secretion of **somatomedins** from the liver. These hormones act with growth hormone to stimulate bone and soft tissue growth. (p. 642)

45. Secretion of growth hormone is controlled by **growth hormone–releasing hormone** (GHRH) and **growth hormone–inhibiting hormone (somatostatin).** (p. 642)

46. Bone has large amounts of calcified extracellular matrix formed when calcium phosphate crystals precipitate and attach to a collagenous lattice support. Bone is a dynamic tissue with living cells that occupy spaces within the calcified matrix. (p. 644)

47. **Osteoblasts** produce collagen, enzymes, and other proteins that make up the organic portion of bone matrix. Linear growth of long bones occurs at the **epiphyseal plates,** where **chondrocytes** produce a layer of cartilage. Osteoblasts lay down bone matrix on top of the original cartilage base. Osteoblasts then revert to a less active form known as **osteocytes.** (p. 644)

48. Nonbone calcium is critical to physiological functioning of the body. It acts as intracellular signals for second messenger pathways, exocytosis, and muscle contraction. (p. 644)

49. Calcium levels are closely regulated by various hormones. If plasma Ca^{2+} falls, Ca^{2+} is moved out of bone. Changes in plasma Ca^{2+} concentrations are monitored by the **parathyroid glands,** which secrete **parathyroid hormone** (parathormone, or PTH). (p. 646)

50. Parathyroid hormone raises calcium concentrations by

mobilizing calcium from bone, by enhancing renal absorption of calcium, and by increasing intestinal absorption of calcium through its effect on **1,25-dihydroxycholecalciferol** (1,25(OH)$_2$D$_3$), or **calcitrol.** (p. 646)

51. **Calcitonin** from the thyroid gland plays only a minor role in daily calcium balance in humans. Calcitonin is released when plasma Ca^{2+} goes up. In animals, calcitonin decreases bone resorption and increases renal calcium excretion. (p. 648)

Questions

LEVEL ONE   Reviewing Facts and Terms

1. Define metabolic, anabolic, and catabolic pathways.
2. List and briefly explain the three forms of biological work.
3. Define a kilocalorie. What is measured with these units? What is direct calorimetry?
4. What is the respiratory quotient (RQ)? What is a typical RQ value for an American diet?
5. Define basal metabolic rate (BMR). Under what conditions is it measured? Why does the average BMR differ in adult males and females? List at least four factors other than gender that may affect BMR in humans.
6. What are the three general fates of biomolecules in the body?
7. What are the main differences between metabolism in the absorptive and postabsorptive states?
8. What is meant by the term *nutrient pool?* What are the three primary nutrient pools of the body?
9. The primary goal during the fasted state is to maintain adequate levels of glucose for which organ? Explain why this organ cannot metabolize proteins and fats.
10. In what forms is excess energy in the body stored?
11. What are the three possible fates for ingested proteins? For ingested fats?
12. Name the two hormones that share primary responsibility for the regulation of glucose metabolism. Name as many other hormones as you can that also influence glucose metabolism. What cell or tissue secretes each hormone?

Explain what effect each hormone has on blood glucose concentrations.

13. What non-carbohydrate molecules can be made into glucose? What are the pathways called through which these molecules are converted to glucose?
14. Under what circumstances are ketone bodies formed? From what biomolecule are ketone bodies formed? How are they used by the body, and why is their formation potentially dangerous?
15. Name two stimuli for insulin secretion and one stimulus that inhibits insulin secretion.
16. Distinguish between the different types of diabetes mellitus, particularly in respect to their basic symptoms.
17. What factors trigger the release of glucagon? What organ is the primary target of glucagon? What effect(s) do(es) glucagon produce?
18. List the four factors that are important in determining that people achieve their full growth. Include five specific hormones known to exert an effect on growth.
19. Name the thyroid hormones. Which one has the highest activity? How and where is most of it produced?
20. Define the following terms and explain their physiological significance:

lipoprotein lipase	osteoporosis
hydroxyapatite	enterostatin
trabecular bone	leptin
epiphyseal plates	adipsin

LEVEL TWO   Reviewing Concepts

21. **Concept map:** Draw a concept map that compares the fed state and the fasted state. For each state, compare metabolism in skeletal muscles, the brain, adipose tissue, and the liver. Indicate which hormones are active in each stage and at what points they exert their influence.
22. Examine the graphs of insulin and glucagon secretion in Figure 21-12 ■. Why have some researchers concluded it is the ratio of these two hormones that determines whether glucose will be stored or removed from storage?
23. Define, compare, and contrast or relate the terms within each set below:
    a. glucose, glycogenolysis, glycogenesis, gluconeogenesis, glucagon, glycolysis
    b. shivering, nonshivering, and diet-induced thermogenesis
    c. lipoproteins, chylomicrons, cholesterol, HDL, LDL, VLDL
    d. cortisol, cortisone, glucocorticoids, ACTH, CRH
    e. thyroid, C cells, follicles, colloid
    f. thyroglobulin, tyrosine, iodide, TBG, deiodinase, TSH, TRH
    g. somatotropin, somatomedins, GHRH, somatostatin, growth hormone binding protein
    h. giantism, acromegaly, dwarfism
    i. hyperplasia, hypertrophy
    j. osteoblasts, osteoclasts, chondrocytes and osteocytes
    k. vitamin D, calcitrol, 1,25-dihydroxycholecalciferol, calcitonin, estrogen
    l. conductive heat loss, radiant heat loss, convective heat loss, evaporative heat loss

24. Explain the physiological processes that lead to the following symptoms in an insulin-dependent diabetic:
    a. hyperglycemia
    b. glucosuria
    c. excretion of copious quantities of urine (polyuria)
    d. ketosis
    e. dehydration
    f. severe thirst

25. Both insulin and glucagon are released following ingestion of a protein meal that raises plasma amino acid levels. Why is the secretion of both hormones necessary?

26. Explain three current theories of the control of food intake. Use the following terms in your explanation: hypothalamus, feeding and satiety centers, appetite, glucostat, CCK, lipostat.

27. Explain temperature regulation in homeotherms. Compare thermoregulation in hot environments to thermoregulation in cold environments.

LEVEL THREE   **Problem Solving**

28. Scott is a bodybuilder who is always seeking ways to enhance his muscle development. He consumes large amounts of amino acid supplements and claims that "you are what you eat." According to Scott, if you eat lots of amino acids, your body will make lots of muscle proteins; if you eat any fat, you will be fat. Scott believes that the amino acids he consumes are stored in his body until he needs them. Is Scott correct? Explain.

29. Billy feels that his 13-year-old son has a better chance of playing college basketball if he is taller. He calls several clinics in an attempt to get his son growth hormone pills. When he speaks to you, what information can you give him about growth hormone and its use?

 Problem Conclusion

In this running problem, you learned about a condition called hyperparathyroidism and its effect on the body. You have also learned how this condition can be treated.

Further check your understanding of this running problem by checking your answers against those in the summary table.

Question	Facts	Integration and Analysis
1 What role does calcium play in the normal function of muscles and neurons?	Calcium triggers neurotransmitter release (∞ p. 226) and is needed to uncover the myosin-binding sites on actin filaments of muscle (∞ p. 334).	Muscle weakness is the opposite effect of what you would predict from knowing the role of calcium in muscles and neurons. However, calcium also affects the sodium permeability of neurons, and it is this effect that leads to muscle weakness and the central nervous system effects.
2 What is the technical term for "elevated levels of calcium in the blood"?	Prefix for elevated levels: hyper- Suffix for "in the blood": -emia	Hypercalcemia is the technical term for elevated levels of calcium in the blood.
3 A significant portion of plasma calcium cannot be filtered into Bowman's capsule. Speculate on why some plasma calcium cannot filter.	Filtration at the glomerulus is a selective process that excludes blood cells and most plasma proteins (∞ p. 525).	A significant amount of calcium in the blood is bound to plasma proteins and therefore does not filter at the glomerulus.
4 What one test could definitively prove that Dr. Bob has hyperparathyroidism?	Hyperparathyroidism is a condition in which excessive amounts of PTH are secreted.	A test for the amount of PTH in the blood would confirm the diagnosis of hyperparathyroidism.
5 Why can't Dr. Bob take replacement PTH by mouth?	PTH is a peptide hormone.	Ingested peptides will be denatured by the acidic environment of the stomach and digested by proteolytic enzymes. Thus PTH taken orally will not be absorbed intact into the body and will not be effective.

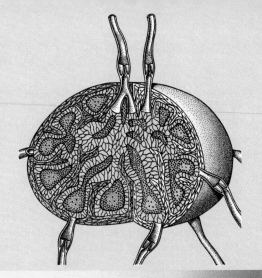

22

The Immune System

Since the 1980s, it seems that human civilization has become afflicted with plagues of infectious diseases. Some of them, such as tuberculosis, rabies, and malaria, are old enemies that we thought we had under control. Others, such as those caused by the human immunodeficiency virus (HIV), the Hanta virus, and the Ebola virus, seem to be cropping up out of nowhere. The conflict between humans and the invaders that cause disease is literally a fight for survival on both parts. The viruses and parasites need a host such as the human body to reproduce, but if they kill off all the hosts, they too will die out. We as hosts must fight off the invaders so that we can continue to survive as a species.

The ability of the body to protect itself from viruses, bacteria, and other disease-causing entities is known as **immunity,** from the Latin word *immunis,* meaning exempt. The human immune system consists of the lymphoid tissues of the body, the immune cells, and chemicals, both intracellular and secreted, that coordinate and execute the immune functions. A key feature of the immune system is plasticity, the ability to change in response to the body's needs.

The immune system serves three major functions:

1. *It protects the body from disease-causing invaders* known as **pathogens.** Microscopic invaders, or **microbes,** include bacteria, viruses, fungi, and one-celled protozoans. Larger pathogens include multicellular parasites

Focus on   The Spleen

■ Figure 22-3

The spleen is the largest lymphoid organ in the body, located in the upper left quadrant of the abdomen close to the stomach.

The outer surface of the spleen is a connective tissue capsule that extends into the interior to create an open framework that supports the blood vessels and lymphoid tissue. Regions of *white pulp* resemble the interior of lymph nodes and are composed mainly of lymphocytes. Darker regions of *red pulp* are closely associated with extensive blood vessels and open venous sinuses.

The red pulp contains many macrophages that act as a filter by trapping and destroying foreign material circulating in the blood. In addition, the macrophages ingest old, damaged, and abnormal red blood cells, breaking down their hemoglobin molecules into amino acids, iron, and bilirubin that is transported to the liver for excretion (∞ Fig. 16-8 ■, p. 464).

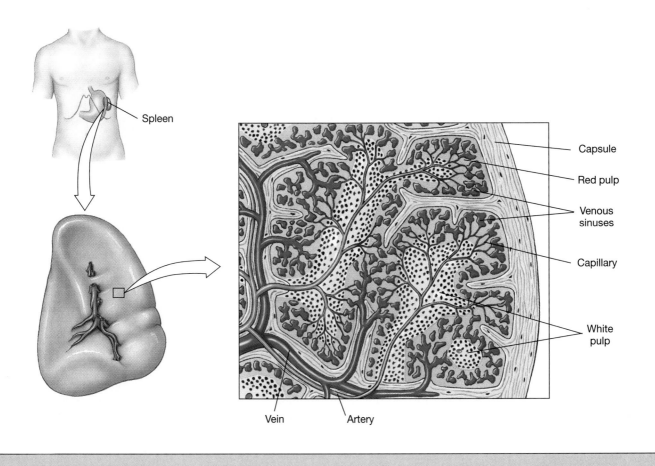

monocytes (which are macrophage precursors), and eosinophils. **Cytotoxic cells** kill the cells they are attacking; this group includes eosinophils and some types of lymphocytes. **Antigen-presenting cells** display fragments of foreign proteins on their cell surface. This group includes lymphocytes, dendritic cells, and macrophages, but not monocytes, the immature form of macrophages.

The terminology associated with macrophages has changed along with the history of histology and immunology. For many years, tissue macrophages were known as the **reticuloendothelial system** and were not

	Basophils and Mast Cells	Neutrophils	Eosinophils	Monocytes and Macrophages	Lymphocytes and Plasma Cells	Dendritic Cells
	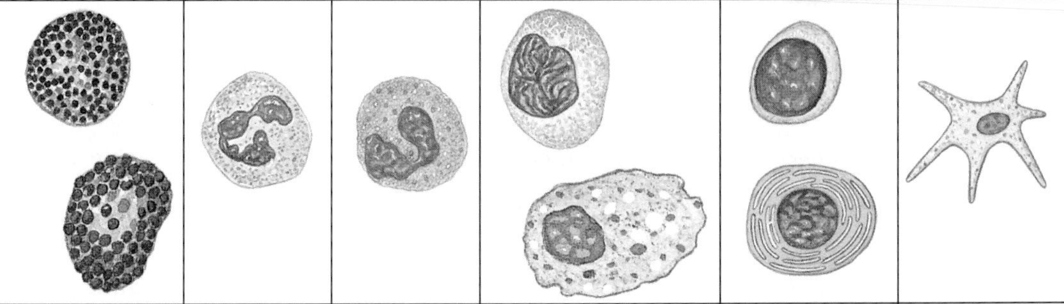					
% in blood	Rare	50–70%	1–3%	1–6%	20–35%	NA
Subtypes and nicknames		Called "polys" or "segs." Immature forms called "bands" or "stabs."		Called the mononuclear phagocyte system.	B lymphocytes, T lymphocytes, memory cells	Also called Langerhans cells, veiled cells
Primary function(s)	Release chemicals that mediate inflammation and allergic responses	Ingest and destroy invaders	Destroy invaders, particularly parasites	Ingest and destroy invaders	Specific responses to invaders, including antibody production	Recognize pathogens and activate other immune cells
Classifications		*Phagocytes*				
		Granulocytes				
			Cytotoxic cells		*Cytotoxic cells (some types)*	
				(Macrophages only)	*Antigen-presenting cells*	

■ Figure 22-4 **Cells of the immune system**

associated with white blood cells. To confuse matters, the cells were named when they were first described in different tissues, before they were all identified as macrophages. Thus, histiocytes in skin, Kupffer cells in the liver, osteoclasts in bone, microglia in brain, and reticuloendothelial cells in spleen are all names for specialized macrophages. The new name for the reticuloendothelial system is the **mononuclear phagocyte system,** a term that refers both to macrophages in the tissues and to their parent monocytes circulating in the blood.

Eosinophils Eosinophils, easily recognized by the bright pink-staining granules in their cytoplasm, are associated with allergic reactions and parasitic diseases. They are known to attach to large parasites such as the blood fluke *Schistosoma* and release substances that damage or kill them. Because eosinophils kill pathogens, they are called cytotoxic cells.

In allergic reactions such as those to pollen, eosinophils congregate near the focus of antigen entry. Although eosinophils have been shown to ingest these particles *in vitro*, there is no evidence that they do so *in vivo*. The granules of eosinophils contain enzymes that generate toxic oxidative substances. When the granules are released, the oxidants attack the invader.

Normally, few eosinophils are found in the peripheral circulation; they account for only 1%–3% of all leukocytes. The life span of an eosinophil in the blood is estimated to be only 6–12 hours. Most functioning eosinophils are found in the digestive tract, lungs, and connective tissue of the skin.

Basophils Basophils are rare in the circulation but are easily recognized when they do appear in a stained blood smear by the large dark blue granules in their cytoplasm. They are very similar to the **mast cells** of tissues. The granules of mast cells and basophils contain histamine, **heparin** (an *anticoagulant* that inhibits blood clotting), and other chemicals involved in the immune response. Mast cells, like eosinophils, are found concentrated in the connective tissue of skin, lungs, and the gastrointestinal tract. In these locations, they are ideally situated to intercept pathogens that are inhaled or ingested or that enter through breaks in the epidermis.

Neutrophils   **Neutrophils** are the most abundant white blood cell (50%–70% of the total) and the most easily identified, with a segmented nucleus of three to five lobes connected by thin strands of nuclear material. Because of the segmented nucleus, neutrophils are also called **polymorphonuclear leukocytes** ("polys") or "segs." Neutrophils, like other blood cells, are formed in the bone marrow. Occasionally, an immature neutrophil makes its way into the circulation, where it can be identified by its horseshoe-shaped nucleus. These immature neutrophils go by the nicknames of "bands" and "stabs."

Neutrophils are phagocytic cells that typically ingest and kill five to twenty bacteria during their programmed life span of one to two days. Most neutrophils spend their lives in the blood, but they can leave the circulation if attracted to an extravascular site of damage or infection. In addition to ingesting bacteria and foreign particles, neutrophils release a variety of chemicals, including fever-causing pyrogens (∞ p. 617) and chemical mediators of the inflammatory response.

Monocytes and macrophages   Until the late 1930s, we did not realize that circulating **monocytes** are the precursor cells of tissue **macrophages.** Monocytes are not very common in the blood (1%–6% of all white blood cells), and it is estimated that they spend only eight hours there in transit from the bone marrow to their permanent positions in various tissues. Once out of the blood, monocytes enlarge and specialize into macrophages. Some tissue macrophages spend their lives patrolling the tissues, creeping along by amoeboid motion, while others find a location and remain fixed in place. In either case, macrophages are the primary scav-

engers for the tissues. They are larger and more effective than neutrophils, ingesting up to 100 bacteria during their life span. Macrophages also remove larger particles such as old red blood cells and dead neutrophils.

Macrophages play an important role in the development of acquired immune responses. When they ingest molecular or cellular antigens and digest them, fragments of the antigen are processed and inserted into the macrophage membrane as part of surface protein complexes (Fig. 22-5 ■). Immune cells that process antigen in this fashion are called **antigen-presenting cells** (APCs). Dendritic cells and some types of lymphocytes also act as antigen-presenting cells.

Lymphocytes and plasma cells   The **lymphocytes** and **plasma cells** are the key cells that mediate the acquired immune response of the body. Only 5% of all lymphocytes are found in the circulation, where they constitute 20%–35% of all white blood cells. Most lymphocytes are found in the lymphoid tissues such as lymph nodes, the spleen, bone marrow, and the submucosal glands of the gastrointestinal tract. These are all locations where lymphocytes are likely to encounter invaders. By one estimate, the adult body contains a trillion (10^{12}) lymphocytes at any one time. Although most of these cells look alike under the microscope, they can have major differences in their function and specificity. The nature of their immune response requires that each lymphocyte bind to a specific ligand; consequently, there are more than 1 million different types of lymphocytes.

Dendritic cells   **Dendritic cells** are antigen-presenting cells characterized by long, thin processes that resemble the dendrites of the neuron. Dendritic cells are found in the skin (where they are called Langerhans cells) and in various organs. When dendritic cells recognize and capture antigens, particularly viruses, they migrate to the lymphoid tissues. There they present antigen fragments

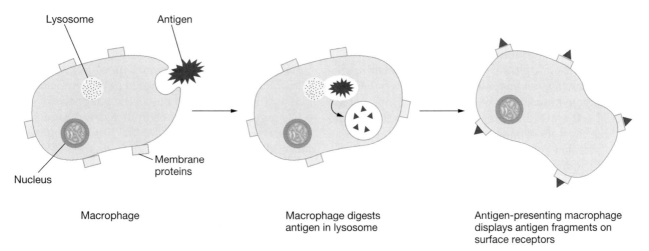

■ Figure 22-5 **Antigen-presenting immune cells**

to lymphocytes to activate them. Dendritic cells in the lymph system are called veiled cells.

INNATE IMMUNITY

The defenses of innate immunity are always present in the body. The body's first line of defense is to exclude pathogens by means of physical and chemical barriers. The second line of defense in the innate immune response consists of patrolling and stationary leukocytes that attack and destroy invaders. These cells respond in the same fashion to any material they identify as foreign, which is why the innate response is considered nonspecific.

Physical and Chemical Barriers Are the Body's First Line of Defense

Physical barriers of the body include the skin (∞ p. 66), the protective mucous linings of the gastrointestinal and genitourinary tracts, and the ciliated epithelium of the respiratory tract. The digestive and respiratory systems are most vulnerable to microbial invasion because these regions have extensive areas of thin epithelium in direct contact with the external environment. In women, the reproductive tract is also vulnerable, but to a lesser degree. The opening to the uterus is normally sealed by a mucus plug that keeps bacteria from ascending from the vagina into the uterine cavity. An infamous example of how this physical barrier can fail was the Dalkon shield, an intrauterine contraceptive device (IUD) with a multi-ply string that extended from the IUD through the mucus plug to the vagina. The string acted as a wick for bacteria, allowing them to bypass the mucus plug that would normally exclude them. Because of bacterial invasions, a number of women with IUDs suffered from pelvic infections and inflammatory reactions that blocked their Fallopian tubes and left them unable to conceive.

In the respiratory system, inhaled particulate matter is trapped in the mucous lining of the upper respiratory system, then transported upward on the mucus escalator to be expelled or swallowed (∞ p. 483). Swallowed pathogens may be disabled by the acidity of the stomach. In addition, respiratory tract secretions and tears contain **lysozyme,** an enzyme with antibacterial activity. Lysozyme attacks components of unencapsulated bacterial cell walls and breaks them down. It cannot digest the capsules of encapsulated bacteria, however.

Pathogens that enter the body past the physical barriers of skin and mucus are dealt with by various immune cells, the second line of defense in the innate immune response. When patrolling and stationary phagocytes in the tissues detect invaders, their response is twofold: Destroy or suppress the invader by ingesting it, and attract additional immune cells. The molecules that attract immune cells are called **chemotaxins.** They include bacterial toxins, products of tissue injury (fibrin

and collagen fragments), and cytokines released by other leukocytes. The latter molecules are also known as **chemokines.**

Phagocytes Recognize Foreign Material and Ingest It

The primary phagocytic cells of the immune system are the tissue macrophages and the neutrophils. If the invader is in the tissue, phagocytes circulating in the blood squeeze out of the capillaries through pores between the endothelial cells, a process known as **diapedesis** [*diapedesis,* leaping through]. Once the phagocytes reach the pathogen, they identify it by chemical cues, then ingest it. Phagocytosis, in which cells engulf and digest particulate matter, is a similar process in macrophages and neutrophils (∞ p. 126). It is a receptor-mediated event to ensure that only unwanted particles are ingested. In the simplest reactions, surface molecules on the pathogen act as ligands that bind directly to receptors on the phagocyte membrane (Fig. 22-6a ■). In a sequence similar to a zipper closing, the ligands and receptors combine sequentially, so that the phagocyte surrounds the unwanted foreign particle. The process is aided by actin filaments that push arms of the cell around the invader.

Phagocytes recognize many different types of foreign particles, both organic and inorganic, by mechanisms that are not well understood. Unencapsulated bacteria, cell fragments, carbon, and asbestos particles are among the materials that can be ingested by phagocytic cells. In the laboratory, scientists use the ingestion rate of tiny polystyrene beads as an indicator of phagocytic activity.

Not all foreign material can be recognized by phagocytes, however. For example, certain bacteria have evolved to hide their surface markers behind a polysaccharide capsule (Fig. 22-6b ■). If pathogens lack markers, they must be "tagged" so that the phagocytes will recognize them as something to be ingested. This tagging is done by coating the pathogen with protein. The tagging proteins are known collectively as **opsonins,** a term derived from the Greek word *opsonin,* meaning to buy provisions. In this situation, opsonins convert unrecognizable particles into "food" for phagocytes. Opsonins act as a bridge between pathogens and phagocyte by binding to receptors on the phagocyte.

The ingested particle ends up in a cytoplasmic vesicle called a **phagosome** (Fig. 22-7 ■). Phagosomes fuse with large intracellular granules that are specialized lysosomes. Enzymes within the granules produce **hypochlorous acid** (HOCl), the active ingredient in household bleach, and **peroxides.** These powerful oxidizing agents destroy ingested pathogens when the phagosomes and granules fuse.

If an area of infection attracts a large number of phagocytes, the result may be the formation of the material known as **pus.** This thick whitish to greenish substance

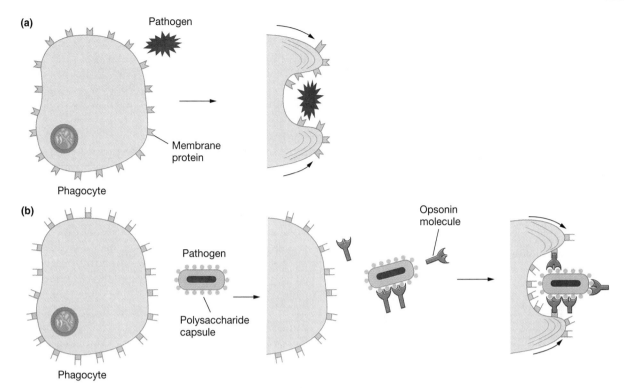

■ Figure 22-6 Phagocytosis (a) Many pathogens can bind directly to receptors on the phagocyte membrane. As the receptors bind sequentially, the phagocyte engulfs the pathogen with the aid of actin filaments in the cytoskeleton. (b) Some bacteria have polysaccharide capsules to which phagocytes cannot bind. These pathogens must first be coated with opsonins before the phagocyte receptors will bind to them.

is a collection of living and dead neutrophils and macrophages, along with tissue fluid, cell debris, and other remnants of the immune process.

Chemical Mediators Create the Inflammatory Response

Inflammation, indicated on the skin by a red, warm, swollen area, is a hallmark reaction of the innate immune pathway. The inflammatory response is created when activated tissue macrophages release cytokines. These chemicals attract other immune cells, increase capillary permeability, and cause fever. Other immune cells attracted to the site in turn release their own cytokines. Some representative examples of the chemicals involved in the immune response are listed in Table 22-2 and described below.

Chemokines and Disease Chemokines are specialized cytokines that attract monocytes, macrophages, neutrophils, and eosinophils, the cells responsible for the nonspecific inflammatory immune response. When chemokines activate these immune cells, the cells respond by ingesting invaders or by releasing various chemicals. However, scientists have discovered that an overenthusiastic immune response triggered by chemokines can cause or contribute to a variety of diseases. For example, neutrophils create the inflammatory response of rheumatoid arthritis, in which the normal connective tissue of the joints is damaged. Eosinophils have been linked to asthma, and monocytes to pneumonia and atherosclerosis. Investigators are now looking for ways to block the chemotaxic effects of various chemokines, hoping that by doing so they may develop new and more specific treatments for these diseases.

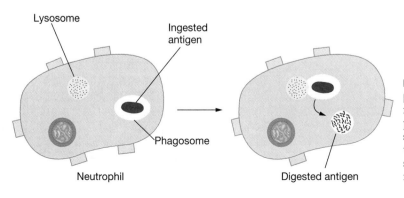

■ Figure 22-7 Digestion in phagocytes Ingested particles move into the phagocyte cytoplasm in vesicles known as phagosomes. When phagosomes fuse with specialized lysosomes, lysosomal enzymes digest the foreign material within the phagosome.

TABLE 22-2   Chemicals of the Immune Response

Functional classes

Acute phase proteins: Liver proteins released during the acute phase that act as opsonins; also enhance the inflammatory response

Chemotaxins: Molecules that attract phagocytes to a site of infection

Cytokines: Molecules released by one cell that affect the growth or activity of another cell

Opsonins: Proteins that coat pathogens so that phagocytes will recognize and ingest them: antibodies, acute phase proteins, and complement proteins

Pyrogens: Fever-producing substances

Specific chemicals and their functions

Antibodies (immunoglobulins, gamma globulins): Fight specific invaders

Bradykinin: Stimulates pain receptors; vasodilator

Complement: Plasma and cell membrane proteins that act as opsonins, cytolytic agents, and mediators of inflammation

Granzymes: Cytotoxic enzymes that trigger cells into committing suicide

Heparin: An anticoagulant

Histamine: Vasodilator and bronchoconstrictor released by mast cells

Hypochlorous acid (HOCl): Powerful oxidant found in lysosomes of phagocytic cells

Interferons: Proteins that inhibit viral reproduction and modulate the immune response

Interleukin-1: Mediates the inflammatory response; secreted primarily by activated macrophages

Kinins: Inactive plasma proteins that activate in a cascade to form bradykinin

Lysozyme: An extracellular enzyme that attacks bacteria

Major histocompatibility complex (MHC): A family of membrane protein complexes used for cell recognition

Membrane attack complex: A membrane pore protein made in the complement cascade; allows ions and water to enter bacteria and lyse them

Perforin: A membrane pore protein made by natural killer and cytolytic T cells and inserted into the membrane of target cells; large enough to allow granzymes to enter the cell

Peroxide: Powerful oxidant found in lysosomes of phagocytic cells

T-cell receptors: Membrane receptors on T lymphocytes that bind to MHC receptors

Acute phase proteins   In the time immediately following an injury or pathogen invasion, the body responds by increasing the concentration of various plasma proteins. Some of these proteins, produced mostly by the liver, are given the general name of **acute phase proteins.** They include molecules that act as opsonins, antiprotease molecules to help prevent tissue damage, and other proteins whose function is less clear. Normally, the levels of these proteins fall back to normal as the immune response proceeds, but in chronic inflammatory diseases such as

rheumatoid arthritis, the elevated levels of acute phase proteins may persist. One acute phase protein that is receiving much attention is **serum amyloid A.** Amyloid deposits have been found in tissues of people suffering from Alzheimer's disease and certain types of diabetes mellitus, so potential links between the deposits and those diseases are being investigated.

Histamine   Histamine is an amine derived from the amino acid histidine (∞ p. 147). Histamine is found primarily in the granules of mast cells and basophils, and it is the active molecule that helps initiate the inflammatory response when mast cells degranulate. Histamine opens pores in capillaries, allowing plasma proteins to escape into the interstitial space. This process causes local edema, or swelling. In addition, histamine dilates blood vessels, increasing blood flow to the area. The result of histamine release is a hot, red, swollen area around a wound or infection site. The purpose of histamine's actions is to bring more leukocytes to the injury site to kill bacteria and remove cellular debris.

Mast cell degranulation is triggered by different cytokines in the immune response. Because mast cells are concentrated under mucous membranes that line the airways and digestive tract, the inhalation or ingestion of certain antigens can trigger histamine release. The resultant edema in the nasal passages leads to one annoying symptom of seasonal pollen allergies, the stuffy nose. Fortunately, pharmacologists have developed a variety of drugs called *antihistamines.* These drugs antagonize histamine by blocking mast cell degranulation or by blocking the action of histamine at its receptor.

Interleukins   Interleukins are cytokines that were initially thought to mediate communication between the different types of leukocytes in the body. However, as scientists learn more about them, we have discovered that many different tissues in the body respond to interleukins.

Interleukin-1 (IL-1) is secreted by activated macrophages and other immune cells. Its main role is to mediate the inflammatory response, but it also has widespread systemic effects on immune function and metabolism. Interleukin-1 mediates the immune response by:

continued from page 655

HIV is a retrovirus that requires an enzyme called reverse transcriptase to reproduce. Two of the drugs in the new AIDS "cocktail," AZT and ddC, are antiretroviral drugs that work by inhibiting reverse transcriptase. The other drug in the cocktail is a protease inhibitor. Protease is an enzyme needed to cleave key proteins during one step of HIV replication.

Question 1: Why can't retroviruses reproduce without reverse transcriptase?

1. Altering blood vessel endothelium to allow easier passage of white cells and proteins during the inflammatory response.
2. Stimulating production of acute-phase proteins by the liver.
3. Inducing fever by acting on the hypothalamic thermostat (∞ p. 615). IL-1 is a known pyrogen of the body.
4. Acting as a stimulator of cytokine and endocrine secretion by a variety of other cells.

Kinins   **Kinins** are a group of inactive plasma proteins that participate in a cascade similar to the coagulation cascade (∞ p. 465). The end product of the kinin cascade is **bradykinin,** a molecule with the same vasodilator effects as histamine. Bradykinin also stimulates pain receptors, creating the tenderness associated with inflammation.

Complement   **Complement** is a collective term for a group of more than 25 plasma and cell membrane proteins. The complement pathway is a cascade of events similar to the blood coagulation cascade. Various intermediates of the complement cascade act as opsonins, chemical attractants for leukocytes, and agents to cause degranulation of mast cells. The terminal step in the complement cascade is the formation of **membrane attack complex,** lipid-soluble molecules that insert themselves into the cell membranes of pathogens and form pores (Fig. 22-8 ■). These pores allow Na^+ to enter the cell, and water follows osmotically. As a result, the cell swells and lyses.

ACQUIRED IMMUNITY

Acquired immune responses are those responses in which the body recognizes a specific foreign substance and selectively reacts to it. Acquired immunity is mediated primarily by lymphocytes, which have membrane receptors that allow them to react to only one type of invader. The process of acquired immunity overlaps with the process of innate immunity. Chemokines released by the inflammatory response attract lymphocytes to the site of an immune reaction, then the lymphocytes release additional cytokines that enhance the inflammatory response.

Lymphocytes Are the Primary Cells Involved in the Acquired Immune Response

If each pathogen that enters the body is to be fought by a specific type of lymphocyte, millions of different types of lymphocytes must be ready to combat millions of different foreign pathogens. On a microscopic level, lymphocytes look alike. At the molecular level, however, the cells can be distinguished by their membrane receptors, which make them specific for a particular ligand (Fig. 22-9 ■). Each unique type of lymphocyte, along with the other cells exactly like it, forms a **clone** [*klon,* a twig].

All lymphocytes go through the same general life cycle. At birth, each clone of lymphocytes is represented by only a few cells, called **naive lymphocytes** [*naif,* natural]. The small number of cells in each naive clone is not enough to fight off foreign invaders, so the first exposure to an antigen activates the appropriate clone and causes it to divide (Fig. 22-10 ■). This process, called **clonal expansion,** creates additional cells in the clone. The newly formed lymphocytes in the clone differentiate into **effector cells,** which create the immune response, or into **memory cells.** Effector cells are short-lived and die within a few days. However, memory cells have longer lives and continue to reproduce themselves. With second and subsequent exposures to the antigen, the memory cells activate, creating a more rapid and stronger secondary response to the antigen.

All lymphocytes secrete cytokines that act on immune cells, on non-immune cells of the body, and, sometimes, on the pathogens themselves. The primary lymphocyte

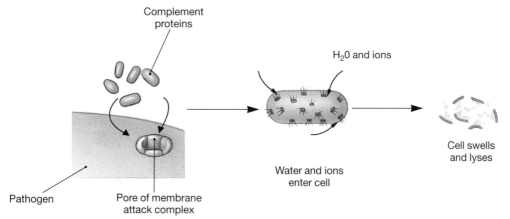

■ **Figure 22-8 Membrane attack complex** Activated complement proteins insert themselves into the membrane of pathogens, creating a pore through which water, Na^+, and other ions can enter the cell. The net entry of ions and water causes the cell to swell and lyse.

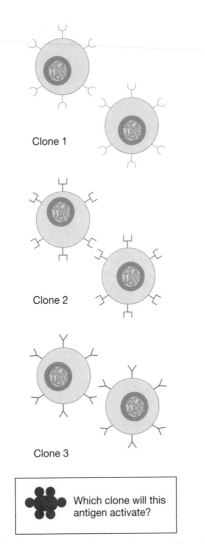

Clone 1

Clone 2

Clone 3

Which clone will this antigen activate?

■ **Figure 22-9 Lymphocyte clones** A group of lymphocytes that are specific to one antigen is known as a clone. This figure shows three pairs of lymphocytes; each pair is from a different clone.

cytokines are the interleukins, discussed earlier, and the interferons. **Interferons** were named for their ability to interfere with viral reproduction in the body. **Alpha-interferon** (interferon-α) and **beta-interferon** (interferon-β) target uninfected cells and activate pathways to prevent viral replication. **Gamma-interferon** (interferon-γ) activates macrophages and other immune cells in addition to stimulating uninfected cells.

There are three main types of lymphocytes: B lymphocytes, T lymphocytes, and natural killer cells. B lymphocytes secrete antibodies; T lymphocytes and natural killer cells attack and destroy infected cells.

B Lymphocytes Secrete Antibodies

B lymphocytes (also called B cells) develop in the bone marrow, a convenient way to remember their name. The primary function of B lymphocytes is **humoral immunity,** in which a special type of B lymphocyte secretes protein molecules called **antibodies,** or immunoglobulins. Antibodies bind to pathogens and help target them for destruction. The term *antibody* describes what these molecules do (work against the foreign body); the equivalent term *immunoglobulin* describes what these molecules are (globular proteins that participate in the humoral immune response).

Mature B lymphocytes insert antibody molecules into their cell membranes so that the antibodies become surface receptors marking the members of each clone (see Fig. 22-9 ■). When a clone of B lymphocytes activates in response to antigen exposure, some differentiate into plasma cells, which do not have antibody proteins bound in their membranes. Plasma cell synthesize additional antibody molecules at incredible rates, estimated to be as high as 2000 molecules of antibody per second!

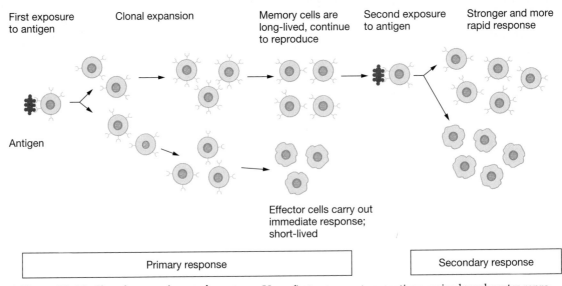

First exposure to antigen

Clonal expansion

Memory cells are long-lived, continue to reproduce

Second exposure to antigen

Stronger and more rapid response

Antigen

Effector cells carry out immediate response; short-lived

Primary response

Secondary response

■ **Figure 22-10 Clonal expansion and memory** Upon first exposure to an antigen, naive lymphocytes reproduce to expand the size of the clone. They then differentiate either into short-lived effector cells that carry out the primary immune response or into long-lived memory cells. When memory cells are reexposed to the appropriate antigen, the clone expands more rapidly to create additional effector and memory cells.

After each invader has been successfully repulsed, a few memory cells of the clone remain behind, so that the next exposure to the same antigen is greeted by a more rapid response. This phenomenon can be demonstrated by measuring the concentration of antibodies in the circulation. Figure 22-11 ■ shows the antibody response after the first and second exposure to an antigen. The primary response is slower and lower in magnitude because the body has not encountered the antigen previously. The secondary response is enhanced by lymphocytes that carry a molecular memory of the first exposure to the antigen.

The existence of the enhanced secondary response is used in immunizations. Patients are given a pathogen that has been altered so that it no longer harms the host cell but can be recognized as foreign by immune cells. The altered pathogen causes the creation of memory cells keyed to that particular pathogen. If the immunized person is later infected by the natural pathogen, a stronger and more rapid immune response is the result.

✓ Why does the action of histamines on capillary permeability result in swelling?

Antibodies Are Glycoproteins Secreted by Plasma Cells

Most antibodies are found in the plasma, where they make up about 20% of the plasma proteins in the healthy individual. Historically, antibodies were among the first aspects of the immune system to be discovered, and traditionally they were named by what they do—agglutinins, precipitins, hemolysins, and so on.

Today, the antibodies are divided into five general classes of immunoglobulins (Ig): IgG, IgA, IgE, IgM, and IgD (pronounced *eye-gee-[letter]*). **IgG** makes up 75% of the serum antibody in adults because these immunoglobulins are produced in secondary immune responses. Maternal IgGs cross the placental membrane and are the antibodies that give infants immunity in the first few months of life.

IgA antibodies are found in the external secretions of the body such as saliva, tears, intestinal and bronchial mucus, and breast milk. In these locations, they disable pathogens before they reach the internal environment. **IgE** is associated with allergic responses. When mast cell receptors combine with IgE and antigen, the mast cells degranulate and release chemical mediators such as histamine. **IgM** is associated with primary immune responses and with the antibodies that react to blood group antigens. **IgD** appears on the surface of B lymphocytes along with IgM, but its physiological role is unclear. When plasma proteins are separated by a biochemical technique called *electrophoresis,* the immunoglobulins move together in a group given the general name of **gamma globulins.**

Antibody structure   Antibodies are made of four linked chains of polypeptides, two light chains and two heavy chains (Fig. 22-12 ■). All the light chains in a given antibody are identical, as are all the heavy chains, but the chains vary widely between different antibodies. The four chains are attached into a Y shape with a hinge region that allows flexible positioning of the arms of the molecule. The two arms, or **Fab regions,** contain antigen-binding sites that confer the antibody's specificity. The stem of the Y-shaped antibody is known as the **Fc region** of the molecule.

Antibody functions   The primary function of antibodies is to bind antigens to B lymphocytes and initiate the production of additional antibodies. Because antibodies are found in the interstitial fluid and plasma, they are most effective against extracellular pathogens: bacteria, some parasites, antigenic macromolecules, and viruses before they have invaded their host cells.

The surface of each B lymphocyte is covered with as many as 100,000 antibody molecules whose Fc ends are inserted into the lymphocyte membrane. Binding of antigen to the Fab regions of the bound antibodies activates the B lymphocyte, converting it into a memory cell or antibody-secreting plasma cell (①, Fig. 22-13 ■).

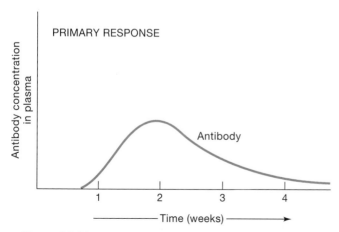
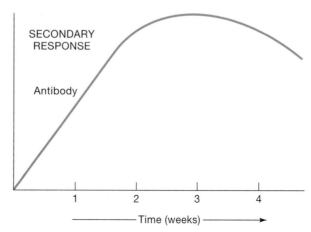

■ **Figure 22-11 Memory in the immune system** Antibody production in response to the first exposure to an antigen is both slower and weaker than the secondary response upon reexposure.

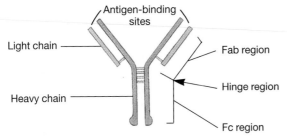

■ **Figure 22-12 Antibody structure** Antibodies are Y-shaped proteins composed of two identical light chains and two identical heavy chains, linked to each other at intervals by disulfide bonds. The tips of the two arms contain variable binding sites for antigens. The positioning of the arms is made flexible by a hinge region where the stem joins the arms.

Antibodies have many secondary functions, some of which are triggered by the Fc end of the molecule rather than by the antigen-binding arms:

1. Antibodies act as opsonins to facilitate recognition and phagocytosis of antigens (②, Fig. 22-13 ■). Antibodies are not toxic by themselves, but they label antigens so that other immune cells will attack them. All antibody-bound substances are eventually ingested by phagocytes.
2. Antibodies cause antigens to clump for easier phagocytosis (③, Fig. 22-13 ■).
3. Antibodies bound to antigens at their Fab regions will bind to other immune cells with the Fc end of the antibody. Through this mechanism, antibody-bound antigen can be recognized by a single Fc receptor, rather than requiring millions of different receptors to recognize millions of antigens. This type of Fc binding occurs, for example, on cytotoxic and phagocytic cells (② and ④, Fig. 22-13 ■).
4. Antibodies activate complement, proteins that play important roles in innate immunity (⑥, Fig. 22-13 ■).

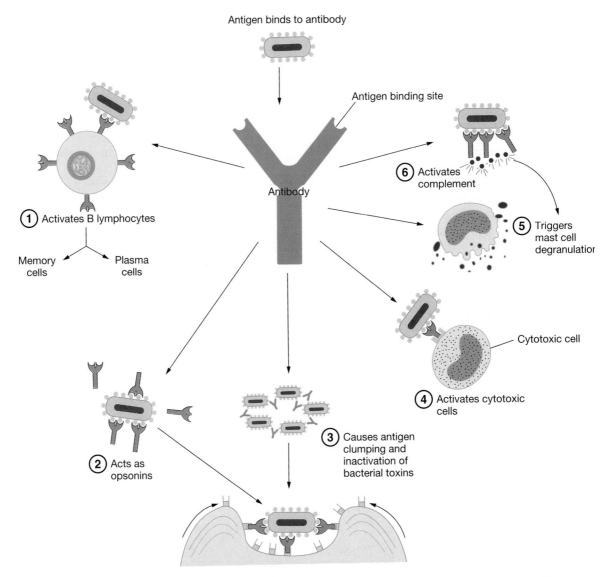

■ **Figure 22-13 Functions of antibodies** The primary function of antibodies is to activate B lymphocytes ①. Activated antibodies also enhance phagocytosis ② and ③, activate cytotoxic cells ④, trigger mast cell degranulation ⑤, and activate complement ⑥.

5. Antibodies activate mast cells, both directly and via complement, causing them to release chemicals that mediate the inflammatory response (⑤, Fig. 22-13 ■).

6. Antibodies may bind to viruses or to toxins produced by bacteria, preventing these substances from affecting host cells. For example, *Corynebacterium diphtheriae* is a bacterium that causes diphtheria, an upper respiratory infection. In this disease, the bacterial toxin kills host cells, leaving ulcers with a characteristic grayish membrane. Immunity to the disease occurs when the host develops antibodies that disable the toxin. By creating an inactivated toxic preparation that did not affect living cells, researchers developed a vaccine for diphtheria. When given to a person, the vaccine triggers antibody production without causing any symptoms of the disease. As a result, diphtheria has been almost eliminated in countries with good immunization programs.

✓ A child is stung by a bee for the first time. Why should the parent be particularly alert when the child is stung a second time?

✓ Because antibodies are proteins, they are too large to cross cell membranes on transport proteins or through channels. How then do IgA and other antibodies become part of external secretions such as saliva, tears, and mucus?

T Lymphocytes Must Make Direct Contact with Their Target Cells

T lymphocytes (T cells) develop in the thymus gland from immature precursor cells that migrate there from the bone marrow. T lymphocytes are responsible for **cell-mediated immunity.** In this process, an activated T lymphocyte must make direct contact with the antigen before the T lymphocyte can destroy or neutralize the antigen.

T lymphocytes bind to their target cells by using membrane receptors known as **T-cell receptors** (Fig. 22-14 ■). T-cell receptors are not antibodies like the B lymphocyte receptors, although their peptide chains are closely related to the chains of antibodies. When the T-cell receptor binds to antigen fragments displayed on the surface of a target cell, the cytoplasmic portion of the receptor initiates signal transduction within the T lymphocyte. This event begins a cascade of intracellular events that ends in activation of T lymphocytes.

Major histocompatibility complex How can T lymphocytes distinguish between cells that need to be eliminated and the normal cells of the body? The key lies with the membrane proteins known as the **major histocompatibility complex,** or MHC, a family of membrane protein complexes encoded by a specific set of genes. These proteins were named when they were first discovered to play a role in the rejection of foreign cells following organ or tissue transplants. We know now that they are present in all nucleated cells of the body, where they are responsible for the extracellular presentation of processed antigen fragments.

All MHC proteins are related, but they vary from person to person because of the huge number of alleles that a person can inherit. There are so many alleles that it is unlikely that any two people other than identical twins will inherit exactly the same set. Major histocompatibility complexes are one reason that tissues cannot be transplanted from one person to another without first establishing compatibility. More recently, it has been discovered that the inheritance of certain MHC alleles is linked to some disorders in which the body fails to recognize body components as "self."

When phagocytes ingest molecular or cellular antigens and digest them, or when normal cells are invaded by viruses, fragments of antigen are inserted back into the cell membrane as part of an MHC complex (see Figs. 22-5, 22-13, and 22-14 ■). These extracellular markers allow lymphocytes to identify antigen-containing cells.

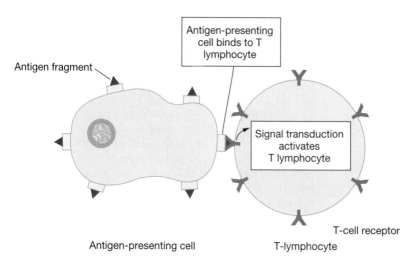

Antigen-presenting cell binds to T lymphocyte

Antigen fragment

Signal transduction activates T lymphocyte

Antigen-presenting cell

T-cell receptor

T-lymphocyte

■ **Figure 22-14 Activation of T lymphocytes** When antigen-presenting cells bind to T-cell receptors on T lymphocytes, they activate the lymphocyte as the result of signal transduction.

Focus on   The Thymus Gland

■ **Figure 22-15**

The thymus gland is a two-lobed organ located in the thorax just above the heart. During embryonic development, undifferentiated lymphocytes are released from the bone marrow and migrate to the thymus, where they differentiate into T lymphocytes and continue to divide. During the maturation process, T lymphocytes synthesize their T-cell receptors and insert them into the membrane. Those cells that would be self-reactive are eliminated, while those that do not react with self tissues multiply to form clones.

The epithelial cells of the thymus secrete peptide factors that influence the development of T lymphocytes, including interleukins, thymosins, thymopoietin, and thymulin. Although these peptides are sometimes classified as hormones, there is little evidence that they are released into the circulation to act on tissues outside the thymus gland.

The thymus gland reaches its greatest size during adolescence, then it shrinks and is largely replaced by adipose tissue as a person ages. New T lymphocyte production in the thymus is low in adults, but the number of T lymphocytes in the blood does not decrease, suggesting that mature T lymphocytes either have long life spans or can reproduce themselves.

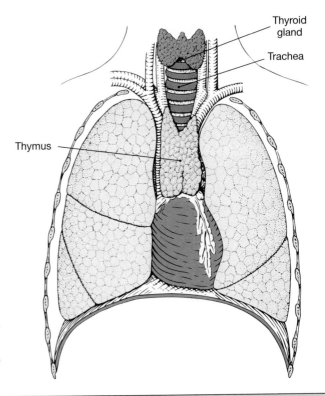

Free antigen in the extracellular fluid will not bind to unoccupied MHC receptors on the surface of a cell.

There are two types of MHC molecules. **MHC class I molecules** are found on all nucleated cells and are used to present peptides from viruses and bacteria that have invaded the cell. T lymphocytes that bind to cells with MHC class I–antigen complexes kill the targeted cell to prevent reproduction of the invading pathogen.

MHC class II molecules are found primarily on antigen-presenting immune cells (macrophages, lymphocytes, and dendritic cells). When these immune cells ingest and process antigen fragments, the fragments are returned to the membrane of the antigen-presenting cell, where they associate with MHC class II proteins. The MHC–antigen complex then binds to and activates other immune cells.

Subtypes of T lymphocytes   T-cell precursors differentiate in the thymus into three major groups: cytotoxic T cells,* helper T cells, and natural killer cells (Fig. 22-16 ■). The existence of a fourth group, suppressor T

cells, has been proposed, but at this time it is unclear whether these cells represent a distinct line.

Cytotoxic T cells attack and destroy cells that display MHC class I–antigen complexes. Although this may be an extreme response, it is the best way to prevent the reproduction of intracellular invaders such as viruses,

Cell Suicide, or Apoptosis   Apoptosis (app-oh-TOE-sis, with the second *p* silent) is a programmed form of cell suicide that the body uses to eliminate cells that it no longer needs. Cells that are undergoing apoptosis do not swell like dying cells that have been attacked by membrane attack complex. Instead, they shrink and fragment into small pieces that are ingested by neighboring phagocytes. Apoptosis is not limited to cells that are attacked by the immune system. Keratinocytes in the skin (∞ p. 66) are programmed to die about three weeks after they migrate toward the surface layers of epidermis. T lymphocytes that would attack self tissues normally die from apoptosis before being released from the thymus gland. Similarly, genetically altered cells often commit suicide before reproducing. One apparent cause of cancer is failure of these abnormal cells to kill themselves.

---

*Cytotoxic T cells are sometimes called "killer T cells," but because that name is very similar to "natural killer cells," it should probably be avoided.

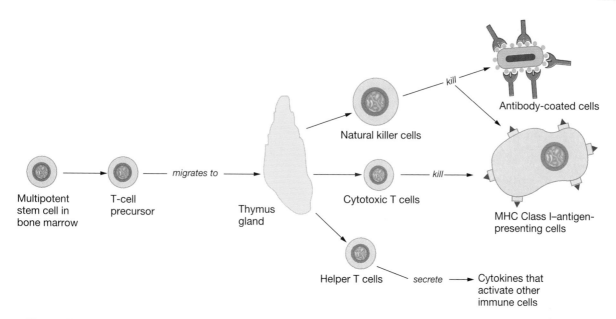

■ Figure 22-16 **Maturation of lymphocytes in the thymus** The immature precursor form of T lymphocytes and natural killer cells migrates from the bone marrow to the thymus, where three cell types differentiate. Helper T cells activate other immune cells by secreting cytokines. Cytotoxic cells and natural killer cells destroy their target cells.

some parasites, and some bacteria. Cytotoxic T cells kill their targets by releasing cytotoxic chemicals: a pore-forming molecule called **perforin** that creates protein channels in the target cell membrane, and enzymes known as **granzymes** that are related to the digestive enzymes trypsin and chymotrypsin. When granzymes enter the target cell through the perforin channels, they trigger the cell into committing suicide (**apoptosis**).

Helper T (T$_H$) cells do not directly attack pathogens and infected cells, but they play an essential role in the immune response by secreting cytokines that influence other cells. Some cytokines secreted by helper T cells include gamma-interferon, which activates macrophages; interleukins that activate antibody production and cytotoxic T lymphocytes; colony-stimulating factors (∞ p. 457) that enhance leukocyte production in the bone marrow; and interleukins that promote growth of mast cells and eosinophils. The HIV virus that causes AIDS preferentially infects and destroys helper T cells, leaving the host unable to respond to pathogens that in other situations could be easily suppressed.

Natural Killer Lymphocytes

A third class of lymphocyte, the **natural killer cells,** or NK cells, was once considered a subset of T lymphocytes because both cell types develop from the same precursor. However, the natural killer cells are now considered a distinct cell line. Sometimes these cells are called *large granular lymphocytes* to distinguish them from the smaller B and T lymphocytes.

In laboratory experiments, natural killer cells attack certain tumor cells and virus-infected cells. They are believed to do the same *in vivo*. For example, in Chédiak-Higashi syndrome, characterized by a lack of

natural killer cells, patients have an increased incidence of lymphoma, a cancer of the lymphoid tissue.

Like cytotoxic T cells, natural killer cells kill their targets by cell lysis and apoptosis. However, natural killer cells do not have to bind to MHC–antigen complexes. They can also bind to the Fc stem of antibody-coated target cells. This nonspecific response of natural killer cells is called **antibody-dependent cell-mediated cytotoxicity** and is similar to the process by which macrophages recognize and interact with antibody-coated pathogens.

IMMUNE RESPONSE PATHWAYS

In the following sections, we will examine the way the body responds to four kinds of immune challenges: an extracellular bacterial infection, a viral infection, an

continued from page 664

Early in the AIDS epidemic, researchers studied the effectiveness of a drug made with CD4. CD4 is a naturally occurring protein found on the surface of helper T cells, the primary target of the HIV virus. When the HIV virus binds to CD4, the virus is taken into the cell and begins replicating. Researchers found that adding high amounts of CD4 to test tubes containing HIV and helper T cells reduced the rate of HIV replication. But in human subjects, the results of clinical trials using the CD4 drug were disappointing. The drug did not slow the rate at which HIV infected the helper T cells.

Question 2: What is the rationale for flooding the body with CD4 protein in order to prevent HIV replication?

allergic response to pollen, and transfusion of incompatible blood. Although we have described the innate and acquired immune responses as if they were two separate entities, in reality they are two interconnected parts of a single process. The innate response is the rapid response, reinforced by the more powerful acquired response. As stated earlier, communication and coordination between the different pathways of immunity is vital for maximum protective effect.

Inflammation Is the Typical Response to Bacterial Invasion

When the passive barricades of the skin and mucous membranes fail, bacteria reach the extracellular fluid. There they usually cause an inflammatory response that represents the combined effects of many cells working to get rid of the invader. Inflammation near the skin's surface is characterized by a red, swollen area that is tender or painful. In addition to the nonspecific inflammatory response, lymphocytes attracted to the area produce antibodies keyed to the specific type of bacterium.

The entry of bacteria sets off a series of reactions (Fig. 22-17 ▇):

1. Components of the bacterial cell wall activate the complement system. Some complement proteins are chemical signals (*chemotaxins*) that attract additional leukocytes from the circulation to help fight the infection.

2. The complement cascade ends with the formation of membrane attack complex molecules that insert themselves into the bacterial wall. The subsequent entry of Na^+ and water lyses the bacteria, aided by the enzyme lysozyme.

3. If the bacteria are not encapsulated, phagocytosis of the bacteria by macrophages can begin immediately.

4. If the bacteria are encapsulated, opsonins must coat the capsule before they can be ingested by phagocytes. Opsonins enhance the process for all bacteria, even those that are not encapsulated. The molecules that act as opsonins include complement, acute phase proteins, and antibodies.

5. Complement causes degranulation of mast cells and basophils. Cytokines secreted by the mast cells act as additional chemotaxins, attracting more immune cells. Vasoactive chemicals such as histamine dilate blood vessels and increase capillary permeability. The enhanced blood supply to the site creates the

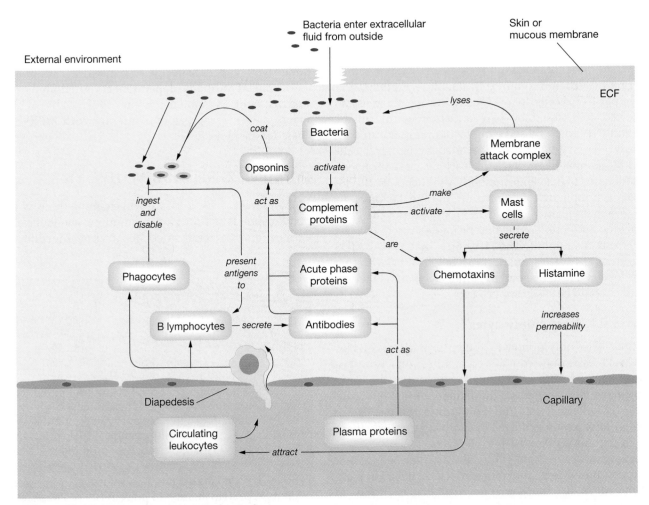

■ **Figure 22-17** **Immune responses to bacteria**

red, warm appearance of inflammation, while plasma proteins that escape into the interstitial space pull water with them, leading to tissue edema.

6. Some elements of the acquired immune response are called into play with bacterial infections. If antibodies to the bacteria are already in the body, they will help the innate response by acting as opsonins and neutralizing bacterial toxins.

7. Memory B cells attracted to the infection site will be activated if they encounter an antigen that they recognize. But this process is much slower than the innate responses described above, and it may be as long as several days before plasma cells begin secreting antibodies against the invader. If the infection is a novel one, phagocytes that ingest the bacteria present bacterial antigen to naive B cells, triggering clonal expansion, antibody production, and the formation of memory cells. This process is even slower, usually requiring at least four days.

8. If the initial wound damaged blood vessels underlying the skin, platelets and the proteins of the coagulation cascade will also be recruited to minimize the damage (∞ p. 465). Once the bacteria are removed by the immune response, repair of the injured site takes place under the control of growth factors and other cytokines.

Intracellular Defense Mechanisms Are Needed to Fight Viral Infections

When viruses first invade the body, they go through an extracellular phase of immune response similar to that described for bacteria. In the early stages of a viral infection, innate immune responses and antibodies can help control the invasion of the virus. However, once the viruses enter the host's cells, humoral immunity in the form of antibodies is no longer effective. The main defense against intracellular viruses is cytotoxic T lymphocytes that bind to infected host cells and destroy them.

For years, cell-mediated immunity mediated by the T lymphocytes and humoral immunity controlled by B lymphocytes were considered independent processes. We now know that the two types of immunity are linked. Figure 22-18 ■ shows the steps of a viral infection and shows how the two types of lymphocytes coordinate to destroy viruses and virus-infected cells.

1. In this figure, we have assumed prior exposure to the virus and preexisting antibodies in the circulation. Antibodies can play an important defensive role in the early extracellular stages of a viral infection.

2. Antibodies act as opsonins, coating the viral particles to make them better targets for macrophages.

3. Antibody-bound viruses are also prevented from entering their target cells. However, once the virus is inside the target host cell, antibodies are no longer useful.

4. Macrophages that ingest viruses insert digested fragments of viral antigen into MHC class II receptors on their membranes. In addition, they secrete a variety of cytokines. Some of these cytokines initiate the inflammatory response. Alpha-interferon causes host cells to make antiviral proteins that keep invading viruses from replicating. Macrophages also secrete cytokines that stimulate natural killer cells.

5. The natural killer cells and macrophages together stimulate the acquired immune response by activating helper T cells. The natural killer cells do this through secretion of gamma-interferon. The macrophages present viral antigen on MHC class II receptors and bind to helper T cells at the T-cell receptors.

6. The activated helper T cells are the primary mediators for additional antibody production by B lymphocytes and for activation of cytotoxic T cells.

7. Cytotoxic T cells and natural killer cells use viral antigen–MHC class I complexes to recognize the infected host cells. When these cytotoxic cells bind to the host cell, they secrete the contents of their granules onto the host cell surface. Perforin molecules insert pores into the host cell membrane so that granzymes can enter the cell, inducing it to commit suicide and undergo apoptosis. Destruction of infected host cells is a key step in halting the reproduction of invading viruses.

Antibodies against viruses   Although the acquired immune response against viruses includes production of antibodies, there is no guarantee that an antibody produced during one infection will be effective against the next invasion by the same virus. Many viruses mutate constantly, and the protein coat forming the primary antigen may change significantly when the virus incorporates sequences of host protein as it leaves an infected cell. The influenza virus is one virus that changes yearly. Consequently, vaccines against influenza must be developed and administered each autumn based on the virologists' predictions of what mutations have occurred.

The rapid mutation of viruses is also one reason that researchers have not yet developed an effective vaccine against HIV, the virus that leads to many cases of acquired immunodeficiency syndrome (AIDS). The HIV virus infects cells of the immune system, particularly T lymphocytes, monocytes, and macrophages. When HIV wipes out the helper T cells, cell-mediated immunity against the virus is lost. The general loss of immune response in AIDS leaves these patients susceptible to a variety of viral, bacterial, fungal, and parasitic infections.

Allergic Responses Are Inflammatory Responses Triggered by Specific Antigens

An allergy is an inflammatory immune response to a nonpathogenic antigen. Left alone, the antigen, called an **allergen,** is not harmful to the body. But if an individual is sensitive to the allergen, the body creates an inflam-

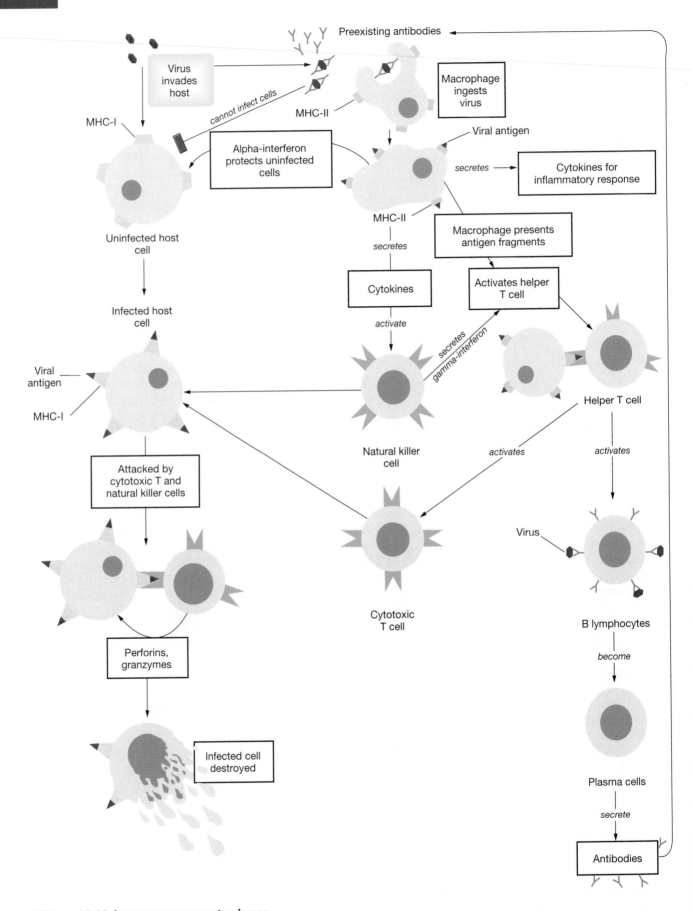

Preexisting antibodies

Virus invades host

cannot infect cells

MHC-I

MHC-II

Macrophage ingests virus

Viral antigen

secretes → Cytokines for inflammatory response

Alpha-interferon protects uninfected cells

MHC-II

Uninfected host cell

secretes

Macrophage presents antigen fragments

Cytokines

Activates helper T cell

activate

secretes gamma-interferon

Infected host cell

Viral antigen

MHC-I

Natural killer cell

activates

Helper T cell

activates

Attacked by cytotoxic T and natural killer cells

Virus

B lymphocytes

Perforins, granzymes

Cytotoxic T cell

become

Infected cell destroyed

Plasma cells

secrete

Antibodies

■ Figure 22-18 **Immune responses to viruses**

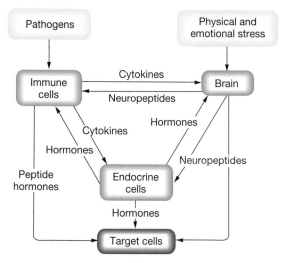

■ Figure 22-21 **Model for interaction between nervous, endocrine, and immune systems**

through bidirectional communication carried out using cytokines, hormones, and neuropeptides. The brain and the immune system are linked by autonomic neurons and neuropeptides from the central nervous system that send information to the immune cells and by cytokines that transfer information from immune cells to the CNS. The nervous system controls various endocrine glands by secretion of hypothalamic releasing hormones, but immune cells also secrete some of the same trophic hormones. The products of various endocrine glands act in a negative feedback manner to control both the nervous and immune systems. The result is a complex network of chemical signals subject to modulation by outside factors. These outside factors include physical and emotional stimuli integrated through the brain, pathogenic stressors integrated through the immune system, and a variety of miscellaneous factors including magnetic fields, chemical factors from brown adipose tissue, and melatonin from the pineal gland. It will probably take years for us to decipher these complex pathways.

The Interaction between Stress and the Immune System

One area of interest is the link between inability to cope with stress and the development of illnesses. The modern study of stress is attributed to Hans Selye, beginning in 1936. He defined **stress** as nonspecific stimuli that disturb homeostasis and elicit an invariable stress response that he termed the General Adaptation Syndrome. Selye's stress response consisted of stimulation of the adrenal glands, followed by suppression of the immune system due to high levels of circulating glucocorticoids (∞ p. 636). For many years, Selye's experiment was the benchmark for defining stress, but since the 1970s, the definition of stress and the stress response has broadened. **Stressors,** the events or items that create stress, are highly variable and difficult to define in experimental settings. Acute stress is different from chronic stress. A person's

reaction to stress is affected by loneliness and by whether the person feels in control of the stressful situation. Many stressors are sensed and interpreted by the brain, leading to modulation of the stressor by experience and expectations. A stressor to one person may not affect another.

Most physical and emotional stressors are integrated through the central nervous system. The two classic stress responses are the fight-or-flight reaction created by massive sympathetic discharge and adrenal catecholamine release, and the elevation of adrenal corticosteroid hormone levels associated with suppression of the immune response. The nervous response is a rapid reaction to acute stress; the hormonal response is a better indicator of chronic or repetitive stress.

One difficulty in studying the body's response to stress is the complexity of integrating information from three levels of response (Fig. 22-22 ■). Scientists have gathered a good deal of information at each of three levels: cellular and molecular factors related to stress, systemic responses to stress, and clinical studies describing the relationships between stress and illness. In some cases, we have linked two different levels together, although in other cases, the experimental evidence is scanty or even contradictory. What we still do not understand is the big picture, the way that all three levels of response overlap.

Experimental progress is slow because many traditional study animals, such as rodents, are not good models for studying stress in humans. At the same time, experiments that create stress in humans are difficult to design and execute for both practical and ethical reasons. Nevertheless, scientists are slowly beginning to piece together the intricate tapestry of mind-body interactions. When we can explain why left-handed people have a higher incidence of autoimmune and allergic diseases than do right-handed people, why students tend to get sick just before exam time, and why relaxation therapy enhances the immune response, we will be well on the way to understanding the complex connections between the brain and the immune system.

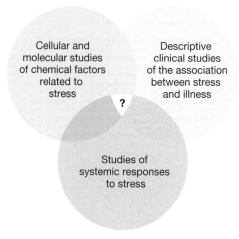

■ Figure 22-22 **Neuroimmunomodulation** The field of neuroimmunomodulation is still in its infancy, and there is much that we do not understand about the mechanisms through which the nervous, immune, and endocrine systems affect health and well-being.

CHAPTER REVIEW

Chapter Summary

1. **Immunity** is the ability of the body to protect itself from disease-causing entities. The human immune system consists of lymphoid tissues, immune cells, and chemicals used to execute the immune functions. (p. 654)

2. The functions of the immune system include protecting the body from **pathogens,** removing dead or damaged tissue and cells, and recognizing and removing abnormal cells, such as cancer cells and cells invaded by viruses. (p. 654)

3. Substances that trigger an immune response from the body and that can react with products of that response are known as **antigens.** (p. 655)

4. Pathogens can be **microbes,** multicellular parasites, or any molecule or cell not of the body. In the United States, the most prevalent infectious diseases are viral and bacterial infections. (p. 655)

Pathogens of the Human Body

5. Bacteria are cells that often have protective outer layers. They can survive and reproduce outside of a living host and can be killed by antibiotics. (p. 655)

6. Viruses are not cells. They consist of a DNA or an RNA core enclosed in protective proteins. Viruses invade host cells and use the intracellular machinery of the host cell to reproduce. To kill a virus, the body's immune system must find and destroy the infected host cell. (p. 655)

7. Viruses that invade a host cell's DNA and turn the host cell into a cancerous cell are called **oncogenic viruses.** (p. 655)

The Immune Response

8. The steps of the immune response include: (1) *detection* and *identification* of the foreign substance, (2) *communication* with other immune cells, (3) *recruitment* of assistance and *coordination* of the response, and (4) *destruction* or *suppression* of the invader. (p. 656)

9. **Innate immunity** consists of the nonspecific immune responses of the body. **Acquired immunity** is directed at specific invaders. With repeated exposure to the pathogen, the acquired response remembers prior exposure and therefore reacts more rapidly. (p. 656)

Anatomy of the Immune System

10. The two primary lymphoid tissues are the thymus gland and the bone marrow, sites where immune cells form and mature. **Encapsulated lymphoid tissues** are the **spleen** and the **lymph nodes. Unencapsulated lymphoid tissues** include the **tonsils,** the **gut-associated lymphoid tissue (GALT),** and clusters of lymphoid tissue associated with the respiratory tract and skin. (p. 657)

11. Immune cells can be found in highest concentrations wherever they are most likely to encounter antigens that penetrate the epithelia. (p. 657)

12. Immune cells are divided into (1) *eosinophils,* (2) *basophils* in the blood and the related *mast cells* in the tissues, (3) *neutrophils,* (4) *monocytes* and their derivative *macrophages,* (5) *lymphocytes* and their derivative *plasma cells,* and (6) *dendritic cells.* (p. 657)

13. **Eosinophils** combat allergic reactions and parasitic diseases. They are **cytotoxic cells** that damage or kill pathogens. **Basophils** and **mast cells** of tissues release histamine, **heparin,** and other chemicals involved in the immune response. (p. 660)

14. **Neutrophils** are phagocytic cells that ingest bacteria and foreign particles. They also release chemicals that cause fever and mediate the inflammatory response. (p. 661)

15. **Monocytes** are the precursors of tissue **macrophages,** the primary scavengers for the tissues. Macrophages are **antigen-presenting cells** that display fragments of antigen on their membrane as part of surface protein complexes. (p. 661)

16. The **lymphocytes** and **plasma cells** mediate the acquired immune response. There are more than 1 million different types of lymphocytes, each of which binds to a specific ligand. (p. 661)

17. **Dendritic cells** capture antigens, particularly viruses, and migrate to the lymphoid tissues, where they present antigen fragments to lymphocytes to activate them (p. 661)

Innate Immunity

18. The first line of defense is to exclude pathogens by means of physical and chemical barriers such as the skin, mucous linings of the digestive system and genitourinary tract, and the ciliated epithelium of the respiratory tract. (p. 662)

19. Pathogens that enter the body are detected by phagocytes in the tissues that destroy or suppress the invader by ingesting it. **Chemotaxins** such as bacterial toxins, fibrin and collagen fragments, and cytokines attract additional immune cells that leave the circulation by **diapedesis.** (p. 662)

20. Phagocytosis is a receptor-mediated event in which surface molecules on the pathogen bind to receptors on the phagocyte membrane. Some foreign material is first coated with tagging proteins known as **opsonins.** (p. 662)

21. Ingested material in **phagosomes** is digested by enzymes from specialized lysosomes. (p. 662)

22. **Inflammation** results when activated tissue macrophages release cytokines that attract other immune cells, increase capillary permeability, and cause fever. (p. 663)

Acquired Immunity

25. Acquired immune responses are mediated primarily by lymphocytes that react to only one type of invader. Each unique type of lymphocyte, along with the other cells exactly like it, forms a **clone.** (p. 665)

26. The first exposure to an antigen activates the appropriate **naive lymphocyte** clone and causes it to divide (**clonal expansion**). Newly formed lymphocytes differentiate into **effector cells,** which create the immune response, or **memory cells.** With second and subsequent exposures to the antigen, the memory cells create a more rapid and stronger secondary response to the antigen. (p. 665)

27. All lymphocytes secrete cytokines. The primary lymphocyte cytokines are **interleukins** and **interferons.** Interferons interfere with viral replication in cells. **Gamma-interferon** also activates macrophages and other immune cells. (p. 665)

28. There are three main types of lymphocytes: B lymphocytes, T lymphocytes, and natural killer cells. **B lymphocytes** secrete antibodies; **T lymphocytes** and **natural killer cells** attack and destroy infected cells. (p. 666)

29. The primary function of B lymphocytes is **humoral immunity,** mediated by protein **antibodies** that bind to pathogens. In mature B lymphocytes, antibodies act as surface receptors marking the members of each clone. Activated B lymphocytes differentiate into plasma cells, which secrete antibodies at very high rates. (p. 666)

30. Antibodies are divided into five general classes of immunoglobulins (Ig): IgG, IgA, IgE, IgM, and IgD. Together, the immunoglobulins are given the general name of **gamma globulins.** (p. 667)

31. Antibodies are made of four polypeptide chains, attached

23. **Chemokines** of the inflammatory response include **acute phase proteins,** histamine, **interleukins, bradykinin,** and **complement proteins.** (p. 663)

24. The terminal step in the complement cascade is the formation of **membrane attack complex,** pore-forming molecules that insert themselves into pathogen cell membranes. Water and Na^+ enter the cell, causing the pathogen to swell and lyse. (p. 665)

to each other in a Y shape. The **Fab regions** of the two arms contain antigen-binding sites that confer the antibody's specificity. The stem of the Y-shaped antibody is known as the **Fc region.** (p. 667)

32. The primary function of antibodies is to bind antigens to B lymphocytes and initiate the production of additional antibodies. Antibodies also act as opsonins, serve as a bridge between antigens and cytotoxic cells, activate complement proteins and mast cells, and disable viruses and toxins. (p. 667)

33. **T lymphocytes** are responsible for **cell-mediated immunity,** in which the lymphocytes bind to their target cells using **T-cell receptors.** (p. 669)

34. The response of the T lymphocyte to binding depends on the type of **major histocompatibility complex** proteins on the target cell. **MHC class I molecules** display antigen peptides and cause T lymphocytes to kill the cell displaying MHC I–peptide complexes. T lymphocytes bound to **MHC class II molecules** on antigen-presenting immune cells are activated to secrete cytokines. (p. 669)

35. **Cytotoxic T cells** kill their targets by releasing **perforin,** a pore-forming molecule that allows **granzymes** to enter the target cell and trigger it into committing suicide (**apoptosis**). (p. 670)

36. **Helper T (T_H) cells** do not directly attack pathogens and infected cells but secrete cytokines that influence other cells. (p. 671)

37. **Natural killer cells** kill certain tumor cells and virus-infected cells. They also bind to and kill antibody-coated target cells. This nonspecific response of natural killer cells is called **antibody-dependent cell-mediated cytotoxicity.** (p. 671)

Immune Response Pathways

38. Bacteria that reach the extracellular fluid usually cause a nonspecific inflammatory response. In addition, lymphocytes attracted to the area produce antibodies keyed to the specific type of bacterium. (p. 672)

39. In the early stages of a viral infection, innate immune responses and antibodies help control the invasion of the virus. Once the viruses enter the host's cells, the main defense is cytotoxic T lymphocytes that bind to infected host cells and destroy them. (p. 673)

40. An allergy is an inflammatory immune response to a nonpathogenic **allergen.** The body gets rid of the allergen by

using an inflammatory response whose symptoms range from mild tissue damage to fatal reactions. (p. 673)

41. **Immediate hypersensitivity reactions** to allergens are mediated by antibodies and occur within minutes of exposure to antigen. **Delayed hypersensitivity reactions** are mediated by T lymphocytes and may take several days to develop. (p. 675)

42. The transfer of tissues between humans is usually restricted by the major histocompatibility complex proteins found on all nucleated cells. Lymphocytes will attack any MHC complexes that they do not recognize as self proteins. (p. 676)

43. Human red blood cells lack MHC proteins but contain other antigenic membrane proteins and glycoproteins such as the ABO and Rh antigens. (p. 676)

44. The ability of the body to distinguish its own cells and not create an immune response is known as **self-tolerance.** It appears to be due to the elimination of self-reactive lymphocytes. When self-tolerance fails, **autoimmune dis-** eases result in which the body makes antibodies against its own components. (p. 677)

45. The theory of **immune surveillance** proposes that cancerous cells develop on a regular basis but are detected and destroyed by the immune system before they can spread. (p. 677)

Integration between the Immune, Nervous, and Endocrine Systems

46. The nervous, endocrine, and immune systems are linked together by signal molecules such as cytokines, neuropeptides, and hormones. The study of brain-immune interactions is now a recognized field known as **neuroimmunomodulation.** (p. 678)

47. The link between inability to cope with stress and the development of illnesses is believed to result from neuroimmunomodulation. (p. 679)

Questions

LEVEL ONE   Reviewing Facts and Terms

1. Define immunity. What is meant by plasticity in the immune system?

2. Name the components of the immune system. What criteria distinguish a primary lymphoid tissue from a secondary one?

3. List and briefly summarize the three main functions of the immune system.

4. Name two ways viruses are released from the host cell. Give three ways in which viruses damage host cells or disrupt their function.

5. If a pathogen has invaded the body, what four steps occur if the body is able to destroy the pathogen? What compromise may be made if the pathogen cannot be destroyed?

6. Define the following terms and explain their significance:
   a. anaphylaxis   b. agglutinate
   c. extravascular   d. degranulation
   e. acute phase protein f. membrane attack complex
   g. clonal expansion   h. immune surveillance

7. How are histiocytes, Kupffer cells, osteoclasts, and microglia related?

8. What is the mononuclear phagocytic system and what role does it play in the immune system?

9. Match the cell type with its description:
   (a) lymphocyte
   (b) neutrophil
   (c) monocyte
   (d) dendritic cell
   (e) eosinophil
   (f) basophil

   1. most abundant leukocyte; phagocytic; lives 1–2 days
   2. cytotoxic; associated with allergic reactions and parasitic infestations
   3. precursor of macrophages; relatively rare in blood smears
   4. related to mast cells; releases chemical mediators such as histamine and heparin
   5. Most of these cells reside in the lymphoid tissues of the body.
   6. These cells capture antigens, then migrate to the lymphoid tissues.

10. Give examples of physical and chemical barriers to infection.

11. Name the three main types of lymphocytes. Explain the functions and interactions of each group.

12. Define self-tolerance. Which cells are responsible? Relate self-tolerance to autoimmune disease.

13. What is meant by the term *neuroimmunomodulation*?

14. Explain the terms *stress, stressor,* and *general adaptation syndrome.*

LEVEL TWO   Reviewing Concepts

15. **Concept map:** Map each of the cell types involved in the immune response. Include information about their physical locations, functions, secretions, and interactions with one another. Construct a table describing each cell's appearance, life span, and normal concentration in the blood.

16. Why do lymph nodes often swell and become tender or even painful when you are sick?

17. Distinguish between the six basic groups of immune cells. Which of these are considered granulocytes? immunocytes? phagocytes? cytotoxic?

18. Summarize the effects of histamine, interleuken-1, acute phase proteins, bradykinin, complement, and gamma interferon. Are these chemicals antagonistic or synergistic to one another?

19. Compare and contrast the following sets of terms:
    a. pathogen, microbe, pyrogen, antigen, antibody, antibiotic
    b. infection, inflammation, allergy, autoimmune disease
    c. virus, bacteria, allergen, retrovirus
    d. chemotaxin, chemokine, opsonin, cytokine, interleukin, bradykinin, interferon

e. diapedesis, phagocytosis
f. acquired immunity, innate immunity, humoral immunity, cell-mediated immunity
g. B lymphocytes, plasma cells, memory cells, helper T cell, natural killer cells, cytotoxic T-cells
h. major histocompatibility complex class I and class II
i. immediate and delayed hypersensitivity

20. Diagram and label the parts of an antibody molecule. What is the significance of each part?
21. Diagram the steps in the process of inflammation.
22. Diagram the steps in the process of fighting viral infections.
23. Diagram the steps in the process of an allergic reaction.

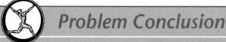

LEVEL THREE   Problem Solving

24. A person with AB blood type is transfused with type O blood. What happens and why? A person with O blood type is transfused with type A blood. What happens? Why?

25. Brian has had a bad semester. His car was wrecked, he may lose his job, his school work is suffering because he has missed so many classes, and he is not much fun to be around. After succumbing to a series of colds, Brian visits his doctor, who tells him his illnesses are stress related. Brian scoffs at this idea. Can you explain to Brian how the doctor may be correct?

26. Barbara is troubled with rheumatoid arthritis, characterized by painful, swollen joints from inflamed connective tissue. She notices that it flares up when she is tired and stressed. This condition is regarded as an autoimmune disease. Outline the physiological reactions that would lead to this condition.

Problem Conclusion

In this running problem, you learned about a new treatment for AIDS. This treatment consists of two antiretroviral drugs and a protease inhibitor. The combination of these drugs seems to stem the tide of HIV replication more effectively than the drugs used alone.

Further check your understanding of this running problem by checking your answers against those in the summary table.

Question	Facts	Integration and Analysis
1 Why can't retroviruses reproduce without reverse transcriptase?	Retroviruses have no DNA to use as a template to make viral mRNA. Reverse transcriptase is an enzyme that allows retroviruses to make complementary DNA from their own viral RNA.	Without reverse transcriptase, the retroviruses cannot make viral DNA. Without viral DNA, viral mRNA cannot be made. mRNA is used to make viral proteins during viral replication. Therefore, without reverse transcriptase, a retrovirus cannot replicate.
2 What is the rationale for flooding the body with CD4 protein in order to prevent HIV replication?	CD4 is the receptor on helper T cells to which HIV binds. Binding causes the virus to be taken into the T cell.	Introducing large amounts of CD4 protein into the body might "trick" HIV into binding to these free proteins instead of to CD4 on helper T cells. This would reduce the number of virus particles that could invade the T cells.
3 George asks, "If I'm making antibodies against HIV, why can't these antibodies kill the virus?" Answer George's question.	Viruses can only be affected by antibodies when they are outside the host cells.	HIV enters helper T cells, making it unavailable for destruction by antibodies. Over time, the HIV infection will destroy the T cells that amplify the immune response against virus-infected cells.
4 What correlation can be made between T-cell count and the viral load of HIV in George's blood?	George's T-cell count increases while the viral load decreases.	Apparently, the drug cocktail is effective in decreasing the rate of T-cell infection by HIV. This has two results. As fewer T cells are destroyed by the virus, the T-cell count rises. And as the drugs suppress the rate of viral replication, the viral load decreases.

23
Integrative Physiology III: Exercise

BACKGROUND BASICS

One of the most common challenges to the body's homeostasis is not a disease but an everyday activity, exercise. The American College of Sports Medicine defines **exercise** as any muscular activity that generates force and disrupts homeostasis. Running is an example of a dynamic endurance exercise, whereas weight lifting and strength training are examples of dynamic resistive exercise. This chapter examines dynamic exercise as a challenge to homeostasis that is met with an integrated response from multiple systems of the body. It does not intend to be a complete description of exercise physiology; for further information, readers should consult an exercise physiology textbook.

Problem

Heat Stroke

"Dangerous heat continues for fifth straight day," reads the morning headline. Had Colleen Warren, 19, taken time to scan the story, she might have avoided the life-threatening health emergency she faced just a few hours later. But Colleen has overslept and is rushing to get to field hockey practice. The weather during this first week of the fall semester is unbearably hot and humid, just as it has been all summer. Today, the temperature is predicted to reach 105° F, with high humidity. In her haste to leave, Colleen forgets her water bottle. "Oh well," she thinks as she hops on her bike to peddle the two miles to the practice field.

continued on page 687

In many ways, exercise is the ideal teaching example of physiological integration. Everyone is familiar with it. It is an activity that does not require special environmental conditions, unlike high-altitude activities or deep-sea diving. Exercise is a normal physiological state rather than a pathological one, although it can be affected by disease and, if excessive, can result in injury. The exercising muscle demands a steady supply of ATP, produced through metabolism, and it is this require-

ment that is the underlying cause of the disruption of homeostasis. For that reason, we begin our discussion by examining muscle metabolism.

METABOLISM AND EXERCISE

Exercise begins with the contraction of skeletal muscles, an active process that requires ATP for energy. But where does the ATP for muscle contraction come from? A small amount is in the muscle fiber cytoplasm when contraction begins (Fig. 23-1 ■). As this ATP is transformed into ADP, another phosphate compound, phosphocreatine (PCr), transfers energy from its high-energy phosphate bond to ADP. This transfer replenishes the muscle's supply of ATP (∞ p. 336). However, the combination of muscle ATP and phosphocreatine is adequate to support only about 15 seconds of intense exercise. Consequently, the muscle fiber must manufacture additional ATP from energy stored in the chemical bonds of other molecules. Some of these molecules are contained within the muscle fiber itself; others must be mobilized from the liver and adipose tissue and transported to the muscle through the circulation.

The primary substrates for energy production are carbohydrates and fats. Recall from Chapter 4 that the most efficient production of ATP occurs through aerobic pathways such as the glycolysis-citric acid cycle pathway (∞ p. 91). If the cell has adequate amounts of oxygen

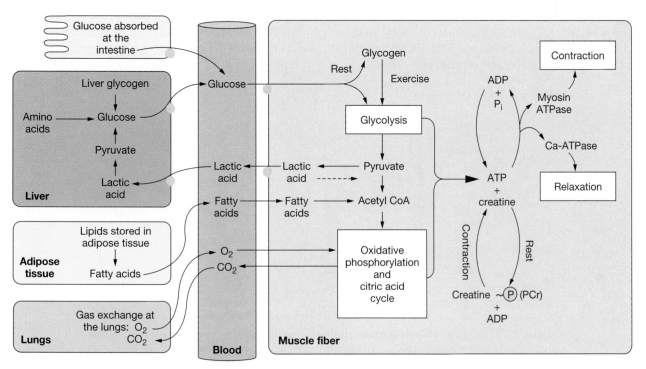

■ **Figure 23-1** **Energy metabolism in skeletal muscle** Glucose for ATP production comes from the diet or from glycogen stored in muscle and liver. Oxygen for aerobic metabolism enters the body at the lungs. Fatty acids from adipose tissue can be used for ATP synthesis only if oxygen is available. If oxygen is not available, glucose is metabolized to lactic acid. The liver is able to convert lactic acid back to glucose. Some energy in the muscle is stored in the form of phosphocreatine (PCr).

for oxidative phosphorylation, then both glucose and fatty acids can be metabolized to provide ATP (Fig. 23-1 ■). If the oxygen requirement of the muscle fiber exceeds its oxygen supply, glucose metabolism shifts to anaerobic glycolysis that ends in lactic acid. With less oxygen available, energy production from fatty acids decreases dramatically.

Anaerobic muscle metabolism is known as **glycolytic metabolism** because anaerobic glycolysis is the primary pathway for ATP production. When the cell has insufficient oxygen, pyruvate, the final product of glycolysis, is converted to lactic acid (∞ p. 90). Anaerobic metabolism produces ATP 2.5 times as rapidly as aerobic pathways do (Fig. 23-2 ■), but it has two distinct disadvantages: (1) It provides only 2 ATP per glucose, compared with the average of 30–32 ATP per glucose for oxidative metabolism; and (2) it contributes to a state of metabolic acidosis by producing lactic acid. (However, the CO_2 generated during metabolism is a more significant source of acid.) In general, exercise that depends on anaerobic metabolism cannot be sustained for an extended period (Fig. 23-2 ■).

The glucose for aerobic and anaerobic ATP production comes from three sources: the plasma glucose pool, intracellular stores of glycogen in muscles and liver, and "new" glucose made in the liver from nonglucose precursors such as amino acids (Fig. 23-1 ■; ∞ p. 97). Muscle and liver glycogen stores provide enough energy substrate for about 2000 kcal (equivalent to about 20 miles of running in the average person), more than adequate for the exercise that most of us do. However, glucose alone cannot provide sufficient ATP for endurance athletes such as marathon runners. To meet their energy demands, they rely on the energy stored in fats, estimated at 70,000 kcal per person.

In reality, aerobic exercise of any duration always uses both fatty acids and glucose as substrates for ATP production. About 30 minutes after aerobic exercise begins, the concentration of free fatty acids in the blood increases significantly, showing that fats are being mobilized from adipose tissue. However, the breakdown of fatty acids through the process of beta oxidation (∞ p. 96) is slower than glycolysis, so muscle fibers use a combination of fatty acids and glucose to meet their energy needs. Aerobic training increases both fat and glycogen stores within the muscle fibers themselves. Training also enhances the ability of exercising muscle to use fatty acids as a source of ATP.

Hormones Regulate Metabolism during Exercise

Several hormones that affect glucose and fat metabolism change their pattern of secretion with exercise. Glucagon, cortisol, the catecholamines (epinephrine and norepinephrine), and growth hormone have all been shown to increase in the plasma during exercise. Cortisol and the catecholamines, along with growth hormone, promote the conversion of triglycerides to glycerol and fatty acids. Glucagon, catecholamines, and cortisol also mobilize liver glycogen and raise plasma glucose levels. A hormonal environment that favors the conversion of glycogen into glucose is desirable, because glucose is the major energy substrate for exercising muscle.

Curiously, although plasma glucose concentrations rise with exercise, the secretion of insulin decreases. This response is contrary to what you would predict, because normally an increase in plasma glucose stimulates insulin release. During exercise, however, insulin secretion is suppressed, probably by sympathetic input onto

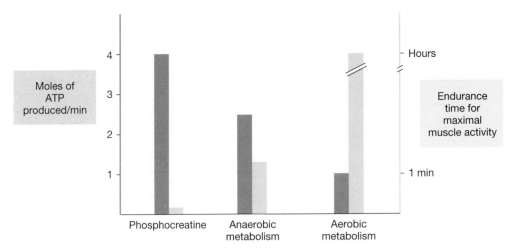

■ Figure 23-2 **Speed of ATP production compared with ability to sustain maximal muscle activity** Phosphocreatine in muscle fibers has the highest rate of energy production. However, a muscle fiber contains only enough phosphocreatine to sustain about 10 seconds of maximal exercise. Anaerobic metabolism can produce ATP at a rate 2.5 times as fast as aerobic metabolism, but it can sustain maximal muscle activity for only about a minute. Aerobic metabolism of glucose and fatty acids is the slowest way for the muscle fiber to produce ATP, but aerobic exercise can be sustained for hours.

the β (beta) cells of the pancreas. Lower blood insulin levels ensure that during exercise, cells other than the muscle fibers will reduce their glucose uptake from blood, sparing glucose for the muscles to use. The muscle cells, on the other hand, are not affected by low levels of insulin: Actively contracting muscle fibers do not require insulin for rapid uptake of glucose. Experiments have shown that a muscle fiber that is electrically stimulated in a glucose solution increases its glucose uptake in proportion to its activity.

Oxygen Consumption Is Related to Exercise Intensity

The activities we call exercise range widely in intensity and duration, from the rapid burst of energy exerted by a sprinter or power lifter to the sustained effort of a marathoner. Physiologists traditionally quantify the intensity of a period of exercise by measuring oxygen consumption ($\dot{V}_{O_2}$). **Oxygen consumption** refers to the fact that oxygen is used up, or consumed, during oxidative phosphorylation, when it combines with hydrogen in the mitochondria to form water (∞ p. 93). Oxygen consumption is a measure of cellular respiration and is usually measured in liters of oxygen consumed per minute.

A person's maximal rate of oxygen consumption ($\dot{V}_{O_2}max$) is an indicator of a person's ability to perform endurance exercise. The greater the $\dot{V}_{O_2}max$, the greater the person's predicted ability to do work.

A metabolic hallmark of exercise is an increase in oxygen consumption that persists even after the activity ceases (Fig. 23-3 ■). When exercise commences, oxygen consumption increases rapidly, but it is not instantaneously matched by the oxygen supply to the muscles. During the lag time, ATP is provided by muscle ATP reserves, phosphocreatine, and aerobic metabolism sup-

continued from page 685

By the time Colleen reaches the practice field, she is already sweating and her face is flushed. At 9 A.M., the air temperature is 80° F, and the humidity is 54%. Colleen takes a quick drink of water from the large water container and runs out to the field.

Question 1: "Humidity" is the percentage of water present in air. Why is the thermoregulatory mechanism of sweating less efficient in humid environments?

ported by oxygen stored on muscle myoglobin and blood hemoglobin (∞ p. 336). The use of these muscle stores creates an *oxygen deficit* because their replacement requires aerobic metabolism and oxygen uptake. Once exercise stops, oxygen consumption is slow to resume its resting level. The excess postexercise oxygen consumption represents oxygen being used to metabolize lactate, restore ATP and phosphocreatine levels, and replenish the oxygen bound to myoglobin. Other factors that play a role in elevating postexercise oxygen consumption include increased body temperature and circulating catecholamines.

Several Factors Limit Exercise

What are the factors that limit exercise capacity? To some extent, the answer depends on the type of exercise. Resistive exercise such as strength training is likely to be limited by the availability of oxygen, so the muscles rely heavily on anaerobic energy production. The picture is more complex with aerobic or endurance exercise. Is the limiting factor for aerobic exercise the ability of the exercising muscle to use oxygen efficiently? The ability of the cardiovascular system to deliver oxygen to the tissues? Or the ability of the respiratory system to provide oxygen to the blood?

One possible limiting factor in exercise is the ability of the muscle fibers to use oxygen. If the muscle mitochondria are limited in number, or if their oxidative enzyme content is insufficient, the muscle fibers are unable to produce ATP rapidly, even if the supply of oxygen and substrates is adequate. Although data suggest that muscle metabolism is not the limiting factor for maximum exercise capacity, it has been shown to influence submaximal exercise capacity—a finding that explains the increase in size and number of muscle mitochondria with endurance training.

The question of whether the pulmonary system or the cardiovascular system limits maximal exercise was resolved when research showed that ventilation is only 65% of its maximum when cardiac output has reached 90% of its maximum. From that information, exercise physiologists concluded that the ability of the cardiovascular system to deliver oxygen and nutrients to the

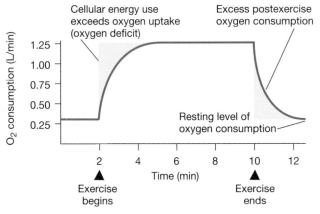

■ **Figure 23-3 Changes in oxygen consumption during and after exercise** The cells' use of energy begins as soon as exercise begins, but oxygen supply as measured by oxygen consumption lags behind. This creates an oxygen deficit that is reflected by elevated oxygen consumption that persists after exercise stops. During the repayment of the oxygen deficit, the muscle fibers replace their stores of ATP and phosphocreatine.

muscle at a rate that supports aerobic metabolism is a major factor in determining maximum oxygen consumption. The next two sections examine the reflexes that integrate breathing and cardiovascular function during exercise.

VENTILATORY RESPONSES TO EXERCISE

Exercise is associated with increased rate and depth of breathing, resulting in enhanced alveolar ventilation (∞ p. 494). **Exercise hyperventilation,** or **hyperpnea,** results from a combination of feedforward signals from the central command neurons in the motor cortex and sensory feedback from peripheral receptors. When exercise begins, muscle mechanoreceptors and proprioreceptors send information to the motor cortex. Descending pathways from the motor cortex to the respiratory control center of the medulla oblongata then immediately increase ventilation. As muscle contraction continues, sensory information from a variety of sources feeds back into the control center to ensure that ventilation and oxygen use by the tissue remain closely matched. The sensory receptors involved in the later response probably include the carotid body chemoreceptors (∞ p. 510), proprioceptors in the joints, and possibly receptors located within the exercising muscle itself. Pulmonary stretch receptors (∞ p. 512) were once thought to play a role, but recipients of heart-lung transplants display a normal ventilatory response to exercise although the neural connections between lung and brain are absent.

Exercise hyperventilation maintains nearly normal arterial P_{O_2} and P_{CO_2} by steadily increasing alveolar ventilation in proportion to the level of exercise. The compensation is so effective that when arterial P_{O_2}, P_{CO_2}, and pH are monitored during mild to moderate exercise, they show no significant change (Fig. 23-4 ■). This observation eliminates the traditional explanation for increased ventilation during exercise. Low arterial P_{O_2}, elevated arterial P_{CO_2}, and decreased plasma pH are the usual stimuli for the carotid and central chemoreceptors. The fact that these stimuli are not present in mild to moderate exercise means that the chemoreceptors or medullary respiratory control center, or both, must be responding to other exercise-induced signals.

Several factors have been postulated as the stimulus for exercise hyperventilation, including sympathetic input on the carotid body and changes in K^+ concentration. During even mild exercise, plasma K^+ levels increase as repeated action potentials in the muscle fibers allow K^+ to move out of the cells and into the extracellular fluid. The carotid chemoreceptors respond to increased K^+ concentration by increasing ventilation. Because the ion concentration changes slowly, however, this mechanism does not explain the sharp initial rise in

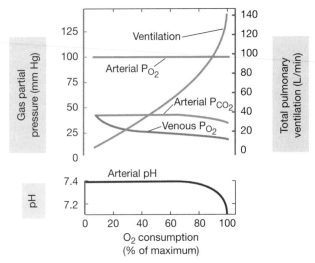

■ Figure 23-4 **Changes in blood gas, partial pressures, and arterial pH with exercise** Pulmonary ventilation increases steadily as exercise increases, but arterial P_{O_2} remains constant, indicating that ventilation is matched to blood flow through the lungs. Venous P_{O_2} drops as exercise increases, showing that the cells are removing more oxygen from hemoglobin as oxygen consumption increases. Arterial P_{CO_2} does not change until oxygen consumption reaches 80% of maximum. At that level, ventilation increases sharply. As the person begins to hyperventilate, arterial (and alveolar) P_{CO_2} declines. Blood pH also remains fairly constant except at near-maximal exercise levels.

ventilation at the onset of activity. It appears likely that sensory input from muscle mechanoreceptors combined with parallel descending pathways from the motor cortex are responsible for the initial change, with sensory input serving to keep ventilation matched to metabolic needs as exercise continues.

✓ If venous P_{O_2} decreases as exercise intensity increases, what do you know about the P_{O_2} of the muscle cells as exercise intensity increases?

CARDIOVASCULAR RESPONSES TO EXERCISE

When exercise begins, mechanosensory input from the working limbs combines with descending pathways from the motor cortex to activate the cardiovascular control center in the medulla oblongata. The center responds with sympathetic discharge that increases cardiac output and causes vasoconstriction in many peripheral arterioles.

Cardiac Output Increases during Exercise

During strenuous exercise, cardiac output rises dramatically. In untrained individuals, cardiac output goes up fourfold, whereas in trained athletes, it may go up six to

continued from page 687

By 10 A.M., the temperature has risen to 93° F, with 50% humidity. Colleen feels dizzy and nauseated, but she presses on. She made the team by the skin of her teeth, and she feels she has to prove herself to her teammates. She takes only a short drink of water at the break and hustles back to the field. At 10:07 A.M. Colleen collapses on the field. Emergency medical services are called. One of Colleen's teammates has received training in first aid. When she feels Colleen's skin, she finds it hot and dry.

Question 2: Individuals with a heat emergency called heat exhaustion have cool, moist skin. Hot, dry skin indicates a more serious emergency called heat stroke. Why is heat exhaustion less serious than heat stroke? (Hint: How does skin temperature relate to the body's ability to regulate body temperature?)

eight times, reaching as much as 35 L/min. As you learned in Chapter 14, cardiac output is influenced by several factors, including (1) heart rate, (2) force of contraction, and (3) venous return (∞ p. 417):

$$\text{Cardiac output (CO)} = \text{heart rate} \times \text{stroke volume}$$

Or, if the factors that influence heart rate and stroke volume are considered,

CO = (SA node rate + influence of autonomic nervous system) × (end-diastolic volume + venous return + force of contraction)

Which of those factors has the greatest effect on cardiac output during exercise in a healthy heart? During exercise, venous return is enhanced by skeletal muscle contraction and deep inspiratory movements (∞ p. 484). It is therefore tempting to postulate that, owing to Starling's law of the heart, the cardiac muscle fibers simply stretch in response to increased venous return, thereby increasing contractility. However, overfilling the ventricles is potentially dangerous, because overstretching may damage the fibers. One way that the body offsets increased venous return is by speeding up the heart rate. If the interval between contractions is less, the heart will not have as much time to fill and therefore is not likely to be damaged by excessive stretch.

The initial change in heart rate at the onset of exercise is due to a decrease in parasympathetic output to the sinoatrial (SA) node (∞ p. 404). As cholinergic inhibition is withdrawn, heart rate rises from its resting rate to around 100 beats per minute, the intrinsic pacemaker rate of the SA node. At that point, sympathetic output from the cardiovascular control center escalates. This escalation has two effects on the heart. First, it increases cardiac contractility so that the heart squeezes out more blood per stroke (increased stroke volume). Second, sympathetic innervation increases heart rate so that the heart does not have much time to relax, protecting it from overfilling. The

combination of faster heart rate and greater stroke volume increases cardiac output during exercise.

Peripheral Blood Flow Redistributes to Muscle during Exercise

At rest, skeletal muscles receive less than a fourth of the cardiac output, or about 1.2 L/min. During exercise, because of local and reflex reactions, a significant shift in peripheral blood flow takes place. With strenuous exercise by highly trained athletes, the combination of increased cardiac output and vasodilation can increase blood flow through exercising muscle to more than 22 L/min! The relative distribution of blood flow to the tissues also shifts. About 88% of the cardiac output is diverted to the exercising muscle, up from 21% at rest (Fig. 23-5 ■).

The redistribution of blood flow during exercise results from a combination of vasodilation in skeletal muscle arterioles and vasoconstriction in other tissues. At the onset of exercise, sympathetic signals from the cardiovascular control center cause vasoconstriction in peripheral tissues. As muscles become active, changes in the microenvironment of the muscle tissue take place. Tissue O_2 concentrations decrease, while temperature, CO_2, and acid in the interstitial fluid around the muscle fibers increase. All of these factors act as paracrines to cause local vasodilation that overrides the sympathetic signal for vasoconstriction. The net result is shunting of blood flow from inactive tissues to the exercising muscles, where it is needed.

Blood Pressure Rises Slightly during Exercise

What happens to blood pressure during exercise? You learned in Chapter 15 that peripheral blood pressure is determined by a combination of cardiac output and peripheral resistance (∞ p. 387). Cardiac output increases during exercise and therefore contributes to increased blood pressure. The changes resulting from peripheral resistance are harder to predict, however, because some peripheral arterioles are constricting while others are dilating:

$$\text{Mean arterial blood pressure} \propto \text{cardiac output} \times \text{peripheral resistance}$$

Skeletal muscle vasodilation causes a decrease in peripheral resistance to blood flow. Sympathetically-induced vasoconstriction in nonexercising tissues offsets the vasodilation, but only partially. Consequently, total peripheral resistance to blood flow falls dramatically as exercise commences, reaching a minimum at about 75% of $\dot{V}_{O_2}$max (Fig. 23-6a ■). If no other compensation occurred, this decrease in peripheral resistance would dramatically lower arterial blood pressure. However, increased cardiac output cancels out the decreased peripheral resistance. When blood pressure is monitored during exercise, mean arterial blood pressure actually increases slightly as exercise intensity increases

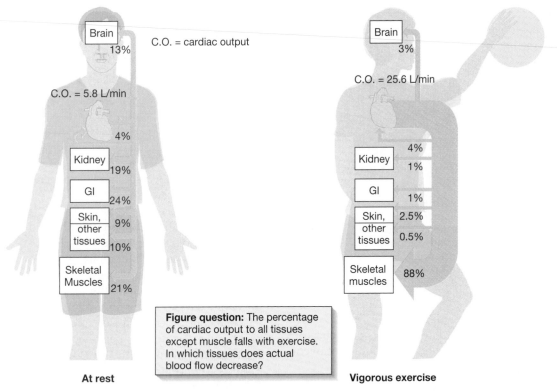

Figure question: The percentage of cardiac output to all tissues except muscle falls with exercise. In which tissues does actual blood flow decrease?

At rest   **Vigorous exercise**

■ Figure 23-5 **Distribution of cardiac output at rest and during exercise** The shunting of blood from nonexercising tissues to muscle is accomplished by sympathetically induced vasoconstriction in the nonexercising tissues combined with vasodilation in muscle.

(Fig. 23-6b ■). The fact that it increases at all, however, suggests that the normal baroreceptor reflexes that control blood pressure are not functioning during exercise.

The Baroreceptor Reflex Adjusts to Exercise

Normally, homeostasis of blood pressure is regulated through the peripheral baroreceptors of the carotid and aortic bodies: An increase in blood pressure initiates responses that return blood pressure to normal. But dur-

✓ Look at Figure 23-6 ■. If peripheral resistance is falling as exercise intensity increases, but peripheral blood pressure is rising only slightly, what must be happening to cardiac output?

✓ In Figure 23-6 ■, why does the line for mean pressure lie closer to the diastolic pressure line instead of being evenly centered between systolic and diastolic pressures?

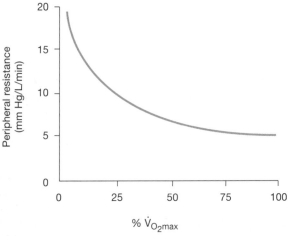

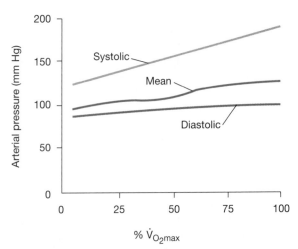

■ Figure 23-6 **Peripheral resistance and arterial blood pressure during exercise** (a) Peripheral resistance falls as the level of exercise increases, primarily because of vasodilation in exercising muscle. (b) Mean arterial blood pressure rises slightly during exercise, despite the drop in resistance. This rise is due to increased cardiac output that results from an increase in heart rate and stroke volume.

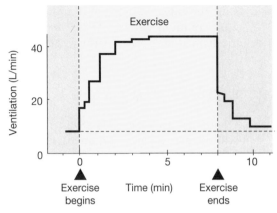

■ **Figure 23-7 Changes in ventilation with exercise**
Ventilation rate jumps as soon as exercise begins, despite the fact that neither arterial P_{CO_2} nor P_{O_2} has changed. This increase in ventilation is a feedforward response initiated by commands from the motor cortex. Modified from P. Dejours, *Handbook of Physiology* (Washington, D.C.: American Physiological Society, 1964).

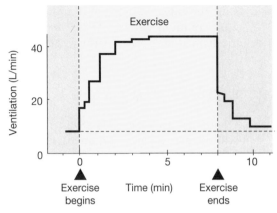

When the paramedics arrive, they take Colleen's blood pressure. It is extremely low. Her body temperature is 103° F. The paramedics quickly transport Colleen to the hospital.

continued from page 689

Question 3: Why is Colleen's blood pressure so low?

ing exercise, blood pressure increases without activating homeostatic compensation. Why is the normal baroreceptor reflex absent during exercise? There are several theories. According to one, the output from the motor cortex during exercise resets the threshold for the arterial baroreceptors to a higher pressure. Blood pressure can then increase slightly during exercise without triggering the homeostatic counterregulatory responses. Another theory suggests that signals in baroreceptor afferent neurons are blocked in the spinal cord by presynaptic inhibition (∞ p. 365) at some point before the afferent neurons synapse with central nervous system neurons. This inhibition makes the baroreceptor reflex ineffective during exercise.

A third theory is based on the postulated existence of muscle chemoreceptors that are sensitive to metabolites, probably H^+, produced during strenuous exercise. When stimulated, these chemoreceptors signal the central nervous system that blood flow in the tissue is not adequate to remove muscle metabolites or to keep the muscle in aerobic metabolism. The chemoreceptor input is reinforced by sensory input from mechanoreceptors in the working limbs. The central nervous system response to this sensory input is to raise blood pressure, overriding the baroreceptor reflex, in order to enhance blood flow to the muscle. The same hypothetical muscle chemoreceptors may play a role in ventilatory responses to exercise.

FEEDFORWARD RESPONSES TO EXERCISE

Interestingly, there is a significant feedforward element in the physiological responses to exercise. It is easy to explain the physiological changes that occur with exercise as reactions to the disruption of homeostasis. However, many of these changes occur in the absence of the normal stimuli or before the stimuli are present. For example, ventilation rates jump as soon as exercise begins, even though experiments have shown that arterial P_{O_2} and P_{CO_2} do not change (Fig. 23-7 ■).

As exercise begins, proprioceptors in the muscles and joints send information to the motor cortex of the brain. Descending signals from the motor cortex go not only to the exercising muscles but also along parallel pathways to the cardiovascular and respiratory control centers and to the limbic system of the brain.

Output from the limbic system and cardiovascular control center triggers generalized sympathetic discharge. As a result, an immediate slight increase in blood pressure marks the beginning of exercise. The sympathetic discharge causes widespread vasoconstriction, increasing blood pressure. Once exercise has begun, this increase in blood pressure compensates for decreases in blood pressure resulting from muscle vasodilation. However, at the onset of exercise, the muscle capillaries are not yet dilated by metabolic paracrines. Consequently, the anticipatory sympathetic vasoconstriction has the potential to elevate blood pressure to potentially lethal levels. To prevent this dangerous elevation of blood pressure, special sympathetic neurons *dilate* (instead of constrict) selected arterioles in the exercising muscles. The peripheral resistance in the muscle falls, and the blood pressure does not rise excessively. The sympathetic fibers to the muscle that control this response are part of a unique system called the **cholinergic vasodilator system.** These sympathetic neurons are notable because they use acetylcholine (not norepinephrine) as their neurotransmitter.

This anticipatory dilation of muscle arterioles by the cholinergic vasodilator system does not improve muscle oxygen and glucose supply. The arterioles involved in the cholinergic vasodilator reflex do not supply the capillaries that provide nutrition to the muscle fibers. Increased oxygen and nutrient supply to the muscle fibers must await the onset of exercise because the muscle arterioles leading to those capillaries are supplied by the more common adrenergic sympathetic neurons.

As exercise proceeds, reactive compensations become superimposed on the feedforward changes. For example, when exercise reaches 50% of aerobic capacity, chemoreceptors in the muscles sense the buildup of lactic acid and metabolites and send this information along to the central command centers in the brain. The command centers then maintain the changes in ventilation and circulation that were initiated in a feedforward manner. Thus, the integration of systems in exercise involves both common reflex pathways and some unique centrally mediated reflex pathways.

TEMPERATURE REGULATION

As exercise continues, heat released through metabolism creates an additional challenge to homeostasis. Most of the energy released during metabolism is not converted into ATP but is released as heat. (Energy conversion from organic substrates to ATP is only 20%–25% efficient.) With continued exercise, heat production exceeds heat loss, and core body temperature rises. In endurance events, body temperature can reach 40°–42° C (104°–108° F), which we would normally call a fever. This rise in body temperature with exercise triggers two thermoregulatory mechanisms: sweating and increased cutaneous blood flow (∞ p. 615). Both mechanisms help regulate body temperature but have the potential to disrupt homeostasis in other ways.

Sweating lowers body temperature through evaporative cooling. At the same time, however, the loss of fluid from the extracellular compartment has the potential to cause dehydration and significantly decrease circulating blood volume. Because sweat is a hypotonic fluid, body osmolarity increases in response to water loss. The combination of decreased extracellular fluid volume and increased osmolarity during extended exercise sets in motion the complex homeostatic pathway for dehydration described in Chapter 19, including thirst and renal conservation of water (∞ p. 560).

The other thermoregulatory mechanism, increased blood flow to the skin, allows loss of body heat to the environment through convective heat loss (∞ p. 614). However, increased sympathetic output during exercise tends to vasoconstrict cutaneous blood vessels and oppose the thermoregulatory response. The primary control of vasodilation in the skin during exercise appears to come from the sympathetic cholinergic vasodilator system. Activation of these neurons as body core temperature rises permits dilation of cutaneous blood vessels without altering sympathetic vasoconstriction in other tissues of the body.

Although vasodilation of cutaneous vessels is essential for thermoregulation, it disrupts homeostasis by decreasing peripheral resistance and diverting blood flow from the muscles. In the face of these contradictory demands, the body initially gives preference to thermoregulation. However, if central venous pressure falls below a critical minimum, the body abandons thermoregulation in the interest of maintaining blood flow to the brain.

The degree to which the body can adjust to both demands depends on the type of exercise being performed and its intensity and duration. Strenuous exercise in hot, humid environments can severely impair the normal thermoregulatory mechanisms and cause heat

continued from page 691

When Colleen arrives in the emergency room, she is quickly diagnosed with heat stroke, a life-threatening condition. She is immersed in a tub of cool water and given intravenous fluids. Heat stroke can occur in athletes who overexert in excessively hot weather. Heat stroke is also common in elderly individuals, whose thermoregulatory mechanisms are less efficient than those in younger people.

Question 4: What steps could Colleen have taken to avoid heat stroke? (Start at 9 A.M., when Colleen set out for the practice field.)

stroke, a potentially fatal condition. Unless prompt measures are taken to cool the body, core temperatures can go as high as 43° C (109° F). It is possible for the body to adapt to repeated exercise in hot environments, however, through a process known as **acclimatization.**

As the body adjusts to exercise in the heat, sweating begins sooner and is more copious, doubling or tripling in volume, enhancing evaporative cooling. With acclimatization, sweat also becomes more dilute, as salt is reabsorbed from the sweat glands under the influence of increased aldosterone. Salt loss in an unacclimatized person exercising in the heat may reach 30 g NaCl per day, but that value drops to as little as 3 g after a month of acclimatization.

EXERCISE AND HEALTH

Physical activity has many positive effects on the human body. Our lifestyles have changed dramatically since humans were hunter-gathers, but our bodies still seem to work best with a certain level of physical activity. Several common pathological conditions—including high blood pressure, strokes, and diabetes mellitus—can be improved by physical activity. Even so, developing regular exercise habits is one lifestyle change that many people find difficult to make. In this section, we look at the effect that exercise has on several common conditions.

Exercise Lowers the Risk of Cardiovascular Disease

As early as the 1950s, scientists showed that men who are physically active have a lower rate of heart attacks than do men who lead a sedentary lifestyle. These studies started many investigations into the exact relationship between cardiovascular disease and exercise. Subsequently, it has been demonstrated that exercise has positive benefits, including lowering blood pressure, decreasing plasma triglyceride levels, and raising plasma high-density lipoprotein (HDL) levels. High blood pressure (hypertension) is a major risk factor for strokes, while elevated triglycerides and LDL-cholesterol levels are associated with development of atherosclerosis and increased risk of heart attack. Overall, exercise reduces the risk of death or illness from a variety of cardiovascular diseases. Even mild exercise, such as walking, has significant health benefits and could help the estimated 40 million American adults with sedentary lifestyles that put them at risk for developing cardiovascular disease, complications of obesity, or diabetes mellitus.

Non-Insulin-Dependent Diabetes Mellitus May Improve with Exercise

The effectiveness of regular exercise in preventing and alleviating non-insulin-dependent, or type II, diabetes mellitus has now been well established. With chronic exercise, skeletal muscle fibers up-regulate both the number of glucose transporters and the number of insulin receptors on their membrane. The addition of transporters decreases the muscle's dependence on insulin for glucose uptake. By increasing the number of insulin receptors, the fibers become more sensitive to insulin: A smaller amount of insulin can achieve a response that previously required more insulin. Because the cells are responding to lower insulin levels, the stress on the endocrine pancreas created by secreting insulin is also reduced, resulting in a lower incidence of type II diabetes mellitus.

Figure 23-8 ■ shows the results of an experiment in which men with impaired glucose tolerance were started on an exercise regimen. Their response to glucose ingestion is compared with that of normal subjects. In each experiment, the individual ingested 100 g of glucose and the changes in his plasma glucose levels were monitored for 120 minutes (this procedure is known as a glucose tolerance test). Simultaneous measurements

were made of plasma insulin. After only seven days of exercise, both the glucose tolerance test and insulin secretion had shifted to a pattern that was more like that of the normal subjects. Because of the beneficial effect of exercise on glucose transport and metabolism, patients with type II diabetes are encouraged to begin a regular exercise program.

Exercise, Stress, and Immunity

Another area that is receiving much attention is the interaction of exercise with the immune system. Epidemiological evidence looking at large populations of people associates exercise with a reduced incidence of disease and with increased longevity. There is also widespread public belief that exercise boosts immunity, prevents cancer, and helps HIV-infected patients combat AIDS. But so far, the results of rigidly controlled research studies have not supported those claims. Indeed, there is good evidence suggesting that strenuous exercise is a form of stress and will suppress the immune response because of corticosteroid release. Some researchers have proposed that the relationship between exercise and decreased immunity forms a J-shaped curve (Fig. 23-9 ■). People who exercise moderately have better immune systems than those who are sedentary, but people who exercise strenuously may see a decrease in immune function because of the stress of the exercise. With so much contradictory information available about the link between exercise and immunity, it is not possible at this time to make any firm conclusions.

Another area of exercise physiology filled with interesting though contradictory results is the effect of exercise on stress, depression, and other psychological

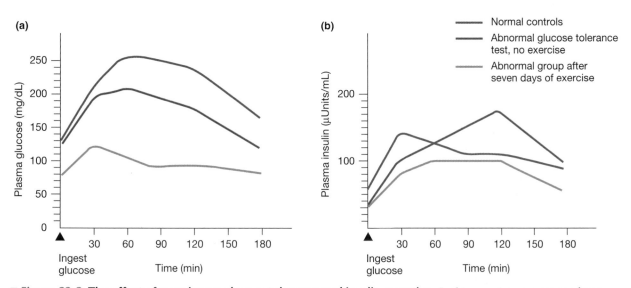

■ **Figure 23-8 The effect of exercise on glucose tolerance and insulin secretion** In this experiment, a group of men with abnormal glucose tolerance tests (red line, part a) exercised for seven days and were tested again. After seven days of exercise, their glucose tolerance test (green line, part a) had shifted in the direction of normal controls, as had their pattern of insulin secretion (green line, part b). Exercise is known to enhance glucose uptake and increase muscle sensitivity to insulin. Data from B. R. Seals et al., "Glucose Tolerance in Older and Younger Athletes and Sedentary Men," *Journal of Applied Physiology* 56(6):1521–1525 (1984); and M. A. Rogers et al., "Improvement in Glucose Tolerance after One Week of Exercise in Patients with Mild NIDDM," *Diabetes Care* 11:613–618 (1988).

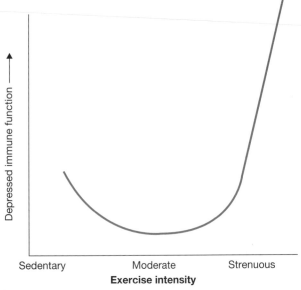

■ **Figure 23-9 Immune function and exercise** This graph shows one theory of the relationship between immune function and exercise. Moderate exercise enhances immunity, but strenuous exercise is a form of stress that depresses immunity.

parameters. The literature shows an inverse relationship between exercise and depression: People who exercise regularly are significantly less likely to be clinically depressed than are people who do not exercise regularly. Although the association exists, assigning cause and effect to the two parameters is difficult. Are the exercisers less depressed because they exercise? Or do the depressed subjects exercise less because they are depressed?

Many published studies appear to show that regular exercise is effective in reducing depression. But a careful analysis of experimental design suggests that the conclusions of those studies may be overstated. The subjects in many of the experiments were being treated concurrently with drugs or psychotherapy, and it is difficult to attribute improvement in their condition solely to exercise. In addition, participation in exercise studies gives subjects a period of social interaction, another factor that might play a role in the reduction of stress and depression. The assertion that exercise reduces depression would be better supported by a demonstration of the effects of exercise on biological parameters such as disturbances of monoamine neurotransmitters (∞ p. 226). To date, however, little work of that type has been done.

CHAPTER REVIEW

Chapter Summary

1. **Exercise** is any muscular activity that generates force and disrupts homeostasis. (p. 684)

Metabolism and Exercise

2. Exercising muscle demands a steady supply of ATP, produced primarily through aerobic metabolism. A small amount of ATP is stored in the muscle fiber. Additional ATP can be rapidly formed from phosphocreatine. (p. 685)

3. Carbohydrates and fats are the primary substrates for energy production. Glucose can be metabolized through both oxidative and anaerobic pathways, but fatty acid metabolism requires adequate amounts of oxygen. (p. 685)

4. Anaerobic, or **glycolytic, metabolism** converts glucose to lactic acid. Glycolytic metabolism is 2.5 times as rapid as aerobic pathways but is not as efficient at ATP production and contributes to a state of metabolic acidosis. (p. 686)

5. Aerobic exercise uses both fatty acids and glucose as substrates for ATP production. (p. 686)

6. Glucagon, cortisol, the catecholamines, and growth hormone influence glucose and fatty acid metabolism during exercise. A hormonal environment that favors the conversion of glycogen into glucose is desirable. (p. 686)

7. Although plasma glucose concentrations rise with exercise, the secretion of insulin decreases. This response reduces glucose uptake by most cells and makes more glucose available in the blood for exercising muscle. (p. 686)

8. The intensity of exercise is indicated by oxygen consumption ($\dot{V}_{O_2}$). A person's maximal rate of oxygen consumption ($\dot{V}_{O_2max}$) is an indicator of that person's ability to perform endurance exercise. (p. 687)

9. Oxygen consumption increases rapidly at the onset of exercise, but the initial energy supply to the muscles is provided by muscle ATP reserves, phosphocreatine, and aerobic metabolism supported by oxygen stored on muscle myoglobin and blood hemoglobin. (p. 687)

10. Postexercise oxygen consumption remains elevated owing to ongoing metabolism, increased body temperature, and circulating catecholamines. (p. 687)

11. Muscle mitochondria increase in size and number with endurance training, suggesting that muscle metabolism may limit submaximal exercise capacity. (p. 687)

12. In maximal exertion, when cardiac output reaches 90% of its maximum, ventilation is only 65% of its maximum. The ability of the cardiovascular system to deliver oxygen and nutrients thus appears to be a major factor limiting maximal exertion. (p. 687)

Ventilatory Responses to Exercise

13. **Exercise hyperventilation** results from a combination of feedforward signals from the central command neurons in the motor cortex and sensory feedback from peripheral chemoreceptors, proprioceptors in the joints, and mechanoreceptors located within the exercising muscle itself. (p. 688)

14. Arterial P_{O_2}, P_{CO_2}, and pH show no significant change during mild to moderate exercise. Exercise hyperventilation appears to be due to some other factor such as increased plasma K^+ levels. (p. 688)

Cardiovascular Responses to Exercise

15. Cardiac output increases with exercise because of increased venous return and sympathetic stimulation of heart rate and contractility. (p. 688)

16. Blood flow through exercising muscle increases dramatically when skeletal muscle arterioles dilate. Arterioles in other tissues constrict. (p. 689)

17. Decreased tissue O_2 and glucose and increased muscle temperature, CO_2, and acid act as paracrines and cause local vasodilation. (p. 689)

18. Mean arterial blood pressure increases slightly as exercise intensity increases. The normal baroreceptor reflexes that control blood pressure do not function during exercise. (p. 689)

Feedforward Responses to Exercise

19. Many of the physiological changes that occur at the beginning of exercise are initiated in the absence of the normal stimuli. For example, a **cholinergic vasodilator** **system** dilates selected muscle arterioles. As exercise proceeds, reactive compensations become superimposed on the anticipatory changes. (p. 691)

Temperature Regulation

20. Heat released during exercise is dissipated by sweating and increased cutaneous blood flow. (p. 692)

Exercise and Health

21. Physical activity has many positive effects and can help prevent or decrease the risk of developing high blood pressure, strokes, and diabetes mellitus. (p. 692)

Questions

LEVEL ONE   Reviewing Facts and Terms

1. Exercise is defined as a _____ activity that generates _____ and disrupts _____.

2. The most efficient ATP production comes through what kind of metabolic pathways? When these pathways are being used, then *glucose/fatty acids/both/neither* can be metabolized to provide ATP.

3. If the skeletal muscle's requirement for _____ is not met, metabolism shifts to _____ that results in the production of lactic acid. This type of muscle metabolism is also known as _____.

4. List three sources of glucose that can be metabolized to ATP, either directly or indirectly.

5. List four hormones that promote the conversion of triglycerides into fatty acids. What effects do these hormones have on plasma glucose levels?

6. What is meant by the term *oxygen deficit*?

7. How does exercise hyperventilation maintain nearly normal P_{O_2} and P_{CO_2}?

8. What role is K^+ hypothesized to play in exercise?

9. In endurance events, body temperature can reach 40–42° C. What is normal body temperature? What two thermoregulatory mechanisms are triggered by this change in temperature during exercise?

10. Sweating is one way that the body thermoregulates. What is a potentially bad result of excessive sweating?

LEVEL TWO   Reviewing Concepts

11. **Concept map:** Map the metabolic, cardiovascular, and respiratory changes that occur with exercise. Include the signals to and from the nervous system and show what specific areas signal and coordinate the exercise response.

12. What causes insulin secretion to decrease during exercise and why is this decrease adaptive?

13. State two advantages and two disadvantages of anaerobic glycolysis.

14. Compare and contrast the following terms, especially as they relate to exercise:
    a. ATP, ADP, PCr
    b. myoglobin, hemoglobin

15. List and discuss five factors that limit exercise capacity.

16. Match the brain area to the response it controls. Answers may be used once, more than once, or not at all. Some responses may be caused by more than one area.
    (a) pons
    (b) medulla oblongata
    (c) midbrain
    (d) motor cortex
    (e) hypothalamus
    (f) cerebellum
    (g) withdrawal of parasympathetic output
    (h) increased sympathetic output
    (i) brain not involved; uses local control

    1. increased cardiac output
    2. vasoconstriction in many peripheral arterioles
    3. exercise hyperventilation
    4. increased stroke volume
    5. increased heart rate
    6. vasodilation of muscle arterioles
    7. coordination of skeletal muscle movement

17. Specify whether each parameter listed below stays the same, increases, or decreases as a person becomes better conditioned for athletic activities:
    a. heart rate during exercise
    b. resting heart rate
    c. cardiac output during exercise
    d. resting cardiac output
    e. breathing rate during exercise
    f. blood flow to muscles during exercise
    g. blood pressure during exercise
    h. total peripheral resistance during exercise

18. Diagram or explain the adjustments the heart makes to increase cardiac output during exercise and avoid damage to itself.

19. Diagram the three theories that have been put forth to explain why blood pressure is allowed to increase during exercise.

20. What is the cholinergic vasodilator system? What is the primary function of this system?

21. List and briefly discuss the benefits of a lifestyle that includes regular exercise.

22. Explain how exercise could improve non-insulin-dependent diabetes mellitus.

LEVEL THREE   Problem Solving

23. You have decided to manufacture a new sports drink that will help athletes, from football players to gymnasts. List at least five different ingredients you would put in your drink and tell why each one is important for the athlete.

24. You are a well-conditioned athlete. At rest, your heart rate is 60 beats per minute and your stroke volume is 70 mL/beat. What is your cardiac output? At one point during exercise, your heart rate goes up to 120 beats/min. Does your cardiac output increase proportionately? Explain.

25. How could a severe sunburn affect an athlete's ability to perform at his or her best?

Problem Conclusion

In this running problem, you learned about a dangerous heat-related emergency called heat stroke. In heat stroke, dehydration seriously impairs the body's thermoregulatory mechanisms and leads to low blood pressure, which in turn prevents sweating and leads to a rise in body temperature.

Further check your understanding of this running problem by checking your answers against those in the summary table.

	Question	Facts	Integration and Analysis
1	Why is the thermo-regulatory mechanism of sweating less efficient in humid environments?	"Humidity" is the percentage of water present in air. Thermoregulation in warm environments includes sweating and evaporative cooling.	Evaporation is slower in humid air; therefore, evaporative cooling is less effective in high humidity.
2	Individuals with a heat emergency called heat exhaustion have cool, moist skin. Hot, dry skin indicates a more serious emergency called heat stroke. Why is heat exhaustion less serious than heat stroke?	Sweating helps the body regulate temperature through evaporative cooling. As evaporation occurs, the skin surface cools.	Cool, moist skin indicates that the sweating mechanism is still functioning. Hot, dry skin indicates that the sweating mechanism has failed. Subjects with hot, dry skin are likely to have higher internal temperatures.
3	Why is Colleen's blood pressure so low?	Blood volume decreases when the body loses large amounts of water through sweating.	Colleen has been sweating but not replacing the fluid she lost. This has caused her blood volume to decrease, with a corresponding decrease in blood pressure.
4	What steps could Colleen have taken to avoid heat stroke? (Start at 9 A.M., when Colleen set out for the practice field.)	Heat stroke is caused by dehydration as a result of excessive sweating. This results in a drop in blood pressure. Peripheral blood vessels constrict in an effort to maintain pressure and blood flow to the brain. Constricted blood vessels in the skin cannot release excess heat. In addition, the sweating response is inhibited to prevent further fluid loss.	Colleen could have avoided heat stroke by taking the following measures: (1) consuming large amounts of fluid (water plus electrolytes) before and during practice; (2) avoiding excessive exercise such as bicycling to practice; (3) stopping exercising at the first signs of heat emergency (dizziness and nausea); (4) seeking shade and drinking large amounts of fluid to replenish lost fluids; (5) using cool wet towels or ice to bring body temperature lower.

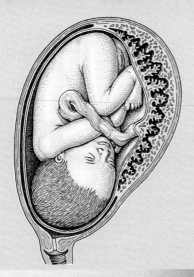

24

Reproduction and Development

Imagine growing up as a girl, then at the age of 12 or so, finding that your voice is deepening and your genitals are developing into those of a man. This scenario actually happens to a small number of men with a condition known as *pseudohermaphroditism* [*pseudes*, false + *hermaphrodites*, the bisexual offspring of Hermes and Aphrodite]. These men have the internal sex organs of a male but inherit a gene that causes a deficiency in one of the male hormones. Consequently, they are born with external genitalia that appear feminine, and they are raised as girls. At puberty, when they begin to secrete male hormones, they develop some, but not all, of the characteristics of men. As you can imagine, a conflict arises: Change gender, or remain a woman? Most choose to continue life as men.

Reproduction is one area of physiology in which we humans like to think of ourselves as significantly advanced over other animals. We mate for pleasure as well as procreation, and the female human is always sexually receptive, not only during her fertile periods. But just how different are we?

Like other terrestrial animals, we have internal fertilization that allows motile flagellated sperm to remain in an

MALE REPRODUCTION

The male reproductive system consists of the testes, the accessory glands and ducts, and the external genitalia. The external genitalia consist of the **penis** and the **scrotum,** a saclike structure that contains the testes (Fig. 24-8 ■). The **urethra** serves as a common passageway for sperm and urine, although not simultaneously. It runs through the ventral aspect of the shaft of the penis, surrounded by a spongy column of tissue known as the **corpus spongiosum** [*corpus*, body; plural *corpora*]. The corpus spongiosum along with two columns of tissue called the **corpora cavernosa** constitute the erectile tissue of the penis. The tip of the penis is enlarged into a region called the **glans** that at birth is covered by a layer of skin called the **foreskin,** or **prepuce.** In some cultures, the foreskin is removed surgically in a procedure called **circumcision.** In the United States, this practice goes through cycles of popularity. Proponents of the procedure claim that it is necessary for good hygiene and they cite evidence suggesting that the incidence of penile cancer, sexually transmitted diseases, and urinary tract infections is lower in circumcised men. Opponents claim that it is cruel to subject newborn boys to an unnecessary and possibly painful procedure.

The scrotum is an external sac into which the testes migrate during fetal development. This location outside the abdominal cavity is necessary because normal sperm development requires a temperature that is 2°–3° F lower than body core temperature. The failure of one or both testes to descend is known as **cryptorchidism** [*crypto*, hidden + *orchis*, testicle] and occurs in 1%–3% of all male births. If left alone, about 80% of cryptorchid testes spontaneously descend later. Those that remain in the abdomen through puberty become sterile and unable to produce sperm. Although they lose their gametogenic potential, cryptorchid testes can produce androgens, indicating that hormone production is not as temperature-sensitive as sperm production. Because undescended testes are prone to become cancerous, authori-

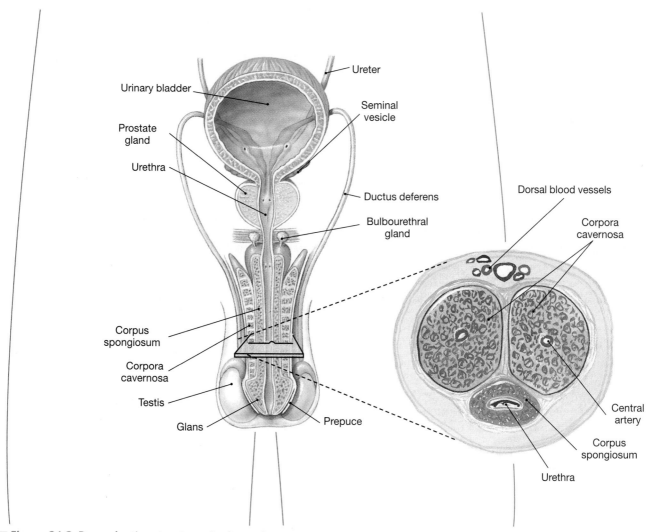

■ Figure 24-8 **Reproductive structures in the male**

ties recommend that they be moved to the scrotum with testosterone treatment or surgery if necessary.

The accessory glands and ducts include the **prostate gland,** the **seminal vesicles,** and the **bulbourethral (Cowper's) glands.** The bulbourethral glands and seminal vesicles empty their secretions into the urethra through ducts. The individual glands of the prostate open directly into the urethral lumen.

The prostate gland is the best-known of the three accessory glands because of its medical significance. Cancer of the prostate is the second most common form of cancer in men, and benign prostatic hypertrophy (enlargement) creates problems for many men after age 50. Because the prostate gland completely encircles the urethra, its enlargement causes difficult urination by narrowing the passageway.

Sperm Production Takes Place in the Testes

The human testes are paired ovoid structures about 5 cm by 2.5 cm (Fig. 24-9 ■). The word *testis* means "witness" in Latin, and its application to the male gonad comes from the fact that in ancient Rome, men taking an oath placed one hand on their genitals.

The testes have a tough outer fibrous capsule that encloses masses of coiled **seminiferous tubules** clustered into 250–300 compartments. Between the tubules is interstitial tissue consisting primarily of blood vessels and testosterone-producing Leydig (interstitial) cells. The seminiferous tubules constitute nearly 80% of the testicular mass in an adult. Each individual tubule is 0.3–1 meter long and, if stretched out and laid end to end, the entire mass would extend for about the length of two and a half football fields.

The seminiferous tubules leave the testis and join the **epididymis** [*epi-,* upon + *didymos,* twin], a single duct that forms a tightly coiled cord on the surface of the testicular capsule (Fig. 24-9b ■). The epididymis becomes the **vas deferens** [*vas,* vessel + *deferre,* to carry away from], also known as the **ductus deferens.** This duct passes into the abdomen, where it eventually empties into the urethra, the passageway from the urinary bladder to the external environment (see Fig. 24-8 ■).

The seminiferous tubules A seminiferous tubule is composed of two types of cells: spermatocytes in various stages of becoming sperm, and Sertoli cells (Fig. 24-9d, e ■). The developing spermatocytes stack in columns from the outer edge of the tubule to the inner lumen. Between each column of spermatocytes is a single Sertoli cell that extends from the outer edge of the tubule to the lumen.

Surrounding the outside of the tubule is a basement membrane. It acts as a barrier, preventing certain large molecules in the interstitial fluid from entering the tubule, yet allowing testosterone to pass easily (Fig. 24-9c ■). Adjacent Sertoli cells in a tubule are linked to each other by tight junctions that form an additional barrier between the lumen of the tubule and the interstitial fluid outside the tubule. The tight junctions are sometimes called the **blood-testis barrier** because functionally they behave much like the impermeable capillaries of the blood-brain barrier, preventing movement of material between two compartments. However, the testes have three functional compartments: (1) the interstitial fluid surrounding the tubules, (2) a basal compartment in the tubules between the basement membrane and basal ends of the Sertoli cells, and (3) an inner compartment that includes the lumen of the tubule. Because of the barriers between these three compartments, fluid in the tubule lumen has a composition different from that of interstitial fluid outside the tubule, with low concentrations of glucose and high concentrations of K^+ and steroid hormones.

Sperm production Spermatogonia, the germ cells that undergo meiotic division to become sperm, are found clustered near the basal ends of the Sertoli cells, just under the basement membrane of the seminiferous tubules. In this basal compartment, they undergo mitotic replication to produce additional spermatogonia. Some of the germ cells remain near the outer edge of the tubule to produce future spermatogonia, while others enter meiosis and become primary spermatocytes. As the spermatocytes differentiate into sperm, they move inward toward the tubule lumen, continuously surrounded by the Sertoli cells. The tight junctions of the blood-testis barrier break and reform around the migrating cells, ensuring that the barrier remains intact. By the time the

continued from page 699

Infertility can be caused by problems in either the man or woman. Sometimes, however, both partners have problems that contribute to their infertility. In general, male infertility is caused by low sperm counts, abnormalities in sperm morphology, or abnormalities in the reproductive structures that carry sperm. Female infertility may be caused by problems in hormonal pathways that govern maturation and release of eggs, or by abnormalities of the reproductive structures (cervix, uterus, ovaries, oviducts). Because tests of male fertility are simple to perform, Dr. Coddington first analyzes Larry's sperm. In this test, a sperm sample is examined under a microscope. Sperm shape is noted and the concentration of sperm in the sample is estimated. The motility of the sperm is also analyzed.

Question 1: Name the male reproductive structures that carry sperm from the testes to the external environment.

Question 2: A new technique for the treatment of male infertility involves retrieval of sperm from the epididymis. The retrieved sperm can then be used to fertilize an egg, which is then implanted in the uterus. What causes of male infertility might make this treatment necessary?

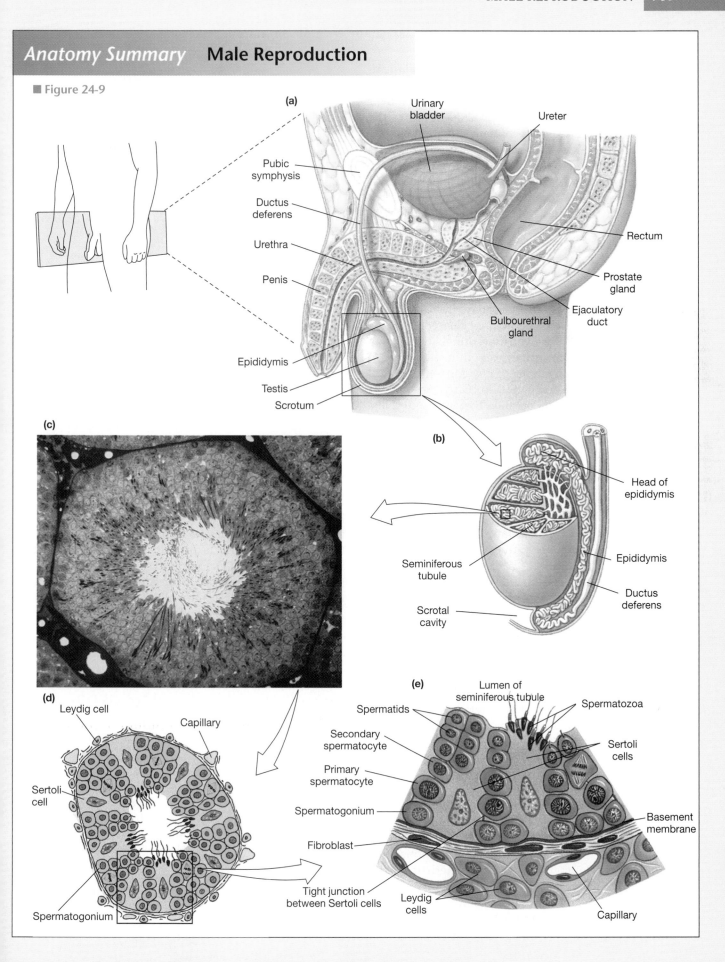

Anatomy Summary   Male Reproduction

■ Figure 24-9

(a)

- Urinary bladder
- Ureter
- Pubic symphysis
- Ductus deferens
- Urethra
- Penis
- Rectum
- Prostate gland
- Ejaculatory duct
- Bulbourethral gland
- Epididymis
- Testis
- Scrotum

(b)

- Head of epididymis
- Seminiferous tubule
- Epididymis
- Ductus deferens
- Scrotal cavity

(c)

(d)

- Leydig cell
- Capillary
- Sertoli cell
- Spermatogonium

(e)

- Lumen of seminiferous tubule
- Spermatids
- Spermatozoa
- Secondary spermatocyte
- Sertoli cells
- Primary spermatocyte
- Spermatogonium
- Basement membrane
- Fibroblast
- Tight junction between Sertoli cells
- Leydig cells
- Capillary

spermatocytes reach the luminal end of the Sertoli cell, they have divided twice and become spermatids.

The spermatids remain embedded in the apical membrane of the Sertoli cells while they complete the transformation into sperm. In this process, they lose most of their cytoplasm and develop a flagellated tail (Fig. 24-10 ■). The chromatin of the nucleus condenses into a dense structure, while a lysosome-like vesicle called an **acrosome** flattens out to cover most of the surface of the nucleus. The acrosome contains enzymes that are essential for fertilization. Mitochondria to produce energy for sperm movement concentrate in the midpiece of the sperm body, along with microtubules that extend into the tail (∞ p. 47). The result is a small, motile gamete that bears little resemblance to the parent spermatid. Finally, the sperm are released into the lumen of the tubule, along with secreted fluid. From there, they are free to move out of the testis.

The entire process takes about 64 days from spermatogonium division until sperm are released into the tubule. However, at any given time, different regions of the tubule contain spermatocytes in different stages of development. The staggering of developmental stages allows sperm production to remain nearly constant at a rate of 200 million sperm per day. That may sound like an extraordinarily high number, but it is about the number of sperm released in a single ejaculation.

Sperm released from the Sertoli cells are not yet mature or capable of swimming. They are pushed through the lumen of the seminiferous tubule by other sperm and by bulk flow of fluid secreted by the Sertoli cells. Sperm entering the epididymis complete their maturation during the 12 or so days of their transit time, aided by protein secretions from the epididymal cells.

The Sertoli cells The function of the Sertoli cells is to regulate the development of spermatogonia into sperm. Another name for Sertoli cells is *sustentacular cells* because they provide sustenance or nourishment for the developing spermatogonia. The Sertoli cells manufacture and secrete a variety of proteins, ranging from the hormones inhibin and activin to growth factors, enzymes, and an **androgen-binding protein.**

Androgen-binding protein is secreted by Sertoli cells into the lumen of the seminiferous tubule. There the protein binds to testosterone secreted by the Leydig cells. Protein binding makes testosterone less lipophilic and concentrates it within the luminal fluid.

Leydig cells The primary function of the Leydig cells, located in the interstitial tissue outside the seminiferous tubules, is the production of testosterone. The Leydig cells are active in the fetus, producing hormones that direct the development of male characteristics. After birth, the cells become inactive until puberty. At that time, they resume their production of testosterone.

The primary androgen in the human body is testosterone. A derivative hormone, dihydrotestosterone, or

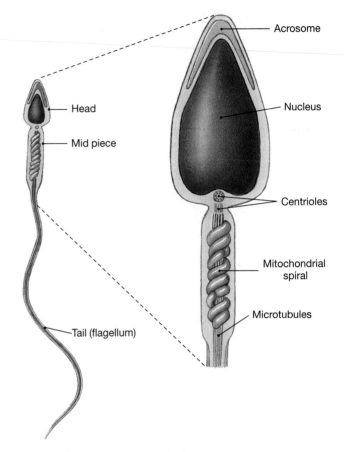

■ **Figure 24-10 Sperm structure**

DHT, is more potent than testosterone and is produced in many peripheral tissues from testosterone. Some of the physiological effects attributed to testosterone are actually the result of DHT activity.

About 5% of testosterone in the adult male comes from the adrenal cortex, but the bulk of the testosterone is

Pseudohermaphrodites and DHT The importance of dihydrotestosterone in male development came to the attention of scientists through studies of the male pseudohermaphrodites described in the opening of this chapter. These men inherit a defective gene for the enzyme *5-alpha-reductase*, which catalyzes the conversion of testosterone to dihydrotestosterone (DHT). The defect, like that of sickle cell anemia, results from the substitution of a single amino acid in the peptide sequence. Because of deficient DHT, the male external genitalia and prostate gland fail to develop fully during fetal development, despite normal testosterone secretion by the testes. At birth, the infants are usually considered female and are raised as such. However, at puberty, the testes again begin to secrete testosterone, causing masculinization of the external genitalia, pubic hair growth (although scanty facial and body hair), and deepening voice. By studying the defect in 5-alpha-reductase in these individuals, we have been able to separate the effects of testosterone from those of DHT.

produced by the Leydig cells of the testes. The Leydig cells convert some androgens to estradiol, the predominant estrogen in females. A small amount of additional estrogen is also made in peripheral tissues. Although males synthesize estrogens, their feminizing effects are usually not obvious.

Spermatogenesis Requires Gonadotropins and Testosterone

The hormonal control of spermatogenesis follows the general pattern described previously. Gonadotropin releasing hormone (GnRH) from the hypothalamus controls the release of LH and FSH from the anterior pituitary. Those hormones in turn stimulate the testes to produce testosterone. The gonadotropins were named originally from their effect on the female ovary, but the same names have been retained in the male.

GnRH release is pulsatile, peaking every 1.5 hours, and LH release follows the same pattern. FSH levels are not as obviously related to GnRH secretion because FSH secretion is also influenced by inhibin and activin. The target of FSH is the Sertoli cells (Fig. 24-11 ■). FSH stimulates production of multiple paracrine factors that are necessary for mitotic replication of spermatogonia and for spermatogenesis. In addition, FSH stimulates the secretion of androgen-binding protein and inhibin. Inhibin feeds back to selectively inhibit FSH release from the pituitary. As far as we can tell, FSH has no direct effect on the developing germ cells.

The primary target of LH is the Leydig cells, which produce testosterone. In turn, testosterone feeds back to inhibit LH release by the anterior pituitary. Testosterone is essential for spermatogenesis, but its actions appear to be mediated by the Sertoli cells. Spermatocytes have no androgen receptors and cannot respond directly to testosterone, but the Sertoli cells do have androgen receptors. Another possibility is that the spermatocytes have receptors for androgen-binding protein rather than for testosterone itself; there is growing evidence to support this theory. Spermatogenesis is a very difficult process to study *in vivo* or *in vitro,* and the available animal models may not accurately reflect the situation in the human testis. Therefore, it may be some time before we can say with certainty how either testosterone or FSH regulates spermatogenesis.

Male Accessory Glands Contribute Secretions to Semen

The male reproductive tract has several accessory glands whose primary function is to secrete various types of fluid. As sperm leave the vas deferens during ejaculation, they are joined by these secretions, resulting in a sperm-fluid mixture known as **semen.** About 99% of the volume of semen is fluid added from the accessory glands, the bulbourethral glands, seminal vesicles, and prostate. The contributions of the accessory glands to the composition of semen are shown in Table 24-3. Semen provides a liquid medium for sperm delivery,

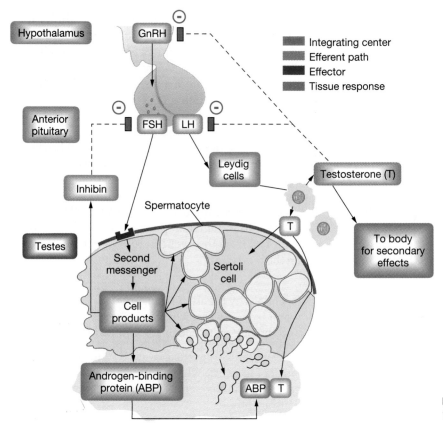

■ Figure 24-11 **Hormonal control of spermatogenesis**

TABLE 24-3	Composition of Semen	
Component	Function	Source
Sperm	Gamete	Seminiferous tubules
Mucus	Lubricant	Bulbourethral glands
Water	Provides liquid medium	All accessory glands
Buffers	Neutralize acidic environment of the vagina	Prostate, bulbourethral glands
Nutrients	Nourish sperm	
Fructose		Seminal vesicles
Citric acid		Prostate
Vitamin C		Seminal vesicles
Carnitine		Epididymis
Enzymes	Clot semen in vagina, then liquefy the clot	Seminal vesicles and prostate
Zinc	Unknown; possible association with fertility	Unknown
Prostaglandins	Smooth muscle contraction; may aid sperm transport	Seminal vesicles

including mucus for lubrication, buffers to neutralize the usually acidic environment of the vagina, and nutrients for sperm metabolism. The seminal vesicles contribute prostaglandins (∞ p. 29), lipid-related molecules that appear to influence sperm motility and transport within both the male and female reproductive tracts. Prostaglandins were originally believed to come from the prostate gland, and the name was well established by the time their true source was discovered.

In addition to providing a medium for sperm, the accessory gland secretions help protect the reproductive tract from pathogens that might ascend the urethra from the external environment. They physically flush out the urethra and supply immunoglobulins, lysozyme, and other compounds with antibacterial action. One interesting component of semen is zinc. Its role in reproduction is unclear, but concentrations of zinc below a certain level are associated with male infertility.

Androgens Influence Primary and Secondary Sex Characteristics

The androgens have a number of effects on the body in addition to gametogenesis. These effects are divided into development of the primary and the secondary sex characteristics. **Primary sex characteristics** are the internal sexual organs and external genitalia that distinguish

DHT and the Prostate   The fetal development of the prostate gland, like that of the external genitalia, is under the control of dihydrotestosterone (DHT). The discovery of the role of DHT in prostate growth led to the development of finasteride, a 5-alpha-reductase inhibitor that blocks production of DHT. This drug provides the first nonsurgical treatment for benign prostatic hypertrophy. Studies are currently under way to determine if lowering (but not eliminating) DHT will decrease the incidence of cancer of the prostate gland.

males from females. As you have already learned, androgens are responsible for their differentiation during embryonic development and for their growth during the maturation period known as puberty.

The **secondary sex characteristics** are other features of the body, such as body shape, that distinguish males from females. The male body shape is sometimes described as an inverted triangle, with broad shoulders and narrow waist and hips; the female body is more pear-shaped. Androgens are responsible for typically male traits such as beard and body hair growth, muscular development, thickening of the vocal chords with subsequent lowering of the voice, and behavioral effects such as the sex drive (**libido**). Because androgens are anabolic hormones that promote protein synthesis, they are sometimes known as anabolic steroids. The illicit use of these drugs by athletes has been widespread despite possible adverse side effects such as liver tumors and infertility. One of the more interesting side effects noted is the apparent addictiveness of anabolic steroids. Withdrawal from taking the drugs may be associated with behavioral changes that include excessive aggression ('roid rage), depression, or psychosis. These psychiatric disturbances suggest that human brain function can be modulated by sex steroids, just as the brain function of other animals can. Fortunately, many side effects of anabolic steroids are reversible once use is stopped.

We now turn to female reproduction, a more complicated topic due to the cyclic nature of gamete production in the ovary.

✓ Explain why the use of exogenous anabolic steroids by males might cause the testes to shrink and the person to become temporarily infertile.

✓ Because GnRH agonists cause down-regulation of GnRH receptors, what would be the advantage and disadvantage of using these drugs as a male contraceptive?

FEMALE REPRODUCTION

The Female Reproductive Tract Consists of the Ovaries, Accessory Structures Such as the Uterus, and the External Genitalia

The female external genitalia are known collectively as the **vulva,** or **pudendum** [*vulva*, womb; *pudere*, to be ashamed]. They are shown in Figure 24-12a ■, the view seen by a health care worker who is about to do a pelvic exam or Pap smear (∞ p. 53). Starting at the periphery are the **labia majora** [*labium*, lip], folds of skin that arise from the same embryonic tissue as the scrotum. Within the labia majora are the **labia minora,** derived from embryonic tissues that in the male give rise to the shaft of the penis (see Fig. 24-3 ■). The **clitoris** is a small bud of erectile, sensory tissue at the anterior end of the vulva, enclosed by the labia minora and an additional fold of tissue equivalent to the foreskin of the penis. In females, the urethra opens to the outside between the clitoris and the vagina [*vagina*, sheath], a cavity that acts as the receptacle for the penis during copulation. At birth, the external opening of the vagina is partially closed by a thin ring of tissue called the **hymen,** or *maidenhead*. Contrary to what some young women believe, the hymen is external to the vagina, not found within it, so the normal use of tampons during menstruation will not rupture the hymen. However, it can be stretched by normal activities such as horseback riding and therefore is not an accurate indicator of a woman's virginity.

Now we will follow the path of sperm deposited in the vagina during intercourse. To continue into the female reproductive tract, the sperm first must pass through the narrow opening of the **cervix,** the neck of the uterus that protrudes slightly into the upper end of the vagina (Fig. 24-12b, c ■). The cervical canal is lined with mucous glands that secrete mucus whose consistency varies with the stage of the menstrual cycle. Sperm that make it through the cervical canal find themselves in the lumen of the uterus, or *womb,* a hollow, muscular organ slightly smaller than a woman's clenched fist.

The uterus is the structure in which fertilized eggs implant and develop during pregnancy. It is composed of three tissue layers: a thin outer connective tissue covering, a thick middle layer of smooth muscle known as the **myometrium** (Fig. 24-12f ■), and a vascular and secretory inner layer known as the **endometrium** that lines the lumen [*metra*, womb]. The thickness and character of the endometrium vary during the menstrual cycle. The cells that constitute the variable innermost layer alternately proliferate and slough off, accompanied by a small amount of bleeding in the process known as **menstruation.**

Sperm swimming upward through the uterus leave its cavity through openings into the two Fallopian tubes. The Fallopian tubes are 20–25 cm long and about the diameter of a drinking straw. Their walls have two layers of smooth muscle, longitudinal and circular, just like the walls of the intestinal tract. The inner lining of the tubes is a ciliated epithelium (Fig. 24-12c ■). The dilated open end of the Fallopian tube is divided into many fingerlike projections called **fimbriae** that form a fringe around the open edge [*fimbriae*, fringe]. The fimbriae are closely associated with the adjacent ovary, both held in position by connective tissue (Fig. 24-12c ■). The close association of the tube and the ovary ensures that eggs released from the surface of the ovary will be swept into the Fallopian tube rather than floating off into the abdominal cavity. Using a combination of muscular contractions and fluid movement created by cilia, the Fallopian tubes transport eggs toward the uterus. Pathological conditions in which ciliary function is absent are associated with infertility or pregnancies in which the embryo implants in the Fallopian tube rather than the uterus.

The Ovary, Like the Testis, Produces Both Gametes and Hormones

The female gonad, the ovary, is responsible for both gamete and hormone production, just as the testis is. The ovary is an elliptical structure, about 2–4 cm long (Fig. 24-12d ■). It has an outer connective tissue layer and an inner connective tissue framework known as the **stroma** [*stroma*, mattress]. The bulk of the ovary is taken up by a thick cortex filled with ovarian follicles in various stages of development or decline. The small central medulla contains nerves and blood vessels.

As discussed earlier, about 7 million oögonia in the embryonic ovary develop into half a million primary oocytes. Each oocyte is contained in a **primary follicle** with an outer layer of granulosa cells (Fig. 24-12e ■). As the primary follicle develops, it gains an additional outer layer of cells known as the **theca** [*theke*, case or cover]. The thecal cells are separated from the granulosa cells by a basal lamina.

Female Reproductive Cycles Last about One Month

Female humans produce gametes in monthly cycles (average 28 days; normal range 24–35 days). These cycles are commonly called **menstrual cycles** because they are marked by a 3–7 day period of bloody uterine discharge known as the **menses** [*menses*, months], or menstruation. The menstrual cycle is described according to changes that occur in follicles of the ovary (**ovarian cycle**) and in the endometrial lining of the uterus (**uterine cycle**).

The ovarian cycle is divided into three phases: the follicular phase, ovulation and the luteal phase. The first half of the ovarian cycle is known as the **follicular phase,** reflecting a period of follicular growth in the ovary. About midway through the cycle, day 14 of a 28-day cycle, an egg is released from its follicle in the process known as **ovulation.** In the period following ovulation,

Anatomy Summary   **Female Reproduction**

■ Figure 24-12

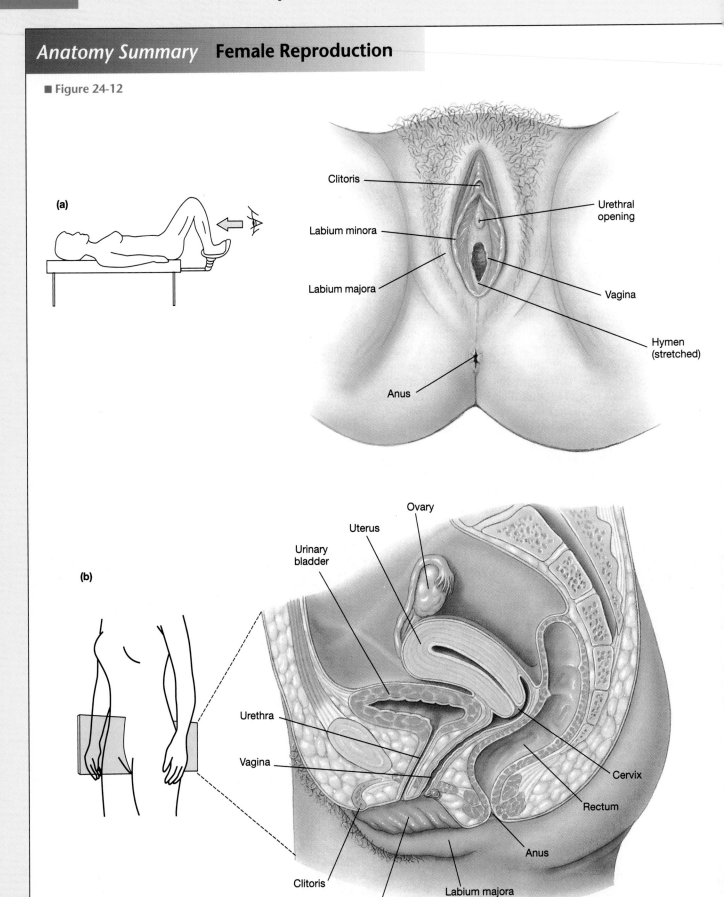

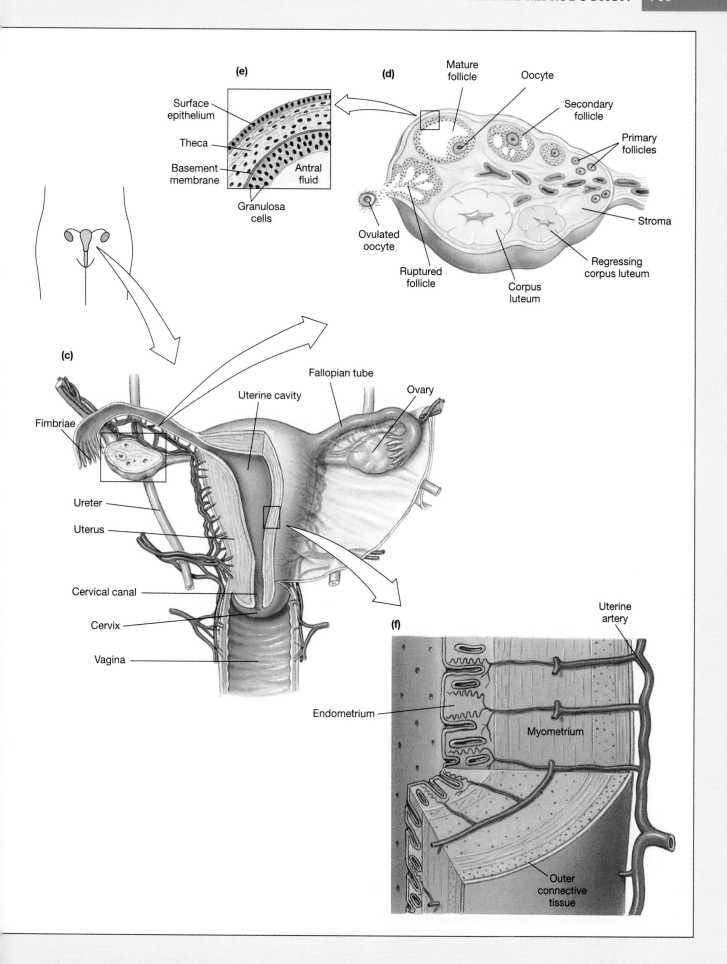

(e)

Surface epithelium

Theca

Basement membrane

Antral fluid

Granulosa cells

(d)

Mature follicle

Oocyte

Secondary follicle

Primary follicles

Stroma

Ovulated oocyte

Ruptured follicle

Corpus luteum

Regressing corpus luteum

(c)

Fimbriae

Ureter

Uterus

Cervical canal

Cervix

Vagina

Uterine cavity

Fallopian tube

Ovary

(f)

Uterine artery

Endometrium

Myometrium

Outer connective tissue

the remains of the ruptured follicle transform into a **corpus luteum** [*corpus,* body + *luteus,* yellow], a secretory structure named from its yellow pigment and lipid deposits. This latter phase of the ovarian cycle is known as the postovulatory, or **luteal phase.** If a pregnancy does not occur, the corpus luteum ceases to function after about two weeks, and the ovarian cycle begins again.

The endometrial lining of the uterus goes through a parallel cycle regulated by hormones secreted during the ovarian cycle. The beginning of the follicular phase in the ovary corresponds to menstrual bleeding from the uterus. This is followed by a **proliferative phase** in the uterus, during which the endometrium begins to build a new layer in anticipation of pregnancy. After ovulation, hormones from the corpus luteum convert the thickened endometrium into a secretory structure. Thus, the luteal phase of the ovary corresponds to the **secretory phase** of the uterus. If no pregnancy occurs, the superficial layers of the endometrium are lost during menstruation as the uterine cycle begins again.

The ovarian and uterine cycles are under the control of GnRH from the hypothalamus and the gonadotropins FSH and LH from the anterior pituitary. Cells of the ovary produce both steroid and protein hormones in response to stimulation by gonadotropins. During the follicular phase, the dominant steroid hormone is estrogen. In the luteal phase, the ovary secretes both estrogen and progesterone. Figure 24-13 ■ is a summary figure showing hormonal and morphological changes that take place during a typical menstrual cycle. In the following sections, we divide the cycle into phases and discuss the events and their hormonal control in sequence.

Early follicular phase   The first day of menstruation is day 1 of a cycle. This point was chosen to start the cycle because the bleeding of menstruation is an easily monitored physical sign. Just before the beginning of each cycle, gonadotropin secretion from the anterior pituitary increases. Under the influence of FSH, several follicles in the ovaries begin to mature (Table 24-4). As the follicles grow, their granulosa cells under the influence of FSH and their thecal cells under the influence of

continued from page 708

The results of Larry's sperm analysis are normal. Dr. Coddington is therefore able to rule out sperm abnormalities as a cause of Peggy's and Larry's infertility. Peggy is instructed to track her body temperature daily and record the results on a chart. This temperature tracking is designed to determine if she is ovulating. Following ovulation, body temperature rises slightly and remains elevated through the remainder of the menstrual cycle.

Question 3: For what causes of female infertility is temperature tracking useful? For what causes is it not useful?

LH start to produce steroid hormones. Thecal cells synthesize androgens that diffuse into the neighboring granulosa cells, where they are converted to estrogens (Fig. 24-14a ■, p. 719).

Gradually increasing estrogen levels in the circulation have several effects. Estrogen exerts a negative feedback effect on pituitary FSH and LH secretion, thus preventing the development of additional follicles in the same cycle. At the same time, estrogen stimulates estrogen production by the granulosa cells. This positive feedback loop allows the follicles to continue estrogen production despite a decrease in gonadotropin stimulation. As the follicles enlarge, the granulosa cells secrete fluid that collects in a cavity known as the **antrum.** The antral fluid contains hormones and enzymes needed for ovulation. At each stage of development, some follicles undergo *atresia* (failure to develop). Only a few follicles reach the final stage, and usually only one dominant follicle continues to ovulation.

In the uterus, menstruation ends during the early follicular phase. Under the influence of estrogens from the ovary, the endometrium begins to grow, or proliferate. This period is characterized by an increase in cell number and by enhanced blood supply to bring nutrients and oxygen to the thickening endometrium.

Late follicular phase   As the follicular phase nears its climax, estrogen secretion from the ovary reaches its peak for the cycle. By this point of the cycle, only one follicle is still developing; the others have degenerated. As the follicular phase ends, the granulosa cells of the dominant follicle begin to secrete a small amount of progesterone in addition to estrogen (Fig. 24-14b ■). Estrogen, which exerted a negative feedback effect on GnRH and gonadotropins earlier in the follicular phase, now exerts a positive feedback effect. Immediately before ovulation, the persistently high levels of estrogen, aided by rising levels of progesterone, create a surge in GnRH secretion by the hypothalamus and enhance the pituitary responsiveness to GnRH. As a result, a surge of LH is secreted by the pituitary. FSH also surges but to a lesser degree, presumably because it is being suppressed by inhibin and estrogen.

The LH surge is an essential part of ovulation. Without it, the final steps of oocyte maturation cannot take place. The meiotic division of the primary oocyte, which had paused after chromosomal duplication, resumes. With this division, the primary oocyte extrudes the first polar body and becomes an egg, or secondary oocyte. As antral fluid collects, the follicle grows to its greatest size, ready to release its egg during ovulation.

The high levels of estrogen in the last part of the follicular phase help prepare the uterus for a possible pregnancy. The endometrium continues to grow to a final thickness of 3–4 mm, while cervical glands begin to produce copious amounts of thin, stringy mucus to facilitate sperm entry. The stage is set for ovulation.

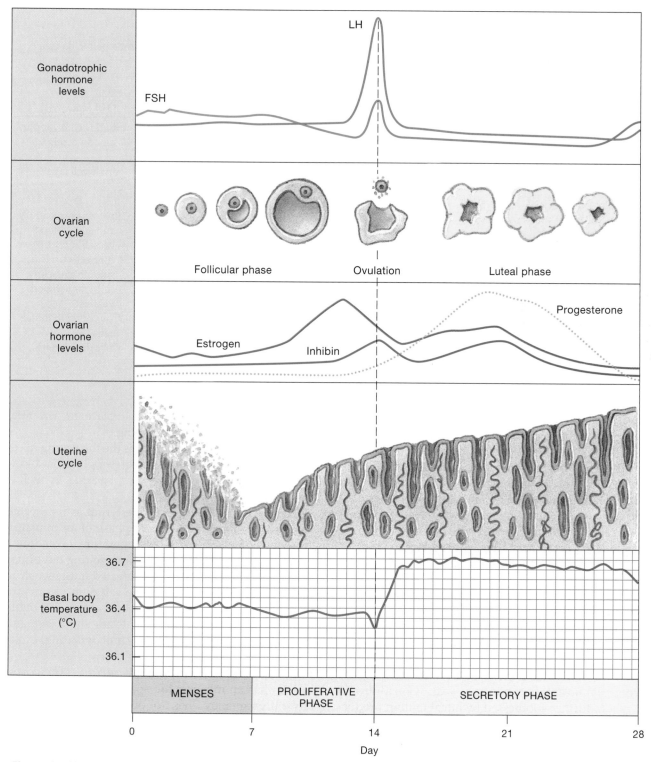

■ Figure 24-13 **The menstrual cycle** The 28-day menstrual cycle is divided into phases based on events in the uterus and ovary.

Ovulation   About 16–24 hours after LH peaks, ovulation occurs. The mature follicle secretes *collagenase*, dissolving collagen in the connective tissue that holds the cells of the follicle together. The breakdown products of collagen create an inflammatory reaction, attracting leukocytes that secrete prostaglandins into the follicle. Our understanding of events from this point is unclear, but it is possible that the prostaglandins cause smooth muscle cells in the outer theca to contract, rupturing the follicle wall at its weakest point. The antral fluid spurts out along with the egg surrounded by two to three layers of granulosa cells. The egg is swept into the Fallopian tubes and carried away to be fertilized or die.

TABLE 24-4 Development of Ovarian Follicles

The contents of the follicle are given starting at the center with the ovum and moving outward. The zona pellucida (see Fig. 24-16) is a glycoprotein coat that protects the ovum.

Stage	Ovum	Zona Pellucida	Granulosa Cells	Antrum	Basal Lamina	Theca
Primary follicle (before FSH stimulation)	Primary oocyte	Minimal	Single layer	None	Separates granulosa and theca	Single cell layer plus blood vessels
Secondary follicle (early follicular phase)	Primary oocyte	Increased in width	2–6 cell layer	None	Present	Single cell layer
Tertiary follicle (late follicular phase)	Primary oocyte, then secondary oocyte with division arrested	Present	3–4 cell layer	Develops within granulosa layer and fills with fluid; swells to 15–20 mm diameter	Present	*Inner layer:* secretory and small blood vessels. *Outer layer:* connective tissue, smooth muscle cells, large blood vessels
Corpus luteum (luteal phase)	None	None	Converted to luteal cells	Fills with migrating cells	Disappears	Converted to luteal cells
Corpus albicans (post-luteal phase)	None	None	Cells degenerate	None	None	Cells degenerate

In addition to promoting rupture of the follicle, the LH surge causes the follicular cells, both granulosa and thecal, to transform into the luteal cells of the corpus luteum. This process, known as *luteinization*, involves biochemical and morphological changes. Thecal cells migrate into the antral space, mingling with the former granulosa cells and filling the cavity. Newly formed luteal cells accumulate lipid droplets and glycogen granules in their cytoplasm and begin to secrete progesterone. Estrogen synthesis diminishes.

Early to mid-luteal phase   After ovulation, the corpus luteum produces steadily increasing amounts of progesterone and estrogen. Progesterone is the dominant hormone of the luteal phase; estrogen levels increase but never reach the peak seen before ovulation. The combination of hormones exerts negative feedback on the hypothalamic-anterior pituitary axis. Gonadotropin secretion, further suppressed by luteal inhibin production, remains shut down throughout most of the luteal phase (Fig. 24-14c ■).

Under the influence of progesterone, the endometrium of the uterus continues its preparation for pregnancy. The inner lining of the uterus is an epithelium with glands that dip into the connective tissue layer below. In the follicular phase, under the influence of estrogens, the endometrium grew and thickened. In the luteal phase, under the influence of progesterone, the endometrium matures and becomes secretory. The endometrial glands coil, and additional blood vessels grow into the connective tissue layer. Endometrial cells deposit lipids and glycogen in their cytoplasm to provide nourishment for

a developing embryo while the placenta, the fetal-maternal connection, is developing.

Another effect of progesterone is the thickening of cervical mucus. Thicker mucus creates a plug that blocks the cervical opening, preventing bacteria as well as sperm from entering the uterus.

One interesting effect of progesterone is its thermogenic ability. During the luteal phase of an ovulatory cycle, a woman's basal body temperature, taken immediately upon awakening and before getting out of bed, jumps 0.3°–0.5° F and remains elevated until menstruation. Because this change in the setpoint occurs after ovulation, it cannot be used effectively to predict ovulation. However, it is a simple way to assess if a woman is having ovulatory or anovulatory (not ovulating) cycles.

Late luteal phase and menstruation   The corpus luteum has an intrinsic life span of approximately 12 days. If pregnancy does not occur, the corpus luteum spontaneously degenerates into an inactive structure called a **corpus albicans** [*albus*, white]. As the luteal cells shut down, production of progesterone and estrogen drops (Fig. 24-14d ■). This drop removes the negative feedback signal to the pituitary and hypothalamus, and secretion of FSH and LH increases.

Maintenance of the secretory endometrium depends on the presence of progesterone. When the corpus luteum degenerates and hormone production falls, blood vessels in the surface layer of the endometrium contract. Without oxygen and nutrients, the surface cells die. About two days after the corpus luteum ceases to function, the endometrium begins to slough its surface

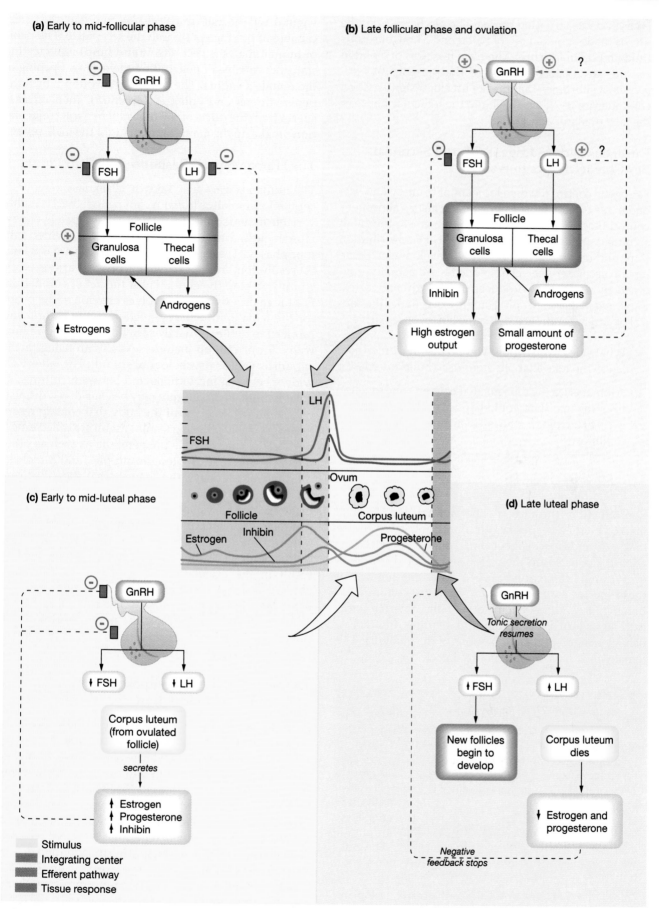

(a) Early to mid-follicular phase

(b) Late follicular phase and ovulation

(c) Early to mid-luteal phase

(d) Late luteal phase

Stimulus
Integrating center
Efferent pathway
Tissue response

■ Figure 24-14 **Hormonal control of the menstrual cycle**

layer, and menstruation begins. The discharge from the uterus totals about 40 mL of blood and 35 mL of serous fluid and cellular debris. There are few clots of blood in the menstrual flow because of the presence of *fibrinolysin*, a substance that breaks up clots. Menstruation continues for 3–7 days, well into the follicular phase of the next ovulatory cycle.

Estrogens and Androgens Control Secondary Sex Characteristics in Women

Estrogens control the development of primary and secondary sex characteristics in females, just as androgens control them in males. In women, the most prominent secondary traits are the female pattern of fat distribution (hips and upper thighs) and breast development. Some female secondary sex characteristics are actually governed by androgens produced in the adrenal cortex. Pubic and axillary (armpit) hair growth is under the control of adrenal androgens, as is the libido, or sex drive.

✓ What side effects would you predict in women athletes who take anabolic steroids to build muscles?

✓ Aromatase is the enzyme that converts androgens to estrogen. What would happen to the ovarian cycle of a woman who is given an aromatase inhibitor?

PROCREATION

Reproduction throughout the animal kingdom is marked by species-specific behaviors designed to ensure that egg and sperm meet. For aquatic animals that release gametes into the water, coordinated timing is everything. Interaction between males and females of these species may be limited to chemical communication by pheromones. On the other hand, internal fertilization in terrestrial vertebrates requires interactive behaviors and special adaptations of the genitalia. The female must have an internal receptacle for sperm (the

continued from page 716

Results of the temperature tracking reveal after several months that Peggy is ovulating regularly. Dr. Coddington therefore believes that her ovaries are functioning normally. Other possible causes for this couple's infertility include abnormalities in Peggy's cervix, Fallopian tubes, or uterus. Dr. Coddington next decides to order a postcoital test. In this test, the couple is instructed to have intercourse 12 hours before the physician visit. The cervical mucus is then analyzed. This test will also analyze the interaction between sperm and mucus.

Question 4: What abnormalities in the cervix, Fallopian tubes, and uterus could cause infertility?

vagina) and the male must possess an organ (the penis) capable of placing sperm into the receptacle. The penis of human males is *flaccid* (soft and limp) in its resting state, not capable of penetrating the narrow opening of the female's vagina. The processes through which the penis stiffens and enlarges (**erection**), then releases sperm from the ducts of the reproductive tract (**ejaculation**), make up the main components of the male sex act.

The Human Sexual Response Has Four Phases

The human sex act, also known as sexual intercourse, copulation, or **coitus** [*coitio*, a coming together] is highly variable in some ways and highly stereotyped in others. Human sexual response in both sexes is divided into four phases: (1) excitement, (2) plateau, (3) orgasm, and (4) resolution. In the excitement phase, various erotic stimuli prepare the genitalia for the act of copulation. For the male, excitement involves erection of the penis; for the female, it includes erection of the clitoris and lubrication of the vagina by secretions from the vaginal walls. Erotic stimuli include sexually arousing tactile stimuli as well as psychological stimuli. The latter vary widely between individuals and between cultures, so what is erotic in one culture may be considered disgusting in another. Regions of the body that contain receptors for sexually arousing tactile stimuli are called **erogenous zones** and include the genitalia as well as other regions such as the lips, tongue, nipples, and ear lobes.

In the plateau phase, the changes that started during excitement intensify, then culminate in an **orgasm** (climax). In both sexes, an orgasm is a series of muscular contractions accompanied by intense pleasurable sensations. In the female, the walls of the vagina and the uterus contract. In the male, the contractions usually result in the ejaculation of semen from the penis. Orgasm is accompanied by increased blood pressure, heart rate, and respiration rate. The final phase of the sexual response is resolution, the period during which the physiological parameters that changed in the first three phases slowly return to normal.

The Male Sex Act Is Composed of Erection and Ejaculation

A key element to successful copulation is the ability of the male to achieve and sustain an erection. Penile erection is a state of vasocongestion in which arterial flow into spongy erectile tissue exceeds venous outflow. In the flaccid penis, the arterioles supplying the erectile tissue are constricted, and blood flow into those areas is minimal. Sexual excitement from either tactile or psychological stimuli triggers the erection reflex, a spinal reflex that is subject to control from higher centers in the brain. The urination and defecation reflexes are similar types of reflexes (∞ pp. 536, 584).

In its simplest form, the erection reflex begins with tactile stimuli sensed by mechanoreceptors in the glans penis or other erogenous zones (Fig. 24-15 ■). Sensory neurons signal the spinal integration center, which inhibits vasoconstrictive sympathetic input on penile arterioles. Simultaneously, increased parasympathetic input mediated primarily by nitric oxide actively vasodilates the penile arterioles. As a result of vasodilation, arterial blood flows into the open spaces of the erectile tissue, passively compressing the veins and trapping blood. The erectile tissue becomes engorged, stiffening and lengthening the penis within 5–10 seconds.

The climax of the male sexual act coincides with emission and ejaculation. **Emission** is the movement of sperm out of the vas deferens and into the urethra, where they are joined by secretions from the accessory glands to make semen. The average semen volume is 3 mL (range 2–6 mL), of which less than 10% is sperm. During ejaculation, semen in the urethra is expelled to the exterior by a series of rapid muscular contractions accompanied by sensations of intense pleasure, the orgasm. A sphincter at the base of the bladder contracts to prevent sperm from entering the bladder and urine from joining the semen.

Both erection and ejaculation can occur in the absence of mechanical stimulation. Sexually arousing thoughts, sights, sounds, emotions, and dreams can all initiate sexual arousal and even lead to orgasm. In addition, nonsexual penile erection accompanies rapid eye movement (REM) sleep.

The inability to obtain or sustain an erection is known as **impotence.** Impotence may result from a variety of physiological and psychological causes. For example, fear of interruption may result in loss of arousal. Alcohol is another factor that inhibits sexual performance, as noted by Shakespeare in *Macbeth* (III, iii, 26–30). When MacDuff asks, "What three things does drink especially provoke?" the porter answers, "Marry, sir, nose-painting, sleep, and urine. Lechery, sir, it provokes and unprovokes: it provokes the desire, but it takes away the performance."

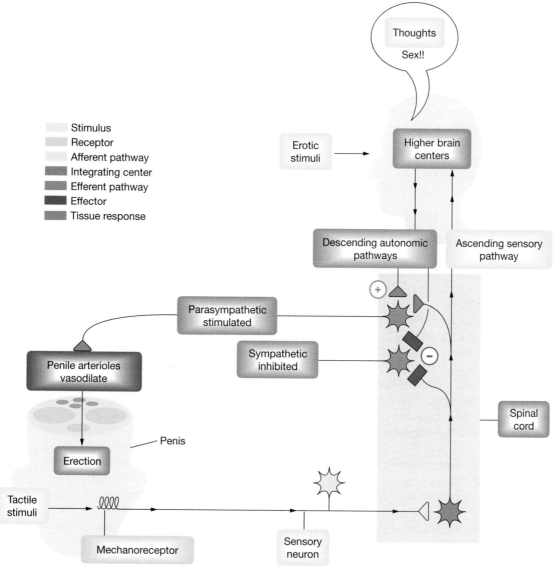

■ Figure 24-15 **The erection reflex** Erection can take place as a simple spinal reflex, without input from higher brain centers. It can also be stimulated (and inhibited) by descending pathways from the cerebral cortex.

Because of the influence of psychological factors on the erection reflex, clinicians often use the unconscious, non-sexual erection of REM sleep to determine whether impotence is physiological or psychological. In its simplest form, the REM erection test consists of gluing a band of postage stamps around the flaccid penis before the subject goes to sleep. If the perforations between stamps have torn by the time the subject awakens, he knows that he had normal erections during REM sleep.

Contraceptive Measures Are Designed to Prevent Pregnancy

One disadvantage of sexual intercourse for pleasure rather than reproduction is the possibility of an unplanned pregnancy. On average, 80% of young women who have intercourse without using any form of birth control will get pregnant within a year. But many women get pregnant after a single unprotected encounter. Couples who hope to avoid unwanted pregnancies generally use some form of birth control, or **contraception.**

Contraceptive practices fall into several broad groups. **Abstinence,** the avoidance of intercourse, is the surest method to avoid pregnancy. Interventional methods include (1) barrier methods that prevent union of eggs and sperm, (2) methods that prevent implantation of the fertilized egg, and (3) hormonal treatment that decreases or stops gamete production.

Barrier methods of contraception are among the earliest recorded means of birth control. Once people made the association between pregnancy and semen, they concocted a variety of substances to block or kill the sperm. An ancient Egyptian papyrus with the earliest known references to birth control describes the use of vaginal plugs made of leaves, feathers, figs, and alum held together with crocodile and elephant dung. Sea sponges soaked in vinegar and disks of oiled silk have also been used at one time or another. Later, women used douches of garlic, turpentine, and rose petals to rinse the vagina after intercourse. As you can imagine, many of these items also caused vaginal or uterine infections. The modern version of the female barrier is the **diaphragm,** introduced into the United States in 1916. These rubber domes and a smaller version called a *cervical cap* are usually filled with a spermicidal cream, then inserted into the top of the vagina so that they cover the cervix. One advantage to the diaphragm is that it is non-hormonal and highly effective (97%–99%) if used properly and regularly. One disadvantage is that it must be inserted close to the time of intercourse, so people depending on the diaphragm may have unprotected intercourse at times.

The male barrier is the **condom,** a closed sheath that fits closely over the penis to catch the ejaculated semen. Males have used condoms made from animal bladders and intestines for centuries. Condoms lost popularity when oral contraceptives came into widespread use in the 1960s and 1970s, but in recent years they have regained favor because they combine pregnancy protection with protection from sexually transmitted diseases.

Sterilization is the most effective contraceptive method for sexually active people, but it is a surgical procedure and is not easily reversed. Female sterilization is called **tubal ligation** and consists of tying off and cutting the Fallopian tubes. A woman with a tubal ligation will still ovulate, but the eggs remain in the abdomen. The male form of sterilization is the **vasectomy,** in which the vas deferens is tied and clipped. Sperm are still made in the seminiferous tubules, but since they cannot leave the reproductive tract, they are reabsorbed.

Some contraceptive methods do not prevent fertilization but do keep a fertilized egg from establishing itself in the endometrium. They include **intrauterine devices (IUDs)** as well as chemicals that change the properties of the endometrium. IUDs are plastic devices that are inserted into the uterine cavity, where they create an inflammatory reaction that prevents implantation. They have low failure rates (0.5% per 100 women per year) but a high incidence of side effects ranging from pain and bleeding to pelvic inflammatory disease can cause infertility due to blockage of the Fallopian tubes.

Techniques for decreasing gamete production depend on altering the hormonal milieu of the body. These methods were tried historically when women would eat or drink various plant concoctions for contraception. Some actually worked because of estrogen-like compounds contained in the plants. Modern pharmacology has improved on this method, and now women can choose between oral contraceptive pills, injections lasting three months, or capsules that are inserted under the skin and release hormones. The oral contraceptives, also known as birth control pills, were first made available in 1960. They rely on various combinations of estrogen and progesterone that inhibit gonadotropin secretion from the pituitary. Without adequate FSH and LH, ovulation is suppressed. In addition, progesterones in the contraceptive pills thicken the cervical mucus and help prevent sperm penetration. These hormonal methods of contraception are highly effective when taken correctly but also carry some risks, including an increased incidence of blood clots and strokes (especially in women who smoke).

Two new forms of hormonal contraception for women may become available soon: the GnRH agonists discussed earlier and mifepristone, also known as RU 486. The GnRH agonists appear to effectively and reversibly block ovulation, as do the estrogen-progesterone combination pills. Mifepristone, a controversial drug because of its use in inducing abortions in Europe, is an antiprogesterone drug that blocks progesterone receptors. In low doses, it blocks ovulation or alters the endometrium so that fertilized eggs do not implant.

Most attempts to devise a male hormonal contraceptive have failed so far because of undesirable side effects. To inhibit sperm production, the contraceptive must block testosterone secretion or action. In doing so, the contraceptive is also likely to decrease the male libido or even cause impotence. Both side effects are unacceptable to the men who would be most interested in using the contraceptive. The search for a male contraceptive pill is concentrating on something that will affect only sperm production in the seminiferous tubules and not testosterone production by the Leydig cells.

The newest methods of contraception being tested are based on development of antibodies to various components of the male and female reproductive systems. These contraceptives would be long-lasting and reversible, and they would be administered as vaccines. One contraceptive vaccine is currently in clinical trials in Sweden.

Infertility Is the Inability to Conceive

While some couples are trying to prevent pregnancy, other couples are spending thousands of dollars trying to get pregnant. Infertility is the inability of a couple to conceive a child after a year of unprotected intercourse. For years, infertile couples had no choice but adoption if they wanted to have a child, but incredible strides have been made in this field since the 1970s. As a result, many infertile couples today are able to have children.

Infertility can arise from a problem in the male, the female, or both. Male infertility usually results from a low sperm count or an abnormally high number of defective sperm. Female infertility can be mechanical (blocked Fallopian tubes or other structural problems), or hormonal, leading to decreased or absent ovulation. One problem involving both partners is production of antibodies to sperm by the female. In addition, not all pregnancies go to a successful conclusion. By some estimates, as many as a third of all pregnancies spontaneously terminate, many within the first weeks, before the woman is even aware that she is pregnant.

Some of the most dramatic advances have been made in the field of *in vitro* fertilization, in which a woman's ovaries are hormonally manipulated to ovulate multiple eggs at one time. The eggs are collected surgically and fertilized outside the body. The developing embryos are then placed in the woman's uterus, which has been primed for pregnancy by hormonal therapy. Because of the expense and complicated nature of the procedure,

multiple embryos are usually placed in the uterus at one time, which may result in multiple births. *In vitro* fertilization has allowed some infertile couples to have children, although the success rate is not very high.

PREGNANCY AND PARTURITION

We now return to a recently ovulated egg and some sperm deposited in the vagina and follow them through fertilization, pregnancy, and **parturition,** the birth process.

Fertilization

Once an egg is released from the ruptured follicle, it is swept into the Fallopian tube on currents created by the beating cilia. Meanwhile, sperm deposited in the vagina must go through their final maturation step, capacitation. **Capacitation** of sperm confers the ability to swim rapidly and fertilize an egg. The process is not well understood, but apparently it involves the removal of a glycoprotein *decapacitation factor* from the outer membrane of the sperm head. Normally, capacitation takes place in the female reproductive tract. This location presents a problem for sperm being used for *in vitro* fertilization. These sperm must be artificially capacitated by physiological salines supplemented with human serum before being exposed to the harvested eggs. Much of what we know about human fertilization has come from infertility research aimed at improving the success rate of *in vitro* fertilization.

Fertilization of an egg by a sperm is the result of a chance encounter, possibly aided by chemical attractants produced by the egg. An egg can be fertilized for about 12–24 hours after ovulation; sperm remain viable for about two days. Fertilization normally takes place in the distal part of the Fallopian tube. Of the millions of sperm in a single ejaculation, only about 100 or so reach this point (Fig. 24-16 ■). To reach the protected egg, a sperm must penetrate several layers: first, an outer layer of loosely connected granulosa cells, then the protective glycoprotein coat of the **zona pellucida.** To get past these barriers, capacitated sperm release powerful enzymes from the acrosome in the sperm head, a process known as the **acrosomal reaction.** The enzymes dissolve cell junctions and the glycoprotein coat, allowing the sperm to wiggle their way toward the egg. The first sperm to reach the egg quickly finds sperm-binding receptors on the membrane and fuses its membrane to the oocyte membrane (Fig. 24-17 ■). The fused section of membrane opens, and the sperm nucleus sinks into the cytoplasm of the egg.

Fusion of the egg and sperm membranes signals the egg to resume meiosis and complete its second division. The final meiotic division creates a second polar body that is discarded. At this point, the 23 chromosomes of the sperm join the 23 chromosomes of the egg, creating a zygote nucleus with a full set of genetic material.

continued from page 720

Analysis of the cervical mucus in Peggy's and Larry's postcoital test shows that sperm are present but not moving. Dr. Coddington explains that it is likely that Peggy's cervical mucus contains antibodies that destroy Larry's sperm.

Question 5: Speculate on how this kind of infertility problem might be treated.

(a)   **(b)**

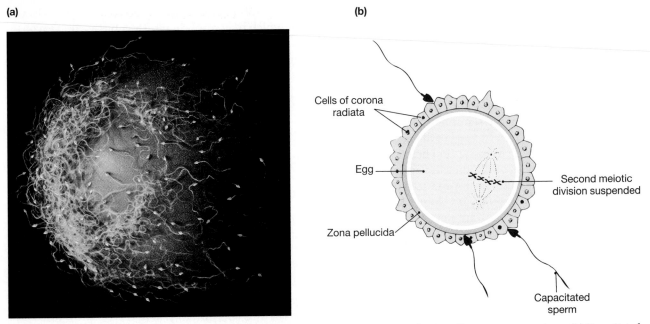

■ **Figure 24-16 Fertilization** (a) This photograph shows the tremendous difference in the size of human sperm and egg. (b) Capacitated sperm release enzymes from their acrosomes in order to penetrate the cells and glycoprotein zona pellucida surrounding the ovum.

Another process initiated by egg-sperm fusion is the **cortical reaction.** This chemical reaction is designed to prevent **polyspermy**, fertilization by more than one sperm. Membrane-bound **cortical granules** in the peripheral cytoplasm of the egg release their contents into the space just outside the oocyte membrane. These chemicals alter the membrane and surrounding zona pellucida so that additional sperm cannot penetrate or bind.

Once the egg is fertilized and becomes a zygote, it begins mitotic division as it slowly makes its way to the uterus, where it will settle for the remainder of the **gestation** period [*gestare*, to carry in the womb].

The Developing Zygote Implants in the Secretory Endometrium

The dividing zygote takes about four days to move from the distal end of the Fallopian tube into the uterine cavity (Fig. 24-18 ■). Under the influence of progesterone, the smooth muscle of the tube relaxes, and transport proceeds slowly. By the time the developing embryo reaches the uterus, it consists of a hollow ball of about 100 cells called a **blastocyst.** Some of the outer layer of blastocyst cells will become the **chorion**, an *extraembryonic membrane* that encloses the entire embryo and forms the placenta

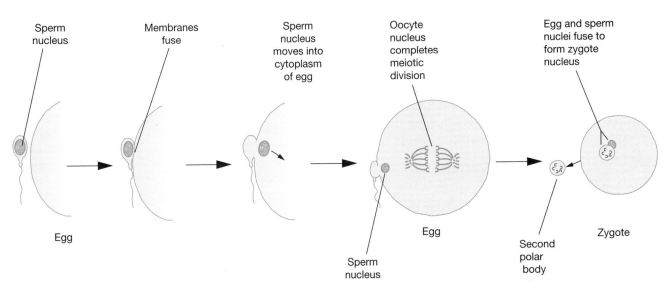

■ **Figure 24-17 Fusion of sperm and egg to form a zygote**

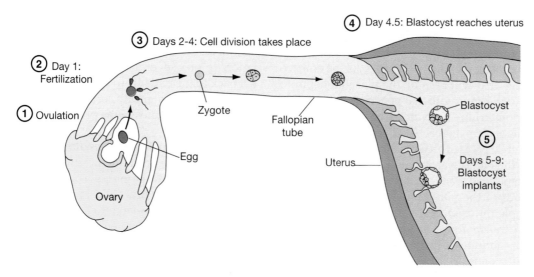

■ **Figure 24-18 Ovulation, fertilization, and implantation of an ovum** ① The ovulated egg is swept into the oviduct by the fimbriae. ② Fertilization occurs within the oviduct. ③ The zygote begins to divide while still in the oviduct. ④ By day 4.5, the embryo has reached the multicelled blastocyst stage. ⑤ Between day 5 and day 9, the blastocyst implants in the endometrial lining of the uterus.

(Fig. 24-19a ■). The inner cell mass of the blastocyst will develop into the embryo itself and other extraembryonic membranes: the **amnion,** which secretes *amniotic fluid* in which the developing embryo floats; the **allantois,** which becomes part of the umbilical cord that links the embryo to the mother; and the **yolk sac,** which degenerates early in human development.

Implantation of the blastocyst into the uterine wall normally takes place about 7 days after fertilization. The outer cells of the blastocyst secrete enzymes that allow

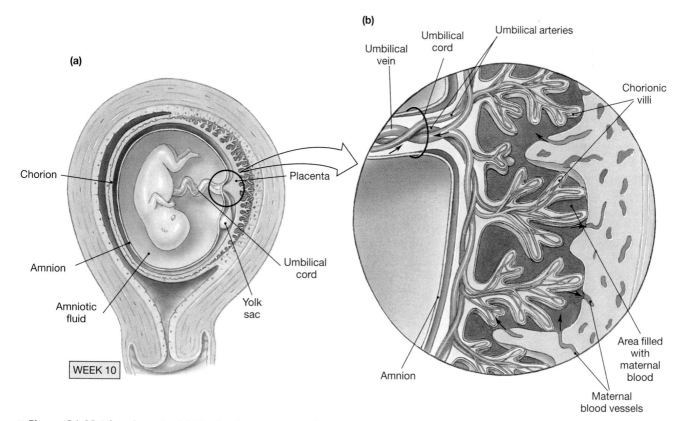

■ **Figure 24-19 The placenta** (a) The developing embryo floats in amniotic fluid contained within the amniotic and chorionic membranes. The embryo obtains oxygen and nutrients from the mother through the placenta and umbilical cord. (b) The maternal blood bathes the fingerlike chorionic villi that contain the embryonic blood vessels. Some material is exchanged by diffusion, but other material must be transported across the membranes.

them to invade the endometrium, like a parasite burrowing into its host. As they do so, the cells of the endometrium grow out around the blastocyst until it is completely engulfed in endometrial cells. As the embryo grows, the cells that will become the placenta begin to form fingerlike **chorionic villi** that penetrate into the vascularized endometrium. Enzymes from the villi break down the walls of the maternal blood vessels until the villi are surrounded by pools of maternal blood (Fig. 24-19b ■). The blood of the embryo and that of the mother do not mix, but nutrients, gases, and wastes are exchanged across the membranes of the villi. Many of these substances move by simple diffusion, but some, such as maternal antibodies, are transported across the membrane.

The placenta continues to grow during pregnancy until by delivery, it is about 20 cm in diameter (the size of a small dinner plate). The placenta receives as much as 10% of the total maternal cardiac output. The tremendous blood flow to the placenta is one reason that the sudden abnormal separation of the placenta from the uterine wall is a medical emergency.

The Placenta Secretes Hormones during Pregnancy

As the embryo implants in the uterine wall and the placenta begins to form, the corpus luteum is coming to the end of its preprogrammed 12-day life span. Unless the embryo sends a hormonal signal, the corpus luteum will disintegrate, progesterone and estrogen levels will drop, and the embryo will be washed from the body along with the surface layers of endometrium during menstruation. The hormones that prevent this from happening come from the cells of the placenta.

Human chorionic gonadotropin The signal that keeps the corpus luteum active during early pregnancy is the hormone **human chorionic gonadotropin** (hCG), secreted by the developing placenta. This peptide is structurally related to LH, and it binds to LH receptors. Under the influence of hCG, the corpus luteum keeps producing progesterone to keep the endometrium intact. However, by the seventh week of development, the placenta has taken over progesterone production and the corpus luteum is no longer needed. At that point, it finally degenerates. Human chorionic gonadotropin production by the placenta peaks at three months of development, then diminishes.

A second function of hCG is stimulation of testosterone production by the developing testes in males fetuses. As you learned in the opening sections of this chapter, fetal testosterone and its metabolite DHT are essential for expression of male characteristics and for descent of the testes into the scrotum before birth.

Human chorionic gonadotropin is the chemical detected by pregnancy tests. Because hCG can induce ovulation in rabbits, years ago the urine from a woman who suspected she was pregnant was injected into a rabbit. The rabbit's ovaries were then inspected for signs of ovulation. It took several days for the results of this test to get back to the patient. Today, with modern biochemical techniques, women can perform their own pregnancy tests in a few minutes in the privacy of their home.

Human chorionic somatomammotropin (hCS) Another peptide hormone produced by the placenta is human **chorionic somatomammotropin** (hCS), also known as **human placental lactogen** (hPL). This hormone, structurally related to growth hormone and prolactin, was initially believed to be necessary for breast development during pregnancy and for milk production (**lactation**). Although hCS probably does contribute to lactation, women who do not make hCS during pregnancy because of a genetic defect have adequate breast development and milk production. The other postulated role for hCS is regulation of the mother's metabolism. During pregnancy, some women develop elevated blood glucose levels (a condition usually associated with diabetes mellitus). This extra glucose is available for fetal needs, moving by facilitated diffusion across the membranes of the placenta into the fetal circulation. After delivery, glucose metabolism in most of these women returns to normal. It seems likely, however, that hCS is not the sole regulator of metabolism during pregnancy, and there are still many unanswered questions about its functions.

Estrogen and progesterone Estrogen and progesterone are produced continuously during pregnancy, first by the corpus luteum under the influence of hCG and then by the placenta. With high circulating levels of the steroid hormones, feedback suppression of the pituitary continues throughout pregnancy, preventing another set of follicles from beginning development.

Estrogen during pregnancy contributes to the development of the milk-secreting ducts of the breasts. Progesterone is essential for maintaining the endometrium and in addition helps suppress uterine contractions, along with **relaxin,** a peptide hormone made by the corpus luteum and placenta. The placenta makes a variety of other hormones including inhibin, prolactin, and prorenin, but at present, the function of most of them is unclear.

Pregnancy Ends with Labor and Delivery

Parturition, the birth process, normally occurs in the 38th to 40th week of gestation. For many years, we developed animal models of the signals that initiate parturition, only to discover in recent years that many of those theories do not apply to humans. Parturition begins with **labor,** the rhythmic contractions of the uterus designed to push the fetus out into the world.

Signals that initiate these contractions can begin with either the mother or the fetus, or they could be a combination of signals from both.

In many nonhuman mammals, a drop in estrogen and progesterone levels marks the beginning of parturition. A drop in progesterone levels is logical, as progesterone inhibits uterine contractions. However, in humans, these hormones do not decrease until labor is well under way. One possible explanation may be that there is inactivation of progesterone receptors. In this fashion, although total progesterone concentration remains elevated through the early part of labor, the *effective* concentration of progesterone decreases.

Another possibility for the induction of labor is that the fetus somehow signals that it has completed development. The ACTH-cortisol axis plays this role in sheep and may be a contributing factor in humans. Human fetuses that develop without most of the brain (*anencephaly*) lack a pituitary gland and functional adrenal cortex, and they also have gestational periods that are highly variable.

Although we do not know what initiates parturition, we do understand the sequence of events. Once the contractions of labor begin, a positive feedback loop consisting of mechanical and hormonal factors is set into motion. The fetus is normally oriented head down at the beginning of labor (Fig. 24-20a ■). In late pregnancy, it repositions itself lower in the abdomen ("the baby has dropped") and begins to push on the softened cervix (Fig. 24-20b ■). Cervical stretch triggers uterine contractions that move in a wave from the top of the uterus down, pushing the fetus farther into the pelvis. The cervix stretches and opens (dilates) even more and starts a positive feedback cycle of escalating contractions (Fig. 24-21 ■). In addition, neurons from the stretching cervix signal the hypothalamus-posterior pituitary to secrete oxytocin (∞ p. 185). This peptide hormone causes uterine muscle contraction and enhances the contractions of labor.

Could oxytocin be the signal that starts parturition? As the pregnancy nears full term, the number of uterine oxytocin receptors increases. However, studies have shown that secretion of the hormone does not increase until labor begins. Synthetic oxytocin is often used to induce labor in pregnant women, but it is not always effective. Apparently, the start of labor requires other conditions in addition to adequate amounts of oxytocin.

Prostaglandins, another chemical factor involved in labor and delivery, are produced in the uterus in response to oxytocin secretion. Prostaglandins are very effective at causing uterine muscle contractions at any time. They are the primary cause of menstrual cramps and have been used to induce abortion in early pregnancy. During labor and delivery, prostaglandins reinforce the uterine contractions induced by oxytocin.

As the contractions of labor intensify, the fetus moves down though the birth canal and out into the world, still attached to the placenta (Fig. 24-20c ■). The placenta then detaches from the uterine wall and is expelled a short time later (Fig. 24-20d ■). The contractions of the uterus clamp the maternal blood vessels and help prevent excessive bleeding, although typically about 240 mL of blood is lost.

The Mammary Glands Secrete Milk during Lactation

When an infant is born, it loses its source of maternal nourishment through the placenta and must rely on an external source of food instead. Primates, who normally have only one or two offspring at a time, have two functional mammary glands; other mammals have functional glands in proportion to their typical litter size. A mammary gland is composed of about 20 milk-secreting lobules, each made of branched hollow ducts of secretory epithelium surrounded by contractile cells (Fig. 24-22 ■). Interestingly, the mammary gland epithelium is closely related to the secretory epithelium of sweat glands, so milk and sweat secretion share some common features.

The breasts first begin to develop under the influence of estrogen during puberty. The milk ducts grow and branch, while fat is deposited behind the glandular tissue. During pregnancy, the glands develop further under the direction of estrogen, aided by growth hormone and cortisol. The final development step also requires progesterone, which converts the duct epithelium into a secretory structure. This process is similar to progesterone's effect on the uterus, in which progesterone makes the endometrium into a secretory tissue during the luteal phase.

Although estrogen and progesterone stimulate mammary development, they inhibit the actual secretion of milk. Milk production is under the control of the hormone prolactin from the anterior pituitary (∞ p. 186). Prolactin is an unusual pituitary hormone in that it is primarily controlled by an inhibiting hormone from the hypothalamus. There is good evidence that the **prolactin inhibiting hormone** (PIH) is actually dopamine, an amine hormone related to epinephrine and norepinephrine. Some experiments suggest that there might be a prolactin releasing hormone as well, but there is no compelling evidence for one in humans.

During the later stages of pregnancy, PIH secretion falls, and prolactin levels in the blood reach levels that are 10 or more times those found in nonpregnant women. Despite the high level of prolactin, the mammary glands produce only small amounts of a thin, low-fat secretion called **colostrum** because the levels of estrogen and progesterone are high. After delivery, when estrogen and progesterone levels decrease, the glands begin to produce greater amounts of milk that contains 4% fat and substantial amounts of calcium. Proteins in colostrum and milk include maternal immunoglobulins, secreted into

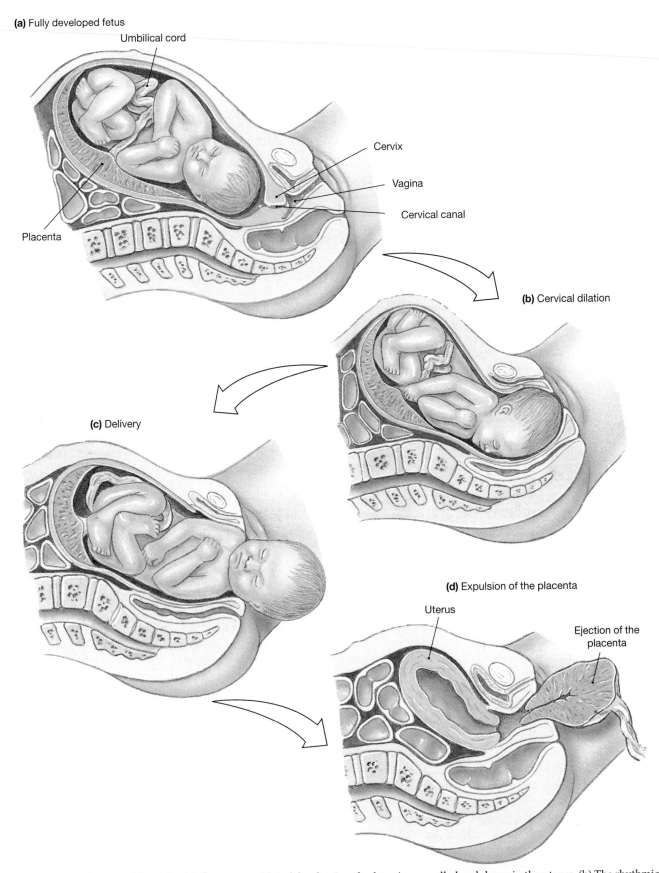

(a) Fully developed fetus

Umbilical cord

Cervix

Vagina

Cervical canal

Placenta

(b) Cervical dilation

(c) Delivery

(d) Expulsion of the placenta

Uterus

Ejection of the placenta

■ **Figure 24-20 Parturition: the birth process** (a) As labor begins, the fetus is normally head down in the uterus. (b) The rhythmic uterine contractions of labor push the head of the fetus against the softened cervix, stretching and dilating it. (c) Once the cervix is fully dilated and stretched, the uterine contractions push the fetus out of the uterus through the vagina. (d) Shortly after the fetus is delivered, the placenta detaches from the uterine wall and is expelled.

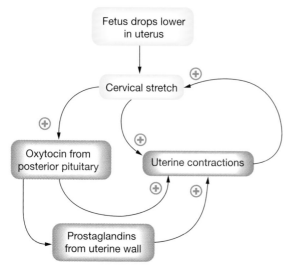

■ **Figure 24-21 The positive feedback loop of parturition** The initiation of this loop is dependent on a uterus that is ready to respond to these stimuli. The positive feedback loop stops when the fetus is delivered and the cervix is no longer being stretched.

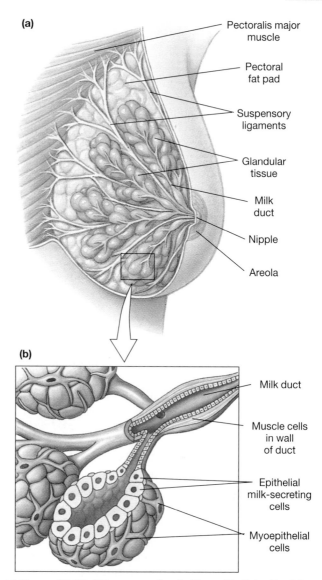

■ **Figure 24-22 Mammary glands** The epithelial cells of the mammary glands secrete milk into the lumen of the gland.

the duct and absorbed intact by the infant's intestinal epithelium (∞ p. 591), so some of the mother's immunity is conferred on the infant during the early weeks of life.

Suckling, the mechanical stimulus of the infant's nursing at the breast, inhibits the release of PIH from the hypothalamus (Fig. 24-23 ■). In the absence of this inhibition, the pituitary increases spontaneous prolactin secretion, resulting in milk production. Pregnancy is not a requirement for lactation, and some women who have adopted babies have been successful in breast-feeding.

The ejection of milk from the glands, known as the **letdown reflex,** requires the presence of a second hormone, oxytocin, from the posterior pituitary. Oxytocin initiates smooth muscle contraction in the breasts and the uterus. In the postpartum (after delivery) uterus, these contractions help return the uterus to a prepregnancy size. In the lactating breast, oxytocin causes contraction of the myoepithelial cells surrounding the mammary glands. This contraction creates high pressure that sends the milk literally squirting into the infant's mouth. Although prolactin release requires the mechanical stimulus of suckling, oxytocin release can be stimulated by various cerebral stimuli, including the thought of the child. Many nursing mothers experience inappropriate milk release triggered by hearing someone else's child cry.

Although we discuss prolactin in the context of nursing mothers, all nonnursing women and men have tonic prolactin secretion that shows a diurnal cycle, peaking during sleep. This fact suggests that the hormone has other roles in the body, although at this point we do not know what they are. Some of the most intriguing research indicates that normal lymphocytes produce a prolactin-like peptide and also have prolactin receptors,

so prolactin may play a role in body immunity, a topic that needs further exploration.

GROWTH AND AGING

Puberty Marks the Beginning of the Reproductive Years

Puberty is the period when a person makes the transition from being nonreproductive to being reproductive. In girls, the onset of puberty is marked by budding breasts and the first menstrual period, or **menarche,** a time of ritual significance in many cultures. In the United States, the average age at menarche is 12 years (normal range is considered 8 to 13 years).

In boys, the signs of puberty are more subtle. They include growth and maturation of the external genitalia, development of secondary sex characteristics such as

continued from page 723

Assisted reproductive technologies (ART) are one treatment option currently available to infertile couples. All ART techniques involve either artificially stimulating the ovaries to produce eggs or using an egg from an egg donor. The eggs are harvested surgically and often fertilized *in vitro*. The zygote may be immediately placed in the Fallopian tube or may be allowed to develop into an early embryo before being returned to the uterus. A different technique used to overcome infertility is intrauterine insemination, a procedure in which sperm that have been washed to remove antigenic material are introduced into the uterus through a tube inserted through the cervix.

Question 6: Based on the results of their infertility workup, which intervention—ART or intrauterine insemination—should be recommended for Peggy and Larry? Why?

pubic and facial hair, change in body shape, and growth in height. The age range for male puberty is 9 to 14 years.

Puberty requires the maturation of the hypothalamic-pituitary axis. Before puberty, the child has low levels of both steroid sex hormones and gonadotropins. Because low sex hormone levels normally enhance gonadotropin release, the combination of low steroids and low gonadotropins indicates that the pituitary is not yet sensitive to steroid levels in the blood. The basis for the onset of puberty, like many other areas in reproductive physiology, is uncertain. One theory says that it is the genetically programmed maturation of hypothalamic neurons. We know that puberty has a genetic base because inherited patterns of maturation are common. If a woman did not start her menstrual periods until she was 16, it is likely that her daughters will also have late menarche.

Menopause and Aging

Several centuries ago, most people died of acute illnesses while still reproductively active. Now modern medicine has overcome most acute illnesses, and we are living well past the time that most of us are likely to have children. In women, the cessation of reproductive cycles is the time known as menopause. After about 40 years of menstrual cycles, a woman's periods become irregular and finally cease. The failure of reproductive cycles is not due to the pituitary but to the ovaries, which can no longer respond to gonadotropins. In the absence of negative feedback, gonadotropin levels increase dramatically in an effort to stimulate the ovaries into maturing more follicles.

The absence of estrogen in postmenopausal women leads to symptoms of varying severity. These may include hot flashes (∞ p. 617), atrophy of genitalia and breasts, and osteoporosis due to loss of calcium from the bones (∞ p. 649). Hormone replacement therapy for women in menopause consists of estrogen or a combination of

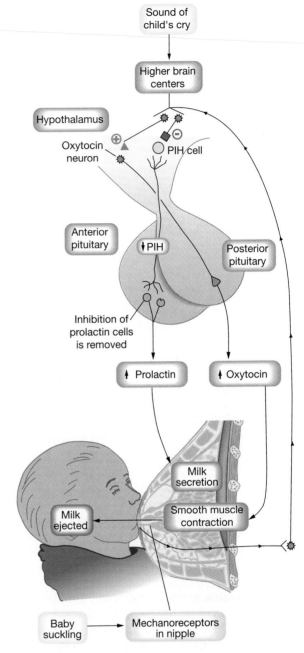

■ Figure 24-23 **The hormonal control of milk secretion and release**

estrogen and progesterone. This treatment continues to be controversial because of a possibility of increased risk of breast and uterine cancer, although studies that contradict one another are published annually.

Males do not go through a dramatic cessation of reproduction as women do, but after age 40 or so, testosterone production declines. Many men remain reproductively active as they age, and it is not uncommon for men in their fifties or sixties to have children with younger women. Postmenopausal women also remain sexually active, although not reproductively active, and some report a more fulfilling sex life once the fear of unwanted pregnancy has been removed.

CHAPTER REVIEW

Chapter Summary

Sex Determination

1. The sex organs consist of the gonads, internal accessory ducts and glands, and the **genitalia.** (p. 699)

2. The male **testes** produce **sperm.** The female **ovaries** produce eggs, or **ova.** Embryonic gonadal cells destined to produce eggs and sperm are called **germ cells.** (p. 699)

3. All nucleated cells of the body except gametes contain 46 chromosomes. Each gamete contains a half set of 23 chromosomes. When egg and sperm unite, the resulting zygote contains a unique set of 46 chromosomes. (p. 699)

4. The two **sex chromosomes,** designated X or Y, direct the development of the internal and external sex organs. The genetic sex of an individual depends on the sex chromosomes: females are XX and males are XY. The Y chromosome is essential for development of the male reproductive organs. In the absence of a Y chromosome, an embryo will develop into a female. (p. 699)

5. In the male embryo, the **SRY gene** produces **testis determining factor,** which binds to DNA and activates a series of genes whose products direct testicular development. (p. 700)

6. **Sertoli cells** in the testes secrete **Müllerian inhibiting substance** that causes the **Müllerian ducts** to regress. **Leydig cells** in the testes secrete **testosterone** that, with its derivative **dihydrotestosterone,** prompts the **Wolffian ducts** to develop into male accessory structures. (p. 700)

7. In female embryos, absence of testosterone and Müllerian inhibiting substance causes the Müllerian ducts to develop into **Fallopian tubes (oviducts), uterus,** and **vagina.** The Wolffian ducts regress and the external genitalia take on the female pattern. (p. 702)

Basic Patterns of Reproduction

8. **Gametogenesis** begins with mitotic divisions of the **spermatogonia** and **oogonia.** The first stage of meiosis creates **primary spermatocytes** and **primary oocytes** that contain twice the usual amount of DNA but have only 46 chromosomes. After the first meiotic division, each primary spermatocyte becomes two identical **secondary spermatocytes.** Each primary oocyte divides into two cells, a large **secondary oocyte,** or egg, and a tiny **first polar body** that degenerates. (p. 703)

9. The second meiotic division creates haploid **spermatids** that mature into sperm. Sperm have a flagellated tail and little cytoplasm. A lysosome-like **acrosome** contains enzymes essential for fertilization. (p. 704)

10. The second meiotic division of the oocyte does not take place unless the egg is fertilized. If fertilization occurs, half of the sister chromatids remain in the zygote, while the other half are discarded in a **second polar body** (p. 704)

11. In both sexes, **gonadotropin releasing hormone** controls the secretion of **follicle stimulating hormone** (FSH) and **luteinizing hormone** (LH) from the anterior pituitary. FSH, along with steroid sex hormones, acts on the gamete-producing cells of the gonads to regulate gametogenesis. LH acts on endocrine cells, stimulating production of the steroid sex hormones. (p. 704)

12. The sex hormones include the steroid **androgens, estrogens,** and **progesterone.** The formation of androgens from estrogens is regulated by **aromatase.** The gonads also produce **inhibin** and **activin,** which inhibit and stimulate secretion of FSH, respectively. (p. 704)

13. Gonadal steroids suppress secretion of GnRH, FSH, and LH in a long-loop response. However, if estrogen secretion increases above a threshold level for at least 36 hours, feedback changes from negative to positive and gonadotropin release (particularly LH) is stimulated. (p. 704)

14. Tonic GnRH release from the hypothalamus occurs in small pulses every 1–3 hours from a region of the hypothalamus called a **pulse generator.** (p. 706)

Male Reproduction

15. The male reproductive system consists of the testes, accessory glands and ducts, and external genitalia. The urethra runs though the **penis.** (p. 707)

16. The **corpus spongiosum** and **corpora cavernosa** comprise the erectile tissue of the penis. The **glans** is covered by the **foreskin.** (p. 707)

17. The scrotum is a sac into which the testes migrate during fetal development. Failure of one or both testes to descend is known as **cryptorchidism.** (p. 707)

18. The testes are composed of **seminiferous tubules** and interstitial tissue with blood vessels and Leydig cells. The seminiferous tubules join the **epididymis,** which becomes the **vas deferens.** The vas deferens empties into the urethra. (p. 708)

19. A seminiferous tubule has spermatocytes in various stages and Sertoli cells. The Sertoli cells form a **blood-testis barrier** between the lumen of the tubule and the interstitial fluid outside the tubule. (p. 708)

20. Spermatogonia in the tubule enter meiosis, becoming primary spermatocytes, spermatids, and finally sperm. This entire process takes about 64 days. (p. 708)

21. The Sertoli cells regulate the development of spermatogonia into sperm. They also secrete a variety of proteins, including inhibin, activin, growth factors, enzymes, and **androgen-binding protein.** (p. 710)

22. The Leydig cells produce testosterone. About 5% of testosterone in the adult male comes from the adrenal cortex. (p. 710)

23. FSH stimulates Sertoli cell production of androgen-binding protein, inhibin, and paracrines necessary for spermatogenesis. The primary target of LH is the Leydig cells, which produce testosterone. (p. 711)

24. The male accessory glands, the **prostate gland,** the **seminal vesicles,** and the **bulbourethral glands,** secrete fluid that, together with sperm, form **semen.** The secretions provide mucus, nutrients for sperm metabolism, and compounds with anti-bacterial action. (p. 711)

25. **Primary sex characteristics** are the internal sexual organs and external genitalia. **Secondary sex characteristics** are other features of the body, such as body shape, that distinguish males from females. (p. 712)

Female Reproduction

26. The female external genitalia consist of the **labia majora, labia minora,** and **clitoris.** The urethral opening is between the clitoris and the vagina, a cavity that acts as the receptacle for the penis during copulation. (p. 713)

27. The uterus is the structure in which fertilized eggs implant and develop during pregnancy. The uterine tissue layers are outer connective tissue, **myometrium,** and inner **endometrium.** (p. 713)

28. The two Fallopian tubes open into the uterus. The Fallopian tubes are lined with ciliated epithelium and are closely associated with the ovaries. The bulk of an ovary consists of ovarian follicles in various stages of development or decline. (p. 713)

29. Gametes are produced in monthly **menstrual cycles.** (p. 713)

30. The first half of the menstrual cycle is the **follicular phase** and reflects a period of follicular growth. **Ovulation** is the release of an egg from its follicle. In the **luteal phase,** the ruptured follicle becomes a **corpus luteum.** (p. 713)

31. The **menses** begin the uterine cycle. This is followed by a **proliferative phase,** during which the endometrium thickens. Following ovulation, the thickened endometrium becomes a secretory structure during the **secretory phase.** (p. 716)

32. During the follicular phase, **granulosa cells** secrete estrogen. As the follicular phase ends, estrogen feedback changes from negative to positive. This creates a surge in LH that is necessary for the final steps of oocyte maturation. (p. 716)

33. Following ovulation, the follicle transforms into a corpus luteum, which secretes progesterone and some estrogen. These hormones exert negative feedback on the hypothalamus-anterior pituitary. (p. 718)

34. Estrogens control the development of primary and secondary sex characteristics in females. (p. 720)

Procreation

35. The human sex act is divided into four phases; (1) excitement, (2) plateau, (3) orgasm, and (4) resolution. (p. 720)

36. The male **erection** reflex is a spinal reflex that can be influenced by higher brain centers. Parasympathetic input mediated by nitric oxide actively vasodilates the penile arterioles. (p. 720)

37. **Emission** is the movement of sperm out of the vas deferens and into the urethra. **Ejaculation** is the expulsion of semen to the exterior by a series of rapid muscular contractions. (p. 721)

38. Contraceptive methods include **abstinence, barrier methods,** prevention of implantation of the fertilized egg, and hormonal treatment that decreases or stops gamete production. (p. 722)

39. Infertility can arise from a problem in the male, the female, or both. *In vitro* fertilization has allowed a number of infertile couples to have children, although the success rates are not very high. (p. 723)

Pregnancy and Parturition

40. Before sperm can fertilize an egg, they must go through **capacitation,** a process that involves the removal of a glycoprotein decapacitation factor from the outer membrane of the sperm head. (p. 723)

41. Fertilization normally takes place in the Fallopian tube. Capacitated sperm release powerful enzymes from the acrosome (the **acrosomal reaction**) to dissolve cell junctions and the **zona pellucida.** The first sperm to reach the egg binds to the oocyte membrane. (p. 723)

42. Fusion of egg and sperm membranes initiates the completion of meiosis as well as the **cortical reaction** designed to prevent **polyspermy.** (p. 724)

43. When the developing embryo reaches the uterus, it consists of a hollow **blastocyst.** Once the embryo implants, it

develops **chorionic, amniotic, and allantoic membranes,** in addition to the embryo itself. (p. 724)

44. The **chorionic villi** of the placenta are surrounded by pools of maternal blood where nutrients, gases, and wastes are exchanged. (p. 726)

45. The corpus luteum remains active during early pregnancy due to **human chorionic gonadotropin** secreted by the developing placenta. By the seventh week of development, the placenta takes over progesterone production. (p. 726)

46. The placenta also secretes **human chorionic somatomam-motropin,** also known as **human placental lactogen.** This hormone plays a role in regulation of maternal metabolism. (p. 726)

47. Estrogen during pregnancy contributes to the development of the milk-secreting ducts of the breasts. Progesterone is essential for maintaining the endometrium and, along with **relaxin,** helps suppress uterine contractions. (p. 726)

48. **Parturition** normally occurs in the 38th–40th week of gestation. It begins with **labor** and ends with delivery of the fetus. A positive feedback loop of oxytocin secretion causes uterine muscle contraction. (p. 726)

49. Following delivery, the mammary glands produce milk under the influence of prolactin. Milk is released during nursing by oxytocin. (p. 727)

Growth and Aging

50. Puberty is the period when a person makes the transition from being nonreproductive to being reproductive. (p. 729)

51. The cessation of reproductive cycles in women is known as the **menopause.** (p. 730)

Questions

LEVEL ONE   Reviewing Facts and Terms

1. Match each item below with all the terms that it applies to:

   (a) X or Y
   (b) inactivated X chromosome
   (c) XX
   (d) XY
   (e) XX or XY
   (f) chromosomes other than sex chromosomes

   1. fertilized egg
   2. sperm or ova
   3. autosomes
   4. sex chromosomes
   5. germ cells
   6. male chromosomes
   7. female chromosomes
   8. Barr body
   9. acrosome

2. The Y chromosome contains a region for male sex determination that is known as the _____ gene. When activated, this gene directs the synthesis of a protein called _____.

3. List the functions of the gonads. How do the products of gonadal function differ in males and females?

4. Trace the anatomical route to the external environment followed by a newly formed sperm and by an ovulated egg. Name all structures that the gametes pass on their way.

5. Define the following terms and give their significance to reproductive physiology:
   a. aromatase
   b. blood-testis barrier
   c. androgen binding protein
   d. first polar body
   e. acrosome

6. Decide if the statements below are true or false, and defend your answer.
   a. All testosterone is produced in the testes.
   b. Each sex hormone is produced only by members of one gender.
   c. Anabolic steroid use appears to be addictive, and withdrawal symptoms include psychological disturbances.
   d. The high levels of estrogen in the late follicular phase help to prepare the uterus for menstruation.
   e. Progesterone is the dominant hormone of the luteal phase of the female cycle.

7. What is semen? What are its main components and where are they produced?

8. List and give a specific example of the various methods of contraception. Which is/are most effective? Least effective?

LEVEL TWO   Reviewing Concepts

9. **Concept maps:** Diagram the hormonal control of gametogenesis in males. Diagram the menstrual cycle in the female, including the ovarian cycle and the uterine cycle. Show long and short feedback loops, the source of all hormones, and the target organs or cells for each hormone.

10. Why are X-linked traits exhibited more frequently by males than females?

11. Define and relate the terms in each group below.
    a. Wolffian ducts, Müllerian ducts, testis determining factor, Müllerian inhibiting substance, testosterone, DHT
    b. Leydig cells and Sertoli cells

    c. spermatogonia, spermatocytes, sperm, spermatids
    d. gamete, zygote, germ cell, embryo
    e. myometrium and endometrium
    f. granulosa cells, corpus luteum, follicle, antrum, thecal cells
    g. coitus, erection, ejaculation, orgasm, emission, erogenous zones
    h. capacitation, decapacitation factor, zona pellucida, acrosomal reaction, cortical reaction, cortical granules
    i. puberty, menarche, menopause
    j. primary and secondary gametes

12. Compare the actions of the hormones below in males and females.
    a. FSH   b. inhibin   c. activin
    d. GnRH   e. LH   f. DHT
    g. estrogen   h. testosterone   i. progesterone

13. Compare and contrast the events of the four phases of sexual intercourse in each gender.

14. Discuss the roles of the following hormones in pregnancy, labor and delivery, and mammary gland development and lactation.

a. human chorionic gonadotropin
b. luteinizing hormone
c. testosterone
d. human placental lactogen
e. estrogen
f. progesterone
g. relaxin
h. prolactin inhibiting hormone

LEVEL THREE   Problem Solving

15. Down syndrome is a chromosomal defect known as "trisomy" (three copies instead of two) of the twenty-first chromosome. Babies born with this condition may be of either gender. Down syndrome has been linked to older mothers, but is not related to the age of the baby's father. Speculate on the link between trisomy and maternal age, using what you have learned about the events surrounding fertilization.

16. Sometimes eggs fail to leave the ovary at ovulation, even though they appear to have gone through all of the correct stages of development. This condition results in benign ovarian cysts, and the unruptured follicles can be palpated as bumps on the surface of the ovary. If the cysts persist, symptoms of this condition often mimic pregnancy, with missed menstrual periods and tender breasts. Can you explain or diagram how these symptoms occur?

Problem Conclusion

In this running problem, you learned how the cause of infertility is diagnosed in a typical couple. You also learned how interventions are used to overcome some causes of infertility.

Further check your understanding of the running problem by checking your answers against those in the summary table.

	Question	*Facts*	*Integration and Analysis*
1	Name the male reproductive structures that carry sperm from the testes to the external environment.	The male reproductive structures include the testes, accessory glandular organs, and a series of ducts.	Sperm leaving the seminiferous tubules pass into the epididymis, vas deferens, and finally the urethra.
2	For what causes of male infertility might retrieval of sperm from the epididymis be necessary?	The epididymis is the first duct the sperm enter upon leaving the seminiferous tubules.	If the infertility problem is due to blockage or congenital defects in the vas deferens or urethra, removal of sperm from the epididymis might be useful. If the problem is due to low sperm count or abnormal sperm morphology, this technique would probably not be useful.
3	For what causes of female infertility is temperature tracking useful? For what causes is it not useful?	Basal body temperature rises slightly following ovulation.	Temperature tracking is a useful way to tell if someone is ovulating but it cannot be used to predict structural problems in the female reproductive tract.
4	What abnormalities in the cervix, Fallopian tubes, and uterus could cause infertility?	The cervix, Fallopian tubes, and uterus are hollow structures through which sperm must pass.	Any blockage of these organs due to disease or congenital defects would prevent normal movement of sperm and cause infertility. Hormonal problems might cause the endometrium to develop incompletely, preventing implantation of the embryo.
5	Speculate on how infertility due to cervical mucus antibodies to sperm might be treated.	Antibodies in the cervical mucus react with antigenic material in the semen or on the sperm, causing the sperm to become immobile.	If the antigenic material can be removed from the semen, this might help the problem of sperm immobilization. A semen sample can be washed to remove nonsperm components. If the antigens are part of the sperm, this method will not work.
6	Should ART or intrauterine insemination be recommended for Peggy and Larry? Why?	ART is used when ovulation is abnormal. Intrauterine insemination is used when ovulation is normal.	Intrauterine insemination can be used to overcome cervical factors, as this technique bypasses the cervix. Peggy can ovulate; therefore, intrauterine insemination should be recommended for Peggy and Larry.

Physics and Math

Richard D. Hill and Daniel Biller, University of Texas

INTRODUCTION

This appendix discusses selected aspects of **biophysics,** the study of physics as it applies to biological systems. Because living systems are in a continual exchange of force and energy, it is necessary to define these important concepts. According to the seventeenth-century scientist Sir Isaac Newton, a body at rest tends to stay at rest, and a body in motion tends to continue moving in a straight line unless the body is acted upon by some force (Newton's First Law). Newton further defined **force** as an influence, measurable in both intensity and direction, that operates on a body in such a manner as to produce an alteration of its state of rest or motion. Put another way, force gives **energy** to a quantity, or mass, thereby enabling it to do work. In general, a driving force multiplied by a quantity yields energy, or work. Some relevant examples of this principle include:

Mechanical force × distance = mechanical energy or mechanical work
Gas pressure × volume of gas = mechanical energy or mechanical work
Osmotic pressure × molar value = osmotic energy or osmotic work
Electrical potential × charge = electrical energy or electrical work
Temperature × entropy = heat energy or mechanical work
Chemical potential × concentration = chemical energy or chemical work

Energy exists in two general forms: kinetic energy and potential energy. **Kinetic energy** [*kinein,* to move] is the energy possessed by a mass in motion. **Potential energy** is energy possessed by a mass because of its position. Kinetic energy (*KE*) is equal to one-half the mass (*m*) of a body in motion multiplied by the square of the velocity (*v*) of the body:

$KE = \frac{1}{2}mv^2$

Potential energy (*PE*) is equal to the mass (*m*) of a body multiplied by acceleration due to gravity (*g*) times the height (*h*) of the body above the earth's surface:

$PE = mgh$ where $g = 10$ m/s^2

Both kinetic and potential energy are measured in joules.

BASIC UNITS OF MEASUREMENT

For physical concepts to be useful in scientific endeavors, they must be measurable and should be expressed in standard units of measurement. Some fundamental units of measure include the following:

Length (*l*): Length is measured in meters (m).
Time (*t*): Time is measured in seconds (s).
Mass (*m*): Mass is measured in kilograms (kg), and is defined as the weight of a body in a gravitational field.
Temperature (*T*): Temperature is measured in degrees Kelvin (°K),
    where °K = degrees Celsius (°C) + 273.15
    and °C = (degrees Fahrenheit − 32)/1.8
Electric current (*I*): Electric current is measured in amperes (A).
Amount of substance (*n*): The amount of a substance is measured in moles (mol).

Using these fundamental units of measure, we can now establish standard units for physical concepts (Table A-1). Although these are the standard units for these concepts at this time, they are not the only units ever used to describe them. For instance, force can also be measured in dynes, energy can be measured in calories, pressure can be measured in torr or mm Hg, and power can be measured in horsepower. However, all of these units can be converted into a standard unit counterpart, and vice versa.

The remainder of this appendix will discuss some biologically relevant applications of physical concepts. This discussion will include topics such as bioelectrical principles, osmotic principles, and behaviors of gases and liquids relevant to living organisms.

BIOELECTRICAL PRINCIPLES

Living systems are composed of different molecules, many of which exist in a charged state. Cells are filled with charged particles such as proteins and organic acids, and ions are in continual flux across the cell membrane. Therefore, electrical forces are important to life.

TABLE A.1 **Standards Units for Physical Concepts**

Measured Concept	Standard (SI*) Unit	Mathematical Derivation/Definition
Force	Newton (N)	$1 \text{ N} = 1 \text{ kg m/s}^2$
Energy/Work/Heat	Joule (J)	$1 \text{ J} = 1 \text{ N} \cdot \text{m}$
Power	Watt (W)	$1 \text{ W} = 1 \text{ J/s}$
Electrical charge	Coulomb (C)	$1 \text{ C} = 1 \text{ A} \cdot \text{s}$
Potential	Volt (V)	$1 \text{ V} = 1 \text{ J/C}$
Resistance	Ohm (Ω)	$1 \Omega = 1 \text{ V/A}$
Capacitance	Farad (F)	$1 \text{F} = 1 \text{ C/V}$
Pressure	Pascal (Pa)	$1 \text{ Pa} = 1 \text{ N/m}^2$

*SI = Système International d'Unites

When molecules gain or lose electrons, they develop positive or negative charges. A basic principle of electricity is that opposite charges attract and like charges repel. A force must act on a charged particle (a mass) to bring about changes in its position. Therefore, there must be a force acting on charged particles to cause attraction or repulsion, and this electrical force can be measured. Electrical force increases as the strength (number) of charges increases, and it decreases as the distance between the charges increases. This observation has been called Coulomb's law, and can be written:

$$F = \frac{q_1 q_2}{\epsilon d^2}$$

where q_1 and q_2 are the electrical charges (coulombs), d is the distance between the charges (meters), ϵ is the dielectric constant, and F is the force of attraction or repulsion, depending on the type of charge on the particles.

When opposite charges are separated, a force acts over a distance to draw them together. As the charges move together, work is being done by the charged particles and energy is being released (Fig. A-1 ■). Conversely, to separate the united charges, energy must be added and work done. If charges are separated and kept apart, they have the potential to do work. This electrical potential is called **voltage.** Voltage is measured in **volts (V).**

If electrical charges are separated and there is a potential difference between them, then the force between the charges will allow electrons to flow. Electron flow is called an electric **current.** The amount of current that flows depends on the nature of the material between the charges. If that material hinders the electron flow, then it is said to offer **resistance (R),** measured in ohms. Current is inversely proportional to resistance, such that current decreases as resistance increases. If a material offers a high resistance, then that material is called an **insulator.** If resistance is low, and current flows relatively freely, then the material is called a **conductor.** Current, voltage, and resistance are related by **Ohm's law,** which states:

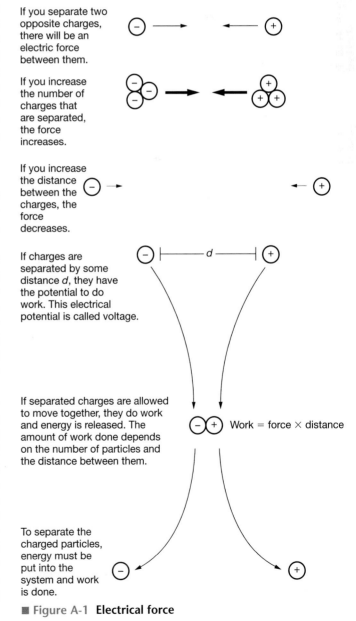

If you separate two opposite charges, there will be an electric force between them.

If you increase the number of charges that are separated, the force increases.

If you increase the distance between the charges, the force decreases.

If charges are separated by some distance d, they have the potential to do work. This electrical potential is called voltage.

If separated charges are allowed to move together, they do work and energy is released. The amount of work done depends on the number of particles and the distance between them.

Work = force × distance

To separate the charged particles, energy must be put into the system and work is done.

■ Figure A-1 **Electrical force**

$$V = IR$$

where V = potential difference in volts
        I = current in amperes
        R = resistance in ohms

In biological systems, pure water is not a good conductor, but water containing dissolved NaCl is a fairly good conductor because ions provide charges to carry the current. In biological membranes, the lipids have few or no charged groups, so they offer high resistance to current flow across them. Thus, different cells can have different electrical properties depending on their membrane lipid composition and the permeability of their membranes to ions.

OSMOTIC PRINCIPLES

Freezing point, vapor pressure, boiling point, and osmotic pressure are properties of solutions collectively called **colligative properties.** These properties depend on the number of solute particles present in a solution. **Osmotic pressure** is the force that drives the diffusion of water across a membrane. Because there are no solutes in pure water, it has no osmotic pressure. However, if one adds a solute like NaCl, the greater the concentration (c) of a solute dissolved in water, the greater the osmotic pressure. The osmotic pressure (π) varies directly with the concentration of solute (number of particles (n) per volume (V)):

$$\pi = (n/V)RT$$

$$\pi = cRT$$

where R is the universal gas constant (8.314 joules/°K • mol) and T is the absolute temperature (in °Kelvin). Osmotic pressure can be measured by determining the mechanical pressure that must be applied to a solution so that osmosis ceases.

Water balance in the body is under the control of osmotic pressure gradients (concentration gradients). Most cellular membranes allow water to pass freely because water is a small, uncharged molecule. Therefore, by making cellular membranes selectively permeable to solutes, the body can control the movement of water by controlling the movement of solutes.

THE RESTING MEMBRANE POTENTIAL

We have now discussed two very important biophysical concepts: electrical force and osmotic force. It is the interaction between these two forces that sets up the membrane potential across a cell membrane. Each living cell membrane effectively acts as a charge separator so that there is a potential difference (**membrane potential, or V_m**) across it. Voltage across a nerve cell membrane is about −70 mV; across a cardiac contractile cell, the membrane potential is about −90 mV. The selective permeability of the cell membrane sets up osmotic gradients that also contribute to the membrane potential by affecting the flow of certain ions either into or out of the cell. The resulting membrane potentials are important because they make phenomena such as nervous impulses and muscle contractions possible.

Two important physical equations describe the relationship between osmotic and electrical forces and membrane potentials. These equations are the Nernst equation and the Goldman equation. The Nernst equation describes the maximum membrane potential that a single ion could produce if the membrane were permeable to only that one ion. In reality, the membrane potential is a result of the many ion concentration gradients and permeabilities; the Goldman equation takes this fact into consideration.

The Nernst Equation

Under resting membrane conditions, the osmotic work (W) required to move an ion across a membrane (assuming a concentration gradient) is equal to the electrical work (W') required to move an ion against a voltage gradient.

Osmotic work required: $W = RT \ln \dfrac{[I]_{out}}{[I]_{in}}$

where R = the gas constant (8.314 joules/°K • mol)
     T = the temperature (in °K)
     ln = natural logarithm
  $[I]_{out}$ = ion concentration outside the cell
  $[I]_{in}$ = ion concentration inside the cell

Electrical work required: $W' = E_{ion}Fz$

where E_{ion} = equilibrium potential (in volts) for the ion (I) in question
     z = the electrical charge on the ion
     F = Faraday constant (96,000 coulombs/mol)

When these two expressions are set equal ($W = W'$) and solved for E_{ion}, you get the membrane potential for a cell that is permeable only to ion I.

Nernst equation: $E_{ion} = \dfrac{RT}{Fz} \ln \dfrac{[I]_{out}}{[I]_{in}}$

The Goldman Equation

The Goldman equation is used to calculate the membrane potential that results from the contribution of all ions that can cross the membrane. The equation takes membrane permeability into account because an ion can contribute to the membrane potential only if the membrane is permeable to it: Its contribution to the membrane potential is proportional to its ability to cross the membrane. The Goldman equation for cells that are permeable to Na^+, K^+, and Cl^- is

$$V_m = \frac{RT}{F} \ln \frac{P_K\,[K^+]_{out} + P_{Na}\,[Na^+]_{out} + P_{Cl}\,[Cl^-]_{in}}{P_K\,[K^+]_{in} + P_{Na}\,[Na^+]_{in} + P_{Cl}\,[Cl^-]_{out}}$$

where V_m = voltage across the membrane
P = relative permeability of the membrane to an ion
R = universal constant (8.314 joules/°K • mol)
T = temperature in °K
[I] = ion concentration (in or out of cell)
F = Faraday constant (96,000 coulombs/mol)

With the Goldman equation, one can calculate membrane potentials based on given ion concentrations and predict the changes expected in the membrane potential if these ion concentrations change.

RELEVANT BEHAVIORS OF GASES AND LIQUIDS

The respiratory and circulatory systems of the human body have developed according to the physical laws that govern the behaviors of gases and liquids. This section will discuss some of the important laws that govern these behaviors and how our body systems utilize these laws.

Gases The **ideal gas law** states:

$PV = nRT$

where P = pressure of gases in the system
V = volume of the system
n = number of moles in gas
T = temperature
R = universal constant (8.314 J/°K • mol)

If n and T are kept constant for all pressures and volumes in a system, then any two pressures and volumes in that system are related to Boyle's Law,

$P_1V_1 = P_2V_2$

where P represents pressure and V represents volume.

This principle is relevant to the human lungs because the concentration of gas in the lungs is relatively equal to that in the atmosphere. In addition, body temperature is maintained at a constant temperature by homeostatic mechanisms. Therefore, if the volume of the lungs is changed, then the pressure in the lungs will change inversely. For example, an increase in pressure causes a decrease in volume, and vice versa.

Liquids **Fluid pressure** (or hydrostatic pressure) is the pressure exerted by a fluid on a real or hypothetical body. In other words, the pressure exists whether or not there is a body submerged in the fluid. Fluid exerts a pressure (P) on an object submerged in it at a certain depth from the surface (h). **Pascal's law** allows us to find the fluid pressure at a specified depth for any given fluid. It states:

$P = \rho g h$

where P = fluid pressure (measured in pascals, Pa)
ρ = density of the fluid
g = acceleration due to gravity (10 m/s²)
h = depth below the surface of the fluid

Fluid pressure is unrelated to the shape of the container in which the fluid is situated.

REVIEW OF LOGARITHMS

Understanding logarithms ("logs") is important in biology because of the definition of pH:

$pH = -\log_{10}[H^+]$

This equation is read as "pH is equal to the negative log to the base 10 of the hydrogen ion concentration." But what is a logarithm?

A logarithm is the exponent to which you would have to raise the base (10) to get the number in which you are interested. For example, to get the number 100, you would have to square the base (10):

$10^2 = 100$

The base 10 was raised to the second power; therefore, the log of 100 is 2:

$\log 100 = 2$

Some other simple examples include:

$10^1 = 10$ The log of 10 is 1.
$10^0 = 1$ The log of 1 is 0.
$10^{-1} = 0.1$ The log of 0.1 is −1.

What about numbers that fall between the powers of 10? If log of 10 is 1 and log of 100 is 2, the log of 70 would be between 1 and 2. The actual value can be looked up on a log table or ascertained with most calculators.

To calculate pH, you need to know another rule of logs that says:

$-\log x = \log (1/x)$

and a rule of exponents that says:

$1/10^x = 10^{-x}$

Suppose you have a solution whose hydrogen ion concentration [H⁺] is 10^{-7} meq/L. What is the pH of this solution?

$pH = -\log [H^+]$

$pH = -\log (10^{-7})$

Using the rule of logs, this can be rewritten as

$pH = \log (1/10^{-7})$

Using the rule of exponents, this can be rewritten as

$pH = \log 10^7$

The log of 10^7 is 7, so the solution has a pH of 7.

Natural logarithms (ln) are logs in the base e. The mathematical constant e is approximately equal to 2.7183.

Genetics

Richard D. Hill, University of Texas

WHAT IS DNA?

Deoxyribonucleic acid (DNA) is the macromolecule that stores the information necessary to build structural and functional cellular components. It also provides the basis for inheritance when DNA is passed from parent to offspring. The union of these concepts about DNA allows us to devise a working definition of a gene. A **gene** (1) is a segment of DNA that codes for the synthesis of a product participating in structural or functional aspects of a cell, and (2) acts as a unit of inheritance that can be transmitted from generation to generation. The genetic organization of an organism's cells determines how those cells work together to create the complete organism. The external appearance (**phenotype**) of an organism is determined to a large extent by the genes it inherits (**genotype**). Thus, one can begin to see how variation at the DNA level can cause variation at the level of the entire organism. These concepts form the basis of **genetics** and evolutionary theory.

NUCLEOTIDES AND BASE-PAIRING

DNA belongs to a group of macromolecules called **nucleic acids. Ribonucleic acid (RNA)** is also a nucleic acid, but it has different functions in the cell that are not discussed here (see Chapter 4, p. 33). Nucleic acids are polymers [*polymeres*, of many parts] made from monomers [*mono-*, one] called **nucleotides**. Each nucleotide consists of a nucleoside (a pentose, or 5-carbon, sugar covalently bound to a nitrogenous base) and a phosphoric acid with at least one phosphate group (Fig. B-1a ■). Nitrogenous bases in nucleic acids are classified as either **purines** or **pyrimidines**. The purine bases are **guanine (G)** and **adenine (A)**; the pyrimidine bases are **cytosine (C)**, **thymine (T)** , found in DNA only, and **uracil (U)**, found in RNA only. To remember which DNA bases are pyrimidines, look at the first syllable. The word "pyrimidine" and names of the DNA pyrimidine bases all have a "y" in the first syllable.

When nucleotides link together to form polymers such as DNA and RNA, the phosphate group of one nucleotide bonds covalently to the sugar group of the adjacent nucleotide (Fig. B-1b, c ■). The end of the polymer that has an unbound sugar is called the 3′ ("three prime") end. The end of the polymer with the unbound phosphate is called the 5′ end.

DNA STRUCTURE

In humans, many millions of nucleotides are joined together to form DNA. Eukaryotic DNA is commonly in the form of a double-stranded double helix (Figure B-2 ■), a structure that looks somewhat like a twisted ladder or twisted zipper. The sugar-phosphate sides, or backbone, are the same for every DNA molecule, but the sequence of the nucleotides is unique for each individual organism.

The backbone of the double helix is formed by covalent **phosphodiester bonds** that link a deoxyribose sugar from one nucleotide to the phosphate group on the adjacent nucleotide. The "rungs" of the double helix are created when the nitrogenous bases on one DNA strand form hydrogen bonds with nitrogenous bases on the adjoining DNA strand.

The structure and orientation of the nitrogenous bases in a DNA molecule allow each type of base to hydrogen bond with only one other type. This phenomenon is called **base-pairing**. The base-pairing rules are as follows:

1. Purines pair only with pyrimidines.
2. Guanine (G) forms three hydrogen bonds with cytosine (C) in both DNA and RNA.
3. Adenine (A) forms with two hydrogen bonds with thymine (T) in DNA or with uracil (U) in RNA.

The number of hydrogen bonds is directly related to the amount of energy necessary to break the base pair. Thus, more energy is required to break G:::C bonds (each ":" represents a hydrogen bond) than A::T bonds. This fact can be useful experimentally to make general conjectures about the similarity of two DNA samples.

The two strands of DNA are bound in **antiparallel** orientation, so that the 3′ end of one strand is bound to the 5′ end of the second strand (see Fig. B-1c ■). This organization has important implications for DNA replication.

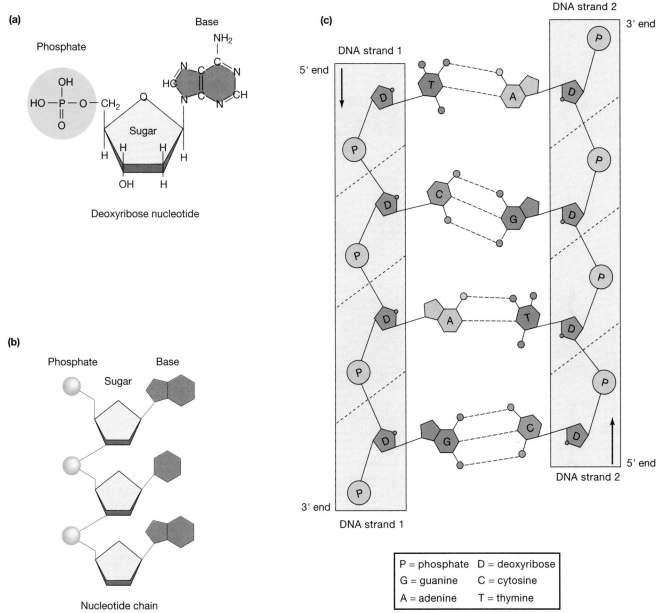

■ **Figure B-1** **Nucleotides and DNA** (a) A nucleotide is composed of a pentose sugar, a nitrogenous base, and a phosphate group. (b) Nucleotides link sugar to phosphate to form nucleic acids. The end of the nucleic acid with an unbound sugar is designated the 3′ end; the end with the unbound phosphate is the 5′ end. (c) Two nucleotide strands link by hydrogen binding between complementary base pairs to form the double helix of DNA. Adenine always pairs to thymine, and guanine pairs to cytosine.

DNA Replication Is Semi-Conservative

In order to be transmitted from one generation to the next, DNA must be replicated. Furthermore, the process of replication must be accurate and fast enough for a living system. The base-pairing rules for nitrogenous bases provide a means for making an appropriate replication system.

In DNA replication, special proteins unzip the DNA double helix and build new DNA by pairing new nucleotide molecules to the two existing DNA strands. The result of this replication is two double-stranded DNA molecules, such that each DNA molecule contains one DNA strand from the template and one newly synthesized DNA strand. This form of replication is called **semi-conservative replication**.

Replication of DNA is bidirectional. A portion of DNA that is "unzipped" and has enzymes performing replication is called a **replication fork** (Fig. B-2 ■). Replication begins at many points (**replicons**), and it continues along both parent strands simultaneously until all the replication forks join.

Nucleotides bond together to form new strands of DNA with the help of an enzyme called **DNA poly-**

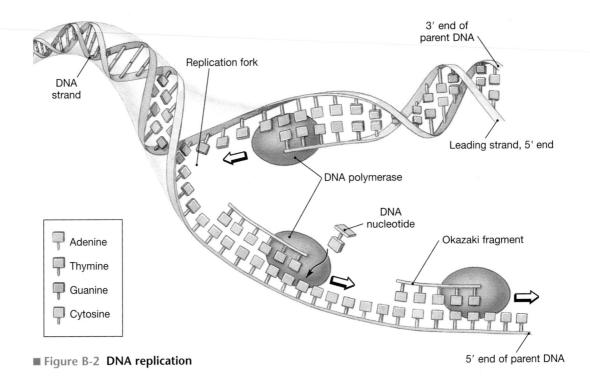

■ **Figure B-2 DNA replication**

merase. DNA polymerase can only add nucleotides to the 3′ end of a growing strand of DNA. For this reason, DNA is said to replicate in a 5′ to 3′ direction.

The antiparallel orientation of the DNA strands and the directionality of DNA polymerase force replication into two different modes: **leading strand replication** and **lagging strand replication**. The DNA polymerase can replicate continuously along only one parent strand of DNA: the parent strand in the 3′ to 5′ orientation. The DNA replicated continuously is called the **leading strand**.

The DNA replication along the other parent strand is discontinuous because of the strand's 5′ to 3′ orientation. DNA replication on this strand occurs in short fragments called **Okazaki fragments** that are synthesized in the direction away from the replication fork. Another enzyme known as **DNA ligase** later connects these fragments into a continuous strand. The DNA replicated in this way is called the **lagging strand**. Because the 5′ ends of the lagging strand of DNA cannot be replicated by DNA polymerase, a specialized enzyme called **telomerase** has arisen to replicate the 5′ ends.

Much of the accuracy of DNA replication comes from base pairing, but on occasion, mistakes in replication happen. However, several quality control mechanisms are in place to keep the error rate at 1 error / 10^9 to 10^{12} base pairs. **Genome** (the entire amount of DNA in an organism) sizes in eukaryotes range from 10^9 to 10^{11} base pairs per genome, so this error rate is low enough to prevent many lethal mutations, yet still allows genetic variation to arise.

FUNCTIONS OF DNA

One of the key functions of DNA is its ability to code for the synthesis of proteins that participate in structural or functional aspects of a cell. The processes by which information coded in DNA is converted into proteins are known as transcription and translation. Both are discussed in Chapter 4. The other key function of DNA is its ability to form genes that act as units of inheritance when transmitted across generations.

Before we begin our discussion of DNA as a unit of inheritance, a few terms need to be introduced. A **chromosome** is one complete molecule of DNA, and each chromosome contains many genes. An **allele** is a form of a gene; interactions between the cell products of alleles determine how that gene will be expressed in the phenotype of an individual. **Somatic** [*soma*, body] **cells** are those cells that comprise the majority of the body (e.g., a skin cell, a liver cell); they are not directly involved with passing on genetic information to future generations. Each somatic cell in a human contains two alleles of each gene, one allele inherited from each parent. For this reason, human somatic cells are called **diploid** ("two chromosome sets"), meaning that they have two complete sets of all their chromosomes. In contrast, **germ cells** pass genetic information directly to the next generation. In human males, the germ cells are the **spermatozoa** (sperm), and in human females, the germ cells are the **oocytes** (eggs). Human germ cells are called **haploid** ("half of the chromosome sets") because each germ cell only contains one chromosome set, which is equal to half of the number of chromosome sets in somatic cells. When a human male germ cell joins with a human female germ cell, the result is a fertilized egg (zygote) containing the diploid number of chromosomes. If this zygote eventually develops into a healthy adult, that adult will have diploid somatic cells and haploid germ cells.

Cells alternate between periods of cell growth and cell division. There are two types of cell division: mitosis and meiosis. **Mitosis** is cell division by somatic cells that results in two daughter cells, each with a diploid set of chromosomes. **Meiosis**, in contrast, is cell division that results in four daughter cells, each with a haploid set of chromosomes. The daughter cells develop into germ cells.

The period of cell growth is called **interphase**. It is divided into three stages: G_1, a period of cell growth, protein synthesis, and organelle production; **S**, the period during which DNA is replicated in preparation for cell division; and G_2, a period of protein synthesis and final preparations for cell division (Fig. B-3 ■). During interphase, the DNA in the nucleus is not visible under the light microscope without dyes because it is uncoiled and diffuse. However, as a cell prepares for division, it must condense all its DNA to form more manageable packages. Each eukaryotic DNA molecule has millions of base pairs, which, if laid end-to-end, could stretch out to about 6 cm. If this DNA did not coil tightly and condense for cell division, imagine how difficult moving it around during cell division would be.

There is a hierarchy of DNA packaging in the cell (Fig. B-4 ■). Each chromosome begins with a linear molecule of DNA about 2 nm in diameter. Then proteins called his-

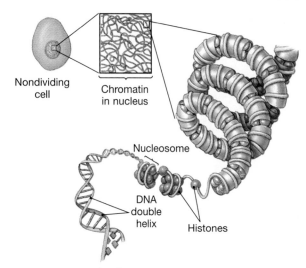

■ Figure B-4 **Levels of organization of DNA**

tones associate with the DNA to form a fiber about 10 nm in diameter that looks like "beads on a string." The "beads" are **nucleosomes** made up of histones wrapped in DNA. The beaded string can twist into a **solenoid** [*solen*, pipe] coil about 30 nm in diameter, consisting of about 6 nucleosomes per turn. The 30-nm solenoid structure can form looped domains, and these looped domains can attach to **nonhistone** (structural) **proteins** to form a 250–300-nm fiber called **chromatid fiber**. The chromatid fiber then coils to form the **chromosome fiber** (about 700 nm in diameter) that is visible during cell division. In this state of packaging, the cell is ready for division.

Mitosis Creates Two Identical Daughter Cells

As stated earlier, mitosis is the cellular division of a somatic cell that results in two diploid daughter cells. The steps of mitosis are **prophase, metaphase, anaphase,** and **telophase** (Fig. B-5 ■). The entire somatic cell cycle can be remembered by the acronym, IPMAT, in which the "I" stands for interphase and the other letters stand for the steps of mitosis that follow.

Prophase   During prophase, chromatin becomes condensed and microscopically visible as duplicate chromosomes. The duplicated chromosomes form **sister chromatids**, which are joined to each other at the **centromere**. The centriole pair (∞ p. 47) duplicates and the centriole pairs move to opposing ends of the cell. The **mitotic spindle,** composed of microtubules, assembles between them. The nuclear membrane begins to break down and disappears by the end of prophase.

Metaphase   In metaphase, mitotic spindle fibers extending from the centrioles attach to the centromere of each chromosome. The forty-six chromosomes, each consisting of a pair of sister chromatids, line up at the "equator" of the cell.

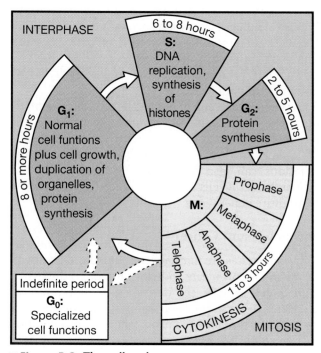

■ Figure B-3 **The cell cycle**

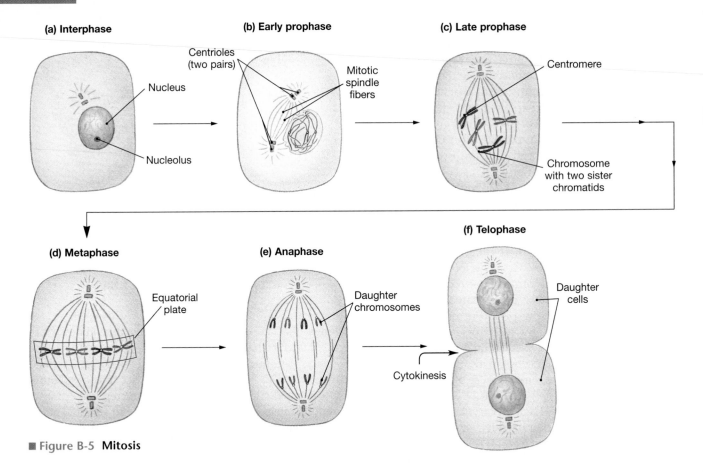

(a) Interphase

Nucleus

Nucleolus

(b) Early prophase

Centrioles
(two pairs)

Mitotic
spindle
fibers

(c) Late prophase

Centromere

Chromosome
with two sister
chromatids

(d) Metaphase

Equatorial
plate

(e) Anaphase

Daughter
chromosomes

Cytokinesis

(f) Telophase

Daughter
cells

■ Figure B-5 **Mitosis**

Anaphase   During anaphase, the spindle fibers pull the sister chromatids apart, so that an identical copy of each chromosome moves toward each pole of the cell. By the end of anaphase, an identical set of forty-six chromosomes is present at each pole. At this point, the cell has a total of ninety-two chromosomes.

Telophase   The actual division of the parent cell into two daughter cells takes place during telophase. In **cytokinesis,** the cytoplasm divides when an actin contractile ring tightens at the midline of the cell. The result is two separate daughter cells, each with a full diploid set of chromosomes. The spindle fibers disintegrate, nuclear envelopes are formed around the chromosomes in each cell, and the chromatin returns to its loosely coiled state.

Mutations Change the Sequence of DNA

Over the course of a lifetime, there are countless opportunities for mistakes to arise in the replication of DNA. A change in a DNA sequence, such as the addition, substitution, or deletion of a base, is a **point mutation.** If a mutation is not corrected, it may cause a change in the gene product. These changes may be relatively minor, or they may result in dysfunctional gene products that could kill the cell or the organism. Only rarely does a

mutation result in a beneficial change in a gene product. Fortunately, our cells contain enzymes that can detect and repair damage to DNA.

Some mutations are caused by **mutagens,** which are factors that increase the rate of mutation. Various chemicals, ionizing radiation such as X-rays and atomic radiation, ultraviolet light, and other factors can behave as mutagens. Mutagens either alter the base code of DNA or interfere with repair enzymes, thereby promoting mutation.

Mutations that occur in body cells are called **somatic mutations.** Somatic mutations are perpetuated in the somatic cells of an individual, but they are not passed on to subsequent generations. However, **germ-line mutations** can also occur. Because these mutations arise in the germ cells of an individual, they are passed on to future generations.

Oncogenes and Cancer

Proto-oncogenes are normal genes in the genome of an organism that primarily code for protein products that regulate cell growth, cell division, and cell adhesion. Mutations in these proto-oncogenes give rise to **oncogenes,** genes that induce uncontrolled cell proliferation and the condition known as **cancer.** The mutations in proto-oncogenes that give rise to cancer-causing oncogenes are often the result of viral activity.

Anatomical Positions of the Body

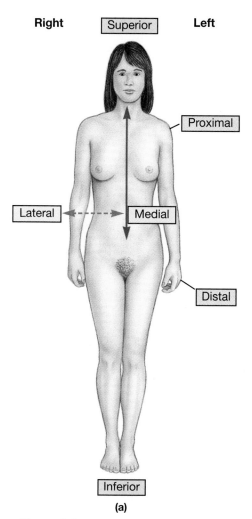

Right **Superior** Left

Proximal

Lateral — Medial

Distal

Inferior

(a)

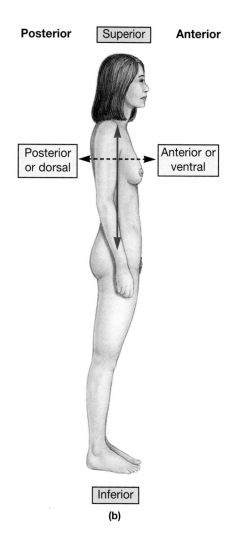

Posterior **Superior** Anterior

Posterior or dorsal — Anterior or ventral

Inferior

(b)

■ **Figure C-1**

ANTERIOR (situated in front of): in humans, toward the front of the body (see VENTRAL)

POSTERIOR (situated behind): in humans, toward the back of the body (see DORSAL)

MEDIAL (middle, as in *median strip*): located nearer to the midline of the body (the line that divides the body into mirror-image halves)

LATERAL (side, as in a *football lateral*): located toward the sides of the body

DISTAL (distant): farther away from the point of reference or from the center of the body

PROXIMAL (closer, as in *proximity*): closer to the center of the body

SUPERIOR (higher): located toward the head or the upper part of the body

INFERIOR (lower): located away from the head or from the upper part of the body

PRONE: lying on the stomach, face downward

SUPINE: lying on the back, face up

DORSAL: refers to the back of the body

VENTRAL: refers to the front of the body

IPSILATERAL: on the same side as

CONTRALATERAL: on the opposite side from

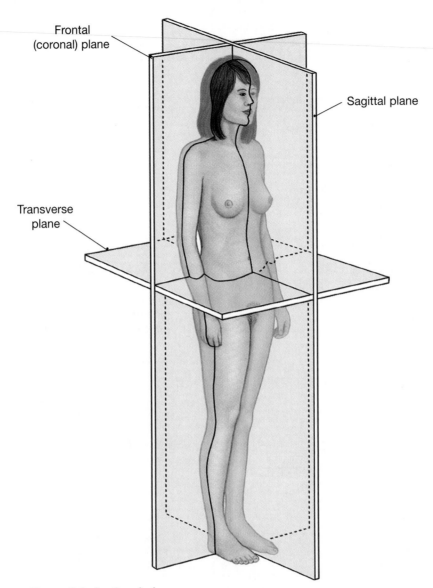

Frontal
(coronal) plane

Sagittal plane

Transverse
plane

■ Figure C-2 Sectional planes

Answers to Concept Check and Graph Questions

CHAPTER 1   Introduction to Physiology
No concept check questions

p. 4

Fig. 1-3 Examples: Humans drive buses; humans eat animals.

CHAPTER 2   Atoms, Ions, and Molecules

p. 19

✓ Zinc has 30 protons.
✓ Sodium has 11 electrons.
✓ One atom of calcium has 20 electrons.
✓ The atomic mass of nitrogen is 14.0067 amu.

p. 20

✓ Hydrogen cannot form a double bond because it has one electron and requires only one more electron to fill its outer shall. A double bond is formed by the sharing of four electrons.

p. 23

✓ $C_6H_8O_7$
✓ Na^+ and Cl^- ions form hydrogen bonds with the polar water molecules. This disrupts the ionic bonds that held the NaCl crystal together.
✓ The outer shell of carbon can hold 8 electrons. The carbon atom contains 4 electrons in that shell, and it obtains the other 4 electrons needed to fill the shell through covalent bonds.
✓ Dishwashing detergents are surfactants that disrupt the hydrogen bonds of water. This disrupts the surface tension of the water and prevents beading.

p. 26

✓ A 1 M solution is equal to a 1000 mM solution, so a 0.1 M solution is equal to a 100 mM solution.
✓ If you add 5 g of glucose to 100 mL of water, the glucose will add volume to the solution and you will no longer have 100 mL of solution. This means you will no longer have a 5% solution. To correctly make the solution, add water to the 5 g of glucose until you have a total solution volume of 100 mL.
✓ Acidity is related to increased H^+ concentration. As H^+ concentration increases, pH decreases. Therefore, increasing acidity is associated with decreasing pH.
✓ Urine, stomach acid, and saliva are all contained in the lumens of hollow organs, where they are not part of the body's internal environment (see Fig. 1-2 on p. 3)

p. 32

✓ Our bodies cannot make the essential amino acids; therefore, we must obtain them through our diet. We can synthesize the other amino acids from nutrients.

✓ Proteins are composed of many different amino acids, which can be linked in an almost infinite number of combinations. These combinations allow proteins to take on different forms, such as the globular and fibrous proteins.

CHAPTER 3   Cells and Tissues

p. 53

✓ Large numbers of mitochondria suggest that the cell has a high energy requirement because mitochondria are the site of greatest energy production in the cell.
✓ Large amounts of smooth endoplasmic reticulum suggest that the tissue synthesizes large amounts of lipids, fatty acids, or steroids.
✓ Without a flagellum, a sperm could not control its movement and would be carried along with the fluid flow.
✓ The rough endoplasmic reticulum is the site of protein synthesis, so pancreatic cells would be more likely to have greater amounts.

p. 60

✓ No, skin would have many layers of cells in order to protect the internal environment. A simple squamous epithelium that is one cell thick with flattered cells would not be a protective epithelium.
✓ The cell is an endocrine cell because it secretes its product into the extracellular space for distribution in the blood.

p. 66

✓ The plasma, or liquid portion of blood, could be considered the matrix.
✓ Cartilage lacks a blood supply, so oxygen and nutrients must reach the cells by diffusion, a slow process.

CHAPTER 4   Cellular Metabolism

p. 75

✓ Potential energy in the body is stored in the chemical bonds of lipids and glycogen.
✓ Kinetic energy is energy in motion: Something is happening. Potential energy is stored energy: Something is waiting to happen.

p. 78

✓ The reactants are baking soda and vinegar; the product is carbon dioxide.
✓ The foaming-up indicates that energy is being released, so this is an exergonic reaction.

p. 80

✓ The substrates are lactose (lactase), peptides (peptidase), lipids (lipase), and sucrose (sucrase).

✓ In the induced-fit model, both reactants (substrates) and products can combine with the active site. In the lock-and-key model, the enzyme fits with the reactants but not the products.

✓ By having isozymes, one reaction could be catalyzed under a variety of conditions.

p. 83

Fig. 4-11: (a) The enzyme is more active at 30°. (b) When pH goes from 8 to 7.4, the activity increases.

p. 85

Fig. 4-14: At enzyme concentration A, the rate is 1 mg product formed/sec. At concentration C, the rate is 2.5 mg product formed/sec.

Fig. 4-15: When substrate concentration is 200 mg/mL, the rate is 4 mg product formed/sec.

p. 87

✓ Modulators may affect the rate of enzymatic reactions by binding to the active site (competitive inhibitors) or by binding to a site on the enzyme protein away from the active site (allosteric and covalent modulators).

✓ As the amount of enzyme present increases, the rate of the reaction increases.

p. 102

✓ The polypeptide will have 149 amino acids.

✓ Cells regulate substrate movement through metabolic pathways by (1) controlling the amount of enzyme, (2) producing allosteric and covalent modulators, (3) using two different enzymes to catalyze reversible reactions, (4) isolating enzymes within intracellular organelles, and (5) altering the ADP:ATP ratio in the cell.

✓ Anaerobic metabolism of glucose can proceed in the absence of oxygen; aerobic metabolism requires oxygen. Anaerobic metabolism produces a maximum yield of 2 ATP/glucose; aerobic metabolism has a maximum yield of 36–38 ATP.

CHAPTER 5   Membrane Dynamics

p. 114

✓ Two types of lipids found in cell membranes are phospholipid and cholesterol.

✓ Integral proteins span the thickness of the membrane and are tightly bound to it. Associated proteins are loosely bound to polar regions of the membrane components.

✓ Membrane proteins serve as structural elements, receptors, enzymes, and transporters.

p. 120

Fig. 5-16: The line is zero from the origin until the point at which glucose is added. It then increases by the same amount that fructose decreases and becomes constant.

p. 125

✓ Over time, both compartments will turn green. Although the total concentrations of dye are equal on both sides of the membrane, there are concentration gradients for the individual dyes. As they move down their individual gradients, yellow dye moves into compartment B and blue dye into compartment A. At equilibrium, both dyes are distributed evenly throughout the system.

✓ The skin's thick extracellular matrix is generally impermeable to oxygen. Also, oxygen needs a moist surface to effectively diffuse across a membrane.

p. 128

✓ In phagocystosis, the cell pushes its membrane out to engulf the particle in a large vesicle. In endocytosis, the membrane surface indents and the vesicle is much smaller.

✓ Most cholesterol in the blood is carried bound to proteins in order to make it more soluble in the aqueous plasma. The membrane carriers keep the bound cholesterol from diffusing into cells.

p. 130

✓ If ouabain is applied to the apical side, nothing will happen because there are no Na^+/K^+-ATPase molecules on that side. If ouabain is applied to the basolateral side, it will stop the pump. Glucose transport will continue for a time until the Na^+ gradient between the cell and the lumen disappears due to Na^+ entry.

✓ Vesicular transport of vesicles from one side of the cell to the other depends on microtubules. If they are not functioning, transcytosis will stop.

p. 136

✓ Both 5% dextrose and 0.9% NaCl are isotonic, so neither would affect body osmolarity. Blood loss is loss of fluid from the extracellular compartment, so the best replacement solution is one that remains in the ECF. NaCl is nonpenetrating, so the NaCl solution meets that requirement. Glucose will slowly enter the cells, taking water with it.

✓ If one liter has been lost, you should replace one liter.

p. 141

✓ Ouabain inhibits the Na^+/K^+-ATPase that maintains the resting membrane potential. Over time, Na^+ would leak into the cell, and the resting membrane potential will become more positive.

CHAPTER 6   Communication, Integration, and Homeostasis

p. 149

✓ Electrical signals pass through gap junctions (#3). The remainder are chemical signals. Some of them, such as neurotransmitters, create electrical signals in their target cells.

✓ Cytokines, hormones, and neurohormones travel through the blood.

✓ The signal to pounce could not have been a paracrine because the eyes are too far away from the legs and because the response was too rapid for it to have taken place by diffusion.

p. 155

✓ Steroid hormones diffuse across the membranes of their target cells and do not need to rely on signal transduction to get their signal inside.

✓ The first messenger is the chemical signal that starts the events leading to a response. Second messengers are intracellular signals that are activated by first messengers that cannot enter the cell.

✓ Insulin can have different effects in different cells by using different second messenger systems.

✓ a. signal ligand → receptor → second messenger → cell response

b. amplifier enzyme → second messenger → protein kinase → phosphorylated protein → cell response

p. 157

Fig. 6-10: (a) 60 beats/min (b) 280 beats/min

p. 168
- ✓ Sense organs are receptors; sensory neurons are afferent pathways; brain and spinal cord are integrating centers; efferent neurons, hormones, and neurohormones are efferent pathways.

CHAPTER 7   Introduction to the Endocrine System

p. 182
- ✓ Steroid-producing cells would have extensive smooth endoplasmic reticulum, and protein-producing cells would have lots of rough endoplasmic reticulum.
- ✓ The shorter half-life suggests that aldosterone is not bound to plasma proteins as much as other steroid hormones.
- ✓ Steroid hormones are slower to act because they cause the synthesis of new proteins rather than the modification of existing proteins.

p. 184
Fig. 7-8: The conversion of tyrosine to dopamine adds a hydroxyl (—OH) group to the 6-carbon ring and changes the carboxyl (—COOH) group to a hydrogen. Norepinephrine is made from dopamine by changing one hydrogen to a hydroxyl group. Epinephrine is made from norepinephrine by changing one hydrogen attached to the nitrogen to a methyl (—CH_3) group.

p. 187
- ✓ The target of these hypothalamic hormones is endocrine cells in the anterior pituitary.
- ✓ A decrease in blood glucose is the negative feedback signal.

p. 195
- ✓ Secondary problem at pituitary: CRH up, ACTH low, cortisol low.
  Primary problem: CRH up, ACTH up, cortisol low.
  Secondary problem at hypothalamus: CRH low, ACTH low, cortisol low.

CHAPTER 8   The Nervous System

p. 209
- ✓ The resting membrane potential of a neuron treated with ouabain will become more positive and eventually reach 0 mV.
- ✓ The ratio would be >11:1.

p. 210
- ✓ If Na^+ leaks into the cell, it will depolarize.

p. 211
Fig. 8-9: The graded potential will be stronger at B.

p. 218
- ✓ If you could disable the inactivation gates, Na^+ would continue to flood in and the membrane potential would stay depolarized.
- ✓ If you artificially depolarize a cell in the middle of the axon, the signal will travel in both directions, toward the cell body and toward the terminal. Normally signals do not go backward because the channels are in their refractory period.

p. 227
- ✓ Some event between the arrival of the action potential at the presynaptic terminal and the depolarization of the postsynaptic cell is dependent on extracellular calcium. We now know that it is the exocytosis of neurotransmitter.
- ✓ Axon terminals convert the electrical signal of the action potential into a chemical signal (neurotransmitter release).

CHAPTER 9   The Central Nervous System

p. 241
- ✓ When H^+ concentration increases, pH decreases.
- ✓ Blood will collect in the space between the membranes, pushing on the soft brain tissue under the skull. This is called a subdural hematoma.
- ✓ Cerebrospinal fluid is more like interstitial fluid because it has little protein and no blood cells.

p. 242
- ✓ Roots are sections of spinal nerves just before they enter the spinal cord. Horns are extensions of gray matter in the spinal cord. Long projections of white matter (axons) that extend up and down the spinal cord are called tracts. Clusters of tracts are called columns.
- ✓ Dorsal roots carry sensory information.

p. 250
- ✓ Neurons cross from one side of the body to the other at the pyramids and along the spinal cord.
- ✓ Spinal cord → medulla → pons, cerebellum → midbrain → diencephalon → cerebrum

CHAPTER 10   Sensory Physiology

p. 272
- ✓ Vision—photoreceptor; hearing—mechanoreceptor; taste—chemoreceptor; smell—chemoreceptor; equilibrium—mechanoreceptor; touch-pressure—mechanoreceptor; temperature—thermoreceptor; pain—nociceptor; proprioreception—mechanoreceptor.
- ✓ Sensory neurons signal intensity of a stimulus by the rate at which they fire action potentials.
- ✓ Nociceptors, which indicate pain, are telling the body that it is being harmed. If possible, the body's response should be something that stops the pain. Therefore it is important that the pain continue as long as the stimulus is present.

p. 276
- ✓ Responses to temperature and pain usually do not require rapid responses as balance and touch do.
- ✓ There are many examples. One would be taste receptors.

p. 280
- ✓ Umami is associated with ingestion of protein foods, a biomolecule necessary for body function.

p. 287
- ✓ The location of somatosensory information is coded according to the side of the brain where the sensory pathway terminates. The location of smell and sound is coded by the time difference in the arrival of the stimulus in each hemisphere.
- ✓ A cochlear implant would not help people with either nerve deafness or conductive hearing loss. It can only help people with sensorineural hearing loss.

p. 289
- ✓ When fluid builds up in the middle ear, the eardrum is unable to move freely and therefore cannot transmit sound through the bones of the middle ear as well.
- ✓ When dancers spot, the endolymph in the ampulla moves with each head rotation, but then stops as the dancer holds his/her head still. This results in less inertia than if the head were continuously turning.

p. 291
- ✓ The aqueous humor acts as a support for the cornea and lens. It also provides a route for bringing nutrients to and

removing wastes from the epithelial layer of the cornea, which has no blood supply.

✓ The aqueous humor is secreted into the anterior chamber of the eye and usually drains into a small duct that joins the venous circulation. If the drainage is impeded, fluid brings up in the eye, causing the condition known as glaucoma. If untreated, the pressure can impair function of the optic nerves and cause blindness.

p. 295

✓ Constriction is mediated by parasympathetic pathways, so an anticholinergic (anti-ACh) drug would cause pupillary dilation.

✓ In myopia, the lens bends light too sharply for the length of the eyeball, and the focal distance falls in front of the retina. Myopia is corrected with a concave lens that scatters the light and increases the focal distance.
Fig. 10-31: (1) Convex lenses focus a beam of light; concave lenses scatter a beam of light passing through them. (2) In hyperopia, the focal point lies behind the retina, so the convex lens focuses it at a shorter distance onto the retina. In myopia, the focal point lies in front of the retina, so scattering the light beam with a concave lens allows the eye's lens to focus the light on the retina.

p. 298

Fig. 10-35: Red cones absorb light over the broadest spectrum, blue cones over the narrowest. At 500 nm, blue and green cones absorb light equally.

p. 301

✓ In both areas, the finest discrimination occurs in the region with the smallest visual or receptive fields.

CHAPTER 11   Efferent Peripheral Nervous System: The Autonomic and Somatic Motor Divisions

p. 316

✓ The preganglionic neurons of sympathetic pathways release acetylcholine onto nicotinic cholinergic receptors. Thus, nicotine will excite the postganglionic neurons of sympathetic pathways, such as those of the pathway that increases heart rate.

p. 319

✓ The monkeys become paralyzed because the nicotinic ACh receptors on skeletal muscles are inactivated, preventing muscle contraction.

✓ The motor end plate has nonspecific monovalent cation (Na^+ and K^+) channels that open when ACh binds to them. The axon contains voltage-gated channels, with separate channels for Na^+ and K^+.

✓ Anticholinesterase drugs block or slow the breakdown of ACh at the motor end plate. This allows ACh to remain active at the motor end plate for a longer duration and helps offset the decrease in active receptors.

CHAPTER 12   Muscles

p. 326

✓ Biceps and triceps in the upper arm; legs: hamstring (flexor) and quadriceps (extensor)

p. 330

✓ Ends of the A bands are darkest because they are where the thick and thin filaments overlap.

✓ T-tubules allow action potentials to travel from the surface of the muscle fiber to its interior.

✓ The banding pattern of organized filaments in the sarcomere forms striations in the muscle.

p. 335

✓ The crossbridges do not all unlink at one time, so while some myosin heads are free and swiveling, others are still tightly bound.

p. 342

✓ Summation in muscle fibers means that the force increases. Temporal summation in neurons means that the membrane potential difference increases.

✓ A marathoner probably has more slow twitch muscle fibers, whereas a sprinter probably has more fast-twitch muscle fibers.

p. 344

✓ An increase in firing of the somatic motor neuron causes the muscle fiber to increase its force of contraction due to summation.

✓ Force in a muscle can be increased by recruiting additional motor units.

p. 347

✓ If the muscle insertion is farther from the joint, the leverage is better and a contraction will create more force for a given increase in distance.

p. 351

✓ In multi-unit smooth muscle, force is increased by recruiting additional muscle fibers. In single-unit smooth muscle, which is electrically coupled, force is increased by increasing the amount of Ca^{2+} that enters the cell.

p. 353

✓ If there is no calcium to enter the smooth muscle, contraction stops.

✓ Tetrodotoxin does not affect action potentials in smooth muscle; therefore, the depolarization of the smooth muscle action potential must not be due to Na^+ entry.

p. 355

✓ Pacemaker potentials are always above threshold and create regular rhythms of contraction. Slow wave potentials are variable in magnitude and are below threshold some of the time.

CHAPTER 13   Integrative Physiology I: Control of Body Movement

p. 367

✓ Sensor (sensory receptor) → afferent pathway (sensory neuron) → integrating center (central nervous system) → efferent pathway (autonomic or somatic motor neuron) → effector (muscles, glands, some adipose tissue)

✓ Hyperpolarization means that the membrane potential of a neuron becomes more negative and moves farther away from threshold.

p. 374

✓ When you pick up a weight, alpha and gamma neurons, spindle afferents, and Golgi tendon organ afferents are all active.

✓ A stretch reflex is initiated by stretch and causes a reflex contraction. An extensor reflex is a postural reflex initiated by withdrawal from a painful stimulus; the extensor muscles contract but the corresponding flexors are inhibited.

CHAPTER 14   Cardiovascular Physiology

p. 390

✓ The bottom tube has the greatest flow because it has the larger pressure gradient (50 mm Hg versus 40 mm Hg).

✓ Tube 3 has the greatest flow because it has the larger radius (less resistance) and the shorter length (less resistance). Tube 4, with the greatest resistance due to length and narrow radius, has the least flow.

✓ If the canals are identical in size and therefore in cross-sectional area, the canal with the faster flow will have the higher rate of flow.

p. 396

✓ Electrical signals within cardiac muscle pass from cell to cell via gap junctions. The connective tissue does not have gap junctions and is not an excitable tissue, so any depolarizations reaching the connective tissue will die out.

✓ Superior vena cava → right atrium → tricuspid (right AV) valve→ right ventricle → pulmonary (right semilunar) valve → pulmonary trunk → right or left pulmonary artery → pulmonary arteriole → pulmonary capillary → pulmonary venule → pulmonary vein → left atrium → mitral (bicuspid, left AV) valve → left ventricle → aortic (left semilunar) valve → aorta

p. 397

✓ From this experiment, it is possible to conclude that myocardial cells require extracellular calcium for contraction, but skeletal muscle cells do not.

p. 400

✓ If all calcium channels in the muscle are blocked, there will be no contraction. If only some are blocked, the force of contraction will be less.

p. 402

✓ The rising phase of the action potential of a myocardial contractile cell is due to Na^+ influx through voltage-gated Na^+ channels, just as in the axon. If those channels are blocked with tetrodotoxin, the cell will not depolarize and will not contract.

p. 403

✓ The Ca^{2+} channels in the autorhythmic cells are not the same as the Ca^{2+} channels in the contractile cells. Those in the autorhythmic cells open rapidly when the membrane potential reaches about -50 mV and close when it reaches about $+20$ mV. The Ca^{2+} channels in the contractile cells are slower and do not open until the membrane has depolarized fully.

✓ If tetrodotoxin is applied to a myocardial autorhythmic cell, nothing will happen as there are no voltage-gated Na^+ channels in those cells.

✓ The vagus nerve contains many parasympathetic pathways to the internal organs. When the nerve was cut, parasympathetic innervation to the heart was eliminated and the rate sped up to the rate set by the autorhythmic cells of the SA node.

p. 407

✓ The AV node provides a pathway for action potentials from the SA node to pass into the ventricles. It also slows down the conduction of those action potentials, allowing completion of atrial contraction before ventricular contraction begins.

✓ The fastest pacemaker sets the heart rate. Therefore, the ectopic pacemaker will speed the heart up to 120 beats/min.

p. 408

✓ In (b), notice that there is no regular association between the P waves and the QRS complexes (the P-R segment varies in length). Notice also that not every P wave has an associated QRS complex. Both sets of waves appear at regular intervals, but the atrial rate (P waves) is faster than the ventricular rate (QRS complexes).
In (c), there are identifiable R waves but no P waves. In (d), there are no recognizable waves at all, showing that the depolarizations are not coordinated.

p. 411

Fig. 14-24: (e) From this ECG, it appears that only one of every three P waves is conducted to the ventricles. This is a condition known as partial heart block.

p. 415

✓ After 10 beats, the pulmonary circulation will have gained 10 mL of blood and the systemic circulation will have lost 10 mL.

✓ (a) One minute. (b) 12 seconds (1/5 of one minute).

p. 416

Fig. 14-29: Maximum stroke volume is 160 mL, first achieved when end-diastolic volume is about 330 mL.

p. 417

Fig. 14-30: At point A, the heart under the influence of epinephrine has a larger stroke volume and therefore is creating more force.

p. 418

Fig. 14-31: Heart rate is the only parameter controlled by ACh. Heart rate and contractility are both controlled by norepinephrine. The SA node has muscarinic receptors; the SA node and contractile myocardium have β_1 receptors.

CHAPTER 15   Blood Flow and the Control of Blood Pressure

p. 429

✓ Veins from the brain to the heart do not require valves because the flow of blood toward the heart is aided by gravity.

✓ The pressure wave at the carotid artery would arrive slightly ahead of the pressure wave to the left wrist because the distance from the heart to the carotid artery is shorter.

p. 430

✓ Pulse pressure = systolic pressure minus diastolic pressure. $(112 - 68) = 44$ mm Hg.
Mean arterial pressure = diastolic pressure + 1/3 pulse pressure. $68 + 1/3 (44) = 82.7$ mm Hg

p. 434

✓ Extracellular K^+ acts as a paracrine that dilates arterioles, thereby increasing blood flow into the tissue.

p. 436

Fig. 15-13: Blood flow through the lungs is 5 L/min.

p. 442

✓ Low protein diets result in a low concentration of plasma proteins, the solute responsible for the osmotic pressure difference between the plasma and the interstitial fluid. With lower osmotic pressure inside the capillaries, reabsorption is reduced while filtration remains constant. This results in a greater net outward movement of fluid and leads to edema.

p. 443

✓ Action potentials are created when a cell depolarizes to threshold. Depolarization results from net influx of positive ions or efflux of negative ions. The most likely ion is Na^+, moving into the receptor cell.

p. 445

✓ The inhibitory neurotransmitter is released on sympathetic neurons in Fig. 15-20.
Fig. 15-21: Arterioles and veins: norepinephrine, α receptors. Ventricles: norepinephrine, β_1 receptors. SA node: norepinephrine, β_1 receptors, and ACh, muscarinic receptors.

CHAPTER 16 Blood

p. 456

✓ Most plasma proteins are synthesized in the liver, so liver degeneration reduces the total plasma protein concentration. This in turn reduces the osmotic pressure gradient that returns fluid into the capillaries. When more fluid is filtered out of the capillary and not reabsorbed, edema results.

p. 465

✓ Some other factor that is essential for red blood cell synthesis must be lacking. These would include iron for hemoglobin production, folic acid, and vitamin B_{12}.

✓ Low atmospheric P_{O_2} at high altitude → low arterial P_{O_2} → sensed by kidney → erythropoietin synthesized and released → acts on bone marrow to increase production of red blood cells

✓ Hematocrit = packed red blood cell volume/total blood volume. A dehydrated person will have a lower plasma volume, but red blood cell volume will remain constant. This results in an elevated hematocrit and the condition known as relative polycythemia.

CHAPTER 17 Respiratory Physiology

p. 480

✓ Increased hydraulic pressure results in greater net filtration out of the capillaries. Accumulation of fluid in the interstitial space of the lungs is the condition known as pulmonary edema.

p. 483

✓ The other factor that affects how much gas will dissolve is the solubility of the gas.

✓ 720 mm Hg × 0.78 N_2 = 561.6 mm Hg

✓ If cilia cannot move mucus up the mucus escalator and out of the lungs, the collected mucus will trigger a cough reflex in an attempt to clear it out.

p. 487

✓ Scarlett will be more successful if she exhales deeply, as this will decrease her thoracic volume and will pull her lower rib cage inward.

✓ Coughing clears the airways of mucus that has captured bacteria and viruses. Inability to cough decreases the ability to expel the potentially harmful material.

p. 488

✓ A hiccup results from a quick contraction of the diaphragm that causes a rapid decrease in both intrapleural and intrapulmonary pressures.

✓ The knife wound would cause the left lung to collapse if it punctured the pleural membrane. The loss of adhesion between lung and chest wall would release the inward pressure exerted on the chest wall, so the rib cage would expand outward. The right side would be unaffected as the right lung is contained within its own pleural sac.

p. 490

✓ Scar tissue will reduce lung compliance.

✓ If surfactant is not present, the first breath will probably not expand the alveoli fully because of the high cohesiveness of the fluid lining the alveoli.

p. 491

✓ Resistance to air flow is inversely proportional to the radius of the tube through which air is flowing. If the bronchiolar lumen becomes more narrow, the resistance will increase.

✓ In a steam room, the water vapor increases the viscosity of the air and therefore increases the resistance of the air to flow.

p. 493

✓ Inspiratory reserve volume will decrease in a lung with lower compliance.

✓ Residual volume will increase in a patient who cannot fully exhale.

✓ Residual volume increases in aging subjects if total lung capacity does not change.

p. 495

✓ An increase in tidal volume will increase alveolar P_{O_2}.

p. 496

✓ If blood flow into one small section of lung decreases, the P_{O_2} in those alveoli will increase because oxygen is not leaving the alveoli and entering the blood. The P_{CO_2} in those alveoli will decrease because new CO_2 is not entering the alveoli from the blood. Bronchioles constrict when P_{CO_2} decreases, shunting blood to areas of the lung with better blood flow. The compensation will not bring ventilation in this section of lung back to normal.

p. 499

✓ Left ventricular failure or mitral valve dysfunction will cause blood to pool in the lungs when the left heart is unable to pump as well as the right heart. The increase in blood volume will result in an increase in blood pressure.

✓ When alveolar ventilation increases, arterial P_{O_2} will increase because more fresh air is entering the alveoli. Arterial P_{CO_2} will decrease because the fresh air in the alveoli has a lower P_{CO_2}, creating a greater pressure gradient for CO_2 to leave the blood. Venous P_{O_2} and P_{CO_2} will not change because these are determined by the oxygen consumption and CO_2 production in the cells.

p. 502

Fig. 17-21: (a) When P_{O_2} is 20 mm Hg, hemoglobin is about 34% saturated with oxygen. (b) Hemoglobin is 50% saturated with oxygen at a P_{O_2} of 23 mm Hg.

p. 503

Fig. 17-22: The P_{O_2} of placental blood is about 30 mm Hg. At a P_{O_2} of 10 mm Hg, maternal blood is only about 5% saturated with oxygen.
Fig. 17-23: When pH falls from 7.4 to 7.2, hemoglobin saturation decreases by 13%, from 37% saturation to 24%. When an exercising muscle cell warms up, it releases more oxygen at any given P_{O_2}.

p. 504

✓ As the P_{O_2} of the exercising muscle falls, more oxygen is released by hemoglobin. The P_{O_2} of the venous blood leaving the muscle will be 25 mm Hg, the same as the P_{O_2} of the muscle cell.

p. 506

✓ An obstruction of the airways would cause CO_2 to be retained in the body. This would increase the H^+ concentration and the pH would fall.

CHAPTER 18   The Kidneys

p. 519

✓ If extracellular K^+ decreases, more K^+ will leave the cell and the membrane potential will hyperpolarize (become more negative, increase).

✓ If plasma Ca^{2+} decreases, the force of contraction will decrease.

p. 525

✓ Filtration and secretion both represent movement of substances from the extracellular fluid into the lumen of the nephron. Filtration is always passive; secretion may be passive or active.

✓ A water molecule goes from the glomerulus → Bowman's capsule → proximal tubule → loop of Henle → distal tubule → collecting duct → renal pelvis → ureter → urinary bladder → urethra.

✓ If reabsorption decreases to half of normal, the body would run out of plasma in under an hour.

p. 530

✓ If the afferent arteriole constricts, the resistance in that arteriole increases, flow through that arteriole is diverted to lower resistance arterioles, and GFR in that nephron will decrease.

✓ The primary driving force for GFP. is blood pressure, opposed by fluid pressure in Bowman's capsule and osmotic pressure due to plasma proteins. If a person has fewer plasma proteins due to liver disease, the plasma will have a lower osmotic pressure. With less osmotic pressure opposing GFR, GFR will increase.

p. 532

✓ Fig. 18-13: The transport rate at 3 mg/mL is 3 mg/min; at 5 and 8 mg/mL, it is 4 mg/min. The transport rate is 4 mg/min at any plasma concentration equal to or higher than 4 mg/mL.

CHAPTER 19   Integrative Physiology II: Fluid and Electrolyte Balance

p. 548

✓ Hyperosmotic NaCl is hypertonic and causes the osmoreceptors to shrink, but hyperosmotic urea is hypotonic and causes them to swell. Thus, they will not fire.

✓ In dehydration, the kidneys try to conserve water. ADH makes the nephron permeable to water and enhances absorption; therefore, ADH levels would increase with dehydration.

✓ If ADH secretion is suppressed, the urine will be dilute.

p. 551

✓ Calcium entry through voltage-gated calcium channels initiates the exocytosis of neurotransmitters by neurons.

✓ Apical membranes on collecting duct cells will have more water pores when ADH is present.

p. 557

✓ Atherosclerotic plaques block blood flow, which decreases GFR and decreases pressure in the afferent arteriole. These are both stimuli for renin release.

✓ Renin secretion begins a cascade that produces angiotensin II. Angiotensin II is a powerful vasoconstrictor, acts on the medullary cardiovascular control center to increase blood pressure, increases ADH and aldosterone secretion and thirst, resulting in fluid volume in the body. These responses may all contribute to increased blood pressure.

p. 567

✓ The muscles of ventilation that are involved include the diaphragm, the external intercostals, and the scalenes.

✓ The bicarbonate is increasing as the reaction shifts to the right and achieves a new equilibrium. At equilibrium, both H^+ and bicarbonate are increased. The bicarbonate cannot act as a buffer when the system is at equilibrium.

p. 569

✓ Both K^+ and H^+ are being moved against their concentration gradients, which requires ATP. In the proximal tubule, the Na^+ is moving down its concentration gradient, providing the energy to push H^+ against its gradient.

✓ When intercalated cells reabsorb K^+, they secrete H^+; therefore pH will increase.

✓ The map for acidosis shows that H^+ is secreted and bicarbonate reabsorbed, whereas the map for alkalosis shows the opposite. Check with your instructor for the details.

CHAPTER 20   Digestion

p. 577

✓ Digestion is the process by which nutrients are broken down into a form that is absorbable by the intestine; it takes place in the lumen of the digestive tract, which is technically outside the body. Metabolism is the sum of all chemical reactions that take place inside the body.

✓ Absorption moves material from the lumen to the extracellular fluid. Secretion moves substances from the cells or extracellular fluid into the lumen.

p. 581

✓ The digestive system has a large and vulnerable surface area facing the external environment; therefore, it needs the immune cells of the lymphoid tissue to combat invaders.

✓ The sphincters are tonically contracted to close off the lumen of the digestive tract from the outside world and to keep material from passing freely between sections of the tract.

Fig. 20-3: The contraction occurs slightly behind the action potential because of the time required for excitation-contraction coupling, which in smooth muscle requires Ca^{2+} entry, complexing to calmodulin, activation of myosin light chain kinase, and phosphorylation of the myosin.

p. 601

✓ Glucose from table sugar enhances the intestinal absorption of Na^+ because the transporter for glucose depends on the movement of Na^+ from the lumen into the intestinal cell.

✓ Bile does not digest triglycerides. It simply emulsifies them into smaller globules so that lipase can digest them.

CHAPTER 21   Endocrine Control of Metabolism and Growth

p. 618

✓ Water has a high specific heat and will draw heat away from the body through conductive heat transfer. If this loss exceeds the person's heat production, they will feel cold.

✓ A person exercising in humid environment loses the benefit of evaporative cooling and is likely to overheat faster.

p. 633

✓ Insulin is a protein and will be digested if administered as a pill.

✓ Although dehydrated patients may have elevated K^+ concentrations in their blood, their total amount of K^+ is below normal. If normal fluid volume is replaced with no K^+ added, the result will be a below-normal K^+ concentration. Thus, these patients are given fluid with K^+ even though they have an initial hyperkalemia.

p. 635

✓ A target cell might not respond to leptin because it has no leptin receptors (or they are defective), or because there is a problem with the signal transduction/second messenger pathway.

p. 638

✓ Anabolic steroids, used to build muscle, do not include cortisol because cortisol is catabolic on muscle proteins.

p. 641

✓ The T_4 given to the patient is converted to T_3 in the peripheral tissues. T_3 is the more active form of the hormone.

✓ When mitochondria are uncoupled, the energy that is normally captured in the high-energy bond of ATP is released as heat instead. This will raise the body temperature of the person and cause heat intolerance.

p. 643

✓ Plasma glucose is regulated by a number of hormones in addition to growth hormone. Cortisol and glucagon both increase plasma glucose concentrations.

p. 649

✓ Adults who hypersecrete growth hormone do not grow taller because the epiphyseal plates of the long bones have closed.

CHAPTER 22   The Immune System

p. 667

✓ When capillary permeability increases, proteins escape from the plasma to the interstitial fluid. This decreases the force opposing capillary filtration, and causes additional fluid to accumulate in the interstitial space (swelling or edema).

p. 669

✓ The first exposure causes a slower and weaker immune response (see Fig. 22-11). The second exposure will cause a stronger response, which may be severe and cause anaphylaxis.

✓ Protein antibodies can be moved across cells by transcytosis or released from cells by exocytosis.

CHAPTER 23   Integrative Physiology III: Exercise

p. 688

✓ If venous P_{O_2} decreases, the P_{O_2} is also decreasing.

p. 690

✓ If resistance falls but blood pressure is increasing, cardiac output must be increasing.

✓ The line for mean blood pressure lies closer to the line for diastolic pressure because the heart spends more time in diastole than systole.

Fig. 23-5: To calculate actual blood flow to an organ, multiply the cardiac output (L/min) times the percentage of flow to that organ. When rest and exercise values are compared, actual blood flow decreases only in the kidneys, GI tract, and other tissues.

CHAPTER 24   Reproduction and Development

p. 702

✓ The male donates the chromosome that determines the sex of the zygote.

✓ In the absence of a Y chromosome, the XO fetus will be a female. She will be infertile, however, as two X chromosomes are required for functioning ovaries.

✓ Developing testes secrete testosterone, which (along with its derivative DHT) prompts the development of the male accessory structures. The external genitalia of the embryo will be female.

p. 703

Fig. 24-5: (1) The gamete in a newborn male is at the spermatogonia stage; it is at the primary oocyte stage in females. (2) The first polar body has twice as much DNA as the second polar body. (3) Each primary oocyte forms one egg; each primary spermatocyte forms four sperm.

p. 712

✓ Exogenous anabolic steroids are usually androgens, which have a negative feedback effect on the anterior pituitary gonadotropins. In the absence of FSH and LH, the testes will shrink and stop producing sperm.

✓ GnRH agonists cause reduced secretion of GnRH and therefore of FSH and LH. In the absence of the gonadotropins, the testes stop producing sperm. However, they also stop producing testosterone. This causes reduced sex drive, which is usually an undesirable side effect.

p. 720

✓ Women who take anabolic steroids may experience growth of facial and body hair, deepening of the voice, increased libido, and irregular menstrual cycles.

✓ A woman given an aromatase inhibitor would have decreased estrogen production during the follicular phase of the menstrual cycle.

vestibular nerve, 287, 289, 302
vestibulocochlear nerve (VIII), 245t
villi   Fingerlike projections of the intestinal surface, 580
VIP. *See* vasoactive intestinal peptide.
viral load, 678
virus, 655, 655t, 673, 674f
visceral nervous system, 202. *See* autonomic division.
visceral reflex. *See* autonomic reflex.
visceral smooth muscle, 351
   control of movement in, 378–79
viscosity   Thickness or resistance to flow of a solution, 389, 490
visible light, 290
vision, 264t, 289–302
vision problems, 291, 295f
visual cortex, 247, 258, 301–2
visual field, 298–301
visual imaging, 249
visual pigment, 296, 298–99f
visual processing, 301–2
visualization technique, in sports, 365
vital capacity   The maximum amount of air that can be voluntarily moved in and out of the respiratory system, 493
vitamin, 81–82, 594–96
vitamin B$_{12}$, 594–96
vitamin D, 635, 647–48
vitamin D$_3$, 176–77f
vitamin K, 468t
vitreous humor, 290
VLDL. *See* very low–density lipoprotein.
VNO. *See* vomeronasal organ.
vocal cords, 475, 477
vocalization, 257
voltage–gated channel   A gated channel that opens or closes in response to a change in membrane potential, 113, 400
voluntary movement, 374t, 375–78
voluntary muscle, 325
vomeronasal organ (VNO), 277
vomiting, 559, 571
vomiting center, 241
von Willebrand factor, 468t
von Willebrand's disease, 454, 456, 460, 466
VRG. *See* ventral respiratory group.
vulva, 713

warm receptor, 275
water, distribution in body, 130–41

estimating body water, 134
   structure of, 21–22
   surface tension of, 22–24
water balance, 596
   disturbances of, 559–60
water balance reflex, 546–47
water excretion, 544–45
Watson, J.D., 33
weight/volume solution, 25
Wernicke's area, 257–58
wheal, 147
white blood cell. *See* leukocyte.
white fat, 63, 64f
white matter   Central nervous system tissue composed primarily of myelinated axons, 242, 243f
"white muscle," 340, 341t
white pulp, 659f
whooping cough, 150t
Wiggers diagram, 413, 414f
wind chill factor, 614
withdrawal reflex, 274
wobbler mouse, 223
Wolffian duct(s)   Embryonic structures that develop into male reproductive structures, 700, 702
womb. *See* uterus.
work, 74, 75f
working memory, 255

X chromosome, 699–700
X-linked disorder, 700
X-rays, 17, 19

Y chromosome, 699–700
yolk sac, 725

Z disk, 327, 331f
Zollinger-Ellison syndrome, 603
zona fasciculata, 636
zona glomerulosa, 636
zona pellucida, 718t, 723
zona reticularis, 636
zonula, 291
zygote   Fertilized egg, 41, 699
zymogen(s)   Inactive proenzymes in the digestive system, 584

Photo Credits

About the Author

Author Dee Silverthorn   **Illustrators** Bill Ober/Claire Garrison   **Clinical Consultant** Dee Silverthorn

Preface

All Dee Silverthorn

Chapter 3

3-1a & c Todd Derksen   **3-1b** David M. Phillips/Visuals Unlimited   **3-6 (left)** Dr. Don Fawcett/Photo Researchers, Inc.   **3-7b** Fawcett/Hirokawa/Heuser/Science Source/Photo Researchers, Inc.   **3-8d** Robert W. Riess   **3-9b** CNRI/Science Source/Photo Researchers, Inc.   **3-10** Robert W. Riess   **3-11a** Robert W. Riess   **3-12** Dr. Don Fawcett/Photo Researchers, Inc. **3-13a** Dr. Don W. Fawcett/Harvard Medical School   **3-13b** Biophoto Associates/Photo Researchers, Inc.   **3-14a** Journal of Cell Biology   **3-14c** J. L. Carson/Custom Medical Stock Photo   **3-15** Todd Derksen **3-18b** Custom Medical Stock Photo   **3-19** Todd Derksen **3-21** Ward's Natural Science Establishment, Inc.   **3-22c** John D. Cunningham/ Visuals Unlimited   **3-23a** Frederic H. Martini   **3-26a & b** Cabisco/Visuals Unlimited

Chapter 7

7-2 Dee Silverthorn

Chapter 8

8-21a Todd Derksen   **8-21b** David M. Phillips/Visuals Unlimited   **8-24a** David Scott/Phototake NYC

Chapter 9

9-13 Mazziotta et al./Science Photo Library/ Photo Researchers, Inc.

Chapter 10

10-14d Todd Derksen   **10-27c** Custom Medical Stock Photo

Chapter 12

12-1a & c Todd Derksen   **12-1b** Phototake NYC   **12-7a & b** J. J. Head/Carolina Biological Supply/Phototake NYC **12-13** J. B. Lippincott Publishers   **12-16** Young-Jin Son **12-25** Biophoto Associates/Photo Researchers, Inc.

Chapter 16

16-6a Todd Derksen   **16-9a & b** Todd Derksen

Chapter 17

17-7b Todd Derksen

Chapter 18

18-1f Todd Derksen   **18-4b** Todd Derksen

Chapter 21

21-10c Ward's Natural Science Establishment, Inc. **21-19c** Ward's Natural Science Establishment, Inc **21-27** Astrid & Hanns-Frieder Michler/Science Photo Library/Photo Researchers, Inc.

Chapter 24

24-1 Photo Researchers, Inc.   **24-9c** David M. Phillips/Visuals Unlimited   **24-16a** Francis Leroy, Biocosmos/Science Photo Library/Custom Stock Medical Photo

MEASUREMENTS AND CONVERSIONS

Prefixes

deci-	(d)	1/10	0.1	1×10^{-1}
centi-	(c)	1/100	0.01	1×10^{-2}
milli-	(m)	1/1000	0.001	1×10^{-3}
micro-	(μ)	1/1,000,000	0.000001	1×10^{-6}
nano-	(n)	1/1,000,000,000	0.000000001	1×10^{-9}
pico-	(p)	1/1,000,000,000,000	0.000000000001	1×10^{-12}
kilo-	(k)		1000	1×10^{3}

Metric system

1 meter (m) = 100 centimeters (cm) = 1000 millimeters (mm)

1 centimeter (cm) = 10 millimeters (mm) = 0.01 meters (m)

1 millimeter (mm) = 1000 micrometers (μm; also called micron, μ)

1 angstrom (Å) = 1/10,000 micrometer = 1×10^{-7} millimeters

1 liter (L) = 1000 milliliters (mL)

1 deciliter (dL) = 100 milliliters (mL) = 0.1 liters (L)

1 cubic centimeter (cc) = 1 milliliter (mL)

1 milliliter (mL) = 1000 microliters (μL)

1 kilogram (kg) = 1000 grams (g)

1 gram (g) = 1000 milligrams (mg)

1 milligram = 1000 micrograms (μg)

Conversions

1 yard (yd) = 0.92 meters	1 meter = 1.09 yards
1 inch (in) = 2.54 centimeters	1 centimeter = 0.39 inches
1 liquid quart (qt) = 946 milliliters	1 liter = 1.05 liquid quarts
1 fluid ounce (oz) = 8 fluid drams = 29.57 milliliters (mL)	
1 pound (lb) = 453.6 grams	1 kilogram = 2.2 pounds

Temperature

Freezing = 0 degrees Celsius (°C) = 32 degrees Fahrenheit (°F) = 273 degrees Kelvin (°K)

To convert degrees Celsius (°C) to degrees Fahrenheit (°F): (°C x 9/5) + 32

To convert degrees Fahrenheit (°F) to degrees Celsius (°C): (°F - 32) x 5/9

Normal values of blood components

Substance or parameter	Normal range	Measured in
Calcium (Ca^{2+})	4.3-5.3 meq/L	Serum
Chloride (Cl^-)	100-108 meq/L	Serum
Potassium (K^+)	3.5-5.0 meq/L	Serum
Sodium (Na^+)	135-145 meq/L	Serum
pH	7.35-7.45	Whole blood
P_{O_2}	75-100 mm Hg	Arterial blood
P_{CO_2}	35-45 mm Hg	Arterial blood
Osmolality	280-296 mosmol/kg water	Serum
Glucose, fasting	70-110 mg/dL	Plasma
Creatinine	0.6-1.5 mg/dL	Serum
Protein, total	6.0-8.0 g/dL	Serum

Modified from W. F. Ganong, *Review of Medical Physiology,* Appleton & Lange, Norwalk, 1995.